Principles of Lasers

FOURTH EDITION

Principles of Lasers

FOURTH EDITION

Orazio Svelto

Polytechnic Institute of Milan
and National Research Council
Milan, Italy

Translated from Italian and edited by

David C. Hanna

Southampton University
Southampton, England

Plenum Press • New York and London

Library of Congress Cataloging in Publication Data:

```
Svelto, Orazio.
   [Principi dei laser. English]
   Principles of lasers / Orazio Svelto ; translated from Italian and
edited by David C. Hanna. -- 4th ed.
      p.   cm.
   Includes bibliographical references and index.
   ISBN 0-306-45748-2
   1. Lasers.   I. Hanna, D. C. (David C.), 1941-   . II. Title.
QC688.S913 1998
621.36'6--dc21                                            98-5077
                                                          CIP
```

Front cover photograph: The propagation of an ultraintense pulse in air results in self-trapping of the laser beam. The rich spectrum of colors produced is the result of the high intensity ($\approx 10^{14}$ W/cm^2) within the self-focused filament, producing nonlinear phenomena such as self-phase modulation, parametric interactions, ionization, and conical emission due to the beam collapse. The rainbowlike display with its sequenced color is due to diffraction of the different colors (copyright 1998 William Pelletier, Photo Services, Inc.).

Back cover photograph: Interaction of an ultraintense ($\approx 10^{20}$ W/cm^2) laser pulse with a target consisting of plastic and aluminum layers. The 450-fs pulse, with peak power of 1200 TW, is produced by the petawatt laser at the Lawrence Livermore National Laboratory. Numerous nonlinear and relativistic phenomena are observable including copious second harmonic generation (green light in photo) (courtesy of M. D. Perry, Lawrence Livermore National Laboratory).

ISBN 0-306-45748-2

© 1998, 1989, 1982, 1976 Plenum Press, New York
A Division of Plenum Publishing Corporation
233 Spring Street, New York, N.Y. 10013

http://www.plenum.com

10 9 8 7 6 5 4 3 2 1

621.366

Printed in the United States of America

To my wife Rosanna
and to my sons Cesare and Giuseppe

Preface to the Fourth Edition

This book is motivated by the very favorable reception given to the previous editions as well as by the considerable range of new developments in the laser field since the publication of the third edition in 1989. These new developments include, among others, quantum-well and multiple-quantum-well lasers, diode-pumped solid-state lasers, new concepts for both stable and unstable resonators, femtosecond lasers, ultra-high-brightness lasers, etc. This edition thus represents a radically revised version of the preceding edition, amounting essentially to a *new book* in its own right. However, the basic aim has remained the same, namely to provide a *broad* and *unified description* of laser behavior at the simplest level which is compatible with a correct physical understanding. The book is therefore intended as a textbook for a senior-level or first-year graduate course and/or as a reference book.

The most relevant *additions* or *changes* to this edition can be summarized as follows:

1. A much-more detailed description of Amplified Spontaneous Emission has been given (Chapter 2) and a novel simplified treatment of this phenomenon, both for homogeneous and inhomogeneous lines, has been introduced (Appendix C).
2. A major fraction of a new chapter (Chapter 3) is dedicated to the interaction of radiation with semiconductor media, either in a bulk form or in a quantum-confined structure (quantum-well, quantum-wire and quantum dot).
3. A modern theory of stable and unstable resonators is introduced, where a more extensive use is made of the ABCD matrix formalism and where the most recent topics of dynamically stable resonators as well as unstable resonators, with mirrors having Gaussian or super-Gaussian transverse reflectivity profiles, are considered (Chapter 5).
4. Diode-pumping of solid-state lasers, both in longitudinal and transverse pumping configurations, are introduced in a unified way and a comparison is made with corresponding lamp-pumping configurations (Chapter 6).
5. Spatially dependent rate equations are introduced for both four-level and quasi-three-level lasers and their implications, for longitudinal and transverse pumping, are also discussed (Chapter 7).

6. Laser mode-locking is considered at much greater length to account for, e.g., new mode-locking methods, such as Kerr-lens mode locking. The effects produced by second-order and third-order dispersion of the laser cavity and the problem of dispersion compensation, to achieve the shortest pulse-durations, are also discussed at some length (Chapter 8).

7. New tunable solid-state lasers, such as Ti: sapphire and Cr: LiSAF, as well as new rare-earth lasers such as Yb^{3+}, Er^{3+}, and Ho^{3+} are also considered in detail (Chapter 9).

8. Semiconductor lasers and their performance are discussed at much greater length (Chapter 9).

9. The divergence properties of a multimode laser beam as well as its propagation through an optical system are considered in terms of the M^2 factor and in terms of the embedded Gaussian beam (Chapters 11 and 12).

10. The production of ultra-high peak intensity laser beams by the technique of chirped-pulse-amplification and the related techniques of pulse expansion and pulse compression are also considered in detail (Chapter 12).

Besides these major additions, the contents of the book have also been greatly *enriched* by numerous examples, treated in detail, as well as several new tables and several new appendixes. The examples either refer to real situations, as found in the literature or encountered through my own laboratory experience, or describe a significant advance in a particular topic. The tables provide data on optical, spectroscopic, and nonlinear-optical properties of laser materials, the data being useful for developing a more quantitative context as well as for solving the problems. The appendixes are introduced to consider some specific topics in more mathematical detail. A great deal of effort has also been devoted to the *logical organization* of the book so as to make its content even more accessible. Lastly, a large fraction of the problems has also been changed to reflect the new topics introduced and the overall shift in emphasis within the laser field.

However, despite these profound changes, the basic philosophy and the basic organization of the book have remained the same. The *basic philosophy* is to resort, wherever appropriate, to an intuitive picture rather than to a detailed mathematical description of the phenomena under consideration. Simple mathematical descriptions, when useful for a better understanding of the physical picture, are included in the text while the discussion of more elaborate analytical models is deferred to the appendixes. The *basic organization* starts from the observation that a laser can be considered to consist of three elements, namely the active medium, the resonator, and the pumping system. Accordingly, after an introductory chapter, Chapters 2–3, 4–5, and 6 describe the most relevant features of these elements, separately. With the combined knowledge about these constituent elements, Chapters 7 and 8 then allow a discussion of continuous-wave and transient laser behavior, respectively. Chapters 9 and 10 then describe the most relevant types of laser exploiting high-density and low-density media, respectively. Lastly, Chapters 11 and 12 consider a laser beam from the user's viewpoint, examining the properties of the output beam as well as some relevant laser beam transformations, such as amplification, frequency conversion, pulse expansion or compression.

The inevitable price paid by the addition of so many new topics, examples, tables, and appendixes has been a considerable increase in book size. Thus, it is clear that the entire content of the book could not be covered in just a one semester-course. However, the organization of the book allows several different learning paths. For instance, one may be more interested in learning the *Principles of Laser Physics*. The emphasis of the study should then be concentrated on the first section of the book (Chapters 2–8 and Chapter 11). If, on the other hand, the reader is more interested in the *Principles of Laser Engineering*, effort should mostly be concentrated on the second part of the book (Chapters 5–12). The *level of understanding* of a given topic may also be suitably *modulated* by, e.g., considering, in more or less detail, the numerous examples, which often represent an extension of a given topic, as well as the numerous appendixes.

Writing a book, albeit a satisfying cultural experience, represents a heavy intellectual and physical effort. This effort has, however, been gladly sustained in the hope that this completely new edition can now better serve the pressing need for a general introductory course to the laser field.

ACKNOWLEDGEMENTS. I wish to acknowledge the following friends and colleagues, whose suggestions and encouragement have certainly contributed to improving the book in a number of ways: Christofer Barty, Vittorio De Giorgio, Emilio Gatti, Dennis Hall, Günther Huber, Gerard Mourou, Nice Terzi, Franck Tittel, Colin Webb, Herbert Welling. I wish also to warmly acknowledge the critical editing of David C. Hanna, who has acted as much more than simply a translator. Lastly I wish to thank, for their useful comments and for their critical reading of the manuscript, my former students: G. Cerullo, S. Longhi, M. Marangoni, M. Nisoli, R. Osellame, S. Stagira, C. Svelto, S. Taccheo, and M. Zavelani.

Milan   Orazio Svelto

Contents

3. Energy Levels, Radiative, and Nonradiative Transitions in Molecules and Semiconductors 81

4. Ray and Wave Propagation through Optical Media 129

Contents   xvii

List of Examples

Chapter 4

Chapter 5

Chapter 6

Chapter 7

Chapter 8

Chapter 9

Chapter 11

Chapter 12

1

Introductory Concepts

In this introductory chapter, the fundamental processes and the main ideas behind laser operation are introduced in a very simple way. The properties of laser beams are also briefly discussed. The main purpose of this chapter is thus to introduce the reader to many of the concepts that will be discussed in the following chapters, and therefore help the reader to appreciate the logical organization of the book.

Following the discussion presented in this chapter, in fact, the organization of the book is based on the observation that a laser can be considered to consist of three elements: an active material, a pumping scheme, and a resonator. Accordingly, after this introductory chapter, Chaps. 2 and 3 deal with the interaction of radiation with matter, starting from the simplest cases, i.e., atoms or ions in an essentially isolated situation (Chap. 2), then going on to the more complicated cases, i.e., molecules and semiconductors (Chap. 3). As an introduction to optical resonators, Chap. 4 considers some topics relating to ray and wave propagation in particular optical elements, such as free space, optical lens-like media, Fabry–Perot interferometers, and multilayer dielectric coatings. Chapter 5 treats the theory of optical resonators, while Chap. 6 discusses pumping processes. Concepts introduced in these chapters are then used in Chaps. 7 and 8, where a theory is developed for continuous wave and transient laser behavior, respectively. The theory is based on the lowest order approximation, i.e., using the rate equation approach. This approach is in fact applicable in describing most laser characteristics. Since lasers based on different types of active media have significantly different characteristics, Chaps. 9 and 10 discuss characteristic properties of a number of laser types: Chapter 9 covers ionic crystal, dye, and semiconductor lasers, which have a number of common features; Chap. 10 considers gas, chemical, and free-electron lasers. By this point the reader should have acquired sufficient understanding of laser behavior to study properties of the output beam (coherence, monochromaticity, brightness, noise), which are considered in Chap. 11. Chapter 12 is then based on the fact that, before being used, a laser beam is generally transformed in some way, which includes: (1) Spatial transformation of the beam due to its propagation through, e.g., a lens system; (2) amplitude transformation as a result of passing through an amplifier; (3) wavelength transformation, or frequency conversion, via a number of nonlinear phenomena (second harmonic generation, parametric processes); (4) time transformation by, e.g., pulse compression or pulse expansion.

1.1. SPONTANEOUS AND STIMULATED EMISSION, ABSORPTION

To describe the phenomenon of spontaneous emission (Fig. 1.1a), let us consider two energy levels, 1 and 2, of some atom or molecule of a given material with energies E_1 and E_2 ($E_1 < E_2$), respectively. In the following discussion, the two levels can be any two of an atom's infinite set of levels. It is convenient however to take level 1 as the ground level. Let us now assume that the atom is initially at level 2. Since $E_2 > E_1$, the atom tends to decay to level 1. The corresponding energy difference $E_2 - E_1$ must therefore be released by the atom. When this energy is delivered in the form of an electromagnetic (em) wave, the process is called *spontaneous* (or *radiative*) *emission*. The frequency v_0 of the radiated wave is then given by the well known expression:

$$v_0 = \frac{(E_2 - E_1)}{h} \qquad (1.1.1)$$

where h is Planck's constant. Spontaneous emission is therefore characterized by the emission of a photon of energy $hv_0 = E_2 - E_1$ when the atom decays from level 2 to level 1 (Fig. 1.1a). Note that radiative emission is just one of two possible ways for the atom to decay. Decay can also occur in a nonradiative way. In this case the energy difference $E_2 - E_1$ is delivered in some form of energy other than em radiation (e.g., it may go into the kinetic or internal energy of the surrounding atoms or molecules). This phenomenon is called *nonradiative decay*.

Let us now suppose that the atom is initially found in level 2 and an em wave of frequency $v = v_0$ (i.e., equal to that of the spontaneously emitted wave) is incident on the material (Fig. 1.1b). Since this wave has the same frequency as the atomic frequency, there is a finite probability that this wave will force the atom to undergo the transition $2 \rightarrow 1$. In this case the energy difference $E_2 - E_1$ is delivered in the form of an em wave that adds to the incident wave. This is the phenomenon of *stimulated emission*. There is a fundamental difference between the spontaneous and stimulated emission processes. In the case of spontaneous emission, atoms emit an em wave that has no definite phase relation to that emitted by another atom. Furthermore the wave can be emitted in any direction. In the case of stimulated emission, since the process is forced by the incident em wave, the emission of any atom adds in phase to that of the incoming wave and in the same direction.

Let us now assume that the atom is initially lying in level 1 (Fig. 1.1c). If this is the ground level, the atom remains in this level unless some external stimulus is applied. We

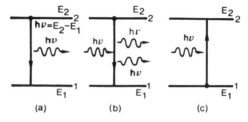

FIG. 1.1. Schematic illustration of the three processes: (a) spontaneous emission, (b) stimulated emission, (c) absorption.

assume that an em wave of frequency $v = v_0$ is incident on the material. In this case there is a finite probability that the atom will be raised to level 2. The energy difference $E_2 - E_1$ required by the atom to undergo the transition is obtained from the energy of the incident em wave. This is the *absorption* process.

To introduce probabilities for these emission and absorption phenomena, let N_i be the number of atoms (or molecules) per unit volume that at time t occupy a given energy level, i. From now on the quantity N_i is called the *population* of the level.

For the case of spontaneous emission, the probability that the process occurs is defined by stating that the rate of decay of the upper state population $(dN_2/dt)_{sp}$ must be proportional to the population N_2. We can therefore write

$$\left(\frac{dN_2}{dt}\right)_{sp} = -AN_2 \tag{1.1.2}$$

where the minus sign accounts for the fact that the time derivative is negative. The coefficient A, introduced in this way, is a positive constant called the rate of spontaneous emission or the Einstein A coefficient. (An expression for A was first obtained by Einstein from thermodynamic considerations.) The quantity $\tau_{sp} = 1/A$ is the spontaneous emission (or radiative) lifetime. Similarly, for nonradiative decay, we can generally write

$$\left(\frac{dN_2}{dt}\right)_{nr} = -\frac{N_2}{\tau_{nr}} \tag{1.1.3}$$

where τ_{nr} is the nonradiative decay lifetime. Note that for spontaneous emission the numerical value of A (and τ_{sp}) depends only on the particular transition considered. For nonradiative decay, on the other hand, τ_{nr} depends not only on the transition but also on characteristics of the surrounding medium.

We can now proceed in a similar way for stimulated processes (emission or absorption). For stimulated emission we can write

$$\left(\frac{dN_2}{dt}\right)_{st} = -W_{21}N_2 \tag{1.1.4}$$

where $(dN_2/dt)_{st}$ is the rate at which transitions $2 \rightarrow 1$ occur as a result of stimulated emission and W_{21} is the rate of stimulated emission. As in the case of the A coefficient defined by Eq. (1.1.2), the coefficient W_{21} also has the dimension of $(\text{time})^{-1}$. Unlike A, however, W_{21} depends not only on the particular transition but also on the intensity of the incident em wave. More precisely, for a plane wave, we can write

$$W_{21} = \sigma_{21}F \tag{1.1.5}$$

where F is the photon flux of the wave and σ_{21} is a quantity having the dimension of an area (the stimulated emission *cross section*) and depending on characteristics of the given transition.

As in Eq. (1.1.4) we can define an absorption rate W_{21} using the equation:

$$\left(\frac{dN_1}{dt}\right)_a = -W_{12}N_1 \tag{1.1.6}$$

where $(dN_1/dt)_a$ is the rate of transitions $1\rightarrow2$ due to absorption and N_1 is the population of level 1. As in Eq. (1.1.5) we can write

$$W_{12} = \sigma_{12}F \tag{1.1.7}$$

where σ_{12} is some characteristic area (the *absorption cross section*), which depends only on the particular transition.

In the preceding discussion the stimulated processes are characterized by the stimulated emission and absorption cross-sections σ_{21} and σ_{12}, respectively. Einstein showed at the beginning of the twentieth century that, if the two levels are nondegenerate, one has $W_{21} = W_{12}$ and thus $\sigma_{21} = \sigma_{12}$. If levels 1 and 2 are g_1-fold and g_2-fold degenerate, respectively, one then has:

$$g_2 W_{21} = g_1 W_{12} \tag{1.1.8}$$

that is

$$g_2 \sigma_{21} = g_1 \sigma_{12} \tag{1.1.9}$$

Note also that the fundamental processes of spontaneous emission, stimulated emission, and absorption can be described in terms of absorbed or emitted photons as follows (see Fig. 1.1): (a) In the spontaneous emission process, the atom decays from level 2 to level 1 through the emission of a photon. (b) In the stimulated emission process, the incident photon stimulates the transition $2\rightarrow1$, so that there are two photons (the stimulating one and the stimulated one). (c) In the absorption process, the incident photon is simply absorbed to produce transition $1\rightarrow2$. Thus each stimulated emission process creates a photon, whereas each absorption process annihilates a photon.

1.2. THE LASER IDEA

Consider two arbitrary energy levels 1 and 2 of a given material, and let N_1 and N_2 be their respective populations. If a plane wave with a photon flux F is traveling in the z-direction in the material (Fig. 1.2), the elemental change dF of this flux along the elemental length dz of the material is due to both stimulated absorption and emission processes occurring in the shaded region of Fig. 1.2. Let S be the cross-sectional area of the beam. The change in number between outgoing and incoming photons in the shaded volume per unit time is thus SdF. Since each stimulated process creates a photon whereas each absorption removes a photon, SdF must equal the difference between stimulated emission and absorption events occurring in the shaded volume per unit time. From Eqs. (1.1.4) and (1.1.6) we can write $SdF = (W_{21}N_2 - W_{12}N_1)(Sdz)$, where Sdz is the volume of the shaded region. With the help of Eqs. (1.1.5), (1.1.7), and (1.1.9), we obtain

$$dF = \sigma_{21}F\left[N_2 - \left(\frac{g_2 N_1}{g_1}\right)\right]dz \tag{1.2.1}$$

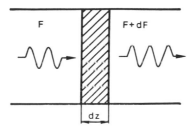

FIG. 1.2. Elemental change dF in the photon flux F for a plane em wave in traveling a distance dz through the material.

Note that, in deriving Eq. (1.2.1), we did not consider radiative and nonradiative decays. In fact nonradiative decay does not add new photons, while photons created by radiative decay are emitted in any direction and thus give negligible contribution to the incoming photon flux F.

Equation (1.2.1) shows that the material behaves as an amplifier (i.e., $dF/dz > 0$) if $N_2 > g_2N_1/g_1$, while it behaves as an absorber if $N_2 < g_2N_1/g_1$. At thermal equilibrium populations are described by Boltzmann statistics. Then if N_1^e and N_2^e are the thermal equilibrium populations of the two levels:

$$\frac{N_2^e}{N_1^e} = \frac{g_2}{g_1}\exp-\left(\frac{E_2 - E_1}{kT}\right) \qquad (1.2.2)$$

where k is Boltzmann's constant and T is the absolute temperature of the material. In thermal equilibrium we thus have $N_2^e < g_2N_1^e/g_1$. According to Eq. (1.2.1) the material then acts as an absorber at frequency v_0. This is what happens under ordinary conditions. However if a nonequilibrium condition is achieved for which $N_2 > g_2N_1/g_1$, then the material acts as an amplifier. In this case we say that there exists a *population inversion* in the material. This means that the population difference $N_2 - (g_2N_1/g_1)$ is opposite in sign to what exists under thermodynamic equilibrium [$N_2 - (g_2N_1/g_1) < 0$]. A material in which this population inversion is produced is referred to as an *active medium*.

If the transition frequency $v_0 = (E_2 - E_1)/kT$ falls in the microwave region, this type of amplifier is called a *maser* amplifier, an acronym for microwave amplification by stimulated emission of radiation. If the transition frequency falls in the optical region, the amplifier is called a *laser* amplifier, an acronym obtained from the preceding one with light substituted for microwave.

To make an oscillator from an amplifier, it is necessary to introduce suitable positive feedback. In the microwave region this is done by placing the active material in a resonant cavity having a resonance at frequency v_0. In the case of a laser, feedback is often obtained by placing the active material between two highly reflecting mirrors, such as the plane parallel mirrors in Fig. 1.3. In this case a plane em wave traveling in a direction perpendicular to the mirrors bounces back and forth between the two mirrors, and is amplified on each passage through the active material. If one of the two mirrors (e.g. mirror 2) is partially transparent, a useful output beam is obtained from that mirror.

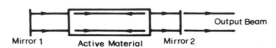

FIG. 1.3. Scheme of a laser.

It is important to realize that, for both masers and lasers, a certain threshold condition must be reached. In the laser case, oscillation begins when the gain of the active material compensates the losses in the laser (e.g., losses due to output coupling). According to Eq. (1.2.1) the gain per pass in the active material (i.e., the ratio between output and input photon flux) is $\exp\{\sigma[N_2 - (g_2 N_1/g_1)]l\}$, where we denote for simplicity $\sigma = \sigma_{21}$, and where l is the length of the active material. Let now R_1 and R_2 be the power reflectivities of the two mirrors (Fig. 1.3), respectively, and let L_i be the internal loss per pass in the laser cavity. If, at a given time, F is the photon flux in the cavity leaving mirror 1 and traveling toward mirror 2, then the photon flux F' leaving mirror 1 after one round trip is $F' = F \exp\{\sigma[N_2 - (g_2 N_1/g_1)]l\} \times (1 - L_i)R_2 \times \exp\{\sigma[N_2 - (g_2 N/g_1)]l\} \times (1 - L_i)R_1$. At threshold we must have $F' = F$ and therefore $R_1 R_2 (1 - L_i)^2 \exp\{2\sigma[N_2 - (g_2 N_1/g_1)]l\} = 1$. This equation shows that threshold is reached when the population inversion $N = N_2 - (g_2 N_1/g_1)$ reaches a critical value, called the *critical inversion*, given by:

$$N_c = - \frac{[\ln R_1 R_2 + 2 \ln(1 - L_i)]}{2\sigma l} \tag{1.2.3}$$

Equation (1.2.3) can be simplified if one defines

$$\gamma_1 = - \ln R_1 = - \ln(1 - T_1) \tag{1.2.4a}$$
$$\gamma_2 = - \ln R_2 = - \ln(1 - T_2) \tag{1.2.4b}$$
$$\gamma_i = - \ln(1 - L_i) \tag{1.2.4c}$$

where T_1 and T_2 are mirror transmissions (for simplicity mirror absorption is neglected). The substitution of Eq. (1.2.4) into Eq. (1.2.3) gives

$$N_c = \frac{\gamma}{\sigma l} \tag{1.2.5}$$

where:

$$\gamma = \gamma_i + \frac{(\gamma_1 + \gamma_2)}{2} \tag{1.2.6}$$

Note that the quantity γ_i, defined by Eq. (1.2.4c), can be called the logarithmic internal loss of the cavity. In fact when $L_i \ll 1$, as usually occurs, one has $\gamma_i \cong L_i$. Similarly, since both T_1 and T_2 represent a loss for the cavity, γ_1 and γ_2, defined by Eqs. (1.2.4a–b), can be called the logarithmic losses of the two cavity mirrors. Thus the quantity γ defined by Eq. (1.2.6) can be called the single-pass loss of the cavity.

Once the critical inversion is reached, oscillation builds up from spontaneous emission. Photons spontaneously emitted along the cavity axis in fact initiate the amplification process.

This is the basis of a laser oscillator, or laser, as it is more simply called. Note that, according to the meaning of the acronym laser, the term should be reserved for lasers emitting visible radiation. However, the same term is commonly applied to any device emitting stimulated radiation, whether in the far or near infrared, ultraviolet, or even in the x-ray region. To specify the kind of radiation emitted, one usually refers to infrared, visible, ultraviolet, or x-ray lasers, respectively.

1.3. PUMPING SCHEMES

We now consider how to produce a population inversion in a given material. At first it may seem possible to achieve this through the interaction of the material with a sufficiently strong em wave, perhaps coming from a sufficiently intense lamp, at the frequency $v = v_0$. Since at thermal equilibrium $(N_1/g_1) > (N_2/g_2)$, absorption in fact predominates over stimulated emission. The incoming wave then produces more transitions $1 \to 2$ than transitions $2 \to 1$, so one would hope in this way to end up with a population inversion. We see immediately however that such a system would not work (at least in the steady state). When in fact $g_2 N_2 = g_1 N_1$, absorption and stimulated emission processes compensate one another, and, according to Eq. (1.2.1), the material becomes transparent. This situation is often referred to as two-level *saturation*.

With just two levels, 1 and 2, it is therefore impossible to produce a population inversion. We then question whether this is possible using more than two levels of the infinite set of levels of a given atomic system. As we shall see the answer in this case is positive, so we accordingly speak of a *three-level* or *four-level laser*, depending on the number of levels used (Fig. 1.4). In a three-level laser (Fig. 1.4a), atoms are in some way raised from level 1 (ground) to level 3. If the material is such that, after an atom is raised to level 3, it decays rapidly to level 2 (perhaps by a rapid nonradiative decay), then a population inversion can be obtained between levels 2 and 1. In a four-level laser (Fig. 1.4b), atoms are again raised from the ground level (for convenience we now call this level 0) to level 3. If the atom then decays rapidly to level 2 (e.g., again by rapid nonradiative decay), a population inversion can again be obtained between levels 2 and 1. Once oscillation starts in such a four-level laser however, atoms are transferred to level 1 through stimulated emission. For continuous wave (cw) operation, it is therefore necessary for the transition $1 \to 0$ also to be very rapid (this again usually occurs by rapid nonradiative decay).

We have just seen how to use three or four levels of a given material to produce population inversion. Whether a system works in a three- or four-level scheme (or whether it

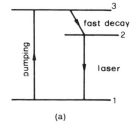 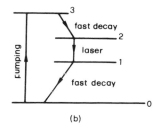

(a)   (b)

FIG. 1.4. (a) Three-level and (b) four-level laser schemes.

works at all) depends on whether preceding conditions are satisfied. We could of course ask why one should bother with a four-level scheme when a three-level scheme already seems to offer a suitable way of producing a population inversion. The answer is that one can, in general, produce a population inversion much more easily in a four-level than in a three-level laser. To see this, we begin by noting that the energy differences between the various levels shown in Fig. 1.4 are usually much greater than kT. According to Boltzmann statistics [see, e.g., Eq. (1.2.2)] we can then say that essentially all atoms are initially (i.e., at equilibrium) at the ground level. If we now let N_t represent the total atom density in the material, these atoms will initially all be in level 1, for the three-level case. Let us now begin raising atoms from level 1 to level 3. They then decay to level 2, and, if this decay is sufficiently rapid, level 3 remains more or less empty. Let us now assume for simplicity that the two levels are either nondegenerate (i.e., $g_1 = g_2 = 1$) or have the same degeneracy. Then, according to Eq. (1.2.1), absorption losses are compensated by the gain when $N_2 = N_1$. From this point on, any atom that is raised contributes to population inversion. In a four-level laser, however, since level 1 is also empty, any atom raised to level 2 immediately produces population inversion.

The preceding discussion shows that whenever possible we should seek a material that can be operated as a four-level rather than a three-level system. It is of course also possible to use more than four levels. It should also be noted that the term four-level laser is used for any laser whose lower laser level is essentially empty by virtue of being above the ground level by many kT. Then if levels 2 and 3 are the same level, we have a level scheme described as four-level in this sense, while having only three levels. Cases based on such a four-level scheme do exist.

Note that, more recently, the so-called *quasi-three-level lasers* have also become a very important laser category. In this case the ground level consists of many sublevels, the lower laser level being one of these sublevels. Therefore the scheme in Fig. 1.4b can still be applied to a quasi-three-level laser with the understanding that level 1 is a sublevel of the ground level and level 0 is the lowest sublevel of the ground level. If all ground-state sublevels are strongly coupled, perhaps by some rapid nonradiative decay process, then populations of these sublevels are always in thermal equilibrium. Let us further assume that the energy separation between levels 1 and 0 (see Fig. 1.4b) is comparable to kT. Then, according to Eq. (1.2.2), there is always some population present in the lower laser level and the laser system behaves in a way that is intermediate between a three- and a four-level laser.

The process by which atoms are raised from level 1 to level 3 (in a three-level scheme), from 0 to 3 (in a four-level scheme), or from the ground level to level 3 (in a quasi-three-level scheme) is known as *pumping*. There are several ways in which this process can be realized in practice, e.g., by some sort of lamp of sufficient intensity or by an electrical discharge in the active medium. We refer to Chap. 6 for a more detailed discussion of the various pumping processes. We note here, however, that, if the upper pump level is empty, the rate at which the upper laser level becomes populated by the pumping, $(dN_2/dt)_p$, can in general be written as $(dN_2/dt)_p = W_p N_g$ where W_p is a suitable rate describing the pumping process and N_g is the population of the ground level for either a three- or four-level laser while, for a quasi-three-level laser, it can be taken to be the total population of all ground state sublevels. In what follows, however, we will concentrate our discussion mostly on four level or quasi-three-level lasers. The most important case of three-level laser, in fact, is the Ruby laser, a historically important laser (it was the first laser ever made to operate) although no longer so widely used. For most four-level and quasi-three-level lasers in common use, the depletion of the ground level, due to the pumping process, can be neglected*. One can then write $N_g = \text{const}$ and the previous equation can be written, more simply, as

$$(dN_2/dt)_p = R_p \qquad (1.3.1)$$

where R_p may be called the pump rate per unit volume or, more briefly, the *pump rate*. To achieve the threshold condition, the pump rate must reach a threshold or critical value, R_{cp}. Specific expressions for R_{cp} will be obtained in Chaps. 6 and 7.

1.4. PROPERTIES OF LASER BEAMS

Laser radiation is characterized by an extremely high degree of monochromaticity, coherence, directionality, and brightness. We can add a fifth property, viz., short duration, which refers to the capability of producing very short light pulses, a less fundamental but nevertheless very important property. We now consider these properties in some detail.

1.4.1. Monochromaticity

This property is due to the following two circumstances: (1) Only an em wave of frequency v given by Eq. (1.1.1) can be amplified. (2) Since a two-mirror arrangement forms a resonant cavity, oscillation can occur only at the resonance frequencies of this cavity. The latter circumstance leads to an often much narrower laser linewidth (by as much as 10 orders of magnitude) than the usual linewidth of the transition $2 \rightarrow 1$, as observed in spontaneous emission.

1.4.2. Coherence

To first order, for any em wave, we can introduce two concepts of coherence, namely, *spatial* and *temporal coherence*. To define spatial coherence, let us consider two points P_1 and P_2 that, at time $t = 0$, lie on the same wave front of some given em wave and let $E_1(t)$ and $E_2(t)$ be the corresponding electric fields at these two points. By definition the difference between phases of the two fields at time $t = 0$ is zero. If this difference remains zero at any time $t > 0$, we say that there is a perfect coherence between the two points. If such coherence occurs for any two points of the em wave front, we then say that the wave has *perfect spatial coherence*. In practice, for any point P_1, point P_2 must lie within some finite area around P_1 to have a good phase correlation. In this case we say that the wave has *partial spatial coherence*, and, for any point P, we can introduce a suitably defined coherence area $S_c(P)$.

To define temporal coherence, we now consider the electric field of the em wave, at a given point P, at times t and $t + \tau$. If, for a given time delay τ, the phase difference between the two field remains the same for any time t, we say that there is a temporal coherence over

---

* Note: As a quasi-three-level laser progressively approaches a pure three-level laser, the assumption that the ground-state population is changed negligibly by the pumping process will eventually not be justified. Also note that in fiber lasers, where very intense pumping is readily achieved, the ground state can be almost completely emptied.

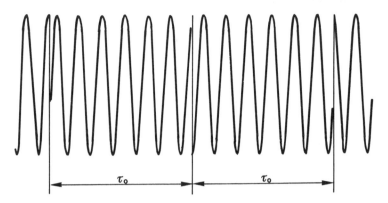

FIG. 1.5. Example of an em wave with a coherence time of approximately τ_0.

a time τ. If this occurs for any value of τ, the em wave is said to have perfect temporal coherence. If this occurs for a time delay τ such that $0 < \tau < \tau_0$, the wave is said to have partial temporal coherence, with a coherence time equal to τ_0. An example of an em wave with a coherence time equal to τ_0 is shown in Fig. 1.5. The figure shows a sinusoidal electric field undergoing phase jumps at time intervals equal to τ_0. We see that the concept of temporal coherence is, at least in this case, directly connected with that of monochromaticity. In fact, we will show in Chap. 11 that any stationary em wave with coherence time τ_0 has a bandwidth $\Delta v \cong 1/\tau_0$. In the same chapter we also show that, for a nonstationary but repetitively reproducing beam (e.g., a repetitively Q-switched or a mode-locked laser beam), coherence time is not determined by the inverse of the oscillation bandwidth Δv and may actually be much greater than $1/\Delta v$.

It is important to point out that the two concepts of temporal and spatial coherence are indeed independent of each other. In fact examples can be given of a wave with perfect spatial coherence but only limited temporal coherence (or vice versa). If the wave in Fig. 1.5 represents electric fields at points P_1 and P_2 considered earlier, spatial coherence between these two points would still be complete, although the wave has limited temporal coherence.

We conclude this section by emphasizing that the concepts of spatial and temporal coherence provide only a first-order description of the laser's coherence. Higher order coherence properties will, in fact, be discussed in Chap. 11. Such a discussion is essential to appreciate fully the difference between an ordinary light source and a laser. We will show in fact that, by virtue of differences between the corresponding higher order coherence properties, a laser beam is fundamentally different from an ordinary light source.

1.4.3. Directionality

This property is a direct consequence of the fact that the active medium is placed in a resonant cavity. For example, in the case of the plane parallel cavity shown in Fig. 1.3, only a wave propagating in a direction orthogonal to the mirrors (or in a direction very near to it) can be sustained in the cavity. To gain a deeper understanding of the directional properties of a laser beam (or in general of any em wave), it is convenient to consider separately the case of a beam with perfect spatial coherence and the case of partial spatial coherence.

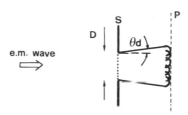

FIG. 1.6. Divergence of a plane em wave due to diffraction.

We first consider the case of perfect spatial coherence. Even for this case a beam of finite aperture has unavoidable divergence due to diffraction. This can be understood with the help of Fig. 1.6, where a beam of uniform intensity and plane wave front is assumed to be incident on a screen S containing an aperture D. According to Huyghens's principle the wave front at some plane P behind the screen can be obtained by the superposition of the elementary waves emitted by each point of the aperture. We thus see that, on account of the finite size D of the aperture, the beam has a finite divergence θ_d. Its value can be obtained from diffraction theory. For an arbitrary amplitude distribution, we obtain

$$\theta_d = \frac{\beta\lambda}{D} \qquad (1.4.1)$$

where λ and D are the wavelength and the diameter of the beam, respectively. The factor β is a numerical coefficient of the order of unity whose value depends on the shape of the amplitude distribution and how both the divergence and the beam diameter are defined. A beam whose divergence can be expressed as in Eq. (1.4.1) is referred to as being *diffraction-limited*.

If the wave has only partial spatial coherence, its divergence is greater than the minimum value set by diffraction. Indeed, for any point P' of the wave front, the Huygens argument in Fig. 1.6 can be applied only for points lying within the coherence area S_c around point P'. The coherence area thus acts as a limiting aperture for the coherent superposition of elementary wavelets. Thus, the beam divergence can now be written as:

$$\theta = \frac{\beta\lambda}{(S_c)^{1/2}} \qquad (1.4.2)$$

where β is a numerical coefficient of the order of unity whose exact value depends on how both the divergence θ and coherence area S_c are defined.

We conclude this general discussion of the directional properties of em waves by pointing out that given suitable operating conditions, the output beam of a laser can be made diffraction limited.

1.4.4. Brightness

We define the brightness of a given source of em waves as the power emitted per unit surface area per unit solid angle. To be more precise let dS be the elemental surface area at point O of the source (Fig. 1.7a). The power dP emitted by dS into a solid angle $d\Omega$ around direction OO' can be written as:

$$dP = B \cos\theta \, dS \, d\Omega \tag{1.4.3}$$

where θ is the angle between OO' and the normal **n** to the surface. Note that the factor $\cos\theta$ occurs because the physically important quantity for emission along the OO' direction is the projection of dS on a plane orthogonal to the OO' direction, i.e., $\cos\theta \, dS$. The quantity B defined through Eq. (1.4.3) is called the *source brightness* at point O in the direction OO'. This quantity generally depends on polar coordinates θ and ϕ of the direction OO' and on point O. When B is a constant, the source is said to be isotropic (or a Lambertian source).

Let us now consider a laser beam of power P, with a circular cross section of diameter D and with a divergence θ (Fig. 1.7b). Since θ is usually very small, we have $\cos\theta \cong 1$. Since the area of the beam is equal to $\pi D^2/4$ and the emission solid angle is $\pi\theta^2$, then, according to Eq. (1.4.3), we obtain the beam brightness as:

$$B = \frac{4P}{(\pi D\theta)^2} \tag{1.4.4}$$

Note that, if the beam is diffraction limited, we have $\theta = \theta_d$, and, with the help of Eq. (1.4.1), we obtain from Eq. (1.4.4):

$$B = \left(\frac{2}{\beta\pi\lambda}\right)^2 P \tag{1.4.5}$$

which is the maximum brightness for a beam of power P.

Brightness is the most important parameter of a laser beam and, in general, of any light source. To illustrate this point we first recall that, if we form an image of any light source through a given optical system and if we assume that the object and image are in the same medium (e.g., air), then the following property holds: The brightness of the image is always less than or equal to that of the source, the equality holding when the optical system provides lossless imaging of the light emitted by the source. To illustrate further the importance of

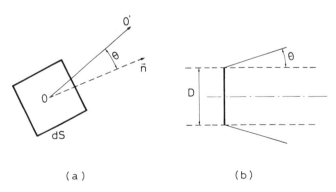

(a)   (b)

FIG. 1.7. (a) Surface brightness at the point O for a general source of em waves. (b) Brightness of a laser beam of diameter D and divergence θ.

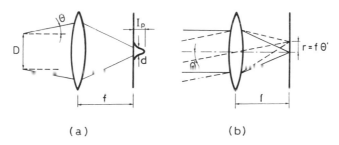

FIG. 1.8. (a) Intensity distribution in the focal plane of a lens for a beam of divergence θ. (b) Plane wave decomposition of the beam in (a).

brightness, let us consider the beam in Fig. 1.7b, with divergence equal to θ, to be focused by a lens of focal length f. We are interested in calculating the peak intensity of the beam in the focal plane of the lens (Fig. 1.8a). To make this calculation we recall that the beam can be decomposed into a continuous set of plane waves with an angular spread of approximately θ around the propagation direction. Two such waves, making an angle θ', are indicated by solid and dashed lines, respectively, in Fig. 1.8b. The two beams are each focused on a distinct spot in the focal plane, and, for a small angle θ', the two spots are transversly separated by a distance $r = f\theta'$. Since the angular spread of the plane waves that make up the beam in Fig. 1.8a equals the beam divergence θ, we conclude that the diameter d of the focal spot in Fig. 1.8a is approximately equal to $d = 2f\theta$. For an ideal lossless lens, the overall power in the focal plane equals the power P of the incoming wave. The peak intensity in the focal plane is thus $I_p = 4P/\pi d^2 = P/\pi(f\theta)^2$. In terms of beam brightness, according to Eq. (1.4.4), we then have $I_p = (\pi/4)B(D/f)^2$. Thus I_p increases with increasing beam diameter D. The maximum value of I_p is then attained when D is made equal to the lens diameter D_L. In this case we obtain

$$I_p = \left(\frac{\pi}{4}\right)(N.A.)^2 B \tag{1.4.6}$$

where $N.A. = \sin[\tan^{-1}(D_L/f)] \cong (D_L/f)$ is the lens numerical aperture. Equation (1.4.6) then shows that, for a given numerical aperture, the peak intensity in the focal plane of a lens depends only on beam brightness.

A laser beam of even moderate power (e.g., a few milliwatts) has a brightness several orders of magnitude greater than that of the brightest conventional sources (see, e.g., Problem 1.7). This is mainly due to the highly directional properties of the laser beam. According to Eq. (1.4.6) this means that the peak intensity produced in the focal plane of a lens can be several orders of magnitude greater for a laser beam compared to that of a conventional source. Thus the intensity of a focused laser beam can reach very large values, a feature exploited in many applications of lasers.

1.4.5. Short Pulse Duration

Without going into detail at this stage, we mention that, by means of a special technique called *mode locking*, it is possible to produce light pulses whose duration is roughly equal to the inverse of the linewidth of the laser transition 2→1. Thus, with gas lasers, whose linewidth is relatively narrow, the pulse width may be ~ 0.1–1 ns. Such pulse durations are not regarded as particularly short, and indeed even some flash lamps can emit light pulses with a duration of somewhat less than 1 ns. On the other hand, the linewidth of some solid-state and liquid lasers can be 10^3–10^5 times greater than that of a gas laser; in this case much shorter pulses may be generated (down to ~ 10 fs). This creates exciting new possibilities for laser research and applications.

Note that the property of short duration, which implies energy concentration in time, can in a sense be considered the counterpart of monochromaticity, which implies energy concentration in wavelength. However, short duration can perhaps be considered a less fundamental property than monochromaticity. In fact, while all lasers can in principle be made extremely monochromatic, only lasers with a broad linewidth, i.e., solid-state and liquid lasers, may produce pulses of very short duration.

1.5. LASER TYPES

The various laser types developed so far display a wide range of physical and operating parameters. Indeed, if lasers are characterized according to the physical state of the active material, we call them *solid-state*, *liquid*, or *gas lasers*. A rather special case is where the active material consists of free electrons at relativistic velocities passing through a spatially periodic magnetic field (*free-electron lasers*). If lasers are characterized by the wavelength of emitted radiation, one refers to *infrared lasers*, *visible lasers*, *ultraviolet* (uv) and *x-ray lasers*. The corresponding wavelength can range from ≈ 1 mm (i.e., millimeter waves) to ≈ 1 nm (i.e., to the upper limit of hard x-rays). Wavelength span can thus be a factor of $\approx 10^6$ (recall that the visible range spans less than a factor 2, roughly from 700–400 nm). Output powers cover an even greater range of values. For cw lasers, typical powers range from a few mW, in lasers used for signal sources (e.g., for optical communications or bar code scanners), to tens of kW, in lasers used for material working, and to a few MW (≈ 5 MW so far), in lasers required in some military applications (e.g., directed energy weapons). In pulsed lasers peak power can be much greater than in cw lasers, and it can reach values as high as 1 PW (10^{15} W)! Again for pulsed lasers, the pulse duration can vary widely from the ms level typical of lasers operating in the so-called *free-running* regime (i.e., without any *Q-switching* or *mode-locking* element in the cavity) to about 10 fs (1 fs $= 10^{-15}$ s) for some mode-locked lasers. Physical dimensions can also vary widely. In terms of cavity length for instance, the length can be as small as ~ 1 μm for the shortest lasers to some km value for the longest (e.g., a 6.5 km long laser, which was set up in a cave for geodetic studies).

This wide range of physical or operating parameters represents both a strength and a weakness. As far as applications are concerned, this wide range of parameters offers enormous potential in several fields of fundamental and applied sciences. On the other hand, in terms of markets, a large variation in terms of devices and systems can be an obstacle to mass production and its associated price reduction.

PROBLEMS

1.1. The part of the em spectrum of interest in the laser field starts from the submillimeter wave region and decreases in wavelength to the x-ray region. This covers the following regions in succession:

far infrared, near infrared, visible, uv, vacuum ultraviolet (vuv), soft x-ray, x-ray: From standard textbooks find the wavelength intervals of these regions. Memorize or record these intervals, since they are frequently used in this book.

1.2. As a particular case of Problem 1.1, memorize or record wavelengths corresponding to blue, green, and red light.

1.3. If levels 1 and 2 in Fig. 1.1 are separated by an energy $E_2 - E_1$ such that the corresponding transition frequency falls in the middle of the visible range, calculate the ratio of the populations of the two levels in thermal equilibrium at room temperature.

1.4. In thermal equilibrium at $T = 300$ K, the ratio of level populations N_2/N_1 for some particular pair of levels is given by $1/e$. Calculate the frequency v for this transition. In what region of the em spectrum does this frequency fall?

1.5. A laser cavity consists of two mirrors with reflectivities $R_1 = 1$ and $R_2 = 0.5$, while the internal loss per pass is $L_i = 1\%$. Calculate total logarithmic losses per pass. If the length of the active material is $l = 7.5$ cm and the transition cross section is $\sigma = 2.8 \times 10^{-19}$ cm^2, calculate the threshold inversion.

1.6. The beam from a ruby laser ($\lambda \cong 694$ nm) is sent to the moon after passing through a telescope of 1-m diameter. Calculate the approximate value of beam diameter on the moon assuming that the beam has perfect spatial coherence. (The distance between earth and moon is approximately 384,000 km.)

1.7. The brightness of probably the brightest lamp so far available (PEK Labs type 107/109™, excited by 100 W of electrical power) is about 95 W/cm^{2}sr in its most intense green line ($\lambda = 546$ nm). Compare this brightness with that of a 1-W argon laser ($\lambda = 514.5$ nm), which can be assumed to be diffraction-limited.

2

Interaction of Radiation with Atoms and Ions

2.1. INTRODUCTION

This chapter discusses the interaction of radiation with atoms and ions that weakly interact with surrounding species, such as atoms or ions in a gas phase or impurity ions in an ionic crystal. The somewhat more complicated case of radiation interacting with molecules or semiconductors is considered in Chap. 3. Since the topic of radiation interacting with matter is very wide, we limit our discussion to phenomena relevant to atoms and ions acting as active media. After an introduction to the theory of blackbody radiation, a milestone for the whole of modern physics, we consider the elementary processes of absorption, stimulated emission, spontaneous emission, and nonradiative decay. These are first considered under the simplifying assumptions of a dilute medium and low intensity. Situations involving high-beam intensity and a nondilute medium (leading to the phenomena of saturation and amplified spontaneous emission) are considered. A number of very important, although perhaps less general topics related to the photophysics of dye lasers, free-electron lasers, and x-ray lasers are briefly considered in Chaps. 9 and 10.

2.2. SUMMARY OF BLACKBODY RADIATION THEORY

Let us consider a cavity containing a homogeneous and isotropic medium. If the cavity walls are kept at a constant temperature T they will continuously emit and receive power in the form of em radiation. When absorption and emission rates become equal, an equilibrium condition is established at the walls of the cavity as well as at each point of the dielectric.[1] This situation can be described by introducing the energy density ρ, which represents the em energy contained in unit volume of the cavity. This energy density can be expressed as a function of the electric field $E(t)$ and magnetic field $H(t)$ according to the equation:

$$\rho = \langle \tfrac{1}{2}\varepsilon E^2 \rangle + \langle \tfrac{1}{2}\mu H^2 \rangle \tag{2.2.1}$$

where ε and μ are the dielectric constant and the magnetic permeability of the medium inside the cavity, respectively, and where the symbol $\langle\ \rangle$ indicates a time average over a cycle of the radiation field. We can then represent the spectral energy distribution of this radiation by the function ρ_v, which is a function of frequency v. This is defined as follows: $\rho_v dv$ represents the energy density of radiation in the frequency range between v and $v + dv$. The relationship between ρ and ρ_v is obviously

$$\rho = \int_0^\infty \rho_v \, dv \tag{2.2.2}$$

Suppose now that a hole is made in the cavity wall. If we let I_v be the spectral intensity of light escaping from the hole, one can show that I_v is proportional to ρ_v, obeying the simple relation:

$$I_v = \left(\frac{c}{4n}\right)\rho_v \tag{2.2.3}$$

where c is the velocity of light in the vacuum and n is the refractive index of the medium inside the cavity. We can now show that I_v, and hence ρ_v, are universal functions, independent of either the nature of the walls or the cavity shape, and depend only on the frequency v and temperature T of the cavity. This property of ρ_v is proven by the following simple thermodynamic argument. Let us suppose we have two cavities of arbitrary shape whose walls are at the same temperature T. To ensure that the temperature remains constant, we imagine that the walls of the two cavities are in thermal contact with two thermostats at temperature T. Let us suppose that at a given frequency v the energy density ρ'_v in the first cavity is greater than the corresponding value ρ''_v in the second cavity. We now optically connect the two cavities by making a hole in each, then imaging with some optical system each hole onto the other. We also insert an ideal filter in the optical system that lets through only a small frequency range around the frequency v. If $\rho'_v > \rho''_v$, then according to Eq. (2.2.3), $I'_v > I''_v$, and there is a net flow of em energy from cavity 1 to cavity 2. Such an energy flow however violates the second law of thermodynamics, since the two cavities are at the same temperature. Therefore one must have $\rho'_v = \rho''_v$ for all frequencies.

The problem of calculating this universal function $\rho_v(v, T)$ proved very challenging for physicists of the time. Its complete solution was provided by Planck, who, to find a correct solution, introduced the so-called hypothesis of *light quanta*. The blackbody theory is therefore one of the fundamental bases of modern physics.[1] Before going further into it, we first need to consider the em modes of a blackbody cavity. Since the function ρ_v is independent of the cavity shape or the nature of the dielectric medium, we choose to consider the relatively simple case of a rectangular cavity uniformly filled with dielectric and with perfectly conducting walls.

2.2.1. Modes of a Rectangular Cavity

Let us consider the rectangular cavity in Fig. 2.1. To calculate ρ_v we begin by calculating the standing em field distributions that can exist in this cavity. According to Maxwell's equations, the electric field $\mathbf{E}(x, y, z, t)$ must satisfy the wave equation:

$$\nabla^2 \mathbf{E} - \frac{1}{c_n^2} \frac{\partial^2 \mathbf{E}}{\partial t^2} = 0 \tag{2.2.4}$$

where ∇^2 is the Laplacian operator and c_n is the velocity of light in the medium considered. In addition the field must satisfy the following boundary condition at each wall:

$$\mathbf{E} \times \mathbf{n} = 0 \tag{2.2.5}$$

where $\mathbf{n}$ is the normal to the particular wall under consideration. This condition expresses the fact that, for perfectly conducting walls, the tangential component of the electric field must vanish on the walls of the cavity.

It can be easily shown that the problem is soluble by separating the variables. Thus, if we set

$$\mathbf{E} = \mathbf{u}(x, y, z)A(t) \tag{2.2.6}$$

and then substitute Eq. (2.2.6) into Eq. (2.2.4), we have

$$\nabla^2 \mathbf{u} = -k^2 \mathbf{u} \tag{2.2.7a}$$

$$\frac{d^2 A}{dt^2} = -(c_n k)^2 A \tag{2.2.7b}$$

where k is a constant. Equation (2.2.7b) has the general solution:

$$A = A_0 \cos(\omega t + \phi) \tag{2.2.8}$$

where A_0 and ϕ are arbitrary constants and where:

$$\omega = c_n k \tag{2.2.9}$$

With $A(t)$ given by Eq. (2.2.8), we see that Eq. (2.2.6) can be written as:

$$\mathbf{E}(x, y, z, t) = E_0 \mathbf{u}(x, y, z) \exp j(\omega t + \phi) \tag{2.2.9a}$$

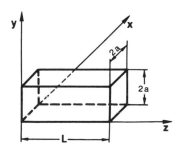

FIG. 2.1. Rectangular cavity with perfectly conducting walls kept at temperature T.

and thus corresponds to a standing wave configuration of the em field within the cavity. In fact the amplitude of oscillation at a given point in the cavity is constant in time. A solution of this type is referred to as an em *mode* of the cavity.

We are now left with the task of solving Eq. (2.2.7a), known as the Helmholtz equation, subject to the boundary condition given by Eq. (2.2.5). It can readily be verified that expressions:

$$u_x = e_x \cos k_x x \sin k_y y \sin k_z z$$
$$u_y = e_y \sin k_x x \cos k_y y \sin k_z z \qquad\qquad (2.2.10)$$
$$u_z = e_z \sin k_x x \sin k_y y \cos k_z z$$

satisfy Eq. (2.2.7a) for any value of e_x, e_y, e_z, provided that:

$$k_x^2 + k_y^2 + k_z^2 = k^2 \qquad\qquad (2.2.11)$$

Furthermore Eq. (2.2.10) already satisfies the boundary condition (2.2.5) on the three planes $x=0$, $y=0$, $z=0$. If we now impose the condition that Eq. (2.2.5) must also be satisfied on the other walls of the cavity, we have

$$k_x = \frac{l\pi}{2a}$$
$$k_y = \frac{m\pi}{2a} \qquad\qquad (2.2.12)$$
$$k_z = \frac{n\pi}{L}$$

where l, m, and n are positive integers. Their physical significance can be seen immediately: They represent the number of nodes that the standing wave mode has along the directions x, y, and z, respectively. For fixed values of l, m, and n, it follows that k_x, k_y, and k_z are also fixed, and, according to Eqs. (2.2.9) and (2.2.11), the angular frequency ω of the mode is also fixed and given by:

$$\omega_{lmn} = c_n \left[\left(\frac{l\pi}{2a}\right)^2 + \left(\frac{m\pi}{2a}\right)^2 + \left(\frac{n\pi}{L}\right)^2 \right]^{1/2} \qquad\qquad (2.2.13)$$

where we have explicitly indicated that the frequency of the mode depends on the indices l, m, and n. The mode is still not completely determined, since e_x, e_y, and e_z are still arbitrary. However Maxwell's equations provide another condition that must be satisfied by the electric field, i.e., $\nabla \cdot \mathbf{u} = 0$, from which with the help of Eq. (2.2.10), we obtain

$$\mathbf{e} \cdot \mathbf{k} = 0 \qquad\qquad (2.2.14)$$

In Eq. (2.2.14) we have introduced the two vectors $\mathbf{e}$ and $\mathbf{k}$ whose components along x-, y-, and z-axes are e_x, e_y, and e_z and k_x, k_y, and k_z, respectively. Equation (2.2.14) therefore shows that, of the three quantities e_x, e_y, and e_z, only two are independent. In fact, once we fix l, m, and n (i.e., once $\mathbf{k}$ is fixed), the vector $\mathbf{e}$ is bound to lie in a plane perpendicular to $\mathbf{k}$. In this plane only two degrees of freedom are left for the choice of vectors **e**, so only two independent modes are present. Any other vector **e** lying in this plane can in fact be obtained as a linear combination of the two previous vectors.

Let us now calculate the number of resonant modes $N(\nu)$ whose frequency lies between 0 and ν. This is the same as the number of modes whose wave vector **k** has a magnitude k between 0 and $2\pi\nu/c$. From Eq. (2.2.12) we see that, in a system coordinate k_x, k_y, k_z, the possible values for **k** are given by vectors connecting the origin with nodal points of the three-dimensional lattice shown in Fig. 2.2. However, since k_x, k_y, and k_z are positive quantities, we must count only those points lying in the positive octant. There is now a one-to-one correspondence between these points and the unit cell of dimensions ($\pi/2a$, $\pi/2a$, π/L). The number of points with k between 0 and ($2\pi\nu/c_n$) can thus be calculated as (1/8) times the volume of the sphere, centered at the origin and of radius ($2\pi\nu/c_n$), divided by the volume of the unit cell of dimensions ($\pi/2a$, $\pi/2a$, π/L). Since, as previously noted, there are two possible modes for each value of k, we have

$$N(\nu) = 2\frac{(1/8)(4/3)\pi(2\pi\nu/c_n)^3}{(\pi/2a)(\pi/2a)(\pi/L)} = \frac{8\pi\nu^3}{3c_n^3}V \qquad (2.2.15)$$

where V is the total volume of the cavity. If we now define p_ν as the number of modes per unit volume and per unit frequency range, we obtain

$$p_\nu = \frac{1}{V}\frac{dN}{d\nu} = \frac{8\pi\nu^2}{c_n^3} \qquad (2.2.16)$$

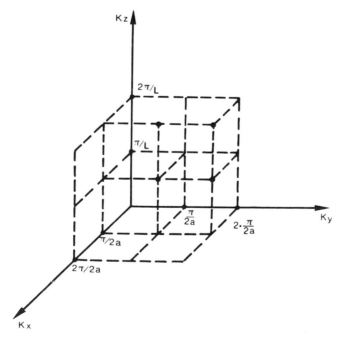

FIG. 2.2. Pictorial illustration of the density of modes in the cavity in Fig. 2.1. Each point of the lattice corresponds to two cavity modes.

2.2.2. Rayleigh–Jeans and Planck Radiation Formula

Having calculated the quantity p_v, we can now proceed to calculate the energy density ρ_v. We begin by writing ρ_v as the product of the number of modes per unit volume per unit frequency range p_v multiplied by the average energy $\langle E \rangle$ contained in each mode, i.e.:

$$\rho_v = p_v \langle E \rangle \tag{2.2.17}$$

To calculate $\langle E \rangle$ we assume that the cavity walls are kept at a constant temperature T. According to Boltzmann's statistics, the probability dp that the energy of a given cavity mode lies between E and $E + dE$ is expressed by $dp = C \exp[-(E/kT)]dE$, where C is a constant to be established by the condition $\int_0^\infty C \exp[-(E/kT)]dE = 1$. The average energy of the mode $\langle E \rangle$ is therefore given by:

$$\langle E \rangle = \frac{\int_0^\infty E \exp[-(E/kT)]dE}{\int_0^\infty \exp[-(E/kT)]dE} = kT \tag{2.2.18}$$

From Eqs. (2.2.16)–(2.2.18) we obtain

$$\rho_v = \left(\frac{8\pi v^2}{c_n^3}\right)kT \tag{2.2.19}$$

This is the well-known Rayleigh–Jeans radiation formula. However it is in complete disagreement with experimental results. Indeed it is immediately obvious that Eq. (2.2.19) must be incorrect, since it implies an infinite total energy density ρ [see Eq. (2.2.2)]. Equation (2.2.19) however represents the inevitable conclusion of the previous classical arguments.

The problem remained unsolved until Planck introduced the hypothesis of light quanta, at the beginning of this century. Planck's fundamental hypothesis was that energy in a given mode could not have an arbitrary value from 0 to ∞, as implicitly assumed in Eq. (2.2.18), but that permitted values of this energy were integral multiples of a fundamental quantity, proportional to the frequency of the mode. In other words, Planck assumed that the energy of the mode could be written as:

$$E = nhv \tag{2.2.20}$$

where n is a positive integer and h a constant (later called Planck's constant). Without entering into too many details about this fundamental hypothesis, we note that essentially this implies that energy exchange between the inside of the cavity and its walls must involve a discrete amount of energy hv. This minimum quantity that can be exchanged is called a *light quantum* or *photon*. According to this hypothesis, the average energy of the mode is now given by:

$$\langle E \rangle = \frac{\sum_n^\infty nhv \exp[-(nhv/kT)]}{\sum_n^\infty \exp[-(nhv/kT)]} = \frac{hv}{\exp(hv/kT) - 1} \tag{2.2.21}$$

This equation is quite different from the classical expression (2.2.18). Obviously, for $hv \to 0$, Eq. (2.2.21) reduces to Eq. (2.2.18). From Eqs. (2.2.16), (2.2.17), and (2.2.21) we obtain the Planck formula:

$$\rho_v = \frac{8\pi v^2}{c_n^3} \frac{hv}{\exp(hv/kT) - 1} \tag{2.2.22}$$

which now agrees perfectly with experimental results if, for h, we choose the value $h = 6.62 \times 10^{-34}$ J × s. For example, Fig. 2.3 shows the behavior predicted by Eq. (2.2.22) for ρ_v versus frequency v for two values of temperature T.

Lastly we note that the ratio:

$$\langle \phi \rangle = \frac{\langle E \rangle}{hv} = \frac{1}{\exp(hv/kT) - 1} \tag{2.2.23}$$

gives the average number of photons $\langle \phi \rangle$ for each mode. If we now consider a frequency v in the optical range ($v \approx 4 \times 10^{14}$ Hz), we obtain $hv \approx 1$ eV. For $T \cong 300$ K we have $kT \cong (1/40)$ eV so that, from Eq. (2.2.23), we obtain $\langle \phi \rangle \cong \exp(-40)$. Thus we see that the average number of photons per mode, for blackbody radiation at room temperature, is much smaller than unity. This value should be compared with the number of photons ϕ_0 that can be obtained in a laser cavity for a single mode laser (see Chap. 7).

2.2.3. Planck's Hypothesis and Field Quantization

The fundamental assumption of Planck given by Eq. (2.2.20) was considered with a degree of caution if not suspicion for some time after it was proposed. Some even considered it as a mere mathematical trick to transform an integral [Eq. (2.2.18)] into a summation [Eq. (2.2.21)] to obtain, by luck, a result in agreement with experiments. However the theory of the photoelectric effect due to Einstein (1904), which was essentially based on Planck's hypothesis, soon provided further evidence that Planck's fundamental assumption was

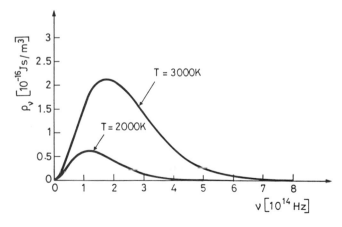

FIG. 2.3. Plot of the function $\rho_v(v, T)$ as a function of frequency v at two values of temperature T.

indeed correct. It was many years later, however, before this assumption received its complete conceptual justification by the quantum field theory of Dirac (1927). Although a detailed description of field quantization is beyond the scope of this book, it is worth devoting a little space to indicate how field quantization arises.[2] This will also help provide a deeper understanding of some topics to be considered later on in the book.

Consider an em mode of the cavity, i.e., characterized by a given standing wave pattern, and let v be its resonance frequency. If $E_x(\mathbf{r}, t)$ and $H_y(\mathbf{r}, t)$ are transverse components of its electric and magnetic fields, respectively, the corresponding energy density ρ is given by Eq. (2.2.1), and its energy equals

$$E = \int \rho \, dV \qquad (2.2.24)$$

where V is the volume of the cavity. To understand the basis of quantum field theory, we must recognize that, likewise the case of a particle, the pair of quantities $E_x(\mathbf{r}, t)$ and $H_y(\mathbf{r}, t)$ cannot be measured simultaneously with arbitrary precision.[2] This means that there is a form of the Heisenberg uncertainty relation between $E_x(\mathbf{r}, t)$ and $H_y(\mathbf{r}, t)$ analogous to that which exists between the position p_x and momentum q_x of a particle moving, e.g., in the x-direction. Note that the Heisenberg uncertainty relation between p_x and q_x can provide the starting point for the quantum theory of a particle. In fact, it indicates that the equations of classical mechanics, which are essentially based on canonical variables p_x and q_x, are no longer valid. Likewise, the uncertainty relations between $E_x(\mathbf{r}, t)$ and $H_y(\mathbf{r}, t)$ can provide the starting point of the quantum theory of radiation in the sense that these show that Maxwell's equations, e.g., Eq. (2.2.4), are no longer valid.

The analogy between the quantum theory of a particle and the quantum theory of radiation can be taken further by considering a particle bound to a given point by an elastic force. This is the case of the harmonic oscillator, one of the fundamental examples of the quantum theory of a bound particle. A harmonic oscillator oscillating, e.g., along the x-direction, is a mechanical oscillator whose total energy is given by:

$$E = \left(\frac{kp_x^2}{2}\right) + \left(\frac{q_x^2}{2m}\right) \qquad (2.2.25)$$

where k is the elastic constant and m is the mass of the particle. In fact this oscillator provides several analogies with a cavity mode. Both of these are, in fact, oscillators in the sense that they are characterized by a resonance frequency. In a mechanical oscillator, oscillation occurs because potential energy, represented by the term $kp_x^2/2$, is periodically transformed into kinetic energy, represented by the term $q_x^2/2m$. In an em oscillator represented by the cavity mode, electric energy, represented by the term $\int (\varepsilon\langle E_x^2\rangle/2)dV$, is periodically transformed into magnetic energy, represented by the term $\int (\mu\langle H_y^2\rangle/2)dV$. Based on this close analogy, one can then look for similar quantization rules. The appropriate quantization procedure leads to the fundamental result that the energy of the given cavity mode is quantized in exactly the same way as the harmonic oscillator. Namely, the eigenvalues for the mode energy are given by:

$$E = \left(\frac{1}{2}\right)hv + nhv \qquad (2.2.26)$$

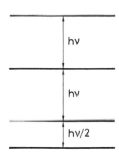

FIG. 2.4. Energy levels of a cavity mode.

where n is an integer value. The first term, zero-point energy, has an origin similar to that of the harmonic oscillator. In the latter case in fact it arises because the energy cannot be zero, since according to Eq. (2.2.25), this requires both p_x and q_x to be zero, which is contrary to the Heisenberg uncertainty principle. Likewise, for the cavity mode, energy cannot be zero because, according to Eq. (2.2.1), this requires both E_x and H_y to be zero, which by the same argument, is again impossible. Thus field quantization predicts that energy levels of a given cavity mode of frequency v are given by Eq. (2.2.26), a conclusion that coincides with Planck's assumption [Eq. (2.2.20)] apart from the zero-point energy term. Thus results of field quantization provide a framework that give Planck's assumption a more fundamental justification. Needless to say, Maxwell's equations (see Sect. 2.2.1) do not impose any condition on the total energy density of a cavity mode. Thus according to these equations, the mode energy can take any value between 0 and ∞, *continuously*.

As a closing comment to this section, we note that, according to Eq. (2.2.26), the energy levels of a cavity mode, like those of the harmonic oscillator, can be displayed as in Fig. 2.4. In the lowest, zero-point, energy level, both $\langle E_x^2 \rangle$ and $\langle H_y^2 \rangle$ are different from zero and referred to as the zero-point fluctuations of the electric and magnetic fields, respectively. Note also that the zero-point energy value of $(hv/2)$ really has no physical significance. If instead of Eq. (2.2.24) we had defined the energy of the mode as:

$$E = \left(\int \rho\, dV \right) - \left(\frac{hv}{2} \right) \qquad (2.2.27)$$

then we would have had a zero value for the lowest energy state. However this state would still include the zero-point field fluctuations of both $\langle E_x^2 \rangle$ and $\langle H_y^2 \rangle$, at the same level as before. Thus these fluctuations are the quantities that actually characterize the zero-point energy state.

2.3. SPONTANEOUS EMISSION

As a first attempt at describing spontaneous emission, we follow a *semiclassical approach* where atoms are treated as quantized (i.e., treated according to quantum mechanics) while fields are treated classically (i.e., described by Maxwell's equations). As we shall see, this attempt fails to describe the phenomenon of spontaneous emission correctly (i.e., in agreement with experimentation); the approach turns out to be very instructive however. Results obtained are then compared to the correct ones, i.e., those predicted by a full *quantum theory*, where both atoms and fields are quantized—the former by quantum mechanics and the latter by quantum field theory. Thus to describe correctly the phenomenon of spontaneous emission, a common phenomenon of everyday experience (the light from the sun or from ordinary lamps arises from spontaneous emission), we must introduce sophisticated concepts of quantum theory.

2.3.1. Semiclassical Approach

Let us assume that a given atom, initially raised to its excited level 2 of energy E_2, is decaying by spontaneous emission to level 1 of energy E_1 (Fig. 1.1a). Assuming that the two levels are nondegenerate, we let

$$\psi_1(\mathbf{r}, t) = u_1(\mathbf{r}) \exp[-j(E_1/\hbar)t] \qquad (2.3.1a)$$

and

$$\psi_2(\mathbf{r}, t) = u_2(\mathbf{r}) \exp[-j(E_2/\hbar)t] \qquad (2.3.1b)$$

be the corresponding wave functions, where $u_{1,2}(\mathbf{r})$ are the eigenfunctions of the two stationary states, $\mathbf{r}$ denotes the coordinate of the electron undergoing the transition, the origin being taken at the nucleus, and $\hbar = h/2\pi$. While the atom undergoes the transition $2 \rightarrow 1$ due to spontaneous emission, its wave function can be expressed as a linear combination of the wave functions of the two states, i.e.:

$$\psi = a_1(t)\psi_1 + a_2(t)\psi_2 \qquad (2.3.2)$$

where a_1 and a_2 are time-dependent complex numbers. Note that, according to quantum mechanics, we have

$$|a_1|^2 + |a_2|^2 = 1 \qquad (2.3.3)$$

and thus $|a_1|^2$ and $|a_2|^2$ represent the probabilities that, at time t, the atom is found in state 1 or 2, respectively.

To understand how spontaneous emission arises, we calculate the electric dipole moment of the atom, $\boldsymbol{\mu}$. According to quantum mechanics we have

$$\boldsymbol{\mu} = -\int e|\psi|^2 \mathbf{r} \, dV \qquad (2.3.4)$$

where e is the magnitude of the electron charge and the integral is taken over the whole volume of the atom. The form of Eq. (2.3.4) can be readily understood by noting that $e|\psi|^2 \, dV$ is the elemental charge expected in the volume dV at position $\mathbf{r}$ and this charge produces an elemental dipole moment $d\boldsymbol{\mu} = -(e|\psi|^2 dV)\mathbf{r}$. The substitution of Eq. (2.3.2) into Eq. (2.3.4) with the help of Eq. (2.3.1) gives

$$\boldsymbol{\mu} = \int e\mathbf{r}|a_1|^2|u_1|^2 \, dV + \int e\mathbf{r}|a_2|^2|u_2|^2 \, dV$$

$$+ \int e\mathbf{r}[a_1 a_2^* u_1 u_2^* \exp j(\omega_0 t) + a_1^* a_2 u_1^* u_2 \exp -j(\omega_0 t)] dV \qquad (2.3.5)$$

where * stands for complex conjugate and $\omega_0 = (E_2 - E_1)/\hbar$. Equation (2.3.5) shows that $\boldsymbol{\mu}$ has a term $\boldsymbol{\mu}_{osc}$, oscillating at the frequency ω_0, which can be written as:

$$\boldsymbol{\mu}_{osc} = \text{Re}[2a_1 a_2^* \boldsymbol{\mu}_{21} \exp j(\omega_0 t)] \qquad (2.3.6)$$

where Re stands for real part and where we have defined a time-independent dipole moment $\boldsymbol{\mu}_{21}$ given by:

$$\boldsymbol{\mu}_{21} = \int u_2^* e\mathbf{r}u_1 \, dV \qquad (2.3.7)$$

Vector $\boldsymbol{\mu}_{21}$ is referred to as the matrix element of the electric dipole moment operator or the electric dipole moment of the atom.

Equation (2.3.6) shows that, during transition $2 \rightarrow 1$, the atom acquires a dipole moment $\boldsymbol{\mu}_{osc}$, which is oscillating at frequency ω_0 and whose amplitude is proportional to the vector $\boldsymbol{\mu}_{21}$ given by Eq. (2.3.7). Now, from classic electrodynamics, we know that an oscillating dipole moment must radiate power into the surrounding space. Accordingly, from a semiclassical standpoint, the process of spontaneous emission can be identified as arising from this radiated power. To be more specific, let us write the oscillating dipole moment as $\boldsymbol{\mu} = \boldsymbol{\mu}_0 \cos(\omega_0 t + \phi) = \text{Re}[\boldsymbol{\mu}_0' \exp(j\omega_0 t)]$, where $\boldsymbol{\mu}_0$ is a real vector describing the amplitude of the dipole moment, Re stands for real part, and $\boldsymbol{\mu}_0'$ is a complex vector given by $\boldsymbol{\mu}_0' = \boldsymbol{\mu}_0 \exp(j\phi)$.[†] According to classical electrodynamics, this oscillating dipole moment radiates into the surrounding space a power P_r given by:

$$P_r = \frac{n\mu^2 \omega_0^4}{12\pi\varepsilon_0 c^3} \qquad (2.3.8)$$

where $\mu = |\boldsymbol{\mu}_0| = |\boldsymbol{\mu}_0'|$ is the amplitude of the electric dipole moment, n is the refractive index of the medium surrounding the dipole, and c is the light velocity in the vacuum. In the present case we can still use Eq. (2.3.8) provided that μ is now taken to be $\mu = 2|a_1 a_2^* \boldsymbol{\mu}_{21}|$; i.e., it is the magnitude of the complex vector $2a_1 a_2^* \boldsymbol{\mu}_{21}$. Thus we see that the radiated power can be written as:

$$P_r = P_r'|a_1|^2|a_2|^2 \qquad (2.3.9)$$

where P_r' is a time-independent quantity given by:

$$P_r' = \frac{16\pi^3 n|\mu|^2 v_0^4}{3\varepsilon_0 c^3} \qquad (2.3.10)$$

and where $|\mu| = |\boldsymbol{\mu}_{21}|$ is the magnitude of the complex vector $\boldsymbol{\mu}_{21}$.

[†] We recall that a complex vector $\mathbf{A}$, is a vector whose components, e.g., A_x, A_y, A_z are complex numbers. The magnitude A of a complex vector is a real quantity given by $A = [\mathbf{A}\mathbf{A}^*]^{1/2}$, where $\mathbf{A}^*$ is the vector conjugate to $\mathbf{A}$ (i.e., with components A_x^*, A_y^*, and A_z^*, which are the complex conjugate of those for $\mathbf{A}$).

To calculate the atom's decay rate we use now an energy balance argument to write

$$\frac{dE}{dt} = -P_r \tag{2.3.11}$$

where the atom energy is given by:

$$E = |a_1|^2 E_1 + |a_2|^2 E_2 \tag{2.3.12}$$

With the help of Eq. (2.3.3), Eq. (2.3.12) can be readily transformed into:

$$E = E_1 + h\nu_0 |a_2|^2 \tag{2.3.13}$$

where $\nu_0 = (E_2 - E_1)/h$ is the transition frequency.

With the help of Eqs. (2.3.9), (2.3.10), and (2.3.13), Eq. (2.3.11) can then be written as:

$$\frac{d|a_2|^2}{dt} = -\frac{1}{\tau_{sp}} |a_1|^2 |a_2|^2 = -\frac{1}{\tau_{sp}} (1 - |a_2|^2)|a_2|^2 \tag{2.3.14}$$

where we have defined a characteristic time $\tau_{sp} = h\nu_0/P_r'$ as:

$$\tau_{sp} = \frac{3h\varepsilon_0 c^3}{16\pi^3 \nu_0^3 n |\mu|^2} \tag{2.3.15}$$

which is called the spontaneous emission (or radiative) lifetime of level 2. The solution of Eq. (2.3.14) can be conveniently written as:

$$|a_2|^2 = \frac{1}{2}\left[1 - \tanh\left(\frac{t - t_0}{2\tau_{sp}}\right)\right] \tag{2.3.16}$$

where t_0 is set by the initial condition, i.e., by the value $|a_2(0)|^2$. Indeed from Eq. (2.3.16) we obtain

$$|a_2(0)|^2 = \frac{1}{2}\left[1 - \tanh\left(\frac{-t_0}{2\tau_{sp}}\right)\right] \tag{2.3.17}$$

which, for a given value of $|a_2(0)|^2$ (provided it is smaller than 1), yields a unique value of t_0. As an example, Fig. 2.5 shows the time behavior of $|a_2(t)|^2$ for the initial condition $|a_2(0)|^2 = 0.96$. Note that, by choosing a different value of $|a_2(0)|^2$, one merely changes the value of t_0 in Eq. (2.3.16), i.e., one changes only the origin of the time axis. Assuming for instance $|a_2(0)|^2 = 0.8$, the curve of $|a_2(t)|^2$ is obtained simply by horizontally shifting the curve in Fig. 2.5 to the left until it crosses the vertical $t=0$ axis at the value 0.8. This shows the advantage of expressing the decay of $|a_2(t)|^2$ in the form of Eq. (2.3.16). Once $|a_2(t)|^2$ has been calculated, the radiated power P_r, according to Eqs. (2.3.11) and (2.3.13), is obtained as $P_r = -h\nu_0 d|a_2|^2/dt$. The time behavior of the normalized radiated power,

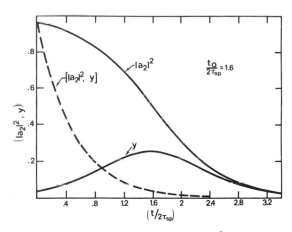

FIG. 2.5. Time behavior of the upper state occupation probability $|a_2|^2$ and of the normalized radiated power $y = \tau_{sp} P_r / h\nu_0$. Continuous lines: semiclassical results. Dashed line: quantum result.

$y = \tau_{sp} P_r / h\nu_0$, is also shown in Fig. 2.5. In the discussion that follows, it is important to notice that the time behavior of $|a_2(t)|^2$ can be approximated by an exponential law, i.e.:

$$|a_2(t)|^2 = |a_2(0)|^2 \exp[-(t/\tau_{sp})] \qquad (2.3.18)$$

only when $|a_2(0)|^2 \ll 1$. In this case, in fact, we can substitute $|a_1|^2 \cong 1$ into Eq. (2.3.14), thus readily obtaining Eq. (2.3.18).

A particularly important case occurs when $|a_2(0)|^2 = 1$. In this case from Eq. (2.3.17) we find that $t_0 = \infty$, which means that, according to this semiclassical theory, the atom should not decay. Indeed when $|a_2(0)|^2 = 1$, then $|a_1(0)|^2 = 0$, and from Eq. (2.3.14) one gets $d|a_2|^2/dt = 0$. Another way of looking at this case is to observe that, when $a_1(0) = 0$, $\boldsymbol{\mu}_{osc}$ given by Eq. (2.3.6) vanishes. In this case, since the atom does not have an oscillating dipole moment, it cannot radiate power; therefore it is in an equilibrium state. Let us now investigate the stability of this equilibrium. To do this, we assume the atom to be perturbed so that $|a_2| \neq 1$ at $t = 0$. Physically this means that, as a result of the perturbation, there is a finite probability $|a_1|^2$ of finding the atom in level 1. Equation (2.3.6) then shows that a dipole moment oscillating at frequency ω_0 is now produced. This moment will radiate into the surrounding space and the atom will tend to decay to level 1. This implies a decrease of $|a_2|^2$, with the atom moving further away from equilibrium. The atom is therefore in unstable equilibrium.

Before going further it is worth summarizing the main results obtained with this semiclassical approach: (1) The time behavior of $|a_2|^2$ can generally be described in terms of an hyperbolic tangent equation (2.3.16), but for very weak excitation (i.e., for $|a_2|^2 \ll 1$) it follows an approximately exponential law [Eq. (2.3.18)]. (2) When the atom is initially in the upper state (i.e., $|a_2(0)|^2 = 1$), the atom is in unstable equilibrium and no radiation occurs.

2.3.2. Quantum Electrodynamics Approach

Although a quantum electrodynamics approach is beyond the scope of this book, it is worthwhile to summarize some of the results obtained from such an approach and compare these with the semiclassical results. The most relevant results from the quantum approach can be summarized as follows:[5,6]

- Unlike the semiclassical case, the time behavior of $|a_2|^2$ is now always described, to a good approximation (Wigner–Weisskopf approximation), by an exponential law. This means that Eq. (2.3.18) is always true no matter what the value of $|a_2(0)|^2$.
- The expression for the spontaneous emission lifetime τ_{sp} is given, in this case too, by Eq. (2.3.15).
- Since the radiated power is given by $P_r = -h\nu_0 d|a_2|^2/dt$, this power also decays exponentially with a time constant τ_{sp}.

We see that the semiclassical and quantum electrodynamics approaches give completely different predictions for the phenomenon of spontaneous emission (see Fig. 2.5). All available experimental results confirm that the quantum electrodynamics approach gives the correct answer to the problem.* From Eq. (2.3.15) we can then write the rate of spontaneous emission $A = 1/\tau_{sp}$ as:

$$A = \frac{16\pi^3 \nu_0^3 n|\mu|^2}{3h\varepsilon_0 c^3} \qquad (2.3.19)$$

The preceding remarks imply that, according to quantum electrodynamics, an atom in the upper level is not in a state of unstable equilibrium; the physical reason for the disappearance of this unstable state on passing from the semiclassical to the quantum electrodynamics approach deserves further discussion. In the semiclassical case, the atom's wave function is generally written as in Eq. (2.3.2), which implies that the atom is not in a stationary state. According to quantum mechanics, this can occur only when some sort of perturbation is applied to the atom. Furthermore, to remove the unstable equilibrium discussed before, we must again assume that the atom is somehow perturbed, and we now look for some cause for this perturbation. At first sight we may be tempted to say that there will always be enough stray radiation around the material to perturb the atom. To be more specific, let us suppose that the material is contained in a blackbody cavity whose walls are kept at temperature T. We might then imagine this stray radiation to be provided by the blackbody radiation within the cavity. This conclusion would be wrong, however, since the radiation produced in this way would actually be due to the process of stimulated emission, i.e., stimulated by the blackbody radiation. The phenomenon of spontaneous emission would then depend upon the wall temperature and would cease at $T = 0$. The correct form of perturbation needed to describe the phenomenon of spontaneous emission is actually provided by the quantum electrodynamics approach. In fact, according to the discussion presented in sect. 2.2.3, the mean square values $\langle E^2 \rangle$ and $\langle H^2 \rangle$ of both the electric and magnetic fields of a given cavity mode are different from zero even at $T = 0$ (*zero-point field fluctuations*). We may therefore consider these fluctuations as the perturbation acting on the atom and which, in particular, upsets the unstable equilibrium predicted by the semiclassical

---

* Of these we mention the very accurate measurements of the so-called Lamb shift, another phenomenon that occurs during spontaneous emission. The center frequency of the spontaneously emitted light does not occur at frequency ν_0 (the transition frequency) but at a slightly different value. Lamb-shift measurements for hydrogen are among the most careful measurements so far made in physics, and they have always agreed exactly (within experimental errors) with predictions by the quantum electrodynamic approach.

treatment. Correspondingly we may think of the spontaneous emission process as originating from these zero-point fluctuations.

2.3.3. Allowed and Forbidden Transitions

Equation (2.3.19) shows that, to have $A \neq 0$, we must have $|\mu| \neq 0$. In this case spontaneous emission arises from the power radiated by the electric dipole of the atom, so the transition is said to be electric-dipole allowed. When $|\mu| = 0$, we have $A = 0$ and the transition is said to be electric-dipole forbidden. In this case the transition may occur via other multipole radiation processes, e.g., through the oscillating magnetic dipole moment of the atom (*magnetic dipole transitions*). This is usually a much weaker process however.

Let us now consider the situation when the transition is electric-dipole forbidden, i.e., when $|\mu| = 0$. Since $|\mu| = |\boldsymbol{\mu}_{21}|$, Eq. (2.3.7) shows that this occurs when the eigenfunctions u_1 and u_2 are either both symmetric or both antisymmetric.* In fact, in this case, the two contributions from the integrand in Eq. (2.3.7) at points $\mathbf{r}$ and $-\mathbf{r}$, respectively, are equal and opposite. It is therefore interesting to see when wave functions $u(\mathbf{r})$ are either symmetric or antisymmetric. This occurs when the Hamiltonian $\mathscr{H}_0(\mathbf{r})$ of the system is unchanged by changing $\mathbf{r}$ into $-\mathbf{r}$, i.e., when:†

$$\mathscr{H}_0(-\mathbf{r}) = \mathscr{H}_0(\mathbf{r}) \qquad (2.3.20)$$

In this case, in fact, for any eigenfunction $u_n(\mathbf{r})$, one has:

$$\mathscr{H}_0(\mathbf{r})u_n(\mathbf{r}) = E_n u_n(\mathbf{r}) \qquad (2.3.21)$$

From Eq. (2.3.21), changing $\mathbf{r}$ into $-\mathbf{r}$ and using Eq. (2.3.20), one gets:

$$\mathscr{H}_0(\mathbf{r})u_n(-\mathbf{r}) = E_n u_n(-\mathbf{r}) \qquad (2.3.22)$$

Equations (2.3.21) and (2.3.22) show that $u_n(\mathbf{r})$ and $u_n(-\mathbf{r})$ are both eigenfunctions of the Hamiltonian $\mathscr{H}_0$ with the same eigenvalue E_n. For nondegenerate levels, there is by definition only one eigenfunction for each eigenvalue, apart from an arbitrary choice of sign, so that:

$$u_n(-\mathbf{r}) = \pm u_n(\mathbf{r}) \qquad (2.3.23)$$

Therefore, if $\mathscr{H}_0(\mathbf{r})$ is symmetric, the eigenfunctions must either be symmetric or antisymmetric. In this case the eigenfunctions are usually said to have well-defined parity.

It now remains to see when the Hamiltonian satisfies Eq. (2.3.20), i.e., when it is invariant under inversion. Obviously this occurs when the system has a center of symmetry. Another important case is that of an isolated atom. In this case the potential energy of the

---

* Recall here that a function $f(\mathbf{r})$ is symmetric (or of even parity) if $f(-\mathbf{r}) = f(\mathbf{r})$, while it is antisymmetric (or of odd parity) if $f(-\mathbf{r}) = -f(\mathbf{r})$.
† If the Hamiltonian $\mathscr{H}_0$ is a function of more than one coordinate $\mathbf{r}_1, \mathbf{r}_2, \ldots$, the inversion operation must be simultaneously applied to all these coordinates.

Example 2.1. *Estimate of* τ_{sp} *and* A *for electric-dipole-allowed-and-forbidden transitions.* For an electric-dipole-allowed transition at a frequency corresponding to the middle of the visible range, an estimate on the order of magnitude of A is obtained from Eq. (2.3.19) by substituting the values $\lambda = c/v = 500$ nm and $|\mu| \approx ea$, where a is the atomic radius ($a \cong 0.1$ nm). We therefore obtain $A \cong 10^8$ s^{-1} (i.e., $\tau_{sp} \cong 10$ ns). For magnetic dipole transitions A is approximately 10^5 times smaller, and therefore $\tau_{sp} \approx 1$ ms. Note: According to Eq. (2.3.19), A increases as the cube of the frequency, so that the importance of spontaneous emmission increases rapidly with frequency. In fact spontaneous emission is often negligible in the middle- to far-infrared where nonradiative decay usually dominates. On the other hand when we consider the x-ray region (say, $\lambda \leq 5$ nm), τ_{sp} becomes exceedingly short (10–100 fs), which constitutes a major problem for achieving a population inversion in x-ray lasers.

k-th electron of the atom is given by the sum of the potential energy due to the nucleus (which is symmetric) and that of all other electrons. For the i-th electron, this energy depends on $|\mathbf{r}_i - \mathbf{r}_k|$, i.e., on the magnitude of the distance between the two electrons. Therefore these terms are also invariant under inversion. An important case where Eq. (2.3.20) is not valid occurs when an atom is placed in an external electric field (e.g., a crystal's electric field) that does not possess a center of inversion. In this case the wave functions do not have definite parity.

To summarize, we say that electric dipole transitions occur only between states of opposite parity and the states have well-defined parity if the Hamiltonian is invariant under inversion.

2.4. ABSORPTION AND STIMULATED EMISSION

In this section we study in some detail absorption and stimulated emission induced in a two-level system for a single atom interacting with a monochromatic em wave. In particular our aim is to calculate the rates of absorption W_{12} and stimulated emission W_{21} [see Eqs. (1.1.4) and (1.1.6)]. We follow the semiclassical approximation, wherein, as already explained, the atom is quantized while the em radiation is treated classically. It can be shown, in fact, that the quantum electrodynamics approach gives the same result as the semiclassical treatment when the number of photons in a given radiation mode is much greater than unity. Since this applies to any situation other than one involving an exceedingly weak em wave, we can dispense with the complication of the full quantum treatment. At first we will assume the two levels to be nondegenerate and treat the case of degenerate levels later in the chapter.

2.4.1. Absorption and Stimulated Emission Rates

Let us first consider the case of absorption and assume that, for time $t \geq 0$, a monochromatic em wave is incident on the atom so that the atomic wave function can be described as in Eq. (2.3.2), where we assume the initial conditions $|a_1(0)|^2 = 1$ and $|a_2(0)|^2 = 0$.

As a result of the interaction with the em wave, the atom acquires an interaction energy H'. In the treatment that follows this energy H' is considered to be due to the interaction of the electric dipole moment of the atom with the electric field $\mathbf{E}(\mathbf{r}, t)$ of the em wave (*electric dipole interaction*), where the origin is taken at the nucleus. The electric field at the nuclear position can then be written as:

$$\mathbf{E}(0, t) = \mathbf{E}_0 \sin(\omega t) \qquad (2.4.1)$$

where ω is the angular frequency of the wave. We also assume that the wavelength of the em radiation is much greater than the atom's dimension, so that the phase shift of the em wave over an atomic dimension is very small. Then Eq. (2.4.1) can be taken to give the value of the electric field for any location in the atom (*electric dipole approximation*). We also assume the frequency ω to be close to the resonant frequency ω_0 of the transition.

Classically, for a given position **r** of the electron within the atom, the atom exhibits an electric dipole moment $\mathbf{\mu} = -e\mathbf{r}$, where e is the magnitude of the electronic charge. The interaction energy H' resulting from the external electric field is then:

$$H' = \mathbf{\mu} \cdot \mathbf{E} - -e\mathbf{r} \cdot \mathbf{E}_0 \sin \omega t \tag{2.4.2}$$

In a quantum mechanical treatment, this sinusoidally time-varying interaction energy is treated as a sinusoidally time-varying interaction Hamiltonian $\mathscr{H}'(t)$, which is then inserted into the time-dependent Schrödinger wave equation. Since $\omega \cong \omega_0$, this interaction Hamiltonian results in the transition of the atom from one level to the other. This implies that, for $t > 0$, $|a_1(t)|^2$ decreases from its initial value $|a_1(0)|^2 = 1$, and $|a_2(t)|^2$ increases correspondingly. To derive an expression for $a_2(t)$ we assume additionally that the transition probability is weak, so that a perturbation analysis can be used (*time-dependent perturbation theory*), and the interaction occurs for a long time after $t = 0$.

Given the preceding assumptions, the time behavior of $|a_2(t)|^2$ is shown in Appendix A to be given by:

$$|a_2(t)|^2 = \frac{\pi^2}{3h^2} |\mu_{21}|^2 E_0^2 \delta(v - v_0) t \tag{2.4.3}$$

where $v = \omega/2\pi$, $v_0 = \omega_0/2\pi$, δ is the Dirac delta function, E_0 is the amplitude of the vector $\mathbf{E}_0$, and $|\mu_{21}|$ is the amplitude of the complex vector $\mathbf{\mu}_{21}$ given by Eq. (2.3.7). Equation (2.4.3) shows that, for $t > 0$, $|a_2(t)|^2$ increases linearly with time. We can then define the transition rate W_{12}^{sa} as:

$$W_{12}^{sa} = \frac{d|a_2|^2}{dt} \tag{2.4.4}$$

and from Eq. (2.4.3), we obtain

$$W_{12}^{sa} = \frac{\pi^2}{3h^2} |\mu_{21}|^2 E_0^2 \delta(v - v_0) \tag{2.4.5}$$

Note that the transition rate defined by Eq. (2.4.4) refers to the case of a single atom interacting with monochromatic radiation, this situation is denoted by the superscript *sa* (single atom) added to W_{12}.

To gain a more physical view of the absorption phenomenon, we note that, for $t > 0$, the wave function can be described as in Eq. (2.3.2). For $t > 0$ the atom thus acquires an oscillating dipole moment $\mathbf{\mu}_{osc}$, given by Eq. (2.3.6). In contrast to the case of spontaneous emission however, since $a_1(t)$ and $a_2(t)$ are now driven by the electric field of the em wave, the phase of $\mathbf{\mu}_{osc}$ turns out to be correlated to that of the wave. In particular for absorption, i.e., when one begins with the initial conditions $a_1(0) = 1$ and $a_2(0) = 0$, the phase of the dipole is such that the dipole absorbs power from the em wave. The interaction phenomenon is thus seen to be very similar to that of a classical oscillating dipole moment driven by an external field.[3]

Equation (2.4.5) can also be expressed in terms of the energy density of the em wave. Since

$$\rho = \frac{n^2 \varepsilon_0 E_0^2}{2} \tag{2.4.6}$$

where n is the refractive index of the medium and ε_0 is the vacuum permittivity, we obtain

$$W_{12}^{sa} = \frac{2\pi^2}{3n^2\varepsilon_0 h^2} |\mu_{21}|^2 \rho \delta(v - v_0) \tag{2.4.7}$$

Note that W_{12}^{sa} is proportional to the Dirac δ function. This implies the unphysical result that $W = 0$ for $v \neq v_0$ and $W_{12} = \infty$ when $v = v_0$, i.e., when the frequency of the em wave is exactly coincident with the frequency of the atomic transition. The reason for this unphysical result can be traced to the assumption that the interaction of the em wave with the atom continues undisturbed for an indefinite time. Indeed, from a classical viewpoint, if a sinusoidal electric field at frequency v drives a (lossless) oscillating dipole moment at frequency v_0, there would be an interaction, i.e., a net energy transfer, only if $v = v_0$. Actually a number of perturbation phenomena (such as collisions with other atoms or lattice phonons) prevent this interaction from continuing undisturbed indefinitely. These phenomena are discussed at some length in a later section, but the general result they lead to can be expressed very simply: Equation (2.4.7) remains valid provided the Dirac δ function—an infinitely sharp function centred at $v = v_0$ and of unit area, i.e., such that $\int \delta(v - v_0)dv = 1$—is replaced by a new function $g(v - v_0)$, symmetric about $v = v_0$, again of unit area, i.e., such that $\int g(v - v_0)dv = 1$, and generally given by:

$$g(v - v_0) = \frac{2}{\pi \Delta v_0} \frac{1}{1 + [2(v - v_0)/\Delta v_0]^2} \tag{2.4.8}$$

where Δv_0 depends on the particular broadening mechanism involved. We can therefore write for W_{12}^{sa} the expression:

$$W_{12}^{sa} = \frac{2\pi^2}{3n^2\varepsilon_0 h^2} |\mu_{21}|^2 \rho g(v - v_0) \tag{2.4.9a}$$

The normalized function $[g(v - v_0)\Delta v_0]$ is plotted in Fig. 2.6 versus the normalized frequency difference $(v - v_0)/(\Delta v_0/2)$. The full width of the curve between the two points having half the maximum value [full width at half-maximum (FWHM)] is simply Δv_0. The maximum of $g(v - v_0)$ occurs for $v = v_0$, and its value is given by:

$$g(0) = \frac{2}{\pi \Delta v_0} = \frac{0.637}{\Delta v_0} \tag{2.4.9b}$$

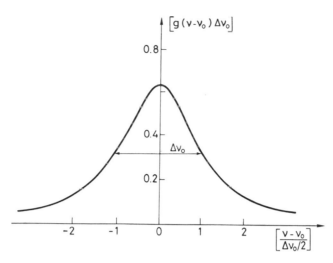

FIG. 2.6. Normalized plot of a Lorentzian line.

A curve of the general form described by Eq. (2.4.8) is referred to as *Lorentzian* after H. E. Lorentz, who first derived it in his theory of the electron oscillator.[3]

For a plane em wave it is often useful to express W_{12}^{sa} in terms of the intensity I of the incident radiation. Since:

$$I = \frac{c\rho}{n} \tag{2.4.10}$$

where n is the refractive index of the medium, we find from Eq. (2.4.9a) that:

$$W_{12}^{sa} = \frac{2\pi^2}{3n\varepsilon_0 ch^2} |\mu_{21}|^2 I g(\nu - \nu_0) \tag{2.4.11}$$

We consider next the case of stimulated emission. The starting points, namely, the wave functions of the two-level system [Eqs. (2.3.1) and (2.3.2)] and the interaction energy H' [Eq. (2.4.2)] remain unchanged. Thus the corresponding pair of equations describing the evolution with time of $|a_2(t)|^2$ and $|a_1(t)|^2$ (see Appendix A) also remain unchanged. The only difference arises from the fact that the initial condition is now given by $|a_2(0)|^2 = 1$ and thus $|a_1(0)|^2 = 0$. It can readily seen that the equations for stimulated emission are then obtained from those corresponding to absorption by simply interchanging the indices 1 and 2. Thus the transition rate W_{21}^{sa} is obtained from Eq. (2.4.5) by interchanging the two indices. From Eq. (2.3.7) we immediately see that $\mu_{21} = \mu_{21}^*$, implying that $|\mu_{12}| = |\mu_{21}|$. Therefore we have

$$W_{12}^{sa} = W_{21}^{sa} \tag{2.4.12}$$

showing that the probabilities of absorption and stimulated emission are equal in this case [compare with Eq. (1.1.8)].

As a conclusion to this section, according to Eqs. (2.4.9) and (2.4.11), the stimulated transition rate can be written as:*

$$W^{sa} = \frac{2\pi^2}{3n^2\varepsilon_0 h^2}|\mu|^2 \rho g(v - v_0) \tag{2.4.13a}$$

$$W^{sa} = \frac{2\pi^2}{3n\varepsilon_0 ch^2}|\mu|^2 Ig(v - v_0) \tag{2.4.13b}$$

where, from Eq. (2.4.12), we have set $W^{sa} = W^{sa}_{12} = W^{sa}_{21}$ and $|\mu| = |\boldsymbol{\mu}_{12}| = |\boldsymbol{\mu}_{21}|$.

2.4.2. Allowed and Forbidden Transitions

Equations (2.4.13a) and (2.3.19) show that the transition rate W^{sa} and the spontaneous emission rate A are proportional to $|v|^2$. This indicates that the two phenomena must obey the same selection rule. Thus the stimulated transition via electric dipole interaction (electric dipole transition) occurs only between states u_1 and u_2 of opposite parity. The transition is then said to be electric-dipole allowed. Conversely, if the parity of two states is the same, then $W^{sa} = 0$ and the transition is said to be electric-dipole-forbidden. This does not mean however that the atom cannot pass from level 1 to level 2 through the influence of an incident em wave. In this case, the transition can occur for instance as a result of the interaction of the magnetic field of the em wave with the magnetic dipole moment of the atom. For the sake of simplicity, we do not consider this case further (*magnetic dipole interaction*) but limit ourselves to observing that the analysis can be carried out in a similar manner to that used to obtain Eq. (2.4.11). We may also point out that a magnetic dipole transition is allowed between states of equal parity (even–even or odd–odd transitions). Therefore a transition forbidden by electric dipole interaction is however allowed for magnetic dipole interaction, and vice versa.

It is now instructive to calculate the order of magnitude of the ratio of the electric dipole transition probability W_e to the magnetic dipole transition probability W_m. Obviously the calculation refers to two different transitions, one allowed for electric dipole and the other for magnetic dipole interaction. We assume that the intensity of the em wave is the same for the two cases. For an allowed electric-dipole transition, according to Eq. (2.4.5) we can write $W_e \propto (\mu_e E_0)^2 \approx (eaE_0)^2$, where E_0 is the electric field amplitude and the electric dipole moment of the atom μ_e is approximated (for an allowed transition) by the product of the electron charge e and the radius a of the atom. For a magnetic dipole interaction, it can

---

* Note: The factor 3 in the denominator of Eqs. (2.4.3), (2.4.5), (2.4.7), (2.4.9a), (2.4.11), (2.4.13a), and (2.4.13b), corresponds to the case of a linearly polarized wave interacting with randomly oriented atoms (such as in a gas). In this case we have

$$W \propto \langle|\boldsymbol{\mu}_{21} \cdot \mathbf{E}_0|^2\rangle = |\mu_{21}|^2 E_0^2 \langle\cos^2\theta\rangle = |\mu_{21}|^2 E_0^2/3$$

where θ is the angle between $\boldsymbol{\mu}_{21}$ and $\mathbf{E}_0$, and the average is taken over all atom field orientations. Indeed for randomly oriented $\boldsymbol{\mu}_{21}$ vectors, $\langle\cos^2\theta\rangle = \frac{1}{3}$, where the average is taken in three-dimensional space, i.e., $\langle\cos^2\theta\rangle = \int \cos^2\theta\, d\Omega/4\pi$. For different cases of atom/field orientation, the factor $|\mu_{21}|^2/3$ must be changed appropriately. Thus, for aligned ions (such as in ionic crystals) and a linearly polarized wave, the factor 3 must be dropped and $|\mu_{21}|$ represents the magnitude of the component of $\boldsymbol{\mu}_{21}$ along $\mathbf{E}_0$.

likewise be shown that $W_m \propto (\mu_m B_0)^2 \approx (\beta B_0)^2$, where B_0 is the magnetic field amplitude of the wave and the magnetic dipole moment of the atom μ_m is approximated (for an allowed transition) by the Bohr magneton β ($\beta = 9.27 \times 10^{-24}$ A $\times$ m^2). Thus we obtain

$$\left(\frac{W_e}{W_m}\right) = \left(\frac{\mu_e E_0}{\beta B_0}\right)^2 = \left(\frac{eac}{\beta}\right)^2 \cong 10^5 \tag{2.4.14}$$

To obtain the numerical result of Eq. (2.4.14) we have made use the fact that, for a plane wave, it is $E_0/B_0 = c$ (where c is the light velocity), and we have assumed that $a \cong 0.05$ nm. The probability of an electric dipole transition is thus much greater than that of a magnetic dipole.

2.4.3. Transition Cross Section, Absorption, and Gain Coefficient

In Section 2.4.1 the transition rate has been calculated for the case of a single atom interacting with an incident em wave and whose linewidth is determined by some line-broadening mechanism. We now consider an ensemble of N_t atoms per unit volume and we want to calculate the corresponding, average, transition rate.

The first case we consider is when both the resonance frequency ν_0 and the lineshape are the same for every atom (the case of *homogeneous broadening*). The transition rate W_h for this homogeneous case is the same for every atom, so that we can simply write

$$W_h(\nu - \nu_0) = W^{sa}(\nu - \nu_0) \tag{2.4.15}$$

If we now let all atoms be in the ground state, the power absorbed per unit volume dP_a/dV is then given by:

$$\left(\frac{dP_a}{dt}\right) = W_h N_t h\nu \tag{2.4.16}$$

Since W_h is proportional to the wave intensity, hence to the photon flux $F = I/h\nu$, we can define an absorption cross section σ_h as:

$$\sigma_h = \frac{W_h}{F} \tag{2.4.17}$$

From Eqs. (2.4.13b) and (2.4.17), σ_h is then given by:

$$\sigma_h = \frac{2\pi^2}{3n\varepsilon_0 ch} |\mu|^2 \nu g(\nu - \nu_0) \tag{2.4.18}$$

With the help of the argument used in connection with Fig. 1.2, we obtain from Eqs. (2.4.16) and (2.4.17) the equation describing the variation of the photon flux along the z-direction as [compare to Eq. (1.2.1)]:

$$dF = -\sigma N_t F \, dz \tag{2.4.19}$$

Examination of Eq. (2.4.19) leads to a simple physical interpretation of this transition cross section. First, let us suppose that we can associate with each atom an effective absorption cross section σ_a in the sense that, if a photon enters this cross section, it is absorbed by the atom (Fig. 2.7). If we let S be the cross-sectional area of the em wave, the number of atoms in the element dz of the material (see also Fig. 1.2) is $N_t S\,dz$, thus giving a total absorption cross section of $\sigma_a N_t S\,dz$. The fractional change (dF/F) of photon flux in the element dz of the material is therefore:

$$\left(\frac{dF}{F}\right) = -\left(\frac{\sigma_a N_t S\,dz}{S}\right) \tag{2.4.20}$$

A comparison between Eqs. (2.4.20) and (2.4.19) shows that $\sigma_h = \sigma_a$, so that the meaning we can attribute to σ_h is that of an effective absorption cross section as just defined.

A somewhat different case occurs when the resonance frequencies v_0' of the atoms are distributed around some central frequency v_0 (a case of *inhomogeneous broadening*). This distribution is described by the function $g^*(v_0' - v_0)$ whose definition is such that $dN_t = N_t g^*(v_0' - v_0)dv_0'$ gives the elemental number of atoms with resonance frequency between v_0' and $v_0' + dv_0'$. According to Eq. (2.4.16), the elemental power absorbed by this elemental number of atoms dN_t is given by $d(dP_a/dV) = (N_t hv)W_h(v - v_0')g^*(v_0' - v_0)dv_0'$, where $W_h(v - v_0')$ is the transition rate for those atoms with resonance frequency v_0'. The total power absorbed per unit volume is then given by:

$$\left(\frac{dP_a}{dV}\right) = N_t hv \int W_h(v - v_0')g^*(v_0' - v_0)dv_0' \tag{2.4.21}$$

A comparison between Eqs. (2.4.21) and (2.4.16) shows that we can define an inhomogeneous transition rate W_{in} as:

$$W_{in} = \int W_h(v - v_0')g^*(v_0' - v_0)dv_0' \tag{2.4.22}$$

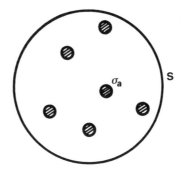

FIG. 2.7. Effective absorption cross section σ_a of atoms in a light beam of cross section S.

According to Eq. (2.4.17) we can now define an inhomogeneous cross section σ_{in} as $\sigma_{in} = W_{in}/F$. Dividing both sides of Eq. (2.4.22) by F and using Eq. (2.4.17), we obtain

$$\sigma_{in} = \int \sigma_h(v - v_0')g^*(v_0' - v_0)dv_0' \qquad (2.4.23)$$

Following the argument presented in connection with Fig. 2.7, we see that σ_{in} is an effective absorption cross section that we can associate with a single atom, so that a photon is absorbed if it enters this cross section. Note, however, that in this case each atom has, in reality, a different cross section $\sigma_h(v - v_0')$ at the frequency of the incoming radiation and σ_{in} is just an effective average cross section. Note also that, according to Eq. (2.4.23), the lineshape and linewidth of σ_{in} depend on the function $g^*(v_0' - v_0)$, i.e., on the distribution of the atomic resonance frequencies. Phenomena leading to this frequency distribution are discussed at some length in a later section. Here we limit ourselves to pointing out that $g^*(v_0' - v_0)$ is generally described by a function of the form:

$$g^*(v_0' - v_0) = \frac{2}{\Delta v_0^*}\left(\frac{\ln 2}{\pi}\right)^{1/2}\exp -\left[\frac{4(v_0' - v_0)^2}{\Delta v_0^{*2}}\ln 2\right] \qquad (2.4.24)$$

where Δv_0^* is the transition linewidth (FWHM), whose value depends on the particular broadening mechanism under consideration.

With the help of Eq. (2.4.18), Eq. (2.4.23) can be tranformed into:

$$\sigma_{in} = \frac{2\pi^2}{3n\varepsilon_0 ch}|\mu|^2 vg_t(v - v_0) \qquad (2.4.25)$$

In Eq. (2.4.25) we have used the symbol $g_t(v - v_0)$ for the total lineshape function, which can be expressed as:

$$g_t = \int_{-\infty}^{+\infty} g^*(x)g[(v - v_0) - x]dx \qquad (2.4.26)$$

where we have set $x = v_0' - v_0$. The expression of the cross section for inhomogeneous broadening σ_{in} is thus obtained from that for homogeneous broadening, given by Eq. (2.4.18), by substituting $g(v - v_0)$ with $g_t(v - v_0)$. Note that, according to Eq. (2.4.26), g_t is the convolution of the functions g and g^*. Since both functions are normalized to unity, it can be shown that g_t is also normalized to unity, i.e., $\int g_t(v - v_0)dv = 1$. Note also that Eq. (2.4.25) provides a generalization of Eq. (2.4.18). Indeed it is immediately seen from Eqs. (2.4.26) and (2.4.25) that σ_{in} reduces to σ_h when $g^*(v_0' - v_0) = \delta(v_0' - v_0)$, i.e., when all atoms have the same resonance frequency. Conversely, if the width of the homogeneous lineshape function $g(v - v_0')$ is much smaller than that due to inhomogeneous broadening $g^*(v_0' - v_0)$, then $g(v - v_0')$ can be approximated by a Dirac δ function in Eq. (2.4.26) to obtain $g_t \cong g^*(v - v_0)$ (case of *pure inhomogeneous broadening*). In this case from Eq. (2.4.24) we obtain

$$g_t = g^*(v - v_0) = \frac{2}{\Delta v_0^*} \left(\frac{\ln 2}{\pi}\right)^{1/2} \exp -\left[\frac{4(v - v_0)^2}{\Delta v_0^{*2}} \ln 2\right] \qquad (2.4.27)$$

The normalized function $[g^*(v - v_0)\Delta v_0^*]$ is plotted in Fig. 2.8 versus the normalized frequency difference $(v - v_0)/(\Delta v_0^*/2)$. According to Eq. (2.4.27) the width of the curve (FWHM) is simply Δv_0^*, the maximum of the function occurs for $v = v_0$, and its value is given by:

$$g^*(0) = \frac{2}{\Delta v_0^*} \left(\frac{\ln 2}{\pi}\right)^{1/2} = \frac{0.939}{\Delta v_0^*} \qquad (2.4.28)$$

A curve of the general form described by Eq. (2.4.27) is referred to as *Gaussian*.

Based on the preceding discussion, from now on we will use the symbol $\sigma = \sigma_{in}$ to indicate the absorption cross section; its general expression can be written as:

$$\sigma = \frac{2\pi^2}{3n\varepsilon_0 ch} |\mu|^2 v g_t(v - v_0) \qquad (2.4.29)$$

The corresponding expression for the absorption rate $W = \sigma F$ can then be written as:

$$W = \frac{2\pi^2}{3n^2\varepsilon_0 h^2} |\mu|^2 \rho g_t(v - v_0) \qquad (2.4.30)$$

where $\rho = (nI/c) = (nFhv/c)$ is the energy density of the em wave.

We can repeat the same arguments for the case of stimulated emission. According to Eq. (2.4.12) one redily sees that, for nondegenerate levels, the general expressions for the

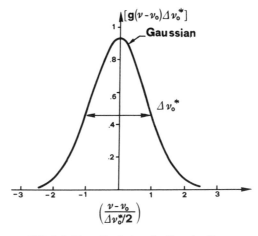

FIG. 2.8. Normalized plot of a Gaussian line.

stimulated emission cross section and the rate of stimulated emission are again given by Eqs. (2.4.29) and (2.4.30), respectively.

We emphasize that, according to Eq. (2.4.29), σ depends only on material parameters ($|\mu|^2$, g_t, and v_0) and the frequency v of the incident wave. A knowledge of σ as a function of v is then all that is needed to describe the interaction process. The transition cross section σ is therefore a very important and widely used transition parameter Note that, when the populations of two levels are N_1 and N_2, Eq. (2.4.19) generalizes to:

$$dF = -\sigma(N_1 - N_2)F\,dz \tag{2.4.31}$$

This has the same form as that originally derived in Chap. 1 [see Eq. (1.2.1) with $g_1 = g_2$]. The discussion presented in this section however provides a deeper understanding of the meaning of the (effective) cross section σ.

Another way of describing the interaction of radiation with matter involves defining a quantity α as:

$$\alpha = \sigma(N_1 - N_2) \tag{2.4.32}$$

If $N_1 > N_2$, then α is positive and referred to as the absorption coefficient of the material. Using Eq. (2.4.29) the following expression is obtained for α:

$$\alpha = \frac{2\pi^2}{3n\varepsilon_0 ch}(N_1 - N_2)|\mu|^2 v g_t(v - v_0) \tag{2.4.33}$$

Since α depends on the populations of the two levels, it is not the most suitable parameter for describing a situation where level populations are changing, such as, for example, in a laser. Its usefulness however lies in the fact that the absorption coefficient α can often be directly measured. From Eqs. (2.4.31) and (2.4.32) we obtain in fact:

$$dF = -\alpha F\,dz \tag{2.4.34}$$

The ratio between the photon flux after traversing a length l of the material and the incident flux is therefore $[F(l)/F(0)] = \exp(-\alpha l)$. Upon experimentally measuring this ratio by a sufficiently monochromatic radiation, we can obtain the value of α for that particular wavelength. The corresponding value of the transition cross section is obtained from Eq. (2.4.32) once N_1 and N_2 are known. If the medium is in thermodynamic equilibrium, N_1 and N_2 can be obtained from Eq. (1.2.2) once the total population $N_t = N_1 + N_2$ and the level's degeneracies are known. The instrument used to measure the absorption coefficient α is known as an absorption spectrophotometer. Note, however, that an absorption measurement obviously cannot be performed for a transition where level 1 is empty. This situation occurs for instance when level 1 is not the ground level and its energy above the ground level is much larger than kT. As a final observation we note that, if $N_2 > N_1$, the absorption coefficient α, defined by Eq. (2.4.32), becomes negative, so the wave is amplified rather than absorbed in the material. In this case it is customary to define the new quantity g as:

$$g = \sigma(N_2 - N_1) \tag{2.4.35}$$

which is positive; it is called the gain coefficient.

2.4.4. Einstein Thermodynamic Treatment

This section describes a treatment given by Einstein[9] for both spontaneous and stimulated transitions (absorption and emission). In this treatment the concept of stimulated emission was first clearly established, and the correct relationship between spontaneous and stimulated transition rates was derived well before the formulation of quantum mechanics and quantum electrodynamics. The calculation uses an elegant thermodynamic argument. Let us assume that the material is placed in a blackbody cavity whose walls are kept at a constant temperature T. Once thermodynamic equilibrium is reached, an em energy density with a spectral distribution ρ_v given by Eq. (2.2.22) is established and the material is immersed in this radiation. In this material, both stimulated emission and absorption processes occur in addition to the spontaneous emission process. Since the system is in thermodynamic equilibrium, the number of transitions per second from level 1 to level 2 must be equal to the number of transitions from level 2 to level 1. We now set

$$W_{21} = B_{21}\rho_{v_0} \tag{2.4.36}$$

$$W_{12} = B_{12}\rho_{v_0} \tag{2.4.37}$$

where B_{21} and B_{12} are constant coefficients (the *Einstein B coefficients*), and let N_1^e and N_2^e be the equilibrium populations of levels 1 and 2, respectively. We can then write

$$AN_2^e + B_{21}\rho_{v_0}N_2^e = B_{12}\rho_{v_0}N_1^e \tag{2.4.38}$$

From Boltzmann statistics we also know that, for nondegenerate levels:

$$\frac{N_2^e}{N_1^e} = \exp(-hv_0/kT) \tag{2.4.39}$$

From Eqs. (2.4.38) and (2.4.39) it then follows that:

$$\rho_{v_0} = \frac{A}{B_{12}\exp(hv_0/kT) - B_{21}} \tag{2.4.40}$$

A comparison between Eqs. (2.4.40) and (2.2.22), when $v = v_0$, leads to the following relations:

$$B_{12} = B_{21} = B \tag{2.4.41}$$

$$\frac{A}{B} = \frac{8\pi h v_0^3 n^3}{c^3} \tag{2.4.42}$$

Equation (2.4.41) shows that the probabilities of absorption and stimulated emission due to blackbody radiation are equal. This relation is therefore analogous to that established, in a completely different way, for monochromatic radiation [see Eq. (2.4.12)]. Equation (2.4.42), on the other hand, allows the calculation of A once B, i.e., the coefficient for stimulated emission due to blackbody radiation, is known. This coefficient can easily be obtained from Eq. (2.4.30) once we remember that this equation was established for monochromatic radiation. For blackbody radiation, we can write $\rho_v\,dv$ for the energy density of radiation whose frequency lies between v and $v + dv$ and simulate this elemental radiation by a monochromatic wave. The corresponding elemental transition probability dW is then obtained from Eq. (2.4.30) by substituting $\rho_v\,dv$ for ρ. Integrating the resulting equation with the assumption that $g_t(v - v_0)$ can be approximated by a Dirac δ function in comparison with ρ_v (see Fig. 2.3), we obtain

$$W = \frac{2\pi^2}{3n^2\varepsilon_0 h^2}|\mu|^2\rho_{v_0} \tag{2.4.43}$$

A comparison between Eqs. (2.4.43) and (2.4.36) or (2.4.37) then gives

$$B = \frac{2\pi^2|\mu|^2}{3n^2\varepsilon_0 h^2} \tag{2.4.44}$$

From Eq. (2.4.42), we obtain

$$A = \frac{16\pi^3 v_0^3 n|\mu|^2}{3h\varepsilon_0 c^3} \tag{2.4.45}$$

Note that the expression just obtained for A is exactly the same as that obtained by a quantum electrodynamics approach [see Eq. (2.3.19)]. Its derivation is in fact based on thermodynamics and the use of Planck's law (which is quantum electrodynamically correct).

2.5. LINE-BROADENING MECHANISMS

This section discusses in some detail various line-broadening mechanisms mentioned in previous sections. According to the earlier discussion, there is an important distinction to be made from the outset between homogeneous and inhomogeneous line broadening. A line-broadening mechanism is referred to as *homogeneous* when it broadens the line of each atom in the same way. In this case the lineshape of the single-atom cross section and that of the overall absorption cross section are identical. Conversely, a line-broadening mechanism is said to be *inhomogeneous* when it distributes the atomic resonance frequencies over some spectral range. Such a mechanism thus broadens the line of the overall system (i.e., that of α) without broadening the lines of individual atoms.

Before proceeding we recall that the shape of the function $g_t(v - v_0)$ can be determined in two ways: (a) In an absorption experiment with the help of a spectrophotometer. In this case one measures the absorption coefficient as a function of frequency v, using the spectrophotometer to select the light frequency. From Eq. (2.4.33) we see that $\alpha \propto vg_t(v - v_0)$. Since the linewidth of the function $g_t(v - v_0)$ is typically much smaller than v_0, we can approximately write $\alpha \propto v_0 g_t(v - v_0)$. Thus, to a very good approximation, the shape of the α versus v curve coincides with that of the function $g_t(v - v_0)$. (b) In an emission experiment one passes the spontaneously emitted light through a spectrometer of sufficiently high resolution and determines $g_t(v - v_0)$ by measuring the shape of the spectral emission. It can be shown that, for any transition, the lineshapes obtained by these two approaches are always the same. Thus, in the discussion that follows, we consider the line-shape function in either absorption or emission, whichever is more convenient.

2.5.1. Homogeneous Broadening

The first homogeneous line-broadening mechanism we consider is due to collisions, and is known as *collision broadening*. In a gas it is due to the collision of an atom with other atoms, ions, free electrons, etc., or with the walls of the container. In a solid it is due to the interaction of the atom with the lattice phonons. After a collision the two-level wave functions ψ_1 and ψ_2 of the atom [see Eq. (2.3.1)] undergo a random phase jump. This means that the phase of the oscillating dipole moment μ_{osc} [see Eq. (2.3.6)] undergoes a random jump relative to that of the incident em wave. These collisions interrupt the process of coherent interaction between the atom and the incident em wave. Since it is the relative phase which is important during the interaction process, an equivalent way of treating this problem assumes that it is the phase of the electric field rather than that of μ_{osc} which undergoes a jump at each collision. The electric field therefore no longer appears sinusoidal but instead appears as shown in Fig. 2.9, where each phase jump occurs at the time of a collision. Under these conditions the atom no longer sees a monochromatic wave. In this case, if we write $d\rho = \rho_{v'} dv'$ for the energy density of the wave in the frequency interval between v' and $v' + dv'$, we can use this elemental energy density in the formula valid for monochromatic radiation, i.e., Eq. (2.4.7), which gives

$$dW_{12} = \frac{2\pi^2}{3n^2\varepsilon_0 h^2} |\mu_{21}|^2 \rho_{v'} \delta(v' - v_0) dv' \tag{2.5.1}$$

The overall transition probability is then obtained by integrating Eq. (2.5.1) over the entire frequency spectrum of the radiation, thus giving

$$W_{12} = \frac{2\pi^2}{3n^2\varepsilon_0 h^2} |\mu_{21}|^2 \int_{-\infty}^{+\infty} \rho_{v'} \delta(v' - v_0) dv' \tag{2.5.2}$$

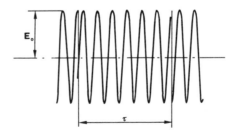

FIG. 2.9. Time behavior of the electric field of an em wave $E(t)$ as seen from an atom undergoing collisions. (In actual cases there may be 10^7 or more cycles during the collision time τ.)

We can now write $\rho_{v'}$ as:

$$\rho_{v'} = \rho g(v' - v) \tag{2.5.3}$$

where ρ is the energy density of the wave [see Eq. (2.4.6)] and $g(v' - v)$ describes the spectral distribution of $\rho_{v'}$. Since $\rho = \int \rho_{v'} \, dv'$, integration of both sides of Eq. (2.5.3) then shows that $g(v' - v)$ must satisfy the normalization condition:

$$\int_{-\infty}^{+\infty} g(v' - v) dv' = 1 \tag{2.5.4}$$

Substituting Eq. (2.5.3) into Eq. (2.5.2) and using a well known mathematical property of the δ function, we obtain

$$W_{12} = \frac{2\pi^2}{3n^2 \varepsilon_0 h^2} |\mu_{21}|^2 \rho g(v - v_0) \tag{2.5.5}$$

As anticipated in Sect. 2.4.1, W_{12} is indeed obtained by substituting $g(v - v_0)$ for $\delta(v - v_0)$ in Eq. (2.4.7). Note that, according to Eq. (2.5.4), we also have

$$\int_{-\infty}^{+\infty} g(v - v_0) dv = 1 \tag{2.5.6}$$

There now remains the problem of calculating the normalized spectral density of the incident radiation $g(v' - v)$. This depends on the time interval τ between collisions (Fig. 2.9), which obviously differs for each collision. We assume that the distribution of the values of τ can be described by a probability density:

$$p_\tau = \frac{[\exp(-\tau/\tau_c)]}{\tau_c} \tag{2.5.7}$$

Here $p_\tau \, d\tau$ is the probability that the time interval between two successive collisions lies between τ and $\tau + d\tau$. Note that τ_c has the physical meaning of the average time $\langle \tau \rangle$ between collisions. It is easy, in fact, to see that:

$$\langle \tau \rangle = \int_0^\infty \tau p_\tau \, d\tau = \tau_c \tag{2.5.8}$$

The mathematical problem to be solved is now well defined. We must obtain the normalized spectral lineshape of a wave as in Fig. 2.9 where the time τ between two successive collisions has the statistical distribution p_τ given by Eq. (2.5.7). Referring to Appendix B for the mathematical details, we quote the final result here. The required normalized spectral lineshape is given by:

$$g(v' - v) = 2\tau_c \frac{1}{[1 + 4\pi^2 \tau_c^2 (v' - v)^2]} \tag{2.5.9}$$

According to Eq. (2.5.5) the lineshape of the transition is obtained from Eq. (2.5.9) by substituting v' for v_0. We then obtain

$$g(v - v_0) = 2\tau_c \frac{1}{[1 + 4\pi^2 \tau_c^2 (v - v_0)^2]} \tag{2.5.10}$$

which is our final result. We thus obtain a function with a Lorentzian lineshape, as generally described by Eq. (2.4.8) [see also Fig. 2.6], where the peak value is now $2\tau_c$ and the linewidth Δv_0 is

$$\Delta v_0 = \frac{1}{\pi \tau_c} \tag{2.5.11}$$

Example 2.2. *Collision broadening of a He–Ne laser.* As a first example of collision broadening, we consider the case of a transition for an atom, or ion, in a gas at pressure p. An estimate of τ_c is in this case given by $\tau_c = l/v_{th}$, where l is the mean free path of the atom in the gas and v_{th} is its average thermal velocity. Since $v_{th} = (3kT/M)^{1/2}$, where M is the atomic mass, and taking l to be given by the expression resulting from the hard-sphere model of a gas, we obtain

$$\tau_c = \left(\frac{2}{3}\right)^{1/2} \frac{1}{8\pi} \frac{(MkT)^{1/2}}{pa^2} \tag{2.5.12}$$

where a is the radius of the atom and p is the gas pressure. For a gas of Neon atoms at room temperature and at a pressure $p \cong 0.5$ Torr (typical pressure in a He–Ne gas laser) using Eq. (2.5.12) with $a \cong 0.1$ nm we obtain $\tau_c \cong 0.5$ μs. We then find from Eq. (2.5.11) that $\Delta v_0 = 0.64$ MHz. Note: τ_c is inversely proportional and hence Δv_0 is directly proportional to p. As a rough rule of thumb we can say that, for any atom, collisions in a gas contribute to line broadening by an amount $(\Delta v_0/p) \cong 1$ MHz/Torr, comparable to that shown in the example of Ne atoms. Note also that, during collision time τ_c, the number of cycles of the em wave equals $m = v\tau_c$. For a wave whose wavelength falls in the middle of the visible range, we have $v = 5 \times 10^{14}$ Hz; thus the number of cycles is 5×10^8. This emphasizes the fact that Fig. 2.9 is not to scale, since the number of cycles in the time τ_c is much greater than the figure suggests.

Example 2.3. *Linewidth of ruby and Nd:YAG.* As a second example of collision broadening, we consider an impurity ion in an ionic crystal. In this case collisions of the ion occur with lattice phonons. Since the number of phonons in a given lattice vibration is a strong function of lattice temperature, we expect the transition linewidth to show a strong dependence on temperature. As a representative example, Fig. 2.10 shows plots of the linewidth versus temperature for both Nd:YAG and ruby; linewidth is expressed in wave numbers (cm^{-1}), a quantity widely used by spectroscopists rather than actual frequency.[*] At 300 K, the laser transition linewidths are seen to be $\Delta v_0 \cong 4$ $cm^{-1} \cong 120$ GHz for Nd:YAG and $\Delta v_0 \cong 11$ $cm^{-1} = 330$ GHz for ruby.

---

[*] For a given wave of frequency v, the corresponding frequency in wave numbers (e.g., in cm^{-1}) is $w = v/c$, where c is the velocity of the wave in a vacuum (expressed in cm/s). Thus the true frequency v is obtained from the frequency in wave numbers by the simple relation $v = cw$. The corresponding wave length is then $\lambda = c/v = 1/w$ (expressed in cm), which illustrates the advantage of wave number notation.

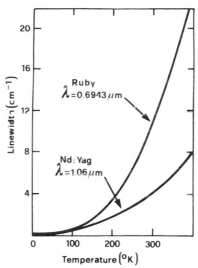

FIG. 2.10. Laser linewidth versus temperature for ruby and Nd:YAG, as determined by phonon broadening.

A second homogeneous line-broadening mechanism originates from spontaneous emission. Since this emission is an inevitable feature of any transition, the corresponding broadening is called *natural* or *intrinsic*. In the case of natural broadening, it is easiest to consider the behavior in terms of the spectrum of the emitted radiation. Note that, as pointed out in Sect. 2.3.2, spontaneous emission is a purely quantum phenomenon; i.e., it can be correctly explained only by quantizing both matter and radiation. It follows therefore that a correct description of the lineshape of emitted radiation also needs a quantum electrodynamics treatment. We therefore limit ourselves to quoting the final result, which happens to be very simple, and to justifying it by some simple arguments. The quantum electrodynamics theory of spontaneous emission shows that the spectrum $g(v - v_0)$ is again described by a Lorentzian line whose shape can be obtained from Eq. (2.5.10) by replacing τ_c by $2\tau_{sp}$, where τ_{sp} is the decay time of spontaneous emission.[10] Thus, in particular, the width of the line (FWHM) is given by:

$$\Delta v_0 = \frac{1}{2\pi\tau_{sp}} \qquad (2.5.13)$$

To justify this result one can note that, since the power emitted by the atom decays in time as $\exp(-t/\tau_{sp})$, the corresponding electric field can be thought of as decaying according to the relationship $E(t) = \exp(-t/2\tau_{sp}) \times \cos\omega_0 t$. The decay of emitted intensity [which is proportional to $\langle E^2(t)\rangle$] would then show the correct temporal behavior, namely, $\exp(-t/\tau_{sp})$. We can now easily calculate the power spectrum corresponding to such a field $E(t)$ and verify that the lineshape is Lorentzian and that its width is given by Eq. (2.5.13).

Example 2.4. *Natural linewidth of an allowed transition.* As a representative example, we can find an order of magnitude estimate for Δv_{nat} for an electric-dipole-allowed transition. Assuming $|v| = ea$ with $a \cong 0.1$ nm and $\lambda = 500$ nm (green light), we found in Example 2.1 that $\tau_{sp} \cong 10$ ns. From Eq. (2.5.13) we then obtain $\Delta v_{nat} \cong 16$ MHz. Note that Δv_{nat}, just as $A = 1/\tau_{sp}$, is expected to increase with frequency as v_0^3. Therefore natural linewidth increases very rapidly for transitions at shorter wavelengths (to the uv or x-ray region).

2.5.2. Inhomogeneous Broadening

We now consider some mechanisms where the broadening arises from a distribution of atomic resonance frequencies (inhomogeneous broadening).

As a first case of inhomogeneous broadening we consider that which occurs for ions in ionic crystals or glasses. Such ions experience a local electric field produced by surrounding atoms of the material. Due to material inhomogeneities that are particularly significant in a glass medium, these fields differ from ion to ion. Via the Stark effect, local field variations then produce local variation of energy levels and thus of the ions' transition frequencies. (The term inhomogeneous broadening originates from this case.) For random local field variations, the corresponding distribution of transition frequencies $g^*(v'_0 - v_0)$ turns out to be given by a Gaussian function, i.e., by the general expression [Eq. (2.4.27)]. The linewidth Δv_0^* (FWHM) depends on the extent of variation of transition frequencies in the material and hence on the amount of field inhomogeneity within the crystal or glass.

> **Example 2.5.** *Linewidth of a Nd:glass laser.* As a representative example we consider the case of Nd^{3+} ions doped into a silicate glass. In this case, due to glass inhomogeneities, the linewidth of the laser transition at $\lambda = 1.05$ μm is $\Delta v_0^* \cong 5.4$ THz, i.e., it is about 40 times broader than that of Nd:YAG at room temperature (see Example 2.3). Note that these inhomogeneities are an unavoidable feature of the glass state.

A second inhomogeneous broadening mechanism, typical of gas, arises from atomic motion and is called Doppler broadening. Assume that an incident em wave of frequency v is propagating in the positive z-direction and let v_z be the component of atomic velocity along this axis. According to the Doppler effect, the frequency of the wave as seen from the rest frame of the atom is $v' = v[1 - (v_z/c)]$, where c is the velocity of light in the medium. Note the well-known result that when $v_z > 0$, we have $v' < v$, and vice versa. Of course absorption by the atom occurs only when the apparent frequency v' of the em wave, as seen from the atom, is equal to the atomic transition frequency v_0, i.e., when $v[1 - (v_z/c)] = v_0$. If we now express this relation as:

$$v = \frac{v_0}{[1 - (v_z/c)]} \qquad (2.5.14)$$

we arrive at a different interpretation of the process: As far as the interaction of the em radiation with the atom is concerned, the result is the same if the atom were not moving but instead had a resonant frequency v'_0 given by:

$$v'_0 = \frac{v_0}{[1 - (v_z/c)]} \qquad (2.5.15)$$

where v_0 is the true transition frequency. Indeed, following this interpretation, absorption is expected to occur when the frequency v of the em wave equals v'_0, i.e., when $v = v'_0$ in agreement with what can be obtained from Eqs. (2.5.14) and (2.5.15). When considered in this way, we see that this broadening mechanism indeed belongs to the inhomogeneous category defined at the beginning of this section.

To calculate the corresponding lineshape $g^*(v_0' - v_0)$, we recall that, if we let $p_v \, dv_z$ equal the probability that an atom of mass M in a gas at temperature T has a velocity component between v_z and $v_z + dv_z$, then p_v is given by the Maxwell distribution:

$$p_v - \left(\frac{M}{2\pi kT} \right)^{1/2} \exp[-(Mv_z^2/2kT)] \qquad (2.5.16)$$

From (2.5.15), since $|v_z| \ll c$, we get $v_0' \cong v_0[1 + (v_z/c)]$ and thus $v_z = c(v_0' - v_0)v_0$. From (2.5.16), one then obtains the desired distribution upon recognizing that one must have $g^*(v_0' - v_0)dv_0' = p_v dv_z$. One then gets

$$g^*(v_0' - v_0) = \frac{1}{v_0} \left(\frac{Mc^2}{2\pi kT} \right)^{1/2} \exp - \left[\frac{Mc^2}{2kT} \frac{(v_0' - v_0)^2}{v_0^2} \right] \qquad (2.5.17)$$

Thus we again obtain a Gaussian function whose FWHM linewidth (Doppler linewidth) is readily found by a comparison of Eqs. (2.5.17) and (2.4.24), giving

$$\Delta v_0^* = 2v_0 \left(\frac{2kT \ln 2}{Mc^2} \right)^{1/2} \qquad (2.5.18)$$

For the purely inhomogeneous case, the lineshape is then given by Eq. (2.4.27), where Δv_0^* is expressed by Eq. (2.5.18).

Example 2.6. *Doppler linewidth of a He–Ne laser.* Consider the Ne line at wavelength $\lambda = 632.8$ nm (the red laser line of a He–Ne laser) and assume $T = 300$ K. Then from Eq. (2.5.18), using the appropriate mass for Ne, we obtain $\Delta v_0^* \cong 1.7$ GHz. Comparing this value with those obtained for collision broadening, see Example 2.2, and natural broadening, see Example 2.4 (the transition is electric-dipole-allowed), shows that Doppler broadening is the predominant line-broadening mechanism in this case.

2.5.3. Concluding Remarks

According to the previous discussion, the shape of a homogeneous line is always Lorentzian, while that of an inhomogeneous line is always Gaussian. When two mechanisms contribute to line broadening, the overall lineshape turns out to be always given by the convolution of the corresponding lineshape functions, as indicated in Eq. (2.5.26) for the case when one line is homogeneously and the other is inhomogeneously broadened. One can show that the convolution of a Lorentzian line of width Δv_1 with another Lorentzian line of width Δv_2 again gives a Lorentzian line whose width is $\Delta v = \Delta v_1 + \Delta v_2$. The convolution of a Gaussian line of width Δv_1 with another Gaussian line of width Δv_2 is again a Gaussian line, this time of width $\Delta v = (\Delta v_1^2 + \Delta v_2^2)^{1/2}$. For any combination of broadening mechanisms, it is therefore always possible to reduce the problem to a convolution of a single Lorentzian line with a single Gaussian line; this integral, known as the Voigt integral,[11] is tabulated. Sometimes however, e.g., as in the previous cases for Ne, one mechanism predominates. In this case it is then possible to speak of a pure Lorentzian or Gaussian line.

TABLE 2.1. Typical magnitude of frequency broadening for various line-broadening mechanisms

	Type	Gas	Liquid	Solid
Homogeneous	Natural	1 kHz–10 MHz	Negligible	
	Negligible			
	Collisions	5–10 MHz/Torr	~ 300 cm^{-1}	—
	Phonons	—	—	~ 10 cm^{-1}
Inhomogeneous	Doppler	50 MHz–1 GHz	Negligible	—
	Local field	—	~ 500 cm^{-1}	1–500 cm^{-1}

We conclude this section by showing in Table 2.1 the actual range of linewidths for various line-broadening mechanisms considered. Note that, in the middle of the visible range, one has $\tau_{sp} \cong 10$ ns and hence $\Delta v_{nat} \cong 10$ MHz for an electric-dipole-allowed transition. For an electric-dipole-forbidden transition, on the other hand, one has $\tau_{sp} \cong 1$ ms and hence $\Delta v_{nat} \cong 1$ kHz. Note also that, in the case of a liquid, collision broadening and local-field inhomogeneous broadening are the predominant broadening mechanisms. In this case the average time between two consecutive collisions is indeed much shorter than in the gas phase ($\tau_c \cong 0.1$ ps); hence one has $\Delta v_c = 1/\pi\tau_c \cong 100$ cm^{-1}. Inhomogeneous broadening arises from local density variations associated with a given temperature; it may produce a value for the linewidth Δv_0^* comparable to that of collision broadening. In a solid, inhomogeneous broadening due to local field variations may be as high as 300 cm^{-1} for a glass and as low as 0.5 cm^{-1}, or even lower, for a good-quality crystal, such as a presently available Nd:YAG crystal.

2.6. NONRADIATIVE DECAY AND ENERGY TRANSFER

In addition to decaying via radiative emission, an excited species can also undergo nonradiative decay. This can occur in a variety of ways, and the detailed description is often complicated. We therefore limit ourselves to a qualitative discussion whose main focus is to elucidate the physical phenomena involved. We then consider the combined effect of radiative and nonradiative decay processes.

2.6.1 Mechanisms of Nonradiative Decay

We first consider a nonradiative decay mechanism that arises from collisions, sometimes called *collisional deactivation*.[12] In this case, for a gas or liquid, the transition energy is released as excitation and/or kinetic energy of the colliding species, or it is transferred to the container walls. In the case of a solid, such as an ionic crystal or glass, the energy of the excited ion is taken up by the lattice phonons or by the glass vibrational modes.

The collisional deactivation process, in the case when the energy of an excited species B^* is released as kinetic energy of a colliding species A, can be expressed in the form:

$$B^* + A \rightarrow B + A + \Delta E \tag{2.6.1}$$

where ΔE is equal to the excitation energy. Since ΔE ends up as kinetic energy of the colliding partners, the process is also referred to as a *superelastic collision* or a *collision of second kind*. For a process of the form shown in Eq. (2.6.1), the rate of change of B^* population N_{B^*} can be written as:

$$\frac{dN_{B^*}}{dt} = -k_{B^*A}N_{B^*}N_A \qquad (2.6.2)$$

where N_A is the population of species A and k_{B^*A} is a coefficient that depends on the transition of species B and on species A. The process is particularly effective, i.e., k_{B^*A} is particularly large, when A has a very small mass (e.g., the He in the gas of a CO_2 laser), so it can more readily take up the surplus energy ΔE from the collision process, as kinetic energy. For the same reason the process can readily occur in a gas discharge when A is a discharge electron (e.g., deactivation of the 2^3S state of He in a He–Ne laser). According to Eq. (2.6.2), we can now define a nonradiative decay rate:

$$W_{nr} = k_{B^*A}N_A \qquad (2.6.3)$$

From Eqs. (2.6.2) and (2.6.3) we then obtain

$$\left(\frac{dN_2}{dt}\right) = -\frac{N_2}{\tau_{nr}} \qquad (2.6.4)$$

where, to conform with previous notations, we have let N_2 be the population of the species undergoing collisional deactivation and we have defined a nonradiative decay time as $\tau_{nr} = (1/W_{nr})$.

Note that, in writing Eq. (2.6.2), we have neglected the reverse process of that given by Eq. (2.6.1), i.e.:

$$B + A \rightarrow B^* + A - \Delta E \qquad (2.6.5)$$

where species B is excited at the expense of the kinetic energy ΔE of the two colliding partners (*thermal activation* or *collision of the first kind*). If this process were taken into account, we would write the following equation instead of Eq. (2.6.2):

$$\left(\frac{dN_{B^*}}{dt}\right) = -k_{B^*A}N_{B^*}N_A + k_{BA}N_B N_A \qquad (2.6.6)$$

where k_{BA} is a coefficient describing the process of thermal activation. To find the relationship between k_{BA} and k_{B^*A}, we consider species B in thermal equilibrium with species A, then apply the so-called *principle of detailed balance*. This principle can be formulated generally by requiring that, in thermodynamic equilibrium, the rate of *any* process must be exactly balanced by the rate of the corresponding reverse process.* In this case, according to Eq. (2.6.6), we require

$$k_{B^*A}N_{B^*}N_A = k_{BA}N_B N_A \qquad (2.6.7)$$

---

* Note: The equation expressing the balance of processes between a two-level atom and the blackbody radiation, established in Section 2.4.4, is another example of the principle of detailed balance.

In thermal equilibrium and for nondegenerate levels, we have $N_{B^*} = N_B \exp(-\Delta E / kT)$ where ΔE is the excitation energy of species B and T is the translational temperature of the ensemble of species B and A. From Eq. (2.6.7) we then obtain

$$k_{B^*A} = k_{BA} \exp(\Delta E / kT) \tag{2.6.8}$$

which shows that the rate coefficient k for the exothermic reaction [Eq. (2.6.1)] is always greater than that of the endothermic reaction [Eq. (2.6.5)]. Actually, for electronic and most vibrational transitions, ΔE is much larger than kT. Thus, according to Eq. (2.6.8), one has $k_{B^*A} \gg k_{BA}$. It is very important to realize that, although Eq. (2.6.8) is derived for thermal equilibrium conditions, the same relation still holds if the population of species B is maintained in a nonequilibrium state of excitation, e.g., by some pumping process, provided that the translational degrees of freedom of both species B and A are still in thermal equilibrium. In fact the quantum mechanical calculation of the rate coefficient k does not depend on the population of B but only on eigenfunctions of the two species involved and on their relative velocities.

For a steady excitation of species B away from thermal equilibrium, i.e., when N_{B^*} is of the same order of N_B, we thus have $k_{B^*A}N_{B^*} \gg k_{BA}N_B$, and Eq. (2.6.6) reduces to Eq. (2.6.2). To conclude, collisional deactivation takes the simple form given by Eq. (2.6.4) only when $\Delta E > kT$, so that thermal activation can be neglected, which is the case for electronic transitions and most vibrational transitions. For deactivation of the lowest energy vibrational levels of some molecules [e.g., the (010) state of CO_2] and for rotational transitions, thermal excitation must however be taken into account.

When the electronic energy of species B is released in the form of internal energy of some other species A, we can represent this with an equation of the form (collision of the second kind*):

$$B^* + A \rightarrow B + A^* + \Delta E \tag{2.6.9}$$

where $\Delta E = E_B - E_A$ is the difference between the internal energies of the two species (see Fig. 2.11a). The quantum mechanical calculation of the corresponding transition rate is beyond the scope of this book, so we refer the reader elsewhere for details.[13] Here we limit ourselves to pointing out that, since ΔE must be added to or removed from the kinetic energy of two colliding partners, the process turns out to be particularly effective when ΔE is appreciably smaller than kT. Therefore the process is also referred to as *near-resonant energy transfer*, and it often plays an important role as a pumping mechanism in gas lasers (e.g., energy transfer between excited He and ground-state Ne in a He–Ne laser or between excited N_2 and ground-state CO_2 in a CO_2 laser). The process also results in an effective deactivation channel for species B. To consider the dynamics of this deactivation process, we must also take into account the reverse process (*back transfer*, see Fig. 2.11b)

$$B + A^* \rightarrow B^* + A - \Delta E \tag{2.6.10}$$

---

* Collisions of the *first kind* involve converting the kinetic energy of one species into internal energy of another species [see Eq. (2.6.5)]. In collisions of the *second kind*, internal energy is converted into some other form of energy (other than radiation), such as kinetic energy [see Eq. (2.6.1)], or transferred into internal energy of another species (same or different species) [see, Eq. (2.6.9)]. Collisions of the second kind thus also include, for instance, the conversion of excitation energy into chemical energy.

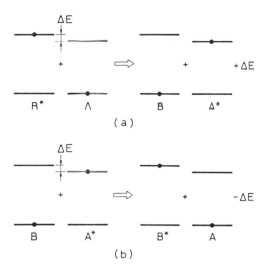

FIG. 2.11. (a) Nonradiative decay of a species B by near-resonant energy transfer to a species A, (b) reverse, back-transfer process.

Again applying the principle of detailed balance we can show that, for the case of exact resonance (i.e., $\Delta E = 0$), we have $k_{B^*A} = k_{BA^*}$, where k_{B^*A} and k_{BA^*} are the rate constants of the two processes described by Eqs. (2.6.9) and (2.6.10), respectively. This indicates that the back-transfer reaction often plays a very important role; however this process can be neglected when the decay of species A from its excited state is very rapid, as it may occur by the onset of stimulated emission. In this case one has $(N_{A^*}/N_A) \ll (N_{B^*}/N_B)$, the back transfer can be neglected, and the rate of decay of the excited species B^* can simply be written as:

$$\left(\frac{dN_{B^*}}{dt}\right) = -k_{B^*A}N_{B^*}N_A \tag{2.6.11}$$

We again obtain an equation of the general form given by Eq. (2.6.4), where now $(1/\tau_{nr}) = k_{B^*A}N_A$.

Finally we consider the case when collisional deactivation of species B (e.g., an active ion in an ionic crystal) occurs through interaction with lattice phonons or glass vibrational modes.* In many cases, except for some nonradiative decay processes occurring in tunable solid-state lasers (see Chap. 9), we are dealing with electronic transitions and thus with transition energies of species B that are many times (typically at least three to four times) greater than that of the most energetic phonon. This means that, to conserve energy, transition energy must be released in the form of many phonons (*multiphonon deactivation*). Thus, in this case, the deactivation process can be represented in the form:

$$B^* \rightarrow B + \sum_1^n {}_i(h\nu_i) \tag{2.6.12}$$

---

* The absence of translational invariance in glass means that, strictly speaking, one should not talk in terms of phonons, in this case, as one does for a crystal. From now on, however, for brevity we will refer to phonons even in this case.

where v_i are the frequencies of phonons involved and the sum is extended over all phonons created in this resonant or near-resonant process. Again we can define a transition rate W_{nr} according to the relation:

$$\frac{dN_{B^*}}{dt} = -W_{nr}N_{B^*} \tag{2.6.13}$$

In this case, since many phonons are involved, the quantum mechanical calculation for the process involves a higher order perturbation theory; it is therefore not considered here in any detail. Instead we limit ourselves to pointing out that, if only a phonon of frequency v is involved, W_{nr} can be written as $W_{nr} = A \exp(-B\Delta E/hv)$, where A and B are host-dependent constants and ΔE is the transition energy of species B. Thus we see that the transition rate rapidly decreases with the increasing number, $n = \Delta E/hv$, of phonons involved, i.e., with increasing order of the multiphonon process. The dominant contribution to the nonradiative process therefore comes from the lattice phonon of the highest energy, since this means that the lowest order process is then involved. The large variation in vibrational spectra shown by different materials then makes W_{nr} extremely host-dependent. By contrast, the rate is found to be relatively independent of the actual electronic state or even the particular active ion involved.

To conclude our discussion on collisional deactivation, we note that, while the process can take a variety of forms, the decay behavior of the excited state can, subject to limits discussed, always be described by the general relation in Eq. (2.6.4), where the value of τ_{nr} depends on the particular process under consideration. It should be noted explicitly, however, that there is a fundamental difference between the nonradiative decay time τ_{nr} discussed here and the collision time τ_c discussed in Section 2.5.1, although they both originate from collisions. In fact a nonradiative decay process requires an inelastic collision, since the decaying species releases its energy to its surroundings. By contrast, τ_c is the average time between two consecutive dephasing collisions, and thus it arises from elastic collisions only. Note that, in general, elastic collisions are more likely than inelastic ones; thus τ_c is smaller, and often much smaller, than τ_{nr}.

A kind of nonradiative decay that does not rely on collisions, arises from *dipole–dipole interaction* between an excited species that we call the donor D and, e.g., a ground-state species that we call the acceptor A. The interaction results in energy transfer between donor and acceptor. This process was extensively studied by Förster for liquids[14] and by Dexter for solids.[15] It plays a very important role, e.g., for active ions in crystals or glasses and for mixtures of organic dyes in solution. Consider the donor undergoing the downward transition at a distance R from the acceptor. During this transition the donor develops a dipole moment μ_D, oscillating at its transition frequency. From the theory of electric dipole radiation,[16] we know that this moment generates, at a distance R, a nonradiating electric field (the so-called *near-zone field*) whose magnitude $E_D(t)$, as for an electrostatic dipole, equals $\mu_D/4\pi\varepsilon_0 R^3$. Under these conditions, nonradiative decay may occur by energy transfer arising from the interaction of the near-zone field $E_D(t, R)$, at the position of the acceptor, with the oscillating dipole moment of the acceptor, μ_A. The interaction energy H can then be written as:

$$H \propto |E_D \cdot \mu_A| \propto \frac{|\mu_D \cdot \mu_A|}{R^3} \tag{2.6.14}$$

Of course, the interaction has a significant strength only if the oscillation frequency of μ_D is nearly resonant with that of μ_A. This means that there must be a good overlap between the

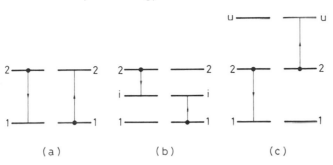

FIG. 2.12. Different forms of energy transfer by dipole–dipole interaction within the same species: (a) Migration of excitation, (b) cross relaxation, (c) cooperative up-conversion.

emission spectrum of the donor and the absorption spectrum of the acceptor from its initial state (not necessarily the ground state). Detailed calculation shows that, for a single donor and acceptor separated by a distance R, the rate of energy transfer can be written as:[14]

$$W_{DA} = \left(\frac{3}{64\pi^5}\right)\left(\frac{1}{R^6}\right)\left[\frac{1}{\tau_{sp}}\int_0^\infty \left(\frac{c}{nv}\right)^4 g_D(v)\sigma_A(v)dv\right] \qquad (2.6.15)$$

where τ_{sp} is the spontaneous lifetime of the donor, n is the refractive index of the surrounding medium, g_D is the lineshape function of the donor, and σ_A is the absorption cross section of the acceptor. Note that since, as usual, the dependence of W_{DA} on the interaction energy H is $W_{DA} \propto H^2$, from Eq. (2.6.14) we expect $W_{DA} \propto |\mu_D|^2|\mu_A|^2/R^6$. From this relation we can now understand the reason for the dependence of W_{DA} on R^{-6} as well as on $(1/\tau_{sp})$ [remember: $1/\tau_{sp} \propto |\mathbf{\mu}_D|^2$, see Eq. (2.3.15)] and on the cross section σ_A of the acceptor [remember: $\sigma_A \propto |\mathbf{\mu}_A|^2$, see Eq. (2.4.29)].

It should be noted finally that, dipole–dipole interactions may take somewhat different forms from the classical donor–acceptor situation just considered. For instance they may occur between members of a single species (referred to as species D) and thus lead for example to energy transfer between an excited and an unexcited atom (Fig. 2.12a). Such a resonant energy transfer usually leads simply to spatial *excitation migration* within the same species D. It may also lead, however, to a nonradiative decay if excitation eventually reaches a D site that is close to a nearly resonant acceptor site that has a rapid decay. Energy transfer may also occur to an intermediate level i, as shown in Fig. 2.12b (*cross relaxation*). This process is particularly effective for near-resonant transfer, i.e., when $\Delta E_{2i} \cong \Delta E_{i1}$. Finally, energy transfer may occur through an upper level, with both donor and acceptor initially excited and the acceptor then excited to a higher level u (Fig. 2.12c). This process, known as *cooperative up-conversion*, is particularly effective for near-resonant energy transfer, i.e., when $\Delta E_{u2} \cong \Delta E_{21}$.

Example 2.7. *Energy transfer in the* Yb^{3+}:Er^{3+}:*glass laser system.*[17] Donor-acceptor type of energy transfer is very effective in the Yb^{3+}:Er^{3+}:glass laser system (see Chap. 9) in transferring excitation from the Yb^3 ion, initially excited to its $^2F_{5/2}$ state, to the $^4I_{11/2}$ excited level of Er^{3+} (Fig. 2.13a). This energy transfer is an effective nonradiative decay mechanism for the Yb ion, and it constitutes a very effective way of pumping the active Er ion. Note that, at high Yb ion concentrations, this energy transfer is assisted by energy migration among Yb ions until excitation reaches a closely spaced Yb–Er pair.

Example 2.8. *Nonradiative decay from the $^4F_{3/2}$ upper laser level of Nd:YAG.* Cross relaxation turns out to be the main nonradiative decay mechanism for the Nd:YAG $^4F_{3/2}$ upper laser level. In this case the intermediate level i in Fig. 2.12b is the $^4I_{15/2}$ level of the Nd ion (Fig. 2.13b). The excitation energy of this level is then rapidly lost via multiphonon relaxation, and the ion passes successively through the $^4I_{13/2}$ and the $^4I_{11/2}$ lower levels (not shown in the figure) to the $^4I_{9/2}$ ground level. The energy difference between these sublevels (e.g., $^4I_{13/2} \rightarrow {}^4I_{11/2}$) is in fact about 2000 cm^{-1} (the Stark sublevels are even closer), i.e., only four times larger than the highest vibrational frequency of YAG crystal (~ 450 cm^{-1}). This mechanism limits the optimal concentration of Nd ions in a YAG crystal to about 1%.

Example 2.9. *Cooperative upconversion in Er^{3+} lasers and amplifiers.*[17] Cooperative up-conversion is believed to be one of the major causes of inefficiency in Er^{3+} lasers or amplifiers (Fig. 2.13c). In this case, of two neighboring Er ions initially excited to the $^4I_{13/2}$ laser level, one decays to the ground $^4I_{15/2}$ level, while the other is raised to the $^4I_{9/2}$ level. From this level, when the ion is in an oxide glass host, it decays rapidly back to the $^4I_{13/2}$ level by multiphonon decay. The net result of this cooperative up-conversion is that one Er ion, initially excited to the laser $^4I_{13/2}$ level, is effectively quenched to the ground level; i.e., one loses 50% of the population.

2.6.2. Combined Effects of Radiative and Nonradiative Processes

First consider the case when nonradiative decay can be described by an equation of the general form [Eq. (2.6.4)]. The time variation of the upper state population N_2 can then be written as:

$$\frac{dN_2}{dt} = -\left(\frac{N_2}{\tau_r} + \frac{N_2}{\tau_{nr}}\right) \tag{2.6.16}$$

Equation (2.6.16) can be put in the simpler form:

$$\frac{dN_2}{dt} = -\left(\frac{N_2}{\tau}\right) \tag{2.6.17}$$

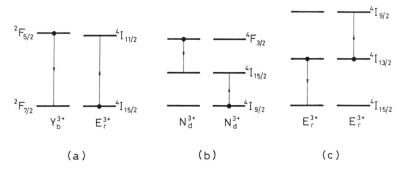

(a)   (b)   (c)

FIG. 2.13. Examples of energy transfer by dipole–dipole interaction: (a) Yb^{3+}–Er^{3+} energy transfer in an Yb:Er laser or amplifier, (b) nonradiative decay of Nd:YAG by cross relaxation, (c) cooperative up-conversion in an Er^{3+} laser or amplifier.

if we define an overall decay time τ given by:

$$\frac{1}{\tau} = \frac{1}{\tau_r} + \frac{1}{\tau_{nr}} \qquad (2.6.18)$$

The population $N_2(t)$ at time t is then obtained by integrating Eq. (2.6.17). We obtain

$$N_2(t) = N_2(0) \exp[-(t/\tau)] \qquad (2.6.19)$$

where $N_2(0)$ is the population at $t = 0$. To calculate the time behavior of the spontaneously emitted light, we note that, according to Eq. (2.6.16), N_2/τ_r gives the number of atoms decaying radiatively per unit volume and unit time. Assuming for simplicity that radiative decay occurs to one lower level only, say level 1, and letting v_0 be the corresponding transition frequency, the spontaneously emitted power at time t is then:

$$P(t) = \frac{N_2(t)hv_0V}{\tau_r} \qquad (2.6.20)$$

where V is the volume of the material. With the help of Eq. (2.6.19), Eq. (2.6.20) gives

$$P(t) = \left[\frac{N_2(0)hv_0V}{\tau_r}\right] \exp[-(t/\tau)] \qquad (2.6.21)$$

Note that the time decay of the emitted light is exponential with a time constant τ rather than τ_r, as perhaps expected at first sight. By monitoring the decay of the spontaneously emitted light from a sample having at $t = 0$ an initial upper state population $N_2(0)$, we thus measure the overall lifetime τ. To obtain τ_r, let us first define the fluorescence quantum yield ϕ as the ratio of the number of emitted photons to the number of atoms initially raised to level 2. Using Eq. (2.6.21) we obtain

$$\phi = \frac{\int (P(t)/hv_0)dt}{N_2(0)V} = \frac{\tau}{\tau_r} \qquad (2.6.22)$$

Note that one can easily show that Eq. (2.6.22) also remains true when the decay is to a number of lower levels provided that, in defining the quantum efficiency ϕ, one includes photons emitted in all of these transitions. The measurement of ϕ thus allows us to calculate τ_r once τ is known from the decay measurement of the emitted radiation. This measurement however is sometimes not easy, especially when τ is very short (picoseconds or even less), i.e., when ϕ is very small.

We now briefly consider the case when nonradiative decay occurs via a dipole–dipole-mediated energy transfer. According to Eq. (2.6.15), the transition rate W_{DA} is strongly dependent on the donor-to-acceptor distance R. For a population of N_D donors and N_A acceptors, due to different distances between donors and acceptors, the decay rate differs for each donor-to-acceptor pair; the resulting overall decay then shows a nonexponential behavior, with the initial faster decay corresponding to sites with the smallest separation R. A particularly important case occurs when there are random values of donor–acceptor spacing and this distance is either fixed, as in a solid, or varies slowly over a spontaneous decay time, as often occurs in liquids (*Förster regime*). In this case, when both radiative and nonradiative decay channels are taken into account, the overall decay turns out to be given by:

$$N_2(t) = N_2(0) \exp\{-[(t/\tau_r) + Ct^{1/2}]\} \tag{2.6.23}$$

where C is a constant proportional to the acceptor concentration, N_A. The time behavior of the radiated power is then obtained by substituting Eq. (2.6.23) into Eq. (2.6.20) and thus follows the same nonexponential decay as that of the population $N_2(t)$.

2.7. DEGENERATE OR STRONGLY COUPLED LEVELS

So far we have considered only the simplest case when both levels 1 and 2 are nondegenerate. We now briefly consider the case when the two levels are degenerate or composed of a number of strongly coupled levels. The situation is depicted in Fig. 2.14, where levels 1 and 2 are assumed to consist of g_1 and g_2 sublevels that are either degenerate (i.e., have the same energy) or are strongly coupled. We let N_1 and N_2 be the total population of levels 1 and 2 and use N_{1i} and N_{2j} to indicate the population of a particular sublevel of the lower and upper manifolds, respectively.

2.7.1. Degenerate Levels

We first look at the degenerate case, and begin by considering the thermal equilibrium situation. In this case the population of each sublevel of both upper and lower state obeys the usual Boltzmann equation; thus:

$$N_{2j}^e = N_{1i}^e \exp[-(E_2 - E_1)/kT] \tag{2.7.1}$$

However, since the sublevels of, e.g., level 1 are also in thermal equilibrium, their populations must all be equal; thus:

$$N_{1i}^e = \frac{N_1^e}{g_1} \tag{2.7.2a}$$

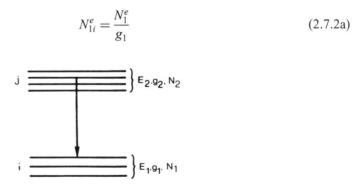

FIG. 2.14. Two-level system where the two levels comprise many sublevels that are either degenerate or strongly coupled.

Similarly we have

$$N_{2j}^e = \frac{N_2^e}{g_2} \tag{2.7.2b}$$

From Eqs. (2.7.1) and (2.7.2) we then obtain

$$N_2^e = N_1^e \left(\frac{g_2}{g_1}\right) \exp\left[-\frac{(E_2 - E_1)}{kT}\right] \tag{2.7.3}$$

Let us now see how to modify expressions for transition cross section, gain, and absorption coefficient in the case of degenerate levels. For this purpose we consider an em wave passing through a material, with given overall populations N_1 and N_2 in the two levels; we calculate the rate of change of the overall population N_2 due to all radiative and nonradiative transitions between sublevels j and i. We therefore write

$$\left(\frac{dN_2}{dt}\right) = -\sum_i^{g_1} \sum_j^{g_2} [W_{ji}N_{2j} - W_{ij}N_{1i} + (N_{2j}/\tau_{ji})] \tag{2.7.4}$$

where W_{ji} is the rate of stimulated transition between j and i sublevels, W_{ij} is the rate of absorption, and $(1/\tau_{ji})$ is the rate of spontaneous decay, radiative and nonradiative, between the same two sublevels. Note that W_{ji} and W_{ij} are obtained from Eq. (2.4.30) by substituting the dipole moments between j and i sublevels, $|\mu_{ij}|^2$ and $|\mu_{ji}|^2$, for $|\mu|^2$. These dipole moments can in turn be readily obtained from Eq. (2.3.7). For instance $|\mu_{ij}|$ is obtained from Eq. (2.3.7) by substituting u_i, the eigenfunction of the i-th lower level, for u_1, and u_j, the eigenfunction of the j-th upper level, for u_2. It then follows that:

$$W_{ji} = W_{ij} \tag{2.7.5}$$

If a rapid relaxation toward thermal equilibrium occurs between sublevels within each level, then all sublevels of the upper level are again equally populated, and the same occurs on lower level sublevels. Therefore:

$$N_{2j} = \frac{N_2}{g_2} \tag{2.7.6a}$$

$$N_{1i} = \frac{N_1}{g_1} \tag{2.7.6b}$$

Substituting Eq. (2.7.6) into (2.7.4) we then obtain

$$\frac{dN_2}{dt} = -W\left(\frac{N_2}{g_2} - \frac{N_1}{g_1}\right) - \frac{N_2}{\tau} \tag{2.7.7}$$

where, with the help of Eq. (2.7.5), we have defined

$$W = \sum_i^{g_1} \sum_j^{g_2} W_{ij} = \sum_i^{g_1} \sum_j^{g_2} W_{ji} \tag{2.7.8}$$

and

$$\frac{1}{\tau} = \frac{\sum_i^{g_1} \sum_j^{g_2} (1/\tau_{ji})}{g_2}$$ (2.7.9)

From Eq. (2.7.7) we observe that WN_2/g_2 represents the rate of change of the total upper state population due to all stimulated emissions processes; likewise WN_1/g_1 represents the population change due to all absorption processes. The change in photon flux dF when the beam travels a distance dz in the material (see Fig. 1.2) can then be written as:

$$dF = W\left(\frac{N_2}{g_2} - \frac{N_1}{g_1}\right)dz$$ (2.7.10)

We can now define a stimulated emission cross section σ_{21} and absorption cross section σ_{12} as:

$$\sigma_{21} = \frac{W}{(g_2 F)}$$ (2.7.11a)

$$\sigma_{12} = \frac{W}{(g_1 F)}$$ (2.7.11b)

from which we obviously have

$$g_2 \sigma_{21} = g_1 \sigma_{12}$$ (2.7.12)

When $(N_1/g_1) > (N_2/g_2)$, Eq. (2.7.10), with the help of Eq. (2.7.11b), can be put in the familiar form $dF = -\alpha F\, dz$ if we define the absorption coefficient α as:

$$\alpha = \sigma_{12}[N_1 - N_2(g_1/g_2)]$$ (2.7.13)

Similarly, when $(N_2/g_2) > (N_1/g_1)$, Eq. (2.7.10), with the help of Eq. (2.7.11a), can be put in the familiar form $dF = gF\, dz$ if we define the gain coefficient g as:

$$g = \sigma_{21}[N_2 - N_1(g_2/g_1)]$$ (2.7.14)

The reasons for defining σ_{21} and σ_{12}, respectively as in Eqs. (2.7.11a) and (2.7.11b), is now apparent. When in fact $N_1 \gg N_2$ (as usually applies to absorption measurements involving optical transitions) Eq. (2.7.13) simply reduces to $\alpha = \sigma_{12}N_1$. Conversely, when $N_2 \gg N_1$ (as in a four-level laser), Eq. (2.7.14) simply reduces to $g = \sigma_{21}N_2$.

2.7.2. Strongly Coupled Levels

We now turn to the case where the upper level, 2, and lower level, 1, actually consist of g_2 and g_1 sublevels, respectively, with different energies but with very rapid relaxation among sublevels belonging to each particular level (strongly coupled levels). Each sublevel of both upper and lower levels may also consist of many degenerate levels. In this case, thermalization among sublevels of either the lower or upper level occurs rapidly, so that we can assume Boltzmann's statistics always to be obeyed. Instead of Eq. (2.7.6) we write now:

$$N_{2j} = f_{2j}N_2 \tag{2.7.15a}$$

$$N_{1i} = f_{1i}N_1 \tag{2.7.15b}$$

where $f_{2j}(f_{1i})$ is the fraction of total population of level 2 (level 1) that is found in sublevel $j(i)$ at thermal equilibrium. According to Boltzmann's statistics, we then have

$$f_{2j} = \frac{g_{2j}\exp[-(E_{2j}/kT)]}{\sum_{m}^{g_2} g_{2m}\exp[-(E_{2m}/kT)]} \tag{2.7.16a}$$

$$f_{1i} = \frac{g_{1i}\exp[-(E_{1i}/kT)]}{\sum_{l}^{g_1} g_{1l}\exp[-(E_{1l}/kT)]} \tag{2.7.16b}$$

where E_{2m} and E_{1l} are sublevel energies in the upper and lower level, respectively, and g_{2m} and g_{1l} are their corresponding degeneracies.

Let us now assume that the stimulated transition occurs between a given sublevel (say, l) of level 1 to a given sublevel (say, m) of level 2. Equation (2.7.4) then simplifies to:

$$\left(\frac{dN_2}{dt}\right) = -W_{ml}N_{2m} + W_{lm}N_{1l} - \sum_{i}^{g_1}\sum_{j}^{g_2}\left(\frac{N_{2j}}{\tau_{ji}}\right) \tag{2.7.17}$$

With the help of Eq. (2.7.15), Eq. (2.7.17) can be written as:

$$\left(\frac{dN_2}{dt}\right) = -W_{ml}^e N_2 + W_{lm}^e N_1 - \left(\frac{N_2}{\tau}\right) \tag{2.7.18}$$

where we have defined the effective rates of stimulated emission W_{ml}^e, stimulated absorption W_{lm}^a, and spontaneous decay $(1/\tau)$, respectively, as:

$$W_{ml}^e = f_{2m}W_{ml} \tag{2.7.19a}$$

$$W_{lm}^a = f_{1l}W_{lm} \tag{2.7.19b}$$

$$\left(\frac{1}{\tau}\right) = \sum_{i}^{g_1}\sum_{j}^{g_2}\left(\frac{f_{2j}}{\tau_{ji}}\right) \tag{2.7.19c}$$

According to Eq. (2.7.18), the change in photon flux dF when the beam travels a distance dz in the material is now given by

$$dF = (W_{ml}^e N_2 - W_{lm}^a N_1)dz \tag{2.7.20}$$

We can then define an effective stimulated emission cross section σ^e_{ml} and an effective absorption cross section σ^a_{lm} as:

$$\sigma^e_{ml} = \frac{W^e_{ml}}{F} = f_{2m}\sigma_{ml} \tag{2.7.21a}$$

$$\sigma^a_{lm} = \frac{W^a_{lm}}{F} = f_{1l}\sigma_{lm} \tag{2.7.21b}$$

where Eqs. (2.7.19a–b) are used and $\sigma_{lm} = W_{lm}/F$ and $\sigma_{ml} = W_{ml}/F$ are, respectively, the actual cross sections of absorption and stimulated emission for the *l*-to-*m* transition. Note that, if the two sublevels *l* and *m* are nondegenerate (or have the same degeneracy), one has $\sigma_{lm} = \sigma_{ml}$. Note also that, according to Eqs. (2.7.20) and (2.7.21), the absorption coefficient for the propagating photon flux can be written as:

$$\alpha_{lm} = \sigma^a_{lm}N_1 - \sigma^e_{ml}N_2 \tag{2.7.22}$$

This shows the usefulness of the concept of effective cross section: The absorption coefficient, or the gain coefficient when $N_2 > N_1$, is simply obtained by multiplying the effective cross section with the *total* population of the upper and lower levels. In particular, at thermal equilibrium, one has $N_2 \cong 0$ and $N_1 \cong N_t$, where N_t is the total population, and Eq. (2.7.22) gives

$$\alpha_{lm} = \sigma^a_{lm}N_t \tag{2.7.23}$$

Example 2.10. *Effective stimulated-emission cross section for the $\lambda = 1.064$-μm laser transition of Nd:YAG.* The scheme of relevant energy levels for the Nd:YAG laser is shown in Fig. 2.15. Laser action can occur on the $^4F_{3/2} \rightarrow {}^4I_{11/2}$ transition ($\lambda = 1.064$ μm), which is the most popular, as well as on $^4F_{3/2} \rightarrow {}^4I_{13/2}$ ($\lambda = 1.32$ μm) and $^4F_{3/2} \rightarrow {}^4I_{9/2}$ transitions ($\lambda = 0.94$ μm). The 1.064-μm transition occurs between one sublevel, $m = 2$, of the $^4F_{3/2}$ level to one sublevel, $l = 3$, of the $^4I_{11/2}$ level ($R_2 \rightarrow Y_3$ transition). We let $f_{22} = N_{22}/N_2 = N_{22}/(N_{21} + N_{22})$ represent the fraction of the total population found in the upper laser level, where N_{22} and N_{21} are populations of the two sublevels of the $^4F_{3/2}$ level and N_2 is the total population of this level. Since the two sublevels are each doubly degenerate, then, according to Eq. (2.7.3), one has $N_{22} = N_{21}\exp[-(\Delta E/kT)]$, where ΔE is the energy separation between the two sublevels. From the previous expression for N_{22}, we then obtain $f_{22} = 1/[1 + \exp(\Delta E/kT)]$. For $\Delta E = 84$ cm^{-1} and $kT = 208$ cm^{-1} ($T = 300$ K), we obtain $f_{22} = 0.4$. From measured spectroscopic data on the $R_2 \rightarrow Y_3$ transition, the actual peak cross section of the transition is deduced as $\sigma_{23} = 6.5 \times 10^{-19}$ cm^2.[21] The effective cross section of the $R_2 \rightarrow Y_3$ transition σ^e_{23} is then obtained from Eq. (2.7.21a) as $\sigma^e_{23} = f_{22}\sigma_{23} \cong 2.8 \times 10^{-19}$ cm^2.

Example 2.11. *Effective stimulated-emission cross section and radiative lifetime in alexandrite.* The relevant energy levels of alexandrite are shown in Fig. 2.16. The upper laser level is the $^4T_2$ state; the laser transition occurs to a vibronic level of the $^4A_2$ ground state ($\lambda \cong 730 \div 800$ nm). Since the $^4T_2$ level is strongly coupled to the ^2E level, the fraction of the total population found in the $^4T_2$ state, f_{2T}, is given by $f_{2T} = N_{2T}/(N_{2E} + N_{2T})$, where N_{2E} and N_{2T} are populations of the two levels. At thermal equilibrium we also have $N_{2T} = N_{2E}\exp[-(\Delta E/kT)]$, where ΔE is the energy separation between the two levels. From the previous expressions we obtain $f_{2T} = \exp-(\Delta E/kT)/[1 + \exp-(\Delta E/kT)]$. Assuming

$\Delta E = 800$ cm^{-1}, $kT = 208$ cm^{-1}, and $\sigma_{TA} = 4 \times 10^{-19}$ cm^2 at $\lambda = 704$ nm,[22] we obtain $\sigma_{TA}^e \cong 0.8 \times 10^{-20}$ cm^2. Note the small value of f_{2T}, i.e., the small fractional value of the upper laser level population, which results in a strong reduction of the effective stimulated emission cross section. Note also that this cross section increases with increasing temperature because f_{2T} increases with temperature. To calculate the effective lifetime, τ, of the upper laser level we note that the rate of spontaneous decay $1/\tau_T$ of the $^4T_2 \rightarrow {}^4A_2$ laser transition is $(1/\tau_T) = 1.5 \times 10^5$ sec^{-1} ($\tau_T \cong 6.6$ μs), while the rate of the $^2E \rightarrow {}^4A_2$ transition is $(1/\tau_E) = 666.6$ sec^{-1} ($\iota_E = 1.5$ ms). From Eq. (2.7.19c) we then obtain $(1/\tau) = (f_{2E}/\tau_E) + (f_{2T}/\tau_T)$ where $f_{2E} = N_{2E}/(N_{2E} + N_{2T}) = 1 - f_{2T}$ is the fraction of the total population found in the ^2E level. Substituting the appropriate values into the previous expression for $(1/\tau)$, we obtain $\iota = 200$ μs at $T = 300$ K. Thus the effective lifetime is considerably lengthened (from 6.6 to 200 μs) due to the presence of the strongly coupled and long-lived ^2E level, which then acts as a storage level or reservoir. Note that the effective lifetime, like the effective cross section, is temperature dependent.

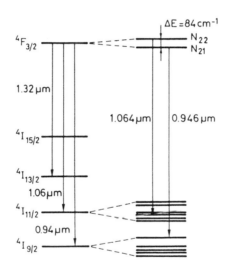

FIG. 2.15. Relevant energy levels for the $\lambda = 1.064$-μm laser transition of Nd:YAG laser.

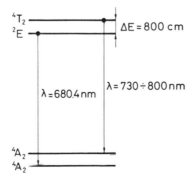

FIG. 2.16. Relevant energy levels of the alexandrite laser.

This equation indicates that σ_{lm}^a can be readily obtained from an absorption measurement.

2.8. SATURATION

This section examines the absorption and emission behavior of a transition (of frequency ν_0) in the presence of a strong monochromatic em wave of intensity I and frequency $\nu \cong \nu_0$. For simplicity we assume the levels to be nondegenerate. Consider first the case when I is sufficiently weak, so that the populations of the two levels N_1 and N_2 do not differ significantly from their thermal equilibrium values. Then $N_1 > N_2$ (often $N_1 \gg N_2$) and the absorption processes, of rate WN_1, dominate the stimulated emission rate WN_2; i.e., more atoms undergo the transition $1\rightarrow2$ than the transition $2\rightarrow1$. Consequently, at sufficiently high values of intensity I, the two populations tend to equalize. This phenomenon is referred to as *saturation*.

2.8.1. Saturation of Absorption: Homogeneous Line

We first consider an absorbing transition ($N_1 > N_2$) and assume the line to be homogeneously broadened. The rate of change of the upper state population N_2 due to the combined effects of absorption, stimulated emission, and spontaneous decay (radiative and nonradiative) (Fig. 2.17) can be written as:

$$\frac{dN_2}{dt} = -W(N_2 - N_1) - \frac{N_2}{\tau} \tag{2.8.1}$$

where N_1 is the population of level 1. We can also write

$$N_1 + N_2 = N_t \tag{2.8.2}$$

where N_t is the total population. Equation (2.8.1) can be put into a simpler form by defining:

$$\Delta N = N_1 - N_2 \tag{2.8.3}$$

Equations (2.8.2) and (2.8.3) then give N_1 and N_2 as a function of ΔN and N_t, so Eq. (2.8.1) becomes

$$\frac{d\Delta N}{dt} = -\Delta N\left(\frac{1}{\tau} + 2W\right) + \frac{1}{\tau}N_t \tag{2.8.4}$$

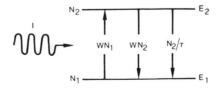

FIG. 2.17. Two-level system interacting with an em wave of high intensity I.

When $(d\Delta N/dt) = 0$, i.e., in the steady state, we obtain

$$\Delta N = \frac{N_t}{1 + 2W\tau} \qquad (2.8.5)$$

To maintain a given population difference ΔN the material must absorb a power per unit volume (dP/dV) from the incident radiation given by:

$$\frac{dP}{dV} = (hv)W\Delta N = (hv)\frac{N_t W}{1 + 2W\tau} \qquad (2.8.6)$$

which at saturation, i.e., for $W\tau \gg 1$, becomes

$$\left(\frac{dP}{dV}\right)_s = (hv)\frac{N_t}{2\tau} \qquad (2.8.7)$$

Equation (2.8.7) shows that the power that must be absorbed by the system to keep it in saturation $(dP/dV)_s$, as expected, equals the power lost by the material due to spontaneous decay of the upper state population $(N_t/2)$.

It is sometimes useful to rewrite Eqs. (2.8.5) and (2.8.6) in a more convenient form. To do this we first note that, according to Eq. (2.4.17), W can be expressed as:

$$W = \frac{\sigma I}{hv} \qquad (2.8.8)$$

where σ is the absorption cross section of the transition. Equations (2.8.5) and (2.8.6) with the help of Eq. (2.8.8) can be recast in the following forms:

$$\frac{\Delta N}{N_t} = \frac{1}{1 + (I/I_s)} \qquad (2.8.9)$$

$$\frac{dP/dV}{(dP/dV)_s} = \frac{I/I_s}{1 + (I/I_s)} \qquad (2.8.10)$$

where:

$$I_s = \frac{hv}{2\sigma\tau} \qquad (2.8.11)$$

is a parameter that depends on the given material and frequency of the incident wave. Its physical meaning is obvious from Eq. (2.8.9). In fact for $I = I_s$, we get $\Delta N = N_t/2$. When $v = v_0$, the quantity I_s has a value that depends only on transition parameters. This quantity is called the *saturation intensity*.

Let us now see how the shape of an absorption line changes with increasing intensity I of the saturating beam. To do this, let us consider the idealized experimental situation shown in Fig. 2.18, where absorption measurements are made using a probe beam, of variable frequency v', whose intensity I' is small enough not to perturb the system appreciably. In practice the beams must be more or less collinear to ensure that the probe beam interacts

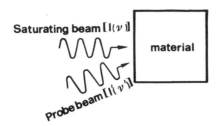

FIG. 2.18. Measurement of the absorption or gain coefficient at frequency v' by a probe beam of intensity $I'(v')$ in the presence of a saturating beam $I(v)$ of intensity I and frequency $v[I(v) \gg I'(v')]$.

only with the saturated region. Under these conditions the absorption coefficient seen by the probe beam is obtained from Eq. (2.4.33) by substituting the homogeneous lineshape $g(v' - v_0)$ for the total lineshape $g_t(v - v_0)$, where v' is substituted for v. Since $N_1 - N_2 = \Delta N$ is now given by Eq. (2.8.9), we can write

$$\alpha = \frac{\alpha_0}{1 + (I/I_s)} \tag{2.8.12}$$

where:

$$\alpha_0 = \frac{2\pi^2}{3n\varepsilon_0 ch} |\mu|^2 v' g(v' - v_0) \tag{2.8.13}$$

is the absorption coefficient when the saturating wave at frequency v is absent (*unsaturated absorption coefficient*). Equations (2.8.12) and (2.8.13) show that, when the intensity I of the saturating beam is increased, the absorption coefficient is reduced. The lineshape however remains the same, since it is always described by the function $g(v' - v_0)$. Figure 2.19 plots the absorption coefficient α versus v' at three different values of I/I_s.

We next consider the case when the saturating em wave consists of a light pulse with intensity $I = I(t)$ rather than a cw beam. For simplicity we compare two limiting cases where pulse duration is either very long or very short compared to the upper state lifetime, τ. If pulse duration is very long compared to the lifetime, the time evolution of the resulting

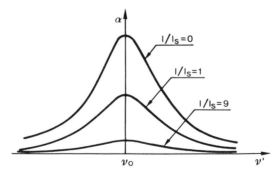

FIG. 2.19. Saturation behavior of the absorption coefficient α versus frequency v' for increasing values of intensity I of the saturating beam (homogeneous line).

population difference ΔN occurs at a very slow rate, so that we can assume in Eq. (2.8.4) $|d\Delta N/dt| \ll N_t/\tau$. Accordingly, ΔN turns out still to be given by the steady-state equation (2.8.9), where now $I = I(t)$. The saturation behavior in this case is essentially the same as for a cw beam. If, on the other hand, the light pulse is very short compared to the lifetime τ, then the stimulated emission term $2W\Delta N$ in Eq. (2.8.4) dominates the spontaneous decay term $(N_t - \Delta N)/\tau$, i.e., $[(N_t - \Delta N)/\tau] \ll 2W\Delta N$. In this case Eq. (2.8.4) reduces to:

$$\left(\frac{d\Delta N}{dt}\right) = -2W\Delta N - -\left(\frac{2\sigma}{h\nu}\right)I(t)\Delta N \tag{2.8.14}$$

where Eq. (2.8.8) has also been used. Integrating Eq. (2.8.14) with the initial condition $\Delta N(0) = N_t$ gives

$$\Delta N(t) = N_t \exp\left[-(2\sigma/h\nu)\int_0^t I(t)dt\right] \tag{2.8.15}$$

Equation (2.8.15) becomes more intuitive if we define the energy fluence $\Gamma(t)$ as:

$$\Gamma(t) = \int_0^t I(t)dt \tag{2.8.16}$$

and the saturation fluence as:

$$\Gamma_s = \frac{h\nu}{2\sigma} \tag{2.8.17}$$

From Eq. (2.8.15) we then obtain

$$\Delta N(t) = N_t \exp[-\Gamma(t)/\Gamma_s] \tag{2.8.18}$$

We see that in this case the beam energy fluence rather than its intensity determines the saturation behavior. The population difference ΔN_∞ that is left in the material after the pulse has passed is, according to Eq. (2.8.18), given by:

$$\Delta N_\infty = N_t \exp[-(\Gamma_t/\Gamma_s)] \tag{2.8.19}$$

where Γ_t is the total energy fluence of the light pulse. The material saturation fluence Γ_s can therefore be viewed as the fluence that the pulse must have to produce a population difference $\Delta N_\infty = N_t/e$.

Having calculated the population difference resulting from saturation by a light pulse, the corresponding absorption coefficient of the material can then be obtained, for a homogeneous line, again from Eq. (2.4.33) by substituting $g(\nu' - \nu_0)$ for $g_t(\nu - \nu_0)$. For a light pulse that is either long or short compared to τ, the value of α is given, respectively, by Eq. (2.8.12) [with $I = I(t)$] or by:

$$\alpha = \alpha_0 \exp[-\Gamma(t)/\Gamma_s] \tag{2.8.20}$$

where α_0 is the unsaturated absorption coefficient. Note that, in the pulsed regime as in the cw regime, the shape of the absorption line remains unchanged when saturation occurs.

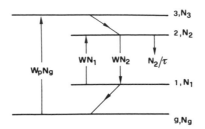

FIG. 2.20. Energy levels and transitions involved in gain saturation of a four-level laser.

2.8.2. Gain Saturation: Homogeneous Line

We now consider the case where the transition $2 \to 1$ exhibits net gain rather than net absorption. We assume that the medium behaves as a four-level system (Fig. 2.20), and the inversion between levels 2 and 1 is produced by some suitable pumping process. We further assume that transitions $3 \to 2$ and $1 \to g$ are so rapid that we can set $N_3 \cong N_1 \cong 0$. With these simplifying assumptions, we can write the following rate equation for the population of level 2:

$$\left(\frac{dN_2}{dt}\right) = R_p - WN_2 - \left(\frac{N_2}{\tau}\right) \tag{2.8.21}$$

where R_p is the pumping rate. In the steady state (i.e., for $dN_2/dt = 0$), we find from Eq. (2.8.21):

$$N_2 = \frac{R_p \tau}{1 + W\tau} \tag{2.8.22}$$

With the help of Eq. (2.8.8), Eq. (2.8.22) can be rewritten as:

$$N_2 = \frac{N_{20}}{1 + (I/I_s)} \tag{2.8.23}$$

where $N_{20} = R_p \tau$ is the population of level 2 in the absence of the saturating beam (i.e., for $I = 0$) and:

$$I_s = \frac{h\nu}{\sigma\tau} \tag{2.8.24}$$

A comparison between Eqs. (2.8.24) and (2.8.11) shows that the expression for the saturation intensity I_s of a four-level system is twice that of the two-level system in Fig. 2.17. The difference arises from the fact that, in a two-level system, a change of population in one level causes an equal and opposite change of population in the other level. Thus ΔN is twice the change in population of each level.

In an experiment such as that shown in Fig. 2.18, the probe beam at frequency ν' now measures gain rather than absorption. From Eq. (2.4.35), with $N_1 = 0$ and also using Eq. (2.8.23), the gain coefficient g can be written as:

$$g = \frac{g_0}{1 + (I/I_s)} \tag{2.8.25}$$

where $g_0 = \sigma N_{20}$ is the gain coefficient for $I = 0$, i.e., when the saturating beam is absent (*unsaturated gain coefficient*). Since the line is homogeneously broadened, this gain coefficient can be obtained, by using Eq. (2.4.18), as:

$$g_0 = \frac{2\pi^2}{3n\varepsilon_0 ch}|\mu|^2 v' N_{20} g(v' - v_0) \qquad (2.8.26)$$

Equations (2.8.25) and (2.8.26) show that, just as in the case of absorption, saturation again leads to a decrease in g as I increases while the gain profile remains unchanged.

We next consider the case when the saturating em wave consists of a light pulse of intensity $I(t)$. If the pulse duration is very long compared to the lifetime τ, we can neglect the time derivative of N_2 in Eq. (2.8.21) compared to the spontaneous decay term N_2/τ. Thus we again obtain Eq. (2.8.23) for the upper state population and Eq. (2.8.25) for the gain coefficient, where I is now a function of time. If the light pulse is very short compared to the lifetime τ, then, during the interaction of the light pulse, the pump term R_p and the spontaneous decay term N_2/τ can be neglected compared to the stimulated term WN_2. Thus we obtain

$$\left(\frac{dN_2}{dt}\right) = -\left(\frac{\sigma I}{hv}\right)N_2 \qquad (2.8.27)$$

where Eq. (2.8.8) is used again. The integration of Eq. (2.8.27) gives

$$N_2(t) = N_{20} \exp\{-[\Gamma(t)/\Gamma_s]\} \qquad (2.8.28)$$

where $N_{20} = R_p\tau$ is the population of level 2 before the arrival of the pulse, $\Gamma(t)$ is the energy fluence of the beam [see Eq. (2.8.16)], and:

$$\Gamma_s = \frac{hv}{\sigma} \qquad (2.8.29)$$

is the amplifier saturation fluence. A comparison between Eqs. (2.8.29) and (2.8.17) shows that the saturation fluence of a four-level amplifier is twice that of an absorber. The saturated gain coefficient is then given by:

$$g = g_0 \exp\{-[\Gamma(t)/\Gamma_s]\} \qquad (2.8.30)$$

where $g_0 = \sigma N_{20}$ is the unsaturated gain coefficient given by Eq. (2.8.26). Thus in the pulsed regime, as for the cw case, the shape of the gain line remains unchanged when saturation occurs.

2.8.3. Inhomogeneously Broadened Line

When the line is inhomogeneously broadened, the saturation phenomenon is more complicated, so we limit ourselves to a qualitative discussion (see Problems 2.16 and 2.17, Chap. 2, for further details). For the sake of generality, we assume that the line is broadened by both homogeneous and inhomogeneous mechanisms so that its shape is expressed by Eq.

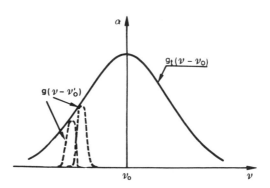

FIG. 2.21. Lineshape of a transition broadened by both homogeneous and inhomogeneous mechanisms. The corresponding lineshape function $g_t(v - v_0)$ is obtained from the convolution of homogeneous lines $g(v - v'_0)$ of individual atoms.

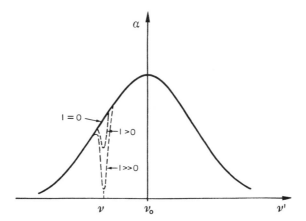

FIG. 2.22. Saturation behavior of the absorption coefficient α versus frequency v' as measured by the weak test beam for increasing values of the intensity $I(v)$ of the saturating beam (inhomogeneous line).

(2.4.26): The overall line $g_t(v - v_0)$ is then given by the convolution of homogeneous contributions $g(v - v'_0)$ of various atoms. Thus, for absorption, the resulting absorption coefficient can be visualized as shown in Fig. 2.21. In this case, for an experiment such as that in Fig. 2.18, the saturating beam of intensity $I(v)$ interacts with only those atoms whose resonance frequency v'_0 is in the neighborhood of the frequency v. Accordingly, only these atoms undergo saturation when $I(v)$ becomes sufficiently large. The modified shape of the absorption line for various values of $I(v)$ is then as shown in Fig. 2.22. In this case, as $I(v)$ is increased, a hole is produced in the absorption line at frequency v. The width of this hole is of the same order as the width of each of the dashed absorption profiles in Fig. 2.21, i.e., the width of the homogeneous line. A similar argument applies if a transition with net gain rather than absorption is considered. The saturating beam, in this case, burns holes in the gain profile rather than in the absorption profile. Note also that a similar argument can be applied when absorption or gain saturation is produced by a light pulse of sufficiently high energy fluence.

2.9. FLUORESCENCE DECAY OF AN OPTICALLY DENSE MEDIUM

In Section 2.3 the decay of an essentially isolated atom or ion was considered. In a real situation an atom is surrounded by many other atoms, some in the ground state and some in the excited state. For an optically dense medium, new phenomena may then occur, since decay may be due to the simultaneous occurrence of both spontaneous and stimulated processes. This section briefly discusses such phenomena.

2.9.1. Radiation Trapping

If the fraction of atoms raised to the upper level is very small and the medium is optically dense, the phenomenon known as *radiation trapping* may play a significant role. In this case instead of escaping from the medium, a photon spontaneously emitted by one atom can be absorbed by another atom, which thereby ends up in the excited state. The process therefore has the effect of reducing the effective rate of spontaneous emission (see Ref. 18 for a detailed discussion of radiation trapping.) We merely limit to point out here that the lifetime increase depends on atomic density, on the cross section of the transition involved, and on the geometry of the medium. Radiation trapping may be particularly important for uv transitions with large cross sections [according to Eq. (2.4.29) $\sigma \propto |\mu|^2 v$, which increases rapidly in going to the uv through the v term and the increase of $|\mu|^2$ with frequency]. This can result in an increase in the effective lifetime of spontaneous emission by as much as a few orders of magnitude.

2.9.2. Amplified Spontaneous Emission

If the fraction of atoms raised to the upper level is very large and the medium is again optically dense, the phenomenon known as amplified spontaneous emission (ASE) may play a very important role.

Consider a cylindrically shaped active medium and let Ω be the solid angle subtended by one face of the cylinder as seen from the center O of the other face (Fig. 2.23a). If the gain of the active material $G = \exp \sigma(N_2 - N_1)/l$ is large enough, the fluorescence power emitted by atoms around point O into the solid angle Ω may be strongly amplified by the active medium (by a factor that may, in some cases, be as high as 10^4 or even higher). Thus, under suitable conditions considered in the following discussion, the active medium preferentially emits its stored energy into the solid angle Ω in Fig. 2.23a and obviously along the opposite direction as well. If a totally reflecting mirror ($R = 1$) is placed at one end of the

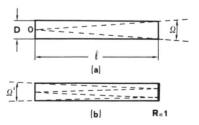

FIG. 2.23. Solid angle of emission in the case of amplified spontaneous emission: (a) Active material without end mirrors, (b) active material with one end mirror.

medium (Fig. 2.23b), then unidirectional output is obtained. This is the basic ASE phenomenon. In contrast to spontaneous emission, ASE possesses distinctive features that show some similarity to laser action: ASE has to some degree the property of directionality; its bandwidth is appreciably narrower than that of spontaneous emission; it exhibits a soft threshold behavior; and the beam of ASE light can be quite intense. We briefly consider these properties here, while we refer the reader to Appendix C for further details.

Directionality is immediately apparent from Fig. 2.23. For $D \ll l$ the emission solid angle Ω in Fig. 2.23a is given by:

$$\Omega = \frac{\pi D^2}{4l^2} \qquad (2.9.1)$$

where D is the diameter and l is the length of the active material, and it is very small. Likewise, in the case of Fig. 2.23b, the emission solid angle is

$$\Omega' = \frac{\pi D^2}{16l^2} \qquad (2.9.2)$$

which is even smaller. Note that, in both cases, due to refraction at the exit face of the active medium, the external solid angle Ω_n (not shown in Fig. 2.23a–b) is obtained from Eqs. (2.9.1) and (2.9.2) by multiplying the right-hand side by n^2, where n is the refractive index of the material. In any case, if $D \ll l$, ASE occurs in a narrow cone (see Example 2.12).

Example 2.12. *Directional property of ASE.* Let us consider the active medium to consist of gaseous nitrogen, for which there is a laser transition occurring at $\lambda \cong 337$ nm (see Chap. 10). We take $D = 2$ cm and $l = 1$ m and assume that a totally reflecting mirror is placed at one end. From Eq. (2.9.2) we obtain $\Omega' \cong 0.8 \times 10^{-4}$ sterad, which shows that the emission solid angle is very much smaller than the 4π-sterad angle into which spontaneous emission occurs. On the other hand beam divergence is much larger than that obtained from the same active medium used in a two-mirror laser resonator. The half-cone divergence angle of the ASE beam θ' is given by $\theta' = (\Omega'/\pi)^{1/2} \cong 5$ mrad. By comparison, in the case of a laser resonator, the minimum attainable divergence, as set by diffraction, is given by $\theta_d \cong (\lambda/D) \cong 20$ μrad; i.e., it is 250 times smaller.

The spectral narrowing of ASE can be understood when we note that the gain experienced by the spontaneously emitted beam is much higher at the peak, i.e., at $v = v_0$ than in the wings of the gain line. This situation is illustrated in Fig. 2.24 for a Lorentzian line. The dashed curve shows the normalized spectral profile $g(v - v_0)/g_p$ of the spontaneously emitted light, while solid lines show the normalized profile I_v/I_{vp} of ASE spectral emission at two different values for the peak gain G. In previous expressions: g_p and I_{vp} are the peak values of the functions g and I_v, respectively; $G = \exp(\sigma_p N_2 l)$, where σ_p is the peak cross section and N_2 is the upper state population (we assume $N_1 \cong 0$). The ASE spectral profile is obtained from the approximate theory shown in Appendix C. The ratio of the ASE linewidth Δv_{ASE} to the spontaneous emission linewidth Δv_0 (FWHM), as obtained from the same approximate theory, is plotted versus $(\sigma_p N_2 l)$ in Fig. 2.25 as a dashed curve. Figure 2.25 also shows, as continuous lines, the corresponding plots, at three values of $(\Omega/4\pi)$, when gain saturation is taken into account (see Ref. 23). Note that, for practical

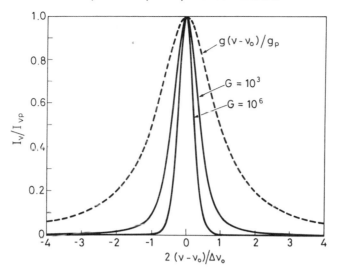

FIG. 2.24. Normalized ASE spectral emission at two different values of the peak, unsaturated, single-pass gain.

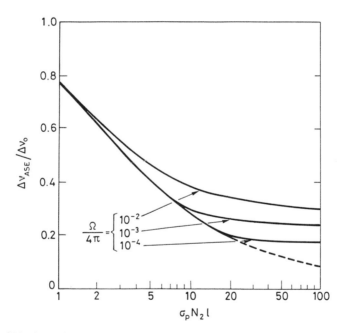

FIG. 2.25. Linewidth of ASE Δv_{ASE} normalized to the linewidth of spontaneous emission Δv_0 as a function of the unsaturated single-pass gain $\sigma_p N_2 l$.

values of the unsaturated gain $10^3 \leq G \leq 10^6$, i.e., for $7 \leq \sigma_p N_2 l \leq 14$ and for practical values of the emission solid angle $10^{-5} \leq (\Omega/4\pi) \leq 10^{-3}$, the reduction in linewidth lies roughly between 3 and 4.

To calculate the apparent threshold, we begin by pointing out that, as shown in Appendix C and ref. 24 the intensity of one of the two ASE beams in Fig. 2.23a is given by:

$$I = \phi I_s \left(\frac{\Omega}{4\pi^{3/2}} \right) \frac{(G-1)^{3/2}}{(G \ln G)^{1/2}} \qquad (2.9.3a)$$

for a Lorentzian line and by:

$$I = \phi I_s \left(\frac{\Omega}{4\pi} \right) \frac{(G-1)^{3/2}}{(G \ln G)^{1/2}} \qquad (2.9.3b)$$

for a Gaussian line. In both the above equations, ϕ is the fluorescence quantum yield and $I_s = h\nu_0/\sigma_p\tau$ is the saturation intensity of the amplifier at the transition peak. We can now define the ASE threshold as the situation where ASE becomes the dominant mechanism depopulating the available inversion. We thus require I to become comparable to the saturation intensity I_s. In this case, in fact, a sizable fraction of the available energy is found in the two ASE cones in Fig. 2.23 rather than in the 4π-sterad solid angle of the spontaneously emitted radiation. For $I = I_s$ and $G \gg 1$, Eqs. (2.9.3a–b) then show that the threshold peak gain must satisfy the relatively simple conditions:

Example 2.13. *ASE threshold for a solid-state laser rod.* We consider a solid-state laser rod, such as Nd:YAG, with $D = 6$ mm, $l = 10$ cm, $n = 1.82$, and first consider the symmetric configuration in Fig. 2.23a. From Eq. (2.9.1) we obtain $(\Omega/4\pi) = 2.25 \times 10^{-4}$. Since the line of Nd:YAG is Lorentzian and $\phi \cong 1$ for this line, from Eq. (2.9.4a) we obtain $G = 2.5 \times 10^4$, i.e., $\sigma_p N_{th} l = \ln G = 10.12$. Taking 2.8×10^{-19} cm^2 as the value of the peak stimulated emission cross section, σ_p, for Nd:YAG (see Example 2.10), we then obtain a threshold inversion for ASE of $N_{th} = 3.6 \times 10^{18}$ cm^{-3}. For the single-end configuration in Fig. 2.23b we obtain from Eq. (2.9.2) $(\Omega'/4\pi) = 5.62 \times 10^{-5}$ and from Eq. (2.9.5a) $G = 6.4 \times 10^2$, i.e., a much smaller value for the threshold peak gain. The threshold inversion for ASE is, in this case, $N_{th} = \ln G/\sigma_p l = 2.3 \times 10^{18}$ cm^3. Note that the emission solid angle is n^2 times larger than the geometrical solid angles, Ω and Ω', calculated above in this example. We thus obtain $\Omega_n = n^2\Omega = 9.36 \times 10^{-3}$ sterad and $\Omega'_n = n^2\Omega_n = 2.33 \times 10^{-3}$ sterad in the two cases, respectively.

$$G = \frac{4\pi^{3/2}}{\phi\Omega}(\ln G)^{1/2} \qquad (2.9.4a)$$

for a Lorentzian line and:

$$G = \frac{4\pi}{\phi\Omega}(\ln G)^{1/2} \qquad (2.9.4b)$$

for a Gaussian line. Note that, if a totally reflecting mirror is placed at one end of the medium (Fig. 2.23b), the resulting threshold condition is again given by Eq. (2.9.4) provided G, the single-pass peak gain, is replaced by G^2, the double-pass peak gain, and Ω is replaced by Ω'. We thus obtain

$$G^2 = \frac{4\pi^{3/2}}{\phi\Omega'}(\ln G^2)^{1/2} \qquad (2.9.5a)$$

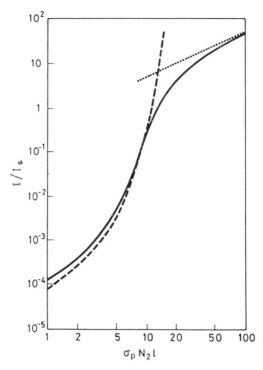

FIG. 2.26. Intensity of ASE emission I normalized with respect to the saturation intensity I_s as a function of the single-pass gain $\sigma_p N_2 l$ for an emission solid angle $\Omega = 4\pi \times 10^{-4}$ sterad.

for a Lorentzian line and:

$$G^2 = \frac{4\pi}{\phi\Omega'}(\ln G^2)^{1/2} \tag{2.9.5b}$$

for a Gaussian line.

The soft threshold characteristics of ASE is apparent from Fig. 2.26, where the normalized intensity of one of the two ASE beams in Fig. 2.23a is plotted against $\sigma_p Nl$ for $(\Omega/4\pi) = 10^{-4}$, assuming a Lorentzian line and $\phi = 1$. The dashed curve is obtained from Eq. (2.9.3a), which is valid in the limit $I \ll I_s$. The dotted line, which applies in the other limit $I \gg I_s$, is obtained from the condition that half of the available fluorescence power is found in the right-propagating ASE beam, i.e., from the equation $(I/I_s) = \sigma_p N_2 l/2$. The solid line is obtained from a more accurate calculation in which the saturation of the upper state population, i.e., gain saturation, is properly taken into account.[24]

ASE is usually used with the configuration in Fig. 2.23b to obtain directional and narrow bandwidth radiation of high intensity from such high-gain media as nitrogen, excimers, or plasmas for x-rays (see Chap. 10). Since either one mirror only or no mirror at all is required, these systems are (incongruously) called *mirrorless lasers*. In fact, although ASE emission has some spatial and temporal coherence, it just consists of amplified spontaneous-emission noise and should therefore not be confused with laser radiation, whose coherent properties are conceptually different, as explained in Chap. 11. In many other situations, ASE is generally a nuisance. For instance it limits the maximum inversion

Table 2.2. Peak transitions cross sections, upper state lifetime, and transition linewidths for some of the most common gas and solid-state lasers

Transition	λ	σ_p (cm^2)	τ (μs)	Δv_0	Remarks
He–Ne	$\lambda = 632.8$ nm	5.8×10^{-13}	30×10^{-3}	1.7 GHz	
Ar$^+$	$\lambda = 514.5$ nm	2.5×10^{-13}	6×10^{-3}	3.5 GHz	
Nd:YAG	$\lambda = 1.064\ \mu m$	2.8×10^{-19}	230	120 GHz	
Nd:Glass	$\lambda = 1.054\ \mu m$	4×10^{-20}	300	5.4 THz	
Rhodamine 6G	$\lambda_p = 570$ nm	3.2×10^{-16}	5.5×10^{-3}	46 THz	
Alexandrite	$\lambda_p = 704$ nm	0.8×10^{-20}	300	60 THz	$T = 300$K
Ti^{3+}:Al$_2$O$_3$	$\lambda_p = 790$ nm	4×10^{-19}	3.9	100 THz	E∥c-axis
Cr^{3+}:LiSAF	$\lambda_p = 845$ nm	5×10^{-20}	67	84 THz	E∥c-axis

that can be stored in high-gain pulsed laser amplifiers. It is also the dominant noise term in fiber amplifiers, such as Er^{3+} doped fiber amplifiers (EDFA), now widely used for optical communications at wavelengths around 1550 nm.

2.10. CONCLUDING REMARKS

This chapter discusses several aspects of the interaction of radiation with matter, mostly relating to atoms or ions. In particular the two most important parameters describing this interaction are the cross section $\sigma = \sigma(v - v_0)$ and the lifetime τ of the upper laser level. In the case of a pure Gaussian or Lorentzian lineshape, we have to know only the peak value σ_p of the cross section and the value of the linewidth (Δv_0 or Δv_0^*). Note also that the lifetime τ refers to the overall upper level lifetime, and, as such, it includes all radiative and nonradiative decay processes that depopulate the upper level. In the case of degenerate or strongly coupled levels, σ_p and τ refer to the effective stimulated emission cross section and upper state lifetime, respectively, as discussed in Section 2.7.

Table 2.2 summarizes the values σ_p, τ, and Δv_0 (or Δv_0^*) for many common laser transitions in gases and ionic crystals. For comparison, the corresponding values for rhodamine 6G, a common dye laser material, are also included. Note that the very high values of $\sigma_p (\approx 10^{-13}$ cm^2) for gas lasers result from the rather small values of Δv_0^* (a few GHz) and the rather short lifetimes (a few ns). The lifetime is short because transitions are electric-dipole-allowed. By contrast, for active ions in ionic crystals or glasses, such as Nd:YAG or Nd:phosphate glass, σ_p is much smaller ($10^{-20} \div 10^{-19}$ cm^2), and the lifetime is much longer (several hundred μs), indicative of a forbidden electric-dipole transition. Note also that the linewidths are much larger (from hundredths to thousands of GHz), which also results in a strong reduction of the peak cross section. Dye laser materials, such as rhodamine 6G, are intermediate between these two cases, showing a fairly high cross section ($\approx 10^{-16}$ cm^2) and also a very short lifetime, a few ns, since again transitions are electric-dipole-allowed. The last three laser materials listed in Table 2.2, namely, alexandrite, Ti^{3+}:Al$_2$O, and Cr^{3+}:LiSAF, belong to the category of tunable solid-state lasers. Indeed, for these materials, laser linewidths are extremely wide (tens to hundred THz). Cross sections are comparable to those of narrower linewidth materials, such as Nd:YAG, while the lifetimes are somewhat shorter.

PROBLEMS

2.1. For a cavity volume $V = 1$ cm^3 calculate the number of modes that fall within a bandwidth $\Delta\lambda = 10$ nm centered at $\lambda = 600$ nm.

2.2. Instead of ρ_ν a spectral energy density ρ_λ can be defined, where ρ_λ is such that $\rho_\lambda\, d\lambda$ gives the energy density for em waves of wavelength between λ and $\lambda + d\lambda$. Find the relationship between ρ_λ and ρ_ν.

2.3. For blackbody radiation find the maximum of ρ_λ versus λ. Show in this way that the wavelength λ_M at which the maximum occurs satisfies the relationship $\lambda_M T = hc/ky$ (Wien's law), where the quantity y satisfies the equation $5[1 - \exp(-y)] = y$. From this equation find an approximate value of y.

2.4. The wavelength λ_M at which the maximum occurs for the distribution in Fig. 2.3 satisfies the relation $\lambda_M T = 2.9 \times 10^{-3}$ m × K (Wien's law). Calculate λ_M for $T = 6000$ K. What is the color corresponding to this wavelength?

2.5. The R_1 laser transition of ruby has to a good approximation a Lorentzian shape of width (FWHM) 330 GHz at room temperature (see Fig. 2.10). The measured peak transition cross section is $\sigma = 2.5 \times 10^{-20}$ cm^2. Calculate the radiative lifetime (the refractive index is $n = 1.76$). Since the observed room temperature lifetime is 3 ms, what is the fluorescence quantum yield?

2.6. Nd:YAG, a typical active laser material, is a crystal of $Y_3Al_5O_{12}$ (yttrium aluminium garnet, YAG) in which some of the Y^{3+} are substituted by Nd^{3+} ions. The typical Nd^{3+} atomic concentration used is 1%, i.e., 1% of Y^{3+} ions are replaced by Nd^{3+}. The YAG density is 4.56 g/cm^3. Calculate the Nd^{3+} concentration in the ground ($^4I_{9/2}$) level. This level is actually made up of five (doubly degenerate) levels (see Fig. 2.15); the four higher levels are spaced from the lowest level by 134, 197, 311, and 848 cm^{-1}, respectively. Calculate the Nd^{3+} concentration in the lowest level of the $^4I_{9/2}$ state.

2.7. The neon laser transition at $\lambda = 1.15$ μm is predominantly Doppler broadened to $\Delta\nu_0^* = 9 \times 10^8$ Hz. The upper state lifetime is $\approx 10^{-7}$ s. Calculate the peak cross section assuming that the laser transition lifetime is equal to the upper state lifetime.

2.8. The quantum yield of the $S_1 \rightarrow S_0$ transition (see Chap. 9) for rhodamine 6G is 0.87, and the corresponding lifetime is ≈ 5 ns. Calculate the radiative and nonradiative lifetimes of the S_1 level.

2.9. Calculate the total homogeneous linewidth of the 633-nm laser transition of Ne if $\Delta\nu_{nat} \cong 20$ MHz and $\Delta\nu_c = 0.64$ MHz. What is the shape of the overall line?

2.10. Find the relationship between the intensity I and the corresponding energy density ρ for a plane wave.

2.11. A cylindrical rod of Nd:YAG with diameter of 6.3 mm and length of 7.5 cm is pumped very hard by a suitable flashlamp. The peak cross section for the 1.064-μm laser transition is $\sigma = 2.8 \times 10^{-19}$ cm^2, and the refractive index of YAG is $n = 1.82$. Calculate the critical inversion for the onset of the ASE process (the two rod end faces are assumed to be perfectly antireflection-coated, i.e., nonreflecting). Also calculate the maximum energy that can be stored in the rod if the ASE process is to be avoided.

2.12. A solution of cryptocyanine (1,1'-diethyl-4,4'-carbocyanine iodide) in methanol is used simultaneously to Q-switch and mode lock (see Chap. 8) a ruby laser. The absorption cross section of cryptocyanine for ruby laser radiation ($\lambda = 694.3$ nm) is 8.1×10^{-16} cm^2. The upper state lifetime is $\tau \cong 22$ ps. Calculate the saturation intensity at this wavelength.

2.13. On applying the principle of detailed balance to the two near-resonant transfer processes, Eqs. (2.6.9) and (2.6.10), show that at exact resonance, i.e., for $\Delta E = 0$, one has $k_{B^*A} = k_{BA^*}$, where k_{B^*A} and k_{BA^*} are the rate constants of the two processes, respectively.

2.14. Instead of observing saturation as in Fig. 2.19, we can use just the beam $I(v)$ and measure the absorption coefficient for this beam at sufficiently high values of intensity $I(v)$. For a homogeneous line, show that the absorption coefficient is in this case:

$$\alpha(v - v_0) = \frac{\alpha_0(0)}{1 + [2(v - v_0)/\Delta v_0]^2 + (I/I_{s0})}$$

where $\alpha_0(0)$ is the unsaturated ($I \ll I_{s0}$) absorption coefficient at $v = v_0$ and I_{s0} is the saturation intensity, as defined by Eq. (2.8.11), at $v = v_0$. Hint: begin by showing that:

$$\alpha(v - v_0) = \frac{\alpha_0(0)}{1 + [2(v - v_0)/\Delta v_0]^2} \frac{1}{1 + (I/I_s)}$$

where I_s is the saturation intensity at frequency v. Continue by expressing I_s in terms of I_{s0}.

2.15. From the expression derived in Problem 2.14, find the behavior of the peak absorption coefficient and the linewidth versus I. How would you measure the saturation intensity I_{s0}?

2.16. Show that, for an inhomogeneous line with lineshape function g, the saturated absorption coefficient for an experiment as in Fig. 2.18 can be written

$$\alpha = \left(\frac{2\pi^2}{3n\varepsilon_0 c_0 h}\right)|\mu|^2 N_t \int \frac{(2/\pi\Delta v_0)v'g^*(v_0' - v_0)}{1 + [2(v' - v_0')/\Delta v_0]^2} \frac{1}{[1 + (I/I_{s0})](1/\{1 + [2(v - v_0')/\Delta v_0]^2\})} dv_0'$$

where the homogeneous contribution is accounted for by a Lorentzian line. Hint: begin by calculating the elemental contribution $d\alpha$ of the saturated absorption coefficient due to the fraction $g^*(v_0' - v_0)dv_0'$ of atoms whose resonant frequencies lie between v_0' and $v_0' + dv_0'$.

2.17. Under the assumptions that the homogeneous linewidth is much smaller than the inhomogeneous linewidth and $I \ll I_{s0}$, show that the expression for α given in Problem 2.16 can be approximated as:

$$\alpha = \left(\frac{2\pi^2}{3n\varepsilon_0 c_0 h}\right)|\mu|^2 N_t v'g^*(v' - v_0)$$
$$\times \left(1 - (2/\pi\Delta v_0)\frac{I}{I_{s0}} \int \frac{dv_0'}{\{1 + [2(v' - v_0')/\Delta v_0]^2\}\{1 + [2(v - v_0')/\Delta v_0]^2\}}\right)$$

Since the integral is now the convolution of two Lorentzian lines, what is the width of the hole in Fig. 2.22?

REFERENCES

1. R. Reiff, *Fundamentals of Statistical and Thermal Physics* (McGraw-Hill, New York, 1965), Chap. 9.
2. W. Heitler, *Quantum Theory of Radiation*, 3rd ed. (Oxford University Press, London, 1953), Sec. II.9.
3. H. A. Lorentz, *Theory of Electrons*, 2nd ed. (Dover, New York, 1952), Chap. 3.

4. J. A. Stratton, *Electromagnetic Theory*, 1st ed. (McGraw-Hill, New York, 1941), pp. 431–38.

5. R. H. Pantell and H. E. Puthoff, *Fundamentals of Quantum Electronics* (Wiley, New York, 1964), Chap. 6.

6. W. Louisell, *Radiation and Noise in Quantum Electronics* (McGraw-Hill, New York, 1964), Chap. 6.

7. R. H. Pantell and H. E. Puthoff, *Fundamentals of Quantum Electronics* (Wiley, New York, 1964), pp. 40–41, 60, 62, and Appendix 4.

8. R. H. Pantell and H. E. Puthoff, *Fundamentals of Quantum Electronics* (Wiley, New York, 1964), Appendix 5.

9. A. Einstein, On the Quantum Theory of Radiation, *Z. Phys.* **18**, 121 (1917).

10. W. Louisell, *Radiation and Noise in Quantum Electronics* (McGraw-Hill, New York, 1964), Chap. 5.

11. H. G. Kuhn, *Atomic Spectra*, 2nd ed. (Longmans, Green, London, 1969), Chap. 7.

12. *Radiationless Transitions* (F. J. Fong, ed.) (Springer-Verlag, Berlin, 1976), Chap. 4.

13. C. K. Rhodes and A. Szoke, Gaseous Lasers: Atomic, Molecular, Ionic in *Laser Handbook* (F. T. Arecchi and E. O. Schultz-DuBois, eds.) (North Holland, Amsterdam, 1972), vol. 1, pp. 265–324.

14. J. B. Birks, *Photophysics of Aromatic Molecules* (Wiley–Interscience, New York, 1970), Sect. II.9.

15. D. L. Dexter, *J. Chem. Phys.* **21**, 836 (1953).

16. J. D. Jackson, *Classical Electrodynamics* (Wiley, New York, 1975), Sect. 9.2.

17. W. J. Miniscalco, Optical and Electronic Properties of Rare Earth Ions in Glasses in *Rare-Earth-Doped Fiber Lasers and Amplifiers* (M. J. F. Digonnet, ed.) (Marcel Dekker, New York, 1993), Chap. 2.

18. T. Holstein, Imprisonment of Resonant Radiation in Gases, *Phys. Rev.* **72**, 1212 (1947).

19. R. Arrathoon, Helium–Neon Lasers and the Positive Column in *Lasers* (A. K. Levine and A. J. DeMaria, eds.) (Marcel Dekker, New York, 1976), Table 2.

20. M. H. Dunn and J. N. Ross, Argon Laser in *Progress in Quantum Electronics*, vol. 4 (J. H. Saunders and S. Stenholm, eds.) (Pergamon, Oxford, 1977), Table 2.

21. W. F. Krupke, M. D. Shinn, J. E. Marion, J. A. Caird, and S. E. Stokowski, Spectroscopic, Optical, and Thermomechanical Properties of Neodymium- and Chromium-Doped Gadolinium Scandium Gallium Garnet, *J. Opt. Soc. Am. B* **3**, 102 (1986).

22. J. C. Walling, O. G. Peterson, J. P. Jennsen, R. C. Morris, and E. W. O'Dell, Tunable Alexandrite Lasers, *IEEE J. Quant. Elect.* **QE-16**, 1302 (1980).

23. L. W. Casperson, Threshold Characteristics of Mirrorless Lasers, *J. Appl. Phys.* **48**, 256 (1977).

24. O. Svelto, S. Taccheo, and C. Svelto, Analysis of Amplified Spontaneous Emission: Some Corrections to the Lyndford Formula, *Optic. Comm.* **149**, 277–282 (1998).

3

Energy Levels, Radiative, and Nonradiative Transitions in Molecules and Semiconductors

Chapter 3 specializes some of the results and considerations from preceding chapters to the somewhat more complicated case of molecules and semiconductors. Particular emphasis is given to semiconductors in either bulk or quantum well form, since these play an increasingly important role as laser media.

3.1. MOLECULES

We first consider energy levels and radiative and nonradiative transitions in molecules, limiting our discussion to a qualitative description of features relevant to correctly understanding laser behavior in active media, such as molecular gases or organic dyes. For a more extensive treatment of the wider subject of molecular physics the reader is referred to specialized texts.[1]

3.1.1. Energy Levels

The total energy of a molecule generally consists of the sum of four contributions: (1) electronic energy E_e due to the motion of electrons about the nuclei, (2) vibrational energy E_v due to the vibrational motion of the nuclei, (3) rotational energy E_r due to the rotational motion of the molecule, and (4) translational energy. We do not consider translational energy here, since it is not usually quantized. The other types of energy are quantized, however, and it is instructive to derive, from simple arguments, the order of magnitude of the energy difference between electronic levels (ΔE_e), vibrational levels (ΔE_v), and rotational levels (ΔE_r). The order of magnitude of ΔE_e is given by:

$$\Delta E_e \cong \frac{\hbar^2}{ma^2} \tag{3.1.1}$$

where $\hbar = h/2\pi$, m is the mass of the electron, and a is the size of molecule. In fact, if we consider an outer electron of the molecule, uncertainty in its position is of the order of a; the uncertainty in momentum, via the uncertainty principle, is $\hbar/a$, and the minimum kinetic energy is therefore $\approx \hbar^2/ma^2$.

For a diatomic molecule consisting of masses M_1 and M_2, we assume that the corresponding potential energy U_p versus internuclear distance R around the equilibrium distance R_0 can be approximated by the parabolic expression $U_p = k_0(R - R_0)^2/2$ (see Fig. 3.1). Then, the energy difference ΔE_v between two consecutive vibrational levels is given by the well-known harmonic oscillator expression:

$$\Delta E_v = h\nu_0 = \hbar \left(\frac{k_0}{\mu} \right)^{1/2} \tag{3.1.2}$$

where $\mu = M_1 M_2/(M_1 + M_2)$ is the reduced mass. For a homonuclear molecule comprising two atoms of mass M, the energy difference between two vibrational levels is then:

$$\Delta E_v = \hbar \left(\frac{2k_0}{M} \right)^{1/2} \tag{3.1.3}$$

We also expect that a displacement of the two atoms from equilibrium by an amount equal to the size of the molecule produces an energy change of about ΔE_e, since this separation results in a considerable distortion of the electronic wavefunctions. Thus we can write

$$\Delta E_e = \frac{k_0 a^2}{2} \tag{3.1.4}$$

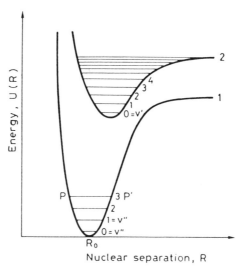

FIG. 3.1. Potential energy curves and vibrational levels of a diatomic molecule.

We can eliminate a^2 and k_0 from Eqs. (3.1.1), (3.1.3), and (3.1.4) to obtain

$$\Delta E_v = 2\left(\frac{m}{M}\right)^{1/2} \Delta E_e \tag{3.1.5}$$

For a homonuclear diatomic molecule, the rotational energy is then given by $E_r = \hbar^2 J(J+1)/Ma^2$, where J is the rotational quantum number. Therefore the difference ΔE_r in rotational energy between, e.g., the $J=0$ and $J=1$ levels is given by $\Delta E_r = 2\hbar^2/Ma^2 = 2(m/M)\Delta E_e$, where Eq. (3.1.1) has been used. From Eq. (3.1.5) we then obtain

$$\Delta E_r = \left(\frac{m}{M}\right)^{1/2} \Delta E_v \tag{3.1.6}$$

Since $m/M \cong 10^{-4}$, it follows that the separation of rotational levels is about one-hundredth that of vibrational levels. The spacing of vibrational levels is in turn about one-hundredth of ΔE_e. In fact, the actual frequency ranges, $(\Delta E_e/h)$, for electronic, $(\Delta E_v/h)$ for vibrational, and $(\Delta E_r/h)$ for rotational transitions are roughly 25–50×10^3 cm^{-1}, 500–3000 cm^{-1}, and 1–20 cm^{-1}, respectively.

After these preliminary considerations, we consider the simplest case of a molecule consisting of two identical atoms. Since, as already stated, rotations and vibrations occur on a much slower time scale than electronic motion, we use the Born–Oppenheimer approximation where two atoms are first considered to be at a fixed nuclear separation R and non-rotating. By solving Schrödinger's equation for this situation, it is then possible to find the dependence of electronic energy levels on the separation R. Even without actually solving the equation (which is usually very complicated), we can appreciate that, for bound states, the energy dependence on R must have the form shown in Fig. 3.1, where the ground state, 1, and the first excited state, 2, are shown as examples. If the atomic separation is very large ($R \rightarrow \infty$), the levels are the same as those of the single atom. If the separation R is finite, then, as a result of interaction between atoms, energy levels are displaced.

To understand the shape of these curves, note that, with the inclusion of a suitable constant, these represent the potential energy of the molecule as a function of the internuclear distance R. In particular, since the minimum energy for curve 1 is set to zero in Fig. 3.1, this curve represents only the potential energy of the ground electronic state. Since the derivative of the potential energy with respect to R gives the force exerted by atoms on each other, the force is attractive at large separations, then it becomes repulsive for small separations. The force is zero for the position corresponding to the minimum of each curve (e.g., R_0), which is therefore the separation that atoms tend to assume (in the absence of vibration). Note that the minimum of the curve for the excited state is generally shifted to larger values of R relative to that of the ground state due to the larger orbit occupied by excited electrons.

Our discussion so far has focused on two atoms held fixed at some nuclear separation R. Let us now assume that the molecule is, e.g., in its electronic state 1 and that the two atoms are released at some value R, with $R \neq R_0$. The internuclear force will then cause the atoms to oscillate around the equilibrium position R_0 and the total energy will be equal to the sum of the potential energy already discussed and vibrational energy. For small oscillations about the position R_0, curve 1 can be approximated by a parabola; the restoring force between the two atoms is then elastic, i.e., proportional to displacement from equilibrium. In this case the problem has well-known solutions, i.e., those of the harmonic oscillator. Energy levels are

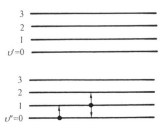

FIG. 3.2. Vibrational energy levels of two consecutive electronic states of a molecule. Arrows indicate allowed transitions starting at $v'' = 0$ and $v'' = 1$ levels.

thus equally spaced by an amount $h\nu_0$ given by Eq. (3.1.2), where the elastic force constant k_0 equals the curvature of the parabola. Therefore, when vibrations are taken into account, energy levels (for each of the two electronic states) are given by levels 1, 2, 3, etc., in Fig. 3.1. Note that the $v = 0$ level does not coincide with the minimum of the curve because of the well-known zero-point energy ($h\nu_0/2$) of a harmonic oscillator. Curves 1 and 2 now no longer represent the energy of the system, since the atoms are no longer fixed, and instead of Fig. 3.1, the simpler scheme in Fig. 3.2 is sometimes used; however the scheme in Fig. 3.1 is more meaningful than that in Fig. 3.2. Suppose, for example, the system is in the $v'' = 3$ vibrational level of ground level 1. From Fig. 3.1 we see that the nuclear distance R oscillates between values corresponding to points P and P' in the figure. At these two points, in fact, vibrational energy coincides with potential energy, which means that kinetic energy must be zero. For large oscillations about the equilibrium position, R_0, the curve for the potential energy cannot be approximated adequately by a parabola, and, in fact, the higher vibrational levels are no longer equally spaced. One can show that level spacing decreases with increasing energy because the restoring force becomes smaller than that predicted by the parabolic approximation.

We next briefly consider the case of a polyatomic molecule. In this case the scheme in Fig. 3.1 can still be used if R is interpreted as some suitable coordinate that describes the

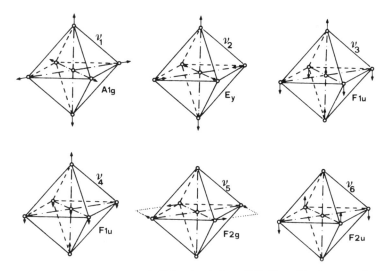

FIG. 3.3. Normal vibration modes of an octahedral molecule (e.g., SF_6). The sulfur atom occupies the center of the octahedron, and the six fluorine atoms are at the corners of the octahedron. (By permission from Ref. 2.)

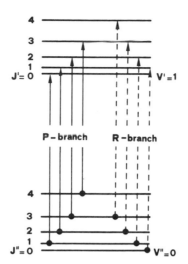

FIG. 3.4. Rotational energy levels belonging to two consecutive vibrational states of a molecule. Arrows indicate allowed transitions belonging to the *P*-branch and *R*-branch.

given mode of vibration. Consider for example the SF_6 molecule, which has an octahedral shape (see Fig. 3.3), with the sulfur atom at the center of the octahedron and each of the six fluorine atoms at an apex. For the symmetric mode of vibration in Fig. 3.3 (mode A_{1g}), the coordinate R may be taken as the distance between the sulfur atom and each of the fluorine atoms. Since Fig. 3.1 shows SF_6 to have six, independent, nondegenerate modes of vibration, the potential energy U for a general state of the molecule will depend on all six vibrational coordinates of the molecule. Therefore the scheme in Fig. 3.1 may be regarded as a section of a seven-dimensional function when only one vibrational coordinate undergoes change.

The description presented so far does not give a complete picture of the molecular system, since the molecule can also rotate. According to quantum mechanics rotational energy is also quantized; for a linear, rigid rotator (e.g., a rigid diatomic or linear triatomic molecule), it can be expressed as:

$$E_r = BJ(J + 1) \qquad (3.1.7)$$

where the rotational constant B is given by $\hbar^2/2I$ with I being the moment of inertia about an axis perpendicular to the internuclear axis and through the center of mass. Thus the total energy of the system is given by the sum of the electronic, vibrational, and rotational energies. Accordingly the energy levels of, say, the $v'' = 0$ and $v' = 1$ vibrational levels of the ground state are indicated in Fig. 3.4. Note that, unlike the case of vibrational levels, spacing between consecutive rotational levels is not constant; in fact it increases linearly with the rotational quantum number J, i.e., $[E_r(J) - E_r(J - 1)] = 2BJ$.

3.1.2. Level Occupation at Thermal Equilibrium

At thermodynamic equilibrium the population $N(E_e, E_v, E_r)$ of a rotational vibrational level belonging to a given electronic state can be written as:

$$N(E_e, E_v, E_r) \propto g_e g_v g_r \exp\{-[(E_e + E_v + E_r)/kT]\} \qquad (3.1.8)$$

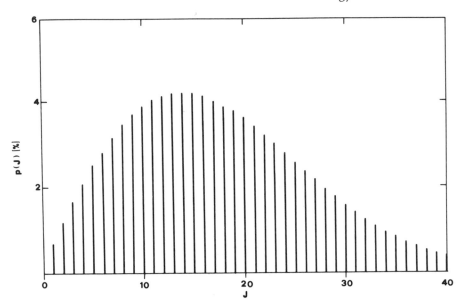

FIG. 3.5. Population distribution among rotational levels of a given vibrational state.

where E_e, E_v, and E_r are the electronic, vibrational, and rotational energies of the level, respectively, and g_e, g_v, and g_r are the corresponding level degeneracies [see Eq. (2.7.3)]. According to estimates in Sect. 3.1.1, the order of magnitude of E_v/hc is 1000 cm^{-1}, while E_e/hc is more than an order of magnitude larger. Since $kT/hc \cong 209$ cm^{-1} (at $T = 300$ K), it then follows that both E_e and E_v are appreciably larger than kT. Accordingly, as a rule of thumb, we can say that, at thermal equilibrium and room temperature, a molecule lies in the lowest vibrational level of the ground electronic state.* The probability of occupation of a given rotational level of this lowest vibrational state can then be written, according to Eqs. (3.1.7) and (3.1.8), as:

$$p(J) \propto (2J + 1) \exp[-BJ(J + 1)/kT] \qquad (3.1.9)$$

The factor $(2J + 1)$ accounts for level degeneracy: A rotational level of quantum number J is in fact $(2J + 1)$-fold degenerate. Taking, as an example, $B = 0.5$ cm^{-1} and assuming $kT = 209$ cm^{-1} (room temperature), Fig. 3.5 shows the population distribution among various rotational levels of a given vibrational state (e.g., the ground state). Note that, as a result of the factor $(2J + 1)$ in Eq. (3.1.9), the most heavily populated level is not the ground (i.e., $J = 0$) level but rather the one whose rotational quantum number J satisfies the relation:

$$(2J + 1)_m = \left(\frac{2kT}{B}\right)^{1/2} \qquad (3.1.10)$$

---

* While this statement is reasonable for diatomic molecules, it is generally not applicable to polyatomic molecules. In the latter case (e.g., the SF_6 molecule) spacing between vibrational levels is often appreciably smaller than 1000 cm^{-1} (down to ~ 100 cm^{-1}); many excited vibrational levels of the ground electronic state may then have a significant population at room temperature.

A conclusion which can be drawn from this section is that, for simple molecules at room temperature, the molecular population is distributed among several rotational levels of the ground vibrational state.

3.1.3. Stimulated Transitions

According to the earlier discussion, transitions among energy levels of a molecule can be divided into three types:

- Transitions between two rotational–vibrational levels of different electronic states— referred to as *vibronic* transitions (a contraction of vibrational and electronic). These generally fall in the near-uv spectral region.
- Transitions between two rotational–vibrational levels of the same electronic state (*rotational–vibrational transitions*). These generally fall in the near- to middle-infrared spectral region.
- Transitions between two rotational levels of the same vibrational state, e.g., $v'' = 0$, of the ground electronic state (*pure rotational transitions*). These generally fall in the far-infrared spectral region.

In the discussion that follows, we briefly consider vibronic and rotational–vibrational transitions, since the most widely used molecular gas lasers are based on these two types of transitions. Lasers based on pure rotational transitions, thus oscillating in the far-infrared, also exist, but their use is relatively limited so far (e.g., for spectroscopic applications). In what follows, the quantum mechanical selection rules for these three types of transitions are briefly considered. (See Appendix D for more details.)

Consider first a vibronic transition and assume that the symmetry of the electronic wavefunctions in the lower and upper electronic states allows an electric dipole transition. Since the electronic motion occurs at much faster speed than nuclear motion, we readily appreciate the so-called *Franck–Condon principle*, which states that nuclear separations do not change during a radiative transition. If we now assume that all molecules are in the $v'' = 0$ level of the ground electronic state,* then, referring to Fig. 3.6, the transition must occur vertically, i.e., somewhere between transitions $A - A'$ and $B - B'$. The Franck–Condon principle can be set in a more quantitative way by saying that the transition probability between a given vibrational level v'' of the ground state and some vibrational level v' of the upper electronic state is given by:

$$W_{12} \propto \left| \int u_{v''} u_{v'} dR \right|^2 \qquad (3.1.11)$$

where $u_{v''}(R)$ and $u_{v'}(R)$ are the vibrational wave functions of the two levels. Within the harmonic approximation these functions are known to be given by the product of a Gaussian function and a Hermite polynomial. Since the $v'' = 0$ wavefunction is known to be a Gaussian function, the transition probability, according to Eq. (3.1.11), is greatest to the

---

* When many vibrational levels of the ground electronic state are occupied, transitions may start from any of these levels. Absorption bands originating from $v'' > 0$ are referred to as *hot bands*.

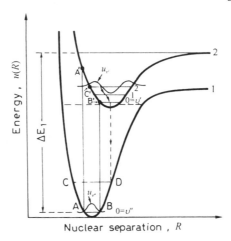

FIG. 3.6. Allowed vibronic transitions for a diatomic molecule.

vibrational state whose wavefunction $u_{v'}$ ensures the best overlap with the function $u_{v''}$. In Fig. 3.6 the most probable transition is therefore to the $v' = 2$ level. A way of understanding this follows from noting that, if we neglect zero-point energy, the molecule in the ground state can be considered at rest with a nuclear separation midway between points A and B. When a photon is absorbed, the molecule passes to an upper vibrational level with the same nuclear separation but still remains at rest (nuclear motion, i.e., position and velocity, cannot change during an electronic transition). This means that the transition occurs toward point C' of vibrational level 2. Since the minimum of the potential energy curve, for the excited state, is shifted toward larger values of the internuclear distance R, the two atoms of the molecule, after absorption, experience a repulsive force, so that the molecule is left in the excited $v' = 2$ vibrational state. To conclude we can say that the transition probability for an electric-dipole-allowed vibronic transition is proportional to $|\int u_{v''} u_{v'}\, dR|^2$; this quantity is referred to as the *Franck–Condon factor.*

Let us now consider a transition between two vibrational levels of the same electronic state (rotational–vibrational transitions) and assume that the symmetry of the molecule allows this transition to occur. In this case the transition is said to be *infrared active.** For such a transition the quantum mechanical selection rule requires that $\Delta v = \pm 1$, where Δv is the change in the vibrational quantum number. Thus starting at the ground state $v'' = 0$, a transition can occur only to the $v'' = 1$ state (see Fig. 3.2). If however we start from the $v'' = 1$ level, then the transition may occur to the $v'' = 2$ (absorption) or $v'' = 0$ (emission) level. This result should be contrasted to that for vibronic transitions where the transition may occur to several vibrational levels with a probability proportional to the corresponding Franck–Condon factor. Note that the $\Delta v = \pm 1$ selection rule holds rigorously within the harmonic potential approximation. Since the electronic energy curves in Fig. 3.6 are not exactly parabolic, it can be shown that transitions obeying the selection rules $\Delta v = \pm 2, \pm 3$, etc., may also occur as a result of this anharmonicity, although with much lower probability (*overtone transitions*).

---

* A simple example of an infrared inactive transition involves homonuclear diatomic molecules (e.g., H_2). Ro-vibrational transitions are not allowed in this case, because, due to symmetry, the molecule does not exhibit an electric dipole moment when it vibrates.

For both vibronic and vibrational-rotational transitions, we have so far ignored the fact that, corresponding to each vibrational level, there actually exists a whole set of closely spaced rotational levels, occupied at thermal equilibrium according to Eq. (3.1.9) (see also Fig. 3.5). Then absorption, e.g., takes place between a given rotational level of the lower vibrational state and some rotational level of the upper vibrational state. For diatomic or linear triatomic molecules, selection rules usually require $\Delta J = \pm 1$, ($\Delta J = J'' - J'$, where J'' and J' are the rotational numbers of lower and upper vibrational states, respectively). In the case of a rotational–vibrational transition, a given vibrational transition (e.g., $v'' = 0 \rightarrow v'' = 1$ in Fig. 3.2), which in the absence of rotation consists of just a single frequency v_0, is in fact made up of two sets of lines (Fig. 3.7). The first set, which has lower frequencies, is called the *P* branch; it corresponds to the $\Delta J = 1$ transition. Transition frequencies in this branch are lower than v_0 because the rotational energy in the upper level is smaller than that in the lower level (see, Fig. 3.4). The second set, which has higher frequencies, is called the *R* branch; it corresponds to $\Delta J = -1$. Using Eq. (3.1.7), we can show that the lines are evenly spaced in frequency by the amount $2B/h$. One also observes from Fig. 3.7 that the amplitudes of lines are not the same due to the different populations in rotational levels of the ground state (see Fig. 3.5). Each line is then assumed to be broadened by some line-broadening mechanism (e.g., Doppler or collision broadening). For more complex molecules, the selection rule $\Delta J = 0$ also holds; in this case transitions from all rotational levels of a given vibrational state produce a single line centered at frequency v_0 (*Q* branch). We observe that, when a population inversion is present between two vibrational levels (such as the $v' = 1$ and $v'' = 0$ levels in Fig. 3.4), the same spectrum as in Fig. 3.7 can be observed in emission rather than in absorption.

Example 3.1. *Emission spectrum of the CO_2 laser transition at $\lambda = 10.6\ \mu m$.* We consider the $00°1 \rightarrow 10°0$ transition (see the section on CO_2 lasers in Chap. 10), whose fundamental frequency v_0, in wave numbers, is at $v_0 = 960.8\ cm^{-1}$.[20] The rotational constant of the CO_2 molecule is $B \cong 0.387\ cm^{-1}$;[20] this value is assumed to be the same for upper ($00°1$) and lower ($10°0$) vibrational levels. From previous considerations, the transition energies of *P*-branch transitions are given by:

$$E = hv_0 + BJ'(J' + 1) - BJ''(J'' + 1) = hv_0 - 2BJ'' \qquad (3.1.12)$$

where, as usual, J'' is the rotational quantum number of the lower vibrational state. The rotational number J'_m of the most populated rotational level of the upper vibrational state is given by Eq. (3.1.10). Assuming a rather hot CO_2 molecule, i.e., $T = 450$ K, we obtain $J_{max} \cong 19.6$. For the CO_2 molecule, symmetry dictates that only $J'(\text{odd}) \rightarrow J''(\text{even})$ transitions can occur. Thus the most populated rotational level in the upper state, available for the transition, is either the $J' = 19$ or the $J' = 21$ level. Assuming that the $J' = 21$ level is the most populated, this level will decay, for a *P*-branch transition, to the $J'' = 22$ level [*P*(22) transition]. According to Eq. (3.1.12), the corresponding transition frequency is $v = v_0 - (2BJ''/h) = 943.8\ cm^{-1}$, corresponding to a wavelength of $\lambda = (1/943.8)\ cm \cong 10.6\ \mu m$. Note that the wavelength corresponding to the fundamental frequency v_0 is $\lambda = c/v_0 \cong 10.4\ \mu m$. Since only even J'' numbers are involved, the separation between two consecutive *P*-branch transitions, according to Eq. (3.1.12), is given by $\Delta v = 2B\Delta J'' = 4B = 1.55\ cm^{-1}$.

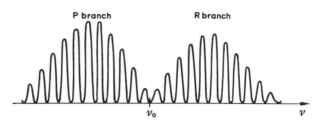

FIG. 3.7. Transitions between two vibrational levels, taking into account rotational splitting. In the absence of rotational energy this transition consists of a single line centered at v_0. Here it consists of two groups of lines: The so-called P-branch, which corresponds to a jump in the rotational quantum number of $\Delta J = +1$, and the so-called R-branch, which corresponds to a jump in rotational quantum number of $\Delta J = -1$.

Example 3.2. *Doppler linewidth of a CO_2 laser.* Consider a CO_2 laser oscillating on the $P(22)$ line at $\lambda = 10.6\ \mu$m and assume $T = 450$ K (see Example 3.1). Using the appropriate mass of CO_2 in Eq. (2.5.18), we obtain $\Delta v_0^* \cong 50$ MHz. Note: Since, according to Eq. (2.5.18), one has $\Delta v_0^* \propto v_0$, the calculated linewidth for a CO_2 molecule is much smaller than that of the He–Ne laser in Example 2.7, essentially because the oscillation frequency v_0 is now approximately 17 times smaller. Note also that the gas is assumed to be hotter in this case because, to obtain the high-output powers typical of CO_2 lasers, higher pump power is used than in the He–Ne case.

Example 3.3. *Collision broadening of a CO_2 laser.* We consider a CO_2 laser containing a gas mixture of H_2, N_2, and CO_2. In this case, the laser linewidth, due to collision broadening, is found experimentally to be given by $\Delta v = 77.58(\psi_{CO_2} + 0.73 - 4\psi_{N_2} + 0.6\psi_{He}) \times p(300/T)^{1/2}$ MHz [compare with Eqs. (2.5.12) and (2.5.11)], where ψ are the fractional partial pressures of the gas mixture, T is the gas temperature, and p is the total pressure (in Torr). Taking, as an example, a typical low-pressure gas mixture ($p \cong 15$ Torr in a 1:1:8 CO_2:N_2:He mixture) at $T = 450$ K, we obtain $\Delta v_c \cong 40$ MHz. A comparison with the result in Example 3.2 shows that, for a low-pressure CO_2 laser, collision broadening is comparable to Doppler broadening. However, for higher pressure CO_2 lasers, e.g., atmospheric pressure lasers (see Chap. 10), collision broadening becomes the dominant line-broadening mechanism.

For pure rotational transitions the selection rule requires the molecule to possess a permanent dipole moment. In fact, for spontaneous emission, emitted radiation can be viewed as originating from the rotation of this dipole moment. For a diatomic or linear triatomic molecule, the selection rule again requires $\Delta J = \pm 1$. Thus, in the case of stimulated emission from a given rotational level J, transitions can occur only to the rotational level with quantum number $J - 1$.

We can now summarize the selection rules that apply to vibronic, rotational–vibrational, and rotational transitions. In an electric-dipole-allowed vibronic transition, one has $\Delta J = \pm 1$ for the change in rotational quantum number, while the change in the vibrational quantum number is not governed by a strict selection rule. In fact, starting from a given vibrational level v'' of the lower electronic state, the transition may occur to several vibrational levels of the upper electronic state with probabilities proportional to the corresponding Franck–Condon factors. In an infrared-active rotational vibrational transition one has $\Delta v = \pm 1$ for the change in vibrational quantum number, within the harmonic approximation, and $\Delta J = \pm 1$ for the change in rotational quantum number. In pure rotational transitions in molecules with a permanent dipole moment, one again has $\Delta J = \pm 1$.

3.1.4. Radiative and Nonradiative Decay

We consider spontaneous emission, assuming first that the molecule is raised to some vibrational level of an excited electronic state (Fig. 3.6). From this state the molecule often decays rapidly by some nonradiative process (e.g., by collision) to the $v' = 0$ vibrational level.* This is particularly true for molecules in the liquid phase, where collisions occur very frequently. From there the molecule may decay radiatively to a vibrational level of the ground state (*fluorescence*; see Fig. 3.6). This transition occurs vertically, and the transition probability from the $v' = 0$ level to some level of the ground state is again proportional to the corresponding Franck–Condon factor. Roughly speaking, the ground-state vibrational levels involved are near the CD level in Fig. 3.6. The molecule then rapidly returns, by nonradiative decay (e.g., by collisions), to the $v'' = 0$ level of the ground electronic state (or, more precisely, thermal equilibrium is again established in the ground electronic state). It is now clear from Fig. 3.6 why the fluorescence wavelength is longer than that of absorption, a phenomenon referred to as *Stoke's law*.

Spontaneous emission may also occur between two rotational–vibrational levels of the ground electronic state, and again, for an infrared-active transition, selection rules $\Delta v = \pm 1$ and $\Delta J = \pm 1$ apply. In pure rotational transitions, spontaneous emission occurs only in molecules having a permanent dipole moment (see Sect. 3.1.3), and the $\Delta J = \pm 1$ selection rule applies. Note however that, for rotational–vibrational transitions and more so in pure rotational transitions, the small value of the transition frequency implies that the spontaneous emission lifetime becomes very long, i.e., from milliseconds to even seconds (recall that $\tau_{sp} \propto 1/v_0^3$). The spontaneous decay of the molecule is then usually dominated by nonradiative processes.

We next briefly consider phenomena that may cause nonradiative decay. Referring to the more general discussion in Sect. 2.6.1, we indicate here the main mechanisms.

Collisional deactivation with another like or unlike species usually occurs in molecules in the liquid phase. In the gas phase this decay route is particularly effective when the transition energy is small (e.g., for a rotational transition) and the colliding species have a small mass [e.g., deactivation of the CO_2 (0,1,0) level by He atoms; see Chap. 10]. Collisional deactivation results in rapid thermalization among the rotational levels of a given vibrational state.

A *near-resonant energy transfer* to another like or unlike species (see Fig. 2.12) is particularly effective when the energy imbalance ΔE is appreciably smaller than kT. A notable example of this nonradiative decay process is again found in a CO_2 molecule for the relaxation of the CO_2 (0,2,0) level to the CO_2 (0,1,0) level (see Chap. 10).

Internal conversion to another vibrational–rotational level of the same molecule may also occur (Fig. 3.8). The process is also referred to as *unimolecular decay*, since it occurs

---

* This rapid decay actually results in *thermalization* of molecules in the upper electronic state. The probability of occupation of a given vibrational level of this state is thus given by Eq. (3.1.8). For simple molecules therefore, the lowest vibrational level has the predominant population.

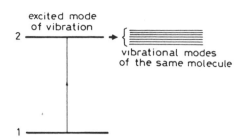

FIG. 3.8. Internal conversion between near-resonant rotational-vibrational modes of the same molecule.

within the same molecule; it is particularly effective when there is a large number of vibrational–rotational modes that are near-resonant with the given transition. These modes may also belong to a different electronic state; for instance, referring again to Fig. 3.6, once the molecule is in the lowest vibrational level of the upper electronic state ($v' = 0$ level), it can decay nonradiatively to a nearly isoenergetic vibrational level of the ground electronic state (*dashed level*). Internal conversion may be particularly effective for large molecules, e.g., dye molecules, which have many vibrational modes. In this case the number of vibrational modes belonging to the ground electronic level in near-resonance with the $v' = 0$ level can be quite large, and the corresponding nonradiative lifetime may even be as short as a few tens of picoseconds.

3.2. BULK SEMICONDUCTORS

This section considers the problem of radiation interacting with matter in a *bulk semiconductor*, i.e., whose physical dimensions are much larger than the de Broglie wavelength of the electrons under consideration. *Quantum-confined semiconductors* (quantum wells, quantum wires, and quantum dots), where one, two, or all three physical dimensions, respectively, are comparable to the de Broglie wavelength, are considered in a following section. These semiconductors play, in fact, an increasingly important role in laser physics. Again we limit our description to the most prominent features of the complex phenomena that occur. For a more extensive treatment of this subject, the reader is referred to Ref. 3.

3.2.1. Electronic States

The outer electrons of the atoms of a semiconductor material are delocalized over the whole crystal; the corresponding wave functions can then be written as *Bloch wave functions*:[4]

$$\psi(\mathbf{r}) = u_k(\mathbf{r}) \exp[j(\mathbf{k} \cdot \mathbf{r})] \tag{3.2.1}$$

where $u_k(\mathbf{r})$ has the periodicity of the crystalline lattice. The substitution of Eq. (3.2.1) into the Schrödinger wave equation shows that the corresponding eigenvalues of the electron energy E are a function of $\mathbf{k}$ and these values fall within allowed bands. From now on we consider only the highest filled band, the *valence band*, and the next higher one, the *conduction band*. Within the *parabolic band approximation*, E versus k relations can be approximated by a parabola. This leads to the view in Fig. 3.9 for valence and conduction bands. The energy E_c in the conduction band, measured from the bottom of the band upwards [Fig. 3.9a], can then be written as

$$E_c = \frac{\hbar^2 k^2}{2m_c} \qquad (3.2.2a)$$

where $m_c - \hbar^2/(d^2E_c/dk^2)_{k=0}$ is the effective mass of the electron at the bottom of the conduction band. Likewise, energy in the valence band, measured from the top of the band downward (Fig. 3.9a), can be written as:

$$E_v = \frac{\hbar^2 k^2}{2m_v} \qquad (3.2.2b)$$

where $m_v = \hbar^2/(d^2E_v/dk^2)_{k=0}$ is the effective mass of the electron at the top of the valence band. When dealing with a given transition, it may be more convenient to refer the energy to the same reference level, e.g., from the top of the valence band upward (Fig. 3.9b). If we let E' be the energy in this coordinate system, energies in the conduction and valence bands are now obviously given by:

$$E'_c = E_g + E_c \qquad (3.2.3a)$$
$$E'_v = -E_v \qquad (3.2.3b)$$

where E_g is the energy gap.

This simple one-dimensional model is generalized to the three-dimensional case if we let k_x, k_y, and k_z be the components of the electron's $\mathbf{k}$ vector and we assume that the

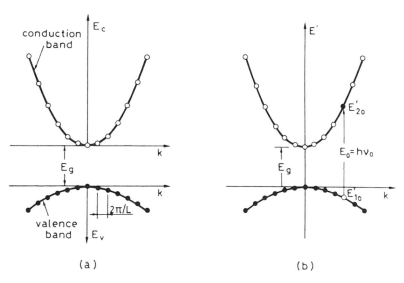

(a)   (b)

FIG. 3.9. Energy versus k relation for a bulk semiconductor: (a) Energy scale starting from the bottom of the conduction band upward for the conduction band and from the top of the valence band downward for the valence band; (b) energy scale starting from the top of the valence band upward for both conduction and valence bands.

effective mass, i.e., the band curvature, is the same along x-, y-, and z-directions. We again obtain Eqs. (3.2.2) and (3.2.3), where now $k^2 = k_x^2 + k_y^2 + k_z^2$.

So far we have assumed that the semiconductor crystal has an infinite extent. For a finite-sized crystal in the form of a rectangular parallelepiped with dimensions L_x, L_y, and L_z, we must impose the boundary condition that the total phase shift, $\mathbf{k} \cdot \mathbf{r}$, across the crystal is some multiple integer of 2π. Thus we obtain

$$k_i = \left(\frac{2\pi l}{L_i}\right) \tag{3.2.4}$$

where $i = x, y, z$ and l is an integer. In the one-dimensional case, available states are then denoted by dots in the valence band or by open circles in the conduction band (see Fig. 3.9).

The existence of valence and conduction bands can also be explained by a simple physical argument. Consider for simplicity the case of sodium, where each atom contains 11 electrons. Ten of these electrons are tightly bound to the nucleus to form an ion of positive charge e. The eleventh electron moves in an orbit around this ion. Let E_1 and E_2 be the energies of this electron in its ground state and first excited state, respectively, and ψ_1, ψ_2 be the corresponding wave functions. Consider now two sodium atoms separated by some distance d. If d is much larger than the atomic dimensions, the two atoms do not interact with each other, and the energy of the two states remains unchanged. That is, considering the two atoms in their energy state E_1, e.g., the one-electron energy level of the two-atom system is still E_1 and this level is doubly degenerate. The overall wave function can, in fact, be expressed as a combination of the two wave functions ψ_{1A} and ψ_{1B}, where the two functions combine either in phase or 180° out of phase (Fig. 3.10). In the absence of an interaction potential, these two states have the same energy E_1. However, when the atomic separation d becomes sufficiently small, the energies of these two states become slightly different owing to the interaction, and the doubly degenerate level is split into two levels. Likewise, in an N-atom system, when atoms are close enough to interact with each other, the N-fold degenerate level of the state of energy E_1 is split into N closely spaced levels. The state of energy E_1 thus gives rise to the valence band, while the state of energy E_2 gives rise to the conduction band (Fig. 3.11). From the previous argument it is apparent that each band consists of N

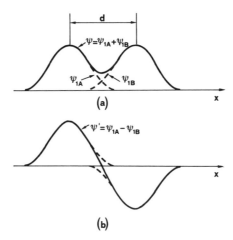

FIG. 3.10. (a) Symmetric and (b) antisymmetric linear combination of the atomic wave functions ψ_{1A} and ψ_{1B} of two identical atoms at separation d.

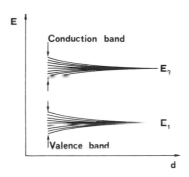

FIG. 3.11. *N*-fold splitting of atomic energy levels as a function of the atomic separation *d* for an *N*-atom system.

closely spaced levels, where N is the total number of atoms in the semiconductor crystal. Since N is usually a very large number, individual energy levels, in each band of a semiconductor, are generally not resolvable.

To summarize we can say that, within the parabolic band approximation, Eqs. (3.2.2) and (3.2.3) and the boundary conditions [Eq. (3.2.4)] describe the allowed energy values in a semiconductor. Within this approximation the electron is then considered as if it were a free particle of momentum $p = \hbar k$ (indeed $E = p^2/2m$ for a free particle) and details of the actual quantum system have been reduced to appearing in the values of the energy gap E_g and effective masses m_c and m_v. Note that, for the three-dimensional case, we can write

$$\mathbf{p} = \hbar\mathbf{k} \tag{3.2.5}$$

for the equation relating momentum $\mathbf{p}$ of the electron to the $\mathbf{k}$-vector of the wave function. Note also that, throughout Eqs. (3.2.2) and (3.2.3), we consider only *direct-gap* semiconductors, where the top of the valence band and the bottom of the conduction band occur at the same k value. Indirect-gap semiconductors, such as Si or Ge, are not considered here, since they are not relevant as laser materials.

Of the various direct-gap semiconductors, we limit our considerations to the III–V compounds, such as GaAs, InGaAs, AlGaAs, or InGaAsP. In particular for GaAs, $m_c = 0.067\, m_0$, where m_0 is the rest mass of a free electron. Note that, for all III–V semiconductors, there are three different types of valence band, namely, the heavy hole (*hh*) ($m_{hh} = 0.46\, m_0$ for GaAs), the light hole (*lh*) ($m_{lh} = 0.08\, m_0$ for GaAs), and the split-off band (see Fig. 3.12). To understand why this is so, based on the previous discussion about sodium atoms, we consider energy bands as originating from the discrete atomic energy levels of isolated atoms that made up the crystal. Accordingly, one can show that there is only one conduction band because the excited state of the corresponding isolated atoms has spherical symmetry like that of the *s*-state atomic orbitals. Likewise, since the lower state (state 1 of energy E_1 in Fig. 3.11) can be shown to have *p*-symmetry, the three valence bands originate from a suitable combination of the p_x, p_y, and p_z orbitals of this state, taking into account the crystal's symmetry. Actually, in a crystal of cubic symmetry, such as in all unstrained III–V compounds, the three bands are expected to have the same energy at $k = 0$. However spin-orbit interaction lowers one of these bands, the split-off band. Since the value of this splitting (e.g., $\Delta E = 0.34$ eV for GaAs) is much larger than kT [$\cong 0.028$ eV], the split-off band is always filled with electrons, so it does not participate in radiative and

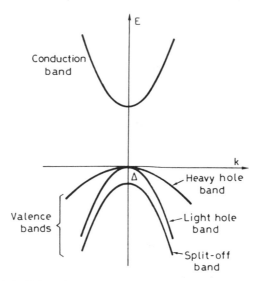

FIG. 3.12. (a) Heavy-hole, light-hole, and split-off valence bands for unstrained III–V semiconductors.

nonradiative transitions. For reasons explained in Sect. 3.2.2, the light hole band also makes little contribution to these transitions. Thus, to first order, the valence band of a III–V semiconductor can be viewed as consisting of only the *hh* band.

3.2.2. Density of States

Following the discussion of cavity modes in Sect. 2.2.1, we calculate the number of energy states $p(k)$ whose k value ranges from 0 to k. Referring to Fig. 2.2, since now both positive and negative values of k_i are allowed*, $p(k)$ is given by the volume of the sphere of radius k, $4\pi k^3/3$, divided by the volume of the unit cell, $(2\pi)^3/L_xL_yL_z$, times a factor 2 to account for the two states arising from the electron spin. Thus:

$$p(k) = \left(\frac{k^3 V}{3\pi^2}\right) \tag{3.2.6}$$

where $V = L_xL_yL_z$ is the crystal volume. Since the number of states is very large, we can calculate the density of states per unit volume $\rho(k)$ as:

$$\rho_{c,v} = \frac{dp}{V\,dk} = \frac{k^2}{\pi^2} \tag{3.2.7}$$

where Eq. (3.2.6) is used. Note that Eq. (3.2.7) is valid for both the valence and conduction bands; to indicate this, the density of states is denoted by both indices c and v. We are also

---

* The treatment so far done could have, as well, been done in terms of positive integer numbers, l. Thus, instead of Eq. (3.2.4), one should have written $k_i = (\pi l/L_i)$. The final result for the density of states would have remained unchanged, however.

interested in calculating the density of states $\rho(E)$ in terms of electron energy. Since $\rho_{c,v}(E)dE = \rho_{c,v}(k)dk$, from Eqs. (3.2.2) we obtain

$$\rho_c(E_c) = \frac{1}{2\pi^2}\left(\frac{2m_c}{\hbar^2}\right)^{3/2} E_c^{1/2} \tag{3.2.8a}$$

$$\rho_v(E_v) = \frac{1}{2\pi^2}\left(\frac{2m_v}{\hbar^2}\right)^{3/2} E_v^{1/2} \tag{3.2.8b}$$

We recall that E_c and E_v are measured from the bottom of the conduction and the top of the valence bands, upward and downward, respectively (Fig. 3.9a). Note that, since for III–V compounds one has $m_c \ll m_v = m_{hh}$, it follows that $\rho_c \ll \rho_v$. Note also that, since $m_{lh} \ll m_{hh}$, the density of states of light holes is only a small fraction of that of heavy holes. Accordingly, light holes are very much in a minority for a III–V semiconductor and their presence can normally be neglected in comparison with heavy holes.

3.2.3. Level Occupation at Thermal Equilibrium

We will first assume that the semiconductor is in overall thermal equilibrium. Since electrons are fermions, i.e., must comply with the Pauli exclusion principle, they must obey Fermi-Dirac statistics rather than Boltzmann statistics. The probability for the electron to occupy a given level of energy E', either in the valence or conduction band, is then given by

$$f(E') = \frac{1}{1 + \exp[(E' - E'_F)/kT]} \tag{3.2.9}$$

where E'_F is the energy of the Fermi level. In this case, the energy of both valence and conduction bands has been referred to the same reference level as in Fig. 3.9b and 3.13a. An interpretation of E'_F is obtained from Eq. (3.2.9) by setting $E' = E'_F$. We get $f(E'_F) = \frac{1}{2}$. Another interpretation of the significance of E'_F is also obtained from Eq. (3.2.9) by letting $T \to 0$. We get $f(E') = 1$ for $E' < E'_F$ and $f(E') = 0$ for $E' > E'_F$. Thus, at $T = 0$, the Fermi level separates the filled region from the empty region in a semiconductor. One should now remember that, for undoped semiconductors, E'_F is situated approximately in the middle of the energy gap. Thus, for $T > 0$, the relation $f(E')$ versus E' will be as shown in Fig. 3.13b. This means that, since $E_g \gg kT$, the level occupancy in the conduction band is very small, i.e., very few electrons are thermally activated to the conduction band. As a consequence of this circumstance, in both Fig. 3.13a and Fig. 3.9a, the available states in the valence band are denoted by a full circle to indicate the presence of an electron. Conversely, the available states in the conduction band are denoted by an open circle to indicate the absence of an electron, i.e., the presence of a hole. For n-type-doping, on the other hand, E'_F must be displaced toward the conduction band in order to accommodate the electrons, in this band, arising from the dopant ions. Similarly, for p-type-doping, E'_F is displaced toward the valence band. Finally, for very heavy doping (doping level of $\sim 10^{18}$ cm^{-3}), E'_F is displaced so much that it actually enters the conduction or the valence band, respectively. The semiconductor is then called degenerate since its conductivity becomes similar to that of a metal.

Suppose now that electrons are raised from the valence to the conduction band by some suitable pumping mechanism. The intraband relaxation (with a typical time constant, τ, as established by electron-phonon collisions, of ≈ 1 ps) is usually much faster than interband

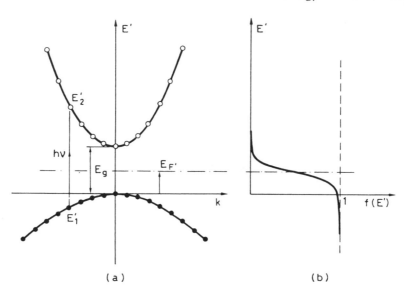

FIG. 3.13. Energy E' versus k relation and (b) level occupation probability $f(E')$ for both conduction and valence bands under thermal equilibrium.

relaxation (with a typical time constant τ of ≈ 1 ns, due to electron-hole recombination). Thus, a thermal equilibrium will be rapidly established within each band even though there is no overall equilibrium in the semiconductor. One can therefore talk of occupation probabilities f_v and f_c for the valence and conduction band separately. This means that f_c and f_v will be given by expressions of the general form of (3.2.9) in each band, respectively. More precisely, referring now to the energy coordinate system of Fig. 3.9a, one can write

$$f_c(E_c) = \frac{1}{1 + \exp[(E_c - E_{F_c})/kT]} \qquad (3.2.10a)$$

and

$$f_v(E_v) = \frac{1}{1 + \exp[(E_{F_v} - E_v)/kT]} \qquad (3.2.10b)$$

where E_{F_c} and E_{F_v} are now the energies of the so-called quasi-Fermi levels of the valence and conduction bands, respectively. Thus, for given values of E_{F_c} and E_{F_v}, the plots of $f_c(E_c)$ versus E_c and of $f_v(E_v)$ versus E_v will be as shown in Fig. 3.14b. Note that, following the previous discussion about the Fermi level, the quasi-Fermi levels indicate, in each band, the boundaries between the zones of fully occupied and completely empty states at $T = 0$ K. Accordingly, for $T = 0$ K, the states occupied by an electron (full circle) and the states occupied by a hole (open circle) will be as shown in Fig. 3.14a. In the same figure, the hatched areas thus correspond to states filled with electrons. Sometimes, it is more convenient to express Eq. (3.2.10) using the energy coordinate of Fig. 3.9b. According to Eq. (3.2.3a) and Eq. (3.2.3b) we then obtain

$$f_c(E_c') = \frac{1}{1 + \exp[(E_c' - E_{F_c}')/kT]} \qquad (3.2.11a)$$

and

$$f_v(E_v') = \frac{1}{1 + \exp[(E_v' - E_{F_v}')/kT]} \tag{3.2.11b}$$

As observed above, the quasi-Fermi levels indicate, in each band, the boundaries between occupied and empty states. Consequently, the values of E_{F_c}' and E_{F_v}' in Eq. (3.2.11) must depend on the number of electrons raised to the conduction band. To obtain this dependence, we calculate the electron density in the conduction band, N_e, as

$$N_e = \int_0^\infty \rho_c(E_c)f_c(E_c)dE_c \tag{3.2.12}$$

To calculate the corresponding hole density, N_h, in the valence band, we notice that $\bar{f}_v(E_v) = 1 - f_v(E_v)$ is the probability that a given state in the valence band is not occupied by an electron and thus filled by a hole. From Eq. (3.2.10b) we then get

$$\bar{f}_v(E_v) = \frac{1}{1 + \exp[(E_v - E_{F_v})/kT]} \tag{3.2.13}$$

Equation (3.2.13) shows that, in the energy coordinate system of Fig. 3.14a, the probability of hole occupation in the valence band takes on the same functional form as that of electron occupation in the conduction band [compare Eqs. (3.2.13) and (3.2.10a)]. This makes the calculation for the valence band completely symmetric to that of the conduction band. Thus,

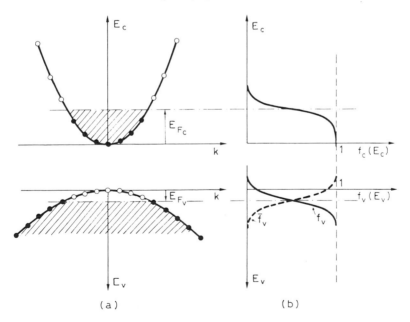

(a)   (b)

FIG. 3.14. (a) Energy E versus k relation and (b) level occupation probability $f_{c,v}(E)$ for conduction and valence bands under thermal equilibrium within each band.

for a given value of the quasi-Fermi level in the valence band, the hole density N_h is obtained as:

$$N_h = \int_0^\infty \rho_v(E_v)\bar{f}_v(E_v)dE_v \tag{3.2.14}$$

Suppose now that a given density of electrons N is raised by a suitable pumping process from the valence to the conduction band. The hole density left in the valence band also equals N, so the quasi-Fermi levels of both the valence and conduction bands can be obtained from Eqs. (3.2.12) and (3.2.14) by setting the condition $N_e = N_h = N$. From Eqs. (3.2.12), (3.2.8a), and (3.2.10a), we obtain

$$N = N_c \frac{2}{\pi^{1/2}} \int_0^\infty \frac{\varepsilon^{1/2}\, d\varepsilon}{1 + \exp(\varepsilon - \varepsilon_F)} \tag{3.2.15}$$

Example 3.4. *Calculation of the quasi-Fermi energies for GaAs.* We take $m_c = 0.067\, m_0$ and $m_v = m_{hh} = 0.46 m_0$ and assume $T = 300$ K. We obtain $N_c = 4.12 \times 10^{17}$ cm^{-3} and $N_v = (m_v/m_c)^{3/2} N_c = 7.41 \times 10^{18}$ cm^{-3}, where N_c is the electron concentration defined in connection with Eq. (3.2.15) and N_v is the corresponding quantity for the holes. At each electron concentration N we can now obtain the quantity N/N_c and, from the general plot of Fig. 3.15a, deduce the corresponding quantity E_{F_c}/kT. A similar calculation can be made for the holes. The values of E_F/kT, calculated in this way for both electrons and holes in GaAs, are plotted in Fig. 3.15b against the carrier concentration N.

where $N_c = 2(2\pi m_c kT/h^2)^{3/2}$, $\varepsilon = E_c/kT$, and $\varepsilon_F = E_{F_c}/kT$. From Eqs. (3.2.14), (3.2.8b), and (3.2.13), we obtain an expression that is the same as Eq. (3.2.15) provided we interchange suffix c with v. Equation (3.2.15) then shows that E_{F_c}/kT is a function of only N/N_c and this function is plotted against N/N_c in Fig. 3.15a. This figure also holds for the valence band provided we interchange suffix c with v.

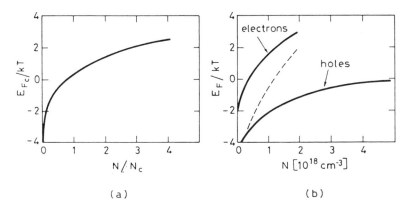

(a)   (b)

FIG. 3.15. (a) Normalized plot of the quasi-Fermi energy of the conduction band E_{F_c} versus the normalized concentration of injected electrons, N. The same normalized relation also holds for holes in the valence band. (b) Normalized plots of the quasi-Fermi levels of both valence and conduction bands E_F/kT versus the concentration of injected carriers N for GaAs.

3.2.4. Stimulated Transitions: Selection Rules

Let us consider the interaction of a monochromatic em wave of frequency v with a bulk semiconductor. As in the case of an atomic system [see Eq. (2.4.2)] the interaction Hamiltonian within the electric dipole approximation can be written as:[*]

$$H' = -e\mathbf{E} \cdot \mathbf{r} \qquad (3.2.16)$$

where $\mathbf{E} = \mathbf{E}(\mathbf{r}, t)$ is the electric field of the em wave at position $\mathbf{r}$ and time t. For a plane wave its expression can be written as:

$$\mathbf{E} = \mathbf{E}_0 \exp j(\mathbf{k}_{opt} \cdot \mathbf{r} - \omega t) \qquad (3.2.17)$$

where $\mathbf{k}_{opt}$ is the field wave vector and $\omega = 2\pi v$. If $v \geq E_g/h$ a transition may occur from a state in the valence band to a state in the conduction band. If we let E'_2 and E'_1 be corresponding energies of the two states, the transition rate W, according to Eq. (A.23) of Appendix A, is

$$W = \frac{\pi^2}{\hbar^2} |H'^0_{12}|^2 \delta(v - v_0) \qquad (3.2.18)$$

where $v_0 = (E'_2 - E'_1)/h$ and:

$$|H'^0_{12}|^2 = \left| \int \psi_c^* (-e\mathbf{r} \cdot \mathbf{E}_0 e^{j\mathbf{k}_{opt} \cdot \mathbf{r}}) \psi_v \, dV \right|^2 \qquad (3.2.19)$$

Note that ψ_v and ψ_c in Eq. (3.2.19) are the Bloch wave functions of levels 1 and 2, as given by Eq. (3.2.1).

From Eqs. (3.2.18) and (3.2.19) we can now obtain the selection rules for the interaction. Noting the Dirac δ-function on the right-hand side of Eq. (3.2.18), we see that $v = v_0$. This means

$$(E'_2 - E'_1) = hv \qquad (3.2.20)$$

which is often referred to as the energy conservation rule for the interaction. Similarly, from Eq. (3.2.19), since $\psi_v \propto \exp(j\mathbf{k}_v \cdot \mathbf{r})$ and $\psi_c \propto \exp(j\mathbf{k}_c \cdot \mathbf{r})$, one can show that the integral is nonvanishing only when:

$$\mathbf{k}_c = \mathbf{k}_{opt} + \mathbf{k}_v \qquad (3.2.21)$$

The proof of Eq. (3.2.21) is somewhat involved, and it requires the periodic properties of $u_c(\mathbf{r})$ and $u_v(\mathbf{r})$ in Eq. (3.2.1) to be properly taken into account.[15] The selection rule, expressed by Eq. (3.2.21), can however be physically understood by noting that an exponential factor of the form $\exp j[(\mathbf{k}_v + \mathbf{k}_{opt} - \mathbf{k}_c) \cdot \mathbf{r}]$ is present in the integrand in Eq. (3.2.19); since this term oscillates rapidly with $\mathbf{r}$, the value of the integral is zero unless $\mathbf{k}_v + \mathbf{k}_{opt} - \mathbf{k}_c = 0$. Since $\hbar\mathbf{k}_{c,v}$ is the electron momentum in the conduction or valence band

---

* To conform with the treatment in Chap. 2, the interaction Hamiltonian is written in terms of an electric dipole interaction rather than the interaction of the vector potential with the electron momentum $\mathbf{p}$, as common in many textbooks on semiconductors. The two Hamiltonians can be shown however to lead to the same final results.

and $\hbar \mathbf{k}_{opt}$ is the photon momentum, Eq. (3.2.21) shows that total momentum must be conserved in the transition. One should note that one has $k_{opt} = 2\pi n/\lambda$ where n is the semiconductor refractive index and λ is the transition wavelength. Thus, with e.g. $n = 3.5$ and $\lambda \cong 1 \ \mu m$, one gets $k_{opt} \cong 10^5 \ cm^{-1}$. On the other hand, one typically has $k_{c,v} = 10^6 \div 10^7 \ cm^{-1}$ for an electron or hole of average thermal energy (see Example 3.5). Thus $k_{opt} \ll k_{c,v}$, so Eq. (3.2.21) simplifies to:

$$k_c = k_v \tag{3.2.22}$$

Example 3.5. *Calculation of typical values of k for a thermal electron.* For an electron in the conduction band having average thermal velocity v_{th}, $m_c v_{th}^2 = 3kT$, where T is the electron temperature. We also have $p = \hbar k_c = m_c v_{th}$. Combining the two previous expressions, we obtain $k_c = (3m_c kT)^{1/2}/\hbar$. If we take $m_c = 0.067m_0$, as for GaAs, and $kT = 0.028$ eV ($T \cong 300$ K), we obtain $k_c = 2.7 \times 10^6 \ cm^{-1}$. Similarly $k_v = (3m_v kT)^{1/2}/\hbar$; thus $k_v = (m_v/m_c)^{1/2}k_c \cong 7 \times 10^6 \ cm^{-1}$ if we take $m_c = m_{hh} = 0.46m_0$ for GaAs.

Equation (3.2.22) is often referred to as the **k**-selection or **k**-conservation rule, and it indicates that stimulated transitions must occur vertically in the E versus k diagram (see Fig. 3.13a). Note lastly that the em wave does not interact with the electron's spin, in other words that spin is not involved in the interaction Hamiltonian (3.3.16). Therefore the spin cannot change in the transition, i.e., the selection rule for the change of the electron spin S is

$$\Delta S = 0 \tag{3.2.23}$$

As in the case of atomic transitions considered in Chap. 2, Eq. (3.2.18) must be modified when line-broadening mechanisms are taken into account. For semiconductors the main broadening mechanism arises from electron-phonon-dephasing collisions. Thus the Dirac δ-function in Eq. (3.2.18) must be replaced by a Lorentzian function $g(v - v_0)$ whose width, according to Eq. (2.5.11), is given by $\Delta v_0 = 1/\pi\tau_c$, where τ_c is the average electron-phonon-dephasing collision time ($\tau_c \cong 0.1$ ps for GaAs). Proceeding as in Sect. 2.4.4 [see also Eq. (2.4.19)], we define a transition cross section of the same form as atomic transitions,* namely:

$$\sigma = \frac{2\pi^2 v}{n\varepsilon_0 ch} \frac{\mu^2}{3} g(v - v_0) \tag{3.2.24}$$

where $\mu = |\boldsymbol{\mu}|$ and

$$\boldsymbol{\mu} = \int u_c e \mathbf{r} u_v \, dV \tag{3.2.25}$$

where $u_c = u_{ck}$ and $u_v = u_{vk}$ are the Bloch wavefunctions in Eq. (3.2.1). Note the factor 3 in Eq. (3.2.24), which arises from averaging the matrix element $\boldsymbol{\mu}$ over all electron $\mathbf{k}$ vector directions for a fixed electric field polarization. [See the footnote to Eqs. (2.4.13a–b)].

---

* The concept of cross section discussed in connection with Fig. 2.7 loses its meaning for a delocalized wavefunction, such as the Bloch wavefunction. We nevertheless retain the same symbol σ for a semiconductor to make an easier comparison with the case of isolated atoms or ions. Here σ means only that the transition rate for a plane wave is $W = \sigma F$, where F is the photon flux of the wave, or alternatively $W = \sigma \rho c/h v$, where ρ is the energy density and v is the frequency of the wave.

3.2.5. Absorption and Gain Coefficients

Consider first two energy levels E_2' and E_1' in the conduction and valence band, respectively, whose energy difference is equal to $E_0 = h\nu_0$, where ν_0 is the frequency of the transition. Under the k-selection rule given by Eq. (3.2.22), energies E_2' and E_1' are uniquely established for a given value of ν_0. From Eqs. (3.2.3) and (3.2.2), we write

$$E_2' = E_g + \left(\frac{\hbar^2 k^2}{2m_c}\right) \tag{3.2.26a}$$

$$E_1' = -\frac{\hbar^2 k^2}{2m_v} \tag{3.2.26b}$$

where we have set $k = k_c = k_v$. Since $E_2' - E_1' = E_0 = h\nu_0$, from Eq. (3.2.26) we obtain

$$h\nu_0 = E_g + \left(\frac{\hbar^2 k^2}{2m_r}\right) \tag{3.2.26c}$$

where m_r is the reduced mass of the semiconductor given by the relation $m_r^{-1} = m_c^{-1} + m_v^{-1}$. Equations (3.2.26) represent a set of three equations in the three unknowns E_2', E_1' and k.

Next we define the *joint density of states* with respect to the energy variable $E_0 = E_2' - E_1'$, so that $\rho_j \, dE_0$ gives the density of transitions with transition energy between E_0 and $E_0 + dE_0$. Under the k-selection and spin selection rules given by Eqs. (3.2.22) and (3.2.23), any state in the valence band, with a given spin, is coupled to only one state in the conduction band with the same spin. The number of transitions thus equals the number of corresponding states in either valence or conduction bands. We thus write $\rho_j \, dE_0 = \rho(k) dk$, where $\rho(k) = \rho_{c,v}(k)$ is given by Eq. (3.2.7), so that we obtain

$$\rho_j(E_0) = \left(\frac{k^2}{\pi^2}\right)\left(\frac{dk}{dE_0}\right) \tag{3.2.27}$$

With the help of Eq. (3.2.26c), Eq. (3.2.27) gives

$$\rho_j(E_0) = \frac{1}{2\pi^2}\left(\frac{2m_r}{\hbar^2}\right)^{3/2}(E_0 - E_g)^{1/2} \tag{3.2.28}$$

For our purposes we also introduce the joint density of states $\rho_j(\nu_0)$ with respect to the transition frequency $\nu_0 = E_0/h$. Since $\rho_j(\nu_0)d\nu_0 = \rho_j(E_0)dE_0$, from Eq. (3.2.28) we obtain

$$\rho_j(\nu_0) = \frac{4\pi}{h^2}(2m_r)^{3/2}(h\nu_0 - E_g)^{1/2} \tag{3.2.29}$$

Consider now the elemental number of transitions $dN = \rho_j(\nu_0)d\nu_0$ whose transition frequency lies between ν_0 and $\nu_0 + d\nu_0$. For absorption to occur, the lower level of energy E_1'

must be occupied by an electron while the upper level of energy E_2' must be empty. The number of transitions available for absorption is therefore:

$$dN_a = (dN)f_v(E_1')[1 - f_c(E_2')] \tag{3.2.30}$$

where $f_v(E_1')$ is the probability that the lower level is full, and $[1 - f_c(E_2')]$ is the probability that the upper level is empty. Note that a general situation of equilibrium within each band is assumed so that $f_v(E_1')$ and $f_c(E_2')$ are obtained from Eqs. (3.2.11b) and (3.2.11a), with E_1' and E_2' substituted for E_v' and E_c', respectively. To calculate the net absorption, we must also take into account stimulated emission between the same two levels. This occurs when the upper state is full while the lower state is empty. The number of transitions available for stimulated emission is then:

$$dN_{se} = (dN)f_c(E_2')[1 - f_v(E_1')] \tag{3.2.31}$$

Once the elemental numbers of available transitions for absorption and stimulated emission are calculated, the elemental absorption coefficient at frequency v is obtained from Eq. (2.4.32) as $d\alpha = \sigma(v - v_0)(dN_a - dN_{se})$, where $\sigma = \sigma_h$ is the homogeneous cross section for the $E_1' \rightarrow E_2'$ transition. From Eq. (3.2.24) we then obtain

$$d\alpha = \left(\frac{2\pi^2 v}{n\varepsilon_0 ch}\right)\frac{\mu^2}{3}g(v - v_0)\rho_j(v_0)[f_v(E_1') - f_c(E_2')]dv_0 \tag{3.2.32}$$

The overall absorption coefficient at frequency v is obtained from Eq. (3.2.32) by integrating over all transition frequencies v_0. If we assume that $g(v - v_0)$ versus v_0 is a much narrower function than both $\rho_j(v_0)$ and $(f_c - f_v)$, then $g(v - v_0)$ can be approximated by the δ function, $\delta = \delta(v - v_0)$. Upon integration over transition frequencies v_0, we then obtain

$$\alpha = \left(\frac{2\pi^2 v}{n\varepsilon_0 ch}\right)\frac{\mu^2}{3}\rho_j(v)[f_v(E_1') - f_c(E_2')] \tag{3.2.33}$$

where E_2' and E_1' are now the energies of the two levels whose energy difference is hv. They can be readily calculated from Eq. (3.2.26) by substituting hv for hv_0.

According to Eq. (3.2.33), the absorption coefficient $\alpha = \alpha(v)$ can be written as:

$$\alpha = \alpha_0[f_v(E_1') - f_c(E_2')] \tag{3.2.34}$$

where:

$$\alpha_0 = \left(\frac{2\pi^2 v}{n\varepsilon_0 ch}\right)\frac{\mu^2}{3}\rho_j(v) \tag{3.2.35}$$

The meaning of $\alpha_0 = \alpha_0(v)$ is understood when we consider a semiconductor in overall thermal equilibrium at $T = 0$ K. The quasi-Fermi levels coincide in this case with the Fermi level and, if this level is within the energy gap, one has $f_v(E_1') = 1$ and $f_c(E_2') = 0$. Then $\alpha(v) = \alpha_0(v)$, which is the maximum absorption coefficient that the semiconductor can have at frequency v. Note that, for an intrinsic semiconductor and assuming $E_g \gg kT$ as is the case for all III–V semiconductors, we still have $f_v(E_1') \cong 1$ and $f_c(E_2') \cong 0$, i.e., $\alpha \cong \alpha_0$ even at room temperature. From Eqs. (3.2.35), with the help of (3.2.29) with v_0 substituted by v, we obtain

$$\alpha \cong \alpha_0 = \frac{\pi^3 v}{n\varepsilon_0 ch^3} \frac{\mu^2}{3} (2m_r)^{3/2} (hv - E_g)^{1/2} \tag{3.2.36}$$

The frequency behavior of $\alpha(v)$ is then determined, to a good approximation, simply by the frequency behavior of $(hv - E_g)^{1/2}$.

Example 3.6. *Calculation of the absorption coefficient for GaAs.* As an approximation we assume the frequency v, in the first term on the right-hand side of Eq. (3.2.36), to have the value $v \cong E_g/h = 3.43 \times 10^{14}$ Hz, where the energy gap E_g is taken to be 1.424 eV. We also take $m_v = 0.46m_0$ and $m_c = 0.067m_0$, so that $m_r = 0.059\, m_0 = 5.37 \times 10^{-32}$ kg. To calculate the average dipole moment $\mu_{av} = (\mu^2/3)^{1/2}$, we recall that the accurate value of the average electron momentum M_{av} was shown to be such that $M_{av}^2 = 3.38\, m_0 E_g$.[5] The relation between average dipole moment and electron momentum is $M_{av} = m_0\omega|\mu_{av}|/e$,[5] so that

$$\mu_{av} = e(3.38E_g/m_0)^{1/2}/2\pi v \cong 0.68 \times 10^{-25} \text{ C} \times \text{m}.$$

Note that, if we write $r_{av} = \mu_{av}/e$, then $r_{av} \cong 0.426$ nm. Substitution into Eq. (3.2.36) of the values for v and μ_{av} just calculated and $n = 3.64$ for the refractive index gives $\alpha_0 = 19,760(hv - E_g)^{1/2}$, where α_0 is expressed in cm^{-1} and the energy in eV. The absorption coefficient as calculated from the latter expression is plotted against $E - E_g$ in Fig. 3.16, where $E = hv$. Note that, when hv exceeds the energy gap by only 10 meV, the absorption coefficient already reaches a very large value ($\approx 2,000$ cm^{-1}).

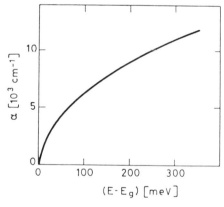

FIG. 3.16. Idealized plot of the absorption coefficient α versus the difference between the photon energy E and gap energy E_g for an intrinsic GaAs bulk semiconductor.

Consider next the case of the gain coefficient of an inverted semiconductor. One can readily see that the previous considerations remain valid provided the subscripts v and c are interchanged. Thus, from Eq. (3.2.34), the gain coefficient is seen to be given by:

$$g = \alpha_0[f_c(E_2') - f_v(E_1')] \tag{3.2.37}$$

It then follows that, at any transition frequency, the maximum gain coefficient is attained at $T = 0$ K, and it equals α_0. Note from Eq. (3.2.37) that, for any temperature, the condition for net gain is $f_c(E_2') > f_v(E_1')$. With the help of Eqs. (3.2.11a–b), we can then show that this implies

$$E_2' - E_1' < E_{F_c}' - E_{F_v}' \tag{3.2.38}$$

This necessary condition for net gain was originally derived by Bernard and Duraffourg.[6] The factor $f_c(E_2') - f_v(E_1')$ in Eq. (3.2.37) originates from the term $f_c(E_2')[1 - f_v(E_1')]$ $-f_v(E_1')[1 - f_c(E_2')]$ which gives the difference in probability between stimulated emission and absorption. Thus the Bernard–Duraffourg condition states that stimulated events must exceed absorption events; in this respect it is equivalent to the $N_2 > N_1$ condition for a simple two-level atomic system. Equation (3.2.38) can also be understood graphically if we consider the simple case of $T = 0$ K. For a given value of electron-hole injection, the position of the quasi-Fermi levels is shown in Fig. 3.17 where dashed zones are filled with electrons and clear zones are unoccupied by electrons (i.e., full of holes). Equation (3.2.38) then simply implies that level 2 must belong to the full zone while level 1 must belong to the empty zone in Fig. 3.17. The derivation of the Bernard–Duraffourg condition shows however that Eq. (3.2.38) is actually valid at any temperature.

It is important now to recall that $E_2' - E_1' = h\nu$ and that one must also have $h\nu > E_g$. Then from Eq. (3.2.38) we obtain

$$E_g \leq h\nu \leq E_{F_c}' - E_{F_v}' \tag{3.2.39}$$

which establishes the gain bandwidth of the semiconductor. According to Eq. (3.2.39), to have gain at any frequency requires $E_{F_c}' - E_{F_v}' > E_g$; the limiting case

$$E_{F_c}' - E_{F_v}' = E_g \tag{3.2.40}$$

is referred to as the *transparency condition*. In this case one has $g = 0$ at $\nu = E_g/h$. To achieve this condition, one must inject a density of electrons into the conduction band (and holes into the valence band) which is called the *transparency density* and indicated as N_{tr}.

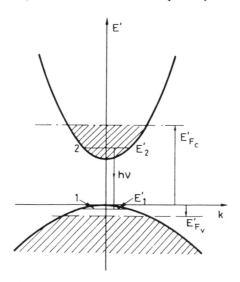

FIG. 3.17. Graphical illustration of the Bernard–Duraffourg condition for achieving net gain in a bulk semiconductor.

Example 3.7. *Calculation of the transparency density for GaAs.* We first transform Eq. (3.2.40) into the unprimed energy axes in Fig. 3.9a. According to Eq. (3.2.3) we can write $E'_{F_c} = E_g + E_{F_c}$ and $E'_{F_v} = -E_{F_v}$, and Eq. (3.2.40) transforms to $E_{F_c} + E_{F_v} = 0$. From Fig. 3.15a we see that E_{F_c}/kT is a function of (N/N_c); i.e., we can write $E_{F_c}/kT = f(N/N_c)$. Similarly we can write $E_{F_v}/kT = f(N/N_v)$, so that the transparency condition becomes

$$f(N_{tr}/N_c) + f(N_{tr}/N_v) = 0 \qquad (3.2.41)$$

To obtain N_{tr} from Eq. (3.2.41) for GaAs, we plot the function $(E_{F_c}/kT) + (E_{F_v}/kT)$ versus N as a dashed line in Fig. 3.15b. At each carrier concentration N, the curve is obtained as the sum of values given by the two continuous curves in the figure. According to Eq. (3.2.41), we can now say that the transparency density N_{tr} is the carrier concentration at which the dashed curve in Fig. 3.15b crosses the zero value of the ordinate. From Fig. 3.15b we obtain $N_{tr} = 1.2 \times 10^{18}$ cm^{-3}.

When the density of injected electrons N exceeds the transparency density, one has $E'_{F_c} - E'_{F_v} > E_g$; according to Eq. (3.2.39) net gain then occurs for photon energy between E_g and $E'_{F_c} - E'_{F_v}$. Plots of the gain coefficient versus photon energy, calculated from Eq. (3.2.37), are shown in Fig. 3.18 for GaAs, using the injected carrier density N as a parameter. One notes that, upon increasing the carrier density, the difference in quasi-Fermi energy, $E'_{F_c} - E'_{F_v}$, increases and this results in a corresponding increase in the gain bandwidth. Even at the highest carrier injection considered in the figure, this bandwidth is however a small

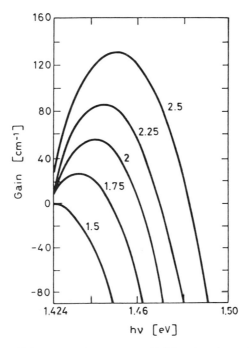

FIG. 3.18. Plot of the gain coefficient versus photon energy with injected carrier density N as a parameter (in units of 10^{18} cm^{-3}) as expected according to Eq. (3.2.37) for GaAs at $T = 300$ K. (By permission from Ref. 15.)

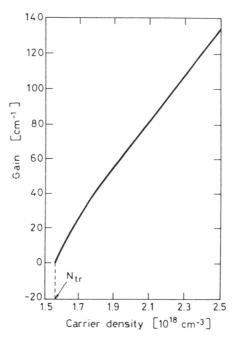

FIG. 3.19. Plot of the peak gain coefficient versus the injected carrier density for GaAs. (Reprinted from Ref 15 by permission of John Wiley & Sons, Inc.)

fraction of the energy gap. We also observe from Fig. 3.18 that the peak gain of each curve increases with increasing N. Again for GaAs, Fig. 3.19 shows the plot of this peak gain coefficient versus the density of injected electrons. For typical gain coefficients of interest in semiconductor lasers ($20 \leq g \leq 80$ cm^{-1}), the plot in Fig. 3.19 can be approximated by a linear relation; i.e., we can write

$$g = \sigma(N - N_{tr}) \tag{3.2.42}$$

where $\sigma \cong 1.5 \times 10^{-16}$ cm^2 for GaAs. Note that the quantity σ, in this way defined, has some analogy to the gain cross section defined for atomic systems [compare Eqs. (3.2.42) and (2.4.35)]. As already mentioned, however, the cross-section concept is not appropriate for a delocalized wavefunction, such as that of an electron in a semiconductor. For this reason, since from Eq. (3.2.42) we have $\sigma = dg/dN$, σ is often referred to as the differential gain coefficient of the semiconductor. We still retain the notation of σ for this differential gain, however, as a reminder of the fact that σ has the dimension of an area.

Most of the examples discussed in this section refer to the particular case of a GaAs semiconductor. However many other materials are also of interest as laser materials; a notable example is the quaternary alloy In$_{1-x}$Ga$_x$As$_y$P$_{1-y}$, which, depending on the composition indices x and y, covers the so-called second and third communication windows of optical fibers (1300 nm $\leq \lambda \leq$ 1600 nm). For the purpose of comparison, Table 3.1 lists the values of E_g, m_c/m_0, m_{hh}/m_0, N_{tr}, σ, and τ for In$_{0.75}$Ga$_{0.25}$As$_{0.55}$P$_{0.45}$ ($\lambda \cong$ 1300 nm) and In$_{0.6}$Ga$_{0.4}$As$_{0.88}$P$_{0.12}$ ($\lambda \cong$ 1550 nm),[7] as well as the corresponding values for GaAs, as discussed in this section. Note that the reported values for N_{tr} and σ fall in a range of values

TABLE 3.1. Values of emission wavelength λ, energy gap E_g, conduction band electron mass m_c, heavy-hole mass m_{hh}, carrier density at transparency N_{tr}, material differential gain σ, and lifetime τ for GaAs ($\lambda \cong 850$ nm) and InGaAsP ($\lambda = 1300$ and $\lambda = 1550$ nm) bulk semiconductors

	GaAs	$In_{0.73}Ga_{0.27}As_{0.6}P_{0.4}$	$In_{0.58}Ga_{0.42}As_{0.9}P_{0.1}$
λ (nm)	840	1310	1550
E_g (eV)	1.424	0.96	0.81
m_c/m_0	0.067	0.058	0.046
m_{hh}/m_0	0.46	0.467	0.44
N_{tr} (10^{18} cm^{-3})	1.2	1	1
σ (10^{-16} cm^2)	1.5	1.2/2.5	1.2/2.5
τ (ns)	3	4.5	4.5

listed for these semiconductors, and these are included in Table 3.1 as indicative numbers. It does seem however that both N_{tr} and σ for InGaAsP are somewhat smaller than corresponding values for GaAs.

3.2.6. Spontaneous Emission and Nonradiative Decay

Let us first consider the spontaneous emission process and define the spectral rate R_ν so that $R_\nu d\nu$ represents the number of spontaneous emission events, per unit time and volume, that result in light emitted with a frequency between ν and $\nu + d\nu$. To calculate R_ν, consider first the elemental transitions, $\rho_j(\nu_0)d\nu_0$, whose transition frequencies lie between ν_0 and $\nu_0 + d\nu_0$. These provide an elemental contribution, dR_ν, to R_ν given by $dR_\nu = A_{21}g(\nu - \nu_0) \times \{f_c(E'_{20})[1 - f_\nu(E'_{10})]\}\rho_j(\nu_0)d\nu_0$ where $A_{21} = A_{21}(\nu_0)$ is the rate of spontaneous emission between the two levels and $g(\nu - \nu_0)$ is the lineshape function of the transition. Note that $\rho_j(\nu_0)$ is multiplied by $f_c(E'_2)[1 - f_\nu(E'_1)]$ since, as in stimulated emission, spontaneous emission can occur only between an occupied upper state and an empty lower state. The total spectral rate R_ν is obtained by integrating the preceding expression over all transition frequencies ν_0:

$$R_\nu = \int A_{21}g(\nu - \nu_0)\{f_c(E'_2)[1 - f_\nu(E'_1)]\}\rho_j(\nu_0)d\nu_0 \qquad (3.2.43)$$

In the limit where $g(\nu - \nu_0)$ can be considered a much narrower function of ν_0 than all other functions in the integrand, $g(\nu - \nu_0)$ can be approximated by a δ function, $\delta(\nu - \nu_0)$, so Eq. (3.2.43) reduces to:

$$R_\nu = A_{21}\{f_c(E'_2)[1 - f_\nu(E'_1)]\}\rho_j(\nu) \qquad (3.2.44)$$

where A_{21} is the spontaneous emission rate for $\nu_0 = \nu$ and E'_2 and E'_1 are now the energy levels corresponding to a transition frequency ν. As an example Fig. 3.20 shows the qualitative behavior of R_ν versus photon energy $h\nu$ for an electron injection rate exceeding the rate for transparency, assuming A_{21} independent of ν. In the same figure the

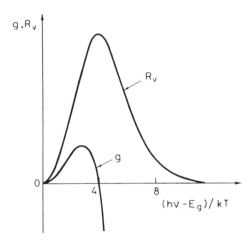

FIG. 3.20. Qualitative behavior of the spontaneous emission spectra R_v and optical gain g at a given value of the injected carrier density. (By permission from Ref. 11.)

corresponding gain coefficient, as calculated by Eq. (3.2.37), is also indicated for comparison. We observe that, unlike the case for atomic systems, the emission spectrum is now different from, and generally wider than, the gain spectrum. This is because R_v is proportional to $f_c(E_2')[1 - f_v(E_1')]$, while g is proportional to $f_c(E_2') - f_v(E_1')$.

Once the spectral rate R_v of spontaneous emission is calculated, the total rate R is obtained by integrating R_v over all emission frequencies:

$$R = \int A_{21} f_c(E_2')[1 - f_v(E_1')] \rho_j(v) dv \tag{3.2.45}$$

In practice, however, one often makes use of the phenomenological relation:

$$R = B N_e N_h \cong B N_e^2 \tag{3.2.46}$$

where B is a suitable constant. Equation (3.2.46) can be justified by assuming that any electron can recombine with any hole, which implies that the k-selection rule does not strictly hold.[9] We will not give any further discussion of this question which is related to the so-called band tails in a semiconductor,[10] and we take Eq. (3.2.46) as a phenomenological relation that holds well at the electron and hole densities of interest. Note that, according to the definition of R, we have $(dN_e/dt) = -R$. We can therefore define a radiative lifetime τ_r so that $R = N_e/\tau_r$ and thus write

$$\tau_r = (BN_e)^{-1} \tag{3.2.47}$$

Let us next consider nonradiative transitions. They generally occur at deep impurity centers where a carrier, electron or hole, is trapped (*deep-trap recombination*). Consider for instance an n-type semiconductor. At sufficiently high doping values, the Fermi level becomes close enough to the conduction band that these traps become filled with electrons.

A nonradiative transition then occurs by recombination of a free hole with this trapped electron, the excess energy being transferred to the lattice. A similar argument applies to p-type doping. For small-gap semiconductors, nonradiative transitions can also occur by direct recombination of untrapped electrons and holes; in this case, the excess energy is transferred to another electron (or hole), which is excited to a higher energy state in the band (*Auger recombination*).[12] Since Auger recombination is a three-body process, the decay of electron density due to this process can be written phenomenologically as $(dN_e/dt) = -CN_eN_hN_e = -CN_e^3$, where C is a suitable constant. Accordingly we can define a nonradiative lifetime due to Auger recombination τ_A as:

$$\tau_A = (CN_e^2)^{-1} \tag{3.2.48}$$

The dominant nonradiative mechanism seems to arise from a deep-trap recombination for GaAs and from Auger recombination in long-wavelength semiconductor-laser materials, such as InGaAsP.

Example 3.8. *Radiative and nonradiative lifetimes in GaAs and InGaAsP.* For GaAs we take $B \cong 1.8 \times 10^{-10}$ cm^3 s^{-1} and $N_e \cong N_{tr} = 1.2 \times 10^{18}$ cm^{-3}. We then get $\tau_r = 1/BN_{tr} \cong 4.6$ ns, to be compared with the measured overall lifetime at transparency of $\tau \cong 3$ ns ($T = 300$ K). Since $\tau^{-1} = \tau_r^{-1} + \tau_{nr}^{-1}$, where τ_{nr} is the lifetime due to the nonradiative process, we infer a nonradiative lifetime τ_{nr}, in this case due to deep trap recombination, of about 9 ns. For InGaAsP at $\lambda = 1300$ nm, we set $B = 2 \times 10^{-10}$ cm^3 s^{-1}, $N_e \cong N_{tr} \cong 1 \times 10^{18}$ cm^3, and $C \cong 3 \times 10^{-29}$ cm^6 s^{-1}. We get from Eq. (3.2.47) $\tau_r \cong 5$ ns and from Eq. (3.2.48) $\tau_A \cong 33.3$ ns, which gives an overall lifetime τ in agreement with the measured value ($\tau \cong 5$ ns at $T = 300$ K).

3.2.7. Concluding Remarks

We have seen in this section that phenomena leading to radiative and nonradiative transitions in a bulk semiconductor are notably more complicated than those, for isolated atoms or ions considered in Chap. 2. From a practical viewpoint, however, the most important physical parameters needed to predict laser behavior are the differential gain σ, the transparency density N_{tr}, and the overall lifetime τ for spontaneous decay (resulting from both radiative and nonradiative processes). For GaAs and InGaAs alloys considered here, these quantities can be obtained from Table 3.1. Note that the lifetime τ depends on carrier concentration and the values reported in Table 3.1 refer to a concentration equal to the transparency density.

We recall lastly that the gain expression given by Eq. (3.2.37) refers to a semiconductor of large dimensions (bulk semiconductor). For this reason, the gain values quoted in Table 3.1 are often referred to as the *material gain*. The actual gain in a double-heterostructure laser is smaller than this, and is determined by the ratio of the transverse dimension of the active layer to that of the cavity mode. This gain, often referred to as the *modal gain*, depends on details of the laser configuration and is thus considered in Chap. 9.

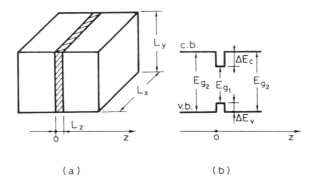

FIG. 3.21. (a) Schematic representation of a quantum well semiconductor and (b) corresponding plot of the energy of the bottom of the conduction band *c.b.* and of the top of valence band *v.b.* as a function of the *z*-coordinate of (a).

3.3. SEMICONDUCTOR QUANTUM WELLS

In a quantum well (QW) semiconductor, a very thin layer ($L_z \cong 5 \div 20$ nm) of a smaller band-gap material E_{g_1} is sandwiched between two layers of a larger band-gap material E_{g_2} (Fig. 3.21a). Technically, this involves the sophisticated techniques of molecular beam epitaxy (MBE) or metallo-organic chemical vapor deposition (MOCVD). Since $E_{g_1} < E_{g_2}$, potential wells are established for electrons, at the top of the valence band *v.b.* and for the holes, at the bottom of the conduction band *c.b.* (Fig. 3.21b). Due to the electron and hole confinement in these potential wells and since the semiconductor dimension is now comparable to the electron and hole De Broglie wavelength, energy levels of electrons and holes show very marked quantum-size effects. Furthermore, due to the small thickness of the layer, we can now allow lattice constants for the two materials to differ significantly; as a result strain develops within the thin quantum layer. The strain changes the quantum properties of the QW semiconductor considerably, and, in particular, it changes the effective masses. The quantum-size effects and, for a strained QW, the change in effective masses cause optical properties of the semiconductor's QW to differ markedly from those of the corresponding bulk material. In particular, the material differential gain increases considerably. The transparency electron density remains comparable to that of the corresponding bulk material, for an unstrained QW, while it shows a sizable decrease for a strained QW. The advantages that these improved properties offer in terms of lowering the laser threshold and increasing the modal gain are discussed in Chap. 9. Here we limit ourselves to pointing out that semiconductor quantum wells, of either strained or unstrained type, have become the most widely used semiconductor-laser materials.

3.3.1. Electronic States

To calculate energy levels of electrons and holes in the potential wells in Fig. 3.21b, we must know how the difference in band gap energy $\Delta E_g = E_{g_2} - E_{g_1}$ is partitioned between the well in the conduction band (ΔE_c) and that in the valence band (ΔE_v). This problem (the so-called band offset) involves complicated details of physics of the semiconductors. Experimentally, for two of the most important types of QW systems, one finds:

- $\Delta E_c = 0.67\ \Delta E_g$, $\Delta E_v = 0.33\ \Delta E_g$ for an AlGaAs/GaAs/AlGaAs QW.
- $\Delta E_c = 0.39\ \Delta E_g$, $\Delta E_v = 0.61\ \Delta E_g$ for a InP/InGaAsP/InP QW.

To calculate the energy levels of both electrons and holes in the corresponding QW, we use the much simplified assumption of infinite well depth [$(\Delta E_c, \Delta E_v) \to \infty$]; potential wells then appear as in Fig. 3.22. We take the z-axis orthogonal to the well with the origin at one well interface. According to Eq. (3.2.1), the Bloch wave functions, in both the conduction and valence bands, can then be written as:

$$\psi_{c,v}(\mathbf{r}) = u(\mathbf{r}_\perp)e^{j\mathbf{k}_\perp \cdot \mathbf{r}_\perp} \sin(n\pi z/L_z) \qquad (3.3.1)$$

where $\mathbf{r}_\perp$ and $\mathbf{k}_\perp$ are the components of $\mathbf{r}$ and $\mathbf{k}$ in the well plane (the x,y plane) and n is a positive integer. Note that, written in this way, $\psi_{c,v}$ already satisfies the boundary conditions $\psi_{c,v} = 0$ for $z = 0$ and $z = L_z$, i.e., at the two well boundaries. If we set similar periodic conditions along the x- and y-axes, we obtain

$$k_x = \left(\frac{l\pi}{L_x}\right) \qquad (3.3.2a)$$

$$k_y = \left(\frac{m\pi}{L_y}\right) \qquad (3.3.2b)$$

where l and m are also positive integers. Note the difference between Eqs. (3.3.2) and (3.2.4) which essentially reflects the fact that, in this case, we are limiting ourselves to positive numbers. Of course, one could also write the boundary condition as in Eq. (3.2.4), i.e., to allow for both positive and negative integers, and still obtain the same final results.

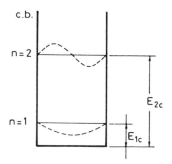

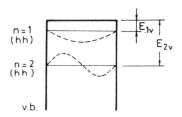

FIG. 3.22. Plots of the $n=1$ and $n=2$ energy levels (*continuous horizontal lines*) and corresponding eigenfunctions (*dashed lines*) in both conduction and valence bands, for infinite well depths.

Within the parabolic band approximation, the energy eigenvalues for either valence or conduction bands can then be written as

$$E_{c,v} = \frac{\hbar^2 k_\perp^2}{2m_{c,v}} + \frac{n^2 \hbar^2 \pi^2}{2m_{c,v} L_z^2} = \frac{\hbar^2 k_\perp^2}{2m_{c,v}} + n^2 E_{1c,v} \qquad (3.3.3)$$

where: $k_\perp^2 = k_x^2 + k_y^2$; $m_{c,v}$ is the electron mass in the conduction band or the hole mass in the valence band (only the heavy hole mass is considered, for simplicity); $E_{1c,v}$ is the energy of the first QW state ($n = 1$), for either conduction or valence bands, and is given by:

$$E_{1c,v} = \frac{\hbar^2 \pi^2}{2m_{c,v} L_z^2} \qquad (3.3.4)$$

One should note that, for both Eqs. (3.3.3) and (3.3.4), energy is measured from the bottom of the conduction band upward for the electrons and from the top of the valence band downward for holes. Note also that, for finite depth of potential wells, electrons are not totally reflected at well interfaces; i.e., the wave function is not zero at interfaces, as assumed in Eq. (3.3.1). The wave function then penetrates into the barrier layer, so that the expressions for wave functions and energy eigenvalues become more complicated.[14] We do not consider this case further, since it produces only quantitative rather than conceptual changes in the results that follow.

To discuss Eqs. (3.3.3) and (3.3.4), we first consider the case of electrons with zero transverse momentum ($k_\perp = 0$). The first two energy levels ($n = 1$ and $n = 2$) for both the conduction and valence bands are shown as solid horizontal lines in Fig. 3.22, while the corresponding eigenfunctions are shown as dashed lines. According to Eq. (3.3.3), we have $E_{2c} = 4E_{1c}$, the same relation also holding for the valence band. If we now consider electrons with $k_\perp > 0$, the energy E versus $k_\perp$ relations, for each of the $n = 1$, $n = 2$, etc., states considered before, will be as shown in Fig. 3.23a. Individual subbands are now introduced into conduction and valence bands. In the same figure, the available states, as obtained via Eq. (3.3.2), are shown as dots in the valence band and as open circles in the conduction band. Note lastly that, when dealing with transitions between valence and conduction subbands, an alternative energy scale, E', starting from the top of the valence band, e.g., and increasing upward may, sometimes, be more convenient (see Fig. 3.23b). The transformation between the primed E' and unprimed E energy scales are again given by Eq. (3.2.3), where E_c and E_v are now expressed by Eq. (3.3.3).

Example 3.9. *Calculation of the first energy levels in a GaAs/AlGaAs QW.* Let us take $L_z = 10$ nm and assume that the electron and hole (heavy hole) masses in the GaAs well are the same as those of the bulk material, i.e., $m_c = 0.067m_0$ and $m_v = m_{hh} = 0.46m_0$. From Eq. (3.3.4) we get $E_{1c} = 56.2$ meV and $E_{1v} = 8$ meV. If the confinement layer at both sides is $Al_{0.2}Ga_{0.8}As$, then $E_{g_2} = 1.674$ eV. Since the band gap of GaAs is $E_{g_1} = 1.424$ eV, we obtain $\Delta E_g = 250$ meV and thus $\Delta E_c = 0.65 E_g = 162.5$ meV and $\Delta E_v = 0.35 \Delta E_g = 87.5$ meV. Since E_{1c} is comparable to ΔE_c, the assumption of an infinite well is not a good approximation in this case. Taking barrier penetration into account, the actual values can be obtained from, e.g., Fig. 9.1 in Ref. 3 as $E_{1c} \cong 28$ meV and $E_{1v} \cong 5$ meV.

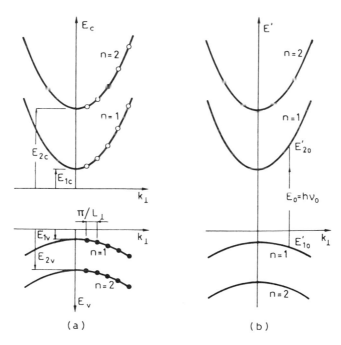

FIG. 3.23. Energy versus $k_\perp$ relations of the $n=1$ and $n=2$ subbands of both valence and conduction bands for a quantum well semiconductor. (a) The origin of the energy axis for conduction subbands is taken at the bottom of the conduction band of the bulk material and energy increases upward. The origin of the energy axis for the valence subbands is taken at the top of the valence band, and energy increases downward. (b) The energy axis is the same for all subbands; the origin is taken at the top of the valence band, and energy increases upward.

3.3.2. Density of States

Let us refer to Fig. 3.24, where the allowed states, as obtained via Eq. (3.3.2), are indicated by dots in the (k_x, k_y) plane (compare with Fig. 2.2). We can observe that only allowed states of the $n=1$ level are indicated. In fact one typically has $L_z = 10$ nm, while L_x and L_y may range between 10 and 100 μm, i.e., they are 10^3–10^4 times larger than L_z. Thus the separation between two successive states along the k_z-axis ($\Delta k_z = \pi/L_z$) is about 10^3–10^4 times larger than the separation between successive states along the k_x or k_y direction. Thus allowed states lie in widely-separated planes orthogonal to the k_z-axis, and it is convenient now to calculate the density of states in each of these planes. Accordingly, let $N(k_\perp)$ be the number of states in each plane, e.g., in the $n=1$ plane in Fig. 3.24, whose transverse vector is between 0 and $k_\perp$. According to the discussion relating to Fig. 2.2, $N(k_\perp)$ is given by one-fourth the area of the circle of radius $k_\perp$ divided by the area $\Delta k_x \Delta k_y$ of the unit cell, then multiplied by 2 to account for the two possible spin orientations in each state; thus:

$$N(k_\perp) = \frac{2(1/4)\pi k_\perp^2}{\Delta k_x \Delta k_y} = \frac{k_\perp^2}{2\pi} A_\perp \qquad (3.3.5)$$

where $A_\perp = L_x L_y$ is the transverse area of the quantum well. The number of states per unit $k_\perp$ and per unit area is then obtained as:

$$\rho_k^{2D} = \frac{dN(k_\perp)}{A_\perp \, dk_\perp} = \frac{k_\perp}{\pi} \qquad (3.3.6)$$

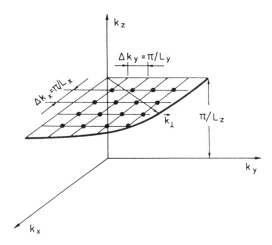

FIG. 3.24. Representation in k_x-, k_y-, k_z-space of the allowed states for the $n = 1$ subband.

Compared to a bulk semiconductor, ρ_k now gives the number of states per unit area rather than per unit volume. Thus we use the superscript $2D$ to remind us that we are in a two-dimensional rather than in a three-dimensional situation. Note also that Eq. (3.3.6) holds for both valence and conduction bands.

To obtain the density of states in energy coordinates, we write $\rho_c^{2D} dE_c = \rho_k^{2D} dk_\perp$ for the conduction band, e.g; from Eq. (3.3.6) we then obtain

$$\rho_c^{2D} = \frac{k_\perp \, dk_\perp}{\pi \, dE_c} \tag{3.3.7}$$

From Eq. (3.3.3), for the $n = 1$ subband, we have

$$k_\perp^2 = \left(\frac{2m_c}{\hbar^2}\right)(E_c - E_{1c}) \tag{3.3.8}$$

The quantity $k_\perp \, dk_\perp$ in Eq. (3.3.7) is readily obtained by differentiating both sides of Eq. (3.3.8). Equation (3.3.7) then gives

$$\rho_c^{2D} = \frac{m_c}{\pi \hbar^2} \tag{3.3.9}$$

Note that ρ_c^{2D} is independent of the value of $k_\perp$, i.e., of the transverse part of the energy $\hbar^2 k_\perp^2 / 2m_c$ [see Eq. (3.3.3)]. This is shown graphically in Fig. 3.25a where the quantity ρ_c^{2D}/L_z is plotted against the electron energy E_c (continuous line). Energy is measured from the bottom of the conduction band, and the function is plotted for $E_{1c} \le E_c \le E_{2c}$, where E_{2c} is the energy of the $n = 2$ subband. In fact, for $E_c \ge E_{2c}$, we must also take into account states lying in the plane $k_z = 2\pi/L_z$ (not shown in Fig. 3.24). The density of these states, however, is the same as that for the $n = 1$ plane; i.e., it is again given by Eq. (3.3.9). For $E_c \ge E_{2c}$ the overall density is then the sum of the densities of both $n = 1$ and $n = 2$ subbands. The corresponding curve is given by the solid-line step-function labeled $n = 1 + 2$

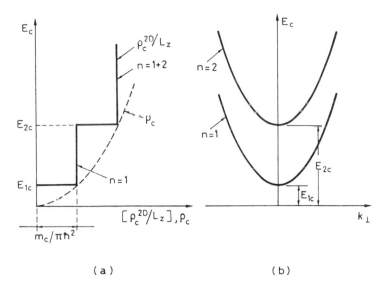

(a)   (b)

FIG. 3.25. (a) Plot of the quantum well density of states in the conduction band ρ_c^{2D} normalized to the well thickness L_z as a function of the state energy E_c (*staircase, solid line*). In the same figure, the plot of the density of states for the corresponding bulk semiconductor ρ_c is also shown as a dashed line. (b) Plots of the E_c versus $k_\perp$ relations for the $n=1$ and $n=2$ conduction subbands.

in Fig. 3.25a. For the sake of comparison we also show, as a dashed curve, the density of states for the same semiconductor material in bulk form, ρ_c, as given by Eq. (3.2.9a). One can readily show that the ρ_c curve touches the ρ_c^{2D}/L_z staircase plot at $E_c = E_{1c}$, $E_c = E_{2c}$, and so on. For completeness, we also show in Fig. 3.25b a plot of E_c versus $k_\perp$ (compare with Fig. 3.23a). Thus, for any value of energy $E_{1c} \leq E_c \leq E_{2c}$, Fig. 3.25b gives the corresponding value of the $k_\perp$ component of the electron **k** vector, directly. Similar considerations also hold for the density of states in the valence band. The corresponding density ρ_v^{2D} is obtained from Eq. (3.3.9) by substituting m_v for m_c; similar plots to those in Fig. 3.25 can then be made for the valence band. Since for GaAs $m_v = m_{hh} \cong 5m_c$, the steps of the staircase in Fig. 3.25a, for the valence band, are five times larger in state density ρ_v^{2D} and five times smaller in energy E_v.

3.3.3. Level Occupation at Thermal Equilibrium

We first consider the case of overall thermal equilibrium. The probability of occupation of a given energy state E' (see Fig. 3.23b), in either conduction or valence subbands, is again given by Fermi–Dirac statistics as in Eq. (3.2.11) where E_F' is the Fermi energy. Suppose now that some electrons are raised to conduction subbands $n=1$, $n=2$, etc., and assume a rapid relaxation among subbands (typically $\tau \cong 0.1$ ps) in both the conduction and valence bands. The equilibrium situation can then be described by introducing two quasi-Fermi levels. The probability of occupation of a given level in conduction or valence subbands is then given by Eqs. (3.2.10a–b) for the unprimed energy axes of Fig. 3.23a or by Eqs. (3.2.11a–b) for the primed energy axes of Fig. 3.23b.

As in a bulk semiconductor, the values of E_{F_c} and E_{F_v} are established by the density of electrons N_e and holes N_h injected into the corresponding bands. We now calculate N_e and N_h by the following relations:

$$N_e = \int (\rho_c^{2D}/L_z) f_c \, dE_c \tag{3.3.10a}$$

$$N_h = \int (\rho_v^{2D}/L_z) \bar{f}_v \, dE_v \tag{3.3.10b}$$

In Eq. (3.3.10a) ρ_c^{2D} is the surface density of states, and is given, for each subband, by Eq. (3.3.9) (see also Fig. 3.25a). In Eq. (3.3.10b) ρ_v^{2D} is the surface density of states for valence subbands, and $\bar{f}_v$ is the occupation probability of holes as given by Eq. (3.2.13). Since ρ^{2D} is constant in each subband, the integrals in Eqs. (3.3.10) can be calculated analytically and the final result written as:

$$N_e = kT \sum_i \left(\frac{m_{ci}}{\pi \hbar^2 L_z} \right) \ln\left[1 + \frac{E_{F_c} - E_{ic}}{kT} \right] \tag{3.3.11a}$$

$$N_h = kT \sum_i \left(\frac{m_{vi}}{\pi \hbar^2 L_z} \right) \ln\left[1 + \frac{E_{F_v} - E_{iv}}{kT} \right] \tag{3.3.11b}$$

Example 3.10. *Calculation of the quasi-Fermi energies for a GaAs/AlGaAs QW.* We take $m_{ci} = m_c = 0.067 \, m_0$ and $m_{vi} = 0.46 m_0$, i.e., we assume that masses are the same as those of the bulk material, and we neglect the contribution from light holes. We also assume $L_z = 10$ nm and $T = 300$ K. From Eq. (3.3.11) we obtain the two plots of concentration N versus $(E_F - E_1)/kT$, for both electrons and holes, shown in Fig. 3.26. From this figure the position of quasi-Fermi levels, for a given injection N of electrons and holes, can be readily obtained.

where the sum is taken over all subbands, m_{ci} and m_{vi} are electron and hole masses in each subband, and E_{ic} and E_{iv} are the minimum energies of each subband. One should note that, by choosing the unprimed energy axes of Fig. 3.23a, expressions for N_e and N_h take exactly the same form.

3.3.4. Stimulated Transitions: Selection Rules

Consider a stimulated transition (absorption or stimulated emission) between two given levels 1 and 2 belonging to a valence and a conduction subband, respectively. Within the electric dipole approximation, the corresponding transition probability W is proportional to $|H_{12}'^0|^2$, given by:

$$|H_{12}'^0|^2 = \left| \int \psi_c^* [-e\mathbf{r} \cdot \mathbf{E}(\mathbf{r})] \psi_v \, dV \right|^2 \tag{3.3.12}$$

where ψ_c and ψ_v are now given by Eq. (3.3.1) and $\mathbf{E}(\mathbf{r})$ is the electric field of the em wave at position $\mathbf{r}$ in the QW [compare to Eq. (3.2.19)]. To simplify our considerations, we take the

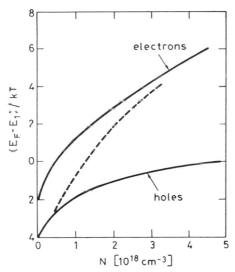

FIG. 3.26. Plots of the normalized difference between the quasi-Fermi energy E_F and the energy of the $n = 1$ subband E_1 versus density of injected carriers, for both electrons and holes, in a 10-nm GaAs/AlGaAs quantum well.

case of E-field polarization in the plane of the well. Then $e\mathbf{r} \cdot \mathbf{E}(\mathbf{r}) = e\mathbf{r}_\perp \cdot \mathbf{E}(\mathbf{r})$, where $\mathbf{r}_\perp$ is the component of the $\mathbf{r}$ vector in the well plane. Since the well thickness is much smaller than the wavelength of light, $\mathbf{E}(\mathbf{r})$ can be taken as a constant in the z-direction orthogonal to the well. We thus write $\mathbf{E} = \mathbf{E}(\mathbf{r}_\perp)$, i.e., a function of only the transverse coordinate $\mathbf{r}_\perp$. It then follows that $e\mathbf{r}_\perp \cdot \mathbf{E}(\mathbf{r})$ reduces to $e\mathbf{r}_\perp \cdot \mathbf{E}(\mathbf{r}_\perp)$, so that Eq. (3.1.12) can be split into two integrals, one over the transverse coordinates x and y and the other over the longitudinal coordinate z, namely:

$$|H_{12}^{\prime 0}|^2 = \left| \int u_c^*(\mathbf{r}_\perp) e^{-j\mathbf{k}_{c\perp} \cdot \mathbf{r}_\perp} (-e\mathbf{r}_\perp \cdot \mathbf{E}_0 e^{j\mathbf{k}_{opt} \cdot \mathbf{r}_\perp}) u_v(\mathbf{r}_\perp) e^{j\mathbf{k}_{v\perp} \cdot \mathbf{r}_\perp} dx\,dy \right|^2$$

$$\times \left| \int \sin(n_c \pi z / L_z) \sin(n_v \pi z / L_z) dz \right|^2 \tag{3.3.13}$$

As in the case of a bulk semiconductor, the integral over the transverse coordinates vanishes unless $\mathbf{k}_{v\perp} + \mathbf{k}_{opt} = \mathbf{k}_{c\perp}$. Since $|\mathbf{k}_{opt}| \ll (|\mathbf{k}_{v\perp}|, |\mathbf{k}_{c\perp}|)$, we obtain the selection rule [compare with Eq. (3.2.22)]:

$$k_{c\perp} = k_{v\perp} \tag{3.3.14}$$

Thus the $\mathbf{k}$-conservation rule still holds for the transverse component $\mathbf{k}_\perp$, which implies that transitions must occur vertically in Fig. 3.23. From the second integral on the right-hand side of Eq. (3.3.13), since n_c and n_v are positive integers, we obtain the selection rule for the quantum number n as:

$$\Delta n = n_c - n_v = 0 \tag{3.3.15}$$

which shows that transitions occur only between two subbands, one in the conduction and the other in the valence band, with the same quantum number n. Note lastly that spin is not involved in the interaction Hamiltonian $e\mathbf{r} \cdot \mathbf{E}$ since the em wave does not interact with spin. This implies that spin cannot change in the transition, i.e.:

$$\Delta S = 0 \qquad (3.3.16)$$

where S is the spin quantum number of the electron involved.

Equations (3.3.14)–(3.3.16) summarize the selection rules holding for transitions in QW. Although they have been derived subject to the assumption that the **E**-field is polarized in the plane of the well, they can be shown to hold in general.[13] Thus, from now on, these results will be used extensively.

3.3.5. Absorption and Gain Coefficients

To calculate absorption obeying the $\mathbf{k}_\perp$-conservation rule, we first introduce the joint density of transitions or joint density of states, ρ_{Jk}^{2D}, such that $\rho_{Jk}^{2D} \, dk_\perp$ gives the number of available transitions or coupled states, per unit area, where $k_\perp$ ranges from $k_\perp$ to $k_\perp + dk_\perp$. Since transitions occur only vertically in Fig. 3.23 and $\Delta S = 0$, this number also equals the number of states in either the valence or conduction band within the same elemental interval $dk_\perp$. Thus we obtain

$$\rho_{Jk}^{2D} = \rho_k^{2D} = \frac{k_\perp}{\pi} \qquad (3.3.17)$$

where Eq. (3.3.6) has been used. Consider now two given levels of energy E_2' and E_1' belonging to the $n = 1$ subbands, e.g., of the conduction and valence bands, respectively. From Fig. 3.23b and Eq. (3.3.3), the energy difference, $E_0 = h\nu_0 = E_2' - E_1'$, is seen to be given by:

$$E_0 = E_g + \frac{\hbar^2 k_\perp^2}{2m_r} + \Delta E_1 \qquad (3.3.18)$$

where m_r is the reduced mass and $\Delta E_1 = E_{1c} + E_{1v}$. If we now introduce the density of states in the E_0 coordinate, $\rho_{jE_0}^{2D}$, we can write

$$\rho_{jE_0}^{2D} \, dE_0 = \rho_{jk}^{2D} \, dk_\perp = \frac{k_\perp \, dk_\perp}{\pi} \qquad (3.3.19)$$

where Eq. (3.3.17) has been used. The quantity $k_\perp \, dk_\perp$ is then obtained by differentiating both sides of Eq. (3.3.18). From Eq. (3.3.19) we obtain

$$\rho_{jE_0}^{2D} = \frac{m_r}{\pi \hbar^2} \qquad (3.3.20)$$

If we now define the density of states with respect to the transition frequency coordinates v_0, ρ_{jv_0}, then, since $\rho_{jv_0}^{2D} \, dv_0 = \rho_{jE_0}^{2D} \, dE_0$, from Eq. (3.3.20) we obtain

$$\rho_{jv_0}^{2D} = \frac{4\pi m_r}{h} \tag{3.3.21}$$

We now calculate the overall absorption at the frequency v of the incoming em wave in the same way as for bulk material, i.e., using Eqs. (3.2.30)–(3.2.32), where $\rho_j(v_0)$ is now replaced by $(\rho_{jv_0}^{2D}/L_z)$, the joint density of states for our case. Again assuming infinitely narrow transitions between any two states, the absorption coefficient for the $(n = 1) \to (n = 1)$ QW transition is readily obtained from Eqs. (3.2.34) and (3.2.35) as:

$$\alpha_{QW} = \left(\frac{2\pi^2 v}{n\varepsilon_0 ch}\right) \frac{\mu^2}{3} \left(\frac{\rho_{jv}^{2D}}{L_z}\right) [f_v(E_1') - f_c(E_2')] \tag{3.3.22}$$

where E_1' and E_2' are now the energies of the two levels whose transition frequency is equal to v. Under the $\mathbf{k}_\perp$ conservation rule, E_2' and E_1' are obtained from Fig. 3.23b and Eq. (3.3.3) by letting $v_0 = v$.

Example 3.11. *Calculation of the absorption coefficient in a GaAs/AlGaAs QW.* We first consider the case of $T = 0$ K. In this case all valence subbands are full, all conduction subbands are empty, and $f_v(E_1') = 1, f_c(E_2') = 0$. The absorption coefficient then attains its maximum value given by:

$$\alpha_{QW}^{\max} = \left(\frac{2\pi^2 v}{n\varepsilon_0 ch}\right) \frac{\mu^2}{3} \frac{\rho_{jv}^{2D}}{L_z} \tag{3.3.23}$$

whose dependence on the photon energy is essentially determined by ρ_{jv}^{2D}. The absorption coefficient versus the difference $(E - E_g)$ between photon energy and the gap energy, as calculated from Eq. (3.3.23) for a $L_z = 10$ nm QW, is shown in Fig. 3.27. According to Fig. 3.25a, ρ_{jv}^{2D} is zero for photon energy $E < E_g + E_{1c} + E_{1v} = E_g + \Delta E_1$. No absorption is thus expected for $(E - E_g) < \Delta E_1$. Assuming $m_c = 0.067 m_0$ and $m_v = 0.46 m_0$, from Example 3.8 we obtain $\Delta E_1 = E_{1c} + E_{1v} \cong 65$ meV. For $\Delta E_1 \leq (E - E_g) \leq \Delta E_2$, where $\Delta E_2 = E_{2c} + E_{2v}$, ρ_{jv}^{2D} is given by Eq. (3.3.21) with $v_0 = v$, and the absorption coefficient has a constant value given by:

$$\alpha_{QW} = \frac{8\pi^3}{n\varepsilon_0 \lambda h^2} \left(\frac{\mu^2}{3}\right) \frac{m_r}{L_z} \tag{3.3.24}$$

where $\lambda = c/v$. According to Example 3.5, we take $(\mu^2/3)^{1/2} = 0.68 \times 10^{-25}$ C × m, $m_r = 5.37 \times 10^{-32}$ kg, $n - 3.64$, and $\lambda = 833$ nm. From Eq. (3.3.24), we obtain $\alpha_{QW} = 5{,}250$ cm^{-1}. For $(E - E_g) \geq \Delta E_2$, transitions between the $n = 2$ subbands also occur, the joint density of states doubles (see also Fig. 3.25a), and the absorption coefficient also doubles. Note that, since $E_{2c} = 4E_{1c}$ and $E_{2v} = 4E_{1v}$, one has $\Delta E_2 = 4\Delta E_1 = 260$ meV. (We can now compare Figs. 3.27 and 3.16.)

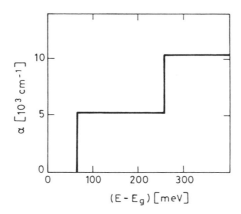

FIG. 3.27. Idealized plot of the absorption coefficient α versus the difference between photon energy and gap energy in a 10-nm GaAs/AlGaAs quantum well.

We can proceed in a similar way for the case of stimulated emission. It can readily be seen that the corresponding formula for gain coefficient can be obtained from Eq. (3.3.22) by interchanging the indices c and v and the indices 1 with 2. We then get

$$g_{QW} = \left(\frac{2\pi^2 v}{n\varepsilon_0 ch}\right)\frac{\mu^2}{3}\left(\frac{\rho_{jv}^{2D}}{L_z}\right)[f_c(E_2') - f_v(E_1')] \tag{3.3.25}$$

The necessary condition for positive net gain is again that $f_c(E_2') \geq f_v(E_1')$, which again implies the Bernard–Duraffourg condition $hv = E_2' - E_1' \leq E_{F_c}' - E_{F_v}'$. On the other hand hv must be larger than $E_g + \Delta E_1$, so that:

$$E_g + \Delta E_1 \leq hv \leq E_{F_c}' - E_{F_v}' \tag{3.3.26}$$

which establishes the gain bandwidth. From Eq. (3.3.26) the transparency condition is

$$E_{F_c}' - E_{F_v}' = E_g + \Delta E_1 \tag{3.3.27}$$

Example 3.12. *Calculation of the transparency density in a GaAs QW.* From Fig. 3.23 [see also Eq. (3.2.3)], $E_{F_c}' = E_{F_c} + E_g$ and $E_{F_v}' = -E_{F_v}$. Using the variables E_{F_c} and E_{F_v}, Eq. (3.3.27) can be simplified as:

$$(E_{F_c} - E_{1c}) + (E_{F_v} - E_{1v}) = 0 \tag{3.3.28}$$

where we have used the relation $\Delta E_1 = E_{1c} + E_{1v}$. To obtain the corresponding value of the transparency density N_{tr} from Eq. (3.3.28), we plot the quantity $[(E_{F_c} - E_{1c}) + (E_{F_v} - E_{1v})]/kT$ in Fig. 3.26 as a dashed curve. The curve is obtained by taking the sum of values given by the two continuous curves in Fig. 3.26 for each carrier concentration N. According to Eq. (3.3.28) the transparency density corresponds to the point where the dashed curve crosses the zero value of the ordinate. From Fig. 3.26 we thus obtain $N_{tr} \cong 1.25 \times 10^{18}$ electrons/cm^3.

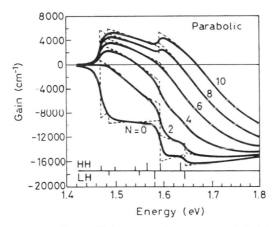

FIG. 3.28. Plots of the gain (or absorption) coefficient versus photon energy with the injected carrier density N as a parameter (in units of 10^{18} cm^{-3}) for an 8-nm GaAs/Al$_{0.2}$Ga$_{0.8}$As quantum well in the parabolic band approximation. (By permission from Ref. 8).

For $N > N_{tr}$ the QW exhibits net gain; its value is obtained from Eq. (3.3.25) once, for a given injection N, the quasi-Fermi levels are calculated (in our example from Fig. 3.26). Typical plots of the gain vs. the photon energy E, as obtained by this procedure for a 8 nm GaAs/Al$_{0.2}$-Ga$_{0.8}$As QW, are shown in Fig. 3.28, as solid curves, for several values of N (in units of 10^{18}cm^{-3}). The case labeled $N = 0$ corresponds to QW absorption and it should be compared to Fig. 3.27. The steps are not sharp here because spectral broadening of individual transitions is also included. Note that all possible transitions to heavy-hole and light-hole subbands are taken into account, in this case, and transparency occurs for $N_{tr} \cong 2 \times 10^{18}$ cm^{-3}. For $N \geq N_{tr}$, the peak gain coefficient can again be approximated by an expression similar to Eq. (3.2.42), namely:

$$g_p = \sigma_{QW}(N - N_{tr}) \qquad (3.3.29)$$

with $\sigma_{QW} \cong 7 \times 10^{-16}$ cm^2. Comparing these results to those for the bulk material shows that, while N_{tr} is almost the same for the two cases, the differential gain σ_{QW} for a QW is considerably larger (about twice) than that of the bulk semiconductor. The same situation also occurs for In$_{1-x}$Ga$_x$As$_y$P$_{1-y}$/InP QW lasers,[14] and is basically related to the different form of the density of states for the two cases (see Fig. 3.25a).[16] Note also that, for both GaAs/AlGaAs and In$_{1-x}$Ga$_x$As$_y$P$_{1-y}$/InP quantum wells, a linear relation between g_p and N holds less accurately than for the corresponding bulk materials. In fact, a plot of g_p versus N, at a given temperature, shows that a saturation of g_p occurs at sufficiently high values of N.[14] This effect is also related to the different form of the density of states for the two cases. We lastly recall that Eq. (3.3.29) gives the material gain coefficient of the QW. How this translates into the modal gain coefficient in a QW semiconductor laser and the real advantages of QW lasers will be discussed in Section 9.4.4.

3.3.6. Strained Quantum Wells

In a GaAs/Al$_{0.2}$Ga$_{0.8}$As QW, the lattice constant of GaAs is well matched (to better than 0.1%) to that of Al$_{0.2}$Ga$_{0.8}$As. (All III–V materials have cubic symmetry.) The same

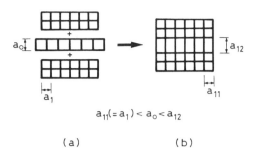

$$a_{11}(=a_1) < a_0 < a_{12}$$

(a)   (b)

FIG. 3.29. Crystal lattice deformation resulting from the epitaxial growth of a thin quantum well layer of III–V material with an original lattice constant a_0 (e.g., $In_xGa_{1-x}As$) between two thick layers of a material with a lattice constant $a_1 < a_0$ (e.g., $Al_{0.2}Ga_{0.8}As$).

situation occurs for the $In_{1-x}Ga_xAs_yP_{1-y}/InP$ QW if we choose $x \cong 0.45\ y$. Consider now for, example, the case of an $In_xGa_{1-x}As/Al_{0.2}Ga_{0.8}As$ QW, where $In_xGa_{1-x}As$ is the well material, because substitution of In for Al lowers the bandgap. For $0 \leq x \leq 0.5$, one has $1.424\ \text{eV} \geq E_g \geq 0.9\ \text{eV}$ and the emitted light covers the important wavelength range $840\ \text{nm} \leq \lambda \leq 1330\ \text{nm}$. The lattice constant of InGaAs is now larger than that of AlGaAs (by as much as 3.6% at $x = 0.5$); before the well is formed, the situation of the two materials will be as shown in Fig. 3.29a. In forming the QW, the two lattice constants must become equal in the QW well plane. This produces a biaxial compression of InGaAs in this plane and a uniaxial tension along the direction orthogonal to the plane (Fig. 3.29b). The InGaAs QW then loses its cubic symmetry, which changes the values of the valence band effective masses and the band gap.*

What needs to concern us mostly is the *hh* mass in the plane of the QW as it enters into the expression for the density of states in the valence band (see Sect. 3.3.2). Under compressive strain this mass is greatly reduced (by as much as a factor of 2 for $x = 0.2$), approaching the value of the electron mass in the conduction band. This makes the density of states in the valence band, ρ_v^{2D}, comparable to that in the conduction band, ρ_c^{2D}. The reduction of *hh* mass and the corresponding reduction in state density ρ_v^{2D} results in two very important advantages compared to a typical unstrained QW: The transparency density N_{tr} is greatly reduced by an amount as large as a factor of 2 [to $N_{tr} \cong (0.5 - 1) \times 10^{18}\ \text{cm}^{-3}$] and the differential gain dg/dN is greatly increased by an amount as large as a factor of 2 [to $(15 - 30) \times 10^{-16}\ \text{cm}^2$]. The reasons for both circumstances are fundamentally related to the reduced value of ρ_v^{2D} and to the shift in position of the quasi-Fermi levels at transparency, on reducing the hole mass.[17] Indeed the lowest value of N_{tr} and the highest value of dg/dN can be shown to be attained in the completely symmetrical case, $m_v = m_c$.

In conclusion there are three main beneficial effects of strained QW lasers:

- A consistent reduction of N_{tr}, which results in a consistent reduction of the threshold current density J_{th}, since (see Section 9.4.4) J_{th} is fundamentally related to N_{tr}.
- An increase of electron-hole recombination time τ, because both the radiative decay rate, $(1/\tau_r) = BN$, and the Auger rate, $(1/\tau_A) = CN^2$, are reduced as a consequence

---

* We recall that, within the parabolic band approximation, all quantum details are essentially hidden in values of the effective masses and energy gap.

of the reduction of N_{tr}. This effect also results in a further decrease in J_{th}, since $J_{th} \propto 1/\tau$ (see Section 9.4.4).

• A considerable increase in the differential material gain and hence the differential modal gain; Chap. 9 shows that this effect not only decreases the threshold current density but also increases the laser efficiency.

For these reasons, strained QW semiconductors are becoming increasingly important as laser media.

3.4. QUANTUM WIRES AND QUANTUM DOTS

According to Sect. 3.3, the improvement in optical properties obtained on going from bulk material to the corresponding QW material is essentially due to a quantum confinement effect arising from the fact that one dimension of the semiconductor has become comparable to the De Broglie wavelength. It is therefore natural to expand this idea to consider the other two possible cases of quantum confinement, namely, quantum wires and quantum dots, where two or all three dimensions become comparable to the De Broglie wavelength (Fig. 3.30a).

As for a QW, the fundamental difference between these quantum confined structures and bulk material arises from the different forms of the density of states. (For more details see Ref. 18.) Figure 3.30b shows, in fact, the qualitative behavior of the density of states for quantum wires and quantum dots as compared to bulk and QW materials. Using these different forms of state density, one can proceed as for a QW to calculate the expected gain. We do not pursue this here, and we limit ourselves to showing in Fig. 3.31, as a representative example, the expected material gain versus photon energy for a $Ga_{0.47}In_{0.51}As/InP$ system. ($Ga_{0.47}In_{0.51}As$ now constitutes the quantum confinement material.) In the figure, curves of predicted material gain versus photon energy in the bulk case for a 10-nm QW, 10 nm × 10 nm quantum wire, and 10 nm × 10 nm × 10 nm quantum dot are plotted for the same electron injection ($N = 3 \times 10^{18}$ cm^{-3}).[19] The calculated transparency density is about the same for bulk, QW and quantum wire ($N_{tr} \cong 1.3 \times 10^{18}$ cm^{-3}), while it is somewhat higher for quantum dot ($N_{tr} \cong 1.8 \times 10^{18}$ cm^{-3}). In agreement with our earlier discussion of QW structures, the peak gain increases on going from bulk to QW, from QW to quantum wire, and from

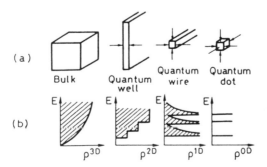

FIG. 3.30. (a) Different configurations and (b) corresponding forms of the density of states for bulk, quantum well, quantum wire, and quantum dot semiconductors. (By permission from Ref. 19.)

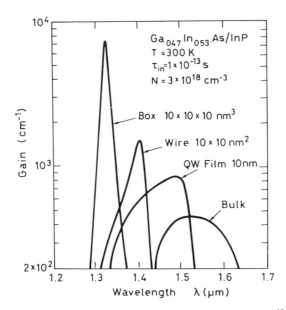

FIG. 3.31. Plot of calculated gain coefficient versus emission wavelength at $N = 3 \times 10^{18}$ cm^{-3} electron injection for a Ga$_{0.47}$In$_{0.53}$. As bulk semiconductor, Ga$_{0.47}$In$_{0.53}$As/InP 10-nm quantum well, 10 nm × 10 nm quantum wire, and 10 nm × 10 nm × 10 nm quantum dot (box). (By permission from Ref. 19.)

quantum wire to quantum dot. The gain bandwidth, on the other hand, decreases from QW to quantum wire and from quantum wire to quantum dot.

As laser media, quantum wires and dots will perhaps be used in the form of an array, such as the planar array in Fig. 3.32. Considerable technological difficulties (due to the need of high packing density, low size fluctuations, and low defect density) plague the fabrication of quantum wires and quantum dots with good optical properties. If these difficulties can be overcome, semiconductor laser materials of still lower threshold, higher differential gain, and narrower bandwidth will become available.

3.5. CONCLUDING REMARKS

Chapter 3, compared to the preceeding one, has progressed from the simple case of atoms to the more complicated cases of molecules and semiconductors, analyzing these in

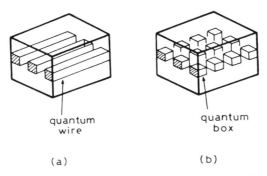

FIG. 3.32. (a) Planar array of multiple quantum wires and (b) multiple quantum dots.

some detail. As our discussion shows, a physical understanding of the optical properties of these materials requires a detailed description of their physical behavior; we have limited ourselves here to the most elementary aspects. From a phenomenological viewpoint, however, (see Chaps. 7 and 8), only a few physical parameters are required to predict laser behavior, namely:

- The wavelengths and bandwidths of the gain transitions.
- The transition cross section or, for a semiconductor, the differential gain and transparency density.
- The lifetime of the upper state or, for a semiconductor, the electron-hole recombination time.

PROBLEMS

3.1. Show that the vibrational frequency of a homonuclear diatomic molecule is $v = (1/2\pi)(2k_0/M)^{1/2}$, where M is the mass of each atom and k_0 is the constant of the elastic restoring force.

3.2. The vibrational frequency of a N_2 molecule is about $\tilde{v} = 2,360$ cm^{-1}. Calculate the value of the force constant k_0. Then calculate the potential energy for a nuclear distance R displaced from equilibrium by $R - R_0 = 0.3$ Å. (Compare this energy to that shown in Chap. 10 for the potential energy curve of a N_2 molecule.)

3.3. The equilibrium internuclear distance of a N_2 molecule is $R_0 \cong 0.11$ nm. Calculate the rotational constant B, the transition frequency, and corresponding transition wavelength for the $J = 0 \rightarrow J = 1$ rotational transition.

3.4. Using the result obtained from Problem 3.3 for the rotational constant B of the N_2 molecule, calculate the frequency separation between two consecutive lines of the P-branch of the $v'' = 0 \rightarrow v' = 1$ transition. Also calculate the quantum number of the most populated rotational level of the $v'' = 0$ state.

3.5. The rotational constant of a CO_2 molecule in its $00°1$ vibrational level is $B = 0.37$ cm^{-1}. Assuming the same value for the rotational constant of the $10°0$ level, calculate the P-branch and R-branch spectrum at $T = 450$ K of the $00°1 \rightarrow 10°0$ transition. (Remember: Only rotational levels of the $00°1$ upper state of an even J-number can participate in the transition.)

3.6. Assuming that the rotational constant of the ground vibrational state of a CO_2 molecule is $B = 0.37$ cm^{-1}, i.e., equal to that of the $(00°1)$ state, calculate the equilibrium distance R_0 between carbon and oxygen atoms.

3.7. From the condition $f_c(E_2') \geq f_v(E_1')$ prove the Bernard–Duraffourg relation $E_{F_c}' - E_{F_v}' \geq h\nu$.

3.8. With the help of Fig. 3.15b calculate for GaAs: The values of E_{F_c} and E_{F_v} at $N = 1.6 \times 10^{18}$ cm^{-3} carrier injection and the overall gain bandwidth at the same injection level.

3.9. In the energy reference system in Fig. 3.9a calculate for GaAs the energies E_2 and E_1 of the upper and lower laser levels for a transition energy exceeding the band gap energy by $0.45kT$.

3.10. With the help of Fig. 3.16 for a bulk GaAs semiconductor, calculate the expected gain at a photon energy exceeding the band gap energy by $0.45kT$ and for a carrier injection of $N = 1.6 \times 10^{18}$ cm^{-3}.

3.11. Assuming that the peak gain in a bulk GaAs semiconductor at a carrier injection $N = 1.6 \times 10^{18}$ cm^{-3} occurs at a photon energy exceeding the gap energy by $0.45kT$ and using results obtained in Problems 3.9–3.11, calculate the differential gain $\sigma = dg/dN$.

3.12. With the help of Fig. 3.15a, plot the quantities E_{F_c}/kT, E_{F_v}/kT, and $(E_{F_c} + E_{F_v})/kT$ versus the concentration N of electrons and holes for bulk InGaAsP at $\lambda = 1300$ nm. From these plots then calculate the transparency density N_{tr}. Compare this figure with Fig. 3.15b to explain why N_{tr} is somewhat smaller than in the GaAs case. From the same plots also calculate the overall gain bandwidth and the values of E_{F_c} and E_{F_v} at $N = 1.6 \times 10^{18}$ cm^{-3} carrier injection. Assuming that maximum gain occurs at an energy of $\Delta E = 0.45kT$ above the band-gap energy, calculate the corresponding wavelength.

3.13. For a 10-nm GaAs quantum well calculate from Fig. 3.26 the overall bandwidth of the gain curve and the energy of the quasi-Fermi levels for an injected carrier density of $N = 2 \times 10^{18}$ cm^{-3}. For the $(E - E_g)$ reference axis in Fig. 3.27 find the energy interval where positive gain occurs.

3.14. Calculate how to modify the first step in Fig. 3.27 at $(E - E_g) \cong 65$ meV if a Lorentzian line shape with dephasing time $\tau_c = 0.1$ ps is assumed for each transition from an upper level in the first conduction subband to a lower level in the first valence subband.

REFERENCES

1. G. Herzberg, *Spectra of Diatomic Molecules* (D. Van Nostrand, Princeton, NJ, 1950).
2. G. Herzberg, *Molecular Spectra and Molecular Structure: Infrared and Raman Spectra of Polyatomic Molecules* (D. Van Nostrand, Princeton, NJ, 1968), p. 122, Fig. 51.
3. G. B. Agrawal and N. K. Dutta, *Long Wavelength Semiconductor Lasers* (Chapman and Hall, New York, 1986).
4. C. Kittel, *Introduction to Solid-State Physics*, 6th ed. (Wiley, New York, 1986).
5. R. H. Yan, S. W. Corzine, L. A. Coldren and I. Suemune, Correction for the Expression for Gain in GaAs, *IEEE J. Quant. Electr.* **QE-26**, 213 (1990).
6. M. G. Bernard and G. Duraffourg, Laser Conditions in Semiconductors, *Phys. Status Solidi* **1**, 699 (1961).
7. Ref. 3, Chap. 3.
8. S. W. Corzine, R. H. Yan and L. A. Coldren, Optical Gain in III–V Bulk and Quantum Well Semiconductors, in *Quantum Well Lasers* (P. S. Zory, ed.) (Academic Press, San Diego, CA, 1993), Chap. 1.
9. G. H. B. Thompson, *Physics of Semiconductor Lasers Devices* (Wiley, New York, 1980), Sect. 2.5.2.
10. Ref. 9, Sect. 2.4.2.
11. Ref. 8, Fig. 2.14.
12. Ref. 3, Sect. 3.3.
13. Ref. 8, Sect. 3.1.
14. Ref. 3, Chap. 9.
15. A. Yariv, *Quantum Electronics*, 3rd ed. (Wiley, New York, 1989), Sect. 11.2.
16. Ref. 8, Sect. 4.1.
17. Ref. 8, Fig. 6.
18. E. Kapon, Quantum Wire Semiconductor Lasers, in *Quantum Well Lasers* (P. S. Zory, ed.) (Academic Press, San Diego, CA, 1983), Chap. 10.
19. M. Asada, Y. Miyamoto and Y. Suematsu, Gain and Threshold of Three-Dimensional Quantum Box Lasers, *IEEE J. Quant. Electr.* **QE-22**, 1915 (1986).
20. P. K. Cheo, CO$_2$ Lasers, in *Lasers* vol. 3 (A. K. Levine and A. DeMaria, eds.) (Marcel Dekker, New York, 1971).

4

Ray and Wave Propagation through Optical Media

4.1. INTRODUCTION

Before entering into a detailed discussion of optical resonators, the subject of Chap. 5, we introduce, in this chapter, a few topics from geometrical and wave optics that are not usually covered in elementary optics texts but provide a useful background. In particular, the matrix formulation of geometrical optics, within the paraxial-ray approximation, and wave propagation, within the paraxial-wave approximation, leading to a treatment of Gaussian beam propagation, will be discussed. In addition, multiple interference, such as in a multilayer dielectric coating or in a Fabry–Perot interferometer, will also be considered.

4.2. MATRIX FORMULATION OF GEOMETRIC OPTICS

Consider a ray of light that is either transmitted by or reflected from an optical element with reciprocal and polarization-independent behavior (e.g., a lens or a mirror).[1] Let z be the optical axis of this element (e.g., the line passing through the centers of curvature of the two spherical surfaces of the lens). Assume that the ray is traveling approximately in the z-direction in a plane containing the optical axis. The ray vector $\mathbf{r}_1$ at a given input plane $z = z_1$ of the optical element (Fig. 4.1) can be characterized by two parameters, namely, its radial displacement $r(z_1)$ from the z-axis and its angular displacement θ_1. Likewise the ray vector $\mathbf{r}_2$ at a given output plane $z = z_2$ can be characterized by its radial, $r_2(z_2)$, and angular, θ_2, displacements. Note that the r-axis is taken to be the same for both input and output rays and is oriented as in Fig. 4.1. The sign convention for angles is that the angle is positive if the $\mathbf{r}$-vector must be rotated clockwise to make it coincide with the positive direction of the z-axis. Thus, for example, θ_1 is positive while θ_2 is negative in Fig. 4.1.

Within the *paraxial-ray approximation* angular displacements θ are assumed to be small enough to allow the approximation to be made, $\sin \theta \cong \tan \theta \cong \theta$. In this case, output,

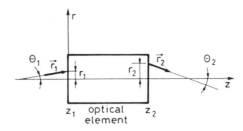

FIG. 4.1. Matrix formulation for the propagation of a ray through a general optical element.

(r_2, θ_2), and input, (r_1, θ_1), variables are related by a linear transformation. If we set $\theta_1 \cong (dr_1/dz_1)_{z_1} = r_1'$ and $\theta_2 \cong (dr_2/dz_2)_{z_2} = r_2'$, we can write

$$r_2 = Ar_1 + Br_1' \tag{4.2.1a}$$
$$r_2' = Cr_1 + Dr_1' \tag{4.2.1b}$$

where A, B, C, and D are constants characteristic of the given optical element. In a matrix formulation, it is therefore natural to write Eqs. (4.2.1) as:

$$\begin{vmatrix} r_2 \\ r_2' \end{vmatrix} = \begin{vmatrix} A & B \\ C & D \end{vmatrix} \begin{vmatrix} r_1 \\ r_1' \end{vmatrix} \tag{4.2.2}$$

where the *ABCD* matrix completely characterizes the given optical element within the paraxial-ray approximation.

As a first and simplest example, we consider the free-space propagation of a ray along a length $\Delta z = L$ of a material with refractive index n (Fig. 4.2a). If input and output planes lie just outside the medium, in a medium of refractive index equal to unity, using Snell's law in the paraxial approximation we get:

$$r_2 = r_1 + \frac{Lr_1'}{n} \tag{4.2.3a}$$
$$r_2' = r_1' \tag{4.2.3b}$$

The corresponding *ABCD* matrix is therefore:

$$\begin{vmatrix} 1 & L/n \\ 0 & 1 \end{vmatrix} \tag{4.2.4}$$

In the next example we consider ray propagation through a lens of focal length f (f is taken to be positive for a converging lens). In a thin lens, we obviously have (Fig. 4.2b)

$$r_2 = r_1 \tag{4.2.5a}$$

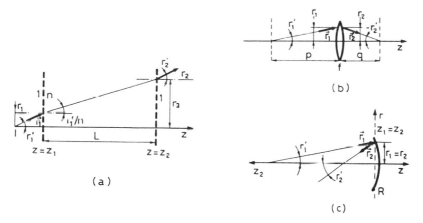

FIG. 4.2. Calculation of the *ABCD* matrix for (a) free-space propagation, (b) propagation through a thin lens, (c) reflection from a spherical mirror.

The second relation is obtained from the well-known law of geometrical optics, viz., $(1/p) + (1/q) = (1/f)$, using the fact that $p = r_1/r'_1$ and $q = -r_2/r'_2$. By also using Eq. (4.2.5a), we obtain

$$r'_2 = -\left(\frac{1}{f}\right)r_1 + r'_1 \tag{4.2.5b}$$

According to Eqs. (4.2.5), the *ABCD* matrix is, in this case:

$$\begin{vmatrix} 1 & 0 \\ -1/f & 1 \end{vmatrix} \tag{4.2.6}$$

As a third example, we consider reflection of a ray by a spherical mirror of radius of curvature R (R is taken to be positive for a concave mirror). In this case the z_1- and z_2-planes are taken to be coincident and placed just in front of the mirror, and the positive direction of the r-axis is taken to be the same for incident and reflected rays (Fig. 4.2c). The positive direction of the z-axis is taken to be from left to right for the incident vector and from right to left for the reflected vector. The angle for the incident ray is positive if the $\mathbf{r}_1$-vector must be rotated clockwise to make it coincide with the positive z_1-direction, while the angle for the reflected ray is positive if the $\mathbf{r}_2$-vector must be rotated anticlockwise to make it coincide with the positive z_2-direction of the z-axis; for example r'_1 is positive, while r'_2 is negative in Fig. 4.2c. Given these conventions, the ray matrix of a concave mirror of curvature R and hence focal length $f = R/2$ can readily be shown to be identical to that of a positive lens of focal length $f = R/2$. The ray matrix is therefore equal to:

$$\begin{vmatrix} 1 & 0 \\ -2/R & 1 \end{vmatrix} \tag{4.2.7}$$

TABLE 4.1. Ray matrices for some common cases

Free-space propagation		$\begin{bmatrix} 1 & \frac{L}{n} \\ 0 & 1 \end{bmatrix}$
Thin lens		$\begin{bmatrix} 1 & 0 \\ \frac{-1}{f} & 1 \end{bmatrix}$
Spherical mirror		$\begin{bmatrix} 1 & 0 \\ \frac{-2}{R} & 1 \end{bmatrix}$
Spherical dielectric interface		$\begin{bmatrix} 1 & 0 \\ \frac{n_2-n_1}{n_2}\frac{1}{R} & \frac{n_1}{n_2} \end{bmatrix}$

Table 4.1 lists ray matrices for the optical elements considered so far, as well as for a spherical dielectric interface. Note that the determinant of the *ABCD* matrix is unitary, i.e.:

$$AD - BC = 1 \qquad (4.2.8)$$

provided input and output planes lie in media of the same refractive index. In fact this situation holds for the first three cases considered in Table 4.1.

Once the matrices of the elementary optical elements are known, one can readily obtain the overall matrix of a more complex optical element by subdividing it into these elementary components. Suppose that, within a given optical element, we can consider an intermediate plane of coordinate z_i (Fig. 4.3) such that the two *ABCD* matrices, between planes $z = z_1$ and $z = z_i$ and planes $z = z_i$ and $z = z_2$, are known. If we now let r_i and r_i' be the coordinates of the ray vector at plane $z = z_i$, we can obviously write

$$\begin{vmatrix} r_i \\ r_i' \end{vmatrix} = \begin{vmatrix} A_1 & B_1 \\ C_1 & D_1 \end{vmatrix} \begin{vmatrix} r_1 \\ r_1' \end{vmatrix} \qquad (4.2.9)$$

$$\begin{vmatrix} r_2 \\ r_2' \end{vmatrix} = \begin{vmatrix} A_2 & B_2 \\ C_2 & D_2 \end{vmatrix} \begin{vmatrix} r_i \\ r_i' \end{vmatrix} \qquad (4.2.10)$$

If we substitute Eq. (4.2.9) for the vector $\mathbf{r}_i$ on the right-hand side of Eq. (4.2.10), we obtain

$$\begin{vmatrix} r_2 \\ r_2' \end{vmatrix} = \begin{vmatrix} A_2 & B_2 \\ C_2 & D_2 \end{vmatrix} \begin{vmatrix} A_1 & B_1 \\ C_1 & D_1 \end{vmatrix} \begin{vmatrix} r_1 \\ r_1' \end{vmatrix} \qquad (4.2.11)$$

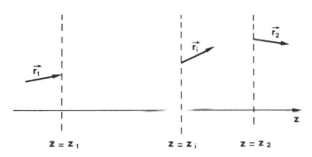

FIG. 4.3. Ray propagation through three distinct planes when the two matrices between planes $z = z_1$ and $z = z_i$ and between $z = z_i$ and $z = z_2$ are known.

The overall *ABCD* matrix can thus be obtained by multiplying the *ABCD* matrices of the elementary components. Note, however, that the order in which matrices appear in the product is the opposite of the order in which corresponding optical elements are traversed by the light ray.

As a first and perhaps somewhat trivial example of using the preceding result, we consider free-space propagation through a length L_1 followed again by free space propagation through a second length L_2, in a medium with refractive index n. According to Eq. (4.2.4) the overall matrix equation can be written as:

$$\begin{vmatrix} r_2 \\ r_2' \end{vmatrix} = \begin{vmatrix} 1 & L_2/n \\ 0 & 1 \end{vmatrix} \begin{vmatrix} 1 & L_1/n \\ 0 & 1 \end{vmatrix} \begin{vmatrix} r_1 \\ r_1' \end{vmatrix} \tag{4.2.12}$$

Using well-known rules of matrix multiplication, the product of the two square matrices gives an overall matrix:

$$\begin{vmatrix} 1 & (L_1 + L_2)/n \\ 0 & 1 \end{vmatrix} \tag{4.2.13}$$

This calculation confirms the obvious result that overall propagation is equivalent to free-space propagation over a total length $L = L_1 + L_2$.

A less trivial and more useful example involves free propagation over a length L (in a medium with refractive index $n = 1$) followed by reflection from a mirror of radius of curvature R. According to Eqs. (4.2.4), (4.2.7), and (4.2.11), the overall *ABCD* matrix is given by:

$$\begin{vmatrix} A & B \\ C & D \end{vmatrix} = \begin{vmatrix} 1 & 0 \\ -(2/R) & 1 \end{vmatrix} \begin{vmatrix} 1 & L \\ 0 & 1 \end{vmatrix} = \begin{vmatrix} 1 & L \\ -(2/R) & 1 - (2L/R) \end{vmatrix} \tag{4.2.14}$$

Note that the determinants of matrices (4.2.13) and (4.2.14) are unitary, and this result holds for any arbitrary cascade of optical elements, since the determinant of a matrix product is the product of their determinants.

We now address the question of finding the ray matrix elements A', B', C', D' for reverse propagation through an optical system in terms of the given matrix elements A, B, C, D for forward propagation. Referring to Fig. 4.1, if we take $-\mathbf{r}_2$ as the input vector, i.e., if we reverse the propagation direction of the $\mathbf{r}_2$ vector, then the output vector must be $-\mathbf{r}_1$.

For backward propagation we use the same sign conventions as used for the ray reflected from a spherical mirror (Fig. 4.2c), namely: The z-axis is reversed, while the r-axis remains unchanged, and the angle between the $\mathbf{r}$-vector and the z-axis is positive if the $\mathbf{r}$-vector must be rotated anticlockwise to coincide with the z-axis. Given these conventions, the rays $-\mathbf{r}_1$ and $-\mathbf{r}_2$ are described by coordinates $(r_1, -r_1')$ and $(r_2, -r_2')$, respectively. Thus one must have:

$$\left| \begin{matrix} r_1 \\ -r_1' \end{matrix} \right| = \left| \begin{matrix} A' & B' \\ C' & D' \end{matrix} \right| \left| \begin{matrix} r_2 \\ -r_2' \end{matrix} \right| \tag{4.2.15}$$

From Eq. (4.2.15) we can obtain r_2 and r_2' as a function of r_1 and r_1'. Since the determinant of the $A'B'C'D'$ matrix is also unitary, we get

$$r_2 = D'r_1 + B'r_1' \tag{4.2.16a}$$
$$r_2' = C'r_1 + A'r_1' \tag{4.2.16b}$$

A comparison between Eqs. (4.2.16) and (4.2.1) then shows that $A' = D$, $B' = B$, $C' = C$, and $D' = A$, so that the overall $A'B'C'D'$ matrix is

$$\left| \begin{matrix} A' & B' \\ C' & D' \end{matrix} \right| = \left| \begin{matrix} D & B \\ C & A \end{matrix} \right| \tag{4.2.17}$$

Equation (4.2.17) thus shows that the matrix for backward propagation is obtained from that for forward propagation simply by interchanging matrix elements A and D.

The matrix formulation is not only useful for describing the behavior of a ray as it passes through an optical system, but it can also be used to describe the propagation of a spherical wave. Consider in fact a spherical wave originating at point P_1 of Fig. 4.4 and propagating in the positive z-direction. After traversing an optical element described by a given $ABCD$ matrix, this wave is generally transformed into a new spherical wave whose center is point P_2. Consider now two conjugate rays $\mathbf{r}_1$ and $\mathbf{r}_2$ of the two waves, which means that the optical element transforms the incident (or input) ray $\mathbf{r}_1$ into the output ray $\mathbf{r}_2$. The radii of curvature R_1 and R_2 of the two waves at the input plane z_1 and output plane z_2 of the optical element are readily obtained as

$$R_1 = \frac{r_1}{r_1'} \tag{4.2.18a}$$

$$R_2 = \frac{r_2}{r_2'} \tag{4.2.18b}$$

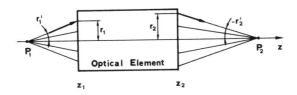

FIG. 4.4. Propagation of a spherical wave emitted from point P_1 through a general optical element described by a given $ABCD$ matrix.

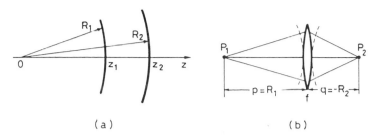

(a)   (b)

FIG. 4.5. Propagation of a spherical wave (a) through free space and (b) through a thin lens.

Note that, in Eqs. (4.2.18), we have used the sign convention that R is positive if the center of curvature is to the left of the wave front. From Eqs. (4.2.1) and (4.2.18) we obtain

$$R_2 = \frac{AR_1 + B}{CR_1 + D} \tag{4.2.19}$$

Equation (4.2.19) is a very important result, since it relates, in simple terms, the radius of curvature R_2 of the output wave to the radius of curvature R_1 of the input wave via the *ABCD* matrix elements of the given optical component.

As a first elementary example using this result, consider the free space propagation of a spherical wave between points having coordinates z_1 and z_2 in Fig. 4.5a. From Eq. (4.2.4), with $n = 1$ and $L = z_2 - z_1$, and Eq. (4.2.19) we obtain $R_2 = R_1 + (z_2 - z_1)$, which of course is an obvious result. Consider next the propagation of a spherical wave through a thin lens (Fig. 4.5b). From Eqs. (4.2.6) and (4.2.19), we obtain

$$\frac{1}{R_2} = \frac{1}{R_1} - \frac{1}{f} \tag{4.2.20}$$

which simply corresponds to the familiar law of geometrical optics $p^{-1} + q^{-1} = f^{-1}$.

Although the two examples in Fig. 4.5 are both rather elementary applications of Eq. (4.2.19), the usefulness of this equation can be appreciated when dealing with a more complicated optical system made up, e.g., of a sequence of lenses and spaces between them. In this case, the overall *ABCD* matrix is given by the product of the matrices of each optical component and the radius of curvature of the output wave is readily obtained by Eq. (4.2.19).

4.3. WAVE REFLECTION AND TRANSMISSION AT A DIELECTRIC INTERFACE

Consider a wave that is incident on the plane interface between two media of refractive indices n_1 and n_2.[2] If the wave is initially in the medium of refractive index n_1 and normally incident on the surface, the electric field reflectivity is

$$r_{12} = \frac{(n_1 - n_2)}{(n_1 + n_2)} \tag{4.3.1}$$

while the field transmission is

$$t_{12} = \frac{2n_1}{(n_1 + n_2)} \tag{4.3.2}$$

Note that, if $n_1 < n_2$, then one has $r_{12} < 0$; this means that the reflected field has a π phase shift compared to the incident field. Of course, if $n_1 > n_2$, then $r_{12} > 0$, so there is no phase shift on reflection. We also observe that, according to Eq. (4.3.2), one always has $t_{12} > 0$, i.e., there is no phase shift on transmission.

For non-normal incidence, the expressions for electric field reflectivity and transmission are more complicated, and they depend also on field polarization. As a representative example, Fig. 4.6 plots the intensity reflectivity, or reflectance, $R = (r_{12})^2$ against the incidence angle θ for a p-polarized wave (E-field in the plane of incidence), and and a s-polarized wave (E-field orthogonal to the plane of incidence), for $n_1 = 1$ and $n_2 = 1.52$. For $\theta = 0$ the two reflectances are obviously equal, and, according to Eq. (4.3.1), have the value $R = 4.26\%$. Note also that, for a p-polarized wave, there is a particular angle ($\theta_B = 56.7°$ in Fig. 4.6) at which $R = 0$.

The situation in this case can be described with the help of Fig. 4.7. Suppose that the incidence angle θ_B is such that the refracted beam is orthogonal to the direction of the reflected beam. The E-field in the optical material and hence its polarization vector are therefore parallel to the direction of reflection. Since the reflected beam can be considered to be produced by radiation emitted by the polarization vector of the medium where refraction occurs, this reflected beam is zero in this case, since an electric dipole does not radiate along its own direction. A straightforward calculation based on geometrical optics can now give the expression for the incidence angle θ_B, referred to as the *Brewster angle* or polarizing angle. According to the previous discussion, we have

$$\theta'_B + \theta_B = \frac{\pi}{2} \tag{4.3.3}$$

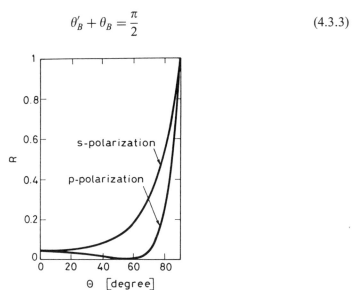

FIG. 4.6. Power reflectivity R versus the angle of incidence θ at an interface between air and a medium of refractive index $n = 1.5$. The two curves refer to cases of E-field polarization in the plane of incidence (p-polarization) and orthogonal to this plane (s-polarization).

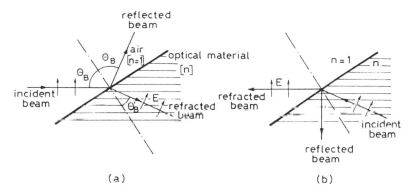

FIG. 4.7. Reflected and refracted beams for incidence at Brewster's angle; (a) incidence from the less dense medium and (b) incidence from the denser medium.

where θ_B' is the angle of the refracted beam. From Snell's law we also have

$$n \sin \theta_B' = \sin \theta_B \qquad (4.3.4)$$

Since, according to Eq. (4.3.3), one has $\sin \theta_B' = \cos \theta_B$, from Eq. (4.3.4) we obtain the following expression for the Brewster angle:

$$\tan \theta_B = n \qquad (4.3.5)$$

Note that, if the direction of the rays is reversed (Fig. 4.7b), the reflected beam is again zero, since refracted and reflected beams are orthogonal. Then, if a plane-parallel plate of a given optical material is inserted at Brewster's angle into a beam polarized in the plane of Fig. 4.7, no reflection occurs at the two surfaces of the plate. Let us now assume that a plane-parallel plate of refractive index $n = 1.52$, e.g., is inserted at the Brewster angle within an optical cavity. According to Fig. 4.6, the reflectance of an *s*-polarized beam at each of the two interfaces is $R \cong 15\%$. Thus an *s*-polarized beam suffers around a 30% loss due to reflection at the two interfaces. If the laser gain per pass is smaller than 30%, *s*-polarization does not oscillate, so the laser beam is linearly polarized in the plane of incidence to the plate.

4.4. MULTILAYER DIELECTRIC COATINGS

Mirror surfaces, used as high-reflectivity laser mirrors or beam splitters, are commonly fabricated by depositing a multilayer dielectric stack on the optical surface, plane or curved, of a substrate material, such as glass.[3,4] The same technique can also be used to reduce the surface reflectivity of optical components (antireflection coating) or to produce such optical elements as interference filters or polarizers. The coating is usually produced in a vacuum chamber by evaporating the required dielectric materials, which then condense in a layer on the substrate. The widespread use of multilayer dielectric coatings for laser optical components arises from the fact that these layers are made of transparent materials that can withstand the high intensity of a laser beam. This is in contrast to the behavior of thin metal layers (e.g., Ag or Au), also produced by vacuum deposition and often used for conventional optical components. In fact, metals and metal layers have a large absorption (5–10%) in the near-infrared to the ultraviolet region and they are not commonly used as materials for laser mirrors. It should be noted however that absorption losses, for these materials, are much less in the middle-to far-infrared, e.g., at the 10.6-μm wavelength of a CO_2 laser. Thus, high-reflectivity gold-coated copper mirrors or, more simply, polished copper mirrors are often used in this wavelength range.

Consider an optical substrate, such as glass, coated with a number of layers having alternately high, n_H, and low, n_L, refractive indices compared to that of the substrate n_s. If the thickness of the layers l_H and l_L are such that $n_H l_H = n_L l_L = \lambda_0/4$, where λ_0 is a specified wavelength, electric field reflections at all layer interfaces, for an incident beam of wavelength $\lambda = \lambda_0$, add in phase. Consider for instance the two interfaces of a high-index layer (Fig. 4.8a). According to Eq. (4.3.1), electric field reflectivity at the low-to-high index interface has a negative sign, and the electric field undergoes a phase shift of $\phi_1 = \pi$ on reflection. Conversely, reflectivity at the high-to-low index interface is positive and no phase shift of the reflected wave occurs there. If the optical thickness $n_H l_H$ of the layer equals $\lambda_0/4$, the phase shift after the round trip in the high-refractive-index layer is

$$\phi_2 = 2kl_H = \left(\frac{2\pi n_H}{\lambda}\right)l_H = \pi.$$

Example 4.1. *Peak reflectivity calculation in multilayer dielectric coatings.* We consider TiO_2 and SiO_2 for the high- and low-index materials, respectively. At the Nd:YAG laser wavelength of $\lambda_0 = 1.064$ μm, one can take $n_H = 2.28$ and $n_L = 1.45$. Assuming BK-7 glass as substrate material, one has $n_s \cong 1.54$. From Eq. (4.4.1) we obtain $R \cong 61.8\%$ for $J = 3$ and $R = 99.8\%$ for $J = 15$. Note that the reflectance at a single interface, according to Eq. (4.3.1), is in our example $[(n_H - n_L)/(n_H + n_L)]^2 = 4.9\%$.

This means that the two reflected waves have the same phase, and the corresponding fields will add. We can easily show that the same conclusion applies to the two interfaces of a low-index layer. It then follows that all reflected beams, in a multilayer dielectric coating, as well as their multiple reflections add in phase. Therefore, if a sufficient number of ($\lambda/4$) layers of alternating low and high indices are deposited, the overall reflectivity due

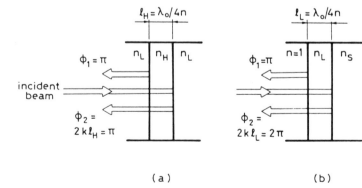

(a)   (b)

FIG. 4.8. (a) First two reflections at the two interfaces of a high-index layer in a multilayer dielectric coating. (b) First two reflections at the two interfaces of a low-index layer in a single-layer antireflection coating. Multiple reflections also occur (see, e.g., the case of a Fabry–Perot interferometer), but these are not shown in the figures.

to all multiple reflections can reach a very high value. If the multilayer stack starts and ends with a high-index layer, so that there is an odd number J of layers, the resulting power reflectivity (at $\lambda = \lambda_0$) turns out to be

$$R(\lambda_0) = \left(\frac{n_H^{J+1} - n_L^{J-1}n_s}{n_H^{J+1} + n_L^{J-1}n_s}\right)^2 \tag{4.4.1}$$

If the wavelength λ of the incident wave is different from λ_0, reflectivity is of course lower than the value given by Eq. (4.4.1). As representative examples, Fig. 4.9 shows curves of reflectivity versus wavelength for $J=15$ and $J=3$. One notices that, as the number of layers increases, the peak reflectivity value obviously increases and the high-reflectivity region becomes wider and has steeper edges. From the high reflectivity curve, one can also observe that high reflectivity is maintained over a wavelength range of $\Delta\lambda = \lambda - \lambda_0 \cong \pm(10\%)\lambda_0$.

To reduce the reflectivity of a given optical surface, a single-layer coating of a material with a refractive index lower than that of the substrate can be used. Since $n_L < n_s$, one can see from Fig. 4.8b that the first two reflections now have opposite phases if $n_L l_L = \lambda_0/4$. Overall reflectivity is thus reduced, and, after taking into account all multiple reflections, one can show that reflectivity at $\lambda = \lambda_0$ is given by:

$$R = \left[\frac{(n_s - n_L^2)}{(n_s + n_L^2)}\right]^2 \tag{4.4.2}$$

From this equation we see that zero reflection is obtained when $n_L = (n_s)^{1/2}$, a condition that is difficult to achieve in practice due to the limited number of available materials with a low enough refractive index.

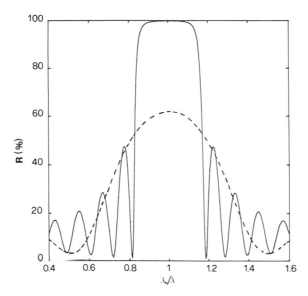

FIG. 4.9. Reflectivity versus wavelength curves of a $\lambda/4$ multilayer dielectric stack made of TiO$_2$ and SiO$_2$ for a total layer number of 3 (dashed curve) and 15 (continuous curve) (substrate material BK7-glass).

Example 4.2. *Single-layer antireflection coating of laser materials.* Consider a BK-7 glass substrate for which $n_s = 1.54$ at $\lambda = 1.06\ \mu$m. To achieve zero reflectivity with a single layer, the refractive index of the layer material must be $n_L = (n_s)^{1/2} \cong 1.24$. The lowest refractive-index material available as a stable film is provided by MgF$_2$ (fluorite) with $n_L = 1.38$. From Eq. (4.4.2), we obtain $R \cong 1.1\%$. Although not zero, this is still significantly smaller than the reflectivity of a bare surface, which, according to Eq. (4.3.1), is given by $R = [(n_s - 1)/(n_s + 1)]^2 = 4.5\%$. Fluorite is a rather soft material however, and it can easily be scratched. Consider next a Nd:YAG rod ($n_s = 1.82$) and a $\lambda/4$ layer of SiO$_2$, a rather hard and durable material ($n_L = 1.45$), for an antireflection coating. From Eq. (4.4.2) we have in this case $R = 3.4\%$, which while far from perfect is significantly less than the reflectivity of the bare surface ($R \cong 8.5\%$). Note that fluorite would provide an almost perfect match, in this case, the reflectivity according to Eq. (4.4.2) being reduced to $R \cong 4 \times 10^{-4}$.

The minimum reflectivity value given by Eq. (4.4.2) applies for $\lambda = \lambda_0$. The width of the low-reflectivity region for a single-layer coating is however very large. For example, if λ_0 corresponds to the center of the visible range, reflectivity is reduced below that of the bare surface for the whole visible range.

Quite often, for laser applications, even lower reflectivities than those considered in Example 4.2 are required (as low as perhaps 0.1%). This is achieved by using more than one layer in the antireflection coating. A coating consisting of two $\lambda/4$ layers of low- and high-refractive-index material with the sequence $n_s/n_L/n_H$ is often used for glass. A very hard and durable two-layer coating often used is ZrO$_2$ ($n_H = 2.1$)-MgF$_2$ ($n_L = 1.38$). The region of low reflectivity is reduced for this type of coating; in fact, the reflectivity versus wavelength curve shows a sharp V-shaped minimum. Such a coating is commonly referred to as a V-coating.

4.5. FABRY–PEROT INTERFEROMETER

We now consider a second example of multiple interference, the case of a Fabry–Perot (FP) interferometer.[5] This interferometer, a common spectroscopic tool since its introduction in 1899, plays a very important role in laser physics for at least three reasons:

- On a fundamental level, its physical behavior forms a basis for the behavior of optical resonators.
- It is often used as a frequency-selective element in a laser cavity.
- It is often used as a spectrometer for analyzing the spectrum of light emitted by a laser.

4.5.1. Properties of a Fabry–Perot Interferometer

The FP interferometer consists of two plane or spherical mirrors with power reflectivities R_1 and R_2 separated by a distance L and containing a medium of refractive index n_r. Although, for the ultimate performance, interferometers use spherical mirrors, for simplicity we consider here the case of two plane and parallel mirrors. We assume a plane wave, of frequency ν, incident on the interferometer in a direction making an angle θ' with the normal to the two mirrors (Fig. 4.10). This wave is indicated schematically by ray 0 in Fig. 4.10. The output beam leaving the interferometer consists of a superposition of the beam resulting from a single pass through the two mirrors (ray 1 in Fig. 4.10) with the beams arising from all multiple reflections (two of which are rays 2 and 3 in the figure). Thus the electric field amplitude of the output beam, E_t, is obtained by summing amplitudes E_l of all these beams, taking into account their corresponding phase shifts. To illustrate this, electric fields of the first three beams are also indicated in the figure. If all multiple reflections are taken into account, we obtain

$$E_t = \sum_{1}^{\infty} {}_l E_l = [E_0 t_1 t_2 \exp(j\phi')]\sum_{0}^{\infty} {}_m (r_1 r_2)^m \exp(2mj\phi) \tag{4.5.1}$$

In Eq. (4.5.1) and Fig. 4.10, E_0 is the amplitude of the beam incident on the interferometer; t_1 and t_2 are electric field transmissions of the two mirrors, and r_1 and r_2 are, respectively, the corresponding electric field reflectivities; ϕ' is the phase shift for a single pass, which also includes any phase shift due to passage through the two mirrors; 2ϕ is the phase difference between successive multiple reflections, given by $2\phi = kL_s = 2kL\cos\theta = (4\pi n_r v/c)L\cos\theta$, where L_s is the sum of the lengths of segments AB and BC in Fig. 4.10, and the angle θ is related to the incidence angle θ' by the Snell's law ($n_r \sin\theta = \sin\theta'$). Note that the previous expression can, for simplicity, be transformed into:

$$\phi = \frac{2\pi L' v}{c} \tag{4.5.2}$$

where:

$$L' = n_r L \cos\theta \tag{4.5.3}$$

The geometrical series in Eq. (4.5.1) can be readily summed to give

$$E_t = E_0 e^{j\phi'} \frac{t_1 t_2}{1 - (r_1 r_2)\exp(2j\phi)} \tag{4.5.4}$$

The power transmission T of the interferometer is given by $T = |E_t|^2/|E_0|^2$; from Eq. (4.5.4) we obtain

$$T = \frac{t_1^2 t_2^2}{1 - 2r_1 r_2 \cos(2\phi) + r_1^2 r_2^2} \tag{4.5.5}$$

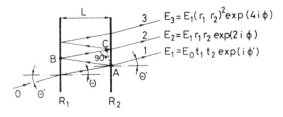

FIG. 4.10. Multiple-beam interference in a Fabry–Perot interferometer.

Since $R_1 = r_1^2$, $R_2 = r_2^2$, and, for a lossless mirror, $t_1^2 = 1 - r_1^2 = 1 - R_1$ and $t_2^2 = 1 - r_2^2 = 1 - R_2$, Eq. (4.5.5) can be transformed into:

$$T = \frac{(1 - R_1)(1 - R_2)}{[1 - (R_1R_2)^{1/2}]^2 + 4(R_1R_2)^{1/2} \sin^2 \phi} \tag{4.5.6}$$

which is the final result of our calculation.

To illustrate the properties of the FP interferometer, Fig. 4.11 shows a plot of transmission T versus frequency v of the incident wave, for $R_1 = R_2 = 64\%$. This plot is obtained from Eq. (4.5.6) with Eq. (4.5.2) used for ϕ. The curve consists of a series of evenly spaced maxima. These maxima occur when $\sin^2 \phi = 0$ in Eq. (4.5.6), i.e., when $\phi = m\pi$, where m is a positive integer. With the help of Eq. (4.5.2), the frequencies v_m of these maxima are obtained as:

$$v_m = \frac{mc}{2L'} \tag{4.5.7}$$

The frequency difference between two consecutive maxima (for reasons that will become clear at the end of this section) is called the free spectral range of the interferometer Δv_{fsr}. From Eq. (4.5.7) we obtain

$$\Delta v_{fsr} = \frac{c}{2L'} \tag{4.5.8}$$

At a transmission maximum, one has $\sin \phi = 0$ and the value of the transmission is seen from Eq. (4.5.6) to be

$$T_{\max} = \frac{(1 - R_1)(1 - R_2)}{[1 - (R_1R_2)^{1/2}]^2} \tag{4.5.9}$$

Note that, if $R_1 = R_2 = R$, then $T_{\max} = 1$ irrespective of the value of the mirror reflectivity R. This result holds only if the mirrors have no absorption, as assumed in our analysis so far.

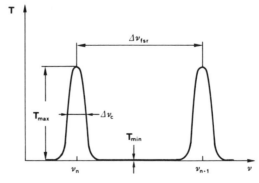

FIG. 4.11. Intensity transmission of a Fabry–Perot interferometer versus the frequency of the incident wave.

The transmission minima occur when $\sin^2 \phi = 1$, i.e., midway between maxima. The transmission at this minimum point is obtained from Eq. (4.5.6) as:

$$T_{\min} = \frac{(1 - R_1)(1 - R_2)}{[1 + (R_1 R_2)^{1/2}]^2} \tag{4.5.10}$$

Note that, under normal circumstances, the value of $T_{\min}$ is very small (see Example 4.3).

To calculate the width Δv_c of a transmission peak, we recall that, according to Eq. (4.5.6), the transmission falls to half of its maximum value for a displacement $\Delta \phi$ from the value $\phi = n\pi$ such that $4(R_1 R_2)^{1/2} \sin^2 \Delta \phi = [1 - (R_1 R_2)^{1/2}]^2$. Assuming that $\Delta \phi$ is much smaller than π, we make the approximation $\sin \Delta \phi \cong \Delta \phi$, which gives $\Delta \phi = \pm [1 - (R_1 R_2)^{1/2}]/2[R_1 R_2]^{1/4}$. The latter equation shows that the two half-intensity points corresponding to $\Delta \phi_+$ and $\Delta \phi_-$ are symmetrically situated at either side of the maximum. If we let $\Delta \phi_c = \Delta \phi_+ - \Delta \phi_-$, then we obtain

$$\Delta \phi_c = \frac{1 - (R_1 R_2)^{1/2}}{(R_1 R_2)^{1/4}} \tag{4.5.11}$$

and, from Eq. (4.5.2):

$$\Delta v_c = \frac{c}{2L'} \frac{1 - (R_1 R_2)^{1/2}}{\pi (R_1 R_2)^{1/4}} \tag{4.5.12}$$

We now define the *finesse* F of the interferometer as:

$$F = \frac{\Delta v_{fsr}}{\Delta v_c} \tag{4.5.13}$$

Example 4.3. *Free-spectral range, finesse, and transmission of a Fabry–Perot etalon.* Consider an FP interferometer made of a piece of glass with two plane parallel surfaces coated for high reflectivity (often called an FP etalon). If we assume $L = 1$ cm and $n_r = 1.54$, the free spectral range for near-normal incidence, i.e., for $\theta \cong 0$, is $\Delta v_{fsr} = c/2n_r L = 9.7$ GHz. If we now take $R_1 = R_2 = 0.98$, we obtain from Eq. (4.5.14a) a finesse $F \cong 150$, so that $\Delta v_c = \Delta v_{fsr}/F = 65$ MHz. According to Eq. (4.5.9), for a lossless coating, the peak transmission is $T_{\max} = 1$; the minimum transmission, according to Eq. (4.5.10), is $T_{\min} \cong 10^{-4}$. Note the very small value of $T_{\min}$.

From Eq. (4.5.8) and (4.5.12) we obtain

$$F = \frac{\pi (R_1 R_2)^{1/4}}{1 - (R_1 R_2)^{1/2}} \tag{4.5.14a}$$

The finesse indicates how much narrower the transmission peak is compared to the free spectral range; typically it is much greater than 1.

The preceding expressions and considerations hold for perfectly lossless mirrors. If we now let A represent the fraction of incident power absorbed by the mirror (mirror absorbance) and assume for simplicity the same reflectivity for the two mirrors, i.e., take $R_1 = R_2 = R$, then from Eq. (4.5.6) one readily gets

$$T_{FP} = \left(\frac{T}{1 - R}\right)^2 \frac{(1 - R)^2}{(1 - R)^2 + 4R \sin^2 \phi} \tag{4.5.14b}$$

where $T = t^2$ is the mirror transmission ($T = 1 - R - A$).

4.5.2. Fabry–Perot Interferometer as a Spectrometer

We now describe the use of a FP interferometer as a spectrum analyzer. We consider the simplest case where the direction of the incident light is normal to the interferometer mirrors (i.e., $\cos\theta = 1$) and the medium inside the interferometer is air ($n_r \cong 1$). We assume that the length L can be changed by a few wavelengths by, e.g., attaching one of the two FP plates to a piezoelectric transducer (scanning FP interferometer). To understand what happens in this case, we first consider a monochromatic wave at frequency v (wavelength λ). According to the preceding discussion, the transmitted light exhibits peaks when $\phi = n\pi$, i.e., when the interferometer length is equal to $L = n\lambda/2$ (see Fig. 4.12a), where n is a positive integer. The change in L required to shift from one transmission peak to the next one is then:

$$\Delta L_{fsr} = \frac{\lambda}{2} \qquad (4.5.15)$$

The width of each transmission peak ΔL_c is such that $(2\pi v/c_0)\Delta L_c = \Delta\phi_c$, where $\Delta\phi_c$ is given by Eq. (4.5.11). With the help of Eq. (4.5.14a), we obtain $\Delta L_c = \lambda/2F$. We therefore have

$$\Delta L_c = \frac{\Delta L_{fsr}}{F} \qquad (4.5.16)$$

i.e., the analogous relation to Eq. (4.5.13).

We now consider the case where two waves at frequencies v and $v + \Delta v$ are incident on the interferometer. The wave at frequency $(v + \Delta v)$ produces a set of transmission peaks displaced by an amount ΔL from those corresponding to frequency v (Fig. 4.12b). Since $2\pi L v/c = n\pi$, the displacement ΔL must be such that $2\pi(L + \Delta L)(v + \Delta v)/c = n\pi$. That is, $\Delta L = -(\Delta v/v)L$. The two frequencies v and $v + \Delta v$ are resolved by the spectrometer if

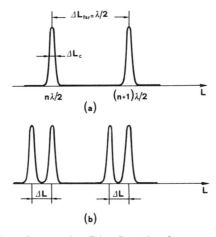

FIG. 4.12. Intensity transmission of a scanning Fabry–Perot interferometer when the incident wave is (a) monochromatic and (b) consists of two closely spaced frequencies.

$|\Delta L| \geq \Delta L_c$. The equality sign in this expression corresponds to the minimum frequency interval Δv_m that can be resolved, which implies $(\Delta v_m/v)L = \Delta L_c$. With the help of Eqs. (4.5.15) and (4.5.16), we then obtain $(\Delta v_m/v)L = \lambda/2F$. Using Eq. (4.5.8) with $L' = L$, we obtain

$$\Delta v_m = \frac{\Delta v_{fsr}}{F} \qquad (4.5.17)$$

Thus the finesse of the interferometer specifies its resolving power in terms of the free spectral range.

Note that when $|\Delta L| = \Delta L_{fsr}$, i.e., when $\Delta v = \Delta v_{fsr} = c/2L$, transmission peaks at frequencies $v + \Delta v$ and v become coincident, although shifted by one order relative to each other. Therefore, when $\Delta v > \Delta v_{fsr}$, an ambiguity by a multiple of Δv_{fsr} occurs in measuring Δv. Thus, when using the interferometer to measure a frequency difference, a simple and unambiguous result is obtained only when $\Delta v < \Delta v_{fsr}$. This explains why Δv_{fsr} is referred to as the free spectral range of the interferometer. We can readily generalize this result to say that, if Δv_{osc} is the spectral bandwidth of the incident light, then to avoid frequency ambiguity, $\Delta v_{osc} \leq \Delta v_{fsr}$. If the equality is assumed to hold in this relation, then from Eq. (4.5.17), we obtain

$$\Delta v_m = \frac{\Delta v_{osc}}{F} \qquad (4.5.18)$$

Thus the finesse F also provides a measure of how *finely* we can discriminate frequencies within the total spectral bandwidth Δv_{osc}.

Example 4.4. *Spectral measurement of an Ar$^+$-laser output beam.* We consider an Ar-ion laser oscillating on its green line at a wavelength $\lambda = 514.5$-nm. We assume that the laser is oscillating on many longitudinal modes encompassing the full Doppler width of the laser line ($\Delta v_0^* = 3.5$ GHz). Thus we have $\Delta v_{osc} = \Delta v_0^* = 3.5$ GHz. To avoid frequency ambiguity, we must have $\Delta v_{fsr} = (c/2L) \geq 3.5$ GHz i.e., $L \leq 4.28$ cm. If we now assume a finesse $F = 150$ and take $L = 4.28$ cm, according to Eq. (4.5.18) we have, for the interferometer resolution, $\Delta v_m = \Delta v_{osc}/F \cong 23$ MHz. If, for example, the length of the laser cavity is $L_1 = 1.5$ m, consecutive longitudinal modes are separated (see Chap. 5) by $\Delta v = c/2L_1 = 100$ MHz. Since $\Delta v_m < \Delta v$, the FP interferometer is able to resolve these longitudinal modes. Note that, since the frequency of the laser light is $v = c/\lambda \cong 5.83 \times 10^{14}$ Hz, the corresponding resolving power of the interferometer is $v/\Delta v_m = 2.54 \times 10^7$. This is a very high resolving power compared, e.g., to the best that can be obtained with a grating spectrometer ($v/\Delta v_m < 10^6$).

4.6. DIFFRACTION OPTICS IN THE PARAXIAL APPROXIMATION

We consider a monochromatic wave under the so-called scalar approximation where em fields are uniformly (e.g., linearly or circularly) polarized.[6] The electric field of the wave can be described by a scalar quantity, viz:

$$E(x, y, z, t) = \tilde{E}(x, y, z) \exp(j\omega t) \qquad (4.6.1)$$

where the complex amplitude $\tilde{E}$ must satisfy the wave equation in scalar form, i.e.:

$$(\nabla^2 + k^2)\tilde{E}(x, y, z) = 0 \qquad (4.6.2)$$

with $k = \omega/c$.

An integral solution for the field amplitude can be obtained by using the Fresnel–Kirchhoff integral. For a given field distribution $\tilde{E}(x_1, y_1, z_1)$ in the $z = z_1$ plane, the resulting field distribution $\tilde{E}(x, y, z)$, at a general plane at coordinate z in the propagation direction, is then given by:

$$\tilde{E}(x, y, z) = \frac{j}{\lambda} \int\!\!\int_s \tilde{E}(x_1, y_1, z_1) \frac{\exp[-(jkr)]}{r} \cos\theta \, dx_1 \, dy_1 \qquad (4.6.3)$$

In Eq. (4.6.3) r is the distance between point P_1 of coordinates (x_1, y_1) and point P of coordinates (x, y) (see Fig. 4.13), θ is the angle that the line P_1P makes with the normal to the plane $z = z_1$, the double integral is taken over the coordinates x_1, y_1 in the $z = z_1$ plane, and the limits are defined by some general aperture S located in that plane. We see that Eq. (4.6.3) is really the expression of the Huygens principle in mathematical form. Indeed $[\tilde{E}(x_1, y_1, z_1)dx_1 \, dy_1][\exp -(jkr)]/r$ represents the Huygens wavelet originating from the elemental area $dx_1 \, dy_1$ around P_1; the field at point P is then obtained by summing wavelets coming from all points in the plane $z = z_1$. The term $\cos\theta$ is the so-called obliquity factor, whose necessity was recognized by Fresnel. The (j/λ) term is a normalization factor arising from a detailed treatment of the theory. It indicates that the Huygens wavelets have a $\pi/2$ phase shift compared to the beam that is incident at the $z = z_1$ plane.

We now consider the E-field solutions, in either differential [Eq. (4.6.2)] or integral forms [Eq. (4.6.3)], within the paraxial wave approximation where the wave is assumed to be propagating at a small angle θ in the z-direction. In this case we can write

$$\tilde{E}(x, y, z) = u(x, y, z) \exp[-(jkz)] \qquad (4.6.4)$$

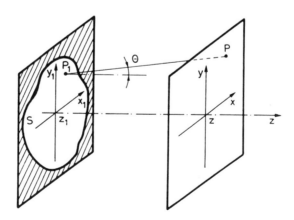

FIG. 4.13. Field calculation $u(P)$ at plane $z > z_1$ when the field profile $u(P_1)$ at plane $z = z_1$ is known.

where u is a slowly varying function, i.e., varying little on a wavelength scale in the z-coordinate. Under the paraxial approximation, the substitution of Eq. (4.6.4) into Eq. (4.6.2) gives

$$\nabla_\perp^2 u - 2jk \frac{\partial u}{\partial z} = 0 \tag{4.6.5}$$

where $\nabla_\perp^2 = (\partial^2/\partial x^2) + (\partial^2/\partial y^2)$. Equation (4.6.5) is the paraxial wave equation.

To obtain an approximate form of Eq. (4.6.3), under the paraxial wave approximation, we write $\cos\theta \simeq 1$ and $r = z - z_1$ in the amplitude factor of the spherical wavelet. To approximate the phase factor $-kr$, we must be more careful however; in fact, consider for instance a distance $r \cong 1$ m and assume that this distance is evaluated with an accuracy of $\Delta r = 1$ μm. For the amplitude factor, this yields a very good relative accuracy of $\Delta r / r = 10^{-6}$. The phase accuracy however is $\Delta\phi = k\Delta r = 2\pi\Delta r/\lambda$ and for $\lambda = 1$ μm it yields $\Delta\phi = 2\pi$. This is, of course, an unacceptable level of accuracy, since for example a phase change $\Delta\phi = \pi$ changes the sign of the entire phase term in the integral. Greater accuracy is thus needed for the phase term in Eq. (4.6.3). To this purpose we write the distance r between points P_1 and P in Fig. 4.13 as $r = [(z - z_1)^2 + (x - x_1)^2 + (y - y_1)^2]^{1/2}$. Under the paraxial wave approximation one has $[|x - x_1|, |y - y_1|] \ll |z - z_1|$. We can therefore write

$$r = (z - z_1) \left[1 + \frac{(x - x_1)^2 + (y - y_1)^2}{(z - z_1)^2} \right]^{1/2}$$

$$\cong (z - z_1) + \frac{(x - x_1)^2 + (y - y_1)^2}{2(z - z_1)} \tag{4.6.6}$$

The substitution of Eq. (4.6.6) into the phase term of Eq. (4.6.3) then gives

$$\tilde{E}(x, y, z) = \frac{j \exp[-jk(z - z_1)]}{\lambda(z - z_1)} \int\int \tilde{E}(x_1, y_1, z_1) \exp\left\{ -jk \left[\frac{(x - x_1)^2 + (y - y_1)^2}{2(z - z_1)} \right] \right\} dx_1 \, dy_1 \tag{4.6.7}$$

which is the Huygens–Fresnel–Kirchhoff integral in the so-called Fresnel approximation. Substituting Eq. (4.6.4) into Eq. (4.6.7) then gives

$$u(x, y, z) = \frac{j}{\lambda L} \int\int_s u(x_1, y_1, z_1) \exp\left\{ -jk \left[\frac{(x - x_1)^2 + (y - y_1)^2}{2L} \right] \right\} dx_1 \, dy_1 \tag{4.6.8}$$

where we have set $L = z - z_1$. Equation (4.6.8) provides a solution for the E-field in integral form within the paraxial wave approximation, while Eq. (4.6.5) gives the same solution in differential form. The two forms are, however, completely equivalent.

We next consider wave propagation within the paraxial approximation in a general optical system described by the $ABCD$ matrix, as in Sect. 4.2. Referring to Fig. 4.14, we let $u(x_1, y_1, z_1)$ and $u(x, y, z)$ be field amplitudes at planes $z = z_1$ and $z = z$ just before and after the optical system, respectively. We also assume that the Huygens principle applies to the

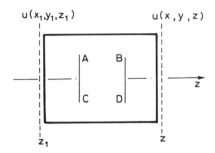

FIG. 4.14. Field calculation $u(x, y, z)$ at plane z after an optical system described by the *ABCD* matrix when the field profile $u(x_1, y_1, z_1)$ at plane $z = z_1$ is known.

general optical system in Fig. 4.14 provided that no field-limiting apertures are present in the optical system. This implies for instance that any lens or mirror within the optical system has an infinite aperture, i.e., an aperture much wider than the transverse dimensions of the field.* According to this extension of the Huygens principle to a general optical system, the field $u(x, y, z)$ is obtained by superposing individual wavelets emitted from plane $z = z_1$ and transmitted through this system. One then obtains[7]

$$u(x, y, z) = \frac{j}{B\lambda} \int\int_s u(x_1, y_1, z_1) \exp\left\{-jk\left[\frac{A(x_1^2 + y_1^2) + D(x^2 + y^2) - 2x_1 x - 2y_1 y}{2B}\right]\right\} dx_1\, dy_1$$

(4.6.9)

which constitutes a generalization of Eq. (4.6.8). Obviously, for free space-propagation, we have $A = D = 1$ and $B = L$ (see Table 4.1), and Eq. (4.6.9) reduces to Eq. (4.6.8).

4.7. GAUSSIAN BEAMS

We now discuss a very important class of *E*-field solutions, commonly called *Gaussian beams*. The properties of these beams, in the paraxial wave approximation, can be derived either from the paraxial wave equation (4.6.5) or the Fresnel–Kirchhoff integral in the Fresnel approximation [see Eqs. (4.6.8) and (4.6.9)]. We will use the integral approach, since it proves to be more useful for describing the properties of optical resonators, discussed in Chap. 5.

4.7.1. Lowest Order Mode

Consider a general optical system described by its corresponding *ABCD* matrix (see Fig. 4.14). We ask the question, "Is there a solution of Eq. (4.6.9) that retains its functional form as it propagates"? In other words is there an eigensolution of Eq. (4.6.9)? An answer is readily obtained if we assume that there is no limiting aperture in the $z = z_1$ plane, so that

---

* For finite apertures of the optical system, diffraction effects are produced at these apertures, thus sizably changing the transmitted field.

the double integral in Eq. (4.6.9) can be taken between $-\infty$ and $+\infty$ for both variables x_1 and y_1. In this case we can show by direct substitution into Eq. (4.6.9) that:

$$u(x, y, z) \propto \exp\{-jk[(x^2 + y^2)/2q]\} \tag{4.7.1}$$

where $q = q(z)$ is a complex parameter (often called the complex beam parameter of a Gaussian beam), is an eigensolution of Eq. (4.6.9). If we write, in fact,

$$u(x_1, y_1, z_1) \propto \exp\{-jk[(x_1^2 + y_1^2)/2q_1]\} \tag{4.7.2}$$

we obtain from Eq. (4.6.9):

$$u(x, y, z) = \frac{1}{A + (B/q_1)} \exp\left[-jk\left(\frac{x^2 + y^2}{2q}\right)\right] \tag{4.7.3}$$

where q is related to q_1 by the very simple law:

$$q = \frac{Aq_1 + B}{Cq_1 + D} \tag{4.7.4}$$

Equation (4.7.4) is a very important relation, known as the *ABCD* law of Gaussian beam propagation. It bears an obvious similarity to Eq. (4.2.19), which shows how the radius of curvature of a spherical wave is transformed by an optical system. We will return to this equation in Sect. 4.7.3 for further discussion.

We now discuss a physical interpretation of the Gaussian solution [Eq. (4.7.1)]. For this we use Eqs. (4.7.1) and (4.6.4) to write

$$\tilde{E} \propto \exp\left\{-jk\left[z + \left(\frac{x^2 + y^2}{2q}\right)\right]\right\} \tag{4.7.5}$$

Consider now a spherical wave with center at coordinates $x = y = z = 0$. Its field at point $P(x, y, z)$ can be written as $\tilde{E} \propto [\exp(-jkR)]/R$, where R is the wave's radius of curvature. Within the paraxial approximation, following an argument similar to that in Eq. (4.6.6), we write

$$R \cong z + \left(\frac{x^2 + y^2}{2R}\right) \tag{4.7.6}$$

The field of the spherical wave is then transformed to:

$$\tilde{E} \propto \exp\left\{-jk\left[z + \left(\frac{x^2 + y^2}{2R}\right)\right]\right\} \tag{4.7.7}$$

A comparison between Eqs. (4.7.7) and (4.7.5) then shows that the Gaussian beam can be viewed as a spherical wave of complex radius of curvature q. To understand the meaning of this complex beam parameter, we separate the real and imaginary parts of $1/q$; i.e., we set

$$\frac{1}{q} = \frac{1}{R} - j\left(\frac{\lambda}{\pi w^2}\right) \tag{4.7.8}$$

Substitution of Eq. (4.7.8) into Eq. (4.7.5) then gives

$$\tilde{E}(x, y, z) \propto \exp\left[-\left(\frac{x^2 + y^2}{w^2}\right)\right] \times \exp\left\{-jk\left[z + \left(\frac{x^2 + y^2}{2R}\right)\right]\right\} \tag{4.7.9}$$

The amplitude factor on the right-hand side of Eq. (4.7.9), i.e., $u_0 = \exp-[(x^2 + y^2)/w^2]$ is plotted in Fig. 4.15a versus r/w, where $r = [x^2 + y^2]^{1/2}$ is the radial beam coordinate. The maximum value is reached at $r = 0$, and, for $r = w$, one has $u_0 = 1/e$. The quantity w therefore defines the transverse scale of the beam; it is called the beam spot size (at the z-position considered). Note that, since the beam intensity is given by $I \propto |\tilde{E}|^2$, we have $I = I_{\max} \exp-[2(x^2 + y^2)/w^2]$. If we define the spot size of the intensity profile, w_I as the value at which $I = I_{\max}/e$, we have $w_I = w/\sqrt{2}$. Note that the intensity I reduces to $1/e^2$ of its maximum at a radial distance of one field spot size.

We now turn to the phase factor in Eq. (4.7.9). A comparison with Eq. (4.7.7), which applies to a spherical wave, shows that the two equations are identical. This leads us to identify R in Eq. (4.7.8) as the radius of curvature of the spherical wavefront of the Gaussian beam. To see this more clearly, consider the equiphase surface of the Gaussian beam that

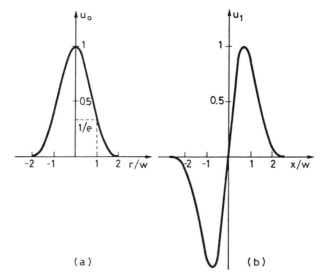

FIG. 4.15. Field profile of (a) the lowest order and (b) next-order Gaussian mode.

intercepts the z-axis at a given position z'. The x, y, z coordinates of this surface must then satisfy the relation $kz + k(x^2 + y^2)/2R = kz'$, which gives

$$z = z' - \left(\frac{x^2 + y^2}{2R}\right) \tag{4.7.10}$$

Equation (4.7.10) thus shows that the equiphase surface is a paraboloid of revolution around the z-axis. It can be shown further that the radius of curvature of this paraboloid at $x = y = 0$, i.e., on the beam axis, is just equal to R. This demonstrates rather clearly why, within the paraxial wave approximation, the phase terms of the spherical wave, Eq. (4.7.7), and Gaussian beam, Eq. (4.7.9), are the same.

4.7.2. Free-Space Propagation

Consider the propagation of the Gaussian beam in Eq. (4.7.1) along the positive z-direction without any restricting aperture in the x- or y-direction (i.e., in free space). From Eq. (4.7.4) with $A = D = 1$ and $B = z$, we obtain

$$q = q_1 + z \tag{4.7.11}$$

Assuming $R = \infty$ at $z = 0$, we then write

$$\left(\frac{1}{q_1}\right) = -j\left(\frac{\lambda}{\pi w_0^2}\right) \tag{4.7.12}$$

where w_0 is the spot size at $z = 0$. We now write Eq. (4.7.11) as $(1/q) = 1/(q_1 + z)$, substitute $1/q$ from Eq. (4.7.8) and $1/q_1$ from Eq. (4.7.12), then separate the real and imaginary parts of the resulting equation. After some straightforward algebraic manipulation, we arrive at expressions for the spot size w and radius of curvature R of the equiphase surfaces as:

$$w^2(z) = w_0^2\left[1 + \left(\frac{\lambda z}{\pi w_0^2}\right)^2\right] \tag{4.7.13a}$$

$$R(z) = z\left[1 + \left(\frac{\pi w_0^2}{\lambda z}\right)^2\right] \tag{4.7.13b}$$

From Eqs. (4.7.3) and (4.7.12) we also write

$$u(x, y, z) = \left[\frac{1}{1 - j(\lambda z/\pi w_0^2)}\right]\exp\left[-jk\left(\frac{x^2 + y^2}{2q}\right)\right] \tag{4.7.14}$$

The complex factor in the first brackets in Eq. (4.7.14) can now be expressed in terms of its amplitude and phase. Also using Eq. (4.7.8) for $(1/q)$, we obtain the expression for the field amplitude as:

$$u(x, y, z) = \frac{w_0}{w} \exp\left[-\left(\frac{x^2 + y^2}{w^2}\right)\right] \exp\left[-jk\left(\frac{x^2 + y^2}{2R}\right)\right] \exp j\phi \qquad (4.7.15a)$$

where:

$$\phi = \tan^{-1}\left(\frac{\lambda z}{\pi w_0^2}\right) \qquad (4.7.15b)$$

Equation (4.7.15a) and the expression for $w(z)$, $R(z)$, and $\phi(z)$ given by Eqs. (4.7.13) and (4.7.15b), solve our problem completely. We see from Eqs. (4.7.13) and (4.7.15b) that w, R, and ϕ (and hence the field distribution) depend only on w_0 (for given λ and z). This can be readily understood when we observe that, once w_0 is known, the field distribution at $z = 0$ is known. In fact we know its amplitude, since the field distribution is a Gaussian function with spot size w_0, and its phase, since we have assumed $R = \infty$ for $z = 0$. Once the field at $z = 0$ is known, the corresponding field at $z > 0$ is uniquely established, since it can be calculated by means, for instance, of the Fresnel–Kirchhoff integral (4.6.8). Again using Eq. (4.7.11) we can show that Eq. (4.7.13) and (4.7.15b) hold also for negative z-values, i.e., for forward propagation toward rather than from the $z = 0$ plane. Note that, if we define

$$z_R = \frac{\pi w_0^2}{\lambda} \qquad (4.7.16)$$

where z_R is called the *Rayleigh range* (whose significance is discussed later), Eqs. (4.7.13) and (4.7.15b) can be written more intuitively as:

$$w^2(z) = w_0^2\left[1 + \left(\frac{z}{z_R}\right)^2\right] \qquad (4.7.17a)$$

$$R(z) = z\left[1 + \left(\frac{z_R}{z}\right)^2\right] \qquad (4.7.17b)$$

$$\phi(z) = \tan^{-1}\left(\frac{z}{z_R}\right) \qquad (4.7.17c)$$

Equations (4.7.15a) and (4.7.17) are the final results of our calculations. We see that $u(x, y, z)$ is made of the product of an amplitude factor, $(w_0/w) \exp - [(x^2 + y^2)/w^2]$, with a transverse phase factor, $\exp -jk[(x^2 + y^2)/2R]$, and a longitudinal phase factor $\exp j\phi$. We discuss the physical meaning of these factors in some detail in the following paragraphs.

The amplitude factor in Eq. (4.7.15a) shows that, while propagating, the beam (for both $z > 0$ and $z < 0$) retains its Gaussian shape but its spot size changes according to Eq. (4.7.17a). Thus we can write $w^2(z)$ as the sum of w_0^2 and $(\lambda z/\pi w_0)^2$, a term arising from beam diffraction. A plot of the normalized spot size w/w_0 versus the normalized propagation

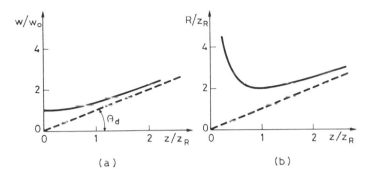

FIG. 4.16. (a) Normalized values of the beam spot size w and (b) radius of curvature of the equiphase surface R versus normalized values of the propagation length z.

length z/z_R is shown as a solid line in Fig. 4.16a for $z > 0$. For $z < 0$ the spot size is readily obtained from the same figure, since $w(z)$ is a symmetric function of z. Thus the minimum spot size occurs at $z = 0$ (hence referred to as the beam waist); for $z = z_R$, one has $w = \sqrt{2} w_0$. The Rayleigh range is thus the distance from the beam waist to where the spot has increased by a factor of $\sqrt{2}$. For $z \to \infty$ (i.e., for $z \gg z_R$), we can write

$$w \approx w_0 \frac{z}{z_R} = \frac{\lambda z}{\pi w_0} \qquad (4.7.18)$$

Equation (4.7.18) is also plotted as a dashed line in Fig. 4.16a. At a large distance, w increases linearly with z; hence we can define a beam divergence due to diffraction as $\theta_d = w/z$. We obtain

$$\theta_d = \frac{\lambda}{\pi w_0} \qquad (4.7.19)$$

The physical reason for the presence of the quantity w_0/w in the amplitude factor in Eq. (4.7.15a) is readily understood when we observe that, since the medium is assumed to be lossless, total beam power must be the same at any plane z. This requires $\iint |u|^2 \, dx \, dy$ to be independent of z. The presence of the quantity $w_0/w(z)$ ensures that this condition holds; in fact, using Eq. (4.7.15a), we can write

$$\iint |u|^2 \, dx \, dy = \left(\frac{w_0^2}{2}\right) \int_{-\infty}^{+\infty} \exp(-\xi^2) d\xi \int_{-\infty}^{+\infty} \exp(-\eta^2) d\eta \qquad (4.7.20)$$

where $\xi = \sqrt{2} x/w$ and $\eta = \sqrt{2} y/w$. By inspection we confirm that $\iint |u|^2 \, dx \, dy$ is independent of z.

We now consider the transverse phase factor in Eq. (4.7.15a). According to Sect. 4.7.1 it indicates that for $z > 0$, due to propagation, the beam acquires an approximately spherical wavefront with a radius of curvature R. A plot of the normalized radius of curvature R/z_R versus the normalized variable z/z_R is shown in Fig. 4.16b for $z > 0$. For $z < 0$ the radius of curvature is readily obtained from the same figure, since $R(z)$ is an antisymmetric function of

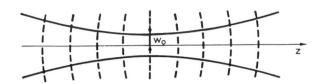

FIG. 4.17. Beam profile, continuous curves, and equiphase surfaces, dashed curves, for a TEM$_{00}$ Gaussian mode.

z. We see that $R \rightarrow \infty$ for $z = 0$, while R reaches its minimum value at $z = z_R$. For $z \gg z_R$, $R \approx z$, and the equation $R = z$ is plotted as a dashed line in Fig. 4.16b. Thus the wavefront is plane at $z = 0$ and, at large distances, it increases linearly with z just as for a spherical wave.

Lastly we consider the longitudinal phase factor in Eq. (4.7.15a). Using Eq. (4.6.4) and (4.7.15b) we see that, besides the phase shift $-kz$ of a plane wave, the Gaussian beam has an additional term $\phi(z)$ that changes from $-(\pi/2)$ to $(\pi/2)$ on going from $z \ll -z_R$ to $z \gg z_R$.

The results of Fig. 4.16 are combined in Fig. 4.17, where the dimensions of the beam profile $2w(z)$ are shown as solid curves and the equiphase surfaces as dashed lines. The beam has a minimum dimension in the form of a waist at $z = 0$; the corresponding spot size w_0 is usually referred to as the spot size at the beam waist or waist spot size. Note that, according to the convention used for the sign of wavefront curvature, since $R > 0$ for $z > 0$ and $R < 0$ for $z < 0$, the center of curvature is to the left of the wavefront for $z > 0$ and to the right of the wavefront for $z < 0$.

4.7.3. Gaussian Beams and the ABCD Law

The propagation of a Gaussian beam through a general medium described by an *ABCD* matrix is given by Eq. (4.7.3).[8] For a given *ABCD* matrix, the solution, depends only on the complex beam parameter q whose expression in terms of matrix elements is given by Eq. (4.7.4). This is a very important law of Gaussian beam propagation, often referred to as the *ABCD* law of Gaussian beams. Its usefulness was already demonstrated in the case of free space propagation in Sect. 4.7.2. In this section we further illustrate the importance of this law with somewhat more complex examples.

Example 4.5. *Gaussian beam propagation through a thin lens.* Consider a thin lens of focal length f. According to Eq. (4.7.4) complex beam parameters q_1, preceding, and q_2, following, the lens, are related by:

$$\frac{1}{q_2} = \frac{C + (D/q_1)}{A + (B/q_1)} \tag{4.7.21}$$

With the help of the matrix elements of a lens given in Table 4.1, we then obtain

$$\frac{1}{q_2} = -\left(\frac{1}{f} + \frac{1}{q_1}\right) \tag{4.7.22}$$

Using Eq. (4.7.8) to express both $1/q_1$ and $1/q_2$, we can separately equate the real and imaginary parts of Eq. (4.7.22) to obtain the following relations between spot sizes and radii of curvature before and after the lens:

$$w_2 = w_1 \tag{4.7.23a}$$

$$\frac{1}{R_2} = \frac{1}{R_1} - \frac{1}{f} \tag{4.7.23b}$$

The physical meaning of Eqs. (4.7.23) can now be discussed in connection with Fig. 4.18. Considering first Eq. (4.7.23a), we see that its physical meaning is obvious, since for a thin lens, beam amplitude distributions immediately before and after the lens must be the same; i.e., there cannot be a discontinuous change of spot size (see Fig. 4.18a). To understand the meaning of Eq. (4.7.23b), consider first the propagation of a spherical wave through the same lens (Fig. 4.18b). Here a spherical wave originating from a point source P_1 is focused by the lens on the image point P_2. The radii of curvature R_1 and R_2 just before and after the lens are, in this case, related by Eq. (4.2.20). A spherical lens can then be seen to tranform the radius of curvature R_1 of an incoming wave into a radius R_2 of the outgoing wave according to Eq. (4.2.20). Since this is expected to occur irrespective of the transverse amplitude distribution, Eq. (4.2.20) is expected to hold also for a Gaussian beam, as indeed Eq. (4.7.23b) indicates.

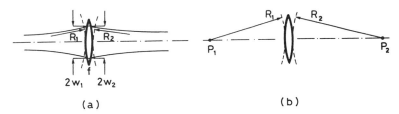

FIG. 4.18. Propagation through a lens of (a) a Gaussian beam and (b) a spherical wave.

Example 4.6. *Gaussian beam focusing by a thin lens.* Consider now a Gaussian beam with spot size w_{01} and plane wave front entering a lens of focal length f (i.e., the beam waist is located at the lens). We are interested in calculating the beam waist position after the lens and its spot size value w_{02}. According to Eqs. (4.2.4) and (4.2.6), the transmission matrix for a lens of focal length f followed by a free-space length z is given by:

$$\begin{vmatrix} 1 - z/f & z \\ -1/f & 1 \end{vmatrix} \tag{4.7.24}$$

The complex beam parameter q_2, after this lens and free-space combination, can again be obtained from Eq. (4.7.21), where the A, B, C, D elements are obtained from Eq. (4.7.24) and $(1/q_1)$ is given by:

$$\left(\frac{1}{q_1}\right) = -\frac{j\lambda}{\pi w_{01}^2} = -\frac{j}{z_{R_1}} \tag{4.7.25}$$

with z_{R_1} the Rayleigh range corresponding to the spot size w_{01}. If now the coordinate z_m after the lens corresponds to the position where the beam waist occurs, then, according to Eq. (4.7.8), $1/q_2$ must also be purely imaginary. This means that the real part of the right-hand side of Eq. (4.7.21) must be zero. With the help of Eqs. (4.7.24) and (4.7.25), we find that z_m is given by:

$$z_m = \frac{f}{[1 + (f/z_{R_1})^2]} \tag{4.7.26}$$

Thus we see, perhaps with some surprise, that the distance z_m from the lens where the minimum spot size occurs is always smaller than the focal distance f. Not however that, under typical conditions, one usually has $z_{R_1} \gg f$ so that $z_m \approx f$. By equating the imaginary parts of both sides of Eq. (4.7.21) and using Eqs. (4.7.24) and (4.7.25) again, the spot size at the focal plane w_{02} is obtained as:

$$w_{02} = \frac{\lambda f}{\pi w_{01}[1 + (f/z_{R_1})^2]^{1/2}} \tag{4.7.27}$$

Again for $z_{R_1} \gg f$ we obtain from Eq. (4.7.27):

$$w_{02} \cong \frac{\lambda f}{\pi w_{01}} \tag{4.7.28}$$

4.7.4. *Higher Order Modes*

We now return to the problem considered in Sect. 4.7.1 to ask whether there are other eigen solutions of Eq. (4.6.8) for, free-space, or of Eq. (4.6.9) for a general optical system. The answer is again positive and we can show that a particularly useful set of eigensolutions can be written as the product of a Hermite polynomial with a Gaussian function. In fact, referring to Fig. 4.14, let us assume that:

$$u(x_1, y_1, z_1) = H_l\left(\frac{\sqrt{2}x_1}{w_1}\right)H_m\left(\frac{\sqrt{2}y_1}{w_1}\right)\exp[-jk(x_1^2 + y_1^2)/2q_1] \tag{4.7.29}$$

where H_l and H_m are Hermite polynomials of order l and m, q_1 is the complex beam parameter at $z = z_1$, and w_1 is the corresponding spot size. The substitution of Eq. (4.7.29) into the right-hand side of Eq. (4.6.9) gives

$$u(x, y, z) = \left[\frac{1}{A + (B/q_1)}\right]^{1+l+m} H_l\left(\frac{\sqrt{2}x}{w}\right)H_m\left(\frac{\sqrt{2}y}{w}\right)\exp[-jk(x^2 + y^2)/2q)] \tag{4.7.30}$$

where q is the complex beam parameter after the optical system in Fig. 4.14, as given by Eq. (4.7.4), and w is the corresponding spot size.

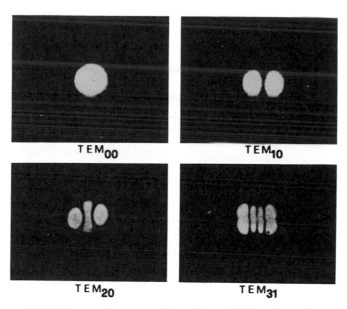

FIG. 4.19. *Intensity patterns of some low-order Gaussian modes.*

For free-space propagation, if we let the z_1-plane be the waist plane, then $q_1 = j\pi w_0^2/\lambda$, where w_0 is the spot size at the beam waist. Substituting the preceeding expression for q_1 into Eq. (4.7.30) and using Eq. (4.7.8), we obtain

$$u_{l,m}(x, y, z) = \left(\frac{w_0}{w}\right) H_l\left(\sqrt{2}\frac{x}{w}\right) H_m\left(\sqrt{2}\frac{y}{w}\right) \exp[-(x^2 + y^2)/w^2]$$
$$\times \exp\{-j[k(x^2 + y^2)/2R] + j(1 + l + m)\phi\} \quad (4.7.31)$$

where ϕ is given by Eq. (4.7.15b) and, using Eq. (4.7.11) to obtain $q = q(z)$, we see that w and R are given by Eqs. (4.7.13a) and (4.7.13b).

The lowest order mode is obtained from Eq. (4.7.31) on setting $l = m = 0$. Since the Hermite polynomial of zero order is a constant, Eq. (4.7.31) reduces to the Gaussian solution discussed in Sect. 4.7.1 [see Eq. (4.7.15a)]. This solution is called the TEM_{00} mode, where TEM stands for Transverse Electric and Magnetic (within the paraxial approximation, both the electric and magnetic fields of the em wave are, in fact, approximately tranverse to the z-direction). The subscripts 00 indicate zero order polynomials for both H_l and H_m in Eq. (4.7.31). The radial intensity profile of a TEM_{00} Gaussian mode, at any z-coordinate, is $I_{00}(x, y) \propto |u_{00}|^2 \propto \exp[-2(x^2 + y^2)/w^2]$. It depends only on the radial coordinate $r = (x^2 + y^2)^{1/2}$. The mode thus corresponds to a circular spot (Fig. 4.19).

The next higher order mode is obtained from Eq. (4.7.31) by setting $l = 1$ and $m = 0$ (or $l = 0$ and $m = 1$). Since $H_1(x) \propto x$, the field amplitude is now given by $|u_{10}| \propto x \times \exp-[(x^2 + y^2)/w^2]$. Thus, at a given x, the field profile is described by a Gaussian function (see Fig. 4.15a) along the y-coordinate, while, at a given y, it is described by the function $x \exp[-(x^2/w^2)]$ along the x-coordinate. This function, normalized to its peak value, is plotted against x/w in Fig. 4.15b. This mode is called TEM_{10} and Fig. 4.19 shows a picture of the corresponding intensity profile. The TEM_{01} ($l=0$ and $m=1$) is obtained by rotating the picture of the TEM_{10} mode in Fig. 4.19 by $90°$. Two pictures of higher order modes are also indicated in this figure. Note that subscripts l and m give, in general, the number of zeros of the field (other than the zeros occurring at $x = \pm\infty$ and $y = \pm\infty$) along the x- and y-axes, respectively.

4.8. CONCLUSIONS

In this chapter, a few topics from geometrical and wave optics, which provide useful background for topics on optical resonators considered in Chap. 5, have been discussed. We have seen in particular that the transformation effected on a ray of an optical element (such as an isotropic material, a thin lens, a spherical mirror, etc.) can be described by a simple 2×2 matrix. The same matrix also describes the propagation of a Gaussian beam. Our discussion also includes a basic description of multilayer dielectric coatings and a somewhat more detailed discussion of a Fabry–Perot interferometer.

PROBLEMS

4.1. Show that the *ABCD* matrix for a ray entering a spherical dielectric interface from a medium of refractive index n_1 to a medium of refractive index n_2 is

$$\begin{vmatrix} 1 & 0 \\ \dfrac{n_2 - n_1}{n_2}\dfrac{1}{R} & \dfrac{n_1}{n_2} \end{vmatrix}$$

where R is the radius of curvature of the spherical surface. ($R > 0$ if the center is to the left of the surface.)

4.2. Considering a thin lens of refractive index n_2 as a sequence of two closely spaced spherical dielectric interfaces of radii R_1 and R_2 and using the result from Problem 4.1, show that its focal length is given by the relation:

$$\frac{1}{f} = \left(\frac{n_2 - n_1}{n_1}\right)\left(-\frac{1}{R_1} + \frac{1}{R_2}\right)$$

where n_1 is the refractive index of the medium surrounding the lens.

4.3. A Fabry–Perot interferometer consists of two identical mirrors with the same power reflectivity $R = 0.99$ and same fractional internal power loss $A = 0.005$. Calculate the peak transmission and the finesse of the interferometer.

4.4. A Fabry–Perot interferometer consisting of two identical mirrors, air-spaced by a distance L, is illuminated by a monochromatic em wave of tunable frequency. From a measurement of the transmitted intensity versus the frequency of the input wave we find that the free spectral range of the interferometer is 3×10^9 Hz and its resolution is 60 MHz. Calculate the spacing L of the interferometer, its finesse, and the mirror reflectivity. If the peak transmission is 50% calculate also the mirror loss.

4.5. A Fabry–Perot interferometer consisting of two identical mirrors, air spaced by a distance L, is illuminated by a 1-ps pulse from an external source at wavelemgth $\lambda \cong 600$ nm. The output beam is observed to consist of a regular sequence of 1-ps pulses spaced by 10 ns. The energy of the pulses decreases exponentially with time with a time constant of 100 ns. Calculate the cavity length and mirror reflectivity.

4.6. Directly substituting Eq. (4.7.2) into the right-hand side of Eq. (4.6.9) show that the double integral in Eq. (4.6.9), taken between $-\infty$ and $+\infty$, gives Eq. (4.7.3), where q is related to q_0 by Eq. (4.7.4).

4.7. A positive lens of focal length f is placed at a distance d from the waist of a TEM_{00} beam of waist spot size w_0. Derive an expression for the focal length f (in terms of w_0 and d) required so that the beam leaving the lens has a plane wave front.

4.8. Show that the power contained in a TEM_{00} Gaussian beam of spot size w is given by $P = (\pi w^2/2)I_p$, where I_p is the peak ($r=0$) intensity of the beam.

4.9. A given He–Ne laser oscillating in a pure Gaussian TEM_{00} mode at $\lambda = 632.8$ nm with an output power of $P = 5$ mW is advertised as having a far-field divergence angle of 1 mrad. Calculate spot size, peak intensity, and peak electric field at the waist position.

4.10. The beam of an Ar laser oscillating in a pure Gaussian TEM_{00} mode at $\lambda = 514.5$ nm with an output power of 1 W is sent to a target at a distance of 100 m from the beam waist. If the spot size at the beam waist is $w_0 = 2$ mm, calculate spot size, radius of curvature of the phase front, and peak intensity at the target position.

4.11. Consider a TEM_{00} Gaussian beam of spot size w_1 entering a lens of diameter D and focal length f. To avoid excessive diffraction effects at the lens edge due to truncation of the Gaussian field by the lens, we usually choose the lens diameter according to the criterion $D \geq 2.25\ w_1$. Assume the following; that the equality holds in the previous expression; that the waist of the incident beam is located at the lens, i.e., $w_1 = w_{01}$; that $f \ll z_{R_1} = \pi w_{01}^2/\lambda$; that Eq. (4.7.27) is still valid. Under these conditions express the minimum spot size after the lens as a function of the lens numerical aperture NA [NA $= \sin \theta$, where $\theta = \tan^{-1}(D/f)$, so that, for small θ, NA $\cong (D/f)$].

4.12. Suppose that a TEM_{00} Gaussian beam from a ruby laser ($\lambda = 694.3$ nm) is transmitted through a 1-m diameter diffraction-limited telescope to illuminate a spot on the surface of the moon. Assuming an earth-to-moon distance of $z \cong 348,000$ km and using the relation $D = 2.25\ w_{01}$ between the telescope objective diameter and beam spot size (see Problem 4.11) calculate the beam spot size on the moon. (Distortion effects from the atmosphere can be important, but these are neglected here.)

4.13. A Gaussian beam of waist spot size w_0 is passed through a solid plate of transparent material of length L and refractive index n. The plate is placed just in front of the beam waist. Using the $ABCD$ law of Gaussian beam propagation show that the spot size and radius of curvature of the phase front after the plate are the same as for propagation in a vacuum over a distance $L' = L/n$. According to this result, is the far-field divergence angle affected by the insertion of the plate?

4.14. From Eq. (4.7.26) show that a Gaussian beam of waist spot size w_{01} cannot be focused at a distance larger than $z_{R_1}/2$, where $z_{R_1} = \pi w_{01}^2/\lambda$. What is the focal length corresponding to this maximum focusing condition?

REFERENCES

1. H. Kogelnik, Propagation of Laser Beams, in *Applied Optics and Optical Engineering* (R. Shannon and J. C. Wynant, eds.) (Academic, New York, 1979), vol. 2, pp. 156–190.
2. M. Born and E. Wolf, *Principles of Optics*, 6th ed. (Pergamon, Oxford, 1980), Sect. 1.5.
3. E. Ritter, Coatings and Thin-Film Techniques, in *Laser Handbook* (F. T. Arecchi and E. O. Schultz–Dubois, eds.) (North-Holland, Amsterdam, 1972), vol. 1, pp. 897–921.
4. *Thin Films for Optical Systems*, F. R. Flory, ed. (Marcel Dekker, New York, 1995).
5. Ref. 2, Sect. 7.6.
6. A. E. Siegman, *Lasers* (Cambridge University, Oxford, 1986), Chap. 16.
7. Ref. 6, Chap. 20.
8. H. Kogelnik and T. Li, Laser Beams and Resonators, *Appl. Opt.* **5**, 1550 (1966).

5

Passive Optical Resonators

5.1. INTRODUCTION

Chapter 5 deals with the theory of passive optical resonators, i.e., where no active medium is present within the cavity. The most widely used laser resonators have either plane or spherical mirrors of rectangular (or more often circular) shape, separated by some distance L. Typically L ranges from a few centimeters to a few tens of centimeters, while mirror dimensions range from a fraction of a centimeter to a few centimeters. Laser resonators thus differ from those used in the microwave field (see, e.g., Sect. 2.2.1.) in two main respects: Resonator dimensions are much greater than the laser wavelength and resonators are usually open, i.e., no lateral surfaces are used. Resonator length is usually much greater than the laser wavelength because this wavelength usually ranges from a fraction of a micrometer to a few tens of micrometers. A laser cavity with length comparable to the wavelength would then generally have too low a gain to allow laser oscillation. Laser resonators are usually open to reduce drastically the number of modes that can oscillate with low loss. In fact in Example 5.1 we see that even a narrow linewidth laser, such as a He-Ne laser, has a large number of modes ($\approx 10^9$) if the resonator is closed. In contrast, on removing the lateral surfaces, the number of low-loss modes is reduced to just a few (≈ 6 in Example 5.1). In these open resonators, in fact, only the few modes corresponding to a superposition of waves traveling nearly parallel to the resonator axis have low enough losses to allow laser oscillation.

According to the preceding discussion, open resonators inevitably have some losses due to diffraction of the em field, which leads to some fraction of the energy leaving the sides of the cavity (*diffraction losses*). Strictly speaking therefore the mode definition given in Sect. 2.2.1. cannot be applied to an open resonator, and true modes (i.e., stationary configurations) do not exist for such a resonator. In what follows however we see that standing wave configurations with very small losses do exist in open resonators. We therefore define a mode as an em configuration whose electric field can be written as:

$$\mathbf{E}(\mathbf{r}, t) = E_0 \mathbf{u}(\mathbf{r}) \exp[(-t/2\tau_c) + j\omega t] \qquad (5.1.1)$$

Here τ_c (decay time of the square of the electric field amplitude) is called the *cavity photon decay time*.

We discuss various possible resonators in the following sections.

5.1.1. Plane Parallel (or Fabry–Perot) Resonator

This consists of two plane mirrors set parallel to one another. To a first approximation, the modes of this resonator can be considered as the superposition of two plane em waves propagating in opposite directions along the cavity axis, as shown in Fig. 5.1a. Within this approximation, resonant frequencies can be readily obtained by imposing the condition that cavity length L must be an integral number of half-wavelengths, i.e., $L = n(\lambda/2)$, where n is a positive integer. This is a necessary condition for the electric field of the em standing wave to be zero on the two mirrors. It then follows that resonant frequencies are given by:

$$v = n\left(\frac{c}{2L}\right) \tag{5.1.2}$$

Note that Eq. (5.1.2) can also be obtained by imposing the condition that the phase shift of a plane wave due to one round-trip through the cavity must equal an integral number times 2π, i.e., $2kL = 2n\pi$. This condition is readily obtained by a self-consistency argument. If the frequency of the plane wave is equal to that of a cavity mode, the phase shift after one round-trip must be zero (apart from an integral number of 2π), since only in this case do the amplitudes at any arbitrary point, due to successive reflections, add up in phase to give an appreciable total field.

According to Eq. (5.1.2) the frequency difference between two consecutive modes, i.e., modes whose integers differ by 1, is given by:

$$\Delta v = \frac{c}{2L} \tag{5.1.3}$$

This difference is referred to as the *frequency difference* between two consecutive longitudinal modes; the word longitudinal is used because the number n indicates the number of half-wavelengths of the mode along the laser resonator, i.e., longitudinally.

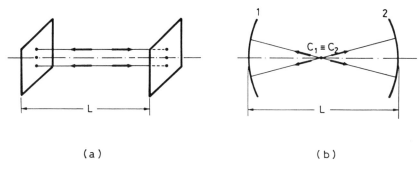

(a)   (b)

FIG. 5.1. (a) Plane parallel resonator and (b) concentric resonator.

5.1.2. *Concentric (Spherical) Resonator*

This consists of two spherical mirrors with the same radius R separated by a distance L such that the mirror centers of curvature C_1 and C_2 are coincident, i.e., $L = 2R$ (Fig. 5.1b). The geometrical optics picture of the modes of this resonator is also shown in the figure. In this case the modes can be approximated by the superposition of two oppositely traveling spherical waves originating from point C. Application of the above self-consistency argument again leads to Eq. (5.1.2) as the expression for the resonant frequencies and to Eq. (5.1.3) for the frequency difference between consecutive longitudinal modes.

5.1.3. *Confocal Resonator*

This consists of two spherical mirrors of the same radius of curvature R separated by a distance L such that mirror foci F_1 and F_2 are coincident (Fig. 5.2). It then follows that the center of curvature C of one mirror lies on the surface of the second mirror (i.e., $L = R$). From a geometrical-optics point of view, we can draw any number of closed optical paths of the type shown in Fig. 5.2 by changing the distance of the two parallel rays from the resonator axis $C_1 C_2$. Note that the direction of the rays can be reversed in Fig. 5.2. This geometrical optics description, however, does not give any indication of what the mode configuration will be. We shall see, in fact, that this configuration cannot be described by either a plane or a spherical wave. For the same reason, resonant frequencies cannot be readily obtained from geometrical optics considerations.

5.1.4. *Generalized Spherical Resonator*

Resonators formed by two spherical mirrors of the same radius of curvature R and separated by a distance L such that $R < L < 2R$ (i.e., somewhere between confocal and concentric conditions) are also often used. In addition we can have $L > R$. For these cases it is not generally possible to use a ray description in which a ray retraces itself after one or a few passes.

All of these resonators can be considered as particular examples of a general resonator consisting of either two concave ($R > 0$) or convex ($R < 0$) spherical mirrors of different radius of curvature at some arbitrary distance L. These resonators can be divided into two categories, namely, *stable* resonators and *unstable* resonators. A resonator is described as unstable when, in bouncing back and forth between the two mirrors, an arbitrary ray

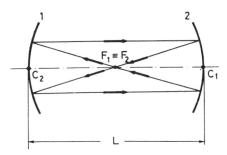

FIG. 5.2. Confocal resonator.

FIG. 5.3. Example of an unstable resonator.

diverges indefinitely away from the resonator axis. An obvious example of an unstable resonator is shown in Fig. 5.3. Conversely a resonator whose ray remains bounded is described as stable.

5.1.5. Ring Resonator

A particularly important class of laser resonator is the *ring resonator*, where the path of the optical rays is arranged in a ring configuration (Fig. 5.4a) or a more complicated configuration, such as the folded configuration in Fig. 5.4b. In both cases resonance frequencies can be obtained by imposing the condition that the total phase shift along the ring path in Fig. 5.4a or the closed loop path in Fig. 5.4b (continuous paths) must equal an integral number of 2π. We then readily obtain the expression for resonance frequencies:

$$v = \frac{nc}{L_p} \tag{5.1.4}$$

where L_p is the perimeter of the ring or the length of the closed loop path in Fig. 5.4b and n is an integer. Note that the arrows of the continuous paths in Fig. 5.4 can in general be reversed, which means that the beam in this figure can propagate either clockwise or counter-clockwise. Thus, in general, a standing wave pattern forms in a ring resonator. However, a unidirectional device can be used, allowing the passage of, e.g., only the right-to-left beam in Fig. 5.4a (optical diode; see Sect. 7.8.2.2. for more details). Then, only the clockwise-propagating beam can exist in the cavity. Therefore the concepts of a cavity mode and cavity resonance frequency are not confined to standing wave configurations. Note also that ring resonators can have either a stable (such as in Fig. 5.4) or unstable configuration.

Example 5.1. *Number of modes in closed and open resonators.* Consider a He-Ne laser oscillating at the wavelength of $\lambda = 633$ nm with a Doppler-broadened gain linewidth of $\Delta v_0^* = 1.7 \times 10^9$ Hz. Assume a resonator length $L = 50$ cm and consider first an open resonator. According to Eq. (5.1.3) the number of longitudinal modes that fall within the laser linewidth is $N_{open} = 2L\Delta v_0^*/c \cong 6$. Assume now that the resonator is closed by a cylindrical lateral surface with a cylinder diameter of $2a = 3$ mm. According to Eq. (2.2.16), the number of modes of this closed resonator that fall within the laser linewidth Δv_0^* is $N_{closed} = 8\pi v^2 V \Delta v_0^*/c^3$, where $v = c/\lambda$ is the laser frequency and $V = \pi a^2 L$ is the resonator volume. From the preceding expressions and data we readily obtain $N_{closed} = (2\pi a/\lambda)^2 N_{open} \cong 1.2 \times 10^9$ modes.

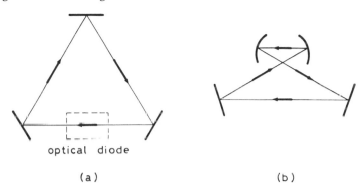

FIG. 5.4. (a) Simplest three-mirror ring resonator and (b) folded ring resonator.

5.2. EIGENMODES AND EIGENVALUES

Consider a general two-mirror resonator (Fig. 5.5a) consisting of two spherical mirrors of different radius of curvature (either positive or negative) separated by a distance L; the resonators may be either stable or unstable.[1] Assume that a beam of general shape is launched in the cavity from, e.g., mirror 1 and consider its propagation back and forth in the cavity. This propagation can be regarded as equivalent to that occurring in the periodic lens-guide structure in Fig. 5.5b, with the same beam traveling in one direction, e.g., along the positive direction of the z-axis. Note that the focal lengths f_1 and f_2 in Fig. 5.5b are related to the radii of curvature R_1 and R_2 in Fig. 5.5a by the well-known relations $f_1 = R_1/2$ and $f_2 = R_2/2$. Note also that diaphragms 1 and 2 of diameter $2a_1$ and $2a_2$, respectively situated after the corresponding lenses in Fig. 5.5b, simulate the apertures of the two mirrors in Fig. 5.5a. Now let $\tilde{E}(x_1, y_1, 0)$ be the complex field amplitude of the beam at some given point having transverse coordinates x_1 and y_1 at diaphragm 1, whose longitudinal coordinate is $z = 0$. The field amplitude $\tilde{E}(x, y, 2L)$ after one lens-guide period, i.e., at $z = 2L$, can be calculated once $\tilde{E}(x_1, y_1, 0)$ and the lens-guide geometry (i.e., quantities f_1, f_2, a_1, a_2, and L) are specified. For this calculation we can use, e.g., the Huygens–Fresnel propagation equation (see Sect. 4.6). The calculation can be somewhat involved for finite values of apertures $2a_1$ and $2a_2$, as we see in Sect. 5.5.2. The calculation can be even more involved if we must consider additional optical elements (e.g., a sequence of lenses) located within the

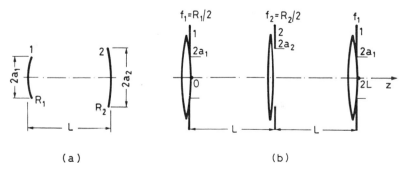

FIG. 5.5. (a) General two-mirror resonator and (b) lens-guide structure equivalent to the resonator in a.

cavity in Fig. 5.5a. In general, as a consequence of the linearity of the Huygens–Fresnel equation with respect to the field amplitudes, we can write

$$\tilde{E}(x, y, 2L) = [\exp(-2jkL)] \int\int_1 K(x, y; x_1, y_1)\tilde{E}(x_1, y_1, 0)dx_1\, dy_1 \qquad (5.2.1)$$

where the double integral is taken over the aperture 1 at the input plane ($z = 0$) and K is a function of the transverse coordinates of both input ($z = 0$) and output ($z = 2L$) planes (known as the *propagation kernel*). A few examples of this kernel are considered in Sect. 5.5.2. One can see, however, from Eq. (5.2.1) that, if $\tilde{E}(x_1, y_1, 0)$ were a bidimensional Dirac δ-function centered at coordinates x_1', y_1', i.e., if $\tilde{E}(x_1, y_1, 0) = \delta(x_1 - x_1', y_1 - y_1')$, then from Eq. (5.2.1) one would have $\tilde{E}(x, y, 2L) = \exp(-2jkL)K(x, y; x_1', y_1')$. Apart from the phase factor $\exp(-2jkL)$, the kernel $K(x, y; x_1', y_1')$ would then represent the field at the output plane generated by a pointlike source located at coordinates x_1', y_1' in the input plane.

Instead of considering a general beam propagating in the lens-guide structure in Fig. 5.5b, we now consider a beam whose transverse structure corresponds to that of a cavity mode in Fig. 5.5a. In this case, for self-consistency, the field must reproduce its shape after one lens-guide period. More precisely we require

$$\tilde{E}(x, y, 2L) = \tilde{\sigma}\exp(-2jkL)\tilde{E}(x, y, 0) \qquad (5.2.2)$$

where the constant $\tilde{\sigma}$ is generally complex since the propagation kernel K is itself a complex function. We can therefore write

$$\tilde{\sigma} = |\tilde{\sigma}|\exp j\phi \qquad (5.2.3)$$

where the amplitude $|\tilde{\sigma}|$ is expected to be less than 1 as a result of beam attenuation due to diffraction losses. The phase ϕ then makes an additional contribution to the round-trip (or single-period of the lens guide) phase shift besides the obvious one, i.e., $-2kL$, arising from the free-space propagation of a plane wave over distance $2L$. According to Eqs. (5.2.2) and (5.2.3) the total single-period phase shift is

$$\Delta\phi = 2kL + \phi \qquad (5.2.4)$$

If the left-hand side of Eq. (5.2.1) is now replaced by the right-hand side of Eq. (5.2.2), we obtain

$$\tilde{\sigma}\tilde{E}(x, y, 0) = \int\int K(x, y; x_1, y_1)\tilde{E}(x_1, y_1, 0)dx_1\, dy_1 \qquad (5.2.5)$$

which represents a Fredholm homogeneous integral equation of the second kind. Its eigensolutions $\tilde{E}_{lm}(x, y, 0)$ (if any exists) give the field distributions which are self-reproducing after each period of the lens-guide structure in Fig. 5.5b. Therefore, these also describe the field distributions over the mirror aperture for the cavity modes of Fig. 5.5a. Each solution in the infinite set of eigenstates is distinguished by a pair of integers l and m. Accordingly, the corresponding eigenvalue will be indicated as $\tilde{\sigma}_{lm}$.

From the preceding discussion, the eigenvalues $\tilde{\sigma}_{lm}$ are seen to be such that $|\tilde{\sigma}_{lm}|^2$ gives the factor by which beam intensity is changed as a result of one round-trip. Since this change is due to diffraction losses, we must have $|\tilde{\sigma}_{lm}|^2 < 1$; the quantity:

$$\gamma_{lm} = 1 - |\tilde{\sigma}_{lm}|^2 \qquad (5.2.6)$$

then gives the round-trip fractional power loss due to diffraction. One can see that, according to Eq. (5.2.4), $\Delta\phi_{lm} = -2kL + \phi_{lm}$ is the corresponding round-trip phase shift. If the field is self-reproducing, we must then require $\Delta\phi_{lm} = -2\pi n$, where n is an integer. Thus we get $-2kL + \phi_{lm} = -2\pi n$. Substituting $k = 2\pi v/c$, we obtain the cavity resonance frequencies as:

$$v_{lmn} - \frac{c}{2L}\left[n + \frac{\phi_{lm}}{2\pi}\right] \qquad (5.2.7)$$

Note that we have indicated explicitly that these frequencies depend on the three integer numbers l, m, and n. Integers l and m represent the order of the eigensolution in Eq (5.2.5); integer n specifies the total phase shift of the beam, after one round-trip, in units of 2π (i.e., $n = -\Delta\phi_{lm}/2\pi$).

To conclude: The solution of the integral equation (5.2.5) gives, via its eigensolutions $\tilde{E}_{lm}$, the field of the eigenmodes at all points in a given plane; for each mode $\tilde{E}_{lm}$, the magnitude $|\tilde{\sigma}_{lm}|$ of the eigenvalue then gives, via Eq. (5.2.6), the round-trip diffraction losses, while the phase ϕ_{lm} gives, via Eq. (5.2.7), the corresponding resonance frequency.

5.3. PHOTON LIFETIME AND CAVITY Q

Consider a given mode of a stable or unstable cavity and assume there are also some losses other than diffraction losses. For instance one may have mirror losses which results in mirror reflectivity being always smaller than unity. One may also have scattering losses in some optical element within the cavity. Under these conditions we want to calculate the rate of energy decay in the given cavity mode. To this purpose, let I_0 be the initial intensity corresponding to the field amplitude $\tilde{E}(x_1, y_1, 0)$ at a given transverse coordinate x_1, y_1. Let R_1 and R_2 be the (power) reflectivities of the two mirrors and T_i the fractional internal loss per pass due to diffraction and any other internal loss. The intensity $I(t_1)$ at the same point x_1, y_1 at a time $t_1 = 2L/c$, i.e., after one cavity round-trip, is

$$I(t_1) = R_1 R_2 (1 - T_i)^2 I_0 \qquad (5.3.1)$$

Note that, since T_i is defined here as the fractional internal loss per pass, the intensity is reduced by a factor $(1 - T_i)$ in a single pass and hence by a factor $(1 - T_i)^2$ in a double pass (round trip). The intensity at the same transverse coordinate after m round-trips, i.e., at time:

$$t_m = \frac{2mL}{c} \qquad (5.3.2)$$

is then:

$$I(t_m) = [R_1 R_2 (1 - T_i)^2]^m I_0 \qquad (5.3.3)$$

Let now $\phi(t)$ be the total number of photons in the given cavity mode at time t. Since the mode retains its shape after each round trip, we can set $\phi(t) \propto I(t)$. From Eq. (5.3.3) we then write

$$\phi(t_m) = [R_1 R_2 (1 - T_i)^2]^m \phi_0 \qquad (5.3.4)$$

where ϕ_0 is the number of photons initially present in the cavity. We can also set

$$\phi(t_m) = [\exp(-t_m/\tau_c)]\phi_0 \tag{5.3.5}$$

where τ_c is a suitable constant. In fact, a comparison between Eqs (5.3.5) and (5.3.4) with the help of Eq. (5.3.2) shows that

$$\exp(-2mL/c\tau_c) = [R_1 R_2 (1 - T_i)^2]^m \tag{5.3.6}$$

We then find that τ_c is independent of the number of round-trips m, and is given by:

$$\tau_c = -\frac{2L}{c \ln[R_1 R_2 (1 - T_i)^2]} \tag{5.3.7}$$

Lets us now assume that Eq. (5.3.5) holds not only at times t_m but at any time $t(>0)$. We can then write

$$\phi(t) \cong \exp(-t/\tau_c)\phi_0 \tag{5.3.8}$$

In this way, we justify Eq. (5.1.1) for the mode field and identify (5.3.7) as the expression for the cavity photon lifetime. Note that Eq. (5.3.7), with the help of Eqs. (1.2.4) and (1.2.6), can be readily transformed into:

$$\tau_c = \frac{L}{c\gamma} \tag{5.3.9}$$

We thus see that the cavity photon lifetime equals the transit time $\tau_T = L/c$ of the beam in the laser cavity divided by the (logarithmic) cavity loss γ.

Example 5.2. *Calculation of the cavity photon lifetime.* We assume $R_1 = R_2 = R = 0.98$ and $T_i \cong 0$. From Eq. (5.3.7) we obtain $\tau_c = \tau_T/[- \ln R] = 49.5\tau_T$, where τ_T is the transit time of the photons for a single pass in the cavity. Note that the photon lifetime is much longer than the transit time, a typical result of low-loss cavities. If we now assume $L = 90$ cm, we get $\tau_T = 3$ ns and $\tau_c \cong 150$ ns.

Having calculated the photon lifetime, the time behavior of the electric field at any point inside the resonator, according to Eq. (5.1.1) and within the scalar approximation, can be written as $E(t) = \tilde{E} \exp[(-t/2\tau_c) + j\omega t]$, where ω is the angular frequency of the mode. The same time behavior then applies for the field of the output wave leaving the cavity through one mirror as a result of finite mirror transmission. If we now take the Fourier transform of this field, we find that the power spectrum of the emitted light has a Lorentzian lineshape with line width (FWHM) given by:

$$\Delta\nu_c = \frac{1}{2\pi\tau_c} \tag{5.3.10}$$

Note that the spectrum of the emitted light, obtained in this way, does not exactly agree with the transmission spectrum shown for a Fabry–Perot interferometer in Sect. 4.5, whose shape is not Lorentzian [see Eq. (4.5.6)]. In particular the expression for $\Delta\nu_c$ obtained here [see Eq.

(5.3.10)], when combined with Eq. (5.3.7) where $T_i \cong 0$, does not coincide with that obtained in Sect. 4.5 [see Eq. (4.5.12) with $L' = L$]. This discrepancy can be traced to the approximation made in Eq. (5.3.8). In numerical terms, however, the discrepancy between the two results is quite small, especially for high values of reflectivity, as we see in Example 5.3. From now on we therefore assume that the cavity lineshape is Lorentzian with width given by Eq. (5.3.10) and the cavity photon lifetime is given by Eq. (5.3.7).

Example 5.3. *Linewidth of a cavity resonance.* If we again take $R_1 = R_2 = 0.98$ and $T = 0$, from Eqs. (5.3.10) and (5.3.7) we obtain $\Delta v_c = 6.4307 \times 10^{-3} \times (c/2L)$, while from Eq. (4.5.12) we obtain $\Delta v_c = 6.4308 \times 10^{-3} \times (c/2L)$. For the particular case of $L = 90$ cm, we then obtain $\Delta v_c \cong 1.1$ MHz. Even at the relatively low reflectivity values of $R_1 = R_2 = 0.5$, the discrepancy is not large. In fact from Eqs. (5.3.10) and (5.3.7) we obtain $\Delta v_c = 0.221 \times (c/2L)$, while from Eq. (4.5.12), $\Delta v_c = 0.225 \times (c/2L)$. Again for $L = 90$ cm we then obtain $\Delta v_c \cong 37.5$ MHz. Thus, in typical cases, Δv_c may range from a few to a few tens of MHz.

Having discussed the cavity photon lifetime, we can now introduce the cavity quality factor, or Q-factor, and derive its relation to the photon lifetime. For any resonant system, and in particular a resonant optical cavity, we define the cavity Q-factor (usually abbreviated to cavity Q) as $Q = 2\pi \times$ (energy stored)/(energy lost in one oscillation cycle). Thus a high cavity Q-value implies low losses in the resonant system. Since, in our case, the energy stored is $\phi h v$ and the energy lost in one cycle is $h v (-d\phi/dt)(1/v) = -h d\phi/dt$, we obtain

$$Q = -\frac{2\pi v \phi}{(d\phi/dt)} \qquad (5.3.11)$$

From Eq. (5.3.8) we then get

$$Q = 2\pi v \tau_c \qquad (5.3.12)$$

which, with the help of Eq. (5.3.10), can be transformed into the more suggestive form:

$$Q = \frac{v}{\Delta v_c} \qquad (5.3.13)$$

Thus the cavity Q-factor can be interpreted as the ratio between the resonance frequency v of the given mode and its linewidth Δv_c.

Example 5.4. *Q factor of a laser cavity.* According to Example 5.2 we take $\tau_c \cong 150$ ns and assume $v \cong 5 \times 10^{14}$ Hz (i.e., $\lambda \cong 630$ nm). From Eq. (5.3.12) we obtain $Q = 4.7 \times 10^8$. Thus very high Q-values can be achieved in a laser cavity, which means that a very small fraction of the energy is lost during one oscillation cycle.

5.4. STABILITY CONDITION

Consider first a general two-mirror resonator (Fig. 5.6) and a ray leaving point P_0 of a plane β inside the resonator, e.g., just in front of mirror 1. After reflection from mirrors 2 and 1, this ray will intersect plane β at some point P_1. If we let r_0 and r_1 be the transverse

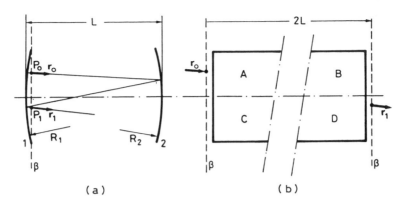

FIG. 5.6. (a) Stability analysis of a two-mirror resonator and (b) stability analysis of a general resonator described by the *ABCD* matrix.

coordinates of P_0 and P_1 with respect to the resonator axis and r'_0 and r'_1 the angles that the corresponding rays make with the axis, then, according to Eq. (4.2.2), we can write

$$\begin{vmatrix} r_1 \\ r'_1 \end{vmatrix} = \begin{vmatrix} A & B \\ C & D \end{vmatrix} \begin{vmatrix} r_0 \\ r'_0 \end{vmatrix} \tag{5.4.1}$$

where the *ABCD* matrix is the cavity round-trip matrix. After one round-trip, the ray leaving point $P_1(r_1, r'_1)$ will intersect plane β at point $P_2(r_2, r'_2)$ given by:

$$\begin{vmatrix} r_2 \\ r'_2 \end{vmatrix} = \begin{vmatrix} A & B \\ C & D \end{vmatrix} \begin{vmatrix} r_1 \\ r'_1 \end{vmatrix} = \begin{vmatrix} A & B \\ C & D \end{vmatrix}^2 \begin{vmatrix} r_0 \\ r'_0 \end{vmatrix} \tag{5.4.2}$$

Therefore, after n round-trips, point $P_n(r_n, r'_n)$ is given by:

$$\begin{vmatrix} r_n \\ r'_n \end{vmatrix} = \begin{vmatrix} A & B \\ C & D \end{vmatrix}^n \begin{vmatrix} r_0 \\ r'_0 \end{vmatrix} \tag{5.4.3}$$

If the resonator is to be stable, we require that, for any initial point (r_0, r'_0), point (r_n, r'_n) does not diverge as n increases. This means that the matrix

$$\begin{vmatrix} A & B \\ C & D \end{vmatrix}^n$$

must not diverge as n increases.

The previous considerations can be readily extended to a general resonator whose round-trip ray transformation is described by a general *ABCD* matrix, e.g., a two-mirror resonator containing other optical elements, such as lenses, telescopes, etc. (see Fig. 5.6b). In this case we again require the nth power of the *ABCD* matrix not to diverge as n increases.

For both resonators in Fig. 5.6, the ray starts from, and arrives at, the same plane β, which means that the refractive index is the same for both rays, input $\mathbf{r}_0$, and output $\mathbf{r}_1$. It then follows that the determinant of the matrix, *AD-BC*, has unit value. A theorem of matrix calculus,[2] sometimes referred to as Sylvester's theorem, then shows that, if we define an angle θ by the relation:

$$\cos \theta = (A + D)/2 \tag{5.4.4}$$

one has

$$\begin{vmatrix} A & B \\ C & D \end{vmatrix}^n = \frac{1}{\sin \theta} \begin{vmatrix} A \sin n\theta - \sin(n-1)\theta & B \sin n\theta \\ C \sin n\theta & D \sin n\theta - \sin(n-1)\theta \end{vmatrix} \tag{5.4.5}$$

Equation (5.4.5) shows that the n-th power matrix does not diverge if θ is a real quantity. Indeed, if θ were complex, say $\theta = a + jb$, the terms proportional to, e.g., $\sin n\theta$ in Eq. (5.4.5) could be written as

$$\sin n\theta = [\exp(jn\theta) + \exp(-jn\theta)]/2j = [\exp(jna - nb) + \exp(-jna + nb)]/2j.$$

The quantity $\sin n\theta$ would then contain a term growing exponentially with n, e.g. $[\exp(-jna + nb)]/2j$ for $b > 0$, and the overall n-th power matrix would thus diverge as n increases. So, for the resonator to be stable, we require θ to be real; according to Eq. (5.4.4), this implies

$$-1 < \left(\frac{A + D}{2}\right) < 1 \tag{5.4.6}$$

Equation (5.4.6) establishes the stability condition for the general resonator in Fig. 5.6b. In the case of the two-mirror resonator in Fig. 5.6a, we can go one step further by explicitly calculating the corresponding $ABCD$ matrix. We recall that a given overall matrix can be obtained by the product of matrices of individual optical elements traversed by the beam, with the matrices written in the reverse order that the ray propagates through the corresponding elements. Thus, in this case, the $ABCD$ matrix is given by the ordered product of the following matrices: Reflection from mirror 1, free-space propagation from mirror 1 to 2, reflection from mirror 2, free-space propagation from mirror 2 to 1. With the help of Table 4.1 we then have

$$\begin{vmatrix} A & B \\ C & D \end{vmatrix} = \begin{vmatrix} 1 & 0 \\ -2/R_1 & 1 \end{vmatrix} \begin{vmatrix} 1 & L \\ 0 & 1 \end{vmatrix} \begin{vmatrix} 1 & 0 \\ -2/R_2 & 1 \end{vmatrix} \begin{vmatrix} 1 & L \\ 0 & 1 \end{vmatrix} \tag{5.4.7}$$

After performing matrix multiplication in Eq. (5.4.7), we obtain

$$\frac{A + D}{2} = 1 - \frac{2L}{R_1} - \frac{2L}{R_2} + \frac{2L^2}{R_1 R_2} \tag{5.4.8}$$

Equation (5.4.8) can be readily transformed to:

$$\frac{A + D}{2} = 2 \left[1 - \left(\frac{L}{R_1}\right)\right]\left[1 - \left(\frac{L}{R_2}\right)\right] - 1 \tag{5.4.9}$$

It is now customary to define two dimensionless quantities for the cavity, referred to as the g_1 and g_2 parameters, defined as:

$$g_1 = 1 - \left(\frac{L}{R_1}\right) \tag{5.4.10a}$$

$$g_2 = 1 - \left(\frac{L}{R_2}\right) \tag{5.4.10b}$$

In terms of these parameters, the stability condition [Eq. (5.4.6)], with the help of Eq. (5.4.9), readily transforms into the very simple relation:

$$0 < g_1 g_2 < 1 \tag{5.4.11}$$

The stability condition (5.4.11) can be conveniently displayed in the g_1, g_2 plane (Fig. 5.7). For this purpose we plot in Fig. 5.7 the two branches of the hyperbola corresponding to the equation $g_1 g_2 = 1$ (*heavy lines*). Since the other limiting condition in Eq. (5.4.11), namely $g_1 g_2 = 0$, implies either $g_1 = 0$ or $g_2 = 0$, the stable regions in the g_1, g_2 plane correspond to the shaded area in the figure. A particularly interesting class of two-mirror resonators is described by points on the straight line AC making an angle of 45° with the g_1 and g_2 axes. This line corresponds to resonators having mirrors of the same radius of curvature (symmetric resonators). As particular examples of these symmetric resonators, we see that those corresponding to points A, B, and C of the figure are the concentric, confocal, and plane resonators, respectively.

It should be noted that the three resonators correponding to points A, B, and C, and, more generally, resonators described by conditions $g_1 g_2 = 0$ or $g_1 g_2 = 1$, lie on the boundary between the stable and unstable regions. For these resonators, only particular rays, e.g., rays

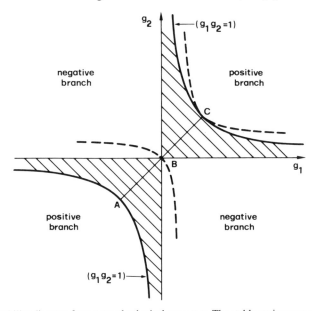

FIG. 5.7. A g_1, g_2 stability diagram for a general spherical resonator. The stable region corresponds to shaded parts in the figure; dashed curves correspond to the possible confocal resonators.

normal to plane mirrors in Fig. 5.1a, do not diverge during propagation. For this reason, these resonators are also said to be marginally stable. Conditions $g_1g_2 = 0$ or $g_1g_2 = 1$ thus correspond to a marginally stable situation.

5.5. STABLE RESONATORS

To simplify our analysis we first consider a resonator with no limiting aperture. We then briefly consider the effects of a finite aperture.

5.5.1. Resonators with Infinite Aperture

Referring to Fig. 5.6b for a general resonator and to Fig. 5.6a for a two-mirror resonator, we assume no limiting aperture; i.e., we take $a_1, a_2 = \infty$ in Fig. 5.5a.[3,4] The field distribution $u(x, y, z)$ after one cavity round-trip in Fig. 5.5a or after one period of the lens-guide system in Fig. 5.5b, i.e., at $z = 2L$, can be obtained from Eq. (4.6.9) with $z_1 = 0$, where the *ABCD* matrix is the one-round-trip (or one-period) matrix. If at the $z_1 = 0$ plane we now take $\tilde{E}(x_1, y_1, 0) = u(x_1, y_1, 0)$, then, within the paraxial wave approximation and according to Eq. (4.6.4), we can write $\tilde{E}(x, y, 2L) = u(x, y, 2L)\exp(-2jkL)$. Inserting the expression for $u(x, y, z)$ given by Eq. (4.6.9) into this relation we obtain

$$\tilde{E}(x, y, 2L) = \exp(-2jkL) \int_{-\infty}^{+\infty} \int_{-\infty}^{+\infty} \left(\frac{j}{B\lambda}\right) \exp -jk\left[\frac{A(x_1^2 + y_1^2) + D(x^2 + y^2) - (2x_1x + 2y_1y)}{2B}\right]$$
$$\times \tilde{E}(x_1, y_1, 0)dx_1\, dy_1 \tag{5.5.1}$$

A comparison between Eqs. (5.5.1) and (5.2.1) shows that the propagation kernel $K(x, y; x_1, y_1)$ is given in this case by:

$$K = \left(\frac{j}{B\lambda}\right) \exp -jk\left[\frac{A(x_1^2 + y_1^2) + D(x^2 + y^2) - (2x_1x + 2y_1y)}{2B}\right] \tag{5.5.1a}$$

As explained in Sect. 4.7, the lowest order Gaussian solution, Eq. (4.7.15), and the general solution for higher order, Eq. (4.7.30), are eigensolutions of the propagation equation (4.6.9) when there is no aperture within the optical system described by the given *ABCD* matrix. For these Hermite–Gaussian eigensolutions to describe the field distribution of cavity eigenmodes, we require the beam to reproduce itself after one round-trip. This means that, if we let q_1 be the complex beam parameter of the Gaussian beam leaving plane β in front of mirror 1, e.g., in Fig. 5.8a, the complex beam parameter q after one round-trip must equal q_1. From the *ABCD* law of Gaussian beam propagation [Eq. (4.7.4)], if we set $q_1 = q$, we obtain

$$q = \frac{Aq + B}{Cq + D} \tag{5.5.2}$$

The q parameter must then satisfy the quadratic equation:

$$Cq^2 + (D - A)q - B = 0 \tag{5.5.3}$$

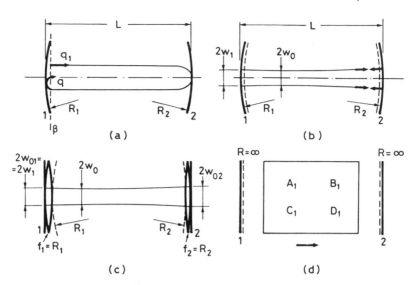

FIG. 5.8. (a) Calculation of the q-parameter for a two-mirror resonator, (b) spot size and equiphase surfaces in a two mirror-resonator, (c) transformation of a two-mirror resonator into a resonator with plane end mirrors, and (d) general resonator with two plane end mirrors.

Since q must be a complex quantity, the standard solution of a quadratic equation shows that the discriminant of Eq. (5.5.3) must be negative, i.e.:

$$(D - A)^2 + 4BC < 0 \qquad\qquad\qquad (5.5.4)$$

However, since $AD - BC = 1$, Eq. (5.5.4) readily gives $(D + A)^2 < 4$, i.e., the same condition as given by Eq. (5.4.6). This means that a Gaussian beam solution can be found only for stable resonators or alternatively all stable resonators with infinite aperture have modes described by the general Hermite–Gaussian solution in Eq. (4.7.30).

5.5.1.1 Eigenmodes

Consider first the two-mirror resonator in Fig. 5.8a. To obtain an expression for the complex amplitude distribution $u(x, y, z)$ at mirror 1, e.g., we must calculate the complex beam parameter q, obtained as a solution to Eq. (5.5.3), for given values of matrix elements A, B, C, and D. Having calculated the q-parameter, we obtain the real and imaginary parts of $1/q$. From these, in accordance with Eq. (4.7.8), we obtain the spot size w and the radius of curvature R of the wavefront at the given position. We proceed similarly to calculate w and R at any position within the resonator including mirror 2 (Fig. 5.8b). For these calculations it is convenient to transform the resonator in Fig. 5.8b into that of Fig. 5.8c, where the spherical mirror of radius R_1, e.g., is substituted by a combination of a plane mirror and a thin lens of focal length $f_1 = R_1$.* The resonator in Fig. 5.8c is then seen to belong to a general class of resonators consisting of two plane mirrors and containing an optical element

---

* The fact that we are considering an equivalent lens of focal length $f_1 = R_1$ when the focal length of the equivalent lens-guide structure was $f_1 = R_1/2$ (see Fig. 5.5b) may generate some confusion. Note however that, due to the reflection at the plane mirror Fig. 5.8c, the lens f_1 in the figure is traversed twice by the beam, so its effect is equivalent to a single lens of overall focal length $f_1/2$.

whose *single-pass* propagation matrix, from mirror 1 to mirror 2, is represented by matrix elements A_1, B_1, C_1, and D_1 (Fig. 5.8d).

To obtain q from Eq. (5.5.3) we must calculate the round-trip matrix for the general resonator in Fig. 5.8d. To this purpose we recall that, according to Eq. (4.2.17), the matrix for a single-pass backward, i.e., for propagation from mirror 2 to mirror 1, is obtained from the $A_1 B_1 C_1 D_1$ matrix by interchanging elements A_1 and D_1. The matrix of a plane mirror is then readily obtained from that of a spherical mirror (see Table 4.1) by letting $R \rightarrow \infty$. The matrix of a plane mirror is thus simply the unit matrix:

$$\begin{vmatrix} 1 & 0 \\ 0 & 1 \end{vmatrix}$$

Starting from mirror 1, the round-trip matrix is then given by:

$$\begin{vmatrix} A & B \\ C & D \end{vmatrix} = \begin{vmatrix} D_1 & B_1 \\ C_1 & A_1 \end{vmatrix} \begin{vmatrix} A_1 & B_1 \\ C_1 & D_1 \end{vmatrix} = \begin{vmatrix} 2A_1 D_1 - 1 & 2B_1 D_1 \\ 2A_1 C_1 & 2A_1 D_1 - 1 \end{vmatrix} \tag{5.5.5}$$

From Eq. (5.5.5) we see immediately that $A = D$; from Eq. (5.5.3) we then obtain

$$q = q_1 = j\sqrt{-\frac{B}{C}} = j\sqrt{-\frac{B_1 D_1}{A_1 C_1}} \tag{5.5.6a}$$

We can show that the stability condition [Eq. (5.4.6)] implies $B_1 D_1 / A_1 C_1 < 0$. This means that q_1 is purely imaginary, i.e., the equiphase surface just in front of mirror 1 (see Fig. 5.8c–d) is plane. We can repeat the same argument starting from mirror 2 to show that:

$$q_2 = j\sqrt{-\frac{A_1 B_1}{C_1 D_1}} \tag{5.5.6b}$$

Since

$$\frac{A_1 B_1}{C_1 D_1} = \left(\frac{A_1}{D_1}\right)^2 \left(\frac{B_1 D_1}{A_1 C_1}\right) < 0,$$

q_2 is also purely imaginary, and the wavefront at mirror 2 is again plane. This means that the wavefront radius of curvature, after lens f_1 in Fig. 5.8c or in front of mirror 1 in Fig. 5.8b, is equal to R_1; a similar argument applies to mirror 2. We thus reach the general conclusion that the equiphase surface on a cavity mirror always coincides with the mirror surface. This result can be understood from Fig. 5.8b, where the field corresponding to the given eigenmode is considered in terms of a superposition of traveling waves. Then the right-traveling wave in Fig. 5.8b (indicated by left-to-right arrows) must be transferred at mirror 2 into the left-traveling wave (indicated by right-to-left arrows) on reflection. In terms of geometrical optics, this implies that propagating rays at mirror 2 must be orthogonal to the mirror surface. Then the wavefront, which is always orthogonal to these rays, must be coincident with the mirror surface at the mirror location.

The general results of Eq. (5.5.6) can now be specialized to the two-mirror resonator. Referring to Fig. 5.8c we see that, in propagating from mirror 1 to mirror 2, the beam passes through lens f_1, then through a free space of length L and then a lens f_2. The A_1, B_1, C_1, D_1 matrix is then simply obtained from the product of the three corresponding matrices, where the written order is the inverse of the propagation order. Using the matrices in Table 4.1, it is a simple matter to show that:

$$\begin{vmatrix} A_1 & B_1 \\ C_1 & D_1 \end{vmatrix} = \begin{vmatrix} g_1 & L \\ -(1 - g_1 g_2)/L & g_2 \end{vmatrix} \tag{5.5.7}$$

where g_1 and g_2 are given by Eq. (4.4.10). From Eq. (5.5.6a), with the help of Eqs. (4.7.8) and (5.5.7), we then obtain

$$w_1 = \left(\frac{L\lambda}{\pi}\right)^{1/2} \left[\frac{g_2}{g_1(1 - g_1 g_2)}\right]^{1/4} \tag{5.5.8a}$$

Similarly, starting from Eq. (5.5.6b), we have

$$w_2 = \left(\frac{L\lambda}{\pi}\right)^{1/2} \left[\frac{g_1}{g_2(1 - g_1 g_2)}\right]^{1/4} \tag{5.5.8b}$$

which could have been obtained straightforwardly from Eq. (5.5.8a) by interchanging subscripts 1 and 2. Starting from the spot size $w_{01} = w_1$ in Fig. 5.8c, we then calculate the spot size w_0 at the beam waist using Eq. (4.7.27) with $f = f_1$ and $w_{02} = w_0$. We obtain

$$w_0 = \left(\frac{L\lambda}{\pi}\right)^{1/2} \left[\frac{g_1 g_2(1 - g_1 g_2)}{(g_1 + g_2 - 2g_1 g_2)^2}\right]^{1/4} \tag{5.5.9}$$

Again knowing the spot size w_1 on mirror 1, we can obtain the waist distance from that mirror by using the expression for z_m given by Eq. (4.7.26) with the substitutions $f = f_1 = R_1$ and $z_{R_1} = \pi w_{01}^2 / \lambda$.

In a symmetric resonator, one has $R_1 = R_2 = R$ and $g_1 = g_2 = g = 1 - (L/R)$; both Eqs. (5.5.8a) and (5.5.8b) then reduce to

$$w = \left(\frac{L\lambda}{\pi}\right)^{1/2} \left(\frac{1}{1 - g^2}\right)^{1/4} \tag{5.5.10a}$$

while Eq. (5.5.9) yields

$$w_0 = \left(\frac{L\lambda}{\pi}\right)^{1/2} \left[\frac{1 + g}{4(1 - g)}\right]^{1/4} \tag{5.5.10b}$$

Example 5.5. *Spot sizes for symmetric resonators.* The first case we consider involves a confocal resonator ($g = 0$). From Eqs. (5.5.10a–b) we obtain respectively,

$$w_c = \left(\frac{L\lambda}{\pi}\right)^{1/2} \qquad w_{oc} = \left(\frac{L\lambda}{2\pi}\right)^{1/2} \tag{5.5.11}$$

where the subscript c stands for confocal. Equation (5.5.11) shows that the spot size at the beam waist is, in this case, $\sqrt{2}$ smaller than that at the mirrors (Fig. 5.9a). For the case of a near-plane resonator, i.e., when $R \gg L$, we can write $g = 1 - \varepsilon$, where ε is a small positive quantity. Neglecting higher order terms in ε, Eq. (5.5.10) gives

$$\left(\frac{w_{np}}{w_c}\right) \cong \left(\frac{w_{onp}}{w_c}\right) \simeq \left(\frac{1}{2\varepsilon}\right)^{1/4} \tag{5.5.12}$$

where the subscript np stands for near-plane and the spot sizes are normalized to the mirror spot size of a confocal resonator. Equation (5.5.12) shows that, to first order, the two spot sizes are equal; thus the spot size is nearly constant over the length of the resonator (Fig. 5.9b). For the case of a near-concentric resonator, i.e., when $L \cong 2R$, we likewise write $g = -1 + \varepsilon$, where again ε is a small positive quantity. Neglecting terms of higher order in ε, Eq. (5.5.10) gives

$$\left(\frac{w_{nc}}{w_c}\right) = \left(\frac{1}{2\varepsilon}\right)^{1/4} \qquad \left(\frac{w_{onc}}{w_c}\right) = \left(\frac{\varepsilon}{8}\right)^{1/4} \tag{5.5.13}$$

where the subscript nc stands for near-concentric. Equation (5.5.13) shows that the mirror spot size, as a function of ε, is given by the same expression as that for a near-plane resonator. The spot size at the beam waist, however, is now much smaller, and it decreases with decreasing value of ε. Spot size behavior along the resonator is shown in Fig. 5.9c. Numerically, if we take $L = 1$ m and $\lambda = 514$ nm (an argon laser wavelength), we get $w_c \cong 0.4$ mm for a confocal resonator. If we now consider a near-plane resonator, still with $L = 1$ m, $\lambda = 514$ nm and with $R = 10$ m, we obtain $g = 0.9$; from Eq. (5.5.10) we get $w_0 \cong 0.59$ mm and $w \simeq 0.61$ mm. Note: Relatively small values of beam spot size are obtained in every case.

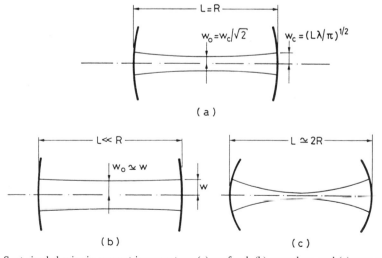

FIG. 5.9. Spot size behavior in symmetric resonators: (a) confocal, (b) near-plane, and (c) near-concentric.

5.5.1.2. Eigenvalues

A comparison of Eq. (4.7.29) with (4.7.30) shows that, if the *ABCD* matrix corresponds to the cavity round-trip matrix and if $q = q_1$, the field amplitude $u(x, y, 2L)$ after one round-trip equals the initial field $u(x_1 = x, y_1 = y, z_1 = 0)$ except for the amplitude factor $1/[A + (B/q)]^{1+l+m}$. According to Eq. (5.2.2) it then follows that:

$$\tilde{\sigma}_{lm} = \frac{1}{[A + (B/q)]^{1+l+m}} \tag{5.5.14}$$

From Eq. (5.5.3) we see that, since $A = D$, then:

$$q = j\sqrt{-B/C} \tag{5.5.15}$$

If we now write

$$\sigma^{-1} = A + \left(\frac{B}{q}\right) \tag{5.5.16}$$

from Eq. (5.5.16) and with the help of Eq. (5.5.15), we obtain $|\sigma^{-1}|^2 = A^2 - BC = AD - BC = 1$. From Eq. (5.5.14) it then follows that the magnitude of $\tilde{\sigma}_{lm}$ is also unity, and, according to Eq. (5.2.6), the diffraction loss γ_{lm} vanishes. This result is actually expected from our analysis, since we stipulated at the onset that there were no limiting apertures (Fig. 5.8d) and we considered in particular a two-mirror resonator with infinite mirror size (Fig. 5.8c).

To obtain an expression for the phase of the eigenvalue, $\tilde{\sigma}_{lm}$, we write

$$\sigma^{-1} = \exp(-j\phi) \tag{5.5.17}$$

From Eqs. (5.5.17) and (5.5.16), with the help of Eqs. (5.5.15) and (5.5.5), we obtain

$$\cos\phi = A = 2A_1 D_1 - 1 \tag{5.5.18a}$$

$$\sin\phi = B\sqrt{-C/B} = 2B_1 D_1\sqrt{-A_1 C_1/B_1 D_1} \tag{5.5.18b}$$

From Eq. (5.5.18b) we see that $0 < \phi < \pi$ for $B_1 D_1 > 0$ and $-\pi < \phi < 0$ for $B_1 D_1 < 0$. From Eq. (5.5.18a) we obtain

$$\cos^2(\phi/2) = \frac{(1 + \cos\phi)}{2} = A_1 D_1$$

and hence:

$$\phi = 2\cos^{-1}(\pm\sqrt{A_1 D_1}) \tag{5.5.19}$$

where the positive or negative signs hold depending on whether $B_1 D_1$ is positive or negative. From Eqs. (5.5.14), (5.5.16), and (5.5.17) we obtain $\tilde{\sigma}_{lm} = \exp j(1 + l + m)\phi = \exp j\phi_{lm}$ i.e., $\phi_{lm} = (1 + l + m)\phi$. From Eqs. (5.5.19) and (5.2.7) we then obtain

$$v_{lmn} = \frac{c}{2L}\left[n + \frac{(1 + l + m)}{\pi}\cos^{-1}(\pm\sqrt{A_1 D_1})\right] \tag{5.5.20}$$

The $+$ or $-$ signs again depend on whether $B_1 D_1$ is positive or negative.

For the particular case of a two mirror resonator, matrix elements A_1 and D_1 are obtained from Eq. (5.5.7). Equation (5.5.20) then transforms into:

$$v_{lmn} = \frac{c}{2L}\left[n + \frac{(1 + l + m)}{\pi}\cos^{-1}(\pm\sqrt{g_1 g_2})\right] \tag{5.5.21}$$

According to Eq. (5.5.7), the $+$ or $-$ signs depends on whether g_2 (and hence g_1) is positive or negative.

Example 5.6. *Frequency spectrum of a confocal resonator.* For a symmetric confocal resonator $(g_1 = g_2 = 0)$, from Eq. (5.5.21) we get

$$v_{lmn} = \frac{c}{4L}[2n + (1 + l + m)] \tag{5.5.22}$$

The corresponding frequency spectrum is shown in Fig. 5.10a. We observe that modes having the same value of $2n + l + m$ have the same resonance frequency, although they correspond to different spatial configurations. These modes are said to be frequency-degenerate. Instead of being given by $c/2L$, as for a plane-parallel resonator [see Eq. (5.1.3)], the frequency spacing between consecutive modes is now given by $c/4L$. Two such consecutive modes, however, must have different (l, m) values; $c/4L$ thus corresponds to the frequency difference between two consecutive *transverse modes* [e.g., $(n, 0, 0) \rightarrow (n, 0, 1)$]. On the other hand, the frequency spacing between two modes with the same (l, m) values (e.g., TEM$_{00}$) and n differing by 1 (i.e., the frequency spacing between adjacent *longitudinal modes*) is still $c/2L$, the same as for the plane-parallel resonator.*

Example 5.7. *Frequency spectrum of a near-planar and symmetric resonator.* In this case, $g_1 = g_2 = 1 - (L/R)$, with $L/R \ll 1$. Thus g is positive and slightly less than unity. Accordingly

$$\cos^{-1} g = \cos^{-1}[1 - (L/R)] \cong (2L/R)^{1/2}$$

---

* The terms longitudinal mode and transverse mode in the laser literature are sometimes confusing and convey the (mistaken) impression that there are two distinct types of modes, viz., longitudinal modes (sometimes called axial modes) and transverse modes. In fact any mode is specified by three numbers, e.g., l, m, n of Eq. (5.5.24). Electric and magnetic fields of the modes are nearly perpendicular to the resonator axis. The variation of these fields in a transverse direction is specified by l, m, while field variation in a longitudinal (i.e., axial) direction is specified by n. When we loosely refer to a (given) transverse mode, it means that we are considering a mode with given values for the transverse indices (l, m), regardless of the value of n. Accordingly, a single transverse mode means a mode with specific values for the transverse indices (l, m). A similar interpretation can be applied to longitudinal modes. Thus two adjacent longitudinal modes mean two modes with consecutive values for the longitudinal index n [i.e., n and $(n + 1)$ or $(n - 1)$].

and Eq. (5.5.21) becomes

$$v_{lmn} = \frac{c}{2L}\left[n + \frac{(1+l+m)}{\pi}\left(\frac{2L}{R}\right)^{1/2}\right] \qquad (5.5.23)$$

The corresponding frequency spectrum is shown in Fig. 5.10b. Note that the frequency spacing between consecutive longitudinal modes is again $c/2L$, while the frequency difference between two consecutive transverse modes is $(c/2L)(2L/\pi^2 R)^{1/2}$.

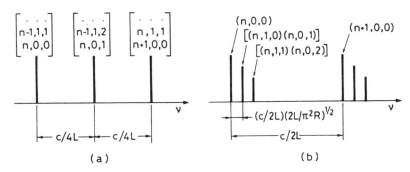

FIG. 5.10. (a) Mode spectrum of a confocal resonator and (b) mode spectrum of a near-plane resonator.

5.5.1.3. Standing and Traveling Waves in a Two-Mirror Resonator

Following the discussion in Sect. 5.5 1.1-2 on spot sizes and resonance frequencies in a general resonator, we now turn to the corresponding behavior of the mode along the laser cavity; we limit our discussion to a two-mirror resonator. According to Eqs. (4.7.31) and (4.6.4), the field inside this resonator can be written as:

$$\tilde{E}_{lmn}(x, y, z) = \frac{w_0}{w}H_l\left[\frac{\sqrt{2}x}{w}\right]H_m\left[\frac{\sqrt{2}y}{w}\right]\exp\left[-\frac{x^2+y^2}{w^2}\right] \qquad (5.5.24a)$$

$$\times \exp[-jkz + j(1+l+m)\phi]\times \qquad (5.5.24b)$$

$$\times \exp[-jk(x^2+y^2)/2R] \qquad (5.5.24c)$$

where $w(z)$, $R(z)$, and $\phi(z)$ are given by Eqs. (4.7.17). They can be calculated once the waist position and the corresponding spot size w_0 are known. Note that the field eigenmode $\tilde{E}$ is explicitly indicated as dependent on subscripts l, m and n. Subscripts l and m come from the order of the Hermite polynomials in Eq. (5.5.24a). Subscript n is also explicitly indicated, since $k = 2\pi v/c$ and the resonant frequency depends on subscripts l, m, and n [see Eq. (5.5.21)]. These subscripts indicate the following:

- Subscripts l and m give the field nulls along the x- and y-axis, respectively, as stated in Sect. 4.7.4.
- Based on the discussion in Sect. 5.1.1, subscript n gives the number of half-wavelengths of the standing wave mode along the resonator; i.e., it gives the number of field nulls along the z-direction.

To conclude this section, we address the question as to whether Eq. (5.5.24) represents a traveling or standing wave pattern for the field eigenmode. The answer depends on the form of the time behavior of the mode. According to Eq. (4.6.1), if we write $E = \tilde{E} \exp(j\omega t)$ where $\omega = \omega_{lmn} = 2\pi v_{lmn}$ is the angular frequency of the mode resonance, then from the longitudinal phase factor in Eq. (5.5.24b), taking the example of a TEM$_{00}$ mode, we have $E \propto \exp j[-kz + \phi + \omega t]$ which corresponds to a wave propagating in the positive z-direction. If, on the other hand, we write $E = \tilde{E} \exp(-j\omega t)$, we obtain a wave propagating in the negative z-direction. The standing wave eigenmode is then obtained from the sum of these two waves, i.e., by writing $E \propto \tilde{E} \cos \omega t$. Following this argument we see that, apart from a proportionality factor given by the mirror's transmission, $E = \tilde{E} \exp(j\omega t)$ also represents the wave escaping through mirror 2 and propagating along the positive z-direction.

5.5.2. Effects of a Finite Aperture

In Sect. 5.5.1.2 we saw that, for a general resonator with no limiting aperture such as that in Fig. 5.8d, diffraction loss vanishes. Indeed, to calculate these losses, we must take into account the actual size of any apertures present in the resonator (often a diaphragm is inserted into the resonator or the aperture is set by the transverse dimension of the active medium). The loss introduced by a finite aperture can in fact be appreciated with the help of Fig. 5.11, where a TEM$_{00}$ mode is considered and we indicate the transverse profile of this mode over the plane containing the aperture of diameter $2a$. The Gaussian TEM$_{00}$ mode is truncated by this aperture, so that the dashed wings of the beam are lost each time the beam passes through the aperture. This description is however approximate because the introduction of a limiting aperture significantly modifies the field distribution, which is then no longer precisely Gaussian.

For a correct and accurate calculation, we must return to the original integral equation (5.2.5) and take into account the finite size of the aperture. We limit the discussion that follows to a two-mirror resonator, assuming the limiting aperture to be set by the finite mirror size.

Consider first a symmetric resonator ($R_1 = R_2 = R$ and $a_1 = a_2 = a$, see Fig. 5.12a) and its equivalent lens-guide structure (Fig. 5.12b). By virtue of the symmetry of the problem, we limit our considerations to one period of length L and require the field to reproduce its shape after this period. We then obtain an integral equation similar to Eq. (5.2.5), namely:

$$\tilde{\sigma}\tilde{E}(x, y, 0) = \int_{-a}^{+a} \int_{-a}^{+a} K(x, y, x_1, y_1)\tilde{E}(x_1, y_1, 0)dx_1\, dy_1 \qquad (5.5.25)$$

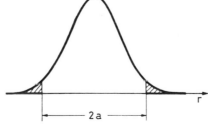

FIG. 5.11. Diffraction losses arising from beam truncation by an aperture of radius a.

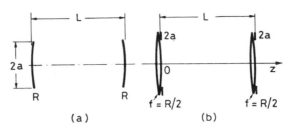

FIG. 5.12. (a) Mode and diffraction loss calculation in a symmetric resonator and (b) equivalent lens-guide configuration.

where the double integral is taken over the limiting aperture and K is the single-pass propagation kernel. Since the beam encounters no other limiting apertures when propagating from the diaphragm at $z = 0$ to that at $z = L$ (Fig. 5.12b), the kernel can be expressed as in Eq. (5.5.1a), where the *ABCD* matrix now refers to one period of length L. The matrix is then given by the product of the matrix of free-space propagation over a length L with the matrix of a lens of focal length $f = R/2$. Since, however, the double integral in Eq. (5.5.25) does not extend between $-\infty$ and $+\infty$, the eigensolutions no longer take the form of a product of a Hermite polynomial and a Gaussian function.

To obtain the eigensolutions one usually solves Eq. (5.2.25) by an iterative procedure, generally with the help of a computer. An often-used approach is the iterative procedure named after Fox and Li,[5] who first applied this method to obtain eigenmodes of a plane-parallel resonator. One first assumes some field expression $\tilde{E}(x, y, 0)$ on the right-hand side of Eq. (5.5.25), then calculates the field $\tilde{E}(x, y, L)$ after one lens-guide period by solving the double integral. This field is then inserted into the right-hand side of Eq. (5.5.25) and a new field $\tilde{E}(x, y, 2L)$ is calculated by performing the double integration, and so on. Although this procedure is rather slow (it usually converges in a few hundred iterations), it eventually leads to a field that does not change on each successive iteration except for an overall

---

Example 5.8. *Diffraction loss of a symmetric resonator.*[6] The diffraction loss per pass for a symmetric two-mirror resonator of finite mirror aperture, as calculated according to the Fox–Li iterative procedure is plotted in Fig. 5.13a (for a TEM_{00} mode) and in Fig. 5.13b (for a TEM_{01} mode) versus the Fresnel number

$$N = \frac{a^2}{L\lambda} \tag{5.5.26}$$

The calculation is performed for a range of symmetric resonators characterized by their corresponding g values. Note that, for a given g value and a given mode (e.g., the TEM_{00} mode), the loss rapidly decreases with an increasing Fresnel number. This is easily understood when, using Eq. (5.5.11), we write the Fresnel number as $N = a^2/\pi w_c^2$ where w_c is the mirror spot size for a confocal resonator of the same length and infinite aperture. Since the mirror spot size does not change strongly with g-value changes (see Example 5.5), the Fresnel number can be interpreted as proportional to the ratio between the area of the mirror cross section (πa^2 for a circular mirror) and the mode cross-section (πw^2) on the mirror. Why loss decreases rapidly when the latter ratio increases is now readily appreciated with the help of Fig. 5.11. Note also that, for a given Fresnel number and g value, the TEM_{00} mode has lower losses than the TEM_{01} mode. The TEM_{00} mode actually has a lower loss than any of the higher order modes. The lowest order mode is thus identified as the lowest loss mode.

> **Example 5.9.** *Limitation on the Fresnel number and resonator aperture in stable resonators.* To obtain oscillation on the TEM_{00} mode only, we must provide a sufficiently high value of diffraction losses γ_{01} for the TEM_{01} mode. On the other hand, to obtain a large value of the spot size, we must design the resonator to operate near the instability boundary $g = 1$ or $g = -1$ (see Example 5.5). Furthermore, if we consider a near-planar resonator, e.g., we cannot operate too close to the instability boundary or the resonator becomes too sensitive to external perturbation (e.g., mirror tilt due to vibrations or temperature changes). We choose then, as an example, $\gamma_{01} = 10\%$ and $g < 0.95$ ($R < 20L$). From Fig. 5.13b we then obtain $N < 2$, which can be considered a typical result. Thus for $L = 2$ m and $\lambda = 1.06$ μm (a Nd:YAG laser wavelength), we obtain $a < 2$ mm. For $L = 2$ m and $\lambda = 10.6$ μm (typical wavelength of a CO_2 laser), we obtain $a < 6.3$ mm.

amplitude reduction due to diffraction loss and a phase factor that accounts for the single-pass phase shift. In this way, we can compute the field amplitude distribution of the lowest order mode and also of higher order modes, as well as their corresponding diffraction losses and resonance frequencies.

We next consider the general two-mirror resonator in Fig. 5.5a and its equivalent lens-guide structure in Fig. 5.5b. If we let $\tilde{E}(x_1, y_1, 0)$ be the field at a general point (x_1, y_1) of the $z = 0$ plane in Fig. 5.5b, the field at point (x_2, y_2) of the $z = L$ plane is readily obtained as $\tilde{E}(x_2, y_2, L) = \exp(-jkL) \iint_1 K_{12}(x_2, y_2; x_1, y_1) \times \tilde{E}(x_1, y_1, 0) dx_1\, dy_1$ where K_{12} is the kernel for beam propagation from $z = 0$ to $z = L$ planes, and where the double integral is taken over aperture 1. Similarly, the field at point (x_3, y_3) of the plane $z = 2L$ is obtained as $\tilde{E}(x_3, y_3, 2L) = \exp(-jkL) \iint_2 K_{21}(x_3, y_3; x_2, y_2) \tilde{E}(x_2, y_2, L) dx_2\, dy_2$ where K_{21} is the kernel for beam propagation from $z = L$ to $z = 2L$ planes, and where the double integral is taken over aperture 2. The combination of the last two equations leads to:

$$\tilde{E}(x_3, y_3, 2L) = \exp(-2jkL) \iint_2 K_{21}(x_3, y_3; x_2, y_2) dx_2\, dy_2$$

$$\times \iint_1 K_{12}(x_2, y_2; x_1, y_1) \tilde{E}(x_1, y_1, 0) dx_1\, dy_1 \qquad (5.5.27)$$

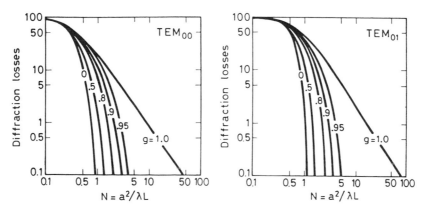

FIG. 5.13. Diffraction loss per transit versus Fresnel number for (a) the TEM_{00} mode and (b) the TEM_{01} mode, for several symmetric resonators. (From Ref. 6, copyright 1965, American Telephone and Telegraph Company. By permission.)

Interchanging the order of integration in Eq. (5.5.27), we readily see that we can write

$$\tilde{E}(x_3, y_3, 2L) = \exp(-2jkL) \iint_1 K(x_3, y_3; x_1, y_1)\tilde{E}(x_1, y_1, 0)dx_1\, dy_1 \qquad (5.5.28)$$

provided we define an overall kernel K (one period in Fig. 5.5b, i.e., one round-trip in Fig. 5.5a) as:

$$K(x_3, y_3; x_1, y_1) = \iint_2 K_{21}(x_3, y_3; x_2, y_2)K_{12}(x_2, y_2, x_1, y_1)dx_2 dy_2 \qquad (5.5.29)$$

This is the appropriate kernel to use in Eq. (5.2.5) for calculating the field eigenmodes and the corresponding eigenvalues.

5.5.3. Dynamically and Mechanically Stable Resonators

A very important problem which arises with stable resonators is to increase the beam spot size within the active medium to a size comparable to the transverse dimensions of the medium. In fact, considering for simplicity a symmetric two-mirror resonator, we can see from Eq. (5.5.10) that, to significantly increase the spot size within the laser cavity beyond the value established for a confocal cavity, we must choose a resonator much closer to the $g = \pm 1$ point (near-plane or near-concentric resonator). The cavity would then be too close to an instability boundary and generally become very sensitive to cavity perturbations such as those arising from variations in pump power. We now consider a laser design which allows us to achieve large spot sizes within the active medium; the design is particularly insensitive to cavity perturbations from either changes in pump power or from mirror tilting (dynamically and mechanically stable resonator).[7]

We first consider a laser resonator consisting of two spherical mirrors, of radii R_1 and R_2, and containing an active medium whose pump-induced thermal effects can be simulated by a thin lens with dioptric power, $1/f$, proportional to the pump power (Fig. 5.14a). This model applies particularly well to solid-state lasers. Some of the ideas that follow, however, can also be applied to the more complex perturbations induced by the pump in a gas medium.

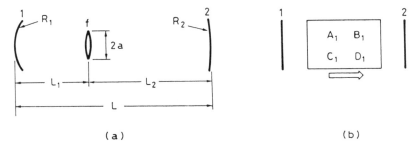

(a)   (b)

FIG. 5.14. (a) General two-mirror spherical resonator that includes a lens of focal length f simulating the thermal lens of the active medium and (b) generalization of the resonator of (a) where the elements A_1, B_1, C_1 and D_1 of the one-way matrix include the matrix of the thermal lens.

A first constraint on the design of the laser cavity in Fig. 5.14a is obtained by the condition that the spot size in the active medium w_a must be insensitive to change in the lens dioptric power. We thus write

$$\frac{dw_a}{d(1/f)} = 0 \qquad (5.5.30)$$

A resonator satisfying this condition is often referred to as *dynamically stable*. A second constraint is obtained from the condition that the spot size w_a must be comparable to the radius of a of the active medium. To avoid introducing excessive diffraction losses due to beam truncation by this finite aperture, we can require, e.g.:[8]

$$2a \cong \pi w_a \qquad (5.5.31)$$

For given values of a and $1/f$, Eqs. (5.5.30) and (5.5.31) provide a pair of equations for the cavity parameters R_1, R_2, L_1, and L_2.

We may now question whether a dynamically stable situation actually exists for the cavity in Fig. 5.14a. To answer this, we show in Fig. 5.15 the general behavior of w_a versus dioptric power $1/f$, as obtained for the preceding cavity at given values of other cavity parameters. From this figure we observe the following general characteristics:

- Two dynamically stable points, i.e., satisfying Eq. (5.5.30), are found when the lens dioptric power is changed.
- Both points correspond to a minimum of w_a, the minimum value w_{am} being the same for the two points.
- The minima belong to two different stability zones; the spot size actually diverges at each zone boundary.
- The width $\Delta(1/f)$ of the two zones is the same; it satisfies a fundamental relation with the minimum spot size as given by the equation:

$$\frac{\pi w_{am}^2}{\lambda} \Delta(1/f) = 2 \qquad (5.5.32)$$

independently of the values of the other cavity parameter.

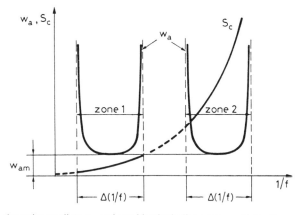

FIG. 5.15. Spot size in the active medium w_a and combined misalignment sensitivity S_c versus dioptric power $1/f$ for the cavity in Fig. 5.14a.

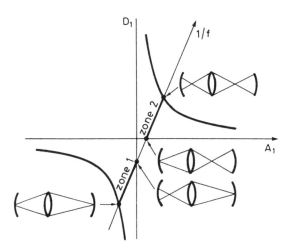

FIG. 5.16. Stability diagram for the general resonator in Fig. 5.14b. The same figure shows the two stability zones discussed in Fig. 5.15 and the corresponding geometrical optics description of the cavities corresponding to stability boundaries.

The existence of two stability zones can generally be understood by referring to Fig. 5.14b (see also Fig. 5.8d), which is a generalization of Fig. 5.14a. In our case, the elements A_1, B_1, C_1, D_1 of the one-way matrix can easily be shown to be linear functions of $1/f$. From Eqs. (5.4.6) and (5.5.5) one sees that, in terms of one-way matrix elements, the cavity stability condition can simply be written as $0 \le A_1 D_1 \le 1$. This condition is represented in Fig. 5.16, where horizontal and vertical axes represent A_1 and D_1, respectively. Since A_1 and D_1 are linear functions of $1/f$, a plot of the values for A_1 versus the corresponding values for D_1, as obtained by changing $1/f$, shows a linear relationship in the $A_1 - D_1$ plane (see Fig. 5.16). This straight line generally intersects the stability boundaries at four distinct points, thus defining two distinct stable zones. The laser beam configurations corresponding to these four limit points can be described by geometrical optics and are also shown in the figure.

Having understood the origin of the two stability zones, we now observe that the dioptric power of an optically pumped rod can be written as:[9]

$$\frac{1}{f} = \frac{k}{\pi a^2} P_a \tag{5.5.33}$$

where P_a is the pump power absorbed in the rod and k is a constant, characteristic of the given material. If the expression for $1/f$, given by Eq. (5.5.33), is substituted into Eq. (5.5.32) and if, according to Eq. (5.5.31), we take $(w_{am}/a) = (2/\pi)$ in the resulting expression, the range of acceptable absorbed power ΔP_a, corresponding to each stability zone, turns out to be a constant for a given laser material (e.g., $\Delta P_a \cong 10\,\text{W}$ for a diode-pumped Nd:YAG).

From the preceding discussion, the optical properties of the two stability zones seem to be identical. A strong distinction between these zones is however revealed when we consider the misalignment properties of the laser cavity. We first define misalignment sensitivities S_1 and S_2 for mirrors 1 and 2 according to the relations $S_1 = \delta r_{c1}/w_a \delta \theta_1$ and $S_2 = \delta r_{c2}/w_a \delta \theta_2$; here δr_{c1} is the displacement of the beam center in the laser rod arising from a tilt $\delta \theta_1$ of mirror 1, and similarly for δr_{c2}. We now define a combined misalignment sensitivity of the two mirrors as $S_c = (S_1^2 + S_2^2)^{1/2}$. A plot of combined sensitivity versus lens dioptric power is also shown in Fig. 5.15. We then see that one of the two zones, henceforth referred to as zone 1, is much less sensitive to mirror misalignment than the other, henceforth referred to as zone 2. The reason for the reduced sensitivity to misalignment in zone 1 is understood by noting that spot sizes at the mirrors are much smaller in zone 1 than in zone 2. Indeed, according to geometrical optics, one of the two stability boundaries of zone 1 corresponds to the beam being focused on both mirrors (see Fig. 5.16) Thus, when close to this boundary, the mirror spot size w_m is very small, so beam divergence $\theta \approx \theta_d = \lambda/\pi w_m$ is very large. Consequently, the mirror tilt required to produce a beam axis rotation comparable to this beam divergence must also be large.

In conclusion, a dynamically and mechanically stable resonator can be designed for a general laser cavity described as in Fig. 5.14b and comprising a variable element, such as the thermally induced lens in the laser rod. The resonator must belong to the more stable zone, zone 1, and it must satisfy Eqs. (5.5.30) and (5.5.31). In practice, instead of satisfying Eq. (5.5.30), the resonator can be designed to correspond to the center of zone 1. If the distance L_1 in Fig. 5.14a is assumed to be the variable parameter, its value can then be taken as the mean of its limiting values L_1' and L_1'' in zone 1. From the geometrical optics description shown in Fig. 5.16, L_1' and L_1'' must accordingly satisfy the conditions $L_1'^{-1} + L_2^{-1} = f^{-1}$ and $(L_1'' - R_1)^{-1} + L_2^{-1} = f^{-1}$, respectively. Once designed for a given focal length f and hence a given absorbed power P_a, the resonator then works for a range of absorbed pump power ΔP_a, whose value, for a given active medium, is independent of cavity parameters.

5.6. UNSTABLE RESONATORS

The stability condition for a generalized spherical resonator was discussed in Sect. 5.4, and the unstable regions were shown to correspond to unshaded regions of the $g_1 - g_2$ plane in Fig. 5.7.[10] Unstable resonators fall into two classes: Positive-branch resonators, which correspond to the case $g_1 g_2 > 1$, and negative-branch resonators, which correspond to the case $g_1 g_2 < 0$.

Before entering into a quantitative discussion of unstable resonators, let us explain why these resonators are of interest in the laser field. First, according to results from Example 5.5, given a stable resonator corresponding to a point in the $g_1 - g_2$ plane that is not close to an instability boundary, the spot size w is typically of the same order as for a confocal resonator; for a wavelength of ≈ 1 μm, it is thus usually smaller than 1 mm. According to the discussion in Example 5.9, a resonator aperture with radius $a < 2$ mm must then be inserted in a laser resonator to limit oscillation to the TEM$_{00}$ mode. When oscillation is confined to a TEM$_{00}$ mode of such a small cross section, the power (or energy) available in the output beam is necessarily limited. For unstable resonators, on the contrary, the field does not tend to be confined to the axis (see for example Fig. 5.3), so a large mode volume in a single transverse mode is expected. With unstable resonators, however, rays tend to walk off out of the cavity. Corresponding modes therefore have substantially greater (geometrical) losses than those in a stable cavity (where losses are due to diffraction). This fact can however be used to advantage if these walk-off losses are converted into useful output coupling.

5.6.1. Geometric Optics Description

To establish the mode configurations of an unstable resonator, we use a geometrical optics approximation, as first done by Siegman.[11] To this purpose we begin by recalling the two main results which were obtained for the eigensolutions of a stable resonator [see Eq. (5.5.24)]: (1) The amplitude is given by the product of a Hermite polynomial with a Gaussian function. (2) The phase distribution gives a spherical wavefront. The presence of the Gaussian function limits the transverse size of the beam; it essentially arises from the focusing properties of a stable spherical resonator. The fact that the wavefront is spherical, on the other hand, is connected to the boundary conditions set by a spherical mirror. In the unstable case there are no Hermite–Gaussian solutions, as discussed in connection with the solution of Eq. (5.5.3). Since the beam is no longer focused toward the resonator axis, but spread out over the whole resonator cross section, it is then natural to assume, as a first approximation, that the solution has constant amplitude over the resonator cross section, while the wavefront is still spherical; i.e., the solution is represented by a spherical wave. More precisely, since the mode can always be considered to be due to the superposition of two counter-propagating waves, we assume that these consist of two counter-propagating *spherical waves*. Note that we can reach the same conclusion by considering the solution of Eq. (5.5.3) in the unstable region. In this case the discriminant of the quadratic equation (5.5.3) is positive, so we generally obtain two real solutions for the parameter q, which just correspond to two spherical waves.

To calculate the mode field, we let P_1 and P_2 be the centers of curvature of the two spherical waves in the general two-mirror unstable resonator in Fig. 5.17a. By symmetry, P_1 and P_2 must lie on the resonator axis. Their positions are then easily calculated by a self-consistency argument: After reflection at mirror 2, the spherical wave originating from point P_1 must produce a spherical wave originating from P_2; conversely, after reflection at mirror 1, the spherical wave originating from P_2 must produce a spherical wave originating from P_1. These two conditions lead to two equations, readily established by straightforward calculations based on geometric optics, in the two unknowns, namely, the positions of points P_1 and P_2. If these positions are expressed in terms of the dimensionless quantities r_1 and r_2 indicated in Fig. 5.17a, these latter quantities can be shown to be functions only of the

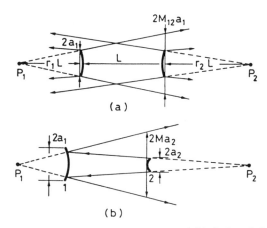

FIG. 5.17. (a) General convex mirror, unstable resonator and (b) single-ended unstable resonator.

resonator g_1, g_2 parameters. In fact, after some lengthy but straightforward calculations, we obtain

$$r_1^{-1} = g_1[1 - (g_1 g_2)^{-1}]^{1/2} + g_1 - 1 \tag{5.6.1a}$$

$$r_2^{-1} = g_2[1 - (g_1 g_2)^{-1}]^{1/2} + g_2 - 1 \tag{5.6.1b}$$

Having calculated r_1 and r_2, we can easily obtain from Fig. 5.17a the so-called single-pass magnification factor, M_{12}, on going from mirror 1 to mirror 2, or, M_{21}, from mirror 2 to mirror 1. For instance M_{12} is defined as the increase in diameter of the spherical wave when propagating from mirror 1 to mirror 2. From simple geometric considerations one obtains from Fig. 5.17a

$$M_{12} = \frac{(1 + r_1)}{r_1} \tag{5.6.2a}$$

Similarly:

$$M_{21} = \frac{(1 + r_2)}{r_2} \tag{5.6.2b}$$

For laser applications, a single-ended resonator, such as that in Fig. 5.17b, is usually of interest. In this case the diameter $2a_1$ of mirror 1 must be greater than the transverse extent, at mirror 1, of the spherical wave originating from point P_2. We thus require $a_1 > M_{21} a_2$. With this condition, the only wave that emerges from the cavity is the spherical wave emitted by point P_1, escaping around mirror 2 (mirrors 1 and 2 are assumed to be 100% reflective). This spherical wave starts from mirror 2 with a diameter $2a_2$ (see Fig. 5.17b), then returns to mirror 2, after one round-trip, magnified by a factor M given by:

$$M = M_{21} M_{12} = (1 + r_1^{-1})(1 + r_2^{-1}) \tag{5.6.3}$$

where Eqs. (5.6.2) have been used. With the help of Eqs. (5.6.1), Eq. (5.6.3) yields

$$M = (2g_1 g_2 - 1) + 2g_1 g_2[1 - (g_1 g_2)^{-1}]^{1/2} \tag{5.6.4}$$

which shows that M, the round-trip magnification factor, depends only on the cavity g parameters. Note that, when $g_1 g_2 < 0$, M becomes negative, the magnitude of this quantity must be considered.

Having calculated the round-trip magnification factor, we easily obtain the expression for the round-trip cavity loss γ, as arising from transmission around the output mirror. In fact, since we assume uniform illumination, the fraction of the beam power coupled from mirror 2 after a round-trip is

$$\gamma = \frac{S_2' - S_2}{S_2'} = \frac{M^2 - 1}{M^2} \tag{5.6.5}$$

where $S_2 = \pi a_2^2$ and $S_2' = \pi M^2 a_2^2$ are, respectively, the cross sections for the beam originating from mirror 2 and after one round trip. Note that the round-trip output-coupling loss γ, like M, is independent of mirror diameter $2a_2$.

Example 5.10. *Unstable confocal resonators.* A particularly important class of unstable resonator is the confocal resonator, which can belong to the negative- or positive-branch, as shown in Figs. 5.18a–b, respectively. In both cases the two mirror foci F_1 and F_2 are coincident. One can show that these resonators are represented, in the $g_1 - g_2$ plane, by the two branches of the hyperbola indicated as dashed curves in Fig. 5.7 [the equation of the hyperbola is $(2g_1 - 1)(2g_2 - 1) = 1$]. Of these various resonators, only the (symmetric) confocal one ($g_1 = g_2 = 0$) and the plane-parallel one ($g_1 = g_2 = 1$) lie on the boundary between the stable and unstable regions. All other confocal resonators are unstable and they may belong to either the negative or positive branch of the instability region. As shown in Fig. 5.18 [and as we can also show from Eq. (5.6.1)], the mode consists of a superposition between a plane wave and a spherical wave originating from the common focus $F_1 = F_2$. The round-trip magnification factor M is simply given by $M = |R_1|/|R_2|$, where R_1 and R_2 are the curvature radii of the two mirrors ($|R_1| > |R_2|$). If the aperture, of diameter $2a_1$, at mirror 1 is made sufficiently large ($2a_1 > 2Ma_2$), only the plane beam escapes from the cavity. Thus, the beam escaping from a single-ended confocal resonator is a plane wave, which constitutes one of the main advantages of unstable confocal resonators. The round-trip loss, or fractional output coupling, of this single-ended resonator is then given by Eq. (5.6.5).

5.6.2. Wave Optics Description

The discussion so far is based on a geometrical-optics approximation. For a more realistic picture of the modes of an unstable resonator, we must use a wave approach, e.g., the integral equation (5.2.5), which arises from the Huygens–Fresnel diffraction equation (5.2.1). For unstable resonators, the limited aperture size of the output mirror constitutes an essential feature, since the beam must exit around this mirror. Consequently, the kernel K used in Eq. (5.2.5) can be obtained, in principle, by essentially the same procedures as those developed, for a stable cavity, in Sect. 5.5.2. Then the integral equation can be solved by an iterative approach, such as the Fox–Li procedure discussed in Sect. 5.5.2. We do not discuss these calculations at any length here, and we limit ourselves to pointing out and commenting on a few relevant results.

A first important result is that the wave optics description shows that eigensolutions, i.e., field profiles that are self-reproducing after one round-trip, also exist for unstable resonators. To show this in some detail, we limit our discussion to a single-ended confocal resonator and define an equivalent Fresnel number as $N_{eq} = [(M - 1)/2] \times (a_2^2/L\lambda)$, for the positive branch, and as $N_{eq} = [(M + 1)/2] \times (a_2^2/L\lambda)$, for the negative-branch, with $2a_2$ being the diameter of the output mirror. A typical example of a computed plot of the radial intensity

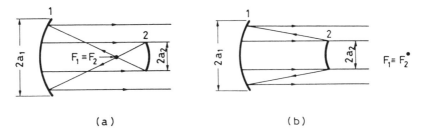

(a)   (b)

FIG. 5.18. Confocal unstable resonators: (a) negative branch and (b) positive branch.

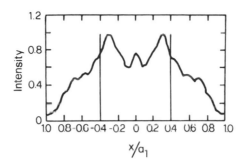

FIG. 5.19. Typical example of the radial behavior of mode intensity distribution in an unstable cavity obtained using a wave-optics calculation. (By permission from Ref. 12.)

profile, which is self-reproducing after one trip, is then shown in Fig. 5.19. The calculation refers to a positive-branch confocal resonator with $M = 2.5$ and $N_{eq} = 0.6$; the intensity profile refers to the field, just in front of mirror 2 (Fig. 5.18b), of a beam propagating to the right inside the resonator. The intensity profile in Fig. 5.19 is plotted against the x (or y) transverse coordinate normalized to the radius a_1 of mirror 1. To ensure a single-ended output, the condition $a_1 = 2.5\, a_2$ is assumed. Consequently, the vertical lines in the figure, occurring at $(x/a_1) = \pm 0.4$, mark the edge of the output mirror.

From Fig 5.19 we can note a peculiar meaning of a round-trip self-reproducing profile for unstable resonators. Starting in fact from mirror 2, the left-propagating spherical wave (see Fig. 5.18b) arises only from that part of the beam in Fig. 5.19 where $-0.4 \leq (x/a_1) \leq 0.4$. In fact the remaining part of the beam escapes around mirror 2 to form the output beam. The part remaining in the resonator, after propagation over a round trip, through the combined effect of spherical divergence and beam diffraction, again reproduces the *whole intensity profile* of Fig. 5.19. The amplitude of the beam profile, after one round-trip, is of course smaller than the original value due to the loss represented by that part of the beam transmitted through mirror 2. We can also note that the beam intensity profile in Fig. 5.19 is quite different from the constant value assumed in the geometrical optic theory; the difference is due to field diffraction, in particular from the edges of mirror 2. We see, in fact, from Fig. 5.19 that, if x is interpreted as the radial distance from the mirror's center, several diffraction rings arising from the sharp edges of mirror 2 are present in the beam. Despite this significant difference between the intensity profile predicted by wave optics and that predicted by geometrical optics, phase variation is found remarkably similar in the two cases. In fact, the wavefront is close to spherical with radius almost equal to that predicted by geometrical optics (i.e., plane in this case).

A second result of the wave optics calculation is that, for unstable resonators as for stable, there are different transverse modes, i.e., different self-reproducing spatial patterns. These modes generally differ from each other in location and strength of the diffraction rings. An example of three such modes for a positive-branch confocal unstable resonator is shown in Fig. 5.20. Unlike the case of stable resonators, it is not possible in this case to distinguish clearly, in terms of these field distributions, between lowest order and higher order modes. Note that the mode labeled $l = 0$ in the figure shows a field amplitude distribution that is more concentrated toward the beam axis. Thus, in this case, this mode has the lowest loss, i.e., it is the 'fundamental' mode.

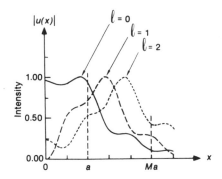

FIG. 5.20. Intensity profile of the three lowest order eigenmodes for a strip unstable resonator with $M = 2.5$ and $N_{eq} = 0.6$. (By permission from Ref. 10.)

A third result is found when changing the equivalent Fresnel number, i.e., changing M or a_2, or L. In fact, at each integer value of the equivalent Fresnel number, a different and distinct mode becomes the lowest-order, i.e., the lowest loss mode. This can be understood with the help of Fig. 5.21, where the magnitude of the eigenvalue σ is plotted vs N_{eq} for the three modes indicated in Fig. 5.20. Note in fact that, since $\gamma = 1 - |\sigma|^2$, the $l = 1$ mode becomes the lowest order mode when N_{eq} becomes larger than 1 (and smaller than 2). This occurs because, as N_{eq} increases starting, e.g., from the value $N_{eq} = 0.6$ in Fig. 5.20, the mode $l = 1$ contracts inward while the mode $l = 0$ spreads outward, so that, at $N_{eq} \cong 1$, the role of the two modes is interchanged. We notice also from Fig. 5.21 that, at each half-integer value of N_{eq}, there is a large difference between losses of the lowest order mode and those of other modes. This seems to suggest that a large transverse mode discrimination is obtained only under these conditions. Note however that, when the loss curves of two modes cross each other (i.e., for integer values of N_{eq} in Fig. 5.21), intensity patterns of these two modes become identical. Thus at $N_{eq} = 1$, e.g., a large difference in loss exists between the $l = 2$ mode and the $l = 0$ and $l = 1$ modes, which, in terms of transverse beam profile, can

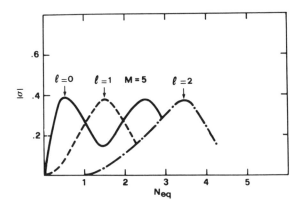

FIG. 5.21. Typical example of the oscillatory behavior of eigenvalue magnitude σ versus the equivalent Fresnel number, N_{eq}, for the three consecutive modes in Fig. 5.20.

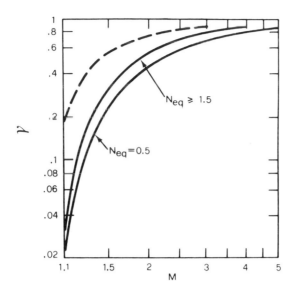

FIG. 5.22. Coupling losses of an unstable resonator versus the magnification factor M. (By permission from Ref. 13).

be considered as effectively corresponding to the same mode.* We conclude that unstable resonators always have a large transverse-mode discrimination, with discrimination being perhaps strongest at half-integer values of N_{eq}.

From the wave optics calculation and for half-integer values of N_{eq}, we obtain a loss of the lowest order mode considerably smaller than the value predicted by geometrical optics. This result is apparent from Fig. 5.22, where the loss γ is plotted against the round-trip magnification factor M. In the figure, the solid curves (which apply to successive half-integer values of N_{eq}) are obtained by wave optics, while the dashed curve corresponds to the geometrical optics results given by Eq. (5.6.5). In any case, the loss of the lowest order mode is seen to be smaller than the value predicted by geometrical optics. This result stems from the fact that, rather than having the constant value predicted by geometrical optics, the intensity distribution of the lowest order mode tends to be more concentrated toward the beam axis (see Fig. 5.20).

5.6.3. Advantages and Disadvantages of Hard-Edge Unstable Resonators

The main advantages of hard-edge unstable resonators as compared to stable resonators can be summarized as follows: (1) Large, controllable mode volume. (2) Good transverse mode discrimination. (3) All reflective optics (which is particularly attractive in the infrared, where metallic mirrors can be used).

The main disadvantages are: (1) The output beam cross section is in the form of a ring (i.e., it has a dark hole at its center). For example, in a confocal resonator (Fig. 5.18), the inner diameter of the ring is $2a_2$, while its outer diameter is $2Ma_2$. Although this hole

---

* The two modes still differ with respect to the total round-trip phase shift, i.e., they still differ in field variation along the longitudinal z-axis and thus in their resonance frequencies.

disappears at the focal plane of a lens which may be used to focus the beam (far-field pattern), the peak intensity in this focal plane decreases with decreasing ring thickness. In fact, for a given total power, the peak intensity for an annular beam is found to be reduced, compared to that of a uniform-intensity beam with diameter equal to the large diameter of the annular beam, by $(M^2 - 1)/M^2$. (2) The intensity distribution in the beam, does not follow a smooth curve but exhibits diffraction rings. (3) An unstable resonator is more sensitive to cavity perturbations compared to a stable resonator.

These advantages and disadvantages mean that unstable resonators can conveniently be used in high-gain lasers (so that M can be relatively large), especially in the infrared, and when high-power (or high-energy) diffraction-limited beams are required.

5.6.4. Unstable Resonators with Variable-Reflectivity Mirrors

Some, if not all, of the disadvantages of hard-edge unstable resonators can be overcome by using a variable-reflectivity unstable resonator. In this case, rather than being equal to 1 for $r < a_2$ and equal to zero for $r > a_2$ as in the hard-edge case, the reflectivity of the output mirror decreases radially from a peak value R_0 to zero over a radial distance comparable to that of the active medium.[14] We let $\rho(r)$ be the field reflectivity of mirror 2 and assume a single-ended resonator with round-trip magnification M. For simplicity we follow an approach based on geometrical optics. In terms of the radial coordinate r, we then say that the field $u_2'(Mr)$, incident after one round-trip at coordinate Mr of mirror 2, comes from the field $u_2(r)$ of the beam incident, at coordinate r of mirror 2, at the start of the round-trip. After taking into account the field reflectivity profile of mirror 2 and the round-trip magnification M, we then write

$$u_2'(Mr) = \frac{\rho(r)u_2(r)}{M} \tag{5.6.6}$$

The term M on the right-hand side of Eq. (5.6.6) is readily understood upon observing that, after magnification by a factor M, the beam area increases by a factor M^2. To conserve the power of the beam, the intensity must therefore decrease by a factor M^2 and the field by a factor M.

If u_2 corresponds to a cavity mode, then one must have $u_2'(r) = \sigma u_2(r)$, where σ is now a real quantity with magnitude smaller than unity to account for cavity losses. From Eq. (5.6.6) we obtain

$$\sigma u_2(Mr) = \frac{\rho(r)u_2(r)}{M} \tag{5.6.7}$$

The eigensolutions $u_2(r) = u_{2l}(r)$ of Eq. (5.6.7) give the field distributions inside the cavity in front of mirror 2. The eigenvalues then give the round-trip losses, due to output coupling, according to the familiar relation [see Eq. (5.2.6)]:

$$\gamma = 1 - \sigma^2 \tag{5.6.8}$$

The first case we consider involves a Gaussian reflectivity profile.[11,12] We therefore write

$$\rho = \rho_0 \exp(-r^2/w_m^2) \tag{5.6.9}$$

where ρ_0 is the peak field reflectivity and w_m sets the transverse scale of the mirror reflectivity profile. Note that, according to Eq. (5.6.9), the reflectivity profile of the intensity, which is the quantity usually measured experimentally, is given by:

$$R = R_0 \exp(-2r^2/w_m^2) \qquad (5.6.10)$$

where $R_0 - \rho_0^2$ is the peak reflectivity. With the help of Eq. (5.6.9), the lowest order solution of Eq. (5.6.7) can be shown by direct substitution to be given by:

$$u_{20}(r) = u_{20}(0) \exp(-r^2/w^2) \qquad (5.6.11)$$

where:

$$w^2 = (M^2 - 1)w_m^2 \qquad (5.6.12)$$

The corresponding eigenvalue σ is

$$\sigma = \frac{\rho_0}{M} \qquad (5.6.13)$$

Then, according to Eq. (5.6.8), the output coupling losses are given by:

$$\gamma = 1 - \left(\frac{R_0}{M^2}\right) \qquad (5.6.14)$$

The radial intensity distribution for the beam incident on mirror 2 is given by:

$$I_{in}(r) = I_{in}(0) \exp(-2r^2/w^2) \qquad (5.6.15)$$

Note that the radial profiles of both the field amplitude u_{20} and beam intensity I_{in} are described by Gaussian functions. On the other hand, the intensity of the output beam I_{out} is given by:

$$I_{out}(r) = I_{in}(r)[1 - R(r)] = I_{in}(0)[\exp(-2r^2/w^2) - R_0 \exp(-2M^2r^2/w^2)] \qquad (5.6.16)$$

where Eqs. (5.6.10), (5.6.12), are (5.6.15) have been used. Thus I_{out} is not described by a Gaussian function and, under appropriate conditions, we can expect an intensity profile with a flat top for $r = 0$, a feature of interest in some applications. This circumstance occurs in fact when $(d^2 I_{out}/dr^2)_{r=0} = 0$. In this case, we find from Eq. (5.6.16) that the central reflectivity R_0 and the cavity magnification M must satisfy the condition:

$$R_0 M^2 = 1 \qquad (5.6.17)$$

For this resonator, the round-trip cavity losses, according to Eqs. (5.6.14) and (5.6.17), are given by:

$$\gamma = 1 - \left(\frac{1}{M^4}\right) \qquad (5.6.18)$$

The preceding equations give the salient results for unstable resonators with mirrors of Gaussian reflectivity profile. Although these results are based on a simple geometrical-optics approach, they are in good agreement with results based on a wave-optics approach for sufficiently large values of the equivalent Fresnel number ($N_{eq} \geq 5$).[15] For Gaussian-reflectivity mirrors, we can also use an elegant wave-optics analysis based on a suitable *ABCD* matrix with complex matrix elements.[16]

Example 5.11. *Design of an unstable resonator with an output mirror having a Gaussian radial reflectivity profile.* We assume $\gamma = 0.5$ as the value that optimizes the output coupling of a given laser (see Chap. 7) and we consider the case where the output beam has its flattest profile. From Eq. (5.6.18) we obtain $M^2 = \sqrt{2}$, and from Eq. (5.6.17):

$$R_0 = \frac{1}{M^2} = \frac{1}{\sqrt{2}} = 0.71$$

From Eq. (5.6.12), we get $w^2 = 0.41\, w_m^2$. The reflectivity profile and the corresponding intensity profiles inside and outside the resonator are shown in Fig. 5.23. If we now let a equal the radius of the active medium and assume the medium be placed in front of mirror 2, the beam intensity profile within the medium is $I_{in}(r)$. To avoid excessive beam truncation by the active medium aperture, i.e., to avoid excessively pronounced diffraction rings arising from this truncation, we can, e.g., impose the condition $I_{in}(a)/I_{in}(0) = 2 \times 10^{-2}$. We then obtain $a \cong 0.9\, w_m$, which, for a given aperture a, establishes the spot size w_m of the Gaussian reflectively profile. As an example, if we take $a = 3.2\,\text{mm}$, we obtain $w_m = 3.5\,\text{mm}$. Thus, to conclude, the Gaussian mirror must have a peak reflectivity of $R_0 \cong 71\%$, a spot size of $w_m = 3.5\,\text{mm}$, and it must be used in an unstable cavity (e.g., a confocal cavity) with a round-trip magnification of $M = (2)^{1/4} = 1.19$.

The second case we consider involves a super-Gaussian reflectivity profile.[17] Instead of Eqs. (5.6.9) and (5.6.10) we write

$$\rho = \rho_0 \exp(-r^n/w_m^n) \tag{5.6.19a}$$

$$R = R_0 \exp(-2r^n/w_m^n) \tag{5.6.19b}$$

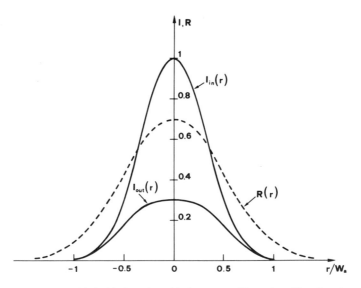

FIG. 5.23. Radial intensity profile inside I_{in} and outside I_{out} an unstable cavity with a Gaussian reflectivity output coupler $R(r)$ (case of the flattest profile for I_{out}).

For $n > 2$, Eqs. (5.6.19) are seen to describe curves with a super-Gaussian reflectivity profile. The substitution of Eq. (5.6.19a) into Eq. (5.6.7) then gives

$$u_{20}(r) = u_{20}(0) \exp(-r^n/w^n) \qquad (5.6.20)$$

where:

$$w = w_m (M^n - 1)^{1/n} \qquad (5.6.21)$$

Again we have $\sigma - \rho_0/M$ and $\gamma = 1 - \sigma^2 = 1 - (R_0/M^2)$. From Eq. (5.6.20) we now obtain

$$I_{in}(r) = I_{in}(0) \exp(-2r^n/w^n) \qquad (5.6.22)$$

Radial profiles of both u_{20} and I_{in} are described by super-Gaussian functions of the same order n as that of the reflectivity profile. The intensity of the output beam I_{out}, on the other hand, is readily available from $I_{out} = I_{in}(r)[1 - R(r)]$ and it is not described by a super-Gaussian function.

To compare the performance of unstable resonators with Gaussian and super-Gaussian reflectivity profiles, in Fig. 5.24a we show intensity profiles I_{in} for $n = 2$ (Gaussian) and $n = 5, 10$ (super-Gaussian). The curves are normalized to their peak values, and the corresponding spot sizes w, in Eqs. (5.6.22) and (5.6.15), are chosen so that $\exp[-(2a^n/w^n)] = 2 \times 10^{-2}$, where a is the radius of the active medium. The comparison is therefore made for the same degree of beam truncation by the active medium. The main advantage of a super-Gaussian mirror compared to a Gaussian mirror is now apparent from Fig. 5.24a: Super-Gaussian mirrors of increasing super-Gaussian order n allow better exploitation of the active medium (i.e., the area of the mode A_m increases as n is increased). One can observe however that the diffraction angle θ_d increases with increasing n, as it can be understood with the help of Fig. 5.24b. This figure shows, in fact, the corresponding radial intensity profiles, predicted by previous equations for $I_{out}(r)$, when $R_0 = 0.45$, and $M = 1.8$. We see that, as n increases, a hole of increasing depth appears in the output beam; this results in an increased beam divergence. As a consequence of these two conflicting

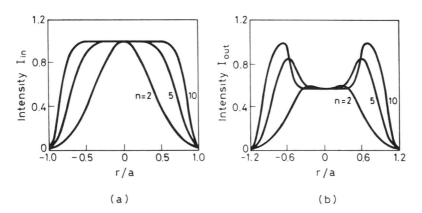

FIG. 5.24. Comparison between a Gaussian and super-Gaussian ($n = 5, n = 10$) reflectivity profile: (a) radial intensity profile inside the resonator and (b) radial intensity profile outside the resonator. (By permission from Ref. 18.)

tendencies, beam brightness, which may be taken to be proportional to A_m/θ_d^2, has an optimum value as a function of n. This optimum value depends on cavity round-trip magnification M and peak mirror reflectivity R_0, but, for all practical cases, it ranges from 5–8.[18] Thus in terms of beam brightness, super-Gaussian mirrors with super-Gaussian order $n = 5 \div 8$ provide the best choice for a variable-reflectivity unstable resonator.

5.7. CONCLUDING REMARKS

Chapter 5 explores some of the most relevant features of stable and unstable resonators. In particular, to obtain single-transverse-mode oscillation, we can use stable resonators provided that the Fresnel number is smaller than 2. For example, this means that the radius of the limiting aperture (e.g., the radius of the active medium) must be smaller than ~ 2 mm at $\lambda = 1.06 \, \mu m$ and ~ 6.5 mm at $\lambda = 10.6 \, \mu m$. For larger values of the active-medium transverse-dimensions, the use of unstable resonators is indicated. In this case, radially-variable-reflectivity output mirrors of Gaussian or, better, super-Gaussian profile, provide the best solution.

PROBLEMS

5.1. A two-mirror resonator is formed by a convex mirror of radius $R_1 = -1$ m and a concave mirror of radius $R_2 = 1.5$ m. What is the maximum possible mirror separation if this is to remain a stable resonator?

5.2. Consider a confocal resonator of length $L = 1$ m used for an Ar^+ laser at wavelength $\lambda = 514.5$ nm. Calculate (a) the spot size at the resonator center and on the mirrors; (b) the frequency difference between consecutive longitudinal modes; (c) the number of non-degenerate modes falling within the Doppler-broadened width of the Ar^+ line ($\Delta \nu_0^* = 3.5$ GHz; see Table 2.2).

5.3. Consider a hemiconfocal resonator (plane-spherical resonator with $L = R/2$) of length $L = 2$ m used for a CO_2 laser at wavelength $\lambda = 10.6 \, \mu m$. Calculate (a) the location of the beam waist, (b) the spot size on each mirror, (c) the frequency difference between two consecutive TEM_{00} modes, and (d) the number of TEM_{00} modes falling within the gain bandwidth. (Consider a typical low-pressure CO_2 laser and thus take $\Delta \nu \simeq 50$ MHz).

5.4. Consider a resonator consisting of two concave spherical mirrors both with radius of curvature 4 m and separated by a distance of 1 m. Calculate the spot size of the TEM_{00} mode at the resonator center and on the mirrors when the laser oscillation is at the Ar^+ laser wavelength $\lambda = 514.5$ nm.

5.5. How is the spot size modified at each mirror if one of the mirrors in Problem 5.4 is replaced by a plane mirror?

5.6. Using Eqs. (4.7.26) and (5.5.8a), show that the beam waist for the two-mirror resonator in Fig. 5.8b–c occurs at a distance z_1 from mirror 1 given by $z_1 = (1 - g_1)g_2L/(g_1 + g_2 - 2g_1g_2)$.

5.7. One of the mirrors in the resonator in Problem 5.4 is replaced by a concave mirror of 1.5-m radius of curvature. Using the results from Problem 5.6, calculate (a) the position of the beam waist and (b) the spot size at the beam waist and on each mirror.

5.8. A resonator consists of two plane mirrors with a positive lens inserted between them. If the focal length of the lens is f, and L_1, L_2 represent the distances of the lens from the two mirrors, calculate (a) the spot size at the lens position and the spot sizes at each mirror and (b) the conditions under which the cavity is stable.

5.9. A triangular ring cavity is made up of three plane mirrors (Fig. 5.4a) with a positive lens inserted between two of the mirrors. If L_p is the length of the ring perimeter, calculate the position of minimum spot size, its value, and the spot size at the lens position. Also find the stability condition for this cavity.

5.10. A laser operating at $\lambda = 630$ nm has a power gain of 2×10^{-2} per pass, and it is provided with a symmetric resonator consisting of two mirrors each of radius $R = 10$ m and separated by $L = 1$ m. Choose an appropriate mirror-aperture size to suppress TEM_{01} mode operation while allowing TEM_{00} mode operation.

5.11. Due to its relatively small sensitivity to mirror misalignment (see Problem 5.16), a nearly hemispherical resonator (i.e., a plane spherical resonator with $R = L + \Delta$ and $\Delta \ll L$) is often used for a He-Ne laser at $\lambda = 630$ nm wavelength. If the cavity length is $L = 30$ cm, calculate (a) the radius of curvature of the spherical mirror so that the spot size at this mirror is $w_m = 0.5$ mm, (b) the location in the $g_1 - g_2$ plane corresponding to this resonator, and (c) the spot size at the plane mirror.

5.12. Consider the nearly hemispherical He-Ne resonator in Problem 5.11 and assume that the aperturing effect produced by the bore of the capillary containing the He-Ne gas mixture (see Chap. 10) can be simulated by a diaphragm of radius a in front of the spherical mirror. If the power gain per pass of the He-Ne laser is taken to be 2×10^{-2}, calculate the diaphragm radius needed to suppress the TEM_{01} mode. (Hint: Show that round-trip loss of this resonator is the same as single-pass loss of a near-concentric symmetric resonator of length $L_{nc} = 2L$ and $R_1 = R_2 = R$. To calculate diffraction loss, use Fig. 5.13, assuming the loss for a resonator of negative g-value to be equal that of the corresponding resonator with corresponding positive g-value).

5.13. Consider the $A_1 B_1 C_1 D_1$ matrix in Fig. 5.8d and show that for a stable cavity, we must have $0 < A_1 D_1 < 1$ and $-1 < B_1 C_1 < 0$. From these results show that $B_1 D_1 / A_1 C_1 < 0$, so that q_1 in Eq. (5.5.6a) is purely imaginary.

5.14. For a stable two-mirror resonator, we can define a misalignment sensitivity δ as the transverse shift of the intersection of the optical axis with a given mirror, normalized to the spot size on that mirror, for a unit angular tilt of one of the two mirrors. In particular, for mirror 1, we can define two misalignment sensitivity factors δ_{11} and δ_{12} as $\delta_{11} = (1/w_1)(dr_1/d\theta_1)$ and $\delta_{12} = (1/w_1)(dr_1/d\theta_2)$, where $dr_1/d\theta_i$ $(i = 1, 2)$ is the transverse change of beam center at mirror 1 for unit angular tilt of either mirror 1 or 2. Show that, for a confocal resonator, one has $(\delta_{11})_c = 0$ and $(\delta_{12})_c = (\pi w_s/\lambda)$.

5.15. Using definitions given in Problem 5.14 show that, for a near-plane symmetric resonator, the misalignment sensitivity is such that $\delta_{11} = \delta_{12} = \delta_{21} = \delta_{22} = (\delta_{12})_c 4w^3/w_s^3$, where $(\delta_{12})_c$ is the misalignment sensitivity of the confocal resonator, w is the spot size on the mirror for the actual resonator, and w_s is the mirror spot size of a confocal resonator of the same length. According to the preceding equation, which of the two resonators is less sensitive to mirror tilt?

5.16. Consider a nearly hemispherical resonator ($R = L + \Delta$ with $\Delta \ll L$) where mirror 1 is the plane mirror. Show that in this case $\delta_{12} = (\delta_{12})_c(w_2/w_s)$ and $\delta_{21} = (\delta_{12})_c(w_s/w_2)$. Comparing this resonator with the long-radius resonator in Problem 5.15, for the same value of mirror spot size, i.e., $w = w_2$, what conclusion can be drawn with regard to the misalignment sensitivity of a nearly hemispherical resonator compared to that of a nearly flat resonator?

5.17. An unstable resonator consists of a plane mirror (mirror 1) and a convex mirror (mirror 2) of radius of curvature $R_2 = 2\,\text{m}$, spaced by a distance $L = 50\,\text{cm}$. Calculate (a) the resonator location in the g_1, g_2 plane, (b) the location of points P_1 and P_2 in Fig. 5.17, (c) the condition under which the resonator is single-ended with beam output only occurring around mirror 2, and (d) the round-trip magnification factor and round-trip losses.

5.18. A confocal unstable resonator is to be used for a CO_2 laser at a wavelength of $\lambda = 10.6\,\mu\text{m}$. The resonator length is chosen to be $L = 1\,\text{m}$. Which branch should be chosen for this resonator if the mode volume is to be maximized? Calculate mirror apertures $2a_1$ and $2a_2$ so that (a) $N_{eq} = 7.5$, (b) single-ended output is achieved, and (c) a 20% round-trip output coupling is obtained. Then find the radii of the two mirrors R_1 and R_2.

5.19. Using a geometric optics approach, calculate the round-trip loss of the resonator designed in Problem 5.18. What are the shape and dimensions of the output beam?

5.20. Consider an unstable resonator consisting of a convex mirror (mirror 1) of radius R_1 and a plane mirror (mirror 2) separated by a distance $L = 50\,\text{cm}$. Assume that the plane mirror has a super-Gaussian reflectivity profile with a super-Gaussian order $n = 6$ and peak power reflectivity $R_0 = 0.5$. Assume also that the active medium consists of a cylindrical rod (e.g., a Nd:YAG rod) with radius $a \cong 3.2\,\text{mm}$ placed just in front of mirror 2. To limit round-trip losses to an acceptable value, assume also a round-trip magnification $M = 1.4$. Calculate (a) the spot size w of the field intensity I_{in} for a 2×10^{-2} intensity truncation by the active medium, (b) the corresponding mirror spot size w_m, (c) the cavity round-trip losses, and (d) the radius of curvature of the convex mirror.

REFERENCES

1. A. E. Siegman, *Lasers* (Cambridge University, Oxford, 1986), Sect. 14.2.
2. M. Born and E. Wolf, *Principles of Optics*, 6th ed. (Pergamon, London, 1980), Sect. 1.6.5.
3. H. Kogelnik and T. Li, Laser Beams and Resonators, *Appl. Opt.* **5**, 1550 (1966).
4. Ref. 1, Chap. 19.
5. A. G. Fox and T. Li, Resonant Modes in a Maser Interferometer, *Bell Syst. Tech. J.* **40**, 453 (1961).
6. T. Li, Diffraction Loss and Selection of Modes in Maser Resonators with Circular mirrors, *Bell System Tech. J.* **44**, 917 (1965).
7. V. Magni, Resonators for Solid-State Lasers with Large-Volume Fundamental Mode and High-Alignment Stability, *Appl. Opt.* **25**, 107–117 (1986). See also *erratum Appl. Opt.* **25**, 2039 (1986).
8. Ref. 1, p. 666.
9. W. Koechner, *Solid-State Laser Engineering*, vol. 1, *Springer Series in Optical Sciences*, 4th ed. (Springer-Verlag, New York, 1996).
10. Ref. 1, Chap. 22.
11. A. E. Siegman, Unstable Optical Resonators for Laser Applications, *Proc. IEEE* **53**, 277–287 (1965).
12. D. B. Rensch and A. N. Chester, Iterative Diffraction Calculations of Transverse Mode Distributions in Confocal Unstable Laser Resonators, *Appl. Opt.* **12**, 997 (1973).
13. A. E. Siegman, Stabilizing Output with Unstable Resonators, *Laser Focus* **7**, 42 (May 1971).
14. H. Zucker, Optical Resonators with Variable Reflectivity mirrors, *Bell System Tech. J.* **49**, 2349 (1970).
15. A. N. Chester, Mode Selectivity and mirror Misalignment Effects in Unstable Laser Resonators, *Appl. Opt.* **11**, 2584 (1972).
16. Ref. 1, Sect. 23.3.
17. S. De Silvestri, P. Laporta, V. Magni, and O. Svelto, Solid-State Unstable Resonators with Tapered Reflectivity mirrors: the Super-Gaussian Approach, *IEEE J. Quantum Electr.* **QE-24**, 1172 (1988).
18. G. Cerullo *et al.*, Diffraction-Limited Solid-State Lasers with Super-Gaussian mirror, in *OSA Proc. On Tunable Solid-State Lasers*, vol. 5 (M. Shand and H. Jenssen, eds.) (Optical Society of America, Washington, DC, 1989), pp. 378–384.

6

Pumping Processes

6.1. INTRODUCTION

We saw in Chap. 1 that the process by which atoms are raised from level 1 to level 3 (for a three-level laser, Fig. 1.4a) or from level 0 to level 3 (for a four-level or a quasi-three-level laser, Fig. 1.4b) is referred to as the pumping process. Usually it is performed in one of the following ways: *Optically*, i.e., by the cw or pulsed light emitted by a powerful lamp or a laser beam; *electrically*, i.e., by a cw, radio-frequency, or pulsed current flowing in a conductive medium, such as an ionized gas or a semiconductor.

In *optical pumping* by an incoherent source, light from a powerful lamp is absorbed by the active medium, so that atoms are pumped into the upper laser level. This method is particularly suited to solid-state or liquid lasers (i.e., dye lasers). Line-broadening mechanisms in solids and liquids produce, in fact, considerable broadening, so that we usually deal with pump bands rather than sharp lines. These bands can therefore absorb a sizable fraction of the, usually broadband, light emitted by the lamp. The availability of efficient and powerful cw or pulsed laser sources at many wavelengths has made *laser pumping* both attractive and practical. In this case the narrow line emitted from a suitable laser source is absorbed by the active medium. This requires the laser wavelength to fall within one of the medium's absorption bands. Note that laser monochromaticity implies that laser pumping needs not be limited to solid-state and liquid lasers; it can also be applied to gas lasers if we can ensure that the line emitted by the pumping laser coincides with an absorption line of the medium to be pumped. This situation occurs for instance in most far-infrared gas lasers (e.g., methyl alcohol or CH_3OH in the vapor state), which are usually pumped by a suitable rotational–vibrational line of a CO_2 laser. For solid-state or liquid lasers, on the other hand, argon ion lasers for cw excitation, nitrogen or excimer lasers for pulsed excitation, and Nd:YAG lasers and their second and third harmonics (either cw or pulsed), are often used. Whenever possible, however, semiconductor diode lasers, due to the inherently high efficiency of these laser sources (overall optical-to-electrical efficiencies larger than 60% have been demonstrated), are now commonly used (*diode laser pumping*).

201

Actually one can foresee that diode laser pumping will become the dominant means of optical pumping, replacing even high-power lamps.

Electrical pumping is usually accomplished by means of a sufficiently intense electrical discharge; it is particularly suited to gas and semiconductor lasers. Gas lasers do not usually lend themselves to lamp pumping because their absorption lines are typically much narrower than the usual broadband emission of a lamp. A notable exception is the optically pumped Cs laser, where Cs vapor is pumped by a lamp containing low-pressure He. In this case the situation is quite favorable for optical pumping, since the strong ~ 390-nm He emission line (which is rather sharp due to the low pressure used) happens to coincide with an absorption line of Cs. This laser however is no longer used and its importance resides mostly in its historic significance as the most notable lamp-pumped gas laser and, particularly, as the earliest proposed laser scheme. Electrical pumping of gas lasers, on the other hand, can be a fairly efficient processs (e.g., for pumping the CO_2 laser) because the linewidth of the excitation cross section of a given transition by electron impact is usually quite large (from a few to a few tens of eV; see Figs. 6.25 and 6.26). This occurs because electron impact excitation, namely, $e + A \rightarrow A^* + e$ is a nonresonant process. Surplus energy beyond that needed to excite species A is in fact left as kinetic energy of the scattered electron. In contrast, the process of optical excitation by an incoming photon of energy $h\nu$, namely, $h\nu + A \rightarrow A^*$, is a resonant process because photon energy must equal excitation energy of species A. As discussed in Chap. 2, some line-broadening processes occur in this case due to energy arising, e.g., from thermal movement of species A (as in Doppler broadening), which can be added to the process. The resulting width of the absorption line, however, is quite small (e.g., $\approx 10^{-5}$ eV for Doppler broadening of Ne atoms); this is the fundamental reason why optical pumping by a broadband source is inefficient for a gas laser. Optical pumping can be used very effectively for semiconductor lasers, since the semiconductor medium has a strong and broad absorption band. Indeed a number of optically pumped semiconductor lasers (particularly by laser pumping) have been operated. Electrical pumping proves to be more convenient, however, since a sufficiently large current density can be made to flow through a semiconductor, usually in the form of a p-n or p-i-n diode.

The two pumping processes just discussed, optical pumping and electrical pumping, are not the only ones available for pumping lasers. A form of pumping somewhat similar to optical pumping involves a medium excited by a beam from an x-ray source (*x-ray pumping*). Likewise, a pumping process somewhat similar to electrical pumping involves a medium excited by a beam of electrons from an electron beam machine (*e-beam pumping*). Although both x-ray and e-beam pumping deliver high pump powers or energies in a large volume of active medium (generally in gaseous form), these pumping mechanisms are not widely used due to the complexity of the x-ray or e-beam apparatus. Note, in this context, that possibly the shortest wavelength so far achieved in a laser ($\lambda \cong 1.4$ nm, i.e., around the boundary between soft and hard x-ray regions) was obtained using intense x-rays produced by a small nuclear detonation. Details of this laser are still classified, but we can readily appreciate that this pumping configuration is not easily duplicated in the typical laboratory!

A conceptually different and rather interesting type of pumping occurs when the required inversion is produced directly from an exothermic chemical reaction (*chemical pumping*). Two such reactions can be used, namely: Associative reactions, i.e., $A + B \rightarrow (AB)^*$, which results in the molecule AB being left in an excited vibrational state; dissociative reactions, where dissociation may be induced by a photon, i.e, $AB + h\nu \rightarrow$

$A + B^*$, which results in species B (atom or molecule) being left in an excited state. Chemical pumping usually applies to materials in the gas phase, and it generally requires highly reactive and often explosive gas mixtures. Energy available in an exothermic reaction is often quite large, and high power (for cw operation) or energies (for pulsed operation) can be available for laser action, if a good fraction of the available energy is converted into laser energy. These features have enabled chemical lasers to produce the largest cw laser power so far available (2.2 MW for the so-called MIRACL laser, an acronym for mid-infrared advanced chemical laser). Given the handling problems associated with reactive and hazardous materials, these lasers have been confined to the military field, for use as directed energy weapons.

Another conceptually different type of pumping mechanism for gas molecules involves supersonic expansion of a gas mixture containing the particular molecule (*gas dynamic pumping*). In this case a suitable mixture, usually involving the CO_2 molecule as the active species (e.g., $CO_2:N_2:H_2O$ in the 6:76:1 partial-pressure ratio), is used. The mixture is raised, in a suitable container, to a high pressure (e.g., ≈ 17 atm) and temperature (e.g., $\approx 1,400$ K) by combustion of appropriate fuels (e.g., benzene, C_6H_6, and nitrous oxide, N_2O, thus automatically supplying hot CO_2 with a CO_2/H_2O ratio of 2:1). The CO_2 molecule in this mixture is of course not inverted, but due to the high temperature, a substantial fraction of molecules is found in the lower laser level ($\approx 25\%$), while a lower but still consistent fraction is found in the upper laser level ($\approx 10\%$). Note, in fact, that the CO_2 laser is a roto-vibrational laser and the lower and upper laser levels of the ground electronic state can be significantly excited thermally, i.e., by having the mixture at a high temperature. The gas mixture is then expanded adiabatically to a very low pressure (e.g., ≈ 0.09 atm) through a row of expansion nozzles. (See the discussion on chemical lasers in Chap. 10.) Due to expansion, the translational temperature of the mixture is reduced to a much lower value (e.g., ≈ 300 K). Consequently, during the expansion process, upper and lower state populations tend to relax to the much lower equilibrium values appropriate to this lower temperature. For a CO_2 laser, however, the lifetime of the upper state is appreciably longer than that of the lower state. This means that relaxation of the lower level occurs at an earlier stage downstream in the expanding beam. Thus, there is a fairly extensive region, downstream from the expansion nozzle, where the population of the lower laser level has decayed while that of the upper level has persisted at its initial value in the container. Thus a population inversion is created in this region by the expansion process. Gas dynamic pumping is mainly applied to CO_2 lasers, and it yields high cw powers (≈ 100 kW). Complications of the system are an obstacle to its use in civilian applications, while its lower power puts it at a disadvantage, when compared to chemical lasers, in military applications.

Whereas the radiation-matter interaction in Chaps. 2 and 3 had the purpose of calculating both stimulated and spontaneous transition rates, the ultimate purpose here would be to calculate the pump rate per unit volume R_p defined by Eq. (1.3.1). When pumping with a broadband light source, e.g., a lamp, the calculation of R_p becomes rather complicated, however.[1] This is also the case when pumping occurs via electron-collision in a gas discharge, where a distribution of electron velocities is involved.[2] Therefore, we limit ourselves here to describing various pumping schemes with some discussion of the underlying physical mechanisms involved in these processes.

6.2. OPTICAL PUMPING BY AN INCOHERENT LIGHT SOURCE

In the case of optical pumping by a powerful incoherent source, i.e. a lamp, the pump light is emitted in all directions and, generally, over a broad spectrum. This light then needs to be transferred into the active medium. The purpose of the next sections is to describe how this transfer can be achieved by a suitable optical system and how light is absorbed in the active medium.

6.2.1. Pumping Systems

Lamps used for laser pumping are often cylindrical; Fig. 6.1 shows two of the most commonly used pumping configurations when a single lamp is used.[3] In both cases the active medium is in the form of a cylindrical rod whose length and diameter are about equal to those of the lamp. The diameter usually ranges from a few millimeters to some tens of millimeters, and the length from a few centimeters to a few tens of centimeters. In Fig. 6.1a the lamp is placed along one of the two focal axes F_1 of a specularly reflecting cylinder of elliptical cross section (labeled 1 in the figure and usually called the *pumping chamber*). The rod is placed along the second focal axis F_2. A well-known property of an ellipse is that a ray F_1P, leaving the first focus F_1, passes, after reflection by the elliptical surface, through the second focus F_2 (ray PF_2). This means that a large fraction of the light emitted by the lamp is conveyed by the pumping chamber to the active rod. High reflectivity in this chamber is achieved by vacuum deposition of a gold or silver layer on the inside surface of the cylinder. Figure 6.1b shows a close-coupled configuration: Rod and lamp are placed as close as possible and closely surrounded by a cylindrical reflector (labeled 1 in the figure). In this case, pumping chambers made of diffusely reflecting materials are often used instead of specular reflectors. For highly diffusing materials, such as compressed $BaSO_4$ powders or white ceramic, which are very efficient scattering media, the efficiency of a close-coupled configuration is usually not much less than that of specularly reflecting cylinders. The pump light distribution within the laser rod is much more uniform however.

Figure 6.2 shows two pumping chambers using two lamps. In Fig. 6.2a, the specularly reflecting cylinder consists of a double-ellipse sharing a common focal axis. The laser rod is placed along this axis while the two lamps lie on the other two focal axes of the ellipses. Figure 6.2b shows two lamps placed as close as possible to the laser rod (close-coupled

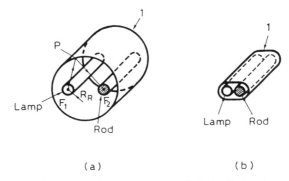

(a)   (b)

FIG. 6.1. Pump configurations using one lamp: (a) elliptical cylinder, (b) close coupling.

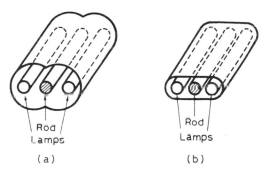

FIG. 6.2. Pump configurations using two lamps: (a) double ellipse; (b) close coupling.

configuration); usually the reflecting cylinder is again made of diffusive material. The efficiencies of these two-lamp configurations are lower than for the corresponding single-lamp configurations in Fig. 6.1. The pump uniformity is however better, and higher pump energies for a given lamp loading can be obtained from a two-lamp as compared to a single-lamp configuration.

For high-power or high-energy systems, multiple-lamp configurations are also used. A widely used configuration involves the active medium arranged in the form of a slab (Fig. 6.3a) or multiple slabs (Fig. 6.3b). In both cases, each lamp may be placed along the focal line of a parabolic reflecting cylinder to ensure uniform illumination of the slab(s). In Fig. 6.3a, laser action occurs by total internal reflections at the two slab faces. The advantage of this zig-zag beam path is that it averages the stress birefringence and thermal focusing induced in the medium by the pump light. Despite its greater complexity, when compared to schemes using a rod-shaped laser medium, this configuration is particularly advantageous when a laser beam of very high optical quality is required. In Fig. 6.3b, laser action occurs in the beam direction indicated by an arrow and the slabs are oriented so that the beam is incident at Brewster's angle. The main advantage of this configuration is that the transverse dimension of the laser medium can be made very large. Furthermore the slabs can be individually cooled by, e.g., a gas refrigerant. This configuration is used in large-aperture (up to 40-cm diameter) Nd:glass amplifiers in laser fusion experiments.

For pulsed lasers, medium-to-high pressure (500–1500 Torr) Xe or Kr flashlamps are used; pump light pulse is produced by discharging, through the lamp, the electrical energy

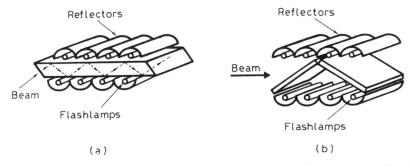

FIG. 6.3. Pumping configuration using many lamps: (a) Active medium in the form of a single slab with the laser beam traversing the slab in a zig-zag path, (b) active medium made of many slabs inclined at Brewster's angle to the laser beam.

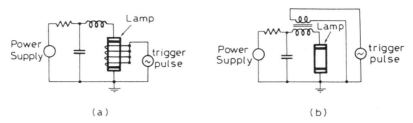

(a)   (b)

FIG. 6.4. Pulsed electrical excitation of a flash lamp using either (a) an external trigger or (b) series trigger configuration.

stored in a capacitor bank, charged by a suitable power supply (Fig. 6.4) A series inductance L is often used in the discharge circuit to limit the current risetime. The discharge may be initiated by ionizing the gas in the lamp through a high-voltage trigger-pulse applied to an auxiliary electrode around the lamp (parallel trigger; see Fig. 6.4a). Alternatively, preionization may be produced by a voltage pulse applied directly between the two main lamp electrodes (series trigger; see Fig. 6.4b). Once ionized, the lamp produces an intense flash of light whose duration is determined by the circuit capacitance and inductance as well as by the lamp's electrical characteristics. (The duration usually ranges from a few microseconds to a few milliseconds.) For cw lasers, high-pressure (1–8 atm) Kr lamps are most often used; the cw current may be delivered by a current-regulated power supply (see Fig. 6.5) where a L/C filter network is used for ripple suppression. In this case, also, an electrical trigger pulse, usually from a series trigger, is needed to provide the required initial ionization. For reliable lamp starting, the voltage across the lamp must be boosted to a sufficiently high value and for a sufficiently long time, during the trigger phase, to ensure a high enough density of ions and electrons in the lamp to stabilize the discharge. This is conveniently done by impulsively charging the filter capacitor C to a high voltage by a low-current booster power supply.

6.2.2. Pump Light Absorption

To illustrate the process of light emission by a lamp, Fig. 6.6a shows the emission spectra, for pulsed excitation, of a Xe flashlamp at two typical current densities. For cw excitation, Fig. 6.6b shows the emission spectrum of a cw Kr lamp at a current density of

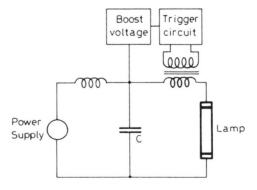

FIG. 6.5. Electrical excitation of a cw lamp.

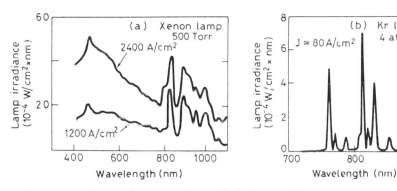

FIG. 6.6. Comparison of the emission spectra of an (a) Xe flashlamp at 500-Torr pressure and of a (b) cw-pumped Kr arc lamp at 4-atm pressure.

$J = 80 \,\mathrm{A/cm^2}$. Typical operating current densities of a Kr lamp are actually somewhat higher than this ($J \cong 150 \,\mathrm{A/cm^2}$) but this difference does not influence the discussion that follows. Note that, at the relatively low current density of a cw lamp, emission is concentrated primarily in various Kr emission lines that are considerably broadened by the high gas pressure. In contrast, at the much higher current densities of a flashlamp, the spectrum also contains a broad continuous component. It arises from electron-ion recombination (*recombination radiation*) and electron deflection by an ion during collision (*bremsstralung radiation*). For both these phenomena, the emission is due to electron-ion interaction. Accordingly, the intensity of the emitted light is expected to be proportional to the product $N_e N_i$, where N_e and N_i are electron and ion densities in the discharge. In a neutral gas discharge, one has $N_e \cong N_i$ while the two densities are proportional to the discharge current density J by the well-known relation $N_e = J/e v_{drift}$, where v_{drift} is the electron drift velocity. It follows that, to a first approximation, the continuous component of the spectrum is expected to grow as J^2. In contrast, to a first approximation, the intensity of the line spectrum in Fig. 6.6b can be taken as proportional to N_e and hence to J. This is why the continuous spectrum dominates the line spectrum at the higher current densities of a pulsed lamp (Fig. 6.6a) while it is not apparent at the much lower current densities of a cw lamp (Fig. 6.6b).

To understand the details of how light emitted by the lamp is absorbed by the active medium, we turn to Fig. 6.7, where the absorption spectrum of Nd:YAG (Nd^{3+} in $Y_3Al_5O_{12}$ crystal) is shown as a solid line and the absorption spectrum of Alexandrite (Cr^{3+} in a $BeAl_2O_4$ crystal) as a dashed line. In both cases the dopant ion, present in the crystal as a trivalent ion impurity, is responsible for the absorption and also acts as the active element. A comparison between Fig. 6.7 and 6.6a shows that the relatively broad spectra of both Nd^{3+} and Cr^{3+} ions allow a reasonably good use of light emitted by a flashlamp. The situation is even more favorable for cw excitation of Nd:YAG by a Kr lamp. The comparison between Fig. 6.7 and 6.6b shows, in fact, that some strong emission lines of Kr, in the 750–900-nm range, coincide with the strongest absorption lines of Nd^{3+} ions. Note that the absorption spectrum of a rare earth element, such as Nd, does not vary much from one host material to another, since absorption arises from electron transitions between inner shells of the ion. Thus the spectrum of Nd:YAG can be taken, to first order, as representative of other Nd-doped materials, such as Nd:YLiF$_4$, Nd:YVO$_4$, and Nd:glass (Nd^{3+} ions in a glass matrix).

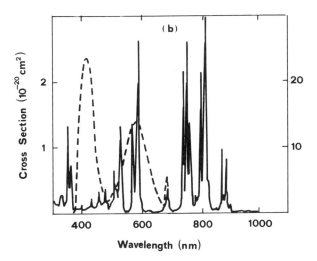

FIG. 6.7. Absorption cross section of Nd^{3+} ion in YAG (*solid line*) and of a Cr^{3+} ion in alexandrite (*dashed line*). The left-hand scale refers to the cross section of Nd:YAG and the right-hand scale to alexandrite. For alexandrite the average of the three values measured for polarization parallel to the *a-*, *b-*, and *c*-axes has been taken.

For a transition metal dopant, such as Cr^{3+}, where the spectrum arises from outermost electrons, the host material has a larger influence on the spectrum. However the spectrum for alexandrite is similar to that of ruby (Cr^{3+} in Al_2O_3 crystal), a historically important and still used material, and to those of more recently developed and now very important laser materials, such as $Cr:LiSrAlF_6$ (LiSAF for short) or $Cr:LiCaAlF_6$ (LiCAF).

6.2.3. Pump Efficiency and Pump Rate

For a cw lamp-pumped laser, we can define the pump efficiency η_p as the ratio between the minimum pump power P_m required to produce a given pump rate R_p and the actual electrical power P_p entering the lamp, i.e.,

$$\eta_p = \frac{P_m}{P_p} \tag{6.2.1}$$

Note that, for a uniform pump rate distribution in the active medium, the minimum pump power can be written as $P_m = (dN_2/dt)_p Vh\nu_{mp} = R_p Vh\nu_{mp}$, where $(dN_2/dt)_p$ is the number of atoms per unit volume raised to the upper laser level by the pump process, V is the volume of the active medium, and ν_{mp} is the frequency difference between the ground level and the upper laser level (see Fig. 6.17). For nonuniform pumping, we can then write

$$P_m = h\nu_{mp}\int_a R_p \, dV = h\nu_{mp}\bar{R}_p V \tag{6.2.2}$$

where the integral is taken over the whole volume of the medium and $\bar{R}_p$ is the average of R_p in the medium. From Eqs. (6.2.1) and (6.2.2) we then obtain

$$\eta_p = \frac{(h\nu_{mp}\bar{R}_p V)}{P_p} \tag{6.2.3}$$

For a pulsed pumping system, we can likewise define η_p as:

$$\eta_p = \frac{(h\nu_{mp} \int R_p \, dV \, dt)}{E_p} \tag{6.2.4}$$

where the integral is also taken over the whole volume of the medium and the whole duration of the pump pulse, and where E_p is the electrical pump energy given to the lamp.

To calculate or simply estimate the pumping efficiency, the pump process can be divided into four distinct steps: (1) Emission of radiation by the lamp; (2) transfer of this radiation to the active medium; (3) absorption in the medium; (4) transfer of the absorbed power to the upper laser level. Consequently, the pumping efficiency can be written as the product of four terms, namely:

$$\eta_p = \eta_r \eta_t \eta_a \eta_{pq} \tag{6.2.5}$$

where: η_r is the efficiency of conversion from electrical input to the lamp to light output in the wavelength range corresponding to pump bands of the laser medium (*radiative efficiency*); η_t is the ratio between power (or energy) actually entering the medium and power emitted by the lamp in the useful pump range (*transfer efficiency*); η_a is the fraction of light entering the medium that is actually absorbed by the material (*absorption efficiency*); η_{pq} is the fraction of absorbed power or energy actually used to populate the upper laser level (*power or energy quantum efficiency*). Note that η_{pq} is given by $\eta_{pq} = R_p V h \nu_{mp} / P_a$, where P_a is the absorbed power. Specific expressions for the preceding efficiency terms can be obtained when the lamp spectral emission, pump geometry, medium absorption coefficient, and geometry are known. We do not consider this topic in depth here and we limit ourselves to a discussion of a few typical results in the example that follows.

Example 6.1. *Pump efficiency in lamp-pumped solid-state lasers.* We take as the active medium a cylindrical rod with a 6.3-mm diameter pumped in a silvered elliptical pumping chamber with major axis $2a = 34$ mm and minor axis $2b = 31.2$ mm. For each laser medium, lamp current density is assumed to have the appropriate value for that laser configuration, generally ranging from 2000–3000 A/cm². Under these pumping conditions, calculated values for efficiency terms $\eta_r, \eta_t, \eta_a, \eta_{pq}$ and overall pump efficiency η_p for ruby, alexandrite, Nd:YAG, and Nd:glass are listed in Table 6.1. In this table we see that

- The lamp radiative efficiency is typically less than 50% in all cases considered.
- Given the larger Nd content in glass and broader absorption bands of Nd:glass material, the overall efficiency of Nd:glass is almost twice that of Nd:YAG.
- The overall efficiency of alexandrite is almost three times higher than for ruby. This is due mainly to the stronger absorption bands in alexandrite owing to the higher Cr^{3+} concentration. Still higher pump efficiency, above the 10% level, is therefore expected in other Cr^{3+}-doped media, such as Cr:LiSAF and Cr:LiCAF because of the even higher (by more than an order of magnitude) Cr concentration.
- In all cases considered, overall efficiency (a product of the four efficiency terms) is quite small (3–8%).

TABLE 6.1. Comparison between computed pumping
efficiency terms for different laser materials

Active Medium	η_r (%)	η_t (%)	η_a (%)	η_{pq} (%)	η_p (%)
Ruby	27	78	31	46	3.0
Alexandrite	36	65	52	66	8.0
Nd:YAG	43	82	17	59	3.5
Nd:Glass (Q-88)	43	82	28	59	5.8

Concluding this section we note that, once overall pump efficiency is calculated or simply estimated, the pump rate can be readily obtained from Eq. (6.2.3) as

$$R_p = \eta_p \left(\frac{P}{Alh\nu_{mp}} \right) \qquad (6.2.6)$$

where A is the cross-sectional area of the active medium and l is its length. This is the simple basic expression for the (average) pump rate often used in the laser literature[4] and which will be used in the following chapters. Note however that, to obtain R_p from Eq. (6.2.6), one must know η_p, implying that detailed calculations, such as those discussed in Ref. 1, must have been performed by someone!

6.3. LASER PUMPING

Laser beams have been often used to pump other lasers since the early days of laser development, for example in the first demonstration of laser action in a dye medium.[5] In particular, Ar ion lasers are now widely used to pump cw dye and $Ti^{3+}:Al_2O_3$ lasers; excimer, nitrogen, and copper vapor lasers are used for pulsed pumping of dye lasers; Nd:YAG and its second harmonic beam are used as pumps for cw and pulsed dye and solid-state lasers (including color center lasers). Laser pumping has become a very important pumping technique, however, since efficient and high-power diode lasers have been developed and made widely available. A particularly interesting case involves using diode lasers to pump other solid-state laser materials, thus providing an all-solid-state laser. The most relevant examples include

- Nd:YAG, Nd:YLF, Nd:YVO$_4$, or Nd:glass pumped by GaAs/AlGaAs quantum well (QW) lasers at $\sim 800\,$nm (typical oscillation wavelengths are around $1\,\mu$m, $1.3\,\mu$m, and $0.95\,\mu$m).
- Yb:YAG, Er:glass or Yb:Er:glass pumped by InGaAs/GaAs strained QW lasers in the 950–980-nm range (oscillation wavelength is around $1\,\mu$m for Yb and $1.54\,\mu$m for Er lasers). Note that, in the case of Er:Yb codoping, pump light is mostly absorbed by Yb^{3+} ions and then transferred to Er^{3+} lasing ions.

- Alexandrite, Cr:LiSAF, or Cr:LiCAF pumped by GaInP/AlGaInP QW lasers in the 640–680-nm range and oscillating in a ~130-nm range around 840 nm.
- Tm:Ho:YAG pumped by AlGaAs QW lasers at 785 nm and oscillating around 2.08 μm. Note that, in this case, pump light is absorbed by Tm^{3+} ions and transferred to Ho^{3+} lasing ions.

As a representative example of Nd-ion lasers, Fig. 6.8a shows the relevant plots of the absorption coefficient versus wavelength for both Nd:YAG (*continuous line*) and Nd:glass (*dashed line*). Note that Nd:YAG is most effectively pumped at a wavelength of $\lambda = 808$ nm, which is obtained by a $Ga_{0.91}Al_{0.09}As/Ga_{0.7}Al_{0.3}As$ QW laser, whose emission bandwidth is typically 1–2 nm wide. Nd:glass on the other hand, due to its broader and featureless absorption profile, can be pumped over a broader range around the 800-nm peak. For the case of Yb-ion lasers, Fig. 6.8b shows the relevant plots of the absorption coefficient versus wavelength for Yb:YAG (*solid line*) and Yb:glass (*dashed line*). Again the absorption coefficient for glass appears broader and featureless compared to that of YAG. The best pumping wavelength is 960 nm for Yb:YAG and 980 nm for glass; these wavelengths are obtained from a InGaAs/GaAs QW laser (e.g., $In_{0.2}Ga_{0.8}As/GaAs$ for $\lambda = 980$ nm). Plots of the absorption coefficient versus wavelength for Cr^{3+} ion lasers (alexandrite, Cr:LiSAF, Cr:LiCAF) show the general structureless shape of the dashed curve in Fig. 6.7. The peak absorption coefficient at 600-nm wavelength is ~0.5 cm^{-1} for alexandrite and up to 50 cm^{-1} for Cr:LiSAF. Note that the higher absorption coefficient in Cr:LiSAF is due to the higher Cr concentration that can be used (~100 times higher than for alexandrite) without incurring in concentration quenching of the upper state. Due to the lack of suitable diode lasers at shorter wavelengths, pumping is achieved in the 640–680 nm wavelength range from GaInP/AlGaInP QW lasers (e.g., $Ga_{0.5}In_{0.5}P/Al_{0.25}Ga_{0.25}In_{0.5}P$ for a 670 nm wavelength) with GaInP being the active QW layer. Table 6.2 summarizes the most relevant pumping data for some of the active media just considered.

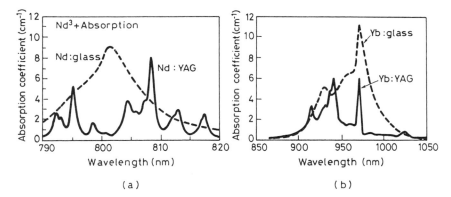

(a)   (b)

FIG. 6.8. Absorption coefficient versus wavelength in the wavelength range of interest for diode laser pumping: (a) Nd:YAG (*solid line*) and Nd:glass (*dashed line*). Neodymium concentration is 1.52×10^{20} cm^{-3} for Nd:YAG (1.1 atomic % doping) and 3.2×10^{20} cm^{-3} for Nd:glass (3.8% by weight of Nd_2O_3). Reprinted with permission from Ref. 15 (b) Yb:YAG (*solid line*) and Yb:glass (dashed line); ytterbium concentration is 8.98×10^{20} cm^{-3} for Yb:YAG (6.5 atomic %) and 1×10^{21} cm^{-3} for Yb:glass. The curves of Yb:YAG and Yb:glass are based on corresponding plots in Refs. 16 and 17, respectively.

TABLE 6.2. Comparison between pumping parameters and laser wavelengths for different laser materials

	Nd:YAG	Yb:YAG	Yb:Er:glass	Cr:LISAF	Tm:Ho:YAG
Concentration	1 at.%	6.5 at.%		1 mol.%	6.5 at.% Tm
					0.36 at.% Ho
Pumping diode	AlGaAs	InGaAs	InGaAs	GaInP	AlGaAs
Diode wavelength (nm)	808	950	980	670	785
Active-ion concentration					
$[10^{20}$ cm$^{-3}]$	1.38	9	10 [Yb]	0.9	8 [Tm]
			1 [Er]		0.5 [Ho]
Pump absorption					
coefficient (cm^{-1})	4	5	16	4.5	6
Oscillation wavelength (μm)	1.06	1.03	1.53	0.72–0.84	2.08
	1.32, 1.34				
	0.947				

6.3.1. Laser-Diode Pumps

There are four types of pumping laser diodes, listed in order of increasing output power as: (1) single stripe; (2) diode array; (3) diode bar; (4) stacked bars.

At the lower end of the output power range ($P < 100$ mW) is the single-stripe semiconductor laser, such as the index-guided laser in Fig. 6.9a. By means of a suitable insulating oxide layer, the diode current is confined to a 3–5-μm-wide stripe extending over the entire length of the diode. The emitted beam has an elliptical shape with a diameter in the direction perpendicular to the laser junction of $d_\perp \cong 1\ \mu$m and a diameter in the junction plane of $d_\| \cong 3$–$6\ \mu$m. With such small spot sizes, the beam is spatially coherent, i.e., it is diffraction-limited. In fact, in a typical situation, the divergence half-angle cone at $1/e^2$ intensity point is $\bar{\theta}_\perp = 20° = 0.35$ rad, perpendicular to the junction. This means that $\theta_\perp \cong 2\lambda/\pi d_\perp$ provided takes, at $\lambda = 800$ nm, $d_\perp \cong 1.4\ \mu$m. In the junction plane one typically has $\theta_\| = 5° = 0.09$ rad and again $\theta_\| = \cong 2\lambda/\pi d_\|$, with $d_\| = 5.8\ \mu$m. [Gaussian distributions in the two planes, with spot sizes $w_{0\perp} \cong d_\perp/2$ and $w_{0\|} \cong d_\|/2$, are assumed so that beam divergence is calculated according to Eq. (4.7.19)]. Note that, in view of this strong difference between beam divergences in the two directions, the beam's major axis direction is rotated by 90° after beam propagation just a few micrometers away from the diode exit face.

To obtain greater output powers, one uses a monolithic array of diode-laser stripes, fabricated on the same semiconductor substrate (Fig. 6.9b). In typical cases, the array may contain 20 stripes, each $5\ \mu$m wide, with their centers spaced by $\sim 10\ \mu$m. The overall dimensions of the emitted beam are $d_\| \cong 200\ \mu$m $\times d_\perp \cong 1\ \mu$m and, for arrays with uncorrelated phases, beam divergences are $\theta_\perp \cong 20°$ and $\theta_\| \cong 5°$, i.e., the same as for a single stripe. The beam divergence parallel to the junction plane, $\theta_\|$, is now about 40 times more than the diffraction limit ($\theta_\| \pi d_\|/2\lambda \cong 34$). Actually, for lower power devices, some phase correlation among various emitters may develop, leading to a characteristic two-lobed angular emission pattern, with the two lobes spaced by $\sim 10°$ and each $\sim 1°$ wide. Output power from such arrays may be up to ~ 2 W.

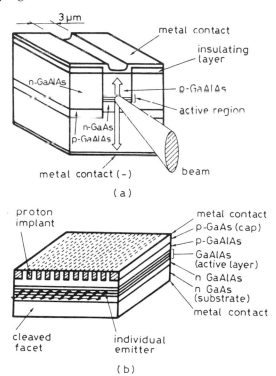

FIG. 6.9. (a) Single-stripe index-guided semiconductor diode laser and (b) monolithic array of many stripes on a single semiconductor chip.

To obtain still greater output powers, the previously described array can be serially repeated in a single substrate to form a monolithic bar structure (Fig. 6.10a). The device shown in this figure consists of 20 arrays whose centers are spaced by 500 μm; each array is 100 μm long and contains 10 laser stripes. The overall length of the bar is thus ≈ 1 cm, the limit being set by processing technology. Again all stripe emitters can be considered phase uncorrelated and output powers up to 10–20 W are usual.

The bar concept can be extended to the case of a stack of bars forming a two-dimensional structure (Fig. 6.10b). In this figure six, 1-cm-long, bars are stacked to form an overall 2 mm $\times$ 1 cm emitting area. These stacked bars are intended for quasi-cw operation with a duty cycle up to 2%. (Peak power density may be up to 1 kw/cm^2 and average power up to 100 W/cm^2.)

To pump laser materials, such as Nd:YAG, having narrow absorption lines, the width of the diode's spectral emission must be considered. The spectral emission bandwidth of a single stripe may be as narrow as 1 nm, which compares favorably with, e.g., the ~ 2 nm bandwidth of the 808-nm absorption peak of Nd:YAG. For the case of arrays and even more so, for bars or stacked bars, spectral emission can be substantially greater than this value due to compositional variation between stripes and due to temperature gradients, both of which leading to different stripe-emission wavelengths. Currently, the best results for a bar may be a spectral width as low as ~ 2 nm. To tune and stabilize the emission wavelength, diode

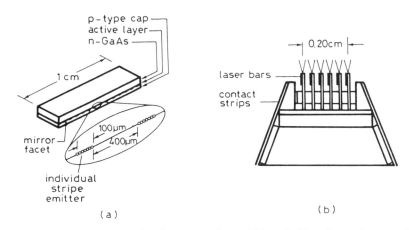

FIG. 6.10. (a) Monolithic 1-cm bar for cw operation and (b) stacked bars for quasi-cw operation.

lasers are normally cooled by a thermoelectric cooler, for low-power devices, and by liquid cooling for the highest powers. (A temperature stability and accuracy of less than 1 °C is usually required.)

6.3.2. Pump Transfer Systems

For efficient pumping, light emitted by the diode laser systems described above must be properly transferred to the active medium. There are basically two types of pump geometry: (1) *Longitudinal pumping*, where the pump beam enters the laser medium along the resonator axis. (2) *Transverse pumping*, where the beam is conveyed to the active medium, generally from one or more directions, transversely to the resonator axis. We consider the two cases separately because the corresponding diode lasers and pump-transfer systems are somewhat different.

6.3.2.1. Longitudinal Pumping

For longitudinal pumping, the beam emitted by the laser diode generally needs to be concentrated in a small (100 μm–1 mm diameter) and possibly circular spot in the active medium. Three of the most common laser configurations are shown in Figs. 6.11a–c. In Fig. 6.11a the laser rod is shown in a plane-concave resonator; the plane mirror is directly deposited on one rod face, with the pump beam focused on this face. In Fig. 6.11b–c two pump beams from two different diode systems are focused, on the rod center, from the two sides of the rod. The laser resonator can consist of either a folded ring configuration (Fig. 6.11b) or a z-shaped folded linear cavity (Fig. 6.11c). For these latter two configurations, the resonator axis is also indicated by a dashed line. Given these resonators, we now address the question of how to transform the pump beam into a possibly circular shape of the appropriate size within the laser rod.

We first consider the single-stripe configuration in Fig. 6.9a, which is used as a pump source for low-power devices. (Output powers up to several tens of mW can be achieved with single-stripe pumping.) The ellipticity of the strongly diverging beam of the diode

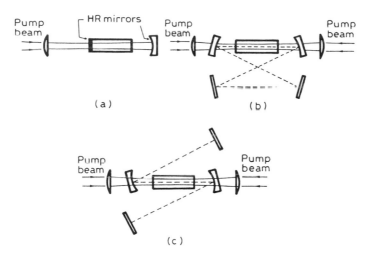

FIG. 6.11. Typical configurations for longitudinal diode laser pumping: (a) single-ended pumping in a simple plane-concave resonator, (b) double-ended pumping for a ring laser in a folded configuration, and (c) double-ended pumping for a *z*-shaped folded linear cavity.

stripe can be compensated by a combination of two spherical lenses and an anamorphic optical system, indicated schematically as a box in Fig. 6.12a. In the figure, the beam indicated by a continuous line corresponds to the beam behavior in the plane parallel to the laser-diode junction; the beam indicated by dashed lines corresponds to the plane perpendicular to the junction. Lens L_1 of focal length f_1 is a spherical lens of short focal length and high numerical aperture to collimate the highly divergent beam from the laser diode. Since $\theta_\perp \cong 4\theta_\parallel$, the beam after the lens has an elliptical shape with dimension $d_\perp = 2f_1 \tan \theta_\perp$ perpendicular to the junction (the so-called fast axis) and $d_\parallel = 2f_1 \tan \theta_\parallel$ parallel to the junction (the slow axis). Thus, in a typical case, $d_\perp/d_\parallel = \tan \theta_\perp/\tan \theta_\parallel \cong 4$. This elliptical beam is then passed through an anamorphic expansion system, i.e., a system providing different beam expansions along the two axes. If, for instance, the system provides a 4:1 beam expansion along the slow axis and no expansion along the fast axis, then a circular spot will result after this expansion. The simplest configuration for such an anamorphic expander is perhaps provided by combining two cylindrical lenses L_3 and L_4 in a confocal (or telescopic) arrangement (Fig. 6.12b). If the two lenses have their focusing action in the plane containing the slow axis, a beam expansion of f_4/f_3 will occur, where f_4 and f_3 are focal lengths of the two lenses, for the beam in this plane (*solid line*). For the other plane however, the two cylindrical lenses behave simply as plane-parallel plates so the beam is unaffected in the fast-axis direction by the beam expander.

The anamorphic system in Fig. 6.12b is not often used in practice, however, because it would require well-corrected lenses of high numerical aperture, which are rather expensive; thus the anamorphic prism pair in Fig. 6.12c is more usually employed.[6] In this figure, we again consider beam behavior only in the slow-axis plane (*solid line*). By simple geometrical considerations, we can show that, after refraction at the front surface of the first prism, the incident beam of diameter D_i is enlarged to a diameter D_r such that $D_r/D_i = \cos \theta_r/\cos \theta_i$, where θ_i and θ_r are, respectively, the angles of incidence and refraction at the prism surface. If the exit face of the first prism is near normal to the beam direction, no refraction occurs at

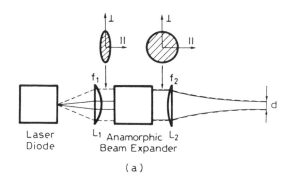

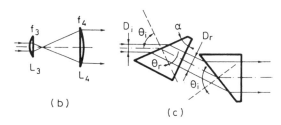

FIG. 6.12. (a) Pump transfer system for compensating the astigmatism of a single-stripe diode laser, (b) simple cylindrical lens combination to realize an anamorphic system, and (c) anamorphic prism pair configuration.

this face so the beam passes through it unchanged. Under these conditions beam magnification M after the first prism is simply given by:

$$M = \frac{D_r}{D_i} = \frac{\cos \theta_r}{\cos \theta_i} \tag{6.3.1}$$

Let us now consider the passage of the beam through the second prism. If the prism is identical to the first one, is oriented as in Fig. 6.12c, and if the angle of incidence at the entrance face is again equal to θ_i, then the beam is again magnified by a factor M on traversing the second prism. The overall beam magnification then equals M^2, and the direction of the output beam is parallel to that of the input beam, although shifted laterally. In the fast-axis plane, the two prisms behave as simple plates, so there is no beam magnification. Thus, for the example considered in Fig. 6.12a, if we choose an anamorphic prism pair with appropriate values for θ_i, for the prism refraction index n, and for the prism apex angle α, we can have $M = 2$, i.e., an overall magnification of $M^2 = 4$. The beam after the prism pair thus has a circular shape. If the collimating lens L_1 in Fig. 6.12a has a sufficiently high numerical aperture to accept the highly diverging beam along the fast axis and the lens is aberration-free, the beam after this lens, and hence after the prism-pair, ideally retains the diffraction-limited quality of the original beam from the diode. Accordingly, since the beam leaving the prism pair has a circular shape, the beam divergence is now equal along the two axes. A spherical lens L_2 of appropriate focal length f_2 can then be used to focus the beam to a round spot of appropriate size in its focal plane (Fig. 6.11a), i.e., where the active medium can conveniently be placed. If lens L_2 is also aberration-free, the beam in the focal plane has circular symmetry and is diffraction-limited.

Example 6.2. *Calculation of an anamorphic prism-pair system to focus the light of a single-stripe diode laser.* We consider the system configuration in Fig. 6.12 and a single-stripe laser with $\theta_\perp = 20°$ and $\theta_\parallel = 5°$, so that, assuming diffraction-limited Gaussian distributions, we can take $d_\perp = 1.4\,\mu m$ and $d_\parallel = 5.8\,\mu m$ ($\lambda \cong 780$ nm). We consider a collimating lens L_1 of focal length $f_1 = 6.5$ mm. After lens L_1, beam diameters along the fast and slow axes are, respectively, $D_\perp = 2f_1 \tan\theta_\perp = 4.73$ mm and $D_\parallel = 2f_1 \tan\theta_\parallel = 1.14$ mm. Each prism must then provide a magnification of $M = (D_\perp/D_\parallel)^{1/2} \cong 2$. Assuming the prisms to be made of fused silica, so that the refractive index at ~ 800 nm wavelength is $n = 1.463$, then θ_i and θ_r are determined from Eq. (6.3.1) and Snell's law ($\sin\theta_i = n\sin\theta_r$). The solution can be readily obtained either graphically or by a fast iterative procedure. For this procedure, we first assume a tentative value of θ_i, then use Snell's law with $n = 1.463$ to calculate a first value for θ_r. This value is substituted in Eq. (6.3.1) with $M = 2$ to calculate a new value for θ_i corresponding to the first iteration, and so on. Starting at $\theta_i = 70°$, e.g., this iterative calculation rapidly converges in a few iterations to $\theta_i = 67.15°$ and $\theta_r \cong 39°$. Since the beam is assumed to exit normal to the second face of the prism, a simple geometrical argument shows that the apex angle of the prism must be $\alpha = \theta_r \cong 39°$. In this way after the second prism, a circular beam with diameter $D = D_\perp = 4.73$ mm is obtained. Let us now take the focal length of lens L_2 to be $f_2 = 26$ mm and assume that the beam is still diffraction-limited after this lens. The beam spot size in the focal plane of this second lens is then $d \cong 4\lambda f_2/\pi D \cong 5.52\,\mu m$ [the expression that applies for Gaussian beam focusing is used again here; see Eq. (4.7.28)]. Note that a very small value for the pump diameter can in principle be achieved. Indeed we see that the effect of the optical system in the fast-axis plane (Fig. 6.12a) is to make a $f_2/f_1 \cong 4$ magnified image of the field distribution at the diode exit face. Since $d_\perp = 1.4\,\mu m$, we expect $d = (f_2/f_1)d_\perp \cong 5.6\,\mu m$. To obtain such a small spot, however, lenses well-corrected for spherical aberration must be used, in particular for the collimator lens L_1. In a typical situation, taking into account the finite resolving powers of lenses L_1 and L_2, the beam diameter in the focal plane of lens L_2 can be 5–10 times larger. In any case, for geometrical reasons, beam divergence in the focal plane of lens L_2 is given by $\theta \cong D/2f_2$, where D is the beam diameter at the lens position. If a rod of refractive index n_R is placed in the focal plane, then, due to beam refraction, the divergence is approximately reduced, in the rod, by a factor n_R. If we then take $n_R = 1.82$, as appropriate for YAG crystals, then $\theta_n \cong D/2f_2 n_R = 0.05$ rad $\cong 3°$.

In the case of a 200-μm-wide array, e.g., since divergence angles $\theta_\perp$ and $\theta_\parallel$ are approximately the same as for a single stripe, the configuration in Figs. 6.12a–c can still be used to produce a circular spot after the anamorphic prism-pair. Since the slow-axis beam divergence is however ~ 40 times larger than the diffraction limit, the spot at the focal plane of lens L_2 is elliptical, with a 40:1 ratio between the two axes. According to Example 6.2, for a well-corrected collimating lens L_1, the elliptical beam could be, e.g., 2.8 μm $\times$ 112 μm. In practice aberrations of the optical system, which are more pronounced for the fast-axis direction, tend to produce a more circular spot, with a spot size of perhaps 150 μm. Another widely used method, in diode arrays, for transferring a pump beam to the active medium involves the use of a multimode optical fiber. For a 200-μm stripe, a fiber with a 200-μm core diameter can be used with the fiber butt-coupled to the diode. With this configuration, however, the fiber numerical aperture NA ($NA = \sin\theta_f$, where θ_f is the acceptance angle of the fiber) must have a sufficiently high value to accept the highly diverging beam of the diode, i.e., $\sin\theta_f > \sin\theta_\perp \cong 0.4$. After propagation in a sufficient length of the fiber, the output beam becomes circular and its divergence is established by the fiber NA, i.e., one has $\theta_{out} = \theta_f$. During propagation along the fiber, the slow axis divergence is thus worsened from $\theta_{in} = \theta_\parallel$ to $\theta_{out} = \theta_f \cong \theta_\perp$. To reduce beam divergence, we can use a cylindrical lens of very short focal length, between array and fiber, to collimate the beam in the fast-axis direction to a diameter equal to the fiber diameter. We can then use a fiber whose NA approximately equals the slow-axis divergence, i.e., take $\theta_f \cong \theta_\parallel$. In this case, as shown in more detail in Example 6.3, the beam of a 200 μm wide array can be focused onto a fiber of perhaps 250–300 μm core diameter and NA of 0.1.

Example 6.3. *Diode-array beam focusing onto a multimode optical fiber.* We consider the simple configuration in Fig. 6.13, where a cylindrical lens of sufficiently short focal lens f is used to collimate the beam along the fast axis (*dashed lines*). The beam diameter after the lens and along this axis is then given by $D_\perp = 2f \tan \theta_\perp$. Along the slow axis, the cylindrical lens behaves like a plane-parallel plate, so the beam (*continuous line*) is essentially unaffected by the lens. (To draw attention to this circumstance, the cylindrical lens is drawn as a dashed line in the figure to indicate that it focuses only on the fast-axis plane.) The beam diameter in the slow-axis plane after the lens is approximately: $D_\parallel \approx L_a + 2f \tan \theta_\parallel$ where L_a is the length of the array. If we now set the condition $D_\parallel = D_\perp$, then we have $f = L_a / 2(\tan \theta_\perp - \tan \theta_\parallel)$. Taking $L_a = 200$ μm, $\theta_\perp = 20°$, and $\theta_\parallel = 5°$, we obtain $f = 350$ μm, a focal length that can be obtained with fiber microlenses. With such a small value of focal length, the beam diameter after the lens is $D = D_\parallel = D_\perp = 2f \tan \theta_\perp = 254$ μm which can easily be accepted into, e.g., a 300-μm diameter multimode fiber butt-coupled to the microlens. For a well-corrected fiber microlens, beam divergence after the lens mostly arises from the uncompensated divergence of the slow-axis beam. The fiber NA must then be $NA = \sin \theta_f \geq \sin \theta_\parallel \cong 0.09$. Beam divergence of light leaving the fiber, for a sufficiently long fiber, is then equal to the fiber NA.

In the case of a 1-cm bar, a single, 1-cm-long, cylindrical microlens can be used to focus each array of the bar onto a separate multimode fiber. Since each array is now typically 100 μm long (see Fig. 6.10a), fibers with a 200-μm core diameter and 0.1 NA can be used for each array (see Fig. 6.13). In this way we can convey the whole beam of the bar into 20 fibers whose ends can then be arranged into a circular fiber bundle of 1–1.5 mm diameter, with overall divergence equal to the NA (0.1) of the fiber. The beam emitted by

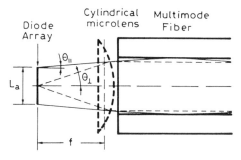

FIG. 6.13. Using a cylindrical microlens to couple the output of a diode array to a multimode optical fiber.

this bundle is then imaged onto the rod along one (Fig. 6.11a) or two longitudinal directions (Figs. 6.11b–c). With this pump configuration, an overall transmission of the transfer system up to 85% has been demonstrated. Output powers up to ~ 15 W in a TEM$_{00}$ mode with an optical-to-optical efficiency of $\sim 50\%$ have been obtained, using a Nd:YVO$_4$ rod pumped by two such fiber-coupled diode bars.

   An interesting and alternative approach was demonstrated that allows the very asymmetric output beam from a diode bar or array to be reshaped so as to produce the same beam dimensions and divergences in the original fast-axis (vertical) and slow-axis (horizontal) directions. The technique involves sending the beam from a diode bar or array, after collimation in the fast direction by a fiber-lens, to a tilted pair of parallel mirrors. This mirror pair effectively divides the beam, by multiple reflections, into several segments in the horizontal direction, then stacks these segments above each other, thus producing an output beam with a rectangular shape.[7] In equalizing beam parameters in the horizontal and vertical directions, the decrease in beam brightness in the vertical direction is compensated for by increased brightness in the horizontal direction and overall brightness is maintained. This shaped beam allows very intense longitudinal pumping, which is particularly effective for the otherwise difficult cases of low-gain, quasi-three-level, lasers.

6.3.2.2. Transverse Pumping

   In the case of transverse pumping, active media in the shape of either slabs or rods can be used. Figure 6.14 shows a transversely pumped slab configuration;[8] pumping occurs

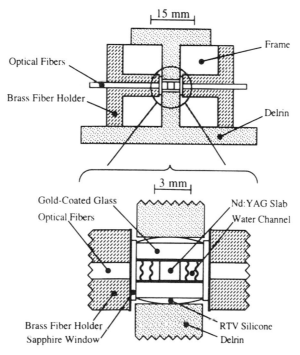

FIG. 6.14. Transverse pumping configuration for a Nd:YAG slab. (By permission from Ref. 8.)

through 25 individual laser arrays, each coupled to a 600-μm core diameter, 0.4 NA, fiber. The power of the beam exiting each fiber is ~ 9.5 W and the total power is 235 W. Fiber ends are spaced along the two sides of a 1.7-mm-thick, 1.8-mm-wide miniature slab. The slab's center-line length is ~ 58.9 mm, which corresponds to 22 total internal reflections at the two slab faces (see Fig. 6.3a). Due to the averaging properties of the resulting zig-zag pattern, the optical quality of the active medium seen by the beam is excellent, so that an output power of 40 W in a TEM$_{00}$ mode, with an optical-to-optical efficiency of $\sim 22\%$, has been achieved.

Figure 6.15 shows an interesting configuration using a Nd:YAG rod.[9] The 4-mm diameter rod, cooled by water flowing in a surrounding tube, is radially pumped by either three (as shown in the figure) or five pump modules placed in a circularly symmetric arrangement. Each pump module consists of 16, 800-μm core diameter 0.22 NA, fibers mounted side by side in a row with 2-mm center-to-center spacing. Into each fiber, the beam of a diode array with a nominal output power of 10 W is injected. The output beam from each fiber directly irradiates the laser rod without additional focusing optics. A pump transfer efficiency of $\sim 80\%$ is estimated for this transverse pump configuration. To help achieving sufficient absorption of the diode laser radiation, pump light reflectors facing each pump module are mounted around the rod. For large enough fiber-to-rod distances, the pump light distribution within the rod, achieved in this way, is rather uniform. Figure 6.16 shows

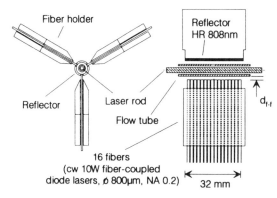

FIG. 6.15. Transverse pumping configuration for a Nd:YAG rod. (By permission from Ref. 9.)

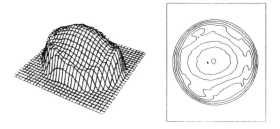

FIG. 6.16. Pump light distribution in the Nd:YAG rod for the transverse pump distribution in Fig. 6.15. (By permission from Ref. 9.)

the pump light distribution for a 13-mm fiber-to-flow-tube distance. Using the configuration in Fig. 6.15, an output power of $\sim 60\,\mathrm{W}$ in a TEM$_{00}$ mode is achieved with an optical-to-optical efficiency of $\sim 25\%$.

6.3.3. Pump Rate and Pump Efficiency

In the case of longitudinal pumping, if we let $I_p(r, z)$ be pump intensity at the location, inside the laser medium, specified by radial coordinate r and longitudinal coordinate z, the pump rate is obtained as

$$R_p(r, z) = \frac{\alpha I_p(r, z)}{h\nu_p} \tag{6.3.2}$$

where α is the absorption coefficient of the laser medium at the frequency ν_p of the pump.

We now assume a Gaussian distribution of the pump beam, i.e., we take

$$I_p(r, z) = I_p(0, 0) \exp\left[-\left(\frac{2r^2}{w_p^2} \right) \right] \exp(-\alpha z) \tag{6.3.3}$$

where $I_p(0, 0)$ is the peak intensity at the entrance face of the rod and w_p is the pump spot size, which is taken for simplicity to be independent of z. Note that α is the absorption coefficient under laser operating conditions. To a good approximation, however, it coincides with the unpumped absorption coefficient, since the population raised to upper levels by the pumping process is usually only a small fraction of the total population. The intensity $I_p(0, 0)$ is obviously related to the incident pump power P_{pi} by the relation:

$$P_{pi} = \int_0^\infty I_p(r, 0) 2\pi r \, dr \tag{6.3.4}$$

From Eqs. (6.3.3) and (6.3.4) we obtain

$$I(0, 0) = \frac{2P_{pi}}{\pi w_p^2} \tag{6.3.5}$$

The incident pump power P_{pi} is then related to the diode laser electrical power P_p by

$$P_{pi} = \eta_r \eta_t P_p \tag{6.3.6}$$

where η_r is the diode radiative efficiency and η_t is the efficiency of the pump transfer system. From Eq. (6.3.2), with the help of Eqs. (6.3.3), (6.3.5), and (6.3.6), we obtain

$$R_p(r, z) = \eta_r \eta_t \left(\frac{P_p}{h\nu_p} \right) \left(\frac{2\alpha}{\pi w_p^2} \right) \exp\left[-\left(\frac{2r^2}{w_p^2} \right) \right] \exp(-\alpha z) \tag{6.3.7}$$

Appendix E shows that, as far as the threshold condition is concerned, the effective pump rate for a given cavity mode is the average, $\langle R_p \rangle$, of R_p taken over the field distribution of the mode. More precisely, if we let $u(r, z)$ be the field amplitude normalized to its peak value, $\langle R_p \rangle$ is given by

$$\langle R_p \rangle = \frac{\int_a R_p |u|^2 \, dV}{\int_a |u|^2 \, dV} \tag{6.3.8}$$

where integrals are taken over the entire volume of the active medium. Let us consider a TEM_{00} single longitudinal mode. If the spot size at the beam waist w_0 is located in the laser rod and the spot size is assumed constant along the rod, then, using Eq. (4.6.4) and Eq. (5.5.24) with $R \to \infty$ and $\phi \cong 0$, we can write:

$$|u|^2 \propto \exp\left[-\left(\frac{2r^2}{w_0^2} \right) \right] \cos^2 kz \tag{6.3.9}$$

Equations (6.3.7)–(6.3.9) then give

$$\langle R_p \rangle = \eta_r \eta_t \left(\frac{P_p}{h\nu_p} \right) \frac{2\{1 - \exp[-(\alpha l)]\}}{\pi(w_0^2 + w_0^2)l} \tag{6.3.10}$$

where l is the length of the laser rod. Note that, in integrating along the z-coordinate in Eq. (6.3.8), we have made the approximation $\int_0^l \exp -(\alpha z) \cos^2 kz \, dz \cong (1/2) \int_0^l \exp -(\alpha z) dz$. This uses the fact that, since $\cos^2 kz$ changes much more rapidly with z than $\exp[-(\alpha z)]$, one can substitute $\cos^2 kz$ with its average value $\langle \cos^2 kz \rangle = (1/2)$. If we now define the absorption efficiency, η_a, as

$$\eta_a = [1 - \exp -(\alpha l)] \tag{6.3.11}$$

Eq. (6.3.10) can be put in the more suggestive form:

$$\langle R_p \rangle = \eta_p \left(\frac{P_p}{h\nu_p} \right) \frac{2}{\pi(w_0^2 + w_p^2)l} \tag{6.3.12}$$

where we have defined $\eta_p = \eta_r \eta_t \eta_a$.

Equation (6.3.12) constitutes the final result of our calculation of the effective pump rate for longitudinal pumping. Note that, for a given value of P_p, $\langle R_p \rangle$ increases as w_p decreases, so that the maximum value of $\langle R_p \rangle$ is attained for $w_p \to 0$. For very small values of pump spot size, however, divergence of the pump beam in the active rod cannot be neglected, so that the beam may actually become larger than the laser beam at the end of the rod. For this reason, and to optimize optical efficiency, the condition $w_p \cong w_0$ is often taken as a rough guide to the optimum case.

In the case of transverse pumping, we begin with the following obvious relation between pump rate and power P_{pi} incident on the rod:

$$\int_a h\nu_p R_p \, dV = \eta_a P_{pi} \tag{6.3.13}$$

where η_a is the fraction of the incident power absorbed in the active medium. Note that, according to Eq. (6.3.11), the absorption efficiency η_a can be written as $\eta_a \cong [1 - \exp(-\alpha D)]$, where D is the relevant transverse dimension of the rod. ($D \cong D_R$, where D_R is the rod diameter, for a single pass, or $D \cong 2D_R$ for a double pass of the pump beam in the rod). Equation (6.3.13) allows us to calculate the pump rate once its spatial variation is known. If, as a simple case, we take $R_p = \text{const}$, then from Eq. (6.3.13) we have $R_p = \eta_r \eta_t \eta_a P_p / h\nu_p Al$ where A is the cross sectional area of the rod and where Eq. (6.3.6) has been used. To calculate $\langle R_p \rangle$ we consider a conceptually simple model of the laser rod, where the active species is assumed to be confined to the central region of the rod, $0 \le r \le a$, while the rod is undoped for $r > 0$ (*cladded rod*). In this case, Eq. (6.3.9) can be taken to hold for any value of r, while one takes $R_p = \text{const}$ for $0 \le r \le a$ and $R_p = 0$ for $r > 0$. Then, from Eqs. (6.3.8) and (6.3.9), we obtain

$$\langle R_p \rangle = \eta_p \left(\frac{P_p}{h\nu_p}\right) \frac{\{1 - \exp[-(2a^2/w_0^2)]\}}{\pi a^2 l} \tag{6.3.14}$$

where again $\eta_p = \eta_r \eta_t \eta_a$. Equation (6.3.14) is the final result of calculating the effective pump rate in transverse pumping.

For the comparison to be performed in Sect. 6.3.5, we must also calculate the effective pump rate for lamp pumping. Assuming the cladded rod model considered above, and again taking R_p constant in the active medium, i.e., for $0 \le r \le a$, we obtain from Eqs. (6.2.6) and (6.3.8):

$$\langle R_p \rangle = \eta_{pl} \left(\frac{P_p}{h\nu_{mp}}\right) \frac{\{1 - \exp[-(2a^2/w_0^2)]\}}{\pi a^2 l} \tag{6.3.15}$$

where η_{pl} is the pumping efficiency for lamp pumping given, according to (6.2.5), by $\eta_{pl} = \eta_r \eta_t \eta_a \eta_{pq}$.

6.3.4. Threshold Pump Power for Four-Level and Quasi-Three-Level Lasers

With the results from Sect. 6.3.3. for the effective pump rate, we can now calculate the expected threshold pump rate and threshold pump power for a given laser. We consider two important cases: (1) An ideal four-level laser where pumped atoms are immediately transferred to the upper laser level while the lower laser level is empty (see Fig. 1.4b). (2) An ideal quasi-three-level laser where pumped atoms are again transferred immediately to the upper laser level and the lower laser level is a sublevel of the ground level. The first case includes such lasers as Nd:YAG at $\lambda = 1.06\ \mu m$ or $\lambda = 1.32\ \mu m$, Ti:Al$_2$O$_3$, and Cr:LiSAF or LiCAF. The most important lasers in the second category are Nd:YAG at $\lambda = 0.946\ \mu m$, Er:glass or Yb:Er:glass at $\lambda \cong 1.54\ \mu m$, Yb:YAG or Yb:glass, and Tm:Ho:YAG.

Let us first consider an ideal four-level laser and assume that the upper laser level actually consists of many strongly coupled sublevels whose *total* combined population is N_2. According to Eq. (1.2.5), the threshold value, N_{2c}, can be written as $N_{2c} = \gamma/\sigma_e l$, where σ_e now indicates the effective stimulated-emission cross section (see Sect. 2.7.2). This expression actually holds only for a spatially uniform model, i.e., when both R_p and the mode configuration $|u|^2$ are considered to be spatially independent. When spatial dependence is taken into account, the preceding expression for the threshold upper-state population is modified as follows (see Appendix E):

$$\langle N_2 \rangle_c = \frac{\gamma}{\sigma_e l} \tag{6.3.16}$$

where $\langle N_2 \rangle$ is the effective value of population, given by

$$\langle N_2 \rangle = \frac{\int_a N_2 |u|^2 \, dV}{\int_a |u|^2 \, dV} \tag{6.3.17}$$

The critical, or threshold, pump rate is now obtained from the condition that the number of atoms raised by the pumping process must equal the number of atoms decaying spontaneously. Thus, we have $R_p = N_{2c}/\tau$, where τ is the effective lifetime of the upper laser level, taking into account the decay of all sublevels (see Sect. 2.7.2). It then follows that

$$\langle R_p \rangle_c = \frac{\langle N_2 \rangle_c}{\tau} \tag{6.3.18}$$

From Eqs. (6.3.16) and (6.3.18) we obtain

$$\langle R_p \rangle_c = \frac{\gamma}{\sigma_e l \tau} \tag{6.3.19}$$

Once the threshold value of the pump rate is calculated, we can readily obtain the corresponding threshold pump power. Substituting Eq. (6.3.12) or (6.3.14) into Eq. (6.3.19), we obtain the following expressions

$$P_{th} = \left(\frac{\gamma}{\eta_p} \right) \left(\frac{h\nu_p}{\tau} \right) \left[\frac{\pi (w_0^2 + w_p^2)}{2\sigma_e} \right] \tag{6.3.20}$$

$$P_{th} = \left(\frac{\gamma}{\eta_p} \right) \left(\frac{h\nu_p}{\tau} \right) \left(\frac{\pi a^2}{\sigma_e \{1 - \exp[-(2a^2/w_0^2)]\}} \right) \tag{6.3.21}$$

which hold for longitudinal and transverse pumping, respectively. The expression for the threshold pump power for longitudinal pumping, given by Eq. (6.3.20), agrees with that given first in Ref. 8. Note that, for longitudinal pumping, the threshold pump power increases as w_0 is increased because, as w_0 increases, the wings of the mode extend further into the less strongly pumped regions of the active medium. Likewise, for transverse pumping and the cladded rod model considered above, the threshold pump power increases as w_0 is increased because, as w_0 increases, the wings of the mode extend further into the cladding, i.e., into the unpumped part of the medium.

Similar considerations apply to the more realistic case of a rod without cladding. In this case, however, the calculation would be more involved because, in general, Eq. (6.3.9) no longer applies, so the true field distribution, taking into account the aperturing effects established by the finite rod diameter, must be used. When w_0 is appreciably smaller than a (say, $w_0 \leq 0.7a$), however, the field distribution is not greatly affected by the presence of this aperture, so Eq. (6.3.21) can be assumed to hold also for a rod without cladding. In this case, of course, $\{1 - \exp[-(2a^2/w_0^2)]\}$ is very close to unity and, in calculating the threshold pump power, this term could even be omitted from Eq. (6.3.21).

For the comparison to be made in Sect. 6.3.5, we must also calculate the threshold pump power for lamp pumping. From Eqs. (6.3.15) and (6.3.19) we obtain

$$P_{th} = \left(\frac{\gamma}{\eta_{pl}}\right)\left(\frac{h\nu_{mp}}{\tau}\right)\left(\frac{\pi a^2}{\sigma_e\{1 - \exp[-(2a^2/w_0^2)]\}}\right) \tag{6.3.22}$$

Let us now consider a quasi-three-level laser where the lower laser level is a sublevel of the ground level (level 1) and assume that the population raised by the pumping process to the pump level(s) is immediately transferred to the upper laser level (ideal quasi-three-level laser). We assume that all ground state sublevels are strongly coupled and hence in thermal equilibrium and let N_1 be the total combined population of level 1. We also assume that the upper laser level, level 2, consists of a number of strongly coupled sublevels and let N_2 be the total combined population of the upper level. Threshold values for the population of the two levels is again established by the condition that total net gain equals losses. For the space-dependent case, according to Eq (6.3.16), we now obtain (see Appendix E)

$$(\sigma_e\langle N_2\rangle_c - \sigma_a\langle N_1\rangle_c)l = \gamma \tag{6.3.23}$$

where $\langle N_2\rangle$ and $\langle N_1\rangle$ again indicate spatially averaged values as in Eq. (6.3.17) and σ_e and σ_a are, respectively, effective values of the stimulated emission and absorption cross sections. Since, for an ideal quasi-three-level laser, $N_1 + N_2 = N_t$, it follows that $\langle N_1\rangle + \langle N_2\rangle = N_t$. Using this expression in Eq. (6.3.23), we can readily calculate $\langle N_2\rangle_c$. The effective value of the threshold pump rate must again satisfy Eq. (6.3.18); using the value of $\langle N_2\rangle$ calculated in this way, we obtain

$$\langle R_p\rangle_c = \frac{(\sigma_a N_t l + \gamma)}{(\sigma_e + \sigma_a)/\tau} \tag{6.3.24}$$

Note that Eq. (6.3.24) obviously reduces to Eq. (6.3.19) if we let $\sigma_a \to 0$.

In calculating the corresponding threshold pump power, we limit our considerations to longitudinal pumping, since this is the only configuration that allows operation with a reasonably low threshold. From Eqs. (6.3.24) and (6.3.12) we obtain

$$P_{th} = \left(\frac{\sigma_a N_t l + \gamma}{\eta_p}\right)\left(\frac{h\nu_p}{\tau}\right)\left[\frac{\pi(w_0^2 + w_p^2)}{2(\sigma_e + \sigma_a)}\right] \tag{6.3.25}$$

which agrees with the expression originally given by Fan and Byer.[20] Note that Eq. (6.3.25) reduces to Eq. (6.3.20) if we let $\sigma_a \to 0$.

6.3.5. Comparison between Diode Pumping and Lamp Pumping

Based on discussions in previous sections, we can now compare lamp pumping and diode pumping. The comparison can be made only for four-level lasers, since quasi-three-level lasers are primarily operated, via longitudinal pumping, by diodes. To compare Eqs. (6.3.22), (6.3.20), and (6.3.21), it is convenient, for diode pumping, to define a pump quantum efficiency η_{pq} as $h\nu_{mp}/h\nu_p$ where ν_p is the actual pump frequency and ν_{mp} is the minimum pump frequency, i.e., the pump frequency that would have been required for direct pumping to the upper laser level (see Fig. 6.17). Equations (6.3.20) and (6.3.21) then readily transform into

$$P_{th} = \left(\frac{\gamma}{\eta_{pd}}\right)\left(\frac{h\nu_{mp}}{\tau}\right)\left[\frac{\pi(w_0^2 + w_p^2)}{2\sigma_e}\right] \tag{6.3.26}$$

$$P_{th} = \left(\frac{\gamma}{\eta_{pd}}\right)\left(\frac{h\nu_{mp}}{\tau}\right)\left(\frac{\pi a^2}{\sigma_e\{1 - \exp[-(2a^2/w_0^2)]\}}\right) \tag{6.3.27}$$

where we have defined $\eta_{pd} = \eta_p\eta_{pq} = \eta_r\eta_t\eta_a\eta_{pq}$ as the overall pump efficiency for diode pumping. Equations (6.3.22), (6.3.26), and (6.3.27) now allow us to compare lamp pumping and diode pumping.

A first comparison can be made in terms of the four efficiency factors $\eta_r, \eta_t, \eta_a, \eta_{pq}$ and hence overall pump efficiency $\eta_p = \eta_r\eta_t\eta_a\eta_{pq}$. Limiting ourselves to the case of Nd:YAG, Table 6.3 shows estimated values of these efficiency factors, where lamp pumping values are taken from Table 6.1. In the case of longitudinal pumping by a diode laser, a 1-cm-long crystal is considered, while for transverse pumping, a 4-mm diameter rod is assumed. Note that, despite the great diversity of pumping systems considered so far, a comparison in terms of the four efficiency factors is very simple and instructive. We see that radiative and transfer efficiencies are approximately the same for lamp and diode pumping. The increase in overall pump efficiency for diode pumping, by almost a factor 10, comes from the very large increase in absorption efficiency (by almost a factor of 6) and a consistent increase in the pump quantum efficiency (by a factor of ~ 1.5). Note also that, in terms of pump efficiency, longitudinal and transverse pumping are roughly equivalent, with a slightly smaller value of absorption efficiency in transverse pumping.

A second comparison can be made in terms of threshold pump power. According to Eqs. (6.3.22) and (6.3.27) and with the same value of rod cross-sectional area, the main difference in pump thresholds between lamp pumping and transverse pumping arises for the

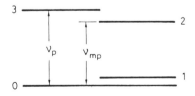

FIG. 6.17. Actual pump frequency ν_p and ideal minimum pump frequency ν_{mp} in a four-level laser.

TABLE 6.3. Comparison between pumping efficiencies of lamp pumping and diode pumping

Pump Configuration	η_r (%)	η_t (%)	η_a (%)	η_{pq} (%)	η_p (%)
Lamp	43	82	17	59	3.5
Diode (longitudinal)	50	80	98	82	32
Diode (transverse)	50	80	90	82	30

almost 10-fold increase in pump efficiency for diode pumping. Comparing longitudinal diode pumping to lamp pumping, we see from Eqs. (6.3.22) and (6.3.26) that the pump threshold for diode pumping, besides being reduced by the increased pump efficiency, is further reduced by a factor $(w_0^2 + w_p^2) \times [1 - \exp -(2a^2/w_{0l}^2)]/2a^2$, where w_{0l} is the laser spot size for lamp pumping. This factor accounts for most of the reduction in threshold pump power when w_0 and w_p are very small. A dramatic case of this type occurs in fiber lasers, where, for single-mode fibers, the value of w_0 as well as that of w_p may be as small as 2–3 μm. If, for example, we take $w_0 = w_p = 2$ μm (for a fiber laser) and $a = 2$ mm and $w_{0l} = 0.5\, a$ (for lamp pumping), the expected reduction arising from the previous geometrical factor is by almost six orders of magnitude! This is why fiber lasers exhibit such small pump thresholds. Of course, to pump single-mode fibers, single-mode, i.e., diffraction-limited, diode lasers must be used. Comparing longitudinal and transverse diode-laser pumping, we note from Eqs. (6.3.26) and (6.3.27) that the pump threshold is lower for longitudinal than transverse pumping by the ratio $(w_0^2 + w_p^2)_l [1 - \exp -(2a^2/w_{0t}^2)]/2a^2$, where suffices l and t stand for longitudinal and transverse pumping, respectively. For the very small values of spot sizes w_0 and w_p that can be used in longitudinal pumping, this ratio may again attain very small values. However, to achieve comparable outputs for the two cases, the two TEM$_{00}$ spot sizes must be made comparable. It is instructive therefore to make this comparison for the same value of spot size, i.e., for $(w_0)_l = (w_0)_t$. To avoid excessive diffraction effects arising from beam truncation at the aperture formed by the rod diameter, the spot size for transverse pumping must then be somewhat smaller than the rod radius a. In practice a value of $(w_0)_t \cong 0.7a$ may be chosen. Assuming the best overlapping condition, i.e., $w_0 = w_p$, for longitudinal pumping, it then follows that $(w_0^2 + w_p^2)_l \times [1 - \exp -(2a^2/w_{0t}^2)]/2a^2 \cong 0.48$; under these conditions, the threshold pump power for longitudinal pumping may be only a factor ~ 0.5 smaller than that for transverse pumping.

In addition to much higher pump efficiency and much lower pump threshold, compared to lamp pumping, diode pumping has the additional advantage of inducing a reduced thermal load in the active medium. In fact, for a given absorbed power P_a in the medium, the fraction $\eta_{pq}P_a$ is available in the upper laser level and, consequently, the fraction $\eta_{pq}(h\nu/h\nu_{mp})P_a$ is available as laser power ($h\nu$ is the energy of the laser photon). The power dissipated as heat is thus $[1 - \eta_{pq}(h\nu/h\nu_{mp})]P_a$. Again for Nd:YAG, since $(h\nu/h\nu_{mp}) \simeq 0.9$, we see from Table 6.3 that the thermal load for lamp pumping is ~ 2 times larger than for diode pumping. This reduced thermal load has two beneficial effects: Reduced thermal lensing and thermally induced birefringence in the rod and reduced thermal fluctuations of the refraction index of the medium, for a given pump power fluctuation. Both of these effects are important in obtaining laser operation on a single transverse and longitudinal mode of high quality.

6.4. ELECTRICAL PUMPING

We recall that this type of pumping is used for gas and semiconductor lasers. We will limit our considerations here to gas lasers and defer discussion of the more straightforward case of semiconductor-laser pumping to Chap. 9.

A gas laser is electrically pumped by allowing a current, continuous (dc current), at radio frequency (rf current), or pulsed, to pass through the gas mixture. Generally, current through the gas passes either along the laser axis direction (longitudinal discharge, Fig. 6.18a) or transversely to it (transverse discharge, Fig. 6.18b). Since the transverse dimension of a laser medium is usually much smaller than its longitudinal dimension, then, for the same gas mixture, the voltage needed in a transverse configuration is significantly less than for a longitudinal configuration. On the other hand, when confined in a dielectric (e.g., a glass) tube, as in Fig. 6.18a, a longitudinal discharge often provides a more uniform and stable pumping configuration. In the discussion that follows we concentrate on the so-called *glow discharge*, where, due to the uniformity of the current density, a uniform bluish glow of light is observed from the discharge. An *arc discharge*, where current is observed to flow in one or more streamers, emitting white light of high intensity (as in lightning), must be avoided.

A series resistance R_B, often called the ballast resistance (Figs. 6.18a–b) is required to stabilize the discharge at the desired operating point. To explain this feature, Fig. 6.19 shows (*solid line*) the voltage versus current characteristic of a gas discharge. Note that, in the operating region, voltage across the discharge remains nearly constant as discharge current increases. A peak voltage V_p, about an order of magnitude larger than this constant operating voltage, is required to induce gas breakdown. Thus the behavior of a discharge tube is very different from that of a simple resistor! The same figure also shows (*dashed line*) the voltage versus current characteristic of a power supply, giving a voltage V_0, in series with a ballast resistance R_B. Note that the current may stabilize at either of the intersections A or C of the two curves. (Intersection B corresponds to an unstable equilibrium situation.) Thus starting with an initially unenergized lamp and applying voltage from the power supply, the lamp stabilizes itself at point C with very little current flowing in the discharge. To reach the other stable point (A, the desired operating point), we can briefly raise the applied voltage to overcome the voltage barrier V_p. This is usually achieved by applying an overvoltage to the high-voltage electrode long enough to produce sufficient gas ionization (as in Figs. 6.4b and 6.5). Alternatively, a high-voltage pulse can be applied to some auxiliary electrode (as in Fig. 6.4a).

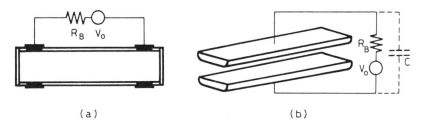

(a)   (b)

FIG. 6.18. Most frequently used pumping configurations for gas discharge lasers: (a) Longitudinal discharge and (b) transverse discharge.

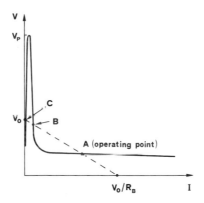

FIG. 6.19. Voltage V versus current I characteristic of a gas discharge (*solid line*) and of a power supply with a series resistance.

Various different electrode structures are used for longitudinal and transverse discharges. In a longitudinal discharge, electrodes often have an annular structure, where the cathode surface is usually much larger than that of the anode to help reduce degradation due to the impact of the heavier ions. In a transverse discharge, electrodes extend over the whole length of the laser material and the opposing surfaces of the two electrodes must have a very smooth curvature (see Fig. 6.18b). In fact, if there is any sharp corner, the high electric field produced there easily results in arc formation rather than a uniform discharge. Longitudinal discharge arrangements are usually used only for cw lasers, while transverse discharges are used with cw, pulsed, or rf lasers. Figure 6.20, shows a particularly interesting case of a transverse discharge using rf excitation where rf electrodes are applied to the outside of the discharge tube, usually made of glass. The presence of a finite thickness of glass tube presents several advantages: (1) It acts as a series capacitor for the discharge whose impedance at the frequency of the rf voltage acts as an effective capacitive ballast for stabilizing the discharge. The loss of pump power in the resistive ballast R_B in Fig. 6.18 is

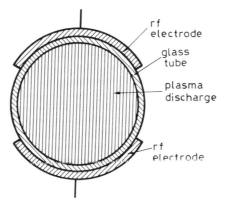

FIG. 6.20. Radio frequency transverse excitation of a gas in a quartz tube.

thus avoided. (2) Since the glass dielectric medium extends over the entire electrode structure, the problem of arc formation is greatly reduced. (3) Since the gas mixture is not in contact with the electrodes, plasma-chemical effects, occurring at the electrode surface and leading to mixture dissociation, are eliminated. When this configuration is applied to a CO_2 laser, for instance, an order of magnitude reduction in electrode maintenance time can be gained and a factor of two decrease in the gas consumption rate.

We now describe the physical phenomena leading to excitation in the gas. First we recall that, in an electrical discharge, both ions and free electrons are produced. Since these charged particles acquire additional kinetic energy from the applied electric field, they are able to excite a neutral atom by collision. Owing to their much greater mass, positive ions are accelerated to much lower velocities than electrons and, therefore, do not play a significant role in the excitation process. Thus electrical pumping of a gas usually occurs via one or both of the following processes: (1) In a gas consisting of only one species, excitation is produced only by electron impact, i.e., the process

$$e + X \rightarrow X^* + e \qquad (6.4.1)$$

where X and X^* represent the atom in the ground state and excited state, respectively. Such a process is referred to as a *collision of the first kind*. (2) In a gas consisting of two species (say, A and B), excitation can also occur from collisions between atoms of different species through *resonant energy transfer* (see Sect. 2.6.1). Referring to Fig. 6.21, let us assume that species B is in the ground state and species A is in the excited state as a result of electron impact. We also assume that the energy difference ΔE between the two transitions is less than kT. In this case there is an appreciable probability that, after collision, species A will be found in its ground state and species B in its excited state. The process can be denoted by

$$A^* + B \rightarrow A + B^* - \Delta E \qquad (6.4.2)$$

where the energy difference ΔE is added to, or subtracted from, the translational energy of the colliding partners depending on its sign; this is why ΔE must be smaller than kT. This process provides a particularly effective way of pumping species B if the upper state of A is metastable (forbidden transition). In this case, once A is excited to its upper level, it remains there for a long time, thus constituting an energy reservoir for excitation of species B. A process of the type indicated in Eq. (6.4.2) is referred to as a *collision of second kind**.

In the discussion that follows we limit our considerations to the electron-impact excitation process, since it is the commonest and simplest excitation mechanism. Electron impact excitation also constitutes the first step in the near-resonant energy transfer process.

---

* Collisions of *the first kind* involve converting kinetic energy of one species into potential energy of another species. In collisions of *the second kind*, potential energy is converted into some other form of energy (other than radiation), such as kinetic energy, or transformed into potential energy (in the form of electronic, vibrational, or rotational energy) of another like or unlike species. Collisions of the second kind therefore include not only the reverse of collisions of the first kind (e.g., $e + X^* \rightarrow e + X$) but also, for instance, the conversion of excitation energy into chemical energy.

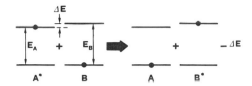

FIG. 6.21. Laser pumping by near-resonant energy transfer.

6.4.1. Electron Impact Excitation

Electron impacts involve both elastic and inelastic collisions. In an inelastic collision, the atom is either excited to a higher state or ionized. Of the various possible excitations, the one we are usually interested in is that which excites the atomic species to the desired upper laser level. To describe the preceding excitation phenomena by appropriate collision cross sections, we first consider the simple case of impact excitation by a collimated beam of mono-energetic electrons. If F_e is the electron flux (number of electrons per unit area per unit time), a total cross section σ_e can be defined similarly to the case of a photon flux [see Eq. (2.4.19)]. Thus if we let dF_e be the change of flux resulting from a beam traveling a distance dz in the material, we can write

$$dF_e = -\sigma_e N_t F_e\, dz \qquad (6.4.3)$$

where N_t is the total population of the atomic species. Collisions that produce electronic excitation account for only a fraction of the total cross section. In fact the cross section for elastic collisions σ_{el} is usually the largest, its order of magnitude being $\sim 10^{-16}$ cm^2. If we now let σ_{e2} be the cross section for electronic excitation from the ground level to the upper laser level, then, according to Eq. (6.4.3), the population rate of the upper state due to the pumping process is

$$\left(\frac{dN_2}{dt}\right)_p = \sigma_{e2} N_g F_e = N_g N_e v \sigma_{e2} \qquad (6.4.4)$$

where v is the electron velocity and N_e is the electron density. The calculation of the pump rate requires knowing the value of σ_{e2}, which is expected to depend on the energy E of the incident electron, i.e., $\sigma_{e2} = \sigma_{e2}(E)$. In fact, in a gas discharge, electrons have a distribution of energies. In this case the population rate of the upper state is obtained from Eq. (6.4.4) by averaging over this distribution, namely:

$$\left(\frac{dN_2}{dt}\right)_p = N_t N_e \langle v\sigma_{e2}\rangle \qquad (6.4.5)$$

In the preceding equation, one has

$$\langle v\sigma\rangle = \int v\sigma(E) f(E) dE \qquad (6.4.6)$$

232

where $f(E)$ describes the energy distribution of the discharge electrons. According to Eqs. (1.3.1) and (6.4.5), the pump rate is then given by

$$R_p = N_t N_e \langle v\sigma_{e2} \rangle \tag{6.4.7}$$

where $\langle v\sigma \rangle$ is given by Eq. (6.4.6). The calculation of R_p thus requires knowledge of the energy dependence of both σ and f. This dependence is considered in the following sections.

6.4.1.1. Electron Impact Cross Section

Figure 6.22 shows the qualitative behavior of σ versus the electron energy E for the three cases: (a) Optically allowed transition. (b) Optically forbidden transition involving no multiplicity change. (c) Optically forbidden transition involving a multiplicity change. In all three cases the peak value of σ is normalized to unity. Note that, in each case, there is a distinct threshold E_{th} for the cross section. As expected, the value of E_{th} is close to the energy of the transition involved. The cross section rises very sharply above threshold, reaches a maximum value, and thereafter decreases slowly. The peak value of σ and the width of the curve depend on the type of transition involved:

- For an optically allowed transition, the peak value of σ can typically be $10^{-16}\,\text{cm}^2$, and the width of the curve can typically be 10 times greater than the threshold energy (curve a in Fig. 6.22).
- For an optically forbidden transition involving no change of multiplicity, the peak cross section is drastically reduced by nearly three orders of magnitude (to about $10^{-19}\,\text{cm}^2$), and the width of the curve may be only 3–4 times the threshold energy (curve b in Fig. 6.22).

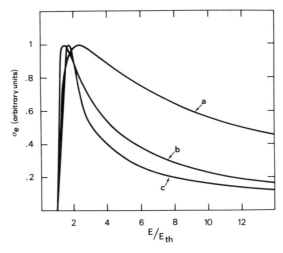

FIG. 6.22. Qualitative behavior of electron impact excitation cross section versus the energy of the incident electron: (a) Optically allowed transitions, (b) optically forbidden transitions involving no change of multiplicity, and (c) optically forbidden transitions involving a change of multiplicity.

- When a change of multiplicity is involved, the peak cross section may actually be larger than in an optically forbidden transition. The width of the curve may now typically be equal to or somewhat smaller than the threshold energy E_{th} (curve *c* in Fig. 6.22).

It should be noted that, in any case, the width of each curve is roughly comparable to the threshold energy, i.e., to the transition energy. In contrast, transition linewidths for photon absorption are much sharper (typically 10^{-4}–10^{-6} of the transition frequency). This very important circumstance arises from the fact that, as explained in Sect. 6.1, electron impact excitation is basically a nonresonant phenomenon. This is why excitation of a gaseous medium is performed much more effectively by a polychromatic source of electrons (such as in a gas discharge) than by a polychromatic light source (such as a lamp).

To provide deeper insight into the mechanisms involved in electron impact excitation, we sketch the procedure for a quantum mechanical calculation of the cross section σ. For optically allowed transitions or optically forbidden transitions involving no multiplicity change, the simplest, and often the most accurate, calculation uses the Born approximation. Before collision, the atom is described by the ground-state wave function u_1 and the incident electron by the plane wave function $\exp(j\mathbf{k}_0 \cdot \mathbf{r})$, where $\mathbf{k}_0$ is the electron wave vector and $\mathbf{r}$ is the vector describing the position of the incident electron with respect to a center situated, e.g., at the nuclear position. After collision, the atom is described by the upper state wave function u_2 and the scattered electron by the plane wave $\exp(j\mathbf{k}_n \cdot \mathbf{r})$, where $\mathbf{k}_n$ is the wave vector of the scattered electron. For the discussion that follows, we recall that $k = 2\pi/\lambda$, where λ is the de Broglie wavelength of the electron. It can be expressed as $\lambda = [1.23/(V)^{1/2}]$ nm, where V is the electron energy in electron volts. The interaction has its origin in the electrostatic repulsion between the incident electron and electrons of the atom. This interaction is assumed to be weak enough for there to be only a very small probability of a transition occurring in the atom during the impact and for the chance of two such transitions to be negligible. In this case the Schrödinger equation for the problem can be linearized. It then turns out that the transition rate, and hence the transition cross section, can be expressed as

$$\sigma_e \propto \left| \int [u_2 \exp(j\mathbf{k}_n \cdot \mathbf{r})]^*[u_1 \exp(j\mathbf{k}_0 \cdot \mathbf{r})]dV \right|^2 \qquad (6.4.8)$$

From the preceding expression for the de Broglie wavelength and assuming an electron energy of only a few eV, the wavelength $\lambda' = 2\pi/|\mathbf{k}_0 - \mathbf{k}_n| = 2\pi/|\Delta\mathbf{k}|$ is seen to be appreciably larger than the atomic dimensions (~ 0.1 nm). This means that $(\Delta\mathbf{k} \cdot \mathbf{r}) \ll 1$ for $|\mathbf{r}| \leq a$, where a is the atomic radius. In this case, the factor $\exp j[(\mathbf{k}_0 - \mathbf{k}_n) \cdot \mathbf{r}] = \exp j(\Delta\mathbf{k} \cdot \mathbf{r})$ in Eq. (6.4.8) can be expanded as a power series of $(\Delta\mathbf{k} \cdot \mathbf{r})$. Since u_1 and u_2 are orthogonal functions, the first term in this expansion, which gives a nonvanishing value for σ_e, is $j(\Delta\mathbf{k} \cdot \mathbf{r})$; we thus obtain

$$\sigma_e \propto \left| \int u_2^* \mathbf{r} u_1 \, dV \right|^2 = |\mathbf{\mu}_{21}|^2 \qquad (6.4.9)$$

where $\mathbf{\mu}_{21}$ is the matrix element of the electric dipole moment of the atom [see Eq. (2.3.7)]. It then follows that, when $\mathbf{\mu}_{21} \neq 0$, i.e., when the transition is optically allowed, the electron impact cross section is proportional to the photon absorption cross section. Thus strong optically-allowed transitions are expected also to show a large cross section for electron impact. For optically-forbidden transitions involving no change of multiplicity ($\Delta S = 0$, e.g., the $1^1S \to 2^1S$ transition in He; see Chap. 10), Eq. (6.4.8) gives a nonvanishing value for the next higher order term in the expansion of $\exp j(\Delta \mathbf{k} \cdot \mathbf{r})$, namely, $-(\Delta \mathbf{k} \cdot \mathbf{r})^2/2$. This means that σ_e can now be written as $\sigma_e \propto |\int u_2^*(\Delta \mathbf{k} \cdot \mathbf{r})^2 u_1 \, dV|^2$. This relation is completely different from the corresponding one that would apply in the case of a photon interaction, i.e., due to a magnetic-dipole interaction. It is therefore no surprise to find that the ratio between the two peak cross sections $\sigma_{\text{forbidden}}/\sigma_{\text{allowed}}$ is, in this case, about 10^{-3}, while the same ratio is $\sim 10^{-5}$ for photon absorption [see Eq. (2.4.14)]. Therefore, optically forbidden transitions are relatively more easily excited by electron impact than by photon impact. This circumstance has some profound consequences for the operating principles of gas lasers, since pumping is often achieved through optically forbidden transitions.

When a change of multiplicity is involved (e.g. the $1^1S \to 2^3S$ transition in He; see Chap. 10), the Born approximation gives a zero cross section in any order of the expansion of $\exp j(\Delta \mathbf{k} \cdot \mathbf{r})$. In fact, such a transition involves a spin change, while, within the Born approximation, the incoming electron, through its electrostatic interaction, can couple only to the orbital motion of the atom rather than to its spin.* The theory in this case is largely due to Wigner, and its starting point is the observation that, in a collision, the total spin of the atom and of the incident electron must be conserved, not necessarily that of the atom alone. Transitions may therefore occur by an electron *exchange collision*, where the incoming electron replaces the electron of the atom involved in the transition and this electron is in turn ejected by the atom. To conserve total spin, the incoming electron must then have its spin opposite to that of the ejected one. To clarify this exchange process, Fig. 6.23 shows the electron impact excitation of the $1^1S \to 2^3S$ transition in He. Note that the process can be visualized as the incident electron, labeled 1, is captured in the $2s$ state of He, while the electron of the atom with opposite spin, labeled 2, is actually ejected; however, this constitutes a very naive way of describing the phenomenon because, during collision, the two electrons are quantum mechanically indistinguishable. Nevertheless, from this simple description, we understand that this exchange mechanism may be even more resonant than those considered in the Born approximation. In fact, there is a high probability that this exchange occurs only if the energy of the incoming electron closely matches the transition energy. In this case, in fact, the electron energy is just what is needed to leave electron 1 after collision in the upper level $2s$, while the second electron, electron 2, is ejected with zero velocity. For higher energies of the incident electron, the exchange process leaves electron 1 in the $2s$ orbital, while electron 2, ejected from the atom, must carry off the corresponding surplus energy. This is definitely less likely to occur. Having established that

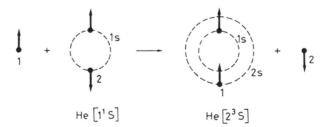

$$\text{He } [1^1S] \qquad\qquad \text{He } [2^3S]$$

FIG. 6.23. The phenomenon of electron exchange in the case of the $1^1S \to 2^3S$ transition in a He atom.

---

* This assumes a negligible spin orbit coupling, which is true for light atoms (e.g., He, Ne), while it is not true for heavy atoms, such as Hg.

this process is somewhat resonant, we now understand why the peak cross section can be even higher than for optically forbidden transitions involving no spin change.

6.4.2. Thermal and Drift Velocities

As mentioned in previous sections, electrons are responsible for the phenomena occurring in a gas discharge. They acquire energy from the applied electric field, then lose or exchange energy through three processes:

- *Inelastic collisions with atoms, or molecules,* of the gas mixture, which either raise the atom to one of its excited states or ionize it: These electron impact excitation or ionization phenomena are perhaps the most important processes in laser pumping, hence the extended discussion in these sections.
- *Elastic collisions with atoms:* If we assume that atoms are at rest before collision (the mean velocity for an atom is indeed much smaller than for an electron), the electron loses energy on collision. Straightforward analysis of the elastic-collision process shows that, for random direction of the scattered electron, the electron loses on average a fraction $2(m/M)$ of its energy, where m is the mass of electron and M is the mass of the atom. Note that this loss is very small, since m/M is small (e.g., $m/M = 1.3 \times 10^{-5}$ for Ar atoms).
- *Electron-electron collisions:* For a gas ionized to even a moderate degree, the frequency of such collisions is usually high, since both particles are charged and exert forces on one another over a considerable distance. Moreover, since both colliding particles have the same mass, the energy exchange in the collision may be considerable.

As a result of such collisions and as a consequence of electrons being accelerated by the electric field of the discharge, the electron gas in the plasma acquires a distribution of velocities. We can describe this by introducing the distribution $f(v_x, v_y, v_z)$ with the meaning that $f(v_x, v_y, v_z)dv_x\, dv_y\, dv_z$ gives the elemental probability that the electron is found with velocity components in a range dv_x, dv_y, dv_z, around v_x, v_y, v_z. Given this distribution, we can define a *thermal velocity* v_{th} so that

$$v_{th}^2 = \langle v^2 \rangle \tag{6.4.10}$$

where the average is taken over the velocity distribution. Similarly, we can define a *drift velocity* v_{drift} as the average velocity along the field direction, i.e.,

$$v_{drift} = \langle v_z \rangle \tag{6.4.11}$$

where the z-axis is taken in the field direction and the average is taken over the electron velocity distribution.

To calculate roughly both v_{th} and v_{drift}, we make the simplifying assumption that, at each collision, some constant fraction δ of the kinetic energy of the electron is lost. A first equation is then obtained from a power balance consideration: The average power lost by the electron must equal the average power delivered to the electron by the external field. Note that the average kinetic energy of the electron is $mv_{th}^2/2$, while v_{th}/l, where l is the electron mean free path, is the average collision rate. The average power lost by the electron is therefore $\delta(v_{th}/l)(mv_{th}^2/2)$; this must equal the power supplied by the electric field $\mathscr{E}$, namely, $e\mathscr{E}v_{drift}$. Hence:

$$e\mathscr{E}v_{drift} = \delta\left(\frac{v_{th}}{l}\right)\left(\frac{mv_{th}^2}{2}\right) \tag{6.4.12}$$

The second equation is obtained from the requirement of an average momentum balance between two consecutive collisions. We assume that after each collision, the electron is scattered in a random direction; hence it loses its preferential drift velocity. Referring to Fig. 6.24, electron velocity at point 1, after the first collision, is thus assumed to have a magnitude equal to the thermal velocity v_{th} and a direction making a general angle θ to the field direction. During its free flight between points 1 and 2, the electron is accelerated by the electric field; at point 2, just before the next collision, it has acquired an additional velocity v_{drift} in the field direction, with a direction opposite to the field. The impulse produced by the corresponding force is $-e\mathscr{E}l/v_{th}$, where l is the distance between points 1 and 2 (which is assumed, on the average, to equal the electron mean free path). This impulse can now be equated to the change of momentum, i.e. $(m\mathbf{v}' - m\mathbf{v}_{th}) = m\mathbf{v}_{drift}$. In terms of their magnitudes we can then write

$$e\mathscr{E}l = mv_{th}v_{drift} \tag{6.4.13}$$

which, together with Eq. (6.4.12), provides the two required equations. From these equations one obtains

$$v_{th} = \left(\frac{2}{\delta}\right)^{1/4}\left(\frac{e\mathscr{E}l}{m}\right)^{1/2} \tag{6.4.14}$$

and

$$v_{drift} = \left(\frac{\delta}{2}\right)^{1/4}\left(\frac{e\mathscr{E}l}{m}\right)^{1/2} \tag{6.4.15}$$

It should be noted that, on taking the ratio of Eq. (6.4.15) and Eq. (6.4.14), we obtain

$$\left(\frac{v_{drift}}{v_{th}}\right) = \left(\frac{\delta}{2}\right)^{1/2} \tag{6.4.16}$$

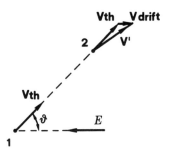

FIG. 6.24. Calculation of drift velocity resulting from electron acceleration by the external electric field between two consecutive collisions.

We mentioned earlier that, after undergoing elastic scattering with an atom, an electron loses a fraction of its kinetic energy equal on the average to $2m/M$. If we then assume $\delta \cong 2m/M$, we obtain from Eq. (6.4.16) $(v_{drift}/v_{th}) \cong (m/M)^{1/2} \cong 10^{-2}$. Thus drift velocity is a very small fraction of the thermal velocity, so that we can consider the movement of electrons in a gas as a slowly drifting swarm of randomly moving particles rather than a particle stream.

Equation (6.4.16) is rather crude, since it assumes that the electron loses a constant fraction δ of its energy in each collision. Although this is true for elastic collisions with atoms, this is obviously not true for inelastic collisions, where energy lost equals the excitation energy of the atom. Note that, although elastic collisions are actually more frequent than inelastic collisions, energy lost in an elastic collision is however very small. Thus if elastic collisions were the dominant process, the discharge would not provide a particularly efficient means of pumping a laser; most of the discharge energy would, in fact, be used to heat rather than to excite atoms. Note that electron-electron collision does not play a role in the energy balance equation expressed in Eq. (6.4.12), since this process simply redistributes electron velocities without changing their average energy.

6.4.3. *Electron Energy Distribution*

We now consider the distribution of electron velocities or electron energies in a gas discharge. If energy redistribution, due to electron-electron collision, is rapid enough compared to energy loss due to both elastic and inelastic collisions with atoms, then the statistical mechanics prediction is that the distribution of electron velocities (or energies) is given by the Maxwell–Boltzmann (MB) distribution function. This can be described for instance by the energy distribution function $f(E)$, where $f(E)dE$ is the elemental probability that electron's kinetic energy lies between E and $E + dE$. We can then write

$$f(E) = \left(\frac{2}{\pi^{1/2}kT_e}\right)\left(\frac{E}{kT_e}\right)^{1/2} \exp[-(E/kT_e)] \tag{6.4.17}$$

where T_e is the electron temperature. Thus, when the distribution can be described by the MB law, the electron temperature is the only parameter that needs to be specified to characterize the distribution.

Once T_e is known, one can calculate v_{th} from Eq. (6.4.10) using the electron energy distribution given by Eq. (6.4.17). Using the standard relation $v^2 = 2E/m$, we readily obtain from Eq. (6.4.10)

$$v_{th} = \left(\frac{3kT_e}{m}\right)^{1/2} \tag{6.4.18}$$

which relates v_{th} to T_e. From Eqs. (6.4.18) and (6.4.14) we then obtain

$$T_e = \left(\frac{2}{\delta}\right)^{1/2}\left(\frac{e}{3k}\right)(\mathscr{E}l) \tag{6.4.19}$$

Since the electron mean free path l is inversely proportional to the gas pressure p, Eq. (6.4.19) shows that, for a given gas mixture, T_e is proportional to the ratio $\mathscr{E}/p$. A more detailed treatment than the simple one leading to Eq. (6.4.14) shows that T_e is a function of $\mathscr{E}/p$ rather than being simply proportional to this ratio, i.e.:

$$T_e = f\left(\frac{\mathscr{E}}{p}\right) \qquad (6.4.20)$$

The $\mathscr{E}/p$ ratio is thus the fundamental quantity involved in establishing a given electron temperature, and it is often used in practice to specify discharge conditions.

We now address the question of whether the electron energy distribution can actually be described by MB statistics. Indeed, an obvious reason why the distribution cannot be Maxwellian is that a MB distribution implies that the velocity distribution in space is isotropic. If this were the case, the drift velocity, as defined by Eq. (6.4.11), would be zero and no current would flow in the discharge! We have seen, however, that the drift velocity is a very small fraction of the thermal velocity; consequently, the role of the drift velocity in altering the MB distribution may be considered negligible. A second important reason for the distribution not being Maxwellian occurs for a weakly ionized gas with high values for the electron impact cross sections, e.g., for CO_2 or CO gas-laser mixtures. In this case, due to the low electron concentration, the energy redistribution process from electron-electron collisions does not proceed at a sufficiently rapid rate compared to that for inelastic collisions. As we discuss in more depth in Example 6.4, we thus expect to find dips in the energy distribution function at energies corresponding to specific transitions of the molecules. In contrast, for neutral atom or ion gas lasers, electron density is much higher because these lasers are relatively inefficient, and, as discussed in Example 6.5, the departure from a Maxwellian distribution is expected to be less significant.

Example 6.4. *Electron energy distribution in a CO_2 laser.* We show in Fig. 6.25 the situation occurring for a CO_2:N_2:He gas mixture with a 1:1:8 ratio between corresponding partial pressures. In the figure the electron impact cross section for N_2 excitation up to the $v = 5$ vibrational level is shown[10] (the main pumping mechanism is in fact by energy transfer from an excited N_2 molecule to the lasing CO_2 molecule). As a result of the very high value for the peak cross section ($\approx 3 \times 10^{-16}$ cm^2) for the N_2 molecule and also as a result of the low value for the current density required in a CO_2 laser (the CO_2 laser is one of the most efficient lasers), the assumption of a Maxwellian distribution is expected to be inadequate in this case. Thus, to calculate the correct electron energy distribution, we must perform an *ab initio* calculation using the appropriate electron transport equation (the Boltzmann transport equation) where all possible electron collision processes leading to excitation (or de-excitation) of vibrational and electronic levels of all gas species are taken into account.[11] The electron distribution $f(E)$, computed in this way for an $\mathscr{E}/p$ ratio of ~ 8 Vcm^{-1} Torr^{-1} corresponding to an average electron energy* of ≈ 1.7 eV, is indicated as a solid line in the same figure.[12] For comparison, the Maxwellian distribution $f'(E)$ for the same average energy is also shown as a dashed line. Note that the depression of the $f(E)$ curve compared to the Maxwellian curve in the figure, for $E > 2$ eV, is due to the very high value of the electron impact cross section for N_2. In fact, when accelerated by the electric field of the discharge, few electrons go beyond the $E = 2$ eV barrier, since they would be immediately involved in N_2 excitation. Consequently, electrons accumulate in the energy range below 2 eV.

---

* Although for a non-Maxwellian distribution the concept of temperature loses its meaning, we can still define an average electron energy, and, for a Maxwellian distribution, this energy turns out to be a function of the $\mathscr{E}/p$ ratio.

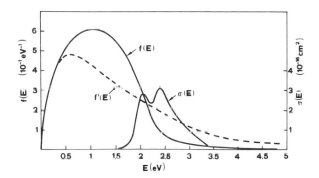

FIG. 6.25. Comparison of the electron energy distribution $f(E)$ for a 1:1:8 CO_2:N_2:He mixture with a Maxwellian distribution $f(E)$ of the same average energy. (Redrawn from Ref. 12.) In the same figure, the electron impact cross section, $\sigma(E)$ for N_2 excitation up to the $v = 5$ vibrational level is also shown. (Redrawn from Ref. 10.) The redrawn curves indicate the physical situation rather than actual original values shown in the cited references.

Example 6.5. *Electron energy distribution in a He-Ne laser.* In contrast to the results of Example 6.4, Fig. 6.26 shows the situation applying to a helium discharge under conditions appropriate to a He-Ne laser. The figure shows two plots of the electron impact cross section to the 2^1S and 2^3S levels of He, versus electron energy. As in Example 6.4, in fact, the main pumping mechanism arises from energy transfer between an excited He atom and the lasing Ne atom. Note however that the peak values of the cross sections are, in this case, about two orders of magnitude smaller than for the N_2 molecule. Since current density and hence electron density are also much higher (the He-Ne laser is a rather inefficient laser) the Maxwellian distribution is expected to hold in this case. Accordingly we show in the same figure a Maxwellian distribution with a mean electron energy of 10 eV, which is the average electron energy in a He-Ne laser corresponding to the optimum excitation condition (see Sect. 6.4.5). Note the much higher value of the average electron energy in this case than in the previous case, a consequence of the fact that we must excite electronic energy levels rather than vibrational energy levels.

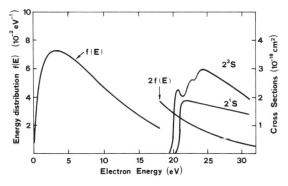

FIG. 6.26. Electron energy distribution $f(E)$ and electron impact cross section for $1^1S \rightarrow 2^1S$ and $1^1S \rightarrow 2^3S$ transitions of He.

Example 6.6. *Thermal and drift velocities in He-Ne and CO$_2$ lasers.* Based on Example 6.5, we assume for a He-Ne laser an average electron energy $\langle E \rangle \cong 10$ eV. This means that $(mv_{th}^2/2) = \langle E \rangle = 10$ eV. Therefore $v_{th} \cong 1.9 \times 10^6$ m/s. Since the electron velocity distribution is assumed in this case to be Maxwellian, then, according to Eq. (6.4.18), the electron temperature can be obtained from the relation $T_e = 2\langle E \rangle/3k$. We obtain $T_e \cong 7.7 \times 10^4$ K. Note that the electron temperature value is much higher than room temperature. To calculate the drift velocity, we use Eq. (6.4.16) and assume that the dominant cooling process for the electrons is by elastic collisions with the lighter He atoms. We then obtain $(v_{drift}/v_{th}) \approx (m/M_{He})^{1/2} \cong 1.16 \times 10^{-2}$, where M_{He} is the mass of the helium, so that $v_{drift} \cong 2.2 \times 10^4$ m/s. In the case of a CO$_2$ laser, based on Example 6.4, we assume an average electron energy of $\langle E \rangle \cong 1.7$ eV. From the relation $(mv_{th}^2/2) = \langle E \rangle$, we obtain $v_{th} \cong 0.78 \times 10^6$ m/s. The drift velocity can then be obtained from Ref. 12, assuming an $\mathscr{E}/p$ ratio of ~ 8 V cm^{-1} Torr^{-1} and a 1:1:8 partial-pressure ratio of the CO$_2$:N$_2$:He mixture. We obtain $v_{drift} \cong 6 \times 10^4$ m/s. Note that, in this case, we cannot speak of electron temperature, since the electron energy distribution departs considerably from a Maxwellian distribution. Note also that, in both cases, the thermal velocity is $\sim 10^6$ m/s, so that the drift velocity is ~ 100 times smaller.

6.4.4. Ionization Balance Equation

In an electrical discharge, electrons and ions are continuously created in the discharge volume by electron impact. Ionization is produced by hot electrons present in the discharge, i.e., those whose energy is larger than the atom's ionization energy. In the steady state, this ionization process must be counterbalanced by some electron-ion recombination process. Radiationless electron-ion recombination cannot occur within the discharge volume, however, because this process cannot conserve both the total momentum and total energy of the particles. Let us consider, for simplicity, head-on collisions. Invoking momentum conservation, the velocity v of the recombined atom is given by $v = (m_1 v_1 + m_2 v_2)/(m_1 + m_2)$, where m_i $(i = 1, 2)$ are the masses and v_i the velocities of the electron and ion before collision. On the other hand, for energy conservation, we require $[(m_1 v_1^2/2) + (m_2 v_2^2)/2] = [(m_1 + m_2)v^2/2] + E_r$ where E_r is the energy released by the electron-ion recombination. For given values of

m_1, m_2, v_1 and v_2, the momentum and energy conservation relations thus supply two equations for the unknown quantity v, the velocity of the recombined atom. Thus, in general, these two equations cannot both be satisfied. Radiative ion-electron recombination, on the other hand, is an unlikely process at the carrier concentrations for a gas laser. The recombination process can thus occur only in the presence of a third partner M, since momentum and energy conservation can be conserved in a three-body collision process. In fact, again assuming head-on collisions, we now have a pair of equations in two unknowns v, the velocity of the recombined atom, and v_M, the velocity of the third partner M after collision. At the low pressures of a gas laser (a few Torr) and if the gas mixture is contained in a cylindrical tube, the necessary third partner M is simply provided by the tube walls. Thus, in a gas laser, electron-ion recombination mostly occurs at the tube walls.

We must now realize that, although electron velocity is much larger than ion velocity, the movement of electrons and ions to the walls must occur together. In fact, if electrons were arriving at the walls more rapidly than ions, a radial field would be established. This field would accelerate the movement of the ions toward the wall and decelerate electrons. For the usual electron and ion concentrations in a gas discharge, this space-charge effect would be quite substantial; consequently, electrons and ions must move to the tube walls at the same rate. This movement can occur by two different mechanisms depending on gas pressure p and tube radius R. If the ion mean free path is much shorter than R, electrons and ions diffuse together to the walls, and recombination occurs via *ambipolar diffusion*. If the ion mean free path is comparable to the tube radius (as happens in the relatively low-pressure ion-gas lasers), electrons and ions reach the wall by free flight rather than by diffusion. The analytical theory of ambipolar diffusion can be obtained from the Schottky theory of a discharge in the so-called positive column.[13] In the low-pressure limit, on the other hand, the *free-fall* model of Tonks–Langmuir for the plasma discharge should be used.[14]

The two theories are rather complicated, and their description goes beyond the scope of this book. In both theories, however, a balance equation must always hold between the number of electron-ion pairs produced and the number of electron-ion pairs recombining at the walls (*ionization balance equation*). So, in the case of the Schottky theory, the balance equation can be written in our notation as

$$\langle v\sigma_i \rangle N_g = \frac{kT_e}{e} \mu_+ \left(\frac{2.405}{R} \right)^2 \tag{6.4.21}$$

where σ_i is the ionization cross section, μ_+ is the ion mobility, and R is the tube radius. We see that, for a given atomic species, i.e., with a given expression for $\sigma_i = \sigma_i(E)$, the average value $\langle v\sigma_i \rangle$ in Eq. (6.4.21) depends only on the electron temperature T_e. Ionization is in fact produced by the most energetic electrons in the energy distribution, and their number depends on the temperature T_e. We also recognize that N_g is proportional to the gas pressure, while the ion mobility is inversely proportional to the pressure. Equation (6.4.21) can then be rearranged as

$$f(T_e) = \frac{C}{(pR)^2} \tag{6.4.22}$$

where we have set $\langle v\sigma_i \rangle / kT_e = f(T_e)$ and where C is a suitable constant. Thus, for a given atomic species, the ionization balance equation leads to a relation between T_e and pR just as the energy and momentum balance equation leads to a relation between T_e and $\mathcal{E}/p$ [see Eq. (6.4.20)]. The functional relation $f = f(T_e)$ in Eq. (6.4.22) is such that T_e increases as pR decreases. In fact, if for a given tube radius we decrease gas pressure, the electron ion recombination due to diffusion to the walls increases. The electron temperature must therefore increase to maintain the balance between ionization and recombination. In the case of the Tonks–Langmuir theory, a similar functional relation again exists between T_e and the pR product.

6.4.5. Scaling Laws for Electrical Discharge Lasers

Equations (6.4.20) and (6.4.22) provide two fundamental relations that can be used to understand a number of aspects of the physical behavior of any gas discharge. For example, we can now explain why, in a stable glow discharge, voltage across a discharge tube is essentially independent of the current flowing (see Fig. 6.19). In fact, if we consider some given gas tube, i.e., with given values of tube radius and gas pressure, then, according to Eq. (6.4.22), the electron temperature is fixed. We then see from Eq. (6.4.20) that the electric field must also be fixed and thus independent of the discharge current.

Let us now see the consequences of Eqs. (6.4.20) and (6.4.2) for a gas-laser discharge. First we observe that, for a given gas medium, an optimum value of electron temperature T_{opt} exists if we want to maximize the pump rate to the upper laser level. Too low a value for the electron temperature, in fact, results in insufficient electron energy to excite the upper laser level. Electron energy is then lost primarily through elastic collisions and excitation of lower levels of the medium, including the lower laser level. Too high a value of the electron temperature, on the other hand, leads to strong excitation of higher levels of the gas mixture (which may not be coupled to the upper laser level) or produces excessive ionization of the gas mixture (which may result in a discharge instability, i.e., a transition from a glow discharge to an arc). If we then set $T_e = T_{opt}$ on the left-hand side of both Eqs. (6.4.20) and (6.4.22), we obtain

$$\left(\frac{\mathscr{E}}{p}\right) = \left(\frac{\mathscr{E}}{p}\right)_{opt} \tag{6.4.23a}$$

$$(pD) = (pD)_{opt} \tag{6.4.23b}$$

where $D = 2R$ is the tube diameter. Thus, for a given gas mixture, some optimum values exist for both pD and $\mathscr{E}/p$ if the mixture is used as the active medium in a gas laser.

Equations (6.4.23) establish the *scaling laws* for any gas laser. To demonstrate application of these laws, let us assume that we start with the optimum operating conditions and that, for some reason, we need to decrease the tube diameter by, e.g., a factor of 2. Equation (6.4.23b) then shows that we must increase the pressure of the gas mixture by the same factor for the laser to still operate with optimum efficiency. If the pressure is doubled, then, according to Eq. (6.4.23a), the electric field $\mathscr{E}$ in the gas discharge and hence the total voltage V across the laser tube must also double. This means that the V versus I characteristic of the given laser tube (*solid line* in Fig. 6.19) will automatically scale up by a factor of 2 in voltage at any given current. The open circuit voltage of the power supply V_0 and the ballast resistance R_B must then be redesigned to have the desired current flowing in the discharge tube.

6.4.6. Pump Rate and Pump Efficiency

To calculate the pump rate, we first recall the standard equation $J = ev_{drift}N_e$, which relates the current density J to the electron density N_e of a discharge. From Eq. (6.4.7) we then obtain

$$R_p = N_t \frac{J}{e} \left(\frac{\langle v\sigma \rangle}{v_{drift}}\right) \tag{6.4.24}$$

If a Maxwellian electron-energy distribution is taken, the term $\langle v\sigma \rangle$ in Eq. (6.4.24) will depend only on the electron temperature T_e. According to Eqs. (6.4.15) and (6.4.19), v_{drift} also depends only on T_e. For given values of the gas pressure p and tube radius R, the electron temperature remains constant, ideally at the optimum operating value. It then follows that the term in parentheses in Eq. (6.4.24) is a constant, i.e., independent of the current density. One thus sees that, in this simple model, R_p increases linearly with the current density. Just as for optical pumping, we can now define pumping efficiency η_p as the ratio between the minimum pump power ideally needed to achieve a given pump rate R_p and the actual electrical pump power P_p to the discharge. We thus write

$$\eta_p = \frac{R_p V_a h\nu_{mp}}{P_p} \tag{6.4.25}$$

where V_a is the volume of the active medium and v_{mp} is the frequency difference between the ground level and the upper laser level. Note that, to a first approximation, η_p can be taken to be independent of the discharge current density, since both R_p and P_p are proportional to J.

It should be noted that the expression for R_p given by Eq. (6.4.24) can be taken only as a qualitative guide to the complex phenomena occurring in gas-laser pumping rather than as an accurate quantitative expression for the actual value of the pump rate. As already mentioned, particularly for the most efficient gas lasers, the electron-energy distribution is significantly different from an MB distribution. Its calculation would then require an *ab initio* treatment of the Boltzmann transport equation with knowledge of all possible electron-collision processes leading to excitation (or de-excitation) of the rotational, vibrational, and electronic levels of all gas species present in the discharge. Furthermore, the number of gas species in the discharge tube may be much larger than the number of species initially introduced. For instance, for a CO_2:N_2:He mixture, various amounts are also found in the discharge of CO, O_2, N_2O, etc., depending on complicated plasma-chemical reactions occurring in the volume of the gas and at electrodes. The calculation of the pump rate therefore becomes quite involved, requiring a computer, and sometimes proving impractical due to a lack of appropriate data on electron-collision cross sections for all components in the gas mixture. Detailed computer calculations have therefore been performed only for gas mixtures of particular importance, such as the CO_2:N_2:He mixture used in high-power CO_2 lasers.[11] One (apparent) way of circumventing the problem is to assume that η_p is known or can be estimated. In this case, we obviously obtain from Eq. (6.4.25)

$$R_p = \eta_p \frac{P}{Alh v_{mp}} \tag{6.4.26}$$

where A is the cross-sectional area of the active medium and l its length. This is the simple pump rate expression often used in the laser literature; we will use it ourselves in the following chapters. As in the case of optical pumping, however, the usefulness of Eq. (6.4.26) relies on someone having already performed the necessary calculations or on reliable estimates of the value of η_p.

Example 6.7. *Pumping efficiency in a CO₂ laser.* As a particularly relevant example of the calculation of η_p, Fig. 6.27 shows computed results for both a 1:2:3 and 1:0.25:3 CO_2:N_2:He gas mixture.[12] The figure gives the percentage of total pump power entering the various excitation channels as a function of either the $\mathcal{E}/p$ ratio or the $\mathcal{E}/N$ ratio, where N is the total concentration of all species in the mixture. Curve I indicates power going into elastic collisions, excitation of ground-state rotational levels of N_2 and CO_2, and excitation of lower vibrational levels of the CO_2 molecule. Curves III and IV give, respectively, power going into electronic excitation and ionization. Curve II gives the power going into excitation of the upper (001) laser level of the CO_2 molecule and the first five vibrational levels of N_2. Assuming a very efficient energy transfer between N_2 and CO_2, this power is all available as useful pump power. Curve II therefore gives the pump efficiency of a CO_2 laser under the stated conditions. Note that, as discussed in Sect. 6.4.5 for electron temperature (which in this case is a meaningless concept, since the electron distribution is far from being Maxwellian), an optimum value of $\mathcal{E}/p$ exists. For too low a value of $\mathcal{E}/p$, pump power is mostly lost in elastic collisions and excitation of lower CO_2 vibrational levels. For too high a value of $\mathcal{E}/p$, electronic excitation becomes the dominant excitation channel. Note also that, at the optimum $\mathcal{E}/p$ value, a very high value of η_p can be obtained ($\sim 80\%$ for the 1:2:3 mixture).

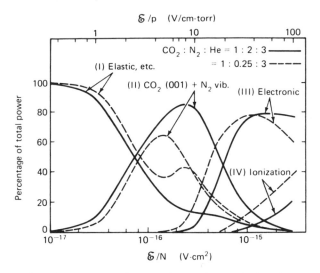

FIG. 6.27. Percentage of total pump power that goes into various excitation channels of a CO_2 laser. (By permission from Ref. 12.)

6.5. CONCLUSIONS

This chapter considers optical and electrical pumping in detail. We see that, for both cases, mechanisms underlying pumping processes involve a variety of physical phenomena. This provides the opportunity to acquire a reasonably in-depth knowledge of, e.g., plasma emission from a lamp, coherent and incoherent emission of diode lasers used for laser pumping, and physical properties of electrical discharges. System configurations used are quite diverse and analogies and similarities among various configurations should help one to devise other workable systems. Despite this diversity, a unified treatment in terms of pump efficiency allows us to compare the effectiveness of the various laser-pumping configurations.

PROBLEMS

6.1. If pump light entering a laser rod is assumed to propagate in a radial direction within the rod, show that the absorption efficiency can be written as:

$$\eta_a = \frac{\int \{1 - \exp[-(2\alpha R)]\} I_{e\lambda}\, d\lambda}{\int I_{e\lambda}\, d\lambda}$$

where R is the radius of the rod, α is the absorption coefficient, and $I_{e\lambda}$ is the spectral intensity of light entering the rod.

6.2. Consider a 6-mm diameter Cr:LiSAF laser rod pumped by a 500-Torr Xe flash-lamp driven at $2400\,\text{A/cm}^2$ current density. Assume for simplicity that the absorption coefficient of Cr:LiSAF versus wavelength consists of two flat bands, each with a peak value of $4\,\text{cm}^{-1}$, the two bands being centered at 420 and 650 nm and having a 80-nm and 120-nm bandwidths, respectively. Using the expression for η_a given in Problem 6.1, calculate the absorption efficiency of the rod for lamp emission in the 400–800 nm band.

6.3. The density of a YAG ($Y_3Al_5O_{12}$) crystal is $4.56 \text{ g} \times \text{cm}^{-3}$. Calculate the density of Nd ions in the crystal when 1% of Yttrium ions are substituted by Neodymium ions (1 atom.% Nd).

6.4. A Nd:YAG rod 6 mm in diameter, 7.5 cm long, with 1 atom.% Nd is cw-pumped by a high-pressure Kr lamp in a close-coupled diffusively-reflecting pumping chamber. Energy separation between the upper laser level and the ground level corresponds to a wavelength of 940 nm. The measured threshold pump power, when the rod is inserted in some given laser cavity, is $P_{th} = 2 \text{ kW}$. Assuming, for this pump configuration, that the rod is uniformly pumped with an overall pump efficiency of $\eta_p = 4.5\%$, calculate the corresponding critical pump rate.

6.5. For a Nd:YAG with 1% atom. Nd, the upper state lifetime is not signifiantly quenched by the mechanism discussed in Example 2.8; it can thus be taken as $\tau = 250 \mu s$. From the pump rate value calculated in Problem 6.4, find the value for critical inversion. According to the Example 2.10, the effective stimulated-emission cross section for the Nd:YAG transition, at $1.064 \mu m$, taking account of the population partition between upper sublevels, can be taken as $\sigma \cong 2.8 \times 10^{-19} \text{ cm}^2$ at T = 300 K. Based on knowledge of the critical inversion, calculate the single-pass cavity loss.

6.6. The laser in Problem 6.4 is pumped by sunlight. Take the average day-time intensity of the sun at the surface of the earth to be $\sim 1 \text{ kW/cm}^2$. Assume that a suitable optical system is used to allow transverse pumping of the laser rod. Assume also that: (a) 10% of the sun's spectrum is absorbed by the rod, (b) the pump quantum efficiency is the same as for flash-lamp pumping (see Table 6.1), (c) transmission of the light-focusing optics is 90%, (d) pump light distribution within the rod is uniform. Given these assumptions, calculate the required area of the collecting optics to allow the laser to be pumped two times above threshold. The focusing system can be made by the admittedly expensive combination of two cylindrical lenses with crossed axes to make a 6 mm × 7.5 cm image of the sun (i.e., suitable for pumping the rod transversely). Knowing that the sun's disk as seen from the earth has a full angle of ~ 9.3 mrad, calculate the focal lengths of the two cylindrical lenses. Can you devise a cheaper focusing scheme?

6.7. A Nd:YAG rod of 5-mm diameter, 5 cm long, with 1 atomic % Nd is pumped by a Xe flash lamp in a close-coupled diffusively-reflecting pumping chamber. Measured threshold pump energy, when the rod is placed in some given laser cavity, is $E_{th} = 3.4 \text{ J}$. Assume that: (a) Overall pumping efficiency is 3.5% (see Table 6.1) and (b) emitted power from the flash lamp lasts $100 \mu s$ and is constant during this time. Given these assumptions, calculate the threshold pump rate R_{cp}. By solving the time-dependent rate equation, which includes effects of both pumping and spontaneous decay, calculate the threshold inversion. If the flash duration is increased to $300 \mu s$ while still remaining constant in time, calculate the new pump rate and pump energy to reach threshold.

6.8. A 1 cm long $Ti^{3+}:Al_2O_3$ rod is longitudinally pumped, from one end, by an argon laser at a 514.5-nm wavelength in a configuration otherwise similar to that in Fig. 6.11c. The absorption coefficient at the pump wavelength for the rod can, in this case, be taken as $\alpha_p \cong 2 \text{ cm}^{-1}$. Transmission at the pump wavelength of the cavity mirror through which the pump beam enters the cavity can be taken as $\eta_t = 0.95$. The wavelength corresponding to the minimum pump frequency ν_{mp} (see Fig. 6.17) for Ti:sapphire is $\lambda_{mp} = 616$ nm. Calculate the overall pumping efficiency. If the pump beam is focused to a spot size of $w_p = 50 \mu m$ in the laser rod, the laser-mode spot size equals the pump spot size, and a single-pass cavity loss $\gamma = 5\%$ is assumed, calculate the optical output power required from the Ar laser at threshold.

6.9. A 2 mm long Nd:glass rod made of LHG-5 glass, with a Nd^{3+} concentration of $3.2 \times 10^{20}\,cm^{-3}$, is longitudinally pumped by a single-stripe AlGaAs QW laser at a 803 nm wavelength in a configuration similar to that in Fig. 6.11a. The pump beam is made circular by the anamorphic system, e.g. as in Fig. 6.12, and focused into the rod to a spot size closely matching the laser-mode spot size $w_0 = 35\,\mu m$. Pump transfer efficiency, including transmission loss at the first cavity mirror in Fig. 6.11a, can be taken as 80%. Assuming an absorption coefficient at the pump wavelength of $9\,cm^{-1}$, an effective stimulated-emission cross section $\sigma_e = 4.1 \times 10^{-20}\,cm^2$, an upper state lifetime of $290\,\mu s$, and a total loss per pass of 0.35%, calculate the threshold pump power. Explain the large difference in pump threshold between this case and that considered in Problem 6.8.

6.10. An Yb:YAG laser rod 1.5 mm long with 6.5 atomic % Yb doping is longitudinally pumped in a laser configuration such as that in Fig. 6.11a by the output of an InGaAs/GaAs QW array at a 940-nm wavelength, focused to a spot size approximately matching the laser-mode spot size $w_0 = 45\,\mu m$. Effective cross sections for stimulated emission and absorption at the $\lambda = 1.03\,\mu m$ lasing wavelength at room temperature can be taken as $\sigma_e \cong 1.9 \times 10^{-20}\,cm^2$ and $\sigma_a \cong 0.11 \times 10^{-20}\,cm^2$, while the effective upper state lifetime is $\tau \cong 1.5\,ms$. Transmission of the output coupling mirror is 3.5%, so that, including other internal losses, single-pass loss can be estimated as $\gamma \cong 2\%$. Calculate the threshold pump power under the stated conditions.

6.11. A Nd:YAG rod with a 4-mm diameter, 6.5 cm long, with 1 atom. % Nd is transversely pumped at a 808-nm wavelength in, e.g., the pump configuration in Fig. 6.15. Assume that 90% of the optical power emitted from the fibers is uniformly absorbed in the rod. To obtain high power from the laser, an output mirror of 15% transmission is used. Including other internal losses, a loss per single pass of $\gamma = 10\%$ is estimated. If the effective stimulated-emission cross section is taken as $\sigma_e = 2.8 \times 10^{-19}\,cm^2$, calculate the optical power required from the fibers to reach laser threshold. Compare this value with that obtained in Problem 6.9 and explain the difference.

6.12. Assuming an MB distribution for electron energy, calculate the electron temperature in eV for a gas of electrons with average kinetic energy of 10 eV.

6.13. Suppose that electrons of mass m collide elastically with an atom of mass M. Assuming the atom is at rest before the collision and electrons are scattered isotropically, show that, as a result of the collision, the electron loses on average a fraction $2m/M$ of its energy.

6.14. A pulsed nitrogen laser operating in the ultraviolet ($\lambda = 337.1\,nm$) requires an optimum electric field of $\sim 10\,kV/cm$ at its typical operating pressure of $p \cong 30$ Torr (for a tube cross section of $5 \times 10\,mm$). A typical length for the nitrogen laser is $\sim 1\,m$. Which of the two pumping configurations shown in Fig. 6.18 would you use for this laser?

6.15. Consider a 1-cm-radius discharge tube filled uniformly with both ions and electrons at a density $N_i = N_e = 10^{13}\,cm^{-3}$. If all electrons were disappearing leaving behind the positive ion charge, what would be the potential of the tube wall V relative to the tube center? (Hence explain ambipolar diffusion.)

6.16. Assume that the ionization cross section is a step function starting at an energy equal to ionization energy E_i and having a constant value σ_i for higher energies. Assuming a Maxwellian distribution for the electron energy, show that the ionization rate can be written as:

$$W_i = N_e \sigma_i \left(\frac{8kT_e}{\pi m}\right)^{1/2} \left(1 + \frac{E_i}{kT_e}\right) \exp\left(-\frac{E_i}{kT_e}\right)$$

6.17. The theory of ambipolar diffusion leads to the following relation between electron temperature T_e and the product pD [compare with Eq. (6.4.22)]:

$$\frac{e^x}{x^{1/2}} = 1.2 \times 10^7 (CpD)^2$$

where C is a constant for a given gas and $x = (E_i/kT_e)$ (E_i is the ionization energy of the gas). Taking values appropriate for helium, $C = 3.2 \times 10^{-4}$ (Torr × mm)$^{-1}$ and $E_i = 24.46$ eV, calculate the required pD value for an electron temperature $T_e = 80{,}000$ K.

6.18. The electron mean free path l can be obtained from the relation $l = 1/N\sigma$, where N is the atomic density and σ is the total electron impact cross section of the atom. Assuming σ to be given by the elastic cross section σ_{el} and taking $\sigma_{el} = 5 \times 10^{-16}$ cm^2 for He, calculate v_{th} and v_{drift} for an average electron energy of $E = 10$ eV, a He pressure of $p = 1.3$ Torr, a temperature $T = 400$ K, and an applied electric field in the discharge of $\mathscr{E} = 30$ V/cm.

6.19. A fluorescent lamp consists of a tube filled with about 3 Torr of Ar and a droplet of Hg, which provides a vapor pressure of ~ 3 mTorr at the normal operating temperature of $T = 300$ K. Thus, as far as discharge parameters are concerned, the tube can be assumed to be filled with Ar gas only. Voltage required across the lamp, for a tube length of 1 m, is about 74 V. Assuming that (a) the fraction lost by electrons per collision is $\delta = 1.4 \times 10^{-4}$, (b) elastic collisions dominate all other collision processes, and (c) $\sigma_{el} = 2 \times 10^{-16}$ cm^2, calculate the electron temperature in the discharge.

REFERENCES

1. P. Laporta, V. Magni, and O. Svelto, Comparative Study of the Optical Pump Efficiency in Solid-State Lasers, *IEEE J. Quantum Electron.* **QE-21**, 1211 (1985).
2. C. S. Willett, *An Introduction to Gas Lasers: Population Inversion Mechanisms* (Pergamon, Oxford, 1974).
3. W. Koechner, *Solid-State Laser Engineering*, vol. 1, *Springer Series in Optical Sciences*, 4th ed. (Springer-Verlag, New York, 1996), Chap. 6, Sect. 6.3.
4. Ref. 3, Chap. 3, Sect. 3.4.
5. P. P. Sorokin and J. R. Lankard, Stimulated Emission Observed by an Organic Dye, Chloro-Aluminum Phtalocyanine, *IBM J. Res. Dev.* **10**, 162 (1966); F. P. Schäfer, F. P. W. Schmidt, and J. Volze, Organic Dye Solution Laser, *Appl. Phys. Lett.* **9**, 306 (1966).
6. J. Berger, D. F. Welch, D. R. Scifres, W. Streifer, and P. S. Cross, High-Power, High-Efficiency Neodymium:Yttrium Aluminum Garnet Laser End Pumped by a Laser Diode Array, *Appl. Phys. Lett.* **51**, 1212 (1987).
7. W. A. Clarkson and D. C. Hanna, Two-Mirror Beam-Shaping Technique for High-Power Diode Bars, *Opt. Lett.* **21**, 375 (1996).
8. R. J. Shine, A. J. Alfrey, and R. L. Byer, 40-W cw, TEM$_{00}$-Mode Diode Pumped, Nd:YAG Miniature Slab Laser, *Opt. Lett.* **20**, 459 (1995).
9. D. Golba, M. Bode, S. Knoke, W. Schöne, and A. Tünnermann, 62-W cw TEM$_{00}$ Nd:YAG Laser Side Pumped by Fiber-Coupled Diode Lasers, *Opt. Lett.* **21**, 210 (1996).
10. G. J. Shulz, Vibrational Excitation of N$_2$, CO and H$_2$ by Electron Impact, *Phys. Rev.* **135A**, 988 (1964).
11. W. L. Nighan, Electron Energy Distribution and Collision Rates in Electrically Excited N$_2$, CO and CO$_2$, *Phys. Rev.* **A2**, 1989 (1970).
12. J. J. Lowke, A. V. Phelps, and B. W. Irwin, Predicted Electron Transport Coefficients and Operating Characteristics of CO$_2$-N$_2$-He Laser Mixtures, *J. Appl. Phys.* **44**, 4664 (1973).
13. Ref. 2, Sect. 3.2.2.
14. C. C. Davis and T. A. King, Gaseous Ion Lasers, in *Advances in Quantum Electronics*, vol. 3 (D. W. Goodwin, ed.) (Academic, New York, 1975), pp. 170–437.

15. T. Y. Fan and R. L. Byer, Diode Laser-Pumped Solid-State Lasers, *IEEE J. Quantum Electr.* **QE-24**, 895 (1988).

16. T. Y. Fan, Diode-Pumped Solid-State Lasers, in *Laser Sources and Applications*, (SUSSP Pubs. And IOP Pubs., 1996), pp. 163–93.

17. S. J. Hamlin, J. D. Myers, and M. J. Myers, *Proc. SPIE* **1419**, 100 (1991).

18. K. Kubodera, K. Otsuka, and S. Miyazawa, Stable $LiNdP_4O_{12}$ Miniature Laser, *Appl. Opt.* **18**, 884 (1979).

19. T. Y. Fan and R. L. Byer, Modeling and CW Operation of a Quasi-Three-Level 946-nm Nd:YAG Laser, *IEEE J. Quantum Electr.* **QE-23**, 605 (1987).

7

Continuous Wave Laser Behavior

7.1. INTRODUCTION

In previous chapters we discussed several features of the components that make up a laser. These are the laser medium itself, whose interaction with an em wave is considered in Chaps. 2 and 3, the passive optical resonator (Chap. 5), and the pumping system (Chap. 6). In this chapter, we will use results from these earlier chapters to develop the theoretical background required to describe cw laser behavior. The case of transient laser behavior is considered in Chap. 8. The theory developed here uses the so-called rate equation approximation and laser equations are derived on the basis of a simple notion that there should be a balance between the total number of atoms undergoing a transition and the total number of photons being created or annihilated.[1,2] This theory provides a rather simple and intuitive picture of laser behavior; furthermore it gives sufficiently accurate results for most practical purposes. For a more refined treatment one should use either the semiclassical approach (where matter is quantized while em radiation is treated classically, i.e., through Maxwell's equations) or the full quantum-electrodynamics approach (where both matter and radiation are quantized). We refer the reader elsewhere for these more advanced treatments.[3]

7.2. RATE EQUATIONS

We consider rate equations for both a four-level laser (Fig. 1.4b) and a quasi-three-level laser, where the lower laser level is a sublevel of the ground state. These laser categories include, in fact, the most important lasers currently in use. In the category of four-level lasers, we have, for instance: (1) Ionic crystal lasers, such as Neodymium lasers, in various hosts, for most of its many possible transitions, Chromium- and Titanium-doped lasers (with the exception of ruby, $Cr^{3+}:Al_2O_3$, the first laser to operate, that involves a pure three-level scheme). (2) Gas lasers, such as CO_2, Ar^+, He-Ne, He-Cd, Cu vapor, HF, and N_2. In the category of quasi-three-level lasers, there are many rare-earth ions in various crystal or glass hosts, such as Yb, Er and Yb:Er, Ho, Tm, and Tm:Ho and, for its shortest wavelength transition, Nd again.

7.2.1. Four-Level Laser

We consider an idealized four-level scheme in which we assume that there is only one pump level or band (band 3 in Fig. 7.1) and relaxation from the pump band to the upper laser level, 2, as well as relaxation from the lower laser level, 1, to the ground level proceeds very rapidly. The following analysis remains unchanged however even if more than one pump band (or level) is involved provided that decay from these bands to the upper laser level is still very rapid. Under these conditions we can make the approximation $N_1 \cong N_3 \cong 0$ for the populations of the lower laser level and pump level(s). Thus we need deal only with two populations, namely, the population N_2 of the upper laser level and the population N_g of the ground level. We assume that the laser is oscillating on only one cavity mode, and we let ϕ represent the corresponding total number of photons in the cavity.

In a first treatment, we consider the case of *space-independent* rate equations, i.e., we assume that the laser is oscillating on a single mode and pumping and mode energy densities are uniform within the laser material. As far as mode energy density is concerned, this means that the mode transverse profile must be uniform and we are neglecting the effects of the standing wave character of the mode. Thus, strictly speaking, the treatment that follows only applies to a unidirectional ring resonator, with uniform transverse beam profile, where pumping is uniformly distributed in the active medium—clearly a rather special and simplified case. This case, although perhaps oversimplified, helps us understand many basic properties of laser behavior. Features arising from space dependence of both pump and mode patterns are discussed at some length later on in this chapter.

For the space-independent case, we can readily write the following equations:

$$\left(\frac{dN_2}{dt}\right) = R_p - B\phi N_2 - \left(\frac{N_2}{\tau}\right) \tag{7.2.1a}$$

$$\left(\frac{d\phi}{dt}\right) = V_a B\phi N_2 - \left(\frac{\phi}{\tau_c}\right) \tag{7.2.1b}$$

In Eq. (7.2.1a) the pumping term R_p [see Eq. (1.3.1)] is based on the assumption of negligible depopulation of the ground state. Explicit expressions for the pumping rate R_p

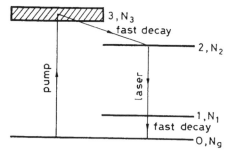

FIG. 7.1. Four-level laser scheme.

were derived in Chap. 6 for both optical and electrical pumping. The term $B\phi N_2$ in Eq. (7.2.1a) accounts for stimulated emission. It was shown in Chaps. 2 and 3 that the stimulated emission rate W is in fact proportional to the square of the magnitude of the electric field of the em wave; hence W can also be taken as proportional to ϕ. The coefficient B is therefore referred to as the stimulated transition rate per photon, per mode. The quantity τ is the lifetime of the upper-laser level; in general it must take into account both radiative and nonradiative processes [see Eq. (2.6.18)]. Note that the upper-laser level often consists of a combination of many strongly coupled sublevels. In this case the lifetime τ is intended to be the effective upper-level lifetime, taking into account the lifetimes of all upper-state sublevels with a weight proportional to the corresponding sublevel population [see Eq. 2.7.19c)]. In (7.2.1.b) the term $V_a B\phi N_2$ represents the growth rate of the photon population due to stimulated emission and is obtained by a simple balance argument. In Eq. (7.2.1.a), in fact, and the term $B\phi N_2$ gives the rate of population decrease due to stimulated emission. Since each stimulated-emission process creates a photon, the photon growth rate must then be $V_a B\phi N_2$, where V_a is the volume of the mode in the active medium. The term ϕ/τ_c, where τ_c is the photon lifetime (see Sect. 5.3), accounts for the removal of photons due to cavity losses.

Before proceeding we point out that, in Eq. (7.2.1b), a term accounting for spontaneous emission is not included. Since, as pointed out in Chap. 1, laser action is actually initiated by spontaneous emission, we do not expect Eq. (7.2.1) to account for the onset of laser oscillation. Indeed if, at time $t = 0$, we set $\phi = 0$ on the right-hand side of Eq. (7.2.1b), then we get $(d\phi/dt) = 0$, implying that laser action does not start. To account for spontaneous emission, we may be tempted to apply similar considerations of balance, starting with the term N_2/τ_r, τ_r being the radiative lifetime of level 2, which is included in the term N_2/τ in Eq. (7.2.1a). One might then think that the appropriate term to include into Eq. (7.2.1.b), to account for spontaneous emission, is $V_a(N_2/\tau_r)$; this would not be true, however. In fact, as seen in Chaps. 2 and 3, spontaneously emitted light is distributed over the entire frequency range corresponding to the gain bandwidth; furthermore emission occurs into a 4π solid angle. The spontaneous emission term, required in Eq. (7.2.1b), must however include only the fraction of the spontaneously emitted light that contributes to the given mode (i.e., that is emitted in the same angular direction and spectral bandwidth of the mode). The correct expression for this term can be obtained only by a quantized treatment of radiation-matter interaction. The result is particularly simple and instructive:[4] In a quantum electrodynamics treatment, Eq. (7.2.1b) is transformed into:

$$\left(\frac{d\phi}{dt}\right) = V_a B(\phi + 1)N_2 - \left(\frac{\phi}{\tau_c}\right) \tag{7.2.2}$$

Thus everything behaves as if there were an *extra photon* in the term describing stimulated emission. When the laser is oscillating however (and unless very close to threshold), the number of photons in the laser cavity may easily range from 10^{10} to 10^{16} for a cw laser (see Example 7.1) and much more than this value for pulsed lasers. In the following analysis, therefore, we do not include this extra term accounting for spontaneous emission; instead we assume that an arbitrarily small number of photons ϕ_i, say $\phi_i = 1$, is initially present in the cavity just to allow laser action to start.

We are now interested in deriving an explicit expression for the quantity B, the stimulated emission coefficient per photon per mode, which is present in Eqs. (7.2.1a–b).

We consider a resonator of length L in which an active medium of length l and refractive index n is inserted. Since, for the time being, we are considering a traveling wave beam, we let I represent the intensity of this beam at a given cavity position and time $t = 0$. Following the argument considered in Sect. 1.2, the intensity I', after one cavity round-trip, is $I' = I \times R_1 R_2 (1 - L_i)^2 \exp(2\sigma N_2 l)$, where R_1 and R_2 are the power reflectivities of the two mirrors, L_i is the single-pass internal loss of the cavity, so that $(1 - L_i)^2$ is the round-trip cavity transmission, and $(2\sigma N_2 l)$ is the round-trip gain of the active medium. Note that, if the upper laser level is either degenerate or consists of many strongly coupled sublevels, one should use here the effective value of the cross section, as discussed in Sect. 2.7. We now write $R_1 = 1 - a_1 - T_1$ and $R_2 = 1 - a_2 - T_2$, where T_1 and T_2 are the power transmissions of the two mirrors and a_1 and a_2 are the corresponding fractional mirror losses. The change of intensity, $\Delta I = I' - I$, for a cavity round-trip is then:

$$\Delta I = [(1 - a_1 - T_1)(1 - a_2 - T_2)(1 - L_i)^2 \exp(2\sigma N_2 l) - 1]I \qquad (7.2.3)$$

We now assume that mirror losses are equal $(a_1 = a_2 = a)$ and so small that we can set $(1 - a - T_1) \cong (1 - a)(1 - T_1)$ and $(1 - a - T_2) \cong (1 - a)(1 - T_2)$. Then Eq. (7.2.3) is obviously transformed into:

$$\Delta I = [(1 - T_1)(1 - T_2)(1 - a)^2 (1 - L_i)^2 \exp(2\sigma N_2 l) - 1]I \qquad (7.2.4)$$

Before proceeding it is convenient to introduce some new quantities γ (see Sect. 1.2), which can be described as the logarithmic loss per pass, namely, [compare with Eqs. (1.2.4)]:

$$\gamma_1 = -\ln(1 - T_1) \qquad (7.2.5)$$

$$\gamma_2 = -\ln(1 - T_2) \qquad (7.2.6)$$

$$\gamma_i = -[\ln(1 - a) + \ln(1 - L_i)] \qquad (7.2.7)$$

As pointed out in Sect. 1.2, γ_1 and γ_2 are the logarithmic losses per pass due to the mirror transmission, and γ_i is the logarithmic internal loss per pass. For brevity however, we call γ_1 and γ_2 the mirror losses and γ_i the internal loss. The logarithmic loss notation proves to be the most convenient way of representing laser losses, given the exponential character of laser gain. Note that, for small transmission values, one has $\gamma = -\ln(1 - T) \cong T$. Likewise, for very small values of a and L_i, from Eq. (7.2.7) one has $\gamma_i \cong a + L_i$, so that the quantities γ really represent loss terms for the cavity. Obviously, the preceding approximations hold only for small values of cavity loss or mirror transmission. For example, if we take $T = 0.1$, then we get $\gamma = 0.104$, i.e., $\gamma \cong T$; however, if we take $T = 0.5$, then $\gamma = 0.695$. With the help of this logarithmic loss notation, we can also define a total (logarithmic) loss per pass γ as:

$$\gamma = \gamma_i + \left[\frac{(\gamma_1 + \gamma_2)}{2}\right] \qquad (7.2.8)$$

We now substitute Eqs. (7.2.5)–(7.2.8) into Eq. (7.2.4) and make the additional assumption:

$$(\sigma N_2 l - \gamma) \ll 1 \qquad (7.2.9)$$

Then the exponential function resulting from Eq. (7.2.4) can be expanded as a power series to yield

$$\Delta I = 2(\sigma N_2 l - \gamma)I \qquad (7.2.10)$$

We then divide both sides of Eq. (7.2.10) by the time Δt taken for the light to make one cavity round-trip, i.e., $\Delta t = 2L_e/c$, where L_e is the optical length of the resonator, given by

$$L_e = L + (n - 1)l \qquad (7.2.11)$$

If the approximation $\Delta I / \Delta t \cong dI/dt$ is also used, we obtain

$$\frac{dI}{dt} = \left(\frac{\sigma l c}{L_e} N_2 - \frac{\gamma c}{L_e} \right)I \qquad (7.2.12)$$

Since the number ϕ of photons in the cavity is proportional to I, a comparison of Eq. (7.2.12) with (7.2.1b) gives

$$B = \frac{\sigma l c}{V_a L_e} = \frac{\sigma c}{V} \qquad (7.2.13)$$

$$\tau_c = \frac{L_e}{\gamma c} \qquad (7.2.14)$$

where

$$V = \left(\frac{L_e}{l} \right) V_a \qquad (7.2.15)$$

is referred to as the mode volume within the laser cavity. (The mode diameter is taken to be independent of the cavity longitudinal coordinate.) Note that Eq. (7.2.14) generalizes the expression for photon lifetime given in Sect. 5.3.

Once explicit expressions for B and τ_c are obtained, then Eq. (7.2.1) provides a description of both static and dynamic behavior of a four-level laser, within the previous limitations and approximations. To simplify notations, we write $N \equiv N_2 - N_1 \cong N_2$ as the population inversion. From Eq. (7.2.1) we then obtain

$$\frac{dN}{dt} = R_p - B\phi N - \frac{N}{\tau} \qquad (7.2.16a)$$

$$\frac{d\phi}{dt} = \left(B V_a N - \frac{1}{\tau_c} \right)\phi \qquad (7.2.16b)$$

These equations, together with the expressions for B, τ_c and V_a given by Eqs. (7.2.13)–(7.2.15), respectively, describe laser behavior in both the cw and transient cases for a four-level laser.

Before proceeding we point out some caveats that apply to Eqs. (7.2.16). One such critical comment was already made at the beginning, namely, that results strictly hold only for a laser oscillating with uniform pump and mode energy distributions in the active medium. This important point indicates a severe limitation to the usefulness of the preceding equations. However results obtained using this simple model are very useful for understanding basic aspects of laser behavior. Furthermore, at least for cw behavior, the much more complicated space-dependent equations lead to rather similar results whose relevance is then better understood by comparison with the space-independent model.

A second critical remark relates to the fact that these rate equations strictly apply only for single-mode oscillation. For n oscillating modes, in fact, we should generally write $2n$ differential equations for the *amplitude and phase* of the field modes to account properly for beating terms among the various modes. In fact, under appropriate locking conditions among mode phases, this leads to the phenomenon of mode locking, described in Chap. 8, which obviously cannot be described by a rate equation model. However, when many modes with random phases are oscillating, the overall beam intensity can, to a first approximation, be taken as the sum of intensities of all modes. For a uniform transverse pump profile, the superposition of the mode transverse profiles thus tends to produce a fairly uniform profile for the overall beam. Futhermore, given the oscillation of many modes with different longitudinal patterns, the corresponding total energy density is expected to not show a pronounced standing wave pattern. In this case, the picture can be greatly simplified by considering just one rate equation for the total number of photons ϕ, i.e., summed over all modes, so Eq. (7.2.16) can still be applied in an approximate fashion.

A third critical remark focuses on the fact that in writing, e.g., Eq. (7.2.3) we have implicitly assumed that population inversion, during laser action, is independent of the longitudinal z-coordinate. Actually, for large values of laser gain, both counter-propagating beams show a strong dependence on the z-coordinate and similarly for the inversion. Under such conditions, laser behavior must be treated on a pass-by-pass basis, as first done by Rigrod (the so-called Rigrod analysis).[5] In the cw case, where Eq. (7.2.9) can be taken to hold, the expression for the output power obtained by Rigrod analysis coincides, however, with that obtained by this far simpler treatment provided we use the γ-notation, as adopted here. For a pulsed laser, Eq. (7.2.9) holds only when the laser is not driven far above threshold. Otherwise Eq. (7.2.16) can no longer be applied, so the dynamic behavior of the laser must be analyzed on a pass-by-pass basis, as in the Rigrod analysis.[5]

A fourth and perhaps more serious remark is that Eq. (7.2.16) really does not apply to an inhomogenously broadened line. To understand this point, let us consider for simplicity an inhomogeneously broadened (but non-Doppler-broadened) transition and assume the laser to be oscillating on a single frequency. The beam then interacts only with that fraction of population whose resonance frequency coincides with the laser frequency, and, at a sufficiently high intensity, the saturated gain profile shows a hole located at this frequency (as in Fig. 2.22 for an absorption profile). Clearly, in this case, the starting point of our analysis on beam amplification, namely Eq. (7.2.3), where N_2 is the *total* upper state population, is no longer valid. The situation is even more complicated for a Doppler-broadened inhomogeneous transition, where, if the laser is oscillating at a frequency sufficiently far from the central frequency of the transition, the right traveling beam and the left traveling beam interact with different sets of atoms or molecules. The cw behavior of an inhomogeneously broadened transition was considered, notably, by Casperson[6] and shown to give results considerably different from those obtained from Eq. (7.2.16).

Within the limitations discussed in the previous paragraphs, we now take Eq. (7.2.16) to be valid for a first-order description of laser behavior. Equation (7.2.16) must then be solved under the appropriate conditions for the case being examined. Thus, to describe cw laser behavior, the case pertinent to this chapter, we merely set the time derivative on the left-hand side of both equations equal to zero. To describe transient laser behavior, we need to specify $R_p = R_p(t)$, and we also need to know the initial conditions. For instance, if pumping is initiated at $t = 0$, initial conditions are $N(0) = 0$ and $\phi(0) = \phi_i$, where ϕ_i is a very small number of initial photons simulating the effect of spontaneous emission (e.g., $\phi_i = 1$). This will be discussed in Chap. 8. For both cw and transient laser behavior, however, once ϕ or $\phi(t)$ is known, the calculation of the output power through one of the cavity mirrors becomes straightforward. In fact, according to Eqs. (7.2.14) and (7.2.8), we can write

$$\frac{1}{\tau_c} = \frac{\gamma_i c}{L_e} + \frac{\gamma_1 c}{2L_e} + \frac{\gamma_2 c}{2L_e} \qquad (7.2.17)$$

If we now substitute Eq. (7.2.17) into the right-hand side of Eq. (7.2.16b), we recognize that the term $(\gamma_2 c/2L_e)\phi$, gives the rate of photon loss due to transmission through mirror 2. The output power through this mirror is therefore given by:

$$P_{out} = \left(\frac{\gamma_2 c}{2L_e}\right)(h\nu)\phi \qquad (7.2.18)$$

Thus the solution of Eq. (7.2.16) allows us to calculate not only the internal laser behavior but also, using Eq. (7.2.18), one of the most pertinent laser parameter, i.e., the output power. If the output power is known, Eq. (7.2.18) can be used to calculate the total number of cavity photons, as shown in Example 7.1.

> **Example 7.1.** *Calculation of the number of cavity photons in typical cw lasers.* As a low-power example, we first consider a 50-cm-long-He Ne laser, oscillating at $\lambda = 630$ nm with an output power of 10 mW. Transmission of the output mirror, for this low-gain laser, may typically be $T_2 = 1\%$, so that $\gamma_2 = -\ln(1 - T_2) \cong 0.01$. From Eq. (7.2.18) we obtain $\phi \cong 1.06 \times 10^{10}$ photons. As a high-power example, we consider a 10-kW CO_2 laser oscillating at a wavelength of 10.6 μm. We take a cavity length of $L_e = 150$ cm and an output mirror transmission, for this higher-gain laser, of $T_2 = 45\%$. We obtain $\gamma_2 = -\ln(1 - T_2) \cong 0.598$ and, from Eq. (7.2.18), $\phi \cong 0.9 \times 10^{16}$ photons.

7.2.2. Quasi-Three-Level Laser

In a quasi-three-level laser, the lower laser level, level 1 in Fig. 7.2, is a sublevel of the ground level, and all ground-state sublevels are assumed to be strongly coupled and hence in thermal equilibrium. Likewise, the upper laser level, level 2 in Fig. 7.2, may be a sublevel of a set of upper state sublevels also assumed to be in thermal equilibrium. In this case, we let N_1 and N_2 represent the total population of all ground-state and all upper-state sublevels, respectively. We again assume a very rapid decay from the pump level(s) to the upper state sub-levels, so that we need to be concerned only with populations N_1 and N_2 (ideal quasi-three-level case). Now let 0 represent the lowest sublevel of the ground state and assume that the energy separation between sublevels 1 and 0 is comparable to kT. Then a non-negligible fraction of the ground-state population N_1 is present in the lower laser level (see Sect. 2.7.2); this results in laser photon absorption.

Following the discussion of Sect. 2.7.2, the rate equations for both upper and lower state laser sublevels can be written in terms of the total populations N_1 and N_2. Rate equations for a quasi-three-level laser can then be written similarly to the four-level case, taking into account the fact that absorption as well as stimulated emission of laser photons now occurs. We thus write

$$N_1 + N_2 = N_t \tag{7.2.19a}$$

$$\left(\frac{dN_2}{dt}\right) = R_p - \phi(B_e N_2 - B_a N_1) - \left(\frac{N_2}{\tau}\right) \tag{7.2.19b}$$

$$\left(\frac{d\phi}{dt}\right) = V_a \phi(B_e N_2 - B_a N_1) - \left(\frac{\phi}{\tau_c}\right) \tag{7.2.19c}$$

where τ is again the effective lifetime of level 2 and B_e and B_a are now given by [compare to Eq. (7.2.13)]:

$$B_e = \frac{\sigma_e c}{V} \tag{7.2.20a}$$

$$B_a = \frac{\sigma_a c}{V} \tag{7.2.20b}$$

where σ_e and σ_a are the effective cross sections for stimulated emission and absorption, respectively (see Sect. 2.7.2). The substitution of Eq. (7.2.20) into Eq. (7.2.19) gives

$$N_1 + N_2 = N_t \tag{7.2.21a}$$

$$\frac{dN_2}{dt} = R_p - \frac{\sigma_e c}{V} \phi(N_2 - f N_1) - \frac{N_2}{\tau} \tag{7.2.21b}$$

$$\frac{d\phi}{dt} = \left[\frac{V_a \sigma_e c}{V}(N_2 - f N_1) - \frac{1}{\tau_c}\right]\phi \tag{7.2.21c}$$

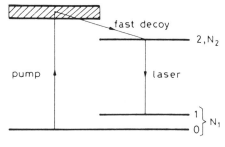

FIG. 7.2. Quasi-three-level laser scheme.

where:

$$f = \frac{\sigma_a}{\sigma_e} \qquad (7.2.22)$$

Equations (7.2.21b–c) suggest that we may now define a population inversion N as:

$$N = N_2 - fN_1 \qquad (7.2.23)$$

Using the pair of equations (7.2.21a) and (7.2.23), we can obtain N_1 and N_2 in terms of N and N_t. Equations (7.2.21a–c) can then be reduced to two equations in the variables ϕ and N. After some straightforward manipulation, we obtain

$$\frac{dN}{dt} = R_p(1+f) - \frac{(\sigma_e + \sigma_a)c}{V}\phi N - \frac{fN_t + N}{\tau} \qquad (7.2.24a)$$

$$\frac{d\phi}{dt} = \left(\frac{V_a\sigma_e c}{V}N - \frac{1}{\tau_c}\right)\phi \qquad (7.2.24b)$$

Within the limits discussed for the validity of Eq. (7.2.16), Eqs. (7.2.24) describe the static and dynamic behavior of a quasi-three-level laser.* In the case of transient laser behavior, if pumping is initiated at $t = 0$, Eqs. (7.2.24) must be solved with the initial conditions $N(0) = -fN_1$ and $\phi(0) = \phi_i \cong 1$. Note that the photon rate equations for four-level (7.2.16b) and quasi-three-level (7.2.24b) lasers are the same. The rate equations for the population inversion are somewhat different, however. In particular, the stimulated term for a quasi-three-level laser is a factor $(\sigma_e + \sigma_a)/\sigma_e$ larger than that of a four-level laser. To understand this result, consider a unit volume, $V_a = 1$, and assume that, in a given time Δt, one photon is created by the stimulated processes in this volume. According to Eq. (7.2.24b), this implies that $(\sigma_e cN\phi/V)\Delta t = 1$. Using this result, from Eq. (7.2.24a) we then see that, correspondingly, the inversion N decreases by an amount equal to $\Delta N = [(\sigma_e + \sigma_a)cN\phi/V]\Delta t = (\sigma_e + \sigma_a)/\sigma_e$. Thus $\Delta N > 1$, which we understand when we notice that, due to this stimulated process, N_2 decreases by 1, while N_1 increases by 1. According to Eq. (7.2.23) the decrease in N must then be larger than 1. By contrast, for a four-level laser, the emission of a photon implies that N_2 decreases by 1 while N_1 remains essentially unchanged (i.e. zero on account of the fast $1\rightarrow0$ decay). Thus, in this case, the decrease of N is simply equal to 1. Note also that, as expected, Eqs. (7.2.24) reduces to (7.2.16) when σ_a, and hence f, are equal to zero.

Within the limits of a space-independent rate equation treatment, Eqs. (7.2.24) thus represent the final result of our calculation for a quasi-three-level laser. For both cw and transient cases, once N and ϕ are obtained by solving these equations with the appropriate boundary conditions, the output power through, e.g., mirror 2 is again obtained from Eq. (7.2.18).

---

* Note: As a quasi-3-level laser becomes progressively closer to a pure 3-level laser, the assumption that the ground-state population is changed negligibly by the pumping process becomes eventually not justified, so the pump rate R_p cannot be taken to be a constant.

7.3. THRESHOLD CONDITIONS AND OUTPUT POWER: FOUR-LEVEL LASER

In this section we investigate, for a four-level laser, the threshold conditions and the output power for the cw case, i.e., when R_p is time independent. We first consider the behavior corresponding to the space-independent rate equations described previously. Results predicted from the space-dependent model are then discussed, and a comparison is made between the two models.

7.3.1. Space-Independent Model

In Sect. 7.2, the rate equations for a four-level laser are derived under the simplifying assumption of a very short lifetime in the lower laser level. Before going into a detailed calculation of cw laser behavior under these conditions, we derive a necessary condition for cw oscillation, when the lifetime of the lower laser level τ_1 has a finite value. To do this, we first observe that the steady-state population in level 1 is of course established by a balance between populations entering and leaving that level. In the absence of oscillation, we thus write $(N_1/\tau_1) = (N_3/\tau_{21})$, where τ_{21} is the lifetime of the $2 \rightarrow 1$ transition. If for simplicity we consider the case where the two levels are actually single levels having the same degeneracy, then to obtain laser action, we require $N_2 > N_1$. Then, the preceding expression implies

$$\tau_1 < \tau_{21} \tag{7.3.1}$$

If this inequality is not satisfied, then laser action is possible only on a pulsed basis provided that the pumping pulse is shorter than or comparable to the lifetime of the upper laser level. Having begun, laser action then continues until the number of atoms accumulated in the lower level, as a result of stimulated emission, is sufficient to destroy the population inversion. These lasers are therefore referred to as *self-terminating*. If, on the other hand, Eq. (7.3.1) is satisfied and R_p is sufficiently strong, then a steady-state oscillation condition is eventually reached. In what follows we examine this situation subject to the assumption that $\tau_1 \ll \tau_{21}$, so that Eqs. (7.2.16) applies.

We begin by considering the *threshold condition* for laser action. Suppose that, at time $t = 0$, an arbitrarily small number ϕ_i of photons (e.g., $\phi_i = 1$) is present in the cavity due to spontaneous emission. From Eq. (7.2.16b) we then see that, to have $(d\phi/dt) > 0$, we must have $BV_aN > 1/\tau_c$. Laser action is therefore initiated when the population inversion N reaches a critical value N_c given by (see also Sect. 1.2):

$$N_c = \left(\frac{1}{BV_a\tau_c}\right) = \left(\frac{\gamma}{\sigma l}\right) \tag{7.3.2}$$

where we have used Eqs. (7.2.13) and (7.2.14). The corresponding critical pump rate R_{cp} is then obtained from Eq. (7.2.16a) by letting $(dN/dt) = 0$, since we are at steady state, $N = N_c$, and $\phi = 0$. The critical pump rate then corresponds to the situation where the rate of pump transitions R_{cp} equals the spontaneous transition rate from level 2, N_c/τ. We thus obtain

$$R_{cp} = \frac{N_c}{\tau} = \left(\frac{\gamma}{\sigma l \tau}\right) \qquad (7.3.3)$$

where we have used Eq. (7.3.2).

If $R_p > R_{cp}$, the photon number ϕ increases from the initial value determined by spontaneous emission, and, if R_p is independent of time, ϕ eventually reaches some steady-state value ϕ_0. This value and the corresponding steady-state value N_0 for the inversion, are obtained from Eq. (7.2.16) by setting $(dN/dt) = (d\phi/dt) = 0$. This gives

$$N_0 = \left(\frac{1}{BV_a\tau_c}\right) = \left(\frac{\gamma}{\sigma l}\right) \qquad (7.3.4a)$$

$$\phi_0 = V_a\tau_c\left[R_p - \left(\frac{N_0}{\tau}\right)\right] \qquad (7.3.4b)$$

Note that, to obtain Eq. (7.3.4b), we have used Eqs. (7.3.4a), (7.2.14), and (7.2.15).

Equations (7.3.4) describe the cw behavior of a four-level laser. We now examine these equations in some detail. Comparing Eqs. (7.3.4a) and (7.3.2) we first observe that, even when $R_p > R_{cp}$, one has $N_0 = N_c$, i.e., the steady-state inversion always equals the critical or threshold inversion. To obtain a better understanding of this result, we show in Fig. 7.3 a plot of both N and ϕ versus the pump rate R_p. When $R_p < R_{cp}$, then $\phi_0 = 0$, and the inversion N increases linearly with R_p. When $R_p = R_{cp}$, obviously one has $N = N_c$ and $\phi_0 = 0$. If R_p is now increased above R_{cp}, Eq. (7.3.4) shows that, while N_0 remains fixed at the critical inversion, ϕ_0 increases linearly with R_p. In other words, the pump rate increases the inversion (i.e., energy stored in the material) below threshold, while it increases the number of photons (i.e., em energy stored in the cavity) above threshold.

We now rewrite Eq. (7.3.4b) in a somewhat simpler form by taking the term $(N_0/\tau) = (N_c/\tau) = R_{cp}$ outside the square brackets. We obtain

$$\phi_0 = (V_aN_0)\frac{\tau_c}{\tau}(x - 1) \qquad (7.3.5)$$

where:

$$x = \frac{R_p}{R_{cp}} \qquad (7.3.6)$$

is the amount by which pump rate exceeds threshold pump rate. We now observe that, for both optical and electrical pumping, we can write

$$x = \frac{P_p}{P_{th}} \qquad (7.3.7)$$

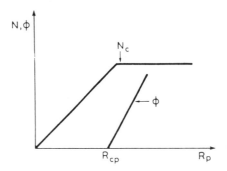

FIG. 7.3. Qualitative behavior of the population inversion N and total number of cavity photons ϕ as a function of the pump rate R_p.

where P_p is the pumping power and P_{th} is its threshold value. Substituting Eqs. (7.3.7) and (7.3.4a) into Eq. (7.3.5), this equation can be transformed into a somewhat more useful form

$$\phi_0 = \frac{A_b \gamma}{\sigma} \frac{\tau_c}{\tau} \left(\frac{P_p}{P_{th}} - 1 \right) \qquad (7.3.8)$$

where $A_b = (V_a/l)$ is the cross-sectional area of the mode (the beam area), which is assumed to be smaller than or equal to the cross-sectional area $A = (V/l)$ of the active medium.

We now go on to derive an expression for the output power. From Eqs. (7.2.18) and (7.3.8) we obtain

$$P_{out} = (A_b I_s) \left(\frac{\gamma_2}{2} \right) \left(\frac{P_p}{P_{th}} - 1 \right) \qquad (7.3.9)$$

where $I_s = h\nu/\sigma\tau$ is the saturation intensity for a four-level system [see Eq. (2.8.24)]. If mirror 1 is totally reflecting, this expression agrees with that of Rigrod,[5] obtained by using a pass-by-pass analysis. Since a plot of P_{out} versus P_p yields a straight line intercepting the P_p-axis at $P_p = P_{th}$, we can define the laser slope efficiency as

$$\eta_s = \frac{dP_{out}}{dP_p} \qquad (7.3.10)$$

where η_s is a constant for a given laser configuration.

With the help of previous expressions and of equations derived in Chap. 6, we can obtain a useful and instructive expression for η_s. First, upon substituting P_{out} from Eq. (7.3.9) into Eq. (7.3.10), we write

$$\eta_s = \frac{A_b h\nu}{\sigma\tau} \frac{\gamma_2}{2} \frac{1}{P_{th}} \qquad (7.3.11)$$

For both lamp pumping and electrical pumping, substituting Eq. (6.2.6) or Eq. (6.4.26) in Eq. (7.3.3), we then obtain

$$P_{th} = \frac{\gamma}{\eta_p} \left(\frac{h\nu_{mp}}{\tau} \right) \left(\frac{A}{\sigma} \right) \tag{7.3.12}$$

where we remind that ν_{mp} is the frequency difference between the upper laser level and the ground level and A is the area of the active medium. Using Eq. (7.3.12) into (7.3.11) we obtain

$$\eta_s = \eta_p \left(\frac{\gamma_2}{2\gamma} \right) \left(\frac{h\nu}{h\nu_{mp}} \right) \left(\frac{A_b}{A} \right) \tag{7.3.13}$$

We can then write

$$\eta_s = \eta_p \eta_c \eta_q \eta_t \tag{7.3.14}$$

where: (i) η_p is the pump efficiency. (ii) The term $\eta_c = \gamma_2/2\gamma$ represents the fraction of generated photons coupled out of the cavity (referred to as the *output coupling efficiency*). Note that η_c is always smaller than 1, and it reaches the value 1 when $\gamma_1 = \gamma_i = 0$. (iii) The term $\eta_q = h\nu/h\nu_{mp}$ gives the fraction of the minimum pump energy transformed into laser energy (referred to as the *laser quantum efficiency*). (iv) The term $\eta_t = A_b/A$ gives the fraction of the active medium cross section used by the beam cross section; we refer to this as the transverse utilization factor of the active medium or the *transverse efficiency*.

Example 7.2. *CW laser behavior of a lamp-pumped high-power Nd:YAG laser.* We consider the laser system in Fig. 7.4, where a 6.35-mm diameter, 7.5 cm long Nd:YAG rod, with 1% atomic concentration of active Nd ions is pumped, in an elliptical pump chamber, by a high-pressure Kr lamp. The laser cavity consists of two plane mirrors separated by 50 cm. The reflectivity of one mirror is $R_1 = 100\%$, while that of the output coupling mirror is $R_2 = 85\%$. A typical curve of the output power P_{out} through mirror 2 (in multimode operation) versus electrical pump power P_p to the Kr lamp is shown in Fig. 7.5.[7] Note that one is dealing with a reasonably high-power cw Nd:YAG laser, with an output power exceeding 200 W. We observe that, since the laser is oscillating on many transverse and longitudinal modes, then, according to the discussion in Sect. 7.2.1, it is reasonable to compare experimental results with theoretical predictions using the preceding space-independent rate equations. In fact, except for input powers just above threshold, experimental points in Fig. 7.5 show a linear relationship between output and input powers, as predicted by Eq. (7.3.9). From the linear part of the curve, an extrapolated threshold of $P_{th} = 2.2$ kW is obtained. Above threshold, the output versus input power relation can be fitted by the equation

$$P_{out} = 53 \left[\left(\frac{P_p}{P_{th}} \right) - 1 \right] \tag{7.3.15}$$

where P_{out} is expressed in watts. The slope efficiency is then easily obtained from Eq. (7.3.15) as $\eta_s = (dP_{out}/dP_p) = 53/P_{th} = 2.4\%$. Equation (7.3.15) can be readily compared to Eq. (7.3.9) once we remember that, as discussed in Example 2.10, the effective values of the cross section and upper level lifetime for the $\lambda = 1.06$ μm transition in Nd:YAG can be taken as $\sigma = 2.8 \times 10^{-19}$ cm^2 and $\tau = 230$ μs, respectively. The energy of the photon at this wavelength is obtained as $h\nu = 3.973 \times 10^{-19} \times (0.5/1.06) = 1.87 \times 10^{-19}$ J, where 3.973×10^{-19} J is the energy of a photon with a wavelength of 0.5 μm (see Appendix I). We then obtain the value of the saturation intensity as $I_s = h\nu/\sigma\tau = 2.9$ kW/cm^2. We now take $R_2 = (1 - a_2 - T_2) \cong (1 - T_2)$ since, for a good multilayer coating, mirror absorption, a_2, may be less than 0.1%. Then $\gamma_2 = -\ln R_2 = 0.162$. Comparing Eqs. (7.3.15) and (7.3.9) yields $A_b \cong 0.23$ cm^2, to be compared with the cross-sectional area of the rod $A \cong 0.317$ cm^2.

To compare measured slope efficiency and extrapolated threshold with values predicted by calculation, we must know γ, i.e., γ_i. Since $\gamma_1 = 0$, Eq. (7.3.12), with the help of Eq. (7.2.8), can be rearranged as:

$$\frac{-\ln R_2}{2} + \gamma_i = \eta_p \left(\frac{\sigma}{A}\right)\left(\frac{P_{th}\tau}{h\nu_{mp}}\right) \tag{7.3.16}$$

Thus, if several measurements are taken of the threshold pump power at different mirror reflectivities R_2, a plot of $\gamma_2 = -\ln R_2$ versus P_{th} should yield a straight line. In fact this is found experimentally, as shown in Fig. 7.6. The intercept of this straight line with the γ_2-axis gives, according to Eq. (7.3.16), the value of internal losses (Findlay and Clay analysis[9]). From Fig. 7.6 we then get $\gamma_i \cong 0.038$, which gives a total loss $\gamma = (\gamma_2/2) + \gamma_i \cong 0.12$.

Once total losses are known, we can use Eq. (7.3.14) to compare the measured slope efficiency $\eta_s = 2.4\%$ with theoretical predictions. We take $\eta_c = \gamma_2/2\gamma \cong 0.68$. We also take $\eta_q = \lambda_{mp}/\lambda = 0.89$, where $\lambda_{mp} = 0.94$ μm is the wavelength corresponding to the transition from the upper laser level to the ground level (see Fig. 2.15) in Nd:YAG; according to the previous calculation we also take $\eta_t = A_b/A \cong 0.72$. From Eq. (7.3.14) we obtain $\eta_p = 5.5\%$, which appears to be a reasonable value for pump efficiency for Kr pumping (see also Table 6.1). The predicted value of P_{th} can now be readily obtained from Eq. (7.3.12) once we take into account that $h\nu_{mp} \cong 2.11 \times 10^{-19}$ J. We obtain $P_{th} \cong 2.26$ kW in good agreement with the experimental result. A knowledge of total losses also allows us to calculate the threshold inversion. From Eq. (7.3.2) we find $N_c \cong 5.7 \times 10^{16}$ ions/cm^3. For a 1% atomic doping, total Nd concentration is $N_t = 1.38 \times 10^{20}$ ions/cm^3. Thus $N_c/N_t = 4.1 \times 10^{-4}$, which shows that population inversion is a very small fraction of the total population.

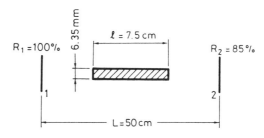

FIG. 7.4. Possible cavity configuration for a lamp-pumped cw Nd:YAG laser.

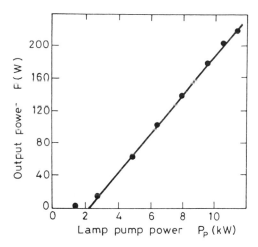

FIG. 7.5. Output power versus lamp input power for a powerful Nd:YAG laser. (By permission from Ref. 7.)

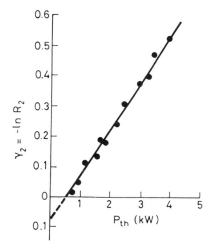

FIG. 7.6. Threshold pump power as a function of mirror reflectivity. (By permission from Ref. 8.)

Example 7.3. *CW laser behavior of a high-power CO_2 laser.* We consider the laser system indicated schematically in Fig. 7.7, where a positive branch unstable resonator is used to obtain a large mode volume and hence high values of output power. The length of the resonator is $L = 175$ cm, while the length of the laser medium is $l = 140$ cm. The active medium consists of a CO_2:N_2:He gas mixture with a 1:1:8 partial pressure ratio and a total pressure of 100 Torr. For cooling reasons, the mixture flows transversely to the resonator axis. Gas excitation is provided by a dc electric discharge between two electrodes, as indicated schematically in the figure (transverse discharge; see also Fig. 6.18b). Typical performance data for the output power P_{out} versus input electrical pump power P_p are shown in Fig. 7.8. Data points can be fitted by the equation

$$P_{out} = 6.66\left[\left(\frac{P_p}{P_{th}}\right) - 1\right] \tag{7.3.17}$$

where P_{out} is given in kilowatts and P_{th} is the extrapolated threshold pump power ($P_{th} \cong 44\,\mathrm{kW}$). Note that we are dealing here with a high-power CO_2 laser producing an output power that exceeds $10\,\mathrm{kW}$.

At a pressure of 100 Torr, the CO_2 laser line is predominantly broadened by collisions. Assuming a gas temperature of $T = 400\,\mathrm{K}$ we find that $\Delta v_c \cong 43\,\mathrm{MHz}$, (see Example 3.3), while Doppler broadening amounts to $\sim 50\,\mathrm{MHz}$ (see Example 3.2). For the given cavity length, the frequency separation between consecutive longitudinal modes is $\Delta v = c/2L = 107\,\mathrm{MHz}$, and, sufficiently far above threshold, a few longitudinal modes are expected to oscillate. Furthermore a few transverse modes are also expected to oscillate. In fact the equivalent Fresnel number (see Sect. 5.6.2) is rather large in our case ($N_{eq} = 7.4$), so that a few transverse modes are expected to have comparable losses (see also Fig. 5.21). Consequently, the transverse beam profile within the laser cavity is expected to be rather uniform. Thus we are dealing with conditions where the previous rate equation treatment provides a reasonable approximation; since the CO_2 laser operates on a four-level scheme, Eq. (7.3.17) can then be compared to Eq. (7.3.9). For this purpose we must know the transmission T_2 of the output mirror. Since the transverse beam profile is assumed to be rather uniform, we use the geometrical optics approximation. Thus T_2, which is equal to the round-trip cavity loss of the unstable resonator, is given by [see Eq. (5.6.5)] $T_2 = (M^2 - 1)/M^2 = 0.45$. In the previous expression M is the round-trip magnification factor of the resonator; it is given by $M = R_1/|R_2| = 1.35$, where R_1 and R_2 are the radii of the two mirrors ($R_2 < 0$, since mirror 2 is concave). Comparing Eqs. (7.3.17) and (7.3.9) and using $\gamma_2 = -\ln(1 - T_2) \cong 0.6$ yields $A_b I_s = 22.3\,\mathrm{kW}$. The beam diameter in the laser cavity is (see also Fig. 5.18b) $D = 2Ma_2 \cong 7.6\,\mathrm{cm}$, where $2a_2 = 5.7\,\mathrm{cm}$ is the diameter of the output coupling mirror (see Fig. 7.7). Thus we get $A_b = \pi D^2/4 \cong 45\,\mathrm{cm}^2$ and $I_s \cong 500\,\mathrm{W/cm}^2$. This value agrees with the best theoretical estimates of the saturation intensity for a CO_2 laser of this kind.[11]

From data in Fig. 7.8 we can now evaluate the unsaturated (i.e., when laser action is prevented) gain coefficient g expected for the laser medium at an input power $P \cong 140\,\mathrm{kW}$. In fact we have

$$g = N_2\sigma = \frac{P_p}{P_{th}}N_{20}\sigma = \frac{P_p}{P_{th}}\frac{\gamma}{l} \tag{7.3.18}$$

where N_2 and N_{20} are the unsaturated upper-state populations at $P_p = 140\,\mathrm{kW}$ and $P_p = P_{th}$, respectively. Note that Eq. (7.3.4a) is used for N_{20}, so the saturated gain coefficient $g_0 = N_{20}\sigma$ is given by $g_0 = \gamma/l$. To calculate g_0, we must know the single-pass cavity loss γ. We assume mirror absorption and scattering losses of 2%. In fact, for this high-power laser oscillating in the 10.6-μm wavelength, polished, water-cooled, copper mirrors are used, which have substantially higher losses than multilayer dielectric mirrors. We then have $\gamma_i \cong 0.02$; since $\gamma_1 = 0$ and $\gamma_2 = 0.6$, we obtain $\gamma \cong 0.32$. Substituting the latter value into Eq. (7.3.18) gives $g_0 = 6.3 \times 10^{-3}\,\mathrm{cm}^{-1}$. The unsaturated gain coefficient can easily be obtained experimentally by measuring the gain coefficient of the laser medium with both mirrors removed. Measured values of gain coefficient for this type of laser are in fairly good agreement with values calculated here.[12]

We now compare the experimental value of slope efficiency in Fig. 7.8 with theoretical predictions. We assume $\eta_p \cong 0.8$ (see Fig. 6.27), $\eta_c = \gamma_2/2\gamma = 0.94$, $\eta_t \cong 1$, and $\eta_q = h\nu/h\nu_{mp} = 0.4$. (See CO_2 laser energy levels in Chap. 10.) From Eq. (7.3.14) we obtain $\eta_s = 0.3$, which is appreciably higher than the experimental value obtained from Fig. 7.8 ($\eta_s \cong 0.21$). This discrepancy can be attributed to at least two distinct causes: (1) The transverse utilization factor η_t may be appreciably smaller than 1. Perhaps by coincidence, if we were using the same value of η_t found in the previous problem, $\eta_t = 0.73$, the theoretical result would almost exactly agree with the experimental one. (2) Data in Fig. 7.8 refer to a partially closed-cycle system, and, in this case, products from the electric discharge (mostly CO and O_2)

are likely to accumulate in the gas mixture, thus reducing pumping efficiency below the theoretical value of 80%. Actually, it is a matter of fact that slope efficiencies larger than ∼20% are seldom found in practice for any CO_2 laser. This discussion clarifies our understanding of how slope efficiency is further reduced from the upper-limit value established by the quantum efficiency ($\eta_q = 40\%$).

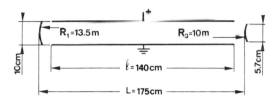

FIG. 7.7. Possible cavity configuration for a powerful cw CO_2 laser.

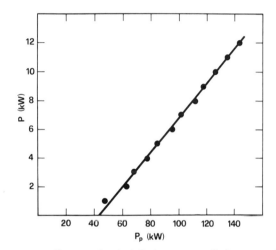

FIG. 7.8. Output power P versus electrical discharge power P_p for a powerful cw CO_2 laser.

7.3.2. Space-Dependent Model

We now consider the case when both the mode distribution and pump rate are spatially dependent. In this case, inversion is also spatially dependent, so the rate equations become more complicated. We therefore limit ourselves to discussing the most relevant results and refer to Appendix E for a detailed treatment. We assume a cylindrical symmetry and let u be the field amplitude of the given mode, normalized to its peak value. For simplicity we take u to be independent of the longitudinal coordinate z along the resonator, while we generally take the pump rate to be dependent on both radial and longitudinal coordinates, i.e., $R_p = R_p(r, z)$.

As far as threshold conditions are concerned, Appendix E shows that Eq. (7.3.2) still holds for the average value of N, i.e.:

$$\langle N \rangle_c = \left(\frac{\gamma}{\sigma l}\right) \tag{7.3.19}$$

where the average is taken over the squared amplitude of the field distribution, namely, [see also Eq. (6.3.17)]:

$$\langle N \rangle = \frac{\int_a N |u|^2 \, dV}{\int_a |u|^2 \, dV} \tag{7.3.20}$$

and the integrals are taken over the volume of the active medium. At each point of the active medium, below or at threshold, an equilibrium must exist between the number of atoms raised by the pumping process and those decaying spontaneously, i.e., $R_p(r, z) = N(r, z)/\tau$. At threshold we then have

$$\langle R_p \rangle_c = \frac{\langle N \rangle_c}{\tau} = \frac{\gamma}{\sigma l \tau} \tag{7.3.21}$$

where $\langle R_p \rangle$ is the average of $R_p(r, z)$ over the squared amplitude of the field distribution [see Eq. (6.3.8)] and where Eq. (7.3.19) has been used.

Above threshold, from the condition $d\phi/dt = 0$, we find from Appendix E that the average gain must equal total losses, i.e.:

$$\sigma l \langle N \rangle_0 = \gamma = \sigma l \langle N \rangle_c \tag{7.3.22}$$

Thus, according to Eq. (7.3.22), it is the average value of the inversion $\langle N \rangle_0$, which actually gets clamped at its threshold value when threshold is exceeded (see Fig. 7.3).

To calculate the threshold pump power P_{th} and the output power P_{out}, we must specify the spatial variation of both $|u|^2$ and R_p. We assume oscillation on a TEM$_{00}$ mode and take

$$|u|^2 = \exp\left[-\left(\frac{2r^2}{w_0^2} \right) \right] \tag{7.3.23}$$

This means that: (1) Spot size is assumed to be independent of z and equal to the spot size w_0 at the beam waist. (2) The standing wave pattern of the mode is neglected [see for comparison Eq. (6.3.9)]. Regarding $R_p(r, z)$ we consider two separate cases; (1) Uniform pumping, i.e., $R_p = $ const. (2) Gaussian pump distribution, as appropriate for longitudinal pumping by, e.g., diode lasers. In this case we take $R_p(r, z) = C \, \exp[-2(r^2/w_p^2)] \exp[-(\alpha z)]$, where C is a constant proportional to the total input pump power [see Eq. (6.3.7)].

We first consider uniform pumping as provided by either electrical or lamp pumping. Then, from either Eq. (6.2.6) or (6.4.26), we obtain

$$R_p = \eta_p \frac{P_p}{\pi a^2 l h \nu_{mp}} \tag{7.3.24}$$

where a cylindrical medium of radius a is considered. We now consider the cladded rod geometry (see Sect. 6.3.3.), where the active species is assumed to be confined to the central region of the rod $0 \le r \le a$, while the rod is undoped for $r > a$. In this case, we need not be concerned with the effects of beam truncation due to the finite aperture of the medium. Thus Eq. (7.3.23) can be taken to hold for $0 \le r \le \infty$ while one has $R_p = $ const. for $0 \le r \le a$ and $R_p = 0$ for $r > 0$. We substitute Eq. (7.3.24) into Eq. (7.3.21) and then use in Eq. (7.3.23) to calculate the average value of R_p. In this way we obtain an expression for P_{th} that is the same as in Eq. (6.3.22) if η_p replaces η_{pl}. The calculation of the output power then proceeds as discussed in Appendix E; we limit ourselves here to quoting and discussing the final result. We define a normalized pump power x as

$$x = \frac{P_p}{P_{mth}} \tag{7.3.25}$$

where P_{mth} is the minimum threshold that occurs when $w_0 \ll a$; according to Eq. (6.3.22) it is given by.

$$P_{mth} = \left(\frac{\gamma}{\eta_p}\right)\left(\frac{h\nu_{mp}}{\tau}\right)\left(\frac{\pi a^2}{\sigma_e}\right) \tag{7.3.26}$$

We also define a normalized value of the output power y as

$$y = \frac{P_{out}}{P_s} \tag{7.3.27}$$

where P_s is a saturation power given by

$$P_s = \frac{\gamma_2}{2}\frac{\pi w_0^2}{2}I_s \tag{7.3.28}$$

The resulting relation between x and y is then

$$x = \frac{y}{\ln[(1+y)/(1+\beta y)]} \tag{7.3.29}$$

where

$$\beta = \exp\left[\left(\frac{2a^2}{w_0^2}\right)\right] \tag{7.3.30}$$

We see that the relation between normalized output power y and the amount by which the threshold is exceeded x is complicated and quite different from the simple one predicted by the space-independent rate equations [see Eq. (7.3.9)].* For comparison Fig. 7.9 shows plots (*solid lines*) of the normalized output power y versus the normalized pump power x, for $w_0 \ll a$, $w_0 = 0.7a$, and $w_0 = \sqrt{2}\, a$. One may note that the relation between y and x is no longer linear, with the derivative dy/dx increasing with x. To understand this behavior it is appropriate to calculate the slope efficiency as

$$\eta_s = \frac{dP_{out}}{dP_p} = \left(\frac{P_s}{P_{mth}}\right)\left(\frac{dy}{dx}\right)$$

where Eqs. (7.3.25) and (7.3.27) have been used. With the help of Eqs. (7.3.26) and (7.3.28) we see that η_s can be expressed as in Eq. (7.3.14), with $\eta_q = h\nu/h\nu_{mp}$, if the transverse efficiency is now defined as:

$$\eta_t = \left[\frac{(\pi w_0^2/2)}{\pi a^2}\frac{dy}{dx}\right] \tag{7.3.31}$$

---

* If we take $P_s = A_b I_s \gamma_2/2$ in Eq. (7.3.9), this equation simply gives $y = (x-1)$, where y and x are again given by Eqs. (7.3.27) and (7.3.25).

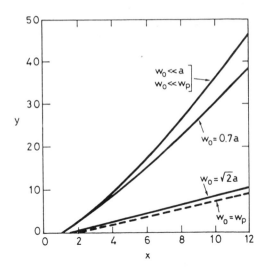

FIG. 7.9. Normalized output power y versus normalized pump power x for a laser oscillating on a TEM_{00} mode. Continuous curves represent uniform pumping in a rod of radius a for several values of the mode spot size w_0. The dashed curve represents a Gaussian distribution of pump light with a spot size w_p such that $w_0 = w_p$.

For transverse pumping by diode lasers (uniform pumping), we obtain the same expression for slope efficiency; the only difference is that now $\eta_q = h\nu/h\nu_p$. Note that, when $\beta \to 0$, i.e., when $w_0 \ll a$, one has $(dy/dx) = 2$ for $y \to 0$ ($x \to 1$), and the transverse efficiency becomes $\eta_t = (\pi w_0^2/\pi a^2)$. Note also that, since (dy/dx) increases with x, the transverse efficiency also increases with x. To understand this point, we consider the case $w_0 = 0.7a$ as an example. The plot of η_t versus x for this case is shown as a solid line in Fig. 7.10. Starting from the value $\eta_t \cong 0.97(w_0/a)^2 \cong 0.473$, at low powers, η_t increases to unity when $x \gg 1$. The increase in transverse efficiency, η_t, with increasing x can now be explained by observing that the energy of an excited atom can be removed either by a stimulated emission process or by a spontaneous decay. Thus at low powers, i.e., for $x \to 1$, stimulated emission

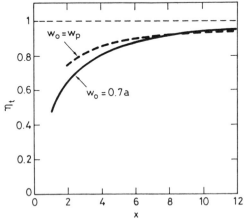

FIG. 7.10. Plot of the transverse efficiency η_t versus normalized pump power x. Continuous and dashed curves represent uniform pumping and Gaussian beam pumping, respectively.

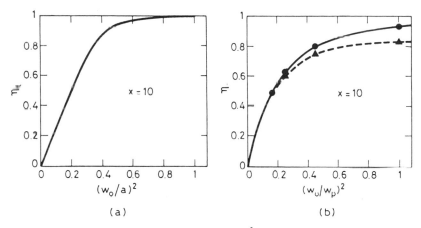

FIG. 7.11. (a) Plot of the transverse efficiency η_t versus $(w_0/a)^2$ at a normalized pump power $x = 10$ for uniform pumping. (w_0 is the mode spot size, and a is the rod radius.) (b) Plot of the transverse efficiency η_t versus $(w_0/w_p)^2$ at a normalized pump power $x = 10$ for Gaussian beam pumping. (w_0 is the mode spot size, and w_p is the spot size of the pump distribution.) Closed circles represent a four-level laser; closed triangles represent a quasi-three-level laser with $B = \sigma_a N_t l/\gamma = 1$.

prevails near the beam axis where beam intensity is high, while spontaneous decay prevails in the wings of the beam. For larger values of x, i.e., with increased beam power, stimulated emission dominates spontaneous decay over a larger portion of the pump profile and hence for a larger fraction of excited atoms. Ultimately, at very large powers, all excited atoms decay by stimulated emission, so that $\eta_t = 1$, and the entire pump profile is used by the laser beam.[6]

For diode pumping we can generally reach pump powers well above threshold and it is instructive to consider the behavior of η_t versus $(w_0/a)^2$ for large values of x. The behavior of η_t versus $(w_0/a)^2$ at $x = 10$ is shown in Fig. 7.11a. The figure shows that, to obtain, e.g., $\eta_t > 90\%$, we must have $w_0 > 0.66a$; for $w_0 = 0.7a$ we obtain the rather high value of 94%. However, for the usual case of a medium without cladding, it is generally not beneficial to increase spot size even further, since this leads to excessive diffraction losses due to the rod aperture.

For a Gaussian distribution of the pump beam, the calculation, also presented in Appendix E, proceeds similarly. We again define a normalized output power y as in Eq. (7.3.27), where P_s is again given by Eq. (7.3.28). We also define a normalized pump power x, as in Eq. (7.3.25), where P_{mth} is now the minimum threshold for Gaussian beam pumping, as obtained from Eq. (6.3.20) when $w_0 \ll w_p$. We then get

$$P_{mth} = \left(\frac{\gamma}{\eta_p}\right)\left(\frac{h\nu_p}{\tau}\right)\left(\frac{\pi w_p^2}{2\sigma_e}\right) \tag{7.3.32}$$

The relation between y and x then turns out to be given by the following expression:

$$\frac{1}{x} = \int_0^1 \frac{t^\delta \, dt}{1 + yt} \tag{7.3.33}$$

where $\delta = (w_0/w_p)^2$. Although Eq. (7.3.33) differs in notation, it agrees with the equation originally given by Moulton.[13] For $w_0 \ll w_p$, one has $\delta \to 0$, so Eq. (7.3.33) gives the same result as Eq. (7.3.29) (when $w_0 \ll a$ i.e., $\beta \to 0$). Thus the plot of y versus x for this case is the same as that for uniform pumping (see Fig. 7.9). For very small values of spot size, in fact, the beam makes no distinction between uniform or Gaussian pump distributions. For $w_0 = w_p$, one has $\delta = 1$; the integration of Eq. (7.3.33) then gives

$$ x = \frac{y}{1 - [\ln(1+y)/y]} \tag{7.3.34} $$

Equation (7.3.34) is plotted in Fig. 7.9 as a dashed line. We can again calculate the slope efficiency as $\eta_s = dP_{out}/dP_p = (P_s/P_{mth})(dy/dx)$ and we again find that η_s can be expressed as in Eq. (7.3.14), where, as for transverse diode pumping, $\eta_q = h\nu/h\nu_p$, and where now:

$$ \eta_t = \frac{\pi w_0^2}{\pi w_p^2} \frac{dy}{dx} \tag{7.3.35} $$

The behavior of η_t versus x for $w_0 = w_p$, as calculated by Eqs. (7.3.34) and (7.3.35), is also plotted in Fig. 7.10 as a dashed line. The increase of η_t with x has the same physical explanation as for uniform pumping. Note that, for sufficiently high values of x ($x > 7$), the transverse efficiency for the two cases is about the same. The behavior of η_t versus $(w_0/w_p)^2$ for $x = 10$ is then shown in Fig. 7.11b. The four data points, indicated by solid dots, are obtained from Ref. 13, and the continuous line represents a suitable interpolation of these points. Comparing Figs. 7.11a–b shows that the increase of η_t with $(w_0/w_p)^2$ is now slower than for uniform pumping. This is due to the lower pump rate in the wings of a Gaussian function compared to a uniform distribution.

It is worthwhile observing from Fig. 7.11b that, to reach the relatively high value of, e.g., 94%, it suffices to have $w_0 \cong w_p$. It is then less advantageous to increase the laser spot size beyond this point. In fact while η_t could increase by only a very small additional amount, the threshold pump power P_{th}, being proportional to $(w_0^2 + w_p^2)$ [see Eq. (6.3.20)], would increase significantly. Thus, at large values of x, the condition $w_0 = w_p$ (also called the mode-matching condition) may be taken to correspond, more or less, to the optimum situation.

So far we have neglected the standing wave character of a mode; i.e., $|u|^2$ was written as in Eq. (7.3.23) rather than as in Eq. (6.3.9). This is correct for a unidirectional ring resonator (see Fig. 5.4a), but for most other cases, e.g., using a two-mirror resonator, a well-defined standing wave pattern is formed when the laser oscillates in a single longitudinal mode.* For a mode with uniform transverse profile, the effect on the output power of a standing wave pattern was considered by Casperson.[6] As far as slope efficiency is concerned, results obtained can be represented in terms of a fifth efficiency factor, introduced in the right-hand side of Eq. (7.3.14); we can refer to this factor as the *longitudinal utilization factor* of the pump distribution η_l, or *longitudinal efficiency*. The value of η_l is $\eta_l = (2/3) = 0.666$ at threshold, and it increases to, e.g., $\eta_l = (8/9) = 0.89$

---

* A notable exception occurs in the twisted-mode technique,[14] where the oppositely traveling beams in the active medium consist of two circularly polarized waves of the same sence (both right or both left) so that, within any plane perpendicular to the beam axis, their fields rotate in opposite directions and no standing wave pattern is produced.

when 10 times above threshold. The physical origin of η_l is similar to that of η_t; namely, just above threshold, only atoms around the peaks of the standing wave decay predominantly by stimulated emission, while atoms near the zeros of the pattern decay spontaneously. At increasing values of x, i.e., at increasing energy densities, more atoms around the field nodes undergo stimulated rather than spontaneous decay, so longitudinal efficiency increases.

Example 7.4. *Threshold and output powers in a longitudinally diode-pumped Nd:YAG laser.* As a representative example of longitudinal diode pumping, we consider the laser configuration in Fig. 7.12, where a 1 cm long Nd:YAG rod is pumped by a 100 μm wide laser array at 805–808 nm wavelength.[15] Coupling optics consists of a 6.5 mm focal-length, 0.615 NA, collecting lens, an anamorphic prism pair providing a 4-times beam magnification, and a 25-mm lens to focus the pump light on the rod (see Fig. 6.12). The Nd:YAG resonant cavity is formed by a plane mirror directly coated on one face of the rod and a 10-cm radius, 95% reflecting mirror separated by approximately 5.5 cm from the plane mirror. About 93% of the pump power is transmitted to the rod through the plane mirror. In this geometry, the TEM_{00} mode waist occurs at the planar reflector, and its spot size can be calculated as $w_0 \cong 130\,\mu$m. (Thermally induced lensing in the rod is neglected). The spot size of the pump beam provides good mode matching with this TEM_{00} laser mode. The laser-operating characteristics are indicated in Fig. 7.13. The threshold pump power is $P_{th} \cong 75$ mW. At an optical pump power of $P_p = 1.4$ W, an output power of $P_{out} = 370$ mW is obtained. At this output power, the measured optical-to-optical slope efficiency is $\eta_s \cong 40\%$.

To compare threshold pump power with the expected value, we assume that the transverse pump beam distribution can be approximated by a Gaussian function and take $w_p \cong w_0 = 130\,\mu$m. From Eq. (6.3.22), with $h\nu_p = 2.45 \times 10^{-19}$ J, $\sigma_e = 2.8 \times 10^{-19}$ cm^2, and $\tau = 230\,\mu$s, we obtain $(\gamma/\eta_p) \cong 3.7 \times 10^{-2}$. For 5% transmission of the output mirror, we get $\gamma_2 \cong 5 \times 10^{-2}$, and assuming an internal loss per pass $\gamma_i = 0.5 \times 10^{-2}$, we obtain $\gamma = \gamma_i + (\gamma_2/2) = 3 \times 10^{-2}$. From the previously obtained value of γ/η_p, we then get $\eta_p \cong 81\%$, which includes overall transmission of the coupling optics and the transmission of the plane mirror at the pump wavelength. Note that the absorption efficiency of the pump radiation in the laser rod in a single pass, $\eta_a = \{1 - \exp -[(\alpha l)]\}$, can be taken as unity for an average absorption coefficient of ~ 6 cm^{-1} in the 805–808 nm band (see Fig. 6.8a) and for a rod length of $l = 1$ cm. We can now compare the measured slope efficiency with the expected value. Since the threshold power for $w_0 \cong w_p$ is 75 mW, the minimum threshold power, which occurs when $w_0 \rightarrow 0$, is expected to be half this value, i.e., $P_{mth} \cong 38$ mW. Thus, at 1.14 W input pump power, we have $x \cong 30$. At this value of x, from Eq. (7.3.34) we obtain $y = 26$, and, from Eq. (7.3.35), $\eta_t \cong 0.97$. We then have $\eta_c = (\gamma_2/\gamma) = 0.83$ and $\eta_q = (807/1060) = 0.76$. Expected overall optical-to-optical slope efficiency is thus $\eta_s = \eta_p \eta_c \eta_t \eta_q = 0.49$, in fair agreement with the measured one. Note that the longitudinal efficiency is not taken into account because the laser is oscillating on many longitudinal modes whose various standing wave patterns add to produce a fairly uniform energy density distribution along the laser rod. According to Eqs. (7.3.27) and (7.3.28), the expected output power at 1.14 W input power is $P_{out} = y P_s = 500$ mW, i.e., somewhat larger than the experimental one. This discrepancy can perhaps be attributed to thermal effects in the laser rod, which, at the highest pump powers, increase losses and decrease spot size w_0.

Note that the quoted 40% efficiency refers to optical-to-optical efficiency. To obtain the overall electrical-to-optical slope efficiency, we must multiply optical efficiency by the radiative efficiency η_r of the array. Again from Fig. 7.13 we obtain $\eta_r \cong 29\%$, so that overall electrical-to-optical slope efficiency is about 11.6%.

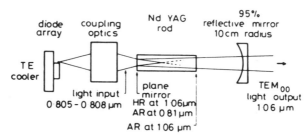

FIG. 7.12. Schematic illustration of the experimental set up of a Nd:YAG laser longitudinally pumped by a diode array. (By permission from Ref. 15).

To conclude this section we observe that, when space dependence is taken into account, the problem becomes somewhat more complicated. However, expressions for threshold inversion and threshold pump rate remain identical to those obtained in a space-independent treatment if appropriate average values $\langle N \rangle_c$ and $\langle R \rangle_c$ are used. Note that, as shown in Sect. 6.3.3, this result also holds when the standing wave pattern of the mode is taken into account. The expression for output power as a function of the amount by which threshold is exceeded becomes more complicated. In terms of slope efficiency however, the results are simple and intuitive and they can be directly related to those obtained for the space-independent case.

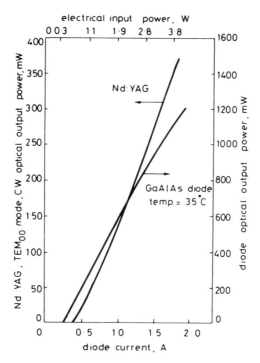

FIG. 7.13. Output power versus diode current for the Nd:YAG laser in Fig. 7.12. The same figure shows output power versus current for the laser diode array. (By permission from Ref. 15.)

7.4. THRESHOLD CONDITION AND OUTPUT POWER: QUASI-THREE-LEVEL LASER

We now investigate the threshold condition and output power for a quasi-three-level laser. Laser behavior is first considered within the space-independent rate-equation model of Sect. 7.2.2. Results predicted by the space-dependent model are then discussed and the two models compared.

7.4.1. Space-Independent Model

The analysis for a quasi-three-level laser proceeds similarly to that for the four-level case, starting with Eqs. (7.2.24).

The threshold inversion is obtained by setting $(d\phi/dt) = 0$ in Eq. (7.2.24b), thus giving

$$N_c = \frac{V}{V_a \sigma_e c_0 \tau_c} = \frac{\gamma}{\sigma_e l} \tag{7.4.1}$$

This is the same expression as for a four-level laser. We obtain the critical pump rate from Eq. (7.2.24a) by setting $(dN/dt) = 0$, $\phi = 0$, and $N = N_c$. We obtain

$$R_{cp} = \frac{f N_t + N_c}{(1+f)\tau} \tag{7.4.2}$$

Since in most cases $f \ll 1$, we can write $(1+f) \cong 1$ in the denominator of Eq. (7.4.2). A comparison between Eqs. (7.4.2) and (7.3.3) shows that the critical pump rate for a quasi-three-level is increased over that for a four-level laser by the addition of the term fN_t in the numerator of Eq. (7.4.2). In typical situations this term is perhaps ~ 5 times larger than N_c. In the case of uniform pumping by a diode laser, according to the discussion in Sect. 6.3.3, we can write $R_p = \eta_p P_p / h\nu_p A l$, where A is the cross-sectional area of the active medium and l its length. Using this expression in Eq. (7.4.2), we obtain the threshold pump power as:

$$P_{th} = \frac{h\nu_p}{\eta_p \tau} \frac{(f N_t + N_c)A l}{(1+f)} \tag{7.4.3}$$

With the help of Eqs. (7.2.22) and (7.4.1), we can write Eq. (7.4.3) in the more intuitive form [compare with Eq. (7.3.12)]:

$$P_{th} = \frac{\gamma(1+B)}{\eta_p} \left(\frac{h\nu_p}{\tau}\right)\left(\frac{A}{\sigma_e + \sigma_a}\right) \tag{7.4.4}$$

where we have set

$$B = \frac{\sigma_a N_t l}{\gamma} \tag{7.4.5}$$

Above threshold the cw inversion N_0 and the cw photon number ϕ_0 are obtained from Eq. (7.2.24) by letting $(dN/dt) = (d\phi/dt) = 0$. As for the four-level laser, N_0 again equals N_c; ϕ_0, as obtained from Eq. (7.2.24a) with the help of Eq. (7.4.2), is then given by

$$\phi_0 = \frac{V}{N_0(\sigma_e + \sigma_a)c} \frac{fN_t + N_0}{\tau}(x - 1) \qquad (7.4.6)$$

where $x = R_p/R_{cp} = P_p/P_{th}$ again represents the amount by which threshold is exceeded. Using Eq. (7.4.1) for N_0 in addition to Eqs (7.2.14), (7.2.15), and (7.2.22), Eq. (7.4.6) becomes

$$\phi_0 = \left[\frac{A_b\gamma(1 + B)}{\sigma_e + \sigma_a}\right]\left(\frac{\tau_c}{\tau}\right)(x - 1) \qquad (7.4.7)$$

where we have set $V_a = A_b l$; A_b is the beam area in the medium, which is assumed to be smaller than, or equal to, the cross-sectional area of the active medium.

The output power through mirror 2, as obtained from Eq. (7.2.18) by using Eqs. (7.4.7) and (7.2.14), is [compare with Eq. (7.3.9)]:

$$P_{out} = \left[\frac{A_b(1 + B)}{\sigma_e + \sigma_a}\right]\left(\frac{h\nu}{\tau}\right)\left(\frac{\gamma_2}{2}\right)\left(\frac{P_p}{P_{th}} - 1\right) \qquad (7.4.8)$$

The laser slope efficiency $\eta_s = (dP_{out}/dP_p)$ is then readily obtained from Eq. (7.4.8) as:

$$\eta_s = \left[\frac{A_b(1 + B)}{\sigma_e + \sigma_a}\right]\left(\frac{h\nu}{\tau}\right)\left(\frac{\gamma_2}{2}\right)\left(\frac{1}{P_{th}}\right) \qquad (7.4.9)$$

With the help of Eq. (7.4.4), Eq. (7.4.9) gives

$$\eta_s = \eta_p\left(\frac{\gamma_2}{2\gamma}\right)\left(\frac{h\nu}{h\nu_p}\right)\left(\frac{A_b}{A}\right) \qquad (7.4.10)$$

Thus, given the same parameters, the slope efficiency of a quasi-three-level laser is predicted to be the same as that of a four-level laser. At first sight this result is unexpected given the increased loss, in a quasi-three-level laser, due to ground-state absorption. However, energy removed by ground-state absorption actually raises atoms to the upper laser level, making these atoms again available to produce stimulated emission.

7.4.2. Space-Dependent Model

We next consider the case where mode distribution and pump rate are assumed to be spatially dependent. We limit our discussion to the most relevant results corresponding to the active medium being longitudinally pumped by a beam with a Gaussian transverse profile. We again refer to Appendix E for a more detailed treatment of this case as well as the case of uniform pumping. We assume that the field intensity profile is again described by

Eq. (7.3.23) and the spatial distribution of the pump rate by Eq. (6.3.7), which is repeated here for convenience:

$$R_p(r,z) = \eta_r\eta_t\left(\frac{P_p}{hv_p}\right)\left(\frac{2\alpha}{\pi w_p^2}\right)\exp\left[-\left(\frac{2r^2}{w_p^2}\right)\right]\exp(-\alpha z) \qquad (7.4.11)$$

As far as the threshold condition is concerned, Eq. (7.4.1) still holds for the average value of N, i.e.,

$$\langle N \rangle_c = \left(\frac{\gamma}{\sigma_e l}\right) \qquad (7.4.12)$$

where the average value is calculated via Eq. (7.3.20). To obtain the threshold pump rate, we notice that, at each point of the active medium, below or at threshold, an equibrium exists between the number of atoms raised by the pumping process and the number of atoms decaying spontaneously. Thus, from Eq. (7.2.24a), we obtain $R_p = [fN_t + N(r,z)]/(1+f)\tau$. Averaging this expression over the mode intensity distribution and using the expression for threshold inversion given by Eq. (7.4.12), we obtain

$$\langle R \rangle_p = \frac{\sigma_a N_t l + \gamma}{(\sigma_e + \sigma_a) l \tau} \qquad (7.4.13)$$

If the pump rate, Eq. (7.4.11), is substituted into the left-hand side of Eq. (7.4.13) and the average of R_p over the field intensity profile is calculated, we end up with the threshold pump power expression given by Eq. (6.3.25), which is repeated here in a slightly different form [compare to Eq. (7.4.4)]:

$$P_{th} = \frac{\gamma(1+B)}{\eta_p}\left(\frac{hv_p}{\tau}\right)\left[\frac{\pi(w_0^2 + w_p^2)}{2(\sigma_e + \sigma_a)}\right] \qquad (7.4.14)$$

where B is again given by Eq. (7.4.5).

Above threshold, using the condition $(d\phi/dt) = 0$, one again finds that average gain must equal losses, thus giving

$$\langle N \rangle_0 = \langle N \rangle_c = \frac{\gamma}{\sigma_e l} \qquad (7.4.15)$$

The calculation of output power is considered in some detail in Appendix E and we limit ourselves here to quoting and discussing the final result. First we define x as the factor by which threshold is exceeded as in Eq. (7.3.25) where the minimum threshold power P_{mth} is now the threshold value that holds for $w_0 \ll w_p$ and $B \ll 1$(i.e.,$\sigma_a N_t l \ll \gamma$). From Eq. (7.4.14) we obtain

$$P_{mth} = \left(\frac{\gamma}{\eta_p}\right)\left(\frac{hv_p}{\tau}\right)\left[\frac{\pi w_p^2}{2(\sigma_e + \sigma_a)}\right] \qquad (7.4.16)$$

We next define normalized output power as in Eq. (7.3.27), where now

$$P_s = \frac{\gamma_2}{2}\left[\frac{\pi w_0^2}{2(\sigma_e + \sigma_a)}\right]\left(\frac{h\nu}{\tau}\right) \tag{7.4.17}$$

The relation between y and x then turns out to be

$$x = \frac{1 + B\dfrac{\ln(1+y)}{y}}{\displaystyle\int_0^1 \frac{t^\delta dt}{1 + yt}} \tag{7.4.18}$$

where $\delta = (w_0/w_p)^2$. Note that Eq. (7.4.18) reduces to Eq. (7.3.33) when $B \to 0$ ($\sigma_a N_t l \ll \gamma$), i.e., for negligible absorption arising from the lower laser level. Apart from the difference in notation, Eq. (7.4.18) is the same as Eq. (25) in Ref. 16.

We can again calculate the slope efficiency as $\eta_s = (dP_{out}/dP_p) = (P_s/P_{mth})(dy/dx)$. With the help of Eqs. (7.4.16) and (7.4.17), we obtain

$$\eta_s = \eta_p \left(\frac{\gamma_2}{2\gamma}\right)\left(\frac{h\nu}{h\nu_p}\right)\left(\frac{\pi w_0^2}{\pi w_p^2}\frac{dy}{dx}\right) \tag{7.4.19}$$

This is the same expression as for a four-level laser. The behavior of the transverse efficiency, $\eta_t = (\pi w_0^2/\pi w_p^2)(dy/dx)$, for $x = 10$ and $B = 1$ is also shown in Fig. 7.11b. The three data points indicated by closed triangles are obtained from the computed results of Ref. 16, while the dashed line indicates a suitable interpolation. Note that, when $(w_0/w_p)^2 \ll 1$, the value of η_t tends to coincide with the corresponding value holding for a four-level laser. In this case, in fact, the population raised to the upper laser level via ground-state absorption is mostly recycled usefully via a stimulated emission process, and ground-state absorption does not degrade laser efficiency. At higher values of $(w_0/w_p)^2$ however, e.g., $(w_0/w_p)^2 = 1$, ground-state absorption in the wings of the inversion profile leads predominantly to spontaneous rather than stimulated decay. Accordingly, the value of η_t becomes smaller than the corresponding value for a four-level laser. It can be shown, however, that, on further increasing the pump power, i.e., the value of x, η_t tends to become equal to the corresponding value for a four-level laser. At sufficiently high values of x in fact, the normalized output power y becomes large enough to that $B[\ln(1+y)]/y \ll 1$. Under this condition Eq. (7.4.18) becomes identical to Eq. (7.3.33).

Example 7.5. *Threshold and output powers in a longitudinally diode-pumped Yb:YAG laser.* As a representative example we consider a $l = 2.5$ mm thick 6.5 atomic % Yb:YAG laser disk longitudinally pumped at $\lambda_p = 941$ nm by a $Ti^{3+}:Al_2O_3$ laser.[18] One face of the disk is plane and coated for high reflectivity at the laser wavelength ($\lambda = 1.03\,\mu m$). The other face is concave with a 1-cm radius of curvature; it is coated to give power reflectivity of $R_2 = 90\%$ at the laser wavelength. Under these conditions, the calculated spot size at the beam waist, i.e., at the plane mirror, is $w_0 = 28\,\mu m$; the spot size can be taken as approximately constant along the resonator. The measured pump spot size in the laser disk is $w_p = 31\,\mu m$. Measured values of output power versus absorbed pump power at $T = 300$ K, are indicated by triangles in Fig. 7.14.

To compare these results with theoretical predictions, we observe that the total Yb concentration is $N_t \cong 9 \times 10^{20}$ cm^{-3} and, from the measured absorption coefficient at $\lambda = 1.03\,\mu m$ (Fig. 6.8b), we obtain

$\sigma_a \cong 1.2 \times 10^{-21}$ cm^2. The effective value of the stimulated emission cross section is then evaluated in Ref. 18 as $\sigma_e \cong 18 \times 10^{-21}$ cm^2. Assuming no other losses except the output coupling loss, we get $\gamma = \gamma_2/2 \cong (1 - R_2)/2 = 5 \times 10^{-2}$. From Eq. (7.4.14) we obtain $P_{th} \cong 83$ mW, to be compared with the experimental value of 70 mW in Fig. 7.14. We now calculate the predicted slope efficiency at the maximum pump power of 180 mW. Note that one has $\gamma = \gamma_2/2$ and that, since the data in Fig. 7.14 are expressed with respect to absorbed pump power, we must set $\eta_p = 1$ in Eq. (7.4.19). The expression for slope efficiency then reduces to $\eta_s = (h\nu/h\nu_p) \times (w_0^2 \, dy/w_p^2 \, dx)$. To calculate the transverse efficiency $\eta_t = (w_0^2 \, dy/w_0^2 \, dx)$, we first observe that, according to Eq. (7.4.16), one has $P_{mth} = 6.5$ mW, so that, at $P_p = 180$ mW, one gets $x = (P_p/P_{mth}) = 27.7$. The transverse efficiency can then be estimated from Fig. 4(e) in Ref. 18 by taking into account that $\eta_t = dS/dF$, $B = \sigma_a N_t l/\gamma \cong 5$, $a = w_p/w_0 \cong 1.1$, $(F/F_{th}) = xa^2/(1 + a^2)(1 + B) \cong 2.16$, where the quantities S, F, and F_{th} are defined in the cited reference. We obtain $\eta_t = dS/dF \cong 70\%$, so that $\eta_s = (h\nu/h\nu_p)\eta_t \cong 63\%$.

A more exact calculation of the output power versus pump power, as obtained directly from Eq. (7.4.18), is also plotted as a solid line in Fig. 7.14. From this calculation we obtain a more precise value of the predicted slope efficiency of $\eta_s = 59\%$ at $P_p = 180$ mW, to be compared with the value of 56% obtained from the four points in the figure corresponding to the highest experimental powers.

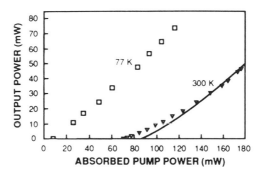

FIG. 7.14. Plots of output power versus absorbed pump power for a Ti:sapphire-pumped Yb:YAG laser for liquid-nitrogen-cooled operation (77 K) and room temperature operation (300 K). (By permission from Ref. 18.)

7.5. OPTIMUM OUTPUT COUPLING

For a fixed rate, there is some value for the transmission, T_2, of the output mirror that maximizes the output power.[19] Physically, the reason for this optimum arises from the fact that, as T_2 is increased, we have the following two contrasting circumstances: (1) The output power tends to increase due to the increased mirror transmission. (2) The output power tends to decrease due to the decreased number of cavity photons, ϕ_0, arising from the increased cavity losses.

To find the optimum output-coupling condition, we limit ourselves to a four-level laser and consider the space- independent model. The optimum transmission is then obtained from Eq. (7.3.9) by imposing the condition $dP_{out}/d\gamma_2 = 0$, for a fixed value of pump power

P_p. Of course, we must take into account the fact that, according to Eq. (7.3.12), P_{th} is also a function of γ_2. From Eq. (7.3.12) we can write

$$P_{th} = P_{mth} \frac{\gamma}{\gamma_i + (\gamma_1/2)} \tag{7.5.1}$$

where P_{mth}, the minimum threshold pump power, is here defined as the threshold pump power for zero output coupling, $\gamma_2 = 0$. Equation (7.3.9) can then be transformed into

$$P_{out} = \left[A_b I_s \left(\gamma_i + \frac{\gamma_1}{2} \right) \right] S \left[\left(\frac{x_m}{S+1} \right) - 1 \right] \tag{7.5.2}$$

where

$$S = \frac{(\gamma_2/2)}{\gamma_i + (\gamma_1/2)} \tag{7.5.3}$$

and

$$x_m = \frac{P_p}{P_{mth}} \tag{7.5.4}$$

The only term in Eq. (7.5.2) that depends on γ_2 is the quantity S, which, according to Eq. (7.5.3), is proportional to γ_2. The optimum-coupling condition can then be obtained by setting $dP_{out}/dS = 0$; the optimum value of S is then readily obtained as:

$$S_{op} = (x_m)^{1/2} - 1 \tag{7.5.5}$$

The corresponding expression for the output power is obtained from Eq. (7.5.2) as:

$$P_{op} = \left[A_b I_s \left(\gamma_i + \frac{\gamma_1}{2} \right) \right] [(x_m)^{1/2} - 1]^2 \tag{7.5.6}$$

The reduction in output power as a result of nonoptimum-operating conditions becomes particularly important when working very close to threshold (i.e., when $x_m \cong 1$). Well above threshold however, on the other hand, the output power becomes rather insensitive to a change in output coupling around the optimum value. As an example, Fig. 7.15 shows a normalized plot of P_{out} versus S for $x_m = 10$. According to Eq. (7.5.2), one has $P_{out} = 0$ for $S = 0$ (i.e., $\gamma_2 = 0$) and $S = 9$; according to Eq. (7.5.5), one then has $S_{op} \cong 2.16$. From Fig. 7.15 we now see that changes in, S, i.e., of the output coupling, around the optimum value by as much as 50% result only in $\sim 5\%$ reduction of the output power.

In the case of a space-dependent model, similar considerations can be developed starting with Eq. (7.3.29) (uniform pumping) or Eq. (7.3.34) (Gaussian pumping). However, at pump powers well above threshold, the relation between output power and pump power tends to become linear (see Fig. 7.11), as predicted by Eq. (7.3.9). Taking into account the relative insensitivity of P_{out} to variation around the optimum output coupling, we can then approximately use the optimum-coupling expression in Eq. (7.5.5) for this case also, where P_{mth} is now obtained from Eq. (7.3.26) (uniform pumping) or Eq. (7.3.32) (Gaussian pumping) by letting $\gamma = \gamma_i + (\gamma_1/2)$.

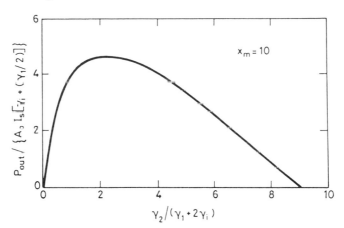

FIG. 7.15. Plot of normalized output power P_{out} versus normalized transmission of the output mirror γ_2 for a pump power, P_p, 10 times larger than the minimum threshold pump power P_{mth}.

Example 7.6. *Optimum output coupling for a lamp-pumped Nd:YAG laser.* We consider the laser configuration discussed in Example 7.2 (see Figs. 7.4 and 7.5) and calculate the optimum transmission of the output mirror when the laser is pumped by a lamp input power of $P_p = 7\,kW$. Since the threshold power P_{th} in Fig. 7.5 was measured to be 2.2 kW, then according to Eq. (7.5.1) with $\gamma_1 = 0$, we obtain $P_{mth} = P_{th}(\gamma_i/\gamma) \cong 697\,W$, where the values $\gamma_i = 0.038$ and $\gamma = 0.12$, obtained in Example 7.2, are used for internal loss and the total loss, respectively. We then get $x_m = P_p/P_{mth} \cong 10$, so that from Eq. (7.5.5), $S_{op} \cong 2.17$. From Eq. (7.5.3) we finally obtain $(\gamma_2)_{op} \cong 0.165$, which corresponds to an optimum transmission of $(T_2)_{op} = 1 - \exp[-(\gamma_2)_{op}] \cong 15\%$, i.e., agreeing with the value actually used in Fig. 7.4.

7.6. LASER TUNING

The gain linewidth of some lasers (e.g., dye lasers or vibronic solid-state lasers) is very wide and, for various applications, we are required to tune the laser output wavelength away from the line center and across the entire available linewidth. In other cases, lasers may exhibit gain on more than one transition (e.g., CO_2 laser or Ar laser); the strongest of these usually oscillates, whereas one may be interested to tune the laser wavelength away from the strongest line. In both of these circumstances, one usually employs a wavelength-selective element within the laser cavity.

In the middle infrared, such as for the CO_2 laser, one generally uses a diffraction grating, aligned in the so-called Littrow configuration (Fig. 7.16a), as one of the cavity mirrors. In this configuration, for a given angular setting of the grating, there is a particular wavelength (labeled λ_1 in the figure) that is reflected exactly back into the resonator; wavelength tuning is achieved by grating rotation.

In the visible or near-infrared spectral region, it is more common to use a dispersive prism, with faces close to Brewster's angle with respect to the laser beam (Fig. 7.16b). Again, for a given angular setting of the prism, a particular wavelength (labeled λ_1 in the figure) is reflected exactly back from mirror 2 into the resonator; then tuning is achieved by prism or mirror rotation.

FIG. 7.16. Laser tuning using wavelength-dispersive behavior of (a) a diffraction grating in the Littrow configuration or (b) a prism.

A third wavelength-selective element, which is becoming increasingly popular in the visible or near-infrared spectral region, uses a birefringent filter within the laser cavity. The filter simply consists of a plate of a suitable birefringent crystal (e.g., quartz or KDP) inclined at Brewster's angle θ_B to the beam direction (Fig. 7.17). Assuming that the optical axis A of the crystal is in a plane parallel to the surface of the plate, we first suppose that the birefringent plate is placed between two polarizers with parallel orientation. This orientation is assumed to be such as to transmit the E-field in the plane of incidence of the plate. Then the input beam does not suffer reflection loss on entering the plate, since this is inclined at Brewster's angle. Provided the optic axis is neither perpendicular nor parallel to the plane of incidence, the input beam contains both ordinary and extraordinary components. These components experience a difference in phase shifts $\Delta\phi = 2\pi(n_e - n_o)L_e$, where n_o and n_e are refractive indices for ordinary and extraordinary beams, respectively, and L_e is the plate thickness along the beam direction within the plate. After passing through the plate, unless $\Delta\phi$ is an integer number of 2π, the two components will combine to form a resultant beam with elliptical polarization; the presence of a second polarizer then leads to loss for this elliptically polarized light. On the other hand, if $\Delta\phi$ is an integral number of 2π, i.e., if

$$\frac{2\pi}{\lambda}(n_e - n_o)L_e = 2l\pi \tag{7.6.1}$$

where l is an integer, beam polarization remains unchanged after passing through the plate. For ideal polarizers, the beam then suffers no loss through the entire system in Fig. 7.17. For a general value of $\Delta\phi$ one can easily calculate the transmission T of the system in Fig. 7.17. Assuming for simplicity that the E-field makes an angle of $45°$ to the optical A-axis, one then has $T = \cos^2(\Delta\phi/2)$.

It should be noted that, consecutive transmission maxima have values of l that differ by unity; their frequency separation is then readily found from Eq. (7.6.1). Assuming that

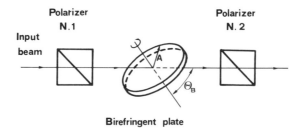

FIG. 7.17. Using a birefringent filter as a wavelength-selective element.

$(n_e - n_o)$ does not change appreciably over the wavelength range of interest, the frequency difference separating two consecutive maxima, i.e., the free spectral range Δv_{fsr} is given by:

$$\Delta v_{fsr} = \frac{c}{(n_e - n_o)L_e} \qquad (7.6.2)$$

Accordingly, the plate thickness, which usually ranges from 0.3–1.5 mm, determines the width of the tuning curve and thus the resolving power. The thinner the plate, the greater the available tuning range and the lower the resolving power. Note also that each transmission peak can be tuned by rotating the plate around the normal to the surface. By doing so, in fact, we change the value of n_e, which depends on the angle between the optical axis and the electric field vector, and hence change the plate birefringence $\Delta n = n_e - n_o$. Note lastly that, in low-gain lasers, such as cw gas or dye lasers, we can dispense with the two polarizers if other polarizing element, such as the active medium or indeed the Brewster's angle surface of the birefringent plate itself, provide sufficient loss discrimination between the two polarizations.

Example 7.7. *Free spectral range and resolving power of a birefringent filter.* Consider a dye laser operating at the wavelength $\lambda = 600$ nm in which a birefringent filter consisting of a $L = 1.5$ mm thick potassium dihydrogen phosphate (KDP) crystal is inserted for laser tuning. The ordinary and extraordinary refractive indices at the laser wavelength are $n_o = 1.47$ and $n_e = 1.51$. The Brewster angle for this plate is $\theta_B \cong \tan^{-1} n \cong 56.13°$, where n is the average of n_o and n_e. The Brewster angle inside the crystal is given by Snell's law as $\theta'_B = 33.9°$, so that $L_e = L/\cos \theta'_B \cong 1.81$ mm. If the orientation of optic axis A in Fig. 7.17 is near-orthogonal to the beam direction, the refractive index of the extraordinary beam to be used in Eq. (7.6.2) is just $n_e = 1.51$ and, from Eq. (7.6.2), we get $\Delta v_{fsr} \cong 4.14 \times 10^{12}$ Hz. The corresponding wavelength interval between two consecutive peaks is $\Delta \lambda = \lambda(\Delta v_{fsr}/v) \cong 5$ nm, where $v = c/\lambda = 5 \times 10^{14}$ Hz is the radiation frequency. Since the transmission of the birefringent filter in Fig. 7.17 is given by $T = \cos^2(\Delta \phi / 2)$, we can readily show that the width of the transmission curve (full width between half-maximum points) is just equal to $\Delta \lambda / 2$, i.e., it is equal to ~ 2.5 nm.

7.7. REASONS FOR MULTIMODE OSCILLATION

Lasers generally tend to oscillate on many modes because the frequency separation of the modes is usually smaller, and often much smaller, than the width of the gain profile. If for example we take $L = 1$ m, the frequency difference between two consecutive longitudinal modes is $\Delta v = c/2L = 150$ MHz. The laser linewidth, on the other hand, may range from ~ 1 GHz, for a Doppler-broadened transition of a visible or near-infrared gas laser, to 300 GHz or more for a transition of a crystal ion in a solid-state material. The number of modes within the laser linewidth can thus range from a few to a few thousands, so the gain difference between these modes, particularly when a few thousand modes are considered, becomes very small. At first sight one might expect a significant fraction of these modes to be excited at a sufficiently high pump rate.

The preceding, seemingly straightforward, conclusion must be examined more carefully, however. In fact, in the early days of laser development, it was argued that in principle lasers always should tend to oscillate on a single mode if the gain line is homogeneously broadened. The argument can be followed with the help of Fig. 7.18, where the laser gain profile is plotted against frequency, for increasing values of the pump rate. For simplicity, one cavity mode is assumed to be coincident with the peak of the gain curve. We further assume that oscillation occurs on the TEM_{00} mode, so that all mode frequencies are separated by $c/2L$ (see Fig. 5.10). The laser gain coefficient is given by Eq. (2.4.35), where the cross section of a homogeneous line is given by Eq. (2.4.18). Oscillation starts in the central mode when the inversion $N = N_2 - N_1$, or the average inversion for the space-dependent model, reaches a critical value N_c, giving a gain equal to cavity losses [see Eq. (7.3.2) or (7.3.19)]. However, even when R_p is increased above the threshold value, the inversion was shown to remains fixed at the critical value N_c in the steady-state. The peak gain, represented by the length OP in Fig. 7.18, therefore remains fixed at the value OP_c when $R_p \geq R_{cp}$. Since the line is homogeneously broadened, its shape cannot change, so the whole gain curve must remain the same for $R_p \geq R_{cp}$, as indicated in Fig. 7.18. The gains of other modes, represented by lengths O'P', O''P'', etc., always remain smaller than the OP_c value for the central mode. If all modes have the same loss, then only the central mode oscillates in the steady state.

The situation is quite different for an inhomogeneous line (Fig. 7.19). In this case the cross section to be used in Eq. (2.4.35) is given by Eq. (2.4.23), i.e., it is given by the superposition of cross sections for the individual atoms whose transition frequencies are

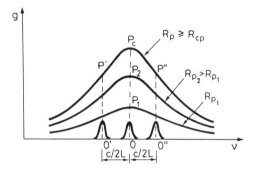

FIG. 7.18. Frequency dependence of laser gain coefficient versus pump rate R_p under saturation condition (*homogeneous line*).

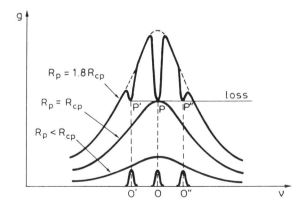

FIG. 7.19. Frequency dependence of laser gain coefficient versus pump rate R_p under saturation conditions (*inhomogeneous line*): frequency hole-burning behavior.

distributed in a given spectrum, as described by the function $g^*(v_0' - v_0)$. Accordingly, it is possible to burn holes in the gain curve, as discussed (for an absorption curve) in Sect. 2.8.3 (see Fig. 2.22). Therefore, when R_p increases above R_{cp}, the gain of the central mode remains fixed at the critical value $\mathrm{OP_c}$; the gains of the other modes $\mathrm{O'P'}$, $\mathrm{O''P''}$, etc., can, however keep on increasing to the corresponding threshold value. In this case, if the laser is operating somewhat above threshold, more than one mode can be expected to oscillate.

Shortly after the invention of the laser, multimode oscillation was actually observed to occur for both inhomogeneous (e.g., He–Ne gas laser) and homogeneous (e.g., ruby laser) lines. This last result appeared to conflict with the preceding argument. This inconsistency was later removed[20] by considering the fact that each mode has a well-defined standing wave pattern in the active medium. For the sake of simplicity, we consider two modes whose standing wave patterns are shifted by $\lambda/4$ in the active medium (Fig. 7.20a).* We assume that mode 1 in Fig. 7.20 is the center mode of Fig. 7.18, so that it is the first to reach threshold. However, when oscillation on mode 1 starts, inversion around those points where the electric field is zero (points A, B, etc.) is mostly left undepleted, so the inversion can continue growing there even when the laser is above threshold. This situation is illustrated in Fig. 7.20b, where the spatial distribution of the population inversion in the laser medium is indicated.† Accordingly, mode 2, which initially has a lower gain, experiences a growth in gain with increased pump rate, since it uses inversion from regions not depleted by mode 1. Therefore, sufficiently far above threshold, mode 2 can also be set into oscillation; this obviously occurs when its gain becomes equal to its losses. Thus, for a homogeneous line, multimode oscillation is not due to holes burned in the gain curve (*spectral hole burning*, as

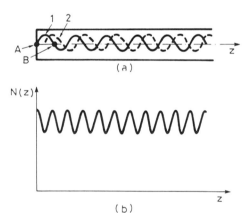

FIG. 7.20. Explanation of multimode oscillation for a homogeneous line: (a) Standing-wave mode-field configurations in the medium for the oscillating mode (continuous line) and for a mode that may oscillate above threshold (dashed line). (b) Spatial hole-burning pattern for the population inversion in the laser medium produced by the oscillating mode.

* We recall that, according to Sect. 5.1, resonant frequencies are due to the condition that cavity length L is an integral number of half-wavelengths, i.e. $L = n(\lambda/2)$, where n is a positive integer. Two consecutive longitudinal modes are thus shifted by $(\lambda/2)$ on going from one mirror to the other. Then, if the active medium is placed at the resonator center, these two modes have their spatial patterns shifted by just $(\lambda/4)$ in the medium, and may thus correspond to the two modes in Fig. 7.20.

† According to the discussion in Sects. 7.3.2 and 7.4.2, the average value $\langle N \rangle$ of the inversion, as defined by Eq. (7.3.20), remains clamped to the threshold value above threshold.

in Fig. 7.19) but to holes burned in the spatial distribution of inversion within the active medium (*spatial hole burning*, Fig. 7.20b). Note that the phenomenon of spatial hole burning does not play a significant role in an inhomogeneous line. In this case, in fact, different modes (with a large enough frequency separation) interact with different sets of atoms, so the hole-burning pattern of one set of atoms is ineffective for the other mode.

In conclusion, a laser always tends to oscillate on many modes. For a homogeneous line, this is due to spatial hole burning, while, for an inhomogeneous line, this is due to spectral hole burning. Note, however, that, in the case of a homogeneous line, when a few modes are oscillating with frequencies around the center of the gain line, the spatial variation of inversion is essentially smeared out due to the presence of the corresponding, spatially shifted, standing wave patterns of these modes. In this case, the homogeneous character of the line prevents other modes, further away from the center of the gain line, from oscillating. Thus, compared to an inhomogeneous line, a homogeneous line restricts oscillation to a smaller number of modes centered around the peak of the gain line.

7.8. SINGLE-MODE SELECTION

Several methods of constraining a laser to oscillate in a single transverse and/or longitudinal mode, for either homogeneous or inhomogeneous lines, are discussed at some length in this section.

7.8.1. Single-Transverse-Mode Selection

For a stable resonator with not too large a value for laser spot size (e.g., less than 0.5 mm for a Nd:YAG or less than 1 cm for a CO_2 laser), it is relatively easy to make the laser oscillate in some particular transverse mode, i.e., one with prescribed values for the transverse mode indices l and m (see Chap. 5). In most applications, oscillation in a TEM_{00} mode is desirable, and, to achieve this, a diaphragm of suitable aperture size may be inserted at some point on the axis of the resonator. If the radius a of this aperture is sufficiently small, it dictates the value of the Fresnel number of the cavity $N = a^2/L\lambda$. As a decreases, the difference in loss between the TEM_{00} mode and the next higher order modes (TEM_{01}, or TEM_{10}) increases, as we can see by comparing Figs. 5.13a–b at the same value of the g parameter. Therefore, by choosing an appropriate aperture size, we can enforce oscillation on the TEM_{00} mode. Note that this mode-selection scheme inevitably introduces some loss for the TEM_{00} mode itself.

For large diameters of the active medium, as discussed in Sect. 5.5.2, it is not possible to obtain a mode spot size comparable to this diameter without incurring serious instability problems in the size of the transverse mode profile. Referring to Example 5.9, we see that the Fresnel number $N = a^2/L\lambda$, where a in this case is the radius of the active medium, must not exceed a value of about 2. For larger values of this radius, we must resort to unstable resonators. As discussed in Sect. 5.6.2, if the equivalent Fresnel number is chosen to be a half-integer value, a large loss discrimination occurs between the lowest order and the higher order modes (see Fig. 5.18). In this case, the output beam has the form of a ring, a shape that is often undesirable. The best way of obtaining oscillation on the lowest-order mode, in this case, is to use an unstable cavity with a radially variable output coupler with a Gaussian or, even better, super-Gaussian profile (Sect. 5.6.4).

7.8.2. Single-Longitudinal-Mode Selection

Even when a laser is oscillating on a single-transverse mode, it can still oscillate on several longitudinal modes (i.e., modes differing in their value of the longitudinal mode index n). These modes are separated in frequency by $\Delta v = c/2L$. Isolation of a single longitudinal mode is achieved, in some cases, by using such a short cavity length that $\Delta v > \Delta v_0/2$, where Δv_0 is the width of the gain curve.* In this case, if a mode is tuned to coincide with the center of the gain curve, the two adjacent longitudinal modes are far enough away from line center that, for a laser not too far above threshold, they cannot oscillate. The requirement for this mode-selecting scheme can then be written as:

$$L \leq \frac{c}{\Delta v_0} \tag{7.8.1}$$

For example, if the equality applies in Eq. (7.8.1) and one mode is coincident with the peak of the gain profile, the two adjacent modes see an unsaturated gain coefficient that, for both a Gaussian or Lorentzian line, is half the peak value. In particular, for a Gaussian line, one can easily understand from Fig. 7.19 that single-longitudinal-mode operation is achieved for $R_{cp} \leq R_p \leq 2R_{cp}$. Note that to tune one mode to coincidence with the line center, we can mount one cavity mirror on a piezo-electric transducer. By applying a voltage to this transducer, we then produce a small, controllable, change in the cavity length. [One can show that the cavity length must be changed by $\lambda/2$ to shift the comb (Fig. 5.10 with $l = m = 0$) of longitudinal modes by one mode spacing].

The method just discussed can be used effectively with a gas laser, notably a He-Ne laser, where gain linewidths are relatively small (a few GHz or smaller). For instance, in the case of an He-Ne laser oscillating on its red transition one has $\Delta v_0^* \cong 1.7\,\text{GHz}$; from Eq. (7.8.1) one then obtains $L \leq 17.5\,\text{cm}$. For solid-state lasers, on the other hand, the gain linewidth is usually much larger (a few hundreds GHz) so, to satisfy Eq. (7.8.1), the equivalent cavity length must typically be appreciably smaller than 1 mm (*microchip lasers*). For lasers with much larger bandwidths (e.g., dye lasers or tunable solid-state lasers), the cavity length to satisfy Eq. (7.8.1) becomes too small to be practical. In this case, and also when longer lengths of the active medium are needed (e.g., for high-power lasers), longitudinal mode selection can be accomplished by a variety of techniques that are the subject of the next two sections.

7.8.2.1. Fabry–Perot Etalons as Mode-Selective Elements

A common way of achieving single-longitudinal-mode oscillation, for both homogeneous or inhomogenous lines, involves inserting one or more Fabry–Perot (FP)

---

* If a tuning element, such as those in Figs. 7.16 and 7.17, is inserted in the laser cavity and the corresponding transmission linewidth, ranging in actual cases from 0.1–1 nm, is smaller than that of the gain medium, the linewidth Δv_0 to be considered in this section is that of the tuning element rather than that of the active medium. Notably, this case occurs for dye lasers or tunable solid-state lasers.

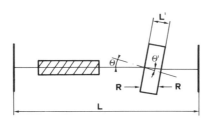

FIG. 7.21. Configuration for longitudinal mode selection using a transmission Fabry–Perot etalon.

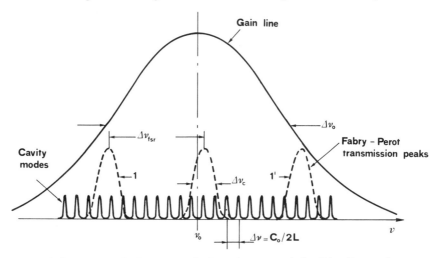

FIG. 7.22. Longitudinal mode selection using a transmission Fabry–Perot etalon.

etalons within the cavity. These consist of a plane-parallel plate of transparent material (fused quartz or glass for visible or near-infrared wavelengths) whose two plane surfaces are coated to a suitably high reflectivity value R.

We first consider the case when a single FP etalon is used, inclined at an angle θ to the resonator axis (Fig. 7.21). According to the discussion in Sect. 4.5.1, the transmission maxima of the etalon occur at frequencies v_n given by:

$$v_n = \frac{nc}{2n_r L_1 \cos \theta'} \qquad (7.8.2)$$

where n is an integer, θ' is the refraction angle of the beam within the etalon, n_r is the etalon refractive index, and L_1 is its length. Since L_1 is much smaller than the cavity length L, only a very small tilt of the angle θ (and hence of θ') from the $\theta = \theta' = 0$ position is needed to tune a transmission maximum of the etalon to coincide with the mode nearest the peak of the laser gain profile (Fig. 7.22). If the frequency separation $\Delta v = c/2L$ between two adjacent longitudinal modes is now $\geq \Delta v_c/2$, where Δv_c is the linewidth of an etalon transmission peak, the etalon selects the mode nearest line center from its neighbors.* According to Eq. (4.5.13), the discrimination between adjacent longitudinal modes requires

$$\frac{\Delta v_c}{2} = \frac{\Delta v'_{fsr}}{2F} \leq \Delta v \qquad (7.8.3)$$

* More precisely single-pass etalon transmission losses of the two neighboring modes are in this case $\geq 50\%$.

where $\Delta v'_{fsr}$ is the free-spectral range and F is the finesse of the etalon. To ensure single-longitudinal-mode operation, we also require the etalon free-spectral range $\Delta v'_{fsr}$ to be larger than or equal to half the gain linewidth Δv_0; otherwise the two neighboring transmission peaks of the etalon would allow corresponding cavity modes to oscillate. Discrimination between adjacent transmission maxima of the etalon then requires

$$\Delta v'_{fsr} \geq \frac{\Delta v_0}{2} \tag{7.8.4}$$

From Eqs. (7.8.3) and (7.8.4) we then find that $(\Delta v_0/2) \leq \Delta v'_{fsr} \leq 2F\Delta v$, which requires as a necessary condition that $(\Delta v_0/2) \leq 2F\Delta v$, i.e., that:

$$L \leq \left(\frac{c}{\Delta v_0}\right) 2F \tag{7.8.5}$$

Comparing this equation with Eq. (7.8.1), we see that compared to a resonator without an etalon, the cavity length can now be increased by a factor of $2F$. Assuming for example $F = 30$ (there are various factors, such as the flatness of etalon surfaces and beam walk-off in the etalon, that limit the achievable finesse to this value), then using a Fabry-Perot etalon allows a factor 60 increase in cavity length while still ensuring single-longitudinal-mode operation.

If the cavity length does not satisfy Eq. (7.8.5), then single-longitudinal-mode operation cannot be achieved using one FP etalon; therefore two or more etalons of different thickness are needed. In the case of two etalons, the thicker etalon is required to discriminate against adjacent longitudinal modes of the cavity. Its free-spectral range $\Delta v'_{fsr}$ must then satisfy Eq. (7.8.3). A second thinner etalon must then discriminate against adjacent transmission maxima from the first etalon; at the same time its free-spectral range $\Delta v''_{fsr}$ must be larger than or equal to the half-width of the gain curve (i.e., $\Delta v''_{fsr} \geq \Delta v_0/2$). To achieve both these conditions, it can be shown that the cavity length must now satisfy the following relation:

$$L \leq \left(\frac{c}{\Delta v_0}\right)(2F)^2 \tag{7.8.6}$$

A comparison between Eqs. (7.8.1), (7.8.5), and (7.8.6) shows that, to achieve single-longitudinal-mode operation without an etalon, with one etalon, or with two etalons, the cavity length must satisfy the respective conditions $L \leq c/\Delta v_0$, $c/\Delta v_0 \leq L \leq (c/\Delta v_0)2F$, or $(c/\Delta v_0)2F \leq L \leq (c/\Delta v_0)(2F)^2$.

Example 7.8. *Single-longitudinal-mode selection in Ar-ion and Nd:YAG lasers.* Consider first an Ar laser oscillating on its $\lambda = 514.5$ nm line, whose gain linewidth is experimentally measured as $\Delta v_0^* = 3.5$ GHz. To achieve single-longitudinal-mode selection without an etalon requires a cavity length $L \leq c/\Delta v_0^* \cong 8.6$ cm; with one etalon cavity length must be $L \leq (c/\Delta v_0^*)(2F) \cong 5.14$ m, where a finesse $F = 30$ has been used. Since the length of an Ar laser is usually smaller than 2 m but larger than a few tens of cm, one FP etalon is required. According to Eq. (7.8.2), for $\cos \theta' \cong 1$, one has $\Delta v'_{fsr} \cong c/2n_r L_1$; taking $n_r = 1.5$ and using Eq. (7.8.3) we get $L_1 \geq L/2Fn_r = 1.66$ cm for a cavity length $L = 1.5$ m. From Eq. (7.8.4) we obtain $L_1 \leq (c/\Delta v_0^* n_r) = 5.71$ cm. The thickness of the etalon can then be chosen between these two values, e.g. $L_1 = 3.7$ cm.

Consider next the case of a Nd:YAG laser, where $\Delta\nu_0 = 120\,\text{GHz}$ (at $T = 300$ K). Single-longitudinal-mode oscillation without an etalon, in this case, requires cavity length $L \leq (c/\Delta\nu_0 n) = 1.4\,\text{mm}$, where the refractive index of YAG is $n = 1.82$. (The two end mirrors are assumed to be directly coated on the two faces of the YAG plate, so that the separation between two consecutive longitudinal modes is now given by $\Delta\nu = c/2nL$). Indeed, single-mode diode-pumped Nd:YAG platelets with a thickness of the order of a few hundred micrometers are now commonly used and even commercially available (*microchip lasers*). When a single-FP etalon is used, then, according to Eq. (7.8.5), the cavity length can be increased to $L \cong 8\,\text{cm}$ (assuming again $F = 30$).

7.8.2.2. *Single-Mode Selection in Unidirectional Ring Resonators*

For a homogeneously broadened transition, single-longitudinal mode operation can automatically be achieved, or at least greatly facilitated, if the laser cavity has the form of a ring and oscillation is constrained to be unidirectional (see Fig. 5.4a). In this case, in fact, the phenomenon of spatial hole burning within the active medium does not occur, and as discussed in Sect. 7.7, the laser tends to oscillate on a single mode. Actually, if the transition is, only partly, homogeneously broadened and the gain profile is very broad, some further bandwidth selecting elements, such as birefringent filters and/or etalons may also be needed. An additional advantage of this unidirectional ring configuration is that higher output power is available, since the entire active medium, rather than just those regions around the maxima of the standing wave pattern, contributes to laser output.

To achieve unidirectional ring operation, a unidirectional device, or *optical diode*, giving preferential transmission for one direction of beam propagation, must be inserted within the cavity. In principle the device can be made as shown in Fig. 7.23. Here the wave going in one direction, e.g., from left to right, passes first through an input polarizer (polarizer 1), then through a rod of suitable transparent material (e.g. glass) to which a longitudinal dc magnetic field is applied (*Faraday rotator*), and then through an output polarizer (polarizer 2) that has the same orientation as the first polarizer.* When a linearly polarized optical beam passes through the Faraday rotator, with its beam axis in the magnetic field direction, the output beam can be shown to still consist of a linearly polarized wave, whose plane of polarization is rotated about the beam axis. The sense of rotation, seen by an observer facing the oncoming beam, depends on the relative direction of magnetic

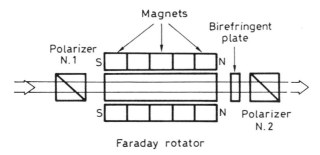

FIG. 7.23. Unidirectional device using a Faraday rotator (optical diode).

---

* To avoid confusion and describe polarization rotation consistently, we will always assume that the observer is facing the oncoming light beam.

field and beam propagation direction. This means that if polarization rotation is seen by an observer facing the oncoming beam to occur, e.g., counterclockwise for a left-to-right propagating beam (Fig. 7.24a), the polarization rotation is seen (the observer is again facing the oncoming beam) to occur clockwise for a right-to-left propagating beam (Fig. 7.24b). For this reason the Faraday rotator is said to represent a non-reciprocal element. The beam is then passed through a birefringent plate, having a $\lambda/2$ optical path difference between the two polarizations. The phase shift between the two polarizations is then equal to π, [i.e., $2\pi(n_o - n_e)l/\lambda = \pi$, where l is the plate length]. If the polarization of the input beam makes an angle of $\alpha/2$ to the extraordinary axis, the plate then rotates the polarization clockwise (seen by the observer facing the beam) by an angle α (Fig. 7.24a). Thus, if the Faraday rotator rotates the polarization counter clockwise by an angle α, the two rotations exactly cancel, so no attenuation is suffered by the beam passing through the output polarizer (polarizer 2 in Fig. 7.23). For a beam traveling in the opposite direction, right to left, as in Fig. 7.24b, however, polarization rotation is again clockwise on passing through the birefringent plate and the two rotations now add to each other. Correspondingly, the beam experiences a loss when passing through the second polarizer (polarizer 1 in Fig. 7.23). Note that this loss may even reach 100% if total rotation is through an angle of $\pi/2$. For a low-gain laser, however, total rotation of only about a few degrees typically introduces enough loss discrimination to ensure unidirectional operation. Note also that, on passing through the birefringent plate, polarization rotation is clockwise independent of the propagation direction, as Fig. 7.24 shows. The birefringent plate thus represents a reciprocal optical element.

Figure 7.25 shows a typical example of a folded ring configuration, including a unidirectional device, used in a commerically available cw dye laser. In this case, pumping is provided by an ion laser and the dye solution is made to flow transversely to the beam in the form of a liquid jet. Single-transverse-mode operation is automatically achieved due to the transverse gain distribution from focused pumping. Laser tuning and gain bandwidth reduction are obtained by combining a birefringent filter with two FP etalons—a thin etalon and a scanning etalon—of different free-spectral ranges. The optical path length of the

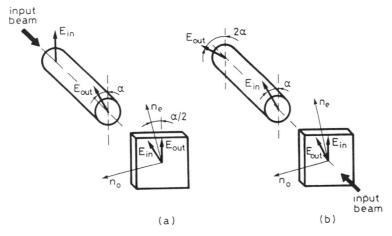

FIG. 7.24. Polarization rotation for a combined Faraday rotator and a $\lambda/2$ birefringement plate for beam propagation from (a) left to right, and (b) right to left.

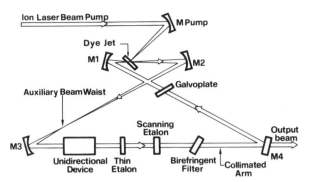

FIG. 7.25. High-power single-longitudinal-mode dye laser using a unidirectional ring cavity.

cavity is conveniently tuned by rotating a tilted, plane-parallel, glass plate inside the resonator (the *galvoplate*). Single-longitudinal-mode operation is then ensured by a unidirectional device consisting of a Faraday rotator and a birefringent plate. Note that no polarizers are used because enough polarization loss is provided in this case by the faces of various optical elements inclined at Brewster's angle.

Another, very interesting, example of a unidirectional ring resonator using a nonplanar cavity, used in a commercially available Nd:YAG laser, is shown in Fig. 7.26.[21] The resonator is made from a small slab (e.g., $3 \times 6 \times 8$ mm) of Nd:YAG whose faces B and D are cut at such an angle that the beam follows the nonplanar path BCD shown in the figure; point C is on the upper surface of the slab. Permanent magnets create a magnetic field in the direction shown in the figure. The beam undergoes total internal reflection at surfaces B, C, and D; it is reflected at surface A by a multilayer dielectric coating that acts as the output coupler. The Nd:YAG slab provides both the active material and Faraday rotator; it is longitudinally pumped by the beam from a semiconductor laser (not shown in the figure). The nonplanarity of the laser path creates an analogous effect to rotation by a half-wave plate. Assume that the planes of incidence at the corner faces B (plane ABC) and D (plane CDA) are at an angle β to the plane of incidence at the front face A (plane DAB). Assume also that the plane of incidence at the top face C (plane BCD) is almost perpendicular to

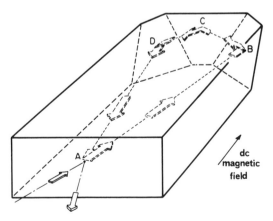

FIG. 7.26. Single-longitudinal-mode Nd:YAG laser using a unidirectional and nonplanar ring cavity [nonplanar ring oscillator (NPRO)]. (By permission from Ref. 21.)

those at the corner faces. Rotations of the planes of incidence then result in a net polarization rotation and image-rotation of 2β after the three reflections at points B, C, and D. The polarization-selective element is simply the multilayer dielectric coating at surface A, whose reflectivity depends on beam polarization. Since the homogeneously broadened linewidth of Nd:YAG is much smaller than that of a dye laser and the frequency separation between longitudinal modes in Fig. 7.26, due to the small cavity dimensions, is much larger than that in Fig. 7.25, no additional frequency-selective elements (such as birefringent filters or FP etalons) are needed in this case. Single-transverse-mode operation is again automatically achieved due to the transverse gain distribution from the focused pumping. A compact and monolithic single-mode device is thus achieved.

7.9. FREQUENCY PULLING AND LIMIT TO MONOCHROMATICITY

Let us assume that oscillation occurs in a cavity mode of frequency v_c that is different from the center frequency v_0 of the transition. We let Δv_c and Δv_0 be the widths of the cavity mode resonance and the laser transition, respectively, and we address the question of finding the laser frequency v_L and the width of the output spectrum Δv_L (Fig. 7.27).

The calculation of v_L can be done within the semiclassical approximation. It can be shown[3,22] that v_L has some intermediate value between v_0 and v_c: i.e., oscillation frequency is pulled toward the transition frequency v_0. To a good approximation for an inhomogeneous line and rigorously for a homogeneous line, the oscillation frequency is given by the weighted average of the two frequencies v_c and v_0, with the weighting factors proportional to the inverse of the corresponding linewidths. Thus we have

$$v_L = \frac{(v_0/\Delta v_0) + (v_c/\Delta v_c)}{(1/\Delta v_0) + (1/\Delta v_c)} \tag{7.9.1}$$

The effect of frequency pulling is generally very small. The value of Δv_0 may, in fact, range from $\sim 1\,\text{GHz}$, for Doppler-broadened transitions in the visible, to as much as $300\,\text{GHz}$ for solid-state lasers (see Table 2.1). On the other hand, for a 1-m-long cavity, $\Delta v_c = 1/2\pi\tau_c = \gamma c/2\pi L_e$ [see Eqs. (7.2.14) and (5.3.10)] may range from $\sim 1\,\text{MHz}$ to a few tens of MHz (for γ ranging from $\sim 1\%$, typical of a low-gain laser medium, such as He-Ne, to values on the order of 50% for high-gain materials). We thus see that the weighting factor $(1/\Delta v_c)$ is more than three order of magnitude larger than $(1/\Delta v_0)$.

We now discuss the width Δv_L of the laser output spectrum when oscillation occurs in this single mode. Its ultimate limit is established by spontaneous emission noise or, more

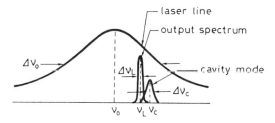

FIG. 7.27. Frequency pulling and spectral output of a single-mode laser.

precisely, by zero point fluctuations in the laser mode field. Since these fluctuations can be treated correctly only by a full quantum electrodynamics approach (see Sect. 2.3.2), derivation of the expression for this limit is beyond the scope of our present treatment. We can understand however that, although both the amplitude and phase of the zero point field fluctuate randomly, spectral broadening of the output must arise predominantly from random-phase fluctuations, while very small amplitude fluctuations are induced by fluctuations of the zero-point field. This is due to the fact that, as discussed earlier in this chapter, the laser cavity photon population, and hence the output power, is quite insensitive to the number q_i of photons considered to be initially present in the cavity to simulate the effect of spontaneous emission. To be more precise we see that, according to Eq. (7.2.2), the rate of increase in cavity photons due to the extra photon from spontaneous emission is given by $(d\phi/dt)_{se} = V_a B N_0$. This term must be compared to the stimulated one, which is given by $V_a B N_0 \phi_0$. Since ϕ_0 may range from 10^{10} to 10^{16} (see Example 7.1), the spontaneous emission term has a negligible effect on the number of cavity photons, i.e., on the field amplitude. According to the preceding discussion, the electric field of the output beam can be written as $E(t) = E_0 \sin[2\pi v_L t + \varphi_n(t)]$, where $\varphi_n(t)$ is a random variable accounting for zero point field fluctuations. It can then be shown that the time behavior of $\varphi_n(t)$ is typical of a diffusion process, i.e., the root-mean-square phase deviation after a time t, $\Delta\phi(t) = \langle[\varphi_n(t) - \varphi_n(0)]^2\rangle^{1/2}$, is proportional to $(t)^{1/2}$. The spectral shape of the emitted light, i.e., the power spectrum of $E(t)$, is then Lorentzian, and, neglecting internal losses, its width (FWHM) is given by[23]

$$\Delta v_L = \left(\frac{N_2}{N_2 - N_1^*}\right) \frac{(2\pi h v_L)(\Delta v_c)^2}{P} \tag{7.9.2}$$

where P is the output power. This is the well-known Schawlow–Townes formula, introduced in their original proposal for the laser,[24] which establishes the *quantum limit* to laser linewidth.

Typically the linewidth predicted by Eq. (7.9.2) is negligibly small compared to that produced by other cavity disturbances (discussed later on) except for the very important case of a semiconductor laser. As shown in Example 7.9, a first reason for this exception is that Δv_c, for a semiconductor laser, is typically about five orders of magnitude larger than for, e.g., a He-Ne laser. A second reason was actually discovered by careful experiments done with GaAs lasers which showed the actual linewidth to be about 50–100 times larger than the value predicted by Eq. (7.9.2). This observation was later on understood in terms of a new phenomenon, peculiar to semiconductor lasers. Fluctuations in electron hole density caused by spontaneous emission produce, in fact, a sizable fluctuation in the refractive index of the laser medium. The resulting

---

Example 7.9. *Limit to laser linewidth in He-Ne and GaAs semiconductor lasers.* Consider first the case of a single-mode He-Ne laser oscillating on its red transition ($\lambda = 632.8$ nm, $v_0 \cong 4.7 \times 10^{14}$ Hz). We take $L_e = 1$ m, $\gamma = 1\%$, and we also assume an output power of $P = 1$ mW. From Eqs. (7.2.14) and (5.3.10) we obtain, respectively, $\tau_c = 3.3 \times 10^{-7}$ s and $\Delta v_c \cong 4.7 \times 10^5$ Hz. Using Eq. (7.9.2) and taking $N_2/(N_2 - N_1) \cong 1$, we obtain $\Delta v_L \cong 0.43$ mHz.

Consider next the case of a single-mode GaAs semiconductor laser ($\lambda = 850$ nm) with cavity length $L = l = 300$ μm and power reflectivity at the two end faces of $R = 0.3$ (i.e., equal to the Fresnel reflectivity of the

fluctuation in optical length of the cavity thereby produces fluctuations in the cavity frequency and hence in the oscillation frequency. Thus, in a semiconductor laser, the right-hand side of Eq. (7.9.2) must be multiplied by a factor, usually denoted α^2, which is considerably larger than 1. The factor α is called the Henry factor, named after the scientist who first provided an explanation of this phenomenon.[25].

uncoated semiconductor surfaces). Neglecting all other cavity losses, we obtain $\gamma = -\ln(R) \cong 1.03$ and hence $\tau_c = nL/c\gamma = 3.4$ ps, where $n = 3.5$ is the refractive index of GaAs. We then get $\Delta v_c = 1/2\pi\tau_c \cong 4.7 \times 10^{10}$ Hz and, from Eq. (7.9.2) assuming $N_2/(N_2 - N_1) = 3$ and $P = 3$ mW, we obtain $\Delta v_L \simeq 3.2$ MHz. Note that the laser linewidth in this case is almost 10 orders of magnitude larger than in the case of a He-Ne laser due to the much shorter cavity decay time and hence the much larger cavity linewidth.

According to the preceding discussion, the linewidth of a typical semiconductor laser arises from *quantum noise*; in practice it is difficult to reduce below 1 MHz. For a He-Ne laser and all other lasers of relevance for obtaining small oscillation linewidths (such as Nd:YAG, CO_2, or Ar lasers), the linewidth determined by the Schawlow–Townes formula, even for modest powers of a few mW, is always well below 1 Hz down to millihertz. Since $v_L = c/\lambda \cong 4.7 \times 10^{14}$ Hz, the relative monochromaticity of this laser, set by zero point fluctuations, would then be $(\Delta v_L/v_L) \cong 2.7 \times 10^{-18}$.

We now examine the cavity length stability requirement to keep the resonator frequency stable to the same degree. Using Eq. (5.1.2) with $n = $ const., we find $(\Delta L/L) \cong -(\Delta v_c/v_c) \cong 2.7 \times 10^{-18}$ so that with $L = 1$ m, e.g., we have $|\Delta L| \cong 2.7 \times 10^{-9}$ nm. This means that a cavity length variation by a quantity $\approx 10^{-8}$ smaller than a typical atomic dimension is sufficient to induce a shift of the cavity frequency v_c, and hence of the oscillation frequency v_L, which is comparable to the oscillation linewidth given by Eq. (7.9.2). Thus, the limit to monochromaticity is in practice set by changes, in cavity length induced by vibrations or thermal drifts, as we see in next section. These changes, arising from noise disturbances of perhaps a less fundamental nature than those previously considered, are often said to be due to *technical noise*.

7.10. LASER FREQUENCY FLUCTUATIONS AND FREQUENCY STABILIZATION

To explore laser frequency fluctuations, let us consider an active medium of refractive index n_m and length l in a laser cavity of length L in air. The effective cavity length is then given by: $L_e = n_a(L - l) + n_m l$, where n_a is the refractive index of air. We divide the mode frequency changes into two parts: (1) *Long-term drifts*, i.e., occurring for time longer than, say, 1 s, of either L or n_a, caused mainly by temperature drifts or slow pressure changes of the ambient air surrounding the laser. (2) *Short-term fluctuations* caused by: Acoustic vibrations of mirrors leading, to cavity length changes; acoustic pressure waves that modulate n_a; short-term fluctuations of n_m, due, e.g., to fluctuations in the discharge current of a gas laser or to air bubbles in the jet flow of a dye laser. In optically pumped solid-state lasers, fluctuations in pump power cause temperature fluctuations that in turn change the refractive index and therefore the optical length of the cavity.

Example 7.10. *Long-term drift of a laser cavity.* Taking for invar $\alpha = 1 \times 10^{-6}$ K^{-1} and considering a frequency in the central part of the visible spectrum, i.e., $v_L \cong 5 \times 10^{14}$ Hz, from previous expressions we obtain that the frequency drift due to a thermal change amounts to $|\Delta v_L|\Delta T = 5 \times 10^8$ HzK^{-1}. This shows that a change in ΔT of only $0.1°$C produces a frequency drift of ~ 50 MHz. To calculate the frequency drift due to a slow pressure change, we recall that for a gas laser, $(L - l)/L_e \cong 0.2$ and, for air, $n_a \cong 1.00027$. Again for $v_L \cong 5 \times 10^{14}$ Hz, we can then write $|\Delta v_L| \cong 2.7 \times 10^{10}|\Delta p/p|$ Hz. Thus, for a relative pressure change of $|\Delta p/p| \cong 3 \times 10^{-3}$, which can readily occur in 1 hour, one has $|\Delta v_L| \cong 80$ MHz.

To illustrate the influence of long-term drifts on cavity length, let α be the thermal expansion coefficient of the elements (e.g. invar rods) that determine mirror separation. Then we have $|\Delta v_L/v_L|$ $\cong |\Delta L/L| = \alpha \Delta T$, where ΔT is the temperature change in the laser environment. Slow pressure changes, on the other hand, contribute to the frequency drift by the amount $|\Delta v_L| = v_L|\Delta n_a|(L - l)/L_e$ $= v_L(n_a - 1)|\Delta p/p|$ $(L - l)/l$, where Δp is the slow change of the ambient pressure p.

According to Example 7.10, to reduce long-term frequency drifts below e.g., 1 MHz requires: (1) Very low-expansion materials for the laser cavity spacer elements, down to an expansion coefficient perhaps smaller than 1×10^{-7} K^{-1}. (2) Stabilization of the ambient temperature below 0.01 °C. We must also enclose the laser in a pressure-stabilized chamber. The reduction of short-term frequency fluctuations is even more difficult, and it requires good vibration isolation optical tables and efficient protection for the whole laser path. It is therefore generally difficult to reduce short-term frequency fluctuations below the 1-MHz level except for monolithic and compact solid-state lasers, such as the NPRO™ laser in Fig. 7.26, where short-term frequency fluctuations of about 10 kHz have been measured.

To characterize laser frequency fluctuations precisely, let us write again the electric field of the output beam as $E(t) = E_0 \sin[2\pi v_L t + \varphi_n(t)]$ where v_L is the central laser frequency and $\varphi_n(t)$ describes the noise phase fluctuations. The instantaneous frequency can then be written as $v(t) = v_L + [d\varphi_n(t)/2\pi \, dt] = v_L + v_n(t)$, where $v_n(t)$ is the frequency noise. This frequency noise is fundamentally related to linewidth or frequency stability, and, consequently, it must be characterized. The measurement used to characterize frequency fluctuations is the spectral power density of frequency noise, represented by $S_v(v_m)$ and described in units of Hz^2/Hz. Here v_m is called the *offset frequency*, which describes the frequency at which the phase $\varphi_n(t)$ is modulated by noise. In practice $S_v(v_m)$ is measured by converting $v_n(t)$ into, e.g., a voltage signal $V_n(t)$ by a frequency-to-voltage converter, then measuring the power spectrum of $V_n(t)$ by an electronic spectrum analyzer. Figure 7.28 shows the square root of the frequency noise spectrum $[S_v(v_m)]^{1/2}$ from a free-running diode-pumped monolithic Nd:YAG laser;[26] the same figure also gives Schawlow–Townes limit for $[S_v(v_m)]^{1/2}$. In fact, for a Lorentzian line as predicted by the Schawlow–Townes theory, one can show that the spectral power density of frequency fluctuations is white; i.e., $S_v(v_m)$ is a constant given by[27]

$$S_v(v_m) = \frac{\Delta v_L}{\pi} \qquad (7.10.1)$$

where Δv_L is the linewidth [given by Eq. (7.9.2) for the Schawlow–Townes theory]. Note from Fig. 7.28 that there is a large increase in the noise spectrum for offset frequencies smaller than 100 kHz due, in this case, to acoustic disturbances and pump power fluctuations.

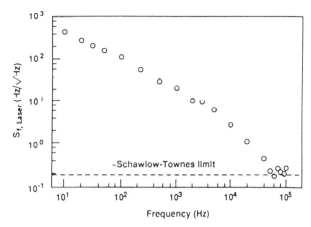

FIG. 7.28. Frequency noise spectrum from a free-running diode-pumped Nd:YAG laser. (By permission from Ref. 26.)

For the most sophisticated applications, e.g., gravitational wave detection, the laser noise spectrum must be strongly reduced. To this purpose one uses techniques to stabilize the cavity length: One of the cavity mirrors is mounted on a piezoelectric transducer, and, for frequency stabilization, a feedback voltage is applied from a suitable electronic circuit to the transducer. By sending a fraction of the laser radiation to a frequency discriminator of sufficiently high resolution and stability, voltage fluctuations at the output yield the required error signal. The sharp transmission (or reflection) lines of a high-finesse FP interferometer, or a sharp absorption line from an atomic or molecular gas kept in a low-pressure cell, are often used as frequency discriminators. Fabry–Perot interferometers with a finesse larger than 10^5 using mirrors with absorption and scattering losses of only a few parts per million have been used. For a 1-m-long FP interferometer, sharp transmission linewidths in the range of a few kilohertz have been obtained. To reduce frequency fluctuations in the FP cavity, mirrors must be mounted using very low-expansion elements (such as a tube made of superinvar or of very low-expansion ceramic) and the interferometer must be placed in a container providing pressure and temperature control. However, long-term and absolute frequency stabilization can be achieved only by using an atomic or molecular absorption line as a frequency reference. A good reference wavelength must be reproducible and essentially independent of such external perturbations as electric or magnetic fields and temperature or pressure changes. Therefore transitions in atoms or molecules without permanent dipole moments, such as CH_4 for the 3.39-μm transition or $^{129}I_2$ for the 633-nm transition in the He-Ne laser, are the most suitable. For a low-pressure gas or vapor, the absorption line is limited by Doppler broadening to a width of ~ 1 GHz (in the visible range). To obtain much narrower linewidths, perhaps in the kilohertz range, some type of Doppler-free nonlinear spectroscopy must be used.[28]

A common method of frequency stabilization relies on the *Pound–Drever technique*,[29] where a small fraction of the output beam is frequency-modulated, then passed through the frequency discriminator, i.e., the FP interferometer or the absorption cell. To understand this technique, we first point out that any element exhibiting a transmission that changes with frequency also induces a phase shift on the incident wave that depends on frequency. For a FP interferometer used in transmission, the phase shift can be obtained from Eq. (4.5.4). For an absorption line on the other hand, the phase shift can be written as $\phi = 2\pi n l/\lambda$, where l is the length of the absorption cell and n is the refractive index of the medium. For a Lorentzian line, the refractive index n can be related to the absorption coefficient α of the medium by the dispersion relation:

$$n(v - v_0) = n_0 + \left(\frac{c}{2\pi v}\right)\left(\frac{v_0 - v}{\Delta v_0}\right)\alpha(v - v_0) \qquad (7.10.2)$$

where n_0 is the refractive index sufficiently far away from the resonance line, v is the frequency of the em wave, v_0 is the transition frequency, and Δv_0 is the transition width. Note that, for $v = v_0$, one has $n = n_0$, so the transition does not contribute to the refractive index. For an inhomogeneous line, we must add at frequency v the phase shifts induced from all atoms, their transition frequencies v_0' being now distributed according to the function $g^*(v_0' - v_0)$. The refractive index of the medium is then obtained from Eq. (7.10.2) upon averaging over the frequency distribution $g^*(v_0' - v)$. This yields

$$n_{eff} = n_o + \frac{cN_t}{2\pi v}\int \frac{v_0' - v}{\Delta v_0}\sigma_h(v - v_0')g^*(v_0' - v_0)dv_0' \qquad (7.10.3)$$

where N_t is the total ground-state population and σ_h is the homogeneous cross section. According to Eq. (7.10.2) or (7.10.3), for a given profile of the absorption coefficient $\alpha = \alpha(\omega - \omega_0)$ (see Fig. 7.29a), the corresponding frequency shift generally takes the form shown in Fig. 7.29b. For simplicity the phase shift at the line center frequency, $\phi_0 = 2\pi n_0 l/\lambda$ where l is the length of the medium, is taken to be zero. A rather similar curve applies for a FP interferometer, so Fig. 7.29b provides a general representation of the phase shift in our frequency discriminator (i.e., the absorption cell or the FP interferometer).

We now consider a frequency-modulated beam and write its electric field as $E(t) = E_0 \exp[j\omega t + j\Gamma \sin \omega_m t]$, where Γ is the phase modulation index and ω_m is the modulation frequency. We expand this field, in terms of Bessel functions, as

$$E(t) = E_0 e^{j\omega t}\sum_{-\infty}^{+\infty}{}_nJ_n(\Gamma)e^{jn\omega_n t} \qquad (7.10.4)$$

where J_n is the Bessel function of order n. If we limit our considerations to the first two side bands at frequency $\pm\omega_m$, we obtain from Eq. (7.10.4):

$$E(t) = E_0 e^{j\omega t}[-J_1(\Gamma)e^{-j\omega_m t} + J_0(\Gamma) + J_1(\Gamma)e^{j\omega_m t}] \qquad (7.10.5)$$

where we have used the property $J_{-1}(\Gamma) = -J_1(\Gamma)$. The electric field of the wave, after leaving the frequency discriminator, is then given by:

$$E(t) = E_0 e^{j\omega t}[-J_1(\Gamma)e^{-j\omega_m t - j\phi_{-1}} + J_0(\Gamma)e^{-j\phi_0} + J_1(\Gamma)e^{j\omega_m t - j\phi_1}] \qquad (7.10.6)$$

where ϕ_0, ϕ_{-1}, and ϕ_1 are the phase shifts associated with the carrier wave and the two sidebands, respectively (Fig. 7.29b). If the beam transmitted by the discriminator is sent to a quadratic detector, the detected photocurrent will be proportional to EE^*, where E^* is the complex conjugate of E. The component of the photocurrent at frequency ω_m is then proportional to:

$$(EE^*)_{\omega_m} = 2|E_0|^2\, \text{Re}\{J_0 J_1 e^{j\omega_m t}[e^{j(\phi_0 - \phi_1)} - e^{j(\phi_{-1} - \phi_0)}]\} \qquad (7.10.7)$$

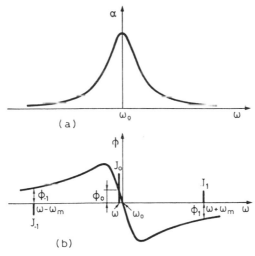

FIG. 7.29. Pound–Drever technique for frequency stabilization to the transmission minimum of an absorption cell (or to the transmission peak of a Fabry–Perot interferometer).

where Re stands for real part. If the carrier frequency ω of the wave coincides with the discriminator central frequency $\omega_0 = 2\pi\nu_0$, then one has $\phi_0 = 0$ and $\phi_{-1} = -\phi_1$ (see Fig. 7.29). From Eq. (7.10.7) we then have $(EE^*)_{\omega_m} = 0$. If, on the other hand, $\omega \neq \omega_0$ and we assume $\omega_m \gg 2\pi\Delta\nu_0$, then we can write $\phi_{-1} \cong -\phi_1$, and from Eq. (7.10.7) we obtain

$$(EE^*)_{\omega_m} = -4|E_0|^2 J_0 J_1 \sin(\phi_0)\sin(\omega_m t - \phi_1) \tag{7.10.8}$$

The sign of the photocurrent component at frequency ω_m therefore depends on the sign of ϕ_0, i.e., on whether ω is above or below ω_0. This component can then be used as the error signal for the electronic feedback loop to force the carrier frequency of the wave to coincide with the central frequency of the discriminator. The precision by which this can be achieved depends on the gain of the feedback loop and on its bandwidth. With very sharply defined frequency discriminators ($\Delta\nu_0 \cong 30\,\text{kHz}$), short-term frequency drifts in the 100-mHz range have been achieved in this way.[30]

7.11. INTENSITY NOISE AND INTENSITY NOISE REDUCTION

Since spontaneous emission and cavity length fluctuations induce only a frequency noise, the field amplitude of the output beam can, to first order, be taken as independent of time. There are, however, other laser perturbations which result in amplitude fluctuations, i.e., produce an *intensity noise*. The most common perturbations of this type can be listed as follow:

- For a gas laser: Fluctuations in the power supply current, instability of the discharge process, and mirror misalignments due to resonator vibrations.
- For dye lasers: Density fluctuations in the dye jet solution and presence of air bubbles in the solution.
- For solid-state lasers: Pump fluctuations (both for lamp pumping and diode pumping) and cavity misalignments.

298 7 • Continuous Wave Laser Behavior

- For semiconductor lasers: Fluctuations in the bias current, amplitude fluctuations due to spontaneous emission and electron-hole recombination noise.

Besides these short-term fluctuations, long-term drift in the output power is also present, generally arising from thermal misalignment in the laser cavity and from degradation of mirrors, windows, or other optical components, including the active medium itself. For well-designed and well-engineered lasers, however, this power degradation occurs only over at least a few thousand hours.

To describe intensity noise let $\delta P(t)$ be the fluctuation of output power around its average value $\langle P \rangle$. We can then define an intensity autocorrelation function $C_{pp}(\tau)$ as

$$C_{pp}(\tau) = \frac{\langle \delta P(t)\delta P(t+\tau) \rangle}{\langle P \rangle^2} \qquad (7.11.1)$$

where $\langle \ \rangle$ stands for (ensemble) average. The Fourier transform of $C_{pp}(\tau)$ is referred to as the *relative intensity noise* (RIN) of the laser source, and it is given by:

$$\text{RIN}(\omega) = \int_{-\infty}^{+\infty} C_{pp}(\tau)\exp(j\omega\tau)d\tau \qquad (7.11.2)$$

Obviously $C_{pp}(\tau)$ is obtained from Eq. (7.11.2) by taking the inverse Fourier transform, i.e.:

$$C_{pp}(\tau) = \frac{1}{2\pi}\int_{-\infty}^{+\infty} \text{RIN}(\omega)\exp(-j\omega\tau)d\omega \qquad (7.11.3)$$

A typical RIN spectrum, as obtained from a single-mode, diode-pumped, NPRO oscillator is shown in Fig. 7.30a (curve 1). Note that the scale is expressed in dB/Hz, a notation that may create some confusion; its meaning is

$$\text{RIN(dB/Hz)} = 10\log[\text{RIN}(v) \times \Delta v] \qquad (7.11.4)$$

where $\text{RIN}(v) = 2\pi\text{RIN}(\omega)$ and $\Delta v = 1\,\text{Hz}$.* The RIN spectrum is strongly peaked at a frequency ($v \cong 300\,\text{kHz}$ in the figure) corresponding to the laser relaxation oscillation frequency (see Chap. 8). The corresponding relative fluctuation of the output power $\delta P(t)/\langle P \rangle$ is shown in Fig. 7.30b (curve 1). In this figure the root-mean-square variation of $\delta P(t)$, i.e., $[\langle \delta P^2(t) \rangle / \langle P \rangle^2]^{1/2}$, is $\sim 2 \times 10^{-3}$. The same result is obtained from Fig. 7.30a if we take into account that the (3-dB) width Δv_p of the relaxation oscillation peak is roughly $2\,\text{kHz}$ and the RIN value at the peak is $\sim -85\,\text{dB/Hz}$. In fact, according to Eq. (7.11.4), we have $\text{RIN}(v) = 10^{-8.5}\,\text{Hz}^{-1} \cong 3.16 \times 10^{-9}\,\text{Hz}^{-1}$. From Eqs. (7.11.3) and (7.11.1) with $\tau = 0$, we then obtain $C_{pp}(0) = \langle \delta P^2(t) \rangle / \langle P \rangle^2 \cong \text{RIN}(v) \times \Delta v_p \cong 6.32 \times 10^{-6}$ and so $[\langle \delta P^2(t) \rangle / \langle P \rangle^2]^{1/2} \cong 2.5 \times 10^{-3}$.

To reduce intensity noise, a negative feedback is often applied to the pump power supply. The time required to establish this feedback is however limited by the discharge response time (pump-rate response time). Accordingly, for a gas laser, this feedback scheme cannot be used to reduce e.g., the intensity noise arising from discharge instabilities. For a diode-pumped solid-state laser, on the other hand, the response time of the diode pump is much shorter than the inverse of the relaxation-oscillation frequency of the solid-state laser.

---

* According to (7.11.4), the dimension of RIN(dB/Hz) is that of a pure number rather than $[\text{Hz}]^{-1}$, as the (dB/Hz) notation could seem to imply. This notation thus simply indicates that one is measuring noise in a bandwidth of $\Delta v = 1\,\text{Hz}$.

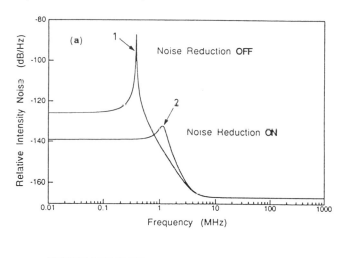

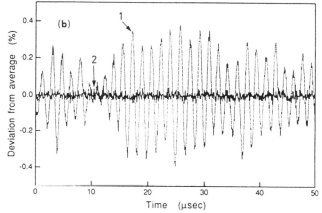

FIG. 7.30. (a) Typical relative intensity noise spectra for a diode-pumped Nd:YAG laser using the NPRO configuration (see Fig. 7.26) without (curve 1) and with (curve 2) active noise reduction. (b) Corresponding plots of typical relative fluctuations in the output power. (By permission from Ref. 31.)

In this case, a negative feedback loop effectively reduces intensity noise up to a frequency higher than the relaxation-oscillation frequency. This is demonstrated in Fig. 7.30a (curve 2), where a feedback loop is seen to reduce the peak value of the RIN by more than 35 dB. The corresponding curve in Fig. 7.30b (curve 2) shows in fact that $[\langle \delta P^2(t)\rangle / \langle P\rangle^2]^{1/2}$ is reduced by more than one order of magnitude.

So far we have considered the intensity noise of a single-mode laser. For multimode oscillation, the situation is much more complicated because, even if total power is kept constant, the power in each mode can fluctuate in time. This phenomenon is known as *mode partition noise*, which often poses a severe problem for intensity noise in each mode. Assume for instance that, besides a main mode, a side-mode with power 20 dB below the main mode is oscillating. The nonlinearity of the corresponding rate equations provides a mechanism that can produce anticorrelation between powers of the two modes.[20] This may result in the power of the side mode varying in time, between its full value and zero, while the power of the main mode shows corresponding fluctuations to keep a constant total power (a phenomenon also referred to as *antiphase dynamic*).[32] The spectral frequency of this

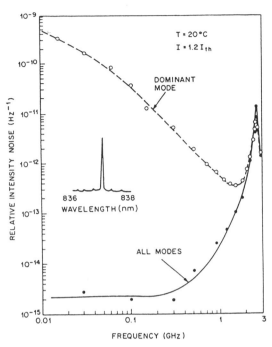

FIG. 7.31. Experimental observation of mode partition noise in a multimode semiconductor laser. (By permission from Ref. 33.)

mode partition noise is then determined by the time behavior of the antiphase-dynamics process. As an example, Fig. 7.31 shows the measured RIN spectra of an AlGaAs Fabry–Perot-type (see Chap. 9) semiconductor laser when the power in all modes (*solid curve*) or in the dominant mode (*dashed curve*) is detected.[33] Note the large increase in RIN that occurs in the dominant mode at frequencies below the relaxation oscillation peak ($\cong 2.5\,\text{GHz}$, the high value, in this case, arising from the laser's short cavity length). It is due to the presence of the other oscillating modes with very low intensity (see inset of Fig. 7.31).

7.12. CONCLUSIONS

In this chapter, a few topics related to the cw behavior of both a four-level and quasi-three-level laser have been considered in some detail. Space-independent rate equations, under the simplest decay-rate conditions, are developed first (ideal four-level and quasi-three-level lasers), and the cw behavior predicted by these equations, including optimum coupling conditions, are considered. Results obtained from the space-dependent equations are discussed at some length.

It should be recalled that the rate equation formulation represents the simplest way of describing cw and transient laser behavior. In order of increasing accuracy and complexity, one should consider the semiclassical and the quantum electrodynamics treatments. For the cw case, however, equations from the semiclassical treatment reduce to the rate equations. The full-quantum treatment, on the other hand, is required to describe correctly the start of laser oscillation as well as the fundamental limit to laser frequency noise. When however the number of photons in a given cavity mode is much greater than 1, (average) results of the quantum treatment coincide with those of the semiclassical treatment. Note also that, in their simplest form, the rate equations, as given here, apply to only relatively few cases. In most cases, there are more than just four levels involved, so the rate equation treatment becomes correspondingly more complicated. In fact it can be said that, in general, each laser has its own particular set of rate equations. The equations considered in this chapter, however, provide a model that can be readily extended to more complicated situations.

Besides topics that can be directly discussed in terms of rate equations, other subjects of fundamental importance for cw laser behavior have also been considered, namely: (1) Reasons for multimode oscillation, methods of single-mode selection, and laser tuning. (2) Limit to monochromaticity for single-mode lasers, as well as field fluctuations in the output beam, in both frequency and amplitude. (3) Methods to activity reduce both frequency and amplitude fluctuations.

The ensemble of these topics constitute the minimum set of knowledge required for a balanced and up-to-date understanding of cw laser behavior.

PROBLEMS

7.1. Calculate the logarithmic loss γ of a mirror with transmission $T = 80\%$ and negligible internal loss.

7.2. Prove Eq. (7.2.11).

7.3. Referring to Fig. 7.4 and Example 7.2 and taking $n = 1.82$ as the refractive index of Nd:YAG, calculate the equivalent resonator length L_e and the cavity photon decay time.

7.4. Consider a four-level laser just at threshold and assume the branching ratio of the transition $2 \rightarrow 1$ compared to the overall spontaneous transition rate as $\beta = 0.51$ and that the overall upper-state lifetime is purely radiative and equal to $\tau = 234\ \mu s$. (Data refer to the 1.064-μm transition of Nd:YAG; see Example 2.10). How short must the lifetime of the lower level (1) be to ensure that, in the steady state, $(N_1/N_2) < 1\%$? Now consider the same laser above threshold; referring to Example 7.2, assume that the laser is oscillating with an output power $P_{out} = 200\ W$. How short must the lifetime of the lower laser level be to ensure that, under these conditions, $(N_1/N_2) < 1\%$?

7.5. Referring to Fig. 7.4 and Example 7.2, suppose that the Nd:YAG rod is replaced by a Nd:YLF rod of the same dimensions (YLF $\equiv$ YLiF$_4$). Oscillation can then occur at either $\lambda = 1.047\ \mu m$ (extraordinary wave or π-transition) or at $\lambda = 1.053\ \mu m$ (ordinary wave or σ-transition). The largest value for the effective stimulated emission cross section is for the π-transition ($\sigma_e \cong 1.8 \times 10^{-19}\ cm^2$). The upper state lifetime is the same for the two transitions, i.e., $\tau = 480\ \mu s$. Assuming that the internal loss in the cavity and lamp-pumping efficiency remain the same as for Nd:YAG, calculate the threshold inversion and the threshold pump power, then compare the results with those of Nd:YAG. Assuming that the energy separation between the $^4F_{3/2}$ upper laser level and the ground level remains the same as for Nd:YAG and taking the same value for the beam area A_b, calculate the slope efficiency.

7.6. Calculate the optimum output-mirror transmission and the corresponding optimum output power when the laser of Problem 7.5 is pumped by a lamp input power $P_p = 7\ kW$.

7.7. For the high-power CO$_2$ laser of Figs. 7.7 and 7.8, and with reference to the data considered in Example 7.3, calculate the optimum output coupling and the optimum power for an input pump power $P_p = 140\ kW$. The resulting optimum value of γ_2 turns out to be substantially smaller than that considered in Example 7.3, i.e., the unstable resonator of Fig. 7.7 is substantially overcoupled. This overcoupling is intentionally made to increase the peak intensity of the output beam when focused by a lens. With reference to the focusing properties of the annular beam of an unstable resonator discussed in Section 5.6.3, show in fact that the ~ 12 kW beam of Fig. 7.8 (at $P_p = 140$ kw) produce a higher intensity, at the focus of a lens, compared to that of the optimum beam considered in this problem.

7.8. Consider the lamp-pumped Nd:YAG laser in Fig. 7.4. The pumping beam induces a thermal lens of focal length f in the rod whose dioptric power $1/f$ is proportional to the pump power P_p. At a pump power of $P_p = 7$ kW and for the rod dimensions in Fig. 7.4, the thermally induced focal length is $f \cong 25$ cm.[34] Now assume that, to calculate the cavity mode, the rod in Fig. 7.4 can be simulated by a thin lens, of focal length $f \cong 25$ cm, placed at the resonator center. In this case calculate the TEM_{00} mode spot size at the lens and at the mirror location.

7.9. One can show that the radial extension of a TEM_{lm} mode of higher order (i.e. for $l \cong m \gg 1$) is approximately given by $w_{lm} \cong l^{1/2}w$, where w is the spot size of the corresponding TEM_{00} mode. The maximum number of Hermite–Gaussian modes that can fit to a rod of radius a is such that $w_{lm} \cong a$.[35] Using this argument and results from Problem 7.8, calculate the approximate number of transverse modes that oscillate for the configuration in Fig. 7.4. The beam divergence of a TEM_{lm} mode ($l \cong m \gg 1$) θ_{lm}, is then approximately given by $\theta_{lm} \cong (l)^{1/2} \lambda/\pi w_0$, where w_0 is the spot size of the TEM_{00} mode at the beam waist.[35] Calculate then the beam divergence for the case in Fig. 7.4, assuming that it equals the divergence of the highest order mode that is oscillating.

7.10. Consider the Nd:YAG laser in Example 7.4 and assume that optimum output coupling can be calculated by the formula established by the space-dependent case in Sect. 7.5. Calculate the optimum output coupling, and using this value for γ_2, calculate with the help of Eq. (7.3.34) the expected value of the output power at a diode laser pump power of $P_p = 1.14$ W.

7.11. Consider a 4-mm diameter, 56 mm long, 0.9 at. %, Nd:YAG rod side pumped, at 807-nm wavelength, by fiber-coupled diode lasers.[36] Assume that the laser is oscillating in a TEM_{00} mode with a constant spot size within the active medium of $w_a \cong 1.4$ mm. Let the transmission of the output mirror be $T_2 = 15\%$ and assume a total internal loss of $\gamma_i = 3.8\%$ (see Example 7.2). Calculate the output power at a laser diode optical pump power of $P_p = 370$ W, corresponding to an absorbed power of $P_{ap} = 340$ W, and the corresponding slope efficiency. Compare calculated values with experimental values in Ref. 36 and try to explain the discrepancy.

7.12. Consider again the diode-pumped laser in Problem 7.11 and assume that, at the stated pump power, the thermally induced lens in the rod has a focal length $f = 21$ cm; take a symmetric flat-flat resonator as in Fig. 7.4. Under the simplifying assumption that the rod can be simulated by a thin lens of focal length f, calculate the distance of the two plane mirrors from this lens to obtain a spot size at the lens position of $w_a = 1.4$ mm. Also calculate the corresponding spot size at the two mirrors.

7.13. Prove Eq. (7.4.7).

7.14. Consider a He-Ne laser oscillating on its red transition, $\lambda = 632.8$-nm in air, and assume a gas-tube length of $l = 20$ m, a tube radius of 1 mm, a 0.1-Torr partial pressure of Ne atoms, an output coupler transmission of 1%, and a single-pass internal loss of 0.5%. The effective cross section and the overall lifetime of the laser transition can be taken as $\sigma_e = 5.8 \times 10^{-13}$ cm^2 and $\tau = 50$ ns, respectively. Assume for simplicity that the lifetime of the lower state is much shorter than that of the upper state. Calculate the threshold inversion, the ratio of this inversion to the total Ne population, and the critical pump rate. The upper laser level is predominantly pumped through the 2^1S state of He that lies ~ 20.5 eV above the ground state. Assuming unit quantum efficiency in this near resonant energy transfer process, calculate the minimum threshold pump power. For an output power of 3 mW, also calculate the ratio of the number of photons emitted by stimulated emission to the number of atoms that decay spontaneously.

7.15. An Ar-ion laser, oscillating on its green $\lambda = 514.5$-nm transition, has a 10% unsaturated gain per pass. The resonator consists of two concave spherical mirrors both of radius of curvature $R = 5$ m and separated by $L = 100$ cm. The output mirror has a $T_2 = 5\%$ transmission; the other mirror is nominally 100% reflecting. Identical apertures are inserted at both ends of the resonator to obtain TEM_{00} mode operation. Neglecting all other types of losses, calculate the required aperture diameter.

7.16. The linewidth, $\Delta v_0^* = 50$ MHz, of a low-pressure CO_2 laser is predominantly established by Doppler broadening. The laser is operating with a pump power twice the threshold value. Assuming one mode coincides with the transition peak and equal losses for all modes, calculate the maximum mirror spacing that still allows single longitudinal operation.

7.17. Consider an Ar-ion laser oscillating on its green $\lambda = 514.5$ nm transition and assume this transition to be Doppler-broadened to a width $\Delta v_0^* = 3.5$ GHz. Assume a cavity length $L_e = 120$ cm, an Ar tube length of 100 cm, and take a single-pass cavity loss of $\gamma = 10\%$. Take an effective value of the stimulated emission cross section and upper state lifetime as $\sigma_e = 2.5 \times 10^{-13}$ cm^2 and $\tau = 5$ ns, respectively. Assume that the lifetime of the lower laser level is much shorter than that of the upper laser level and that one cavity mode coincides with the transition peak. Calculate the threshold inversion for this central mode and the threshold pump rate. At how great a pump rate, above threshold, does oscillation start in the two adjacent longitudinal modes?

7.18. A He-Ne laser is oscillating on three adjacent longitudinal modes, the central one being coincident with the center of the laser transition. The cavity length is 50 cm, and the output coupling is 2%. If the laser linewidth is $\Delta v_0^* = 1.7$ GHz, calculate the mode spacing.

7.19. Assume that one cavity mirror is mounted on a piezoelectric transducer. Show that the comb of longitudinal modes shifts by approximately one comb spacing for a $\lambda/2$ translation of the transducer.

7.20. Consider a single-longitudinal-mode He-Ne laser and assume that the oscillating frequency is made to coincide with the frequency of the transition peak by a piezoelectric transducer attached to a cavity mirror. How far can the mirror be translated before a mode hop (i.e., a switch in oscillation to the next mode) occurs?

7.21. An Ar-ion laser oscillating on its green $\lambda = 514.5$ nm transition has a total loss per pass of 4%, an unsaturated peak gain $G_p = \exp \sigma_p Nl$ of 1.3, and a cavity length of 100 cm. To select a single-longitudinal-mode, a tilted and coated quartz ($n_r = 1.45$) Fabry–Perot etalon with a 2-cm thickness is used inside the resonator. Assuming for simplicity that one cavity mode is coincident with the peak of the transition (whose linewidth is $\Delta v_0^* = 3.5$ GHz), calculate the etalon finesse and the reflectivity of the two etalon faces to ensure single-longitudinal-mode operation.

7.22. Referring to Fig. 7.31, evaluate the relative root-mean-square fluctuation in output power of the semiconductor laser for the dominant mode and for all modes.

REFERENCES

1. H. Statz and G. de Mars, Transients and Oscillation Pulses in Masers, in *Quantum Electronics* (C. H. Townes ed.) (Columbia Univ. Press, New York, 1960), pp. 530–37.
2. R. Dunsmuir, Theory of Relaxation Oscillations in Optical Masers, *J. Electron. Control* **10**, 453 (1961).

3. M. Sargent, M. O. Scully, and W. E. Lamb, *Laser Physics* (Addison-Wesley, London, 1974).

4. R. H. Pantell and H. E. Puthoff, *Fundamentals of Quantum Electronics* (Wiley, New York, 1969), Chap. 6, Sect. 6.4.2.

5. W. W. Rigrod, Saturation Effects in High-Gain Lasers, *J. Appl. Phys.* **36**, 2487 (1965).

6. L. W. Casperson, Laser Power Calculations: Sources of Error, *Appl. Optics* **19**, 422 (1980).

7. W. Koechner, *Solid-State Laser Engineering*, vol. 1, Springer Series in Optical Sciences, 4th ed. (Springer-Verlag, Berlin, 1996), Chap. 3, adapted from Fig. 3.21.

8. Ref. 7, Chap. 3, Fig. 3.22.

9. D. Findlay and R. A. Clay, Measurement of Internal Losses in 4-Level Lasers, *Phys. Lett.* **20**, 277 (1966).

10. Private communication, Istituto di Ricerca per le Tecnologie Meccaniche, Vico Canavese, Torino, Italy.

11. M. C. Fowler, Quantitative Analysis of the Dependence of CO_2 Laser Performance on Electrical Discharge Properties, *Appl. Phys. Letters* **18**, 175 (1971).

12. E. Hoag *et al.*, Performance Characteristics of a 10-kW Industrial CO_2 Laser System, *Appl. Opt.* **13**, 1959 (1974).

13. P. F. Moulton, An Investigation of the $Co:MgF_2$ Laser System, *IEEE J. Quant. Electr.* **QE-21**. 1582 (1985).

14. V. Evtuhov and A. E. Siegman, A Twisted-Mode Technique for Obtaining Axially Uniform Energy Density in a Laser cavity, *Appl. Opt.* **4**, 142 (1965).

15. J. Berger *et al.*, 370-mW, 1.06-μm, cw TEM_{00} Output from a Nd:YAG Laser Rod End-Pumped by a Monolithic Diode Array, *Electr. Letters* **23**, 669 (1987).

16. W. P. Risk, Modeling of Longitudinally Pumped Solid-State Lasers Exhibiting Reabsorption Losses, *J. Opt. Soc. Am. B* **5**, 1412 (1988).

17. T. Y. Fan and R. L. Byer, Modeling and CW Operation of a Quasi-Three-Level 946 nm Nd:YAG Laser, *IEEE J. Quant. Electr.* **QE-23**, 605 (1987).

18. P. Lacovara *et al.*, Room Temperature Diode-Pumped Yb:YAG Laser, *Opt. Letters* **16**, 1089 (1991).

19. A. Yariv, Energy and Power Considerations in Injection and Optically Pumped Lasers, *Proc. IEEE* **51**, 1723 (1963).

20. C. L. Tang, H. Statz, and G. de Mars, Spectral Output and Spiking Behavior of Solid-State Lasers, *J. Appl. Phys.* **34**, 2289 (1963).

21. T. J. Kane and R. L. Byer, Monolithic, Unidirectional, Single-Mode Nd:YAG Ring Laser, *Opt. Lett.* **10**, 65 (1985).

22. A. E. Siegman, *Lasers* (Oxford University, Cambridge, 1986), Chap. 12, Sect. 12.2.

23. A. Yariv, *Optical Electronics* (Saunders College Publ., Fort Worth, TX, 1991), Sect. 10.7.

24. A. L. Schawlow and and C. H. Townes, Infrared and Optical Masers, *Phys. Rev.* **112**, 1940 (1958).

25. C. H. Henry, Theory of Linewidth of Semiconductor Lasers, *IEEE J. Quant. Electr.* **QE-18**, 259 (1982).

26. T. Day, E. K. Gustafson , and R. L. Byer, Subhertz Relative Frequency Stabilization of Two Diode Laser-Pumped Nd:YAG Lasers Locked to a Fabry-Perot Interferometer, *IEEE J. Quant. Electr.* **QE-28**, 1106 (1992).

27. D. K. Owens and R. Weiss, Measurement of the Phase Fluctuation in a He-Ne Zeeman Laser, *Rev. Sci. Instruments* **45**, 1060 (1974).

28. Wolfgang Demtröder, *Laser Spectroscopy*, 2nd ed. (Springer-Verlag, Berlin, 1996), Chap. 7.

29. R. W. T. Drever *et al.*. Laser Phase and Frequency Stabilization Using an Optical Resonator, *Appl. Phys. B* **31**, 97 (1983).

30. N. Uekara and K. Ueda, 193-mHz Beat Line Width of Frequency-Stabilized Laser-Diode-Pumped Nd:YAG Ring Lasers, *Opt. Letters* **18**, 505 (1993).

31. Introduction to Diode-Pumped Solid-State Lasers, LIGHTWAVE Electronics Corp. Techn. Information N. 1 (1993).

32. Kenju Otsuka, Winner-Takes-All and Antiphase States in Multimode Lasers, *Phys. Rev. Letters* **67**, 1090 (1991).

33. G. P. Agrawal and N. K. Dutta, *Long-Wavelength Semiconductor Lasers* (Chapman and Hall, New York, 1986), Fig. 6.11.

34. Ref. 7, Sect. 7.1.1. and Fig. 7.5.

35. Ref. 22, Chap. 17, Sect. 17.5, 17.6.

36. D. Golla *et al.*, 62-W cw TEM_{00} Mode Nd:YAG Laser Side Pumped by Fiber-Coupled Diode Lasers, *Opt. Letters* **21**, 210 (1996).

8

Transient Laser Behavior

8.1. INTRODUCTION

Chapter 8 examines a few cases where the pump rate and/or cavity losses are time-dependent. We also consider situations when a nonlinear optical element, such as a saturable absorber, is inserted in the laser cavity, where nonlinearity causes the laser to depart from stable cw operation. In such cases we are dealing with transient laser behavior. The transient cases we consider can be divided into two categories: (1) Cases, such as relaxation oscillations, Q-switching, gain-switching, and cavity-dumping, where, ideally, a single-mode laser is involved; these can be described by a rate equation treatment. (2) Cases involving many modes, e.g., mode-locking, which requires a different treatment, such as a description in terms of either the fields of all the oscillating modes (frequency domain description) or in terms of a self-consistent circulating pulse within the cavity (time domain description).

8.2. RELAXATION OSCILLATIONS

We first consider a step-function pump rate. We assume that $R_p = 0$ for $t < 0$ and $R_p(t) = R_p$ (independent of time) for $t > 0$. We also assume that the laser is oscillating in a single mode, so that a simple rate equation treatment can be applied. As seen in Chap. 7, rate equations are nonlinear in the variables $N(t)$ and $\phi(t)$ since they involve products of the form ϕN. Consequently, analytical solutions for this case or other cases about to be considered are generally not possible, so one must often resort to numerical computation.[1,2]

As a representative example, Fig. 8.1 shows one of the first computed plots of $N(t)$ and $\phi(t)$ carried out for a three-level laser such as a ruby laser.[2] In this case the initial condition for the population inversion is $N(0) = -N_t$, where N_t is the total population, because at time $t = 0$, the entire population is in the lower laser level (1; see Fig. 1.4a). The initial condition for the total number of cavity photons is then $\phi(0) = \phi_i$, where ϕ_i can be taken as some

305

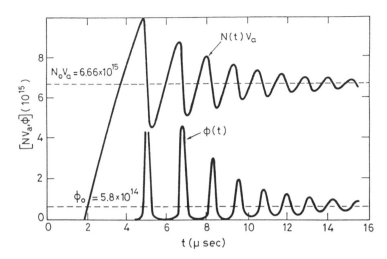

FIG. 8.1. Example of the temporal behavior of the total inversion $V_0 N(t)$, and photon number, $\phi(t)$, for a three-level laser. (By permission from Ref. 8.)

small integer (e.g., $\phi_i = 1$) that just allows the laser action to begin. Note that a similar behavior to that in Fig. 8.1 is also expected for a four-level laser; one of the main differences is that, in this case, the initial condition is $N(0) = 0$. Thus, if the time origin in Fig. 8.1 is shifted approximately to time $t = 2$ μs, where population inversion becomes zero in the figure, curves in Fig. 8.1 can be used to provide a qualitative description for a four-level laser, as well.

Several features of this figure are worth pointing out: (1) After time $t = 2$ μs, the population inversion keeps growing due to the pumping process while the photon number remains at its initial low value, as determined by quantum field fluctuations, until the inversion crosses the threshold value ($N_0 V_a = 6.66 \times 10^{15}$ in the figure). From this time on, roughly for $t > 3.5$ μs, the population exceeds the threshold value, so the number of cavity photons can begin to grow. From either Eq. (7.2.16b) (four-level laser) or Eq. (7.2.24b) (quasi-three-level laser), we find in fact that $d\phi/dt > 0$ when $N > N_c$, where N_c is the critical or threshold inversion. (2) After threshold is exceeded, the photon number requires some time to grow from its initial value ($\phi_i = 1$) to a value, e.g., equal to the steady-state value ($\phi_0 = 5.8 \times 10^{14}$ in the figure); meanwhile the population continues growing due to the pumping process. (3) When the photon number becomes large enough (roughly when $\phi > \phi_0$), the stimulated emission process becomes dominant over the pumping process. The population then begins to decrease and, at the time corresponding to the maximum of $\phi(t)$, $N(t)$ is seen to have decreased to N_c. This can be readily shown from either Eq. (7.2.16b) or Eq. (7.2.24b), since, when $d\phi/dt = 0$, one has $N = N_c$. (4) After this photon peak, the population inversion is driven below N_c by the continuing high rate of stimulated emission. Thus the laser goes below threshold, and the photon number decreases. (5) When this photon number decreases to a sufficiently low value (roughly when $\phi < \phi_0$), the pumping process again becomes dominant over the stimulated-emission process. The population inversion can now begin growing again and the whole series of events considered in features (1)–(5) repeats itself. The photon number $\phi(t)$ is then seen to display a regular sequence of peaks (or *laser spikes*) of decreasing amplitude; consecutive peaks are approximately equally spaced in time. The output power therefore shows a similar time behavior. This aspect of regular oscillation for the output power is usually referred to as a *damped relaxation oscillation*. The time behavior of the population inversion undergoes a similar oscillatory behavior; the oscillation of $N(t)$ precedes that of $\phi(t)$ by about half the oscillation period, since we must first produce an increase in population, $N(t)$, to have a corresponding increase in the photon number $\phi(t)$. Note also that, since a steady-state solution is eventually reached [as given by Eqs. (7.3.4a) and (7.3.4b) for a four-level laser or by Eqs. (7.4.1) and (7.4.6) for a quasi-three-level laser], the computer calculation confirms that the solution corresponds to a stable operating condition.

For small oscillations about the steady-state values (e.g., roughly for $t > 14$ μs in Fig. 8.1), the dynamic behavior can be described analytically. In fact, if we write

$$N(t) = N_0 + \delta N(t) \tag{8.2.1}$$

$$\phi(t) = \phi_0 + \delta\phi(t) \tag{8.2.2}$$

and assume $\delta N \ll N_0$ and $\delta\phi \ll \phi_0$, we can neglect the product $\delta N\delta\phi$ in the expression $N\phi$ in the rate equations; these equations then become linear in the variables δN and $\delta\phi$. Limiting ourselves to the case of a four-level laser, we can substitute Eqs. (8.2.1) and (8.2.2) into Eqs. (7.2.16a) and (7.2.16b). Since N_0 and ϕ_0 must satisfy the same equations with time derivatives set to zero, we readily obtain from Eqs. (7.2.16):

$$\left(\frac{d\delta N}{dt}\right) = -\delta N\left[B\phi_0 + \left(\frac{1}{\tau}\right)\right] - BN_0\delta\phi \tag{8.2.3}$$

$$\left(\frac{d\delta\phi}{dt}\right) = BV_a\phi_0\delta N \tag{8.2.4}$$

Note in particular that Eq. (8.2.4) is obtained from Eq. (7.2.16b) using the fact that $BV_aN_0 - (1/\tau_c) = 0$. The substitution of Eq. (8.2.4) into Eq. (8.2.3) gives

$$\frac{d^2\delta\phi}{dt^2} + \left[B\phi_0 + \left(\frac{1}{\tau}\right)\right]\frac{d\delta\phi}{dt} + (B^2V_aN_0\phi_0)\delta\phi = 0 \tag{8.2.5}$$

We now look for a solution of the form:

$$\delta\phi = \delta\phi_0\exp(pt) \tag{8.2.6}$$

The substitution of Eq. (8.2.6) into Eq. (8.2.5) then shows that p must obey the following equation:

$$p^2 + \frac{2}{t_0}p + \omega^2 = 0 \tag{8.2.7}$$

where:

$$\left(\frac{2}{t_0}\right) = \left[B\phi_0 + \left(\frac{1}{\tau}\right)\right] \tag{8.2.8}$$

and

$$\omega^2 = B^2 V_a N_0 \phi_0 \qquad (8.2.9)$$

The solution of Eq. (8.2.7) is obviously given by:

$$p = -\frac{1}{t_0} \pm \left(\frac{1}{t_0^2} - \omega^2\right)^{1/2} \qquad (8.2.10)$$

The first case we consider is when $(1/t_0) < \omega$. In this case the square root in Eq. (8.2.10) yields an imaginary number, so that we can write $p = -(1/t_0) \pm j\omega'$, where:

$$\omega' = \left[\omega^2 - \left(\frac{1}{t_0}\right)^2\right]^{1/2} \qquad (8.2.11)$$

From Eq. (8.2.6) we see that $\delta\phi$ corresponds to a damped sinusoidal oscillation, i.e.:

$$\delta\phi = C \exp\left(-\frac{t}{t_0}\right) \sin(\omega' t + \varphi) \qquad (8.2.12)$$

where C and φ are established by the initial conditions. If Eq. (8.2.12) is then substituted into Eq. (8.2.4), we find that δN is also described by a damped sinusoidal oscillation. Assuming $(1/t_0) \ll \omega'$, we obtain

$$\delta N \cong \frac{\omega' C}{B V_a \phi_0} \exp\left(-\frac{t}{t_0}\right) \cos(\omega' t + \varphi) \qquad (8.2.13)$$

Note that $\delta N(t)$ leads $\delta\phi(t)$ by half an oscillation period, as discussed previously, since we must first have a growth of inversion $\delta N(t)$ before we can have a growth of $\delta\phi(t)$.

Equations (8.2.8) and (8.2.9) can be recast in a form more convenient for calculation if the explicit expressions for N_0 and ϕ_0 given by Eq. (7.3.4a–b) are used. We readily obtain

$$t_0 = \frac{2\tau}{x} \qquad (8.2.14)$$

$$\omega = \left[\frac{(x-1)}{\tau_c \tau}\right]^{1/2} \qquad (8.2.15)$$

where $x = R_p/R_{cp}$ is the amount by which threshold is exceeded. Note that, while the damping time t_0 for the oscillation is determined by the upper state lifetime, the oscillation period $T = 2\pi/\omega' \cong 2\pi/\omega$ is determined by the geometrical mean of τ and the photon lifetime τ_c.

Example 8.1. *Damped oscillation in a Nd:YAG and a GaAs laser.* We first consider the single-mode Nd:YAG laser in Fig. 7.26 and assume that the preceding space-independent relaxation oscillation theory can be applied to this diode-pumped NPRO laser. Assuming the laser to be $x = 5$ times above threshold, we obtain from Eq. (8.2.14) $t_0 = 92$ μs, where we take $\tau = 230$ μs. We also take $l = 11.5$ mm as the round-trip path length of the NPRO resonator, and we assume a $T = 0.4\%$ output coupling transmission of the laser and a round-trip cavity loss of $L - 0.5\%$. The round-trip loss will then be $\gamma \cong (T + L) = 0.9\%$ and the cavity photon decay time $\tau_c = nl/c\gamma \cong 7.8$ ns, where $n = 1.82$ is the refractive index of the YAG material. From Eq. (8.2.15) we then obtain $v = \omega/2\pi \cong 238$ kHz for the frequency of the relaxation oscillation. Note that, in this case, we have $t_0 \gg 1/\omega$, thus the approximation $\omega' \cong \omega$ is justified. Note also that the spectrum of this damped oscillation is Lorentzian with a width $\Delta v_0 = 1/2\pi t_0 = 1.73$ kHz; this is also the 3-dB width of the relaxation oscillation peak of the RIN spectrum for this laser (see Fig. 7.30a).

Consider next a typical GaAs injection laser with cavity length $L = l = 300$ μm, whose two end faces are cleaved and act as cavity mirrors. According to Eq. (4.3.1) the power reflectivity of both mirrors, in this case, is equal to $R = [(n - 1)/(n + 1)]^2 \cong 0.3$, where $n = 3.35$ is the refractive index of GaAs. Thus $\gamma_1 = \gamma_2 = -\ln R = 1.2$. We also assume a distributed loss coefficient of $\alpha_0 - 60$ cm^{-1} along the semiconductor length, so that we can write $\gamma_i = \alpha_0 L = 1.8$. We thus get $\gamma = \gamma_i + [(\gamma_1 + \gamma_2)/2] = 3$ and $\tau_c = L_e/c\gamma = nL/c\gamma = 1.1$ ps. The upper state lifetime may be taken as $\tau \cong 3$ ns. Assuming $x = 1.5$, we get from Eq. (8.2.14) $t_0 = 4$ ns and, from Eq. (8.2.15), $v = \omega/2\pi \cong 2$ GHz. In this case we also have $t_0 \gg 1/\omega$, so the approximation $\omega' \cong \omega$ is again justified. Note also that, according to this calculation, the relaxation oscillation peak of the RIN spectrum of this laser is expected to be in the range of some GHz (see Fig. 7.31).

If the condition $t_0 > 1/\omega$ is not satisfied, the two solutions for p given by Eq. (8.2.10) are both real and negative. In this case the time behavior of $\delta\phi(t)$ consists of a superposition of two exponential decays. To have $t_0 < 1/\omega$, we must have, according to Eqs. (8.2.14) and (8.2.15):

$$\left(\frac{\tau_c}{\tau}\right) > \frac{4(x - 1)}{x^2} \tag{8.2.16}$$

The right-hand side of Eq. (8.2.16) has a maximum value of 1 when $x = 2$. This means that, if $\tau_c > \tau$, Eq. (8.2.16) is satisfied for any value of x. This situation usually occurs in gas lasers, which, therefore, generally do not exhibit spiking behavior.

Example 8.2. *Transient behavior of a He-Ne laser.* Consider an He-Ne laser oscillating on its red transition ($\lambda = 632.8$ nm). In this case $\tau = 50$ ns. Assuming a cavity length of $L = 50$ cm, an output coupling of 1%, and neglecting all other losses, we obtain $\gamma = \gamma_2/2 = 5 \times 10^{-3}$ and $\tau_c = L/c\gamma = 322$ ns. Thus $\tau_c > \tau$, and Eq. (8.2.16) is satisfied for any value of x. From Eqs. (8.2.14) and (8.2.15) with $x = 1.5$, we obtain $t_0 = 66.6$ ns and $\omega \cong 5.6 \times 10^6$ Hz. From Eq. (8.2.10) we then see that the two lifetimes describing the decays are 1 μs and 33.3 ns.

Before concluding this section we observe that the linearized analysis just considered also applies to a slightly different case, i.e., when we must test the stability of a given steady-state solution by a *linear stability analysis*. We assume in this case that the laser is already operating in the steady state and that a small step perturbation is applied (i.e., $\delta N = \delta N_0$ and $\delta\phi = \delta\phi_0$ at $t = 0$, where δN_0 and $\delta\phi_0$ are two given quantities). According to the preceding discussion, the perturbation introduced at time $t = 0$ decays with time by either a damped sinusoidal oscillation or a bi-exponential law. Steady-state solutions N_0 and ϕ_0, discussed in previous Chap. 7, therefore correspond to a stable equilibrium.

8.3. DYNAMIC INSTABILITIES AND PULSATIONS IN LASERS

The simple results obtained in the previous section appear to conflict with many experimental results observed from the earliest days of lasers, which indicated that many lasers, even when operating cw, tend to exhibit a continuous pulsating behavior, sometimes irregular and sometimes regular in character. A classical example of this type is shown in Fig. 8.2, which indicates pulsations observed in the first cw-excited ruby laser.[3] The output consists of a train of pulses irregularly spaced in time and of random amplitude (*irregular spiking*). Furthermore these pulsations do not tend to a steady-state value, as in Fig. 8.1. This kind of unstable behavior has been the subject of more than 25 years of theoretical and experimental investigations, revealing that this behavior can be attributed to a variety of reasons briefly summarized in the following paragraphs.[4]

In single-mode lasers, one of the main causes of instability arises from external and usually accidental modulation of such laser parameters as pump rate or cavity loss. For random modulation, this simply leads to laser intensity noise, discussed in Sect. 7.11. For sinusoidal modulation, the time behavior can be described in terms of rate equations by writing, e.g., for pump modulation, $R_p = R_{p0} + \delta R_p \exp(j\omega t)$, with $\delta R_p \ll R_{p0}$. According to Eqs. (8.2.1) and (8.2.2), we can then write $N(t) = N_0 + \delta N_0 \exp(j\omega t)$ and $\phi(t) = \phi_0 + \delta\phi_0 \exp(j\omega t)$, with $\delta N_0 \ll N_0$ and $\delta\phi_0 \ll \phi_0$, then solve the corresponding linearized equations. From the previous section, a single-mode laser is then seen to present a natural resonance at its relaxation oscillation frequency, ω_R, given e.g., by Eq. (8.2.11) for a four-level laser. A sinusoidal pump modulation thus forces the laser to exhibit small oscillations at the modulation frequency ω; the oscillation amplitude is a maximum when ω coincides with ω_R. For a white spectrum of pump modulation, an intensity noise spectrum peaking at ω_R is therefore observed (see Fig. 7.30a). Besides this instability of technical origin, single-mode lasers, under special circumstances, can also show a natural dynamic instability leading to pulsations and even chaotic behavior. For instance, for a homogeneously broadened transition, the laser must be driven sufficiently far above threshold (typically more than 10 times) and the cavity linewidth $\Delta\nu_c$ must be sufficiently *larger* than the transition linewidth $\Delta\nu_0$ (the so-called *bad-cavity case*). Conditions of this sort are experimentally realized in, specially prepared, optically pumped far-infrared lasers. Dynamic instabilities of this type can be accounted for only by a semiclassical treatment of laser behavior, i.e., by means of the Maxwell–Bloch equations.[5]

In multimode lasers, a new type of instability may easily set in due to, e.g., a switching in time between one mode and another or from one set of modes to another set, induced by pump modulation.[6] This instability leads to a kind of antiphase motion (antiphase dynamics) among modes, and it is adequately described, for a homogeneous line, by a rate equation treatment that takes into account the cross saturation effect due to spatial hole burning.[7]

FIG. 8.2. Typical time behavior of early cw-pumped solid-state lasers. Time scale is 50 μs/div. (By permission from Ref. 3.)

To summarize, we can say that single-mode lasers do not usually exhibit dynamic instability but instead show some intensity noise due to unavoidable perturbations of laser parameters. On the other hand, multimode lasers may also be subjected to additional instabilities due to a kind of antiphase motion among oscillating modes. Depending on the amplitude of modulation of the laser parameters, the type of laser, and whether the line is homogeneously or inhomogeneously broadened, this instability may lead to either a mode partition noise (see Sect. 7.11) or even to strong laser pulsations.

8.4. Q-SWITCHING

In Chap. 7 we saw that, under cw operation, the population inversion is fixed at its threshold value when oscillation starts. Even under the pulsed operating conditions considered in Sect. 8.2, the population inversion is seen to exceed the threshold value by only a relatively small amount (see Fig. 8.1), due to the onset of stimulated emission. Suppose now that a shutter is introduced into the laser cavity. If the shutter is closed, laser action is prevented, so the value of the population inversion may far exceed the threshold population holding when the shutter is absent. If the shutter is now opened suddenly, the laser will exhibit a gain that greatly exceeds losses; stored energy may then be released in the form of a short and intense light pulse.[8] Since this operation involves switching the cavity Q-factor from a low to a high value, the technique is usually called *Q-switching*. This technique allows one to generate laser pulses with a duration comparable to the photon decay time (i.e., from a few nanoseconds to a few tens of nanoseconds) and high-peak power (in the megawatt range).

8.4.1. Dynamics of the Q-Switching Process

To describe the Q-switching dynamic behavior, we assume that a step pump pulse is applied to the laser starting at time $t = 0$, i.e., $R_p(t) = 0$ for $t < 0$ and $R_p(t) = R_p = $ constant for $0 < t < t_p$; meanwhile the shutter is closed (Fig. 8.3a). For $0 < t < t_p$, the time behavior of the population inversion can be calculated from Eq. (7.2.16a), for a four-level laser, or

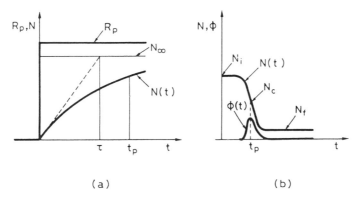

(a)   (b)

FIG. 8.3. Sequence of events in a Q-switched laser: (a) Idealized time behavior of the pump rate, R_p, and of the population inversion, N, before Q-switching. (b) Time behavior of population inversion, N, and photon number, ϕ, after Q-switching (fast switching case.)

from Eq. (7.2.24a), for a quasi-three-level laser, with ϕ set to zero. For instance, for a four-level laser, we obtain

$$N(t) = N_\infty[1 - \exp(-t/\tau)] \tag{8.4.1}$$

where the asymptotic value N_∞ is given by:

$$N_\infty = R_p\tau \tag{8.4.2}$$

as we readily obtain from Eq. (7.2.16a) by setting $dN/dt = 0$. The time behavior of $N(t)$ is also shown in Fig. 8.3a. From Eq. (8.4.1) and Fig. 8.3a we see that the duration t_p of the pump pulse ideally should be comparable to, or shorter than, the upper state lifetime τ. In fact, for $t_p \gg \tau$, $N(t)$ does not undergo any further appreciable increase and the pump power, rather than being accumulated as inversion energy, would be wasted through spontaneous decay. From Eq. (8.4.2) we also see that, to achieve a sufficiently large inversion, a long lifetime τ is required. Thus Q-switching can be used, effectively, with electric-dipole-forbidden laser transitions, where τ generally falls in the millisecond range. This is the case of most solid-state lasers (e.g., Nd, Yb, Er, Ho in different host materials; Cr-doped materials, such as alexandrite, Cr:LiSAF, and ruby) and some gas lasers (e.g., CO_2 or iodine). On the other hand, for semiconductor lasers, dye lasers, and a number of important gas lasers (e.g., He-Ne, Ar, Excimers), the laser transition is electric-dipole allowed and the lifetime is on the order of a few to a few tens of nanoseconds. In this case, with the usual values for pump rates R_p available, the achievable inversion N_∞ is too low to be of interest for Q-switching.

We now assume that the shutter is suddenly opened at time $t = t_p$, so that cavity loss $\gamma(t)$ is switched from a very high value, corresponding to a closed shutter, to the value γ for the same cavity with the shutter open (*fast switching*). We now take the time origin at the instant when switching occurs (Fig. 8.3b). The time behavior of the population inversion $N(t)$ and the number of photons $\phi(t)$ can then be obtained from the rate equations with the simplifying assumption that, during the short time of the Q-switching process, the effect of the decay term N/τ can be neglected. The qualitative behavior of $N(t)$ and $\phi(t)$ can then be depicted as in Fig. 8.3b. The population inversion starts from the initial value N_i, which can be obtained from Eq. (8.4.1) for $t = t_P$, then remains constant for some time, and finally begins to be depleted when the cavity photon number reaches a sufficiently high value. When $N(t)$ eventually falls to the threshold inversion N_c, the photon number reaches its peak value, as discussed earlier for the case of relaxation oscillations in Sect. 8.2. From this time on, the laser exhibits net loss rather than net gain; as a consequence, the photon number decreases to zero. During the same time the population inversion decreases to a final value N_f, which is left in the active medium; its value is established by the dynamics of the Q-switching process (see Sect. 8.4.4). Note that the time scales in Figs. 8.3a–b are very different; in fact, the time scale in Fig. 8.3a is established by the value of the upper state lifetime and thus corresponds to the microsecond range (usually, 100 μs–1 ms). The time scale in Fig. 8.3b, on the other hand, turns out to be of the order of the cavity photon decay time (see Sect. 8.4.4) and hence falls in the nanosecond range (usually, 5–50 ns).

So far we have considered the dynamic behavior corresponding to fast switching, where the switching of cavity loss is treated as instantaneous. In practice, fast switching requires

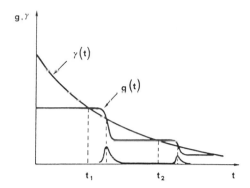

FIG. 8.4. Sequence of events in the slow switching case demonstrating the occurrence of multiple pulses. In the figure, $g(t) = \sigma N(t)l$, where l is the length of the active medium, represents the laser gain.

their peak value (several photon decay times, i.e., typically from a few tens to a few hundreds of a nanosecond). In the case of *slow switching*, the dynamic behavior is somewhat more complicated and multiple pulses may result. Figure 8.4 depicts this behavior, where cavity loss $\gamma(t)$ is assumed to decay from its initially high value to a final value in a relatively long time. The same figure also shows the corresponding time behavior of the single-pass gain $g(t) = \sigma Nl$ and the cavity photon number $\phi(t)$. The first pulse starts at time t_1 when the decreasing loss $\gamma(t)$ equals the instantaneous gain $g(t)$. The pulse then reaches its peak value when the gain, owing to its decrease due to saturation, equals the loss. After this first pulse, the gain is driven below the loss, so further oscillation cannot occur until the switch opens further, thus decreasing the loss below the gain. A second pulse can then be produced (occurring at time t_2 in the figure) whose peak again occurs when gain saturation makes the gain equal to loss.

8.4.2. Q-Switching Methods

Several methods have been developed to achieve switching of the cavity Q; in this section we limit ourselves to discussing the most commonly used, namely:[9] (1) Electro-optical shutters. (2) Rotating prisms. (3) Acoustooptical switches. (4) Saturable absorbers. These devices are generally grouped into two categories, *active* and *passive Q-switches*. In an active Q-switching device, some external active operation must be applied to this device (e.g., a change of the voltage applied to the electrooptical shutter) to produce Q-switching. In a passive Q-switch, the switching operation is automatically produced by the optical nonlinearity of the element used (e.g., a saturable absorber).

8.4.2.1. Electrooptical Q-Switching

These devices use a cell exploiting an electrooptical effect, usually the Pockels effect, to induce Q-switching. A cell based on the Pockels effect (*Pockels cell*) consists of a suitable nonlinear crystal, such as KD*P or lithium niobate for the visible-to-near-infrared region or cadmium telluride for the middle-infrared, in which an applied dc voltage induces a change in the crystal's refractive indices. This induced birefringence turns out to be proportional to the applied voltage. Figure 8.5a shows a Q-switched laser using a suitable combination of a polarizer and a Pockels cell. The Pockels cell is oriented and biased so that the x- and y-axes of the induced birefringence lie in the plane orthogonal to the resonator axis. The polarizer axis then makes a 45° angle to the birefringence axes.

Consider now a laser beam propagating from the active medium toward the polarizer–Pockels cell combination with polarization parallel to the polarizer axis. Ideally this beam is totally transmitted by the polarizer and then incident on the Pockels cell. The E-field of the incoming wave is at 45° to the birefringence x- and y-axes of the Pockels cell and it can be resolved into components E_x and E_y (Fig. 8.5b) with their oscillations in phase. After passing through the Pockels cell, these two components have experienced different phase shifts, giving rise to the phase difference:

$$\Delta\varphi = k\Delta n L' \tag{8.4.3}$$

where $k = 2\pi/\lambda$, $\Delta n = n_x - n_y$ is the value of the induced birefringence, and L' is the crystal length. If the voltage applied to the Pockels cell is such that $\Delta\varphi = \pi/2$, then the two field components leaving the Pockels cell differ in phase by $\pi/2$. This means that when E_x is maximum, E_y is zero and vice versa; i.e., the wave is circularly polarized (Fig. 8.5c). After reflection at the mirror, the wave passes once more through the Pockels cell, so its x- and y-components acquire an additional $\Delta\varphi = \pi/2$ phase difference. The total phase difference then becomes π, so that, when, e.g., E_x is at its maximum (positive) value, E_y is at its maximum (negative) value, as shown in Fig. 8.5d. As a result, the overall field **E** is again linearly polarized but with a polarization axis at 90° to that of the original wave in Fig. 8.5b. This beam is therefore not transmitted by the polarizer, but instead it is reflected out of the cavity (see Fig. 8.5a). This condition corresponds to the Q-switch being closed. The switch is then opened by removing the bias voltage to the Pockels cell. In this case the induced

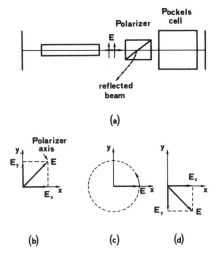

FIG. 8.5. (a) Possible polarizer-Pockels-cell combination for Q-switching. Figures (b), (c), and (d) show the E-field components along the birefringence axes of the Pockels cell in a plane orthogonal to the resonator axis.

is then opened by removing the bias voltage to the Pockels cell. In this case the induced birefringence disappears, so that the incoming light is transmitted without a change of polarization. Note that the required voltage for operation in this arrangement is called the $\lambda/4$ voltage or the *quarter-wave voltage*; in fact, the quantity $\Delta nL'$, i.e., the difference in optical path lengths for the two polarizations, is $\lambda/4$, as we can see from Eq. (8.4.3).

Pockels cell Q-switches are widely used. Depending on the particular nonlinear crystal used in the cell, the particular arrangement of the applied field, the crystal dimensions, and the value of the wavelength involved, the $\lambda/4$ voltage may range from 1 to 5 kV. This voltage must then be switched off in a time, t_s, less than the build-up time of the Q-switched pulse (typically $t_s < 20$ ns).

8.4.2.2. Rotating Prisms

The most common mechanical means of Q-switching involves rotating one of the end mirrors of the laser resonator about an axis perpendicular to the resonator axis. In this case the high Q-condition is reached when the rotating mirror passes through a position parallel to the other cavity mirror. To simplify alignment requirements, a 90° roof-top prism with the roof edge perpendicular to the rotation axis is often used instead of an ordinary mirror (Fig. 8.6). Such a prism has the property that, for light propagating orthogonal to the roof edge (see Fig. 8.6), the reflected beam is always parallel to the incident beam regardless of rotation of the prism about its roof edge. This ensures that alignment between the prism and the other cavity mirror is always achieved in the plane orthogonal to the roof. The effect of rotation is then to bring the prism into alignment in the other direction.

Rotating-prism Q-switches are simple and inexpensive devices that can be used at any wavelength; they are rather noisy however, and, due to the limited speed of the rotating motor, they generally result in slow Q-switching. For a typical multitransverse-mode solid-state laser, for instance, the beam divergence is around a few mrad. The high-Q situation then corresponds to an angular range of ~ 1 mrad around the perfect alignment condition. Thus, even for a motor rotating at the rapid speed of 24,000 rpm (400 Hz), the duration of the high Q-switching condition is about 400 ns. This slow switching time sometimes results in the production of multiple pulses.

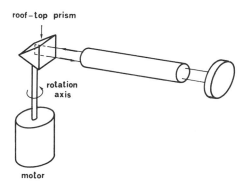

FIG. 8.6. Mechanical Q-switching system using a rotating 90° roof-top prism.

8.4.2.3. Acoustooptic Q-Switches

An acoustooptic modulator consists of a block of transparent optical material (e.g., fused quartz in the visible-to-near infrared and germanium or cadmium selenide in the middle-far infrared) in which an ultrasonic wave is launched by a piezoelectric transducer bonded to one side of the block and driven by a rf oscillator (Fig. 8.7a). The side of the block opposite to the transducer side is cut at an angle, and has an absorber for the acoustic wave placed on its surface (see Fig. 8.7b). With back reflection of the acoustic wave thus suppressed, only a traveling acoustic wave is present in the medium. The strain induced by the ultrasonic wave results in local changes in the material refractive index due to the photoelastic effect. This periodic change in the refractive index acts as a phase grating with a period equal to the acoustic wavelength, amplitude proportional to the sound amplitude, and which is traveling at the sound velocity in the medium (traveling-wave phase grating). Its effect is to diffract a fraction of the incident beam from the incident beam direction.[10] Thus, if an acoustooptic cell is inserted in a laser cavity (Fig. 8.7b), an additional loss occurs due to beam diffraction while the driving voltage to the transducer is applied. If the driving voltage is high enough, this additional loss is sufficient to prevent the laser from oscillating. The laser is then returned to its high-Q condition by switching off the transducer voltage.

To gain a more detailed understanding of the operation of an acoustooptic modulator, we now consider the case where the length L' of the optical medium is sufficiently large so that the grating acts as a thick phase grating. This occurs when the following condition is satisfied

$$\frac{2\pi\lambda L'}{n\lambda_a^2} \gg 1 \tag{8.4.4}$$

where λ is the wavelength of the incident beam, n is the material refractive index, and λ_a is the wavelength of the acoustic wave. In typical cases, Eq. (8.4.4) requires L' to be larger than 1 cm. In this case, known as the Bragg regime, a single beam is diffracted from the cavity at an angle $\theta' = \lambda/\lambda_a$. Note that this angle equals the divergence angle of a beam of wavelength λ diffracted from an aperture of size λ_a. Maximum diffraction efficiency is achieved when the angle of the incident light θ_B satisfies the condition $\theta_B = \lambda/2\lambda_a$ (Fig. 8.7a), originally derived by Bragg for x-ray diffraction from crystallographic planes. In such a case, the diffracted beam can be considered to arise from specular reflection of the incident

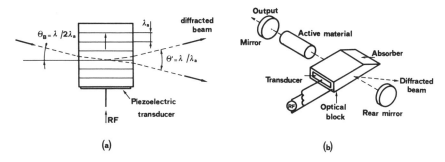

FIG. 8.7. (a) Incident, transmitted, and diffracted beams in an acoustooptic modulator (Bragg regime). (b) *Q*-switched laser arrangement incorporating an acoustooptic modulator.

beam at the phase planes produced by the acoustic wave. For sufficiently high values of the rf drive power to the piezoelectric transducer, a relatively large fraction η of the incident beam can be diffracted from the cavity. (Typical diffraction efficiencies are about 1–2% per watt of rf power.) Note that Eq. (8.4.4) can be approximated as $(\lambda L'/n\lambda_a) \gg \lambda_a$. This can be interpreted by saying that a wavelet diffracted by a given aperture λ_a, at the crystal entrance, spreads out, at the crystal exit, by an amount $\lambda L'/n\lambda_a$ that must be much larger than each aperture λ_a. Under this condition, in fact, each wavelet diffracted at the crystal entrance is summed, before exiting the crystal, with wavelets produced by other apertures λ_a of the crystal; this results in volume diffraction.*

> **Example 8.3.** *Condition for the Bragg regime in a quartz acoustooptic modulator.* We consider an acoustooptic cell driven at a frequency of $v_a = 50$ MHz and take $v = 3.76 \times 10^5$ cm/s as the shear wave velocity in quartz. The acoustic wavelength is then $\lambda_a = v/v_a = 75$ μm. Taking $n = 1.45$ for the refractive index of quartz at $\lambda = 1.06$ μm, from Eq. (8.4.4) we obtain $L' \gg 1.3$ mm. Thus, for a crystal length of about 5 cm, the condition for the Bragg diffraction regime is amply satisfied. Note that, in this example, the beam is diffracted at an angle $\theta' = \lambda/\lambda_a \cong 0.8°$ to the incident beam direction, so the angle of incidence at the modulator must be $\theta_B = \lambda/2\lambda_a \cong 0.4°$.

Acoustooptic modulators have the advantage of low optical insertion losses and, for repetitive Q-switching, they can readily be driven at high-repetition rates (kHz). The loss introduced in the low-Q situation is rather limited, however, and Q-switching time is rather long (primarily established by the time taken for an acoustic wavefront to traverse the laser beam). These modulators are therefore used, primarily, for repetitive Q switching of low-gain cw-pumped lasers (e.g., Nd:YAG or Ar-ion lasers).

8.4.2.4. Saturable Absorber Q-Switch

The three Q-switching devices considered so far are active Q-switches, since they must be driven by an appropriate driving source (Pockels cell voltage power supply, rotating mirror, or rf oscillator). We now consider a case of passive Q-switching exploiting the nonlinearity of a saturable absorber; this is by far the most common passive Q switch used so far.

A saturable absorber consists of a material that absorbs at the laser wavelength and has a low value of saturation intensity. It often assumes the form of a cell containing a solution of a saturable dye in an appropriate solvent (e.g., the dye known as BDN, bis 4-dimethyl-aminodithiobenzil-nickel, dissolved in 1,2-dichloroethane in the case of Nd:YAG). Solid-state (e.g., BDN in a cellulose acetate, F_2:LiF, or Cr^{4+}:YAG, again for a Nd:YAG laser) or gaseous saturable absorbers (e.g., SF_6 for CO_2 lasers) are also used. To a first approximation, a saturable absorber can be treated as a two-level system with a very large peak cross section (10^{-16} cm^2 is typical for a saturable dye). From Eq. (2.8.11) it then follows that the corresponding saturation intensity I_s is comparatively small (1–10 MW/cm^2), so the absorber becomes almost transparent, due to saturation, for a comparatively low-incident-light intensity.

---

* When $(2\pi\lambda L') \ll n\lambda_a^2$, the acoustic grating behaves like a thin phase grating, and the cell is said to be operating in the Raman–Nath regime. This regime is seldom used for acoustooptic Q-switching due to the higher requirement for rf power per unit volume of the cell.

To understand the dynamic behavior of a saturable absorber Q-switch, let us assume that a cell containing this absorber, with a peak absorption wavelength coincident with the laser wavelength, is introduced into the laser cavity. As a typical case, assume that the initial, i.e., unsaturated absorption of the cell is 50%. Then laser action begins only when the gain of the active medium compensates for the loss of the saturable absorber plus the unsaturable cavity losses. Due to the large value of the cell absorption, the required critical population inversion is thus very high. When laser action eventually starts, beam intensity inside the laser cavity, $I(t)$, builds up from the starting noise, I_n, arising from spontaneous emission. To appreciate the full time evolution of $I(t)$, Fig. 8.8a shows a logarithmic plot of $I(t)/I_n$ versus time in a typical situation (see Example 8.5). When laser intensity equals I_s, which occurs at time $t = t_s$ in the figure, the absorber begins to bleach due to saturation. The growth rate of laser intensity is thus increased; this in turn results in an increased rate of absorber bleaching, and so on. The overall result is a much more rapid bleaching of the saturable absorber. Since I_s is comparatively small, the inversion left in the laser medium, after bleaching of the absorber, is essentially the same as the initial inversion, i.e., very large. After the absorber is bleached, the laser will thus have a gain in excess of losses, and a giant pulse is produced. The dynamic behavior, during this last phase, is illustrated in more detail in Fig. 8.8b, where the quantity $\phi(t)/\phi_i = I(t)/I_n$ [where $\phi(t)$ is the cavity photon-number, and $\phi_i \cong 1$ is the initial value due to spontaneous emission] is plotted against time, in a linear scale and in a more restricted time interval. The time behavior of the total inversion NV_a, where V_a is the volume of the mode in the active medium, is also shown in the same figure (see Example 8.5). As in any other Q-switching case, the photon number now increases rapidly until saturation of the inversion and hence of the gain sets in. The pulse thus reaches a maximum value when the inversion equals the critical inversion N_c of the laser without saturable absorber and the pulse thereafter decreases.

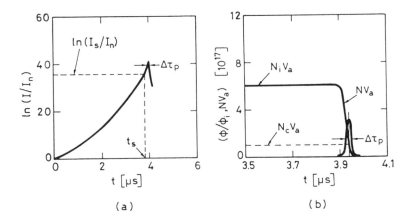

(a)   (b)

FIG. 8.8. Typical time behavior for the laser beam intensity I and cavity photon number ϕ in a 50-cm long Nd:YAG laser which is passively Q-switched by a saturable absorber: (a) Logarithmic plot of I/I_n, where I_n is the noise intensity due to spontaneous emission. It provides the most convenient description of dynamic behavior before saturation of the saturable absorber. (b) Linear plot of $\phi/\phi_i = I/I_n$, where $\phi_i = 1$ is the initial number of photons due to spontaneous emission. It provides the most convenient description of time evolution around the peak of the pulse.

An important feature to note from Fig. 8.8 is that the time taken for the pulse to increase from the noise level to the peak value is very long ($t_p \cong 3.94$ μs). This is essentially due to the fact that, during the unbleached phase of the saturable absorber, i.e., for $t < t_s$, laser gain barely exceeds the high threshold value established by the presence of the, as yet unsaturated, absorber. The growth of laser intensity is thus very slow and a large number of passes is required for the beam to reach its peak value (~ 2370 in Fig. 8.8). This results in a natural selection of cavity modes.[11] Suppose in fact that two modes have single-pass unsaturated gains g_1 and g_2 ($g = \sigma Nl$) and single-pass losses γ_1 and γ_2. Since the two modes start from the same intensity, as established by spontaneous emission, the ratio of the two intensities at time $t = t_s$, i.e., before saturable-absorber saturation, is given by

$$\frac{I_1}{I_2} = \left[\frac{e^{(g_1 - \gamma_1)}}{e^{(g_2 - \gamma_2)}} \right]^n \tag{8.4.5}$$

where n is the number of round trips within the cavity up to a time $t = t_s$ ($n \cong 2310$ in Fig. 8.8). If we now let $\delta = (g_1 - \gamma_1) - (g_2 - \gamma_2)$ be the difference between the two net gains, from Eq. (8.4.5) we can write $(I_1/I_2) = \exp n\delta$. We thus see that, even assuming the very modest value of 0.001 for δ, for $n = 2310$ we obtain $(I_1/I_2) = \exp 2.3 \cong 10$. Thus, even a very modest discrimination of either gain or loss between the two modes results in a large discrimination between their intensities at time $t = t_s$, and hence also at the pulse peak, which occurs shortly afterward (~ 100 ns in the example). As a result, single-mode operation can be achieved rather easily in the case of a saturable absorber Q-switch. Note that, in the case of active Q-switching, this mode selection mechanism is much less effective since laser build-up from noise is much faster and the total number of transits may now be on the order of only 10 or 20.*

Passive Q-switching by saturable absorbers provides the simplest method of Q-switching. Photochemical degradation of the absorber, particularly for dye saturable absorbers, was the main drawback to this type of Q-switch. This situation is now changing with the advent of solid-state absorbers that do not degrade.

8.4.3. Operating Regimes

Q-switched lasers can operate in either one of the following ways: (1) *Pulsed operation* (Fig. 8.9), where the pump rate $R_p(t)$ generally takes the form of a pulse of duration comparable to the upper state lifetime τ (Fig. 8.9a). Without Q-switching the population inversion $N(t)$ would reach a maximum value and decrease thereafter. The cavity Q is switched when $N(t)$ reaches its maximum value ($t = 0$ in Fig. 8.9b). Then, for $t > 0$, the number of photons begins to grow, leading to a pulse whose peak occurs at some time t_d after switching. As a result of the growth of the photon number, the population inversion $N(t)$ decreases from its initial value N_i (at $t = 0$) to the final value N_f left after the pulse is over. Note that, according to a comment made in connection with Fig. 8.3, relevant time scales for $t < 0$ and $t > 0$ are completely different. In fact the time scale for events when $t < 0$

---

* Note: We can operate an active Q-switch in an fashion analogous to the saturable absorber by setting the low-Q (high-loss) condition to a value that permits lasing to start (known as prelasing) before eventually switching to high Q after a long prelase has allowed mode selection to occur.[38]

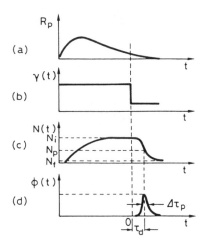

FIG. 8.9. Development of a Q-switched laser pulse in pulsed operation. The figure shows the time behavior of: (a) the pump rate R_p; (b) the resonator loss γ; (c) the population inversion N; (d) the number of photons ϕ.

falls in the millisecond range, while the time scale for events when $t > 0$ falls in the nanosecond range. Q-switched lasers with a pulsed pump can obviously be operated repetitively; the typical repetition rates range from a few to a few tens of Hz. (2) *Continuously pumped, repetitively Q-switched operation* (Fig. 8.10), where a cw pump R_p is applied to the laser (Fig. 8.10a) and cavity losses are periodically switched from a high to a low value (Fig. 8.10b). Laser output then consists of a continuous train of Q-switched pulses (Fig. 8.10c). During each pulse, inversion falls from its initial value N_i (before Q-switching) to a final value N_f (after the Q-switched pulse; see Fig. 8.10d). The population inversion is then restored to its initial value N_i by the pumping process before the next Q-switching event. Since the time taken to restore the inversion is roughly equal to the upper-state

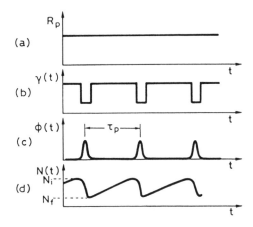

FIG. 8.10. Development of Q-switched laser pulses in a repetitively Q-switched, cw pumped laser. The figure shows the time behavior of: (a) the pump rate R_p; (b) the resonator losses γ; (c) the number of photons ϕ; (d) the population inversion N.

lifetime τ, the time τ_p between two consecutive pulses must equal or be shorter than τ. In fact, if τ_p were much longer than τ, most of the available inversion would be lost by spontaneous decay. Therefore, repetition rates of cw-pumped Q-switched lasers typically range from a few kilohertz to a few tens of kilohertz.

Electrooptical and mechanical shutters as well as saturable absorbers are commonly used for pulsed operation. For repetitively Q switching of continuously-pumped lasers (which have lower gain than pulsed lasers), acoustooptic Q-switches and, sometimes, mechanical shutters are commonly used. With cw lasers, a saturable absorber within the cavity can, under appropriate conditions, lead to a repetitive Q-switching operation. In this case, the repetition rate of the Q-switched pulses is established by the nonlinear dynamics of the absorber rather than by an external control.

8.4.4. Theory of Active Q-Switching

For the sake of simplicity we consider only the case of active Q-switching and we further assume that the switching is instantaneous (fast switching case).[12] The dynamic behavior of the laser can again be obtained from Eqs. (7.2.16) and (7.2.24) for four- and quasi-three-level lasers, respectively.

We first consider a four-level laser operating in a pulsed regime (Fig. 8.9) and assume that, for $t < 0$, losses are large enough for the laser to be below threshold. If Q-switching is performed when $N(t)$ attains its maximum value, the corresponding initial inversion can be obtained from Eq. (7.2.16a) by setting $(dN/dt) = 0$; thus one gets

$$N_i = \tau R_p(0) \tag{8.4.6}$$

where $R_p(0)$ is the pump rate value when Q-switching occurs (i.e., at $t = 0$). We now assume that the time behavior of $R_p(t)$ is always the same whatever the value of $\int R_p \, dt$, i.e., of the pump energy. We then set $R_p(0) \propto \int R_p \, dt$, so that, for example, if $\int R_p \, dt$ is doubled, then $R_p(0)$ is also doubled. Thus, if we let E_p be the pump energy corresponding to the given pump rate, since $E_p \propto \int R_p \, dt$, we then have $E_p \propto R_p(0)$ and, according to Eq. (8.4.6), $E_p \propto N_i$. Therefore, if we let N_{ic} and E_{pc} be the initial inversion and the corresponding pump energy, respectively, when the laser is operated just at threshold, we can write

$$\left(\frac{N_i}{N_{ic}}\right) = \left(\frac{E_p}{E_{pc}}\right) = x \tag{8.4.7}$$

where $x = (E_p/E_{pc})$ is the amount by which threshold is exceeded. Since N_{ic} is the critical inversion for normal laser action (i.e., when the Q-switching element is open), its value is obtained by the usual relation $N_{ic} = N_c = \gamma/\sigma l$, where γ is the cavity loss with the Q-switching element open. If N_{ic} is known, i.e., if γ, σ, and l are known and the ratio x between actual pump energy and threshold pump energy is also known, then Eq. (8.4.7) allows the initial inversion N_i to be calculated.

Once N_i is known, the time evolution of the system, after Q-switching i.e. for $t > 0$, can be obtained from Eq. (7.2.16) with the initial conditions $N(0) = N_i$ and $\phi(0) = \phi_i$. Here again ϕ_i is some small number of photons needed for laser action to begin ($\phi_i \cong 1$). The equations can now be considerably simplified since we expect the evolution of both $N(t)$ and $\phi(t)$ to occur on a time scale so short that the pump term R_p and the spontaneous decay term N/τ in Eq. (7.2.16a) can be neglected. Equations (7.2.16) then reduces to:

$$\frac{dN}{dt} = -B\phi N \tag{8.4.8a}$$

$$\frac{d\phi}{dt} = \left[V_a BN - \left(\frac{1}{\tau_c} \right) \right] \phi \tag{8.4.8b}$$

Note that, according to Eq. (8.4.8b), the population N_p corresponding to the peak of the photon pulse (see Fig. 8.9c), i.e., when $(d\phi/dt) = 0$, is

$$N_p = \frac{1}{V_a B \tau_c} = \frac{\gamma}{\sigma l} \tag{8.4.9}$$

which is the same as the critical inversion N_c. This result, with the help of Eq. (8.4.7), allows us to express the ratio N_i/N_p in a useful form for the discussion that follows, viz.:

$$\left(\frac{N_i}{N_p} \right) = x \tag{8.4.10}$$

After these preliminary considerations, we can proceed to calculate the peak power of the laser output pulse P_p through, e.g., mirror 2. According to Eq. (7.2.18) we write

$$P_p = \left(\frac{\gamma_2 c}{2L_e} \right) h\nu \phi_p \tag{8.4.11}$$

where ϕ_p is the number of photons in the cavity at the peak of the laser pulse. To calculate ϕ_p we take the ratio of Eqs. (8.4.8a–b). Using Eq. (8.4.9) we obtain

$$\frac{d\phi}{dN} = -V_a \left[1 - \left(\frac{N_p}{N} \right) \right] \tag{8.4.12}$$

which can be readily integrated to give

$$\phi = V_a [N_i - N - N_p \ln(N_i/N)] \tag{8.4.13}$$

where, for simplicity, the small number ϕ_i is neglected. At the peak of the pulse, we then obtain

$$\phi_p = V_a N_p \left[\frac{N_i}{N_p} - \ln \left(\frac{N_i}{N_p} \right) - 1 \right] \tag{8.4.14}$$

which readily gives ϕ_p once N_p and the ratio (N_i/N_p) are known through Eqs. (8.4.9) and (8.4.10), respectively. The peak power is then obtained from Eqs. (8.4.9), (8.4.11), and (8.4.14) as

$$P_p = \frac{\gamma_2}{2}\left(\frac{A_b}{\sigma}\right)\left(\frac{h\nu}{\tau_c}\right)\left[\frac{N_i}{N_p} - \ln\left(\frac{N_i}{N_p}\right) - 1\right] \qquad (8.4.15)$$

where $A_b = V_a/l$ is the beam area and the expression for τ_c, given by Eq. (7.2.14), has also been used.

To calculate the output energy E we observe that we can write

$$E = \int_0^\infty P(t)dt = \left(\frac{\gamma_2 c}{2L_e}\right)h\nu \int_0^\infty \phi\, dt \qquad (8.4.16)$$

where $P(t)$ is the time behavior of the output power and Eq. (7.2.18) has again been used. The integration of Eq. (8.4.16) is easily carried out by integrating both sides of Eq. (8.4.8b) and noting that $\phi(0) = \phi(\infty) \cong 0$. We then get $\int_0^\infty \phi\, dt = V_a\tau_c \int_0^\infty B\phi N\, dt$. The quantity $\int_0^\infty B\phi N\, dt$ is then obtained by integrating both sides of Eq. (8.4.8a) to give $\int_0^\infty B\phi N\, dt = (N_i - N_f)$ where N_f is the final inversion (see Fig. 8.3b). Thus $\int_0^\infty \phi\, dt = V_a\tau_c(N_i - N_f)$ so that Eq. (8.4.16) becomes

$$E = \left(\frac{\gamma_2}{2\gamma}\right)(N_i - N_f)(V_a h\nu) \qquad (8.4.17)$$

Note that Eq. (8.4.17) is readily understood by observing that $(N_i - N_f)$ is the available inversion and this inversion produces a number of photons $(N_i - N_f)V_a$. Out of this number of photons emitted by the medium, only the fraction $(\gamma_2/2\gamma)$ is available as output energy.

To calculate E from Eq. (8.4.17) we need to know N_f, which can be obtained from Eq. (8.4.13) by letting $t \to \infty$. Since $\phi(\infty) \cong 0$, we get

$$\frac{N_i - N_f}{N_i} = \frac{N_p}{N_i}\ln\left(\frac{N_i}{N_f}\right) \qquad (8.4.18)$$

which gives N_f/N_i as a function of N_p/N_i. We can now define the quantity $\eta_E = (N_i - N_f)/N_i$ in Eq. (8.4.18) as the inversion- (or energy-) utilization factor. In fact, out of the initial inversion N_i, the inversion actually used is $(N_i - N_f)$. In terms of η_E, Eq. (8.4.18) can be recast in the form:

$$\eta_E\left(\frac{N_i}{N_p}\right) = -\ln(1 - \eta_E) \qquad (8.4.19)$$

Figure 8.11 plots the energy-utilization factor η_E against (N_i/N_p) as obtained from Eq. (8.4.19). Note that, for large values of (N_i/N_p), i.e., for pump energy far exceeding threshold

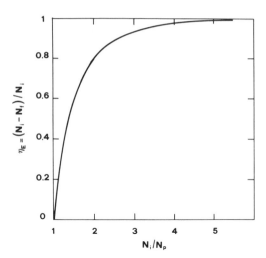

FIG. 8.11. Energy utilization factor η_E versus the ratio, N/N_p, between the initial inversion and the peak inversion.

pump energy, the energy-utilization factor tends to unity. Note also that, in terms of η_E, Eq. (8.4.17), with the help of Eq. (8.4.9), can be written in the simpler and more intuitive form

$$E = \left(\frac{\gamma_2}{2}\frac{N_i}{N_p}\right)\eta_E\left(\frac{A_b}{\sigma}\right)hv \qquad (8.4.20)$$

where $A_b = V_a/l$ is again the beam area.

Once output energy and peak power are known, we can obtain an approximate value, $\Delta\tau_p$, for the width of the output pulse by defining it as $\Delta\tau_p = E/P_p$. From Eqs. (8.4.20) and (8.4.15):

$$\Delta\tau_p = \tau_c \frac{(N_i/N_p)\eta_E}{[(N_i/N_p) - \ln(N_i/N_p) - 1]} \qquad (8.4.21)$$

Note that $\Delta\tau_p/\tau_c$ depends only on the value of $(N_i/N_p) = x$; for (N_i/N_p) ranging, e.g., from 2 to 10, $\Delta\tau_p$ ranges from ~ 5 to 1.5 times the photon decay time τ_c.

We can now calculate the time delay τ_d between the peak of the pulse and the time of Q-switching (see Fig. 8.9). This delay can be approximated by the time required for the photon number to reach some given fraction of its peak value. If for instance we choose this fraction to be $(1/10)$, no appreciable saturation of the inversion is expected to occur up to this point, so we can approximate $N(t) \cong N_i$ in Eq. (8.4.8b). With the help of Eqs. (8.4.9) and (8.4.10), Eq. (8.4.8b) is transformed into $(d\phi/dt) = (x - 1)\phi/\tau_c$, which upon integration gives

$$\phi = \phi_i \exp[(x - 1)t/\tau_c] \qquad (8.4.22)$$

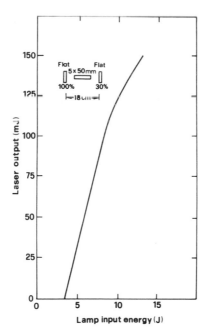

FIG. 8.12. Laser output energy versus input energy to the flashlamp for a Q-switched Nd:YAG laser, whose geometrical dimensions are shown in the inset. (By permission from Ref 13.)

The time delay τ_d is obtained from Eq. (8.4.22) by putting $\phi = \phi_p/10$. Setting $\phi_i = 1$, we obtain

$$\tau_d = \frac{\tau_c}{x - 1} \ln(\phi_p/10) \qquad (8.4.23)$$

where ϕ_p is given by Eq. (8.4.14). Note that, since ϕ_p is a very large number ($\approx 10^{17}$ or more; see Example 8.4) and it appears in the logarithmic term of Eq. (8.4.23), τ_d does not change much if we choose a different fraction of ϕ_p in this logarithmic term, e.g., ($\phi_p/20$).

Example 8.4. *Output energy, pulse duration, and pulse buildup time in a typical Q-switched Nd:YAG laser.* Figure 8.12 shows a typical plot of laser output energy E versus input energy E_p to the flash lamp for a Q-switched Nd:YAG laser. Rod and cavity dimensions are also indicated in the figure inset.[13] The laser is operated in a pulsed regime, and Q-switched by a KD*P (deuterated potassium dihydrogen phosphate, i.e., KD_2PO_4) Pockels cell. From the figure we observe that the laser has a threshold energy $E_{cp} \cong 3.4$ J, and it gives, e.g., an output energy $E \cong 120$ mJ for $E_p \cong 10$ J. At this value of pump energy, the laser pulse width is found experimentally to be ~6 ns.

We now compare these experimental results with those predicted from previous equations. We neglect mirror absorption and so put $\gamma_2 \cong -\ln R_2 = 1.2$ and $\gamma_1 \cong 0$. Internal losses of the polarizer-Pockels-cell combination are estimated to be $L_i \cong 15\%$, while, in comparison, internal losses of the rod can be neglected. We thus get $\gamma_i = -\ln(1 - L_i) \cong 0.162$ and $\gamma = [(\gamma_1 + \gamma_2)/2] + \gamma_i = 0.762$. The predicted value of laser energy at $E_p = 10$ J is obtained from Eq. (8.4.20) once we observe that, for our case, $(N_i/N_p) = (E_p/E_{cp}) = 2.9$. We now assume $A_b \cong A = 0.19$ cm^2, where A is the cross-sectional area of the rod. Since $(N_i/N_p) = 2.9$, we find from Fig. 8.11 that $\eta_E \cong 0.94$. From Eq. (8.4.20), assuming an effective value of the stimulated emission cross section of $\sigma = 2.8 \times 10^{-19}$ cm^2 (see Example 2.10), we obtain $E \cong 200$ mJ. The somewhat larger value predicted by the theory can be attributed to two factors: (1) The area of the beam is smaller than that of the rod. (2) Due to the short cavity length, the condition for fast switching, namely that switching time is much shorter than the buildup time of the laser pulse, may not be well-satisfied in our case. Later in this example we show, in fact, that the predicted buildup time for the Q-switched pulse τ_d is about 20 ns. It is difficult to switch the Pockels cell in a much shorter time than this; as a consequence some energy is lost through the polarizer during the switching process. (In some cases, with pulses of this short a duration, as much as 20% of the output energy is switched out of the cavity by the polarizer during the Q-switching process).

   To calculate the predicted pulse duration we observe that, according to Eq. (7.2.11), the effective resonator length is $L_e = L + (n - 1)l \cong 22$ cm, where $n \cong 1.83$ for Nd:YAG; thus from Eq. (7.2.14) we obtain $\tau_c = L_e/c\gamma \cong 1$ ns. The laser pulse width is obtained from Eq. (8.4.21) as $\Delta\tau_p = \tau_c\eta_E x/[x - \ln(x - 1)] \cong 3.3$ ns, where Fig. 8.11 is used to calculate η_E. The discrepancy between this value and the experimental value $\Delta\tau_p = 6$ ns is attributed to two factors: (1) Multimode oscillation. In fact the buildup time is expected to differ for different modes due to their slightly different gain, and this should appreciably broaden the pulse duration. (2) As already mentioned, the condition for fast switching may not be completely satisfied in our case, so pulse width is expected to be somewhat broadened by slow switching.

   The buildup time for the Q-switched pulse is obtained from Eq. (8.4.23) once ϕ_p is known. If we take $N_p = \gamma/\sigma l \cong 5.44 \times 10^{17}$ cm^{-3} and assume $V_a = A_b l \cong Al \cong 1$ cm^3, from Eq. (8.4.14) we obtain $\phi_p \cong 4.54 \times 10^{17}$ photons. Then from Eq. (8.4.23) with $\tau_c = 1$ ns and $x = 2.9$, we obtain $\tau_d \cong 20$ ns.

Example 8.5. *Dynamical behavior of a passively Q-switched Nd:YAG laser.* We consider a laser cavity with equivalent length $L_e = 50$ cm whose active medium is a Nd:YAG rod of diameter $D = 5$ mm. We assume that the laser is passively Q-switched by a saturable absorber with saturation intensity $I_s = 1$ MW/cm^2; we also assume that the cell containing the saturable absorber solution has an unsaturated loss of $L = 50\%$. We take output mirror reflectivity to be $R_2 = 74\%$, and we neglect all other cavity losses. Thus $\gamma_2 = -\ln R_2 \cong 0.3$ for the loss of the output-coupling mirror and $\gamma_a = -\ln(1 - L) = 0.693$ for the unsaturated loss of the saturable absorber. Total unsaturated loss will then be $\gamma_t = \gamma_a + (\gamma_2/2) = 0.843$. We now assume that the pump rate provides a square pulse lasting for time $t_p = 100$ μs (see Fig. 8.13). According to Eq. (8.4.1), the population inversion, at the end of the pumping pulse and in the absence of laser action, is given by

$$N(t_P) = N_\infty[1 - \exp(-t_P/\tau)] = 0.35N_\infty \qquad (8.4.24)$$

where we have taken $\tau = 230$ μs. Oscillation threshold is reached at time t_{th} (see Fig. 8.13) such that:

$$\sigma N(t_{th})l = \gamma_t \qquad (8.4.25)$$

   We now set the pumping rate to exceed the laser threshold by 10%, so that:

$$N(t_p) = 1.1 \, N(t_{th}) \qquad (8.4.26)$$

where $N(t_p)$ is the inversion present at $t = t_P$ in the absence of laser action (see Fig. 8.13). From Eqs. (8.4.26) and (8.4.24) we obtain $N(t_{th}) \cong 0.32N_\infty$, so that from Eq. (8.4.1), $t_{th} \cong 88$ μs; from Eq. (8.4.25), we get $\sigma N_\infty l = \gamma_t/0.32 = 2.64$.

For $t > t_{th}$ the laser shows a net gain $g_{net}(t')$, which, before appreciable absorber saturation occurs, can be written as $g_{net}(t) = \sigma Nl - \gamma_t \cong \sigma l(dN/dt)_{th}t'$, where we changed to a new reference time axis t' whose origin is $t = t_{th}$. From Eq. (8.4.1) and the previously calculated value of t_{th}, we have $(dN/dt)_{th} = (N_\infty/\tau)\exp(-t_{th}/\tau) = 0.68(N_\infty/\tau)$. Net gain is given by $g_{net}(t') = 0.68(\sigma N_\infty l)(t'/\tau)$ and, using the previously calculated value of $(\sigma N_\infty l)$, we obtain

$$g_{net}(t') \cong 1.8\left(\frac{t'}{\tau}\right) \tag{8.4.27}$$

Once the expression for the net gain is calculated, the growth of the cavity photons in the laser cavity is obtained from the equation

$$\left(\frac{d\phi}{dt}\right) = \left(\frac{g_{net}}{t_T}\right)\phi \tag{8.4.28}$$

where $t_T = L_e/c \cong 1.66$ ns is the single-pass transit time for the laser cavity. Equation (8.4.28) is readily derived from an argument very similar to that used to obtain Eq. (7.2.12). If Eq. (8.4.27) is substituted into Eq. (8.4.28) and the resulting equation integrated, we obtain

$$\phi(t') = \phi_i \exp[0.9(t_T/\tau)(t'/t_T)^2] \tag{8.4.29}$$

Note that, since net gain increases linearly with time [see Eq. (8.4.27)], $\phi(t')$ increases exponentially with t'^2 [see Eq. (8.4.29) and Fig. 8.8a, where time t coincides with time t' in this example]. To calculate the time t'_s at which saturation occurs from Eq. (8.4.29), we must relate $\phi(t')$ to the circulating beam intensity $I(t')$. To do so, we first observe that, if two beams of the same intensity I travel in opposite directions within a laser cavity, the spatially averaged value of energy density within the cavity is $\rho = 2I/c$ [compare to Eq. (2.4.10)]. The relation between the photon number and circulating laser intensity within the cavity is then

$$\phi = \frac{\rho A_b L_e}{h\nu} = \frac{2I A_b L_e}{ch\nu}$$

where A_b is the beam area. From the previous expression, if $A_b = \pi D^2/4 \cong 0.196$ cm^2, the photon number ϕ_s corresponding to saturation intensity $I_s = 1$ MW/cm^2 is $\phi_s \cong 3.49 \times 10^{15}$. From Eq. (8.4.29) if $\phi_i = 1$, then $(t'_s/t_T) \cong 2347$, i.e., $t'_s \cong 3.89$ μs and $t_s \cong 92$ μs (see Fig. 8.13). Therefore starting from the noise level, it takes ~ 2350 transits for light to reach an intensity equal to the absorber saturation intensity. From this point on the saturable absorber bleaches very rapidly.

For $t > t_s$ we approximate the dynamic behavior of the laser by assuming that the saturable absorber is completely bleached. Time evolution is then obtained from equations in this section by assuming, as an initial inversion, $N_i = N(t_s)$. Then, for the initial gain we get $g_i = \sigma N_i l = \sigma N_\infty\{1 - \exp[-(t_s/\tau)]\} = 0.87$, while the gain at the peak of the pulse is $g_p = \sigma N_p l = \gamma = 0.15$. Thus $N_i/N_p = g_i/g_p = 5.8$. The total initial inversion in the mode volume V_a is then given by $N_i V_a = N_i A_b l = (A_b/\sigma)g_i$. Taking $\sigma = 2.8 \times 10^{-19}$ cm^2 (see previous example) and the previously calculated value of A_b and g_i, we obtain $N_i V_a = 6.09 \times 10^{17}$ ions (see Fig. 8.8b). Likewise we have $N_p V_a = N_i V_a/5.8 = 1.05 \times 10^{17}$ ions. The calculation of the peak photon number is then readily obtained from Eq. (8.4.14) as $\phi_p = 3.19 \times 10^{17}$ photons. The calculation of the peak pulse delay time τ_d and pulsewidth $\Delta\tau_p$ are also readily obtained from Eqs. (8.4.23) and (8.4.21), respectively, once the cavity photon decay time is calculated as $\tau_c = t_T/\gamma \cong 11$ ns. We get $\tau_d \cong 88$ ns and $\Delta\tau_p \cong 21$ ns, so that the pulse peak occurs approximately at a time $t'_p \cong t'_s + \tau_d + (\Delta\tau_p/2) = 3.99$ μs (see Fig. 8.8b).

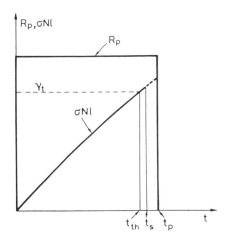

FIG. 8.13. Time evolution of the pump rate and of the laser gain for a square pump of duration $t_p = 100$ μs and for a medium with relaxation time $\tau = 230$ μs.

We now consider the case of a continuously pumped, repetitively Q-switched laser (Fig. 8.10). Note that, after switching and during the evolution of the Q-switched pulse, Eq. (8.4.8) still applies; peak power, output energy, and pulse duration are therefore still given by Eqs. (8.4.15), (8.4.20), and (8.4.21), respectively. However the expression for (N_i/N_p) is no longer given by Eq. (8.4.10), since it is determined by a different pump dynamics. In fact we now require that, in the time τ_p between two consecutive pulses, the pump rate must reestablish the initial inversion, starting from the population N_f which is left after the preceding Q-switching event. From Eq. (7.2.16a), setting $\phi = 0$, we get upon integration

$$N_i = (R_p\tau) - (R_p\tau - N_f)\exp(-\tau_p/\tau) \qquad (8.4.30)$$

From Eqs. (7.3.6), (7.3.3), and (8.4.9), we have $R_p\tau = xN_c = xN_p$; then Eq. (8.4.30) gives

$$x\frac{N_p}{N_i}[1 - \exp(-1/f^*)] = 1 - \frac{N_f}{N_i}\exp(-1/f^*) \qquad (8.4.31)$$

where x is the amount by which the cw pump exceeds its threshold value and $f^* = \tau f$, where $f = 1/\tau_p$ is the laser repetition rate. Equation (8.4.31) together with Eq. (8.4.18), which still hold, provide a pair of equations that can be solved for both (N_i/N_p) and (N_i/N_f) once x and f^* are known. Fig. 8.14 plots the solution for (N_i/N_p) against the amount x by which threshold is exceeded for several values of the normalized frequency f^*. For given values of x and f^*, Fig. 8.14 gives the corresponding value of (N_i/N_p). Once (N_i/N_p) is known, the quantity (N_i/N_f) or, equivalently, the energy-utilization factor η_E can be obtained from Fig. 8.11. When (N_i/N_p) and η_E are calculated, P_p, E, and $\Delta\tau_p$ are readily obtained from Eqs. (8.4.15), (8.4.20), and (8.4.21), respectively. Note that, within the range we consider for variables x and f^*, the relation between (N_i/N_p) and x is close to linear.

Calculations for a quasi-three-level laser proceed similarly from Eq. (7.2.24). Due to space limitations, these calculations are not presented here.

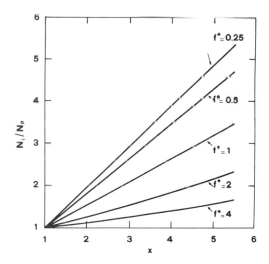

FIG. 8.14. Case of a cw-pumped, repetitively Q-switched laser. Plot of (N/N_p) versus the amount x by which threshold is exceeded, for several values of the normalized pulse repetition rate f^*.

8.5. GAIN SWITCHING

Gain switching like Q-switching is a technique that allows us to generate a laser pulse of short duration (generally from a few tens to a few hundreds of nanoseconds) and high-peak power. Unlike Q-switching however, where losses are rapidly switched to a low value, in the case of gain switching, it is the laser gain that is rapidly switched to a high value. Gain switching is achieved by using a pumping pulse that is so fast that the population inversion and hence the laser gain, reach a value considerably above threshold before the number of cavity photons has had time to build up to a sufficiently high level to deplete the inversion.

The physical phenomena involved can be simply described by referring to the spiking situation depicted in Fig. 8.1. If we assume that the pump rate $R_p = R_p(t)$ takes the form of a square pulse starting at $t = 0$ and ending at $t \cong 5$ μs, light emission would consist of just the first spike of $\phi(t)$, occurring at about $t = 5$ μs. After this spike in fact, inversion is driven by the light pulse to a value well below the threshold value, and it does not grow thereafter since there is no further pumping. Thus we see that gain switching is similar to laser spiking described in Sec. 8.2. The main difference arises from the fact that, to obtain pulses of high peak power and short duration, the peak value of R_p must be much larger than the cw value in Fig. 8.1. Depending on the peak value of this pump rate, the peak inversion may then range from 4 to 10 times the threshold value rather than the value of ~ 1.48 shown in Fig. 8.1. The buildup time of the laser radiation to its peak value may correspondingly range from 5 to 20 times the cavity photon decay time τ_c. The time duration of the pumping pulse must therefore be approximately equal to this buildup time and hence very short.[14]

In an actual situation the time behavior of the pumping process usually takes the form of a bell-shaped pulse rather than a square pulse. In this case we require the peak of the photon spike to occur at an appropriate time in the trailing edge of the pumping pulse. In fact, if this peak occurs at, e.g., the peak of the pumping pulse, there may be enough pump input, after the laser pulse, to allow the inversion to exceed the threshold again, thereby producing a second, although weaker, laser pulse. If, on the other hand, the photon peak occurs much later in the tail of the pulse, this implies insufficient pumping time for the inversion to reach a sufficiently high value. The preceding discussion implies that for a given pump pulse duration, there is an optimum value for the peak pump rate. By decreasing pump duration, the optimum value of the peak pump rate increases, so a more intense and narrower laser pulse is produced.

It should be noted that, given a sufficiently rapid and intense pump pulse, any laser can in principle be gain-switched even if the spontaneous decay of its upper laser level is allowed by electric dipole interaction and the lifetime correspondingly falls in the nanosecond range. In this case the pumping pulse and the cavity photon decay time must be appreciably shorter than this lifetime; correspondingly, very short gain-switched pulses, i.e., with duration shorter than ~ 1 ns can be obtained.

Example 8.6. *Typical cases of gain-switched lasers.* The most common example of a gain-switched laser is the electrically-pulsed, transversely excited at atmospheric pressure (TEA, see Chap. 10) CO_2 laser. Taking a typical cavity length of $L = 1$ m, a 20% transmission from the output mirror, and assuming that internal losses arise only from this mirror's transmission, one has $\gamma \cong 0.1$ and $\tau_c = L/c\gamma \cong 30$ ns. Assuming that laser buildup time is 10 times longer, we see that the pumping pulse should last ~ 300 ns which agrees with experimental findings.

Typical examples of gain-switched lasers using active media with upper state lifetime τ in the nanosecond range include a short-cavity dye laser (e.g., rhodamine 6G dye laser, $\tau \cong 5$ ns) pumped by the fast (~ 0.5-ns) pulse of an atmospheric pressure N_2 laser and a semiconductor laser (e.g., GaAs, $\tau \cong 3$ ns) pumped by a very short (~ 0.5-ns) current pulse. In both cases gain-switched laser pulses of ~ 100 ps duration can be obtained.

8.6. MODE LOCKING

Let us now consider a laser oscillating in a large number of longitudinal modes. Under ordinary circumstances, phases of these modes have random values, and, for cw oscillation beam, intensity shows random time behavior. Figure 8.15 shows the time behavior of the square of the electric field amplitude $|A(t)|^2$ of the output beam for $N = 51$ oscillating modes, each with the same amplitude E_0, and evenly separated in frequency by the frequency difference Δv between consecutive longitudinal modes. The output beam is seen to consist of a random sequence of light pulses. Despite this randomness, since these pulses arise from the sum of N frequency components evenly spaced in frequency, the pulse waveform in Fig. 8.15 has the following features which are characteristics of a Fourier series: (1) The waveform is periodic with a period $\tau_p = 1/\Delta v$. (2) Each light pulse of the random waveform has a duration $\Delta \tau_p$ roughly equal to $1/\Delta v_L$, where $\Delta v_L = N\Delta v$ is the total oscillating bandwidth. Thus, for lasers with relatively large gain bandwidths, such as solid-state, dye, or semiconductor lasers, Δv_L may be comparable to this bandwidth; hence short noise pulses with durations of picoseconds or less can be produced. Note that, since the response time of a conventional photodetector is usually much longer than a few picoseconds, we do not resolve this complex time behavior in the detected output of a

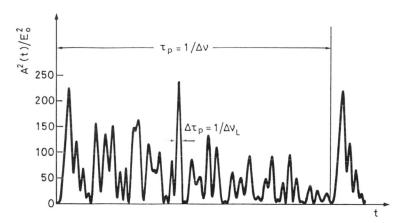

FIG. 8.15. Example of time behavior of the squared amplitude of the total electric field, $|A(t)|^2$, for the case of 51 oscillating modes, all with the same amplitude E_0 and with random phases.

random-phase, multimode laser; instead its average value is monitored. This value is the sum of powers in the modes; hence is proportional to NE_0^2.

Suppose now that the oscillating modes, while still having equal or comparable amplitudes, are somehow made to oscillate with some definite relation between their phases. Such a laser is referred to as mode locked, and the process by which modes are made to adopt a definite phase relation is referred to as *mode locking*.[15] Mode-locked lasers are considered at some length in the following sections.

8.6.1. Frequency-Domain Description

We first describe mode-locking in the frequency domain and consider, as a first example, the case of $2n + 1$ longitudinal modes oscillating with the same amplitude E_0 (Fig. 8.16a). We assume that phases φ_l of the modes in the output beam to be locked according to the relation

$$\varphi_l - \varphi_{l-1} = \varphi \qquad (8.6.1)$$

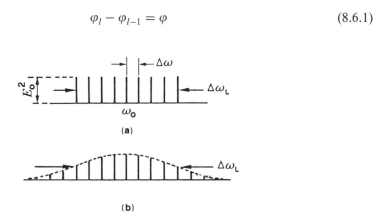

FIG. 8.16. Mode amplitudes (represented by vertical lines) versus frequency for a mode-locked laser. (a) Uniform amplitude distribution, and (b) Gaussian amplitude distribution, over a bandwidth (FWHM) $\Delta\omega_L$.

where φ is a constant. The total electric field $E(t)$ of the em wave at any given point in the output beam can be written as:

$$E(t) = \sum_{-n}^{+n} {}_lE_0 \exp\{ j[(\omega_0 + l\Delta\omega)t + l\varphi]\}$$ (8.6.2)

where ω_0 is the frequency of the central mode, $\Delta\omega$ is the frequency difference between two consecutive modes, and the value of the phase for the central mode has, for simplicity, been taken as zero. According to Eq. (8.6.2), the total electric field of the wave can be written as

$$E(t) = A(t) \exp(j\omega_0 t)$$ (8.6.3)

where

$$A(t) = \sum_{-n}^{+n} {}_lE_0 \exp[jl(\Delta\omega t + \varphi)]$$ (8.6.4)

Equation (8.6.3) shows that $E(t)$ can be represented in terms of a sinusoidal carrier wave, at the center mode frequency ω_0, whose amplitude $A(t)$ is time-dependent.

To calculate the time behavior of $A(t)$, we now change to a new time reference t' such that $\Delta\omega t' = \Delta\omega t + \varphi$. In terms of the new variable t', Eq. (8.6.4) can be transformed into:

$$A(t') = \sum_{-n}^{+n} {}_lE_0 \exp jl(\Delta\omega t')$$ (8.6.5)

The sum on the right-hand side is easily recognized as a geometric progression with a ratio $\exp j(\Delta\omega t')$ between consecutive terms. This progression is then easily summed to obtain

$$A(t') = E_0 \frac{\sin[(2n + 1)\Delta\omega t'/2]}{\sin(\Delta\omega t'/2)}$$ (8.6.6)

To understand the physical significance of Eq. (8.6.6), Fig. 8.17 shows the quantity $A^2(t')/E_0^2$, $A^2(t')$ being proportional to beam intensity, versus time t', for $2n + 1 = 7$ oscillating modes. As a result of the phase-locking condition (8.6.1), oscillating modes are seen to interfere so as to produce a train of evenly spaced light pulses. Pulse maxima occur when the denominator in Eq. (8.6.6) vanishes. In the new time reference t', the first maximum occurs for $t' = 0$. Note that, at this time, the numerator in Eq. (8.6.6) also vanishes. Making the approximation $\sin \alpha \cong \alpha$, which holds for small values of α, we readily see from Eq. (8.6.6) that $A^2(0) = (2n + 1)^2 E_0^2$. The next pulse occurs when the denominator in Eq. (8.6.6) again vanishes; this happens at time t' such that $(\Delta\omega t'/2) = \pi$. Two successive pulses are therefore separated by a time:

$$\tau_p = \frac{2\pi}{\Delta\omega} = \frac{1}{\Delta\nu}$$ (8.6.7)

where $\Delta\nu$ is the frequency separation between two consecutive modes. For $t' > 0$, the first zero for $A^2(t')$ in Fig. 8.17 occurs when the numerator in Eq. (8.6.6) again vanishes. This

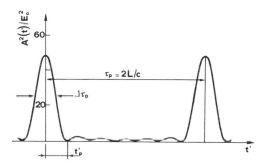

FIG. 8.17. Time behavior of the squared amplitude of the electric field for the case of seven oscillating modes with locked phases and equal amplitudes, E_0.

occurs at time t'_p such that $[(2n + 1)\Delta\omega t'_p/2] = \pi$. Since the width $\Delta\tau_p$ (FWHM) of $A^2(t')$, i.e., of each laser pulse, is approximately equal to t'_p, we have

$$\Delta\tau_p \cong \frac{2\pi}{(2n + 1)\Delta\omega} = \frac{1}{\Delta\nu_L} \qquad (8.6.8)$$

where $\Delta\nu_L = (2n + 1)\Delta\omega/2\pi$ is the total laser bandwidth (see Fig. 8.16a).

The mode-locking behavior in Fig. 8.17 can be readily understood if we represent field components in Eq. (8.6.5) by vectors in the complex plane. The lth amplitude component thus corresponds to a complex vector of amplitude E_0 rotating at the angular velocity $l\Delta\omega$. At time $t' = 0$ these vectors have zero phase [see Eq. (8.6.5)] and, accordingly, they lie in the same direction, which we assume to be the horizontal direction in Fig. 8.18. The total field in this case is $(2n + 1)E_0$. For $t' > 0$, the vector corresponding to the central mode remains fixed; vectors for modes where $l > 0$, i.e., with $\omega > \omega_0$, rotate in one direction (e.g., counterclockwise), while vectors for modes where $\omega < \omega_0$ rotate in the opposite (clockwise) sense. Thus, in the case of, e.g., five modes, the situation at some later time t' will be as indicated in Fig. 8.18a. If now time t' is such that mode 1 has made a counterclockwise

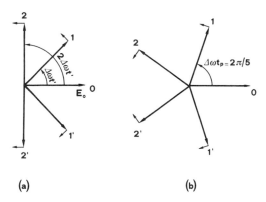

(a)   (b)

FIG. 8.18. Representation of the cavity mode amplitudes in the complex plane for the case of five modes. (a) shows the situation at a general time $t' > 0$, while (b) depicts the time instant at which the sum of the five mode amplitudes is zero.

rotation of 2π (which occurs when $\Delta\omega t' = 2\pi$), mode -1 will also have rotated (clockwise) by 2π, while modes 2 and -2 will have rotated by 4π. Therefore all these vectors are again aligned with the vector at frequency ω_0 and total field amplitude is again $(2n + 1)E_0$. Thus the time interval τ_p between two consecutive pulses must be such that $\Delta\omega\tau_p = 2\pi$, as shown by Eq. (8.6.7). In this picture, the time t'_p at which $A(t')$ vanishes (see Fig. 8.17) corresponds to the situation where all vectors are evenly spaced around the 2π angle (Fig. 8.18b). To achieve this condition, mode 1 must have rotated only by $2\pi/5$ or, more generally for $(2n + 1)$ modes, by $2\pi/(2n + 1)$. The time duration t'_p and hence the pulse duration $\Delta\tau_p$ are thus given by Eq. (8.6.8).

Before proceeding we point out here some peculiar properties of mode-locked lasers. We first recall that, under the mode-locking condition given by Eq. (8.6.1), the output beam consists of a train of mode-locked pulses; the duration of each pulse $\Delta\tau_p$ is about equal to the inverse of the oscillating bandwidth $\Delta\nu_L$. This result is due to a general property of a Fourier series. Since $\Delta\nu_L$ can be of the order of the width of the gain line $\Delta\nu_0$, very short pulses (down to a few picoseconds) can be expected to result from mode-locking in solid-state or semiconductor lasers. For dye or tunable solid-state lasers, the gain linewidth can be at least a factor 100 times larger, so that much shorter pulse widths are possible and indeed have been obtained (e.g., ~ 25 fs for rhodamine 6G dye laser and ~ 7 fs for Ti:sapphire laser). In the case of gas lasers the gain linewidth is much narrower (up to a few GHz) so relatively long pulses are generated (to ~ 100 ps). Note also that the peak power of the pulse is proportional to $(2n + 1)^2 E_0^2$, whereas, for modes with random phases, average power is simply the sum of powers in the modes and hence proportional to $(2n + 1)E_0^2$. Therefore, for the same number of oscillating modes and for the same field amplitudes E_0, the ratio between the peak-pulse power in the mode-locked case and the average power in the non-mode-locked case is equal to the number, $(2n + 1)$, of oscillating modes, which for solid-state or liquid lasers, can be very high ($10^3 - 10^4$). Mode-locking is thus useful not only for producing pulses of very short duration but also for producing high peak powers.

So far we have restricted our considerations to the unrealistic case of an equal-amplitude mode spectrum (Fig. 8.16a). The spectral envelope is generally expected to have a bell shape, however, and, as a simple example, we consider a Gaussian shape (Fig. 8.16b). We can therefore write the amplitude E_l of the lth mode as:

$$E_l^2 = E_0^2 \exp[-(2l\Delta\omega/\Delta\omega_L)^2 \ln 2] \tag{8.6.9}$$

where $\Delta\omega_L$ represents the bandwidth (FWHM) of the spectral intensity. If we assume that phases are locked according to Eq. (8.6.1) and the phase of the central mode is zero, we find that $E(t)$ can be expressed as in Eq. (8.6.3), where the amplitude $A(t')$, in the time reference t', is given by:

$$A(t') = \sum_{-\infty}^{+\infty} {}_l E_l \exp j(l\Delta\omega t') \tag{8.6.10}$$

If the sum is approximated by an integral [i.e., $A(t) \cong \int E_l \exp j(l\Delta\omega t)dl$], the field amplitude $A(t)$ is seen to be proportional to the Fourier transform of the spectral amplitude E_l. We then find that $A^2(t)$, i.e., the pulse intensity, is a Gaussian function of time, which can be written as

$$A^2(t) \propto \exp[-(2t/\Delta\tau_p)^2 \ln 2] \tag{8.6.11}$$

where $\Delta\tau_p$ is the width (FWHM) of the pulse intensity and is given by

$$\Delta\tau_p = \frac{2\ln 2}{\pi\Delta\nu_L} = \frac{0.441}{\Delta\nu_L} \tag{8.6.12}$$

To conclude our discussion of these two examples, we can say that, when the mode locking condition (8.6.1) holds, the field amplitude is given by the Fourier transform of the magnitude of the spectral amplitude. In such a case, the pulsewidth $\Delta\tau_p$ is related to the width of the laser spectrum $\Delta\nu_L$ by the relation $\Delta\tau_p = \beta/\Delta\nu_L$, where β is a numerical factor (of the order of unity) that depends on the particular shape of the spectral intensity distribution. A pulse of this sort is said to be *transform-limited*.

Under locking conditions different from Eq. (8.6.1), the output pulse may be far from being transform-limited. As an example, instead of Eq. (8.6.1) which can be written as $\varphi_l = l\varphi$, we consider the situation when

$$\varphi_l = l\varphi_1 + l^2\varphi_2 \tag{8.6.13}$$

where φ_1 and φ_2 are two constants. If we again assume a Gaussian amplitude distribution, such as in Eq. (8.6.9), the Fourier transform of the spectrum can again be analytically calculated, and $E(t)$ can be written as

$$E(t) \propto \exp(-\alpha t^2)\exp[j(\omega_0 t + \beta t^2)] \tag{8.6.14}$$

where the two constants α and β are related to $\Delta\omega_L$ and φ_2. For brevity these relations are not given here, since they are not needed for what follows. Note here, however, the following three points: (1) The beam intensity, being proportional to $|E(t)|^2$, is still described by a Gaussian function whose pulsewidth $\Delta\tau_p$ (FWHM) in terms of the parameter α is given by

$$\Delta\tau_p = 2\ln(2/\alpha)^{1/2} \tag{8.6.15}$$

(2) The presence of a phase term $l^2\varphi_2$ in Eq. (8.6.13), that is quadratic in the mode index l, results in $E(t)$ having a phase term βt^2 that is quadratic in time. This means that the instantaneous carrier frequency of the wave, $\omega(t) = d(\omega_0 t + \beta t^2)/dt = \omega_0 + 2\beta t$, has a linear frequency sweep (or *frequency chirp*). (3) For large values of φ_2 and hence of β, the product $\Delta\tau_p\Delta\nu_L$ can be much larger than the minimum value 0.441 given by Eq. (8.6.12). To understand this point, we go back to calculate the spectrum of a field such as in Eq. (8.6.14). The spectral intensity again turns out to be given by a Gaussian function whose bandwidth $\Delta\nu_L$ (FWHM) is given by

$$\Delta\nu_L = \frac{0.441}{\Delta\tau_p}\left[1 + \left(\frac{\beta\Delta\tau_p^2}{2\ln 2}\right)^2\right]^{1/2} \tag{8.6.16}$$

where the expression for α given by Eq. (8.6.15) has been used. Equation (8.6.16) shows that, if $\beta \neq 0$, one has $\Delta\tau_p\Delta\nu_L > 0.441$. In fact, spectral broadening now arises both from the pulsed behavior of $|E(t)|^2$, i.e., from the amplitude modulation of $E(t)$ [which accounts for the first term on the right-hand side of Eq. (8.6.16)] and from the frequency chirp term $2\beta t$ of $E(t)$ [which accounts for the second term on the right-hand side of Eq. (8.6.16)]. Thus, for $\beta \Delta \tau_p^2 \gg 1$, i.e., for sufficiently large values of the frequency chirp, the product $\Delta \tau_p \Delta \nu_L$ becomes much larger than 1.

8.6.2. Time-Domain Picture

We recall that, under the mode-locking condition (8.6.1), two consecutive pulses of the output beam were separated by a time τ_p given by Eq. (8.6.7). Since $\Delta \nu = c/2L$, where L is the cavity length, τ_p equals $2L/c$, which is just the cavity-round trip time. Note also that the spatial extent Δz of a typical mode-locked pulse is usually much shorter than the cavity length (e.g., for a pulse of duration $\Delta \tau_p = 1$ ps, $\Delta z = c \Delta \tau_p = 0.3$ mm, while a laser cavity is typically several tens of centimeters long). The oscillating behavior inside the laser cavity can therefore be visualized as consisting of a single ultrashort pulse, of duration $\Delta \tau_p$ given by Eq. (8.6.8), that propagates back and forth within the cavity. In such a case, in fact, the output beam obviously consists of a train of pulses with a time separation between consecutive pulses equal to the cavity-round trip time. This is the so-called *time-domain picture* of mode locking.

According to this picture, we readily understand that the mode-locking condition (8.6.1) can be achieved by placing a suitably fast shutter at one end of the cavity (Fig. 8.19a). In fact, if an initially non-mode-locked beam is present within the cavity, its spatial amplitude distribution can be represented as in Fig. 8.15, with time t replaced by z/c, where z is the longitudinal coordinate along the laser cavity. Let us now assume that the shutter is periodically opened with a period $T = 2L/c$, possibly at the time when the most intense noise pulse in Fig. 8.15 reaches the shutter. If the opening time of the shutter is comparable to the duration of this noise pulse, then only this pulse will survive in the laser cavity and thus produces the mode-locking situation in Fig. 8.19a. Following a similar argument, we can now realise that, if the shutter is placed at the cavity center and periodically opened with a period $T = L/c$, the mode-locking situation described in Fig. 8.19b develops. In this case, two ultrashort pulses are present in the cavity, and they are located and travel in such a way that they cross each other at the shutter position when the shutter is open. By the same argument, if the shutter is placed at a distance $L/3$ from one cavity mirror and periodically opened with a period $T = 2L/3c$, the mode-locking situation described in Fig. 8.19c develops. In this case, three ultrashort pulses are present in the cavity, and they are located and travel in such a way that two pulses always cross at the shutter position when the shutter is open. Note that, for the cases in Figs. 8.19b–c, the repetition rate for the train of output pulses is $2\Delta \nu$ and $3\Delta \nu$, respectively, where $\Delta \nu = c/2L$ is the pulse repetition rate of the mode-locking case considered in Fig. 8.19a. For this reason, the mode-locking situations in Figs. 8.19b–c are referred to as cases of *harmonic mode locking*. In contrast to these cases, the mode-locking situation in Fig. 8.19a is sometimes referred to as mode locking at the fundamental frequency or *fundamental mode locking*. Note also that the frequency-domain description of mode locking for the cases in Figs. 8.19b–c must correspond to a phase-locking condition that differs from Eq. (8.6.1), since this condition was shown to lead to the mode locking in Fig. 8.19a. For instance, the phase-locking condition corresponding to Fig.

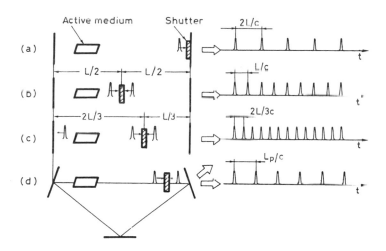

FIG. 8.19. Use of a fast cavity shutter to achieve mode locking: (a) Shutter placed at one cavity end thus leading to an output pulse train with a repetition frequency of $\Delta\nu = c/2L$. (b) Shutter placed at a distance $L/2$, from one cavity mirror. (c) Shutter placed at a distance, $L/3$, from one cavity mirror. The output pulse repetition rates for cases (b) and (c) are $2\Delta\nu$ and $3\Delta\nu$, respectively, these being examples of harmonic mode locking. (d) Represents the case of ring-laser mode-locking where the shutter position required for establishing mode locking becomes irrelevant.

8.19b is $\varphi_{l+1} - \varphi_l = \varphi_l - \varphi_{l-1} + \pi$, instead of $\varphi_{l+1} - \varphi_l = \varphi_l - \varphi_{l-1}$ which is another way of expressing Eq. (8.6.1).

An interesting situation occurs when a ring laser cavity is used in mode-locked operation (Fig. 8.19d). In this case, independently of the shutter position within the cavity, the laser produces fundamental, second harmonic, or third harmonic mode locking depending on whether the shutter repetition rate equals c/L_p, $2c/L_p$, or $3c/L_p$, where L_p is the length of the ring cavity perimeter. For instance, assume that the shutter is opened with a repetition rate c/L_p (Fig. 8.19d). Then if two counterpropagating pulses meet once at the shutter position, they then keep meeting at the same position after each round-trip, independently of shutter position.

8.6.3. Mode-Locking Methods

The methods of mode locking, like those for Q-switching, can be divided into two categories: (1) Active mode locking, in which the mode-locking element is driven by an external source. (2) Passive mode locking, in which the element that induces mode locking is not driven externally but instead exploits some nonlinear optical effect, such as saturation of a saturable absorber or a nonlinear refractive index change in a suitable material.

8.6.3.1. Active Mode Locking

There are three main types of active mode locking: (1) Mode locking induced by an amplitude modulator (*AM mode locking*). (2) Mode locking induced by a phase modulator (*FM mode locking*). (3) Mode locking induced by periodic modulation of the laser gain at a repetition rate equal to the fundamental cavity frequency $\Delta v = c/2L$ (*mode locking by synchronous pumping*). We discuss AM mode locking in most details, since this type is the most popular, then briefly discuss FM mode locking. Mode locking by synchronous pumping is not discussed here, since it is now less widely used. In fact it applies only to active media with nanosecond relaxation time, notably dye media and, to obtain the shortest pulses, requires the modulation rate of the pump to equal (with high precision) the fundamental frequency of the laser cavity. Therefore short pulse durations (shorter than 1 ps) are difficult to achieve from a synchronously pumped dye laser.

To describe AM mode locking, we consider a modulator inserted into the cavity that produces a time-varying loss at frequency ω_m. If $\omega_m \neq \Delta\omega$, where $\Delta\omega = 2\pi\Delta v$, Δv being the frequency difference between longitudinal modes, this loss will simply amplitude modulate the electric field $E_l(t)$ of each cavity mode to give

$$E_l(t) = E_0\{1 - (\delta/2)[1 - \cos(\omega_m t)]\}\cos(\omega_l t + \phi_l) \qquad (8.6.17)$$

where ω_l is the mode frequency, ϕ_l its phase, and δ is the depth of amplitude modulation. This means that the field amplitude is modulated from E_0 to $E_0(1 - \delta)$. Note that the term $E_0(\delta/2)\cos\omega_m t \times \cos(\omega_l t + \phi_l)$ in Eq. (8.6.17) can be written as $(E_0\delta/4) \times \{\cos[(\omega_l + \omega_m)t + \phi_l] + \cos[(\omega_l - \omega_m)t + \phi_l]\}$. Thus $E_l(t)$ actually contains two terms oscillating at the frequencies $\omega_l \pm \omega_m$ (modulation side bands). If now $\omega_m = \Delta\omega$, these modulation side bands coincide with the adjacent mode frequencies of the resonator. These two side bands thus contribute to the field equations of the two adjacent cavity modes. So, equations for cavity modes become coupled; i.e., the field equation of a given cavity mode contains two contributions from the modulation of the adjacent modes.[16] If the modulator is placed very close to one cavity mirror, this mode-coupling mechanism can be shown to lock the mode phases according to Eq. (8.6.1).

Details of AM mode locking are more readily understood in the time-domain rather than in the frequency domain. Thus, Fig. 8.20a shows the time behavior of cavity round-trip power losses*, 2γ, modulated with a modulation period $T = 2\pi/\omega_m$. We assume the modulator is placed at one end of the cavity (see Fig. 8.19a). If now $\omega_m = \Delta\omega$, the modulation period T equals the cavity round-trip time, so the stable steady-state condition corresponds to light pulses, $I(t)$, passing through the modulator at times t_m when a minimum loss of the modulator occurs (Fig. 8.20a). Indeed, if a pulse is assumed to pass through the modulator at a time of minimum loss, it will return to the modulator after a time, $2L/c$, where the loss is again at a minimum. Note that, if the pulse is assumed to pass initially through the modulator at a time, e.g., slightly earlier than t_m (solid-line pulse in Fig. 8.20b), then time-varying loss in the modulator causes the leading edge of the pulse to suffer more attenuation than the trailing edge. Thus, after passing through the modulator, the pulse indicated by the dashed line in Fig. 8.20b will result. Its peak has thus advanced in such a way that, during the next passage, the peak becomes closer to t_m. This shows that, eventually, the steady-state situation in Fig. 8.20a is reached. In this case, the pulse duration is shortened each time the pulse passes through the modulator because both the leading and trailing edges of the pulse are somewhat attenuated while the pulse peak is not attenuated by

---

* In a mode-locked linear cavity, it is simpler to speak in terms of round-trip loss and round-trip gain rather than in terms of the corresponding single-pass values.

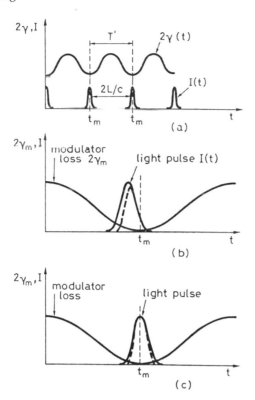

FIG. 8.20. Time-domain description of AM mode locking: (a) steady state condition; (b) light pulse arriving before the time t_m of minimum loss; (c) pulse-shortening occurring when the pulse arrives at time t_m.

the time-varying loss $2\gamma_m(t)$ of the modulator (see Fig. 8.20c). Thus, if it were only for this mechanism, pulse duration would tend to zero with progressive passages through the modulator. This is however prevented by the finite bandwidth of the gain medium. In fact, as the pulse becomes shorter, its spectrum eventually becomes so large to fill the bandwidth of the laser medium. The wings of the pulse spectrum are then no longer amplified; this constitutes the fundamental limitation to the pulse bandwidth and hence to pulse duration.

The finite bandwidth of the active medium influences the steady-state pulse duration in quite a different way, however, for homogeneous or inhomogeneous lines. For an inhomogeneously broadened line, if the laser is sufficiently far above threshold, the oscillating bandwidth $\Delta\nu_L$ tends to cover the whole gain bandwidth $\Delta\nu_0^*$. In this case, in a frequency-domain description, the primary purpose of the modulator is then to lock the phases of these already oscillating modes. Under the synchronism condition $\omega_m = \Delta\omega$ and if the AM modulator is placed at one cavity end, the phase-locking condition (8.6.1) develops. Assuming for simplicity a Gaussian distribution for mode amplitudes, we thus obtain from Eq. (8.6.12):

$$\Delta\tau_p \cong \frac{0.441}{\Delta\nu_0^*} \qquad (8.6.18)$$

By contrast to this situation, for a homogeneous line, the phenomenon of spatial hole burning tends to concentrate the width of the oscillating spectrum in a narrow region around the central frequency v_0 (see Sect. 7.7). Thus, assuming the laser originally to be unlocked, noise light pulses (see Fig. 8.15) tend to be appreciably longer than $1/\Delta v_0$, where Δv_0 is the width of the gain line. In this case the mechanism described in Fig. 8.20c is really effective in shortening the pulse duration, i.e., in broadening its spectrum. This pulse narrowing is however counteracted by pulse broadening that occurs when the pulse passes through the active medium and undergoes spectral narrowing. The theory of active mode locking, for a homogeneously broadened gain medium, was given a detailed and elegant treatment by Kuizenga and Siegman[17] and, later, presented in a more general framework by Haus[18]. We limit ourselves here to quoting the most relevant results and refer the reader to Appendix F for a more detailed treatment. The intensity profile turns out to be well-described by a Gaussian function whose width $\Delta\tau_p$ (FWHM) is approximately given by

$$\Delta\tau_p \cong \frac{0.45}{(v_m \Delta v_0)^{1/2}} \qquad (8.6.19)$$

where v_m is the modulator frequency ($v_m = \omega_m/2\pi = c/2L$ for fundamental harmonic mode locking).

If pulse-width expressions for inhomogeneous [Eq. (8.6.18)] and homogeneous [Eq. (8.6.19)] lines are compared at the same value of the laser linewidth (i.e., for $\Delta v_0^* = \Delta v_0$), we see that, since $(v_m/\Delta v_0) \ll 1$, then $(\Delta\tau_p)_{\text{hom}} \gg (\Delta\tau_p)_{\text{inhom}}$. Note that the pulse-narrowing mechanism depicted in Fig. 8.20c does not play any appreciable role in the case of an inhomogeneous line, although it is still obviously present. In this case, in fact, short-noise pulses with duration about equal to the inverse of the gain bandwidth are already present even under non-mode-locking conditions. The main role of the modulator is then to establish a synchronism between oscillating modes so that, out of the noise pulses in Fig. 8.15, only one pulse survives, i.e., the pulse passing through the modulator at the time of minimum loss (Fig. 8.20a).

Example 8.7. *AM mode-locking for cw Ar and Nd:YAG lasers.* We first consider a mode-locked Ar-ion laser oscillating on its $\lambda = 514.5$-nm green transition; this transition is Doppler-broadened to a width of $\Delta v_0^* = 3.5$ GHz. From Eq. (8.6.18) we then obtain $\Delta\tau_p \cong 126$ ps. We consider next a mode-locked Nd:YAG laser oscillating on its $\lambda = 1.064$-μm transition, whose linewidth is phonon-broadened to $\Delta v_0 \cong 4.3$ cm^{-1} = 129 GHz at $T = 300$ K. We consider a laser cavity with optical length $L_e = 1.5$ m whose AM modulator is located at one cavity end (Fig. 8.19a). We then get $v_m = c/2L_e = 100$ MHz, and, from Eq. (8.6.19), $\Delta\tau_p \cong 125$ ps. Note that, on account of the different expressions for $\Delta\tau_p$ for a homogeneous or inhomogeneous line, the pulsewidths for the two cases are almost the same, although the linewidth of Nd:YAG is almost 30 times wider than that of Ar-ion.

To describe FM mode-locking, we consider a modulator, whose refractive index n is sinusoidally modulated at frequency ω_m, placed at one end of the cavity. Any given cavity mode is therefore subjected to a time-varying phase shift given by $\varphi = (2\pi L'/\lambda) \times n(t)$, where L' is the modulator length. The phase-modulated modes then show sidebands [see Eq. (7.10.5)] whose frequencies, for $\omega_m = \Delta\omega$, coincide with those of neighboring modes. Thus cavity modes become coupled again and their phases locked,[16] although the locking condition turns out to be different from that given by Eq. (8.6.1). In the time-domain, this FM mode-locking produces pulses as indicated in Fig. 8.21. In this case, two stable mode-locking states can occur, i.e., such that a light pulse passes through the modulator either at each minimum of $n(t)$ (solid-line pulses) or at each maximum (dotted-line pulses).

To obtain a physical understanding of what happens in this case, we observe that, since the optical length of the modulator is $L'_e = n(t)L'$, the modulation actually results in variation of the overall optical length l_e of the cavity. In effect, the cavity is equivalent to one without a modulator but with the position of one cavity mirror oscillating at frequency ω_m. Either one of the two stationary situations in Fig. 8.21 thus corresponds to mode-locked pulses striking this moving mirror when it is at either of its extreme positions (i.e., when the mirror is stationary). Note that, after reflection by this moving mirror, the pulse acquires a nearly parabolic phase variation of either a positive sign (for the solid-line pulses) or a negative sign (for the dotted-line pulses), and its spectrum is slightly broadened. The overall phase modulation of the pulse and the corresponding pulse duration are then established by the condition that spectral broadening produced at each reflection from the moving mirror must be compensated by spectral narrowing resulting from passage through the amplifier. A stability analysis can also be performed to understand what happens when the pulse strikes the mirror not exactly at a stationary point. However this analysis is rather complicated, and due to the somewhat limited importance of this type of mode locking, it is not considered here. In fact, this type of locking is less frequently used for two reasons: (1) The pulses are frequency-modulated. (2) Mode locking tends to be somewhat unstable in the sense that switching between the two states in Fig. 8.21 often occurs in practice.

For a pulsed and generally high-gain laser, AM mode locking is commonly achieved by a Pockels cell amplitude modulator. Figure 8.5a shows a possible configuration; the Pockels cell voltage is sinusoidally modulated from zero to some fraction of the $\lambda/4$ voltage. For a cw-pumped and generally low-gain laser, AM mode locking is more commonly achieved by an acoustooptic modulator, which exhibits lower insertion loss than a Pockels cell modulator. However the acoustooptic modulator used for mode locking differs significantly from that used for Q-switching in Fig. 8.7. In fact, the face to which the piezoelectric transducer is bonded and the opposite face of the optical material are now cut parallel to each other. The sound wave launched into the material by the transducer is then reflected back by the opposite face of the material and, if the length of the optical block equals an integral number of half-wavelengths of the sound wave, an acoustic standing wave pattern is produced. Since the standing wave amplitude is sinusoidally modulated in time, the same happens for diffraction losses. It should be noted however that, if the sound wave is oscillating at frequency ω, the diffraction loss is modulated at frequency 2ω. Consider, in fact, an acoustic standing wave of the form $S = S_0(\cos \omega t)(\sin kz)$. Modulator loss reaches a

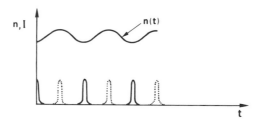

FIG. 8.21. FM mode locking. Time behavior of modulator refractive index n and of output intensity I.

maximum whenever a maximum *amplitude* of the standing wave pattern occurs; this maximum is reached twice in an oscillation period (i.e., at $t = 0$ and $t = \pi/\omega$). Thus modulator loss is modulated at frequency 2ω and, for fundamental mode locking (see Fig. 8.19a), mode locking is achieved when the modulator is located as near as possible to one cavity mirror and the modulation frequency 2ν is set equal to $(c/2L)$. This means that the transducer must be driven at frequency ν equal to $c/4L$ (e.g., $\nu = 50$ MHz for $L = 1.5$ m). In the case of FM mode locking (for both pulsed or cw lasers), a Pockels cell electrooptic phase modulator is commonly used. In this case one of the two axes, e.g., x, of induced birefringence (see Fig. 8.5b) is oriented along the polarizer axis. Therefore the beam does not rotate its polarization when passing through the Pockels cell but acquires a phase shift given by $\phi = (2\pi L'/\lambda)n_x$, where L' is the Pockels cell length and n_x is its refractive index for polarization in the x-direction. If now the voltage to the Pockels is sinusoidally modulated, the refractive index n_x, due to the Pockels effect, is also sinusoidally modulated and the same modulation occurs to the beam phase.

8.6.3.2. Passive Mode Locking

There are four main types of passive mode locking (ML): (1) *Fast saturable absorber* ML, which uses saturation properties of a suitable absorber (e.g., a dye molecule or a semiconductor) with a very short upper state lifetime. (2) *Kerr lens mode-locking* (KLM), which exploits the self-focusing property of a suitable, transparent, nonlinear optical material. (3) *Slow saturable absorber* ML, which exploits the dynamic saturation of the gain medium. (4) *Additive pulse* ML (APM), which exploits the self-phase modulation induced in a suitable nonlinear optical element inserted in an auxiliary cavity and coupled to a main cavity of identical length. In this last type, the pulse-shortening mechanism arises from interference between the main pulse in the laser cavity and the pulse coupled from the auxiliary cavity, which has been phase-modulated by the nonlinear material. Additive-pulse mode locking requires the optical length of the two cavities to be equal with a typical accuracy of a fraction of the laser wavelength. For this reason, this type of mode locking is not so widely used as the other techniques, so it is not discussed further here.

To illustrate mode locking by a fast saturable absorber, we consider an absorber with low saturation intensity and a much shorter relaxation time than the duration of the mode-locked pulses. The theory of mode locking by a fast saturable absorber for a homogeneously broadened gain medium was treated, in detail, notably by Haus[19]. We quote here only the most relevant results and refer the reader to Appendix F for a more detailed treatment.

For low values of intracavity beam intensity, I, compared to the absorber's saturation intensity, I_s, the cavity round-trip power loss* can be written as:

$$2\gamma_t = 2\gamma - 2\gamma'\left(\frac{I}{I_s}\right) \tag{8.6.20}$$

where γ is the unsaturable single-pass loss and γ' is the low-intensity single-pass loss of the saturable absorber.† Suppose now that a very thin absorber is placed in contact with one cavity mirror (Fig. 8.19a), and that the laser initially oscillates with unlocked mode phases.

---

* In a mode-locked linear cavity, it is preferable to think in terms of round-trip loss and gain rather than the corresponding single-pass values.

† According to Eq. (2.8.12) and $I \ll I_s$, the absorbance γ_a of an absorber of length l_a can be written as $\gamma_a = \alpha l_a = \alpha_0 l_a[1 - (I/I_s)]$; the intensity-independent term $\alpha_0 l_a$ can then be included in the total unsaturable loss γ.

The intensity of each of the two traveling waves is then composed of a random sequence of light bursts (see Fig. 8.15); for the initially low peak intensity of these bursts, the saturated round-trip power gain $2g_0$ roughly equals the cavity unsaturable loss. However the most intense pulse in Fig. 8.15, due to absorber saturation, suffers the least attenuation in the saturable absorber. If certain special conditions are met,[19] this pulse then grows more rapidly than the others; after many round-trips, the situation described in Fig. 8.22 is eventually established, where, for simplicity, the gain medium and saturable absorber are assumed to be together at one end of the cavity. In this case a single, intense, mode-locked pulse survives in the cavity and, due to the reduced loss caused by the more pronounced absorber saturation, the average power increases compared to the unlocked case. The round-trip saturated gain $2g_0'$ then becomes smaller than round-trip unsaturable loss in the cavity. Accordingly, a time window of net gain is established during the passage of the pulse (between times t_1 and t_2 in the figure), i.e., the pulse tails see a net loss and the pulse peak a net gain. If it were only for this mechanism, the pulse would be progressively shortened after each pass through the absorber-amplifier combination. However, a steady-state condition is again established by the balance between this pulse-shortening mechanism and pulse broadening arising from the finite gain bandwidth. The steady-state pulse amplitude, in this case, is described by a hyperbolic secant function, namely:

$$E(t) \propto \mathrm{sech}(t/\tau_p) \tag{8.6.21}$$

The duration $\Delta\tau_p$ of the pulse intensity (FWHM) is related to τ_p by $\Delta\tau_p \cong 1.76\tau_p$ and is given by

$$\Delta\tau_p \cong \frac{0.79}{\Delta\nu_0}\left(\frac{g_0'}{\gamma'}\right)^{1/2}\left(\frac{I_s}{I_p}\right)^{1/2} \tag{8.6.22}$$

where $\Delta\nu_0$ is the gain bandwidth (FWHM) and I_p is the peak intensity of the pulse.

It should be noted that the physical picture in Fig. 8.22 applies to long-lifetime (hundreds of microseconds) gain media, such as crystalline or glass solid-state media. In this case no appreciable gain variation occurs during the passage of the pulse, so the saturated gain g_0' is established by the average intracavity laser power. It should also be recalled that, for a simple two-level system, the absorber's saturation intensity is given by $I_s = h\nu/2\sigma\tau$

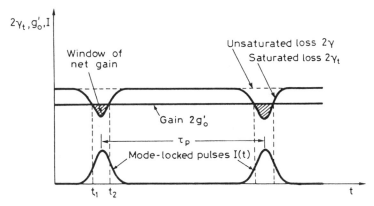

FIG. 8.22. Passive ML by a fast saturable absorber.

[see Eq. (2.8.11)]. Since τ must be very short ($\sim$ a few picoseconds or shorter), the required low value of saturation intensity calls for very large values of absorption cross section σ ($\sim 10^{-16}$ cm^2 or larger). It follows that the most commonly used saturable absorbers are either solutions of fast dye molecules or semiconductors.

In the case of dye solutions, cyanine dyes consisting of a long chain of the form ($-CH=CH$)$_n$, where n is an integer, and terminated by two aromatic end groups, are often used. The upper state relaxation time of cyanine dyes used for mode locking is typically some tens of picoseconds, as established by nonradiative decay via both internal conversion (see Fig. 3.6) and rotation of the aromatic rings. Thus the absorber remains saturated for a time roughly equal to this relaxation time, so mode-locked pulses shorter than a few picoseconds cannot be obtained.

For a semiconductor saturable absorber, absorption recovery typically shows a multicomponent decay, namely: (1) A fast decay ($\tau \approx 100$ fs) due to intraband thermalization of electrons, within the conduction band, arising from electron-electron collisions. (2) A slower relaxation (≈ 1 ps) due to intraband thermalization of conduction band electrons with the lattice, arising from electron-phonon collisions. (3) A still slower relaxation (from a few picoseconds to some nanoseconds) due to interband radiative and nonradiative decay. The longest relaxation time results in the lowest saturation intensity, which is helpful for starting the mode-locking process. The fastest relaxation time provides the rapid saturable absorber mechanism needed to produce short pulses. A particularly interesting solution consists in integrating a multiple-quantum-well saturable absorber between two mirrors whose spacing is such that the resulting Fabry–Perot etalon operates at antiresonance, i.e., at a point where a minimum of transmission or a maximum of reflection occurs (see Fig. 4.11). If the etalon is used as one cavity mirror, the laser intensity within the etalon can be substantially reduced compared to the value in the laser cavity. This offers the considerable advantage of increasing, in a controlled manner, the effective value of saturation intensity, decreasing the effective unsaturable losses, and increasing the damage threshold.[20] The effectiveness of this simple-to-use antire-sonance Fabry–Perot saturable absorber (A-FPSA) has been convincingly proven to generate both pico-second and femtosecond laser pulses from several wide-bandwidth solid-state lasers.

Another fast passive mode-locking technique relies on the lens effect induced in a suitable material by a Kerr-type nonlinearity, and it is thus referred to as Kerr-lens mode locking.[21,22] Consider first an optical material, such as quartz or sapphire, traversed by a light beam of uniform intensity I.

Example 8.8. *Passive mode locking of Nd:YAG and Nd:YLF lasers by a fast saturable absorber.* We consider a cw Nd:YAG laser passively mode-locked by an ~ 0.6 μm thick multiple-quantum-well (~ 50 wells) InGaAs/GaAs A-FPSA.[20] We take $g_0' = 2\%$, $\gamma' = 1\%$, $\Delta\nu_0 = 4.5$ cm^{-1} $\cong 135$ GHz ($T = 300$ K), $I_p = 0.3I_s$. From Eq. (8.6.22) we get $\Delta\tau_p \cong 15$ ps. Note that, in this case, the absorber is heavily saturated, so Eq. (8.6.22) can be taken only as a first-order approximation in calculating the predicted pulse width. For the case of a Nd:YLF laser, we assume the same value for unsaturable and saturable losses as in Nd:YAG. We therefore assume the same value of g_0'. The gain linewidth of Nd:YLF $\Delta\nu_0'$ ($\Delta\nu_0' \cong 13$ cm^{-1}) is about three times larger than that of Nd:YAG and the comparison is made at the same value of the output power, i.e., at the same value of the pulse energy $E \cong I_p\Delta\tau_p$. From Eq. (8.6.22) one readily finds that the pulsewidth $\Delta\tau_p'$ in this case is related to the pulsewidth in the previous case by $\Delta\tau_p' = (\Delta\nu_0/\Delta\nu_0')^2\Delta\tau_p$. For $\Delta\nu_0' = 2.89\Delta\nu_0$, one gets $\Delta\tau_p' \cong 1.8$ ps. Note the strong dependence of laser pulsewidth on gain linewidth under these conditions.

At sufficiently high intensity, the refractive index of the medium is modified, to a readily observable extent, by the field intensity; i.e., we can generally write $n = n(I)$. The first term of a Taylor expansion of n versus I is proportional to I, so we can write

$$n = n_0 + n_2 I \qquad (8.6.23)$$

where n_2 is a positive coefficient depending on the material (e.g., $n_2 \cong 4.5 \times 10^{-16}$ cm^2/W for fused quartz and $n_2 \cong 3.45 \times 10^{-16}$ cm^2/W for sapphire). This phenomenon is known as the *optical Kerr effect*; it is generally due to hyperpolarizability in the medium occurring at high electric fields and arising from either a deformation of atomic or molecular electronic orbitals or from a reorientation of elongated molecules (for a gas or liquid). For a solid, only deformation of the atom's electron cloud can occur, so the optical Kerr effect is very rapid, the response time being on the order of a rotation period of the outermost electrons of the atom (a few femtoseconds).

Assume now that beam intensity, in a medium exhibiting the optical Kerr effect (a *Kerr medium*), has a non-uniform transverse profile, e.g., Gaussian. Intensity at the beam center is then larger than in the wings and, according to Eq. (8.6.23), a nonlinear refractive index change $\delta n = n_2 I$ is induced, which is positive at the beam center and goes to zero in the beam wings. For a Gaussian beam profile one can write $I = I_p \exp[-2(r/w)^2]$, where I_p is the peak intensity and w is the (field) spot size. The nonlinear phase shift acquired by the beam in traversing a length l of the medium will then be $\delta\phi = 2\pi\delta n l/\lambda = (2\pi n_2 I_p l/\lambda)\exp[-2(r/w)^2] \cong (2\pi n_2 I_p l/\lambda) \times [1 - 2(r/w)^2]$. Thus, to first order in $(r/w)^2$, $\delta\phi$ can be taken as a parabolic function of (r/w); this is equivalent to saying that a spherical lens is induced in the medium by the optical Kerr effect. In fact, this induced lens may lead to beam focusing when beam power exceeds a critical value, a phenomenon known as *self-focusing*.

A nonlinear loss element providing a loss of the general form in Eq. (8.6.20) can then be realized as shown in Fig. 8.23. In fact, at higher beam intensities, the beam is focused more strongly at the aperture, so less loss is experienced at this aperture. If the nonlinear loss element in Fig. 8.23 is now correctly located within a laser cavity, passive mode locking can be achieved according to the mechanism described in Fig. 8.22 for a fast saturable absorber. The time response of Kerr lens mode locking is very short, so that, for all practical purposes, it can be taken as instantaneous. By appropriately controlling cavity dispersion, the fastest mode-locked pulses are achieved by this technique using ultrabroadband gain media (bandwidths of ≈ 100 THz).

Although many passively mode-locked lasers use fast saturable absorbers, under special circumstances slow saturable absorbers can also lead to mode locking. This type of mode locking is often referred to as *slow saturable absorber mode locking*. The required

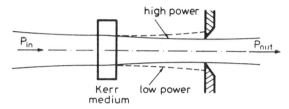

FIG. 8.23. Non-linear loss element exploiting the optical Kerr effect in a suitable nonlinear material.

special circumstances involve the following: (1) The relaxation time of both absorber and amplifier must be comparable to the cavity round-trip time. (2) The saturation fluence of both gain medium [$\Gamma_{sg} = h\nu/\sigma_g$, see Eq. (2.8.29)] and saturable absorber [$\Gamma_{sa} = h\nu/2\sigma_a$; see Eq. (2.8.17)] must be sufficiently low to allow both media to be saturated by the intracavity laser fluence. (3) The saturation fluence of the gain medium must be comparable to, although somewhat larger than, that of the saturable absorber.

The physical phenomena leading to mode locking in this case are rather subtle;[23] we describe them with the help of Fig. 8.24, where for simplicity we suppose that both the saturable absorber and active medium are at one end of the cavity. Before the mode-locked pulse arrives, gain is assumed to be smaller than losses, so that the early part of the leading edge of the pulse suffers a net loss. If the total energy fluence of the pulse has a suitable value, the accumulated energy fluence of the pulse may become comparable to the saturation fluence of the absorber during the leading edge of the pulse. Saturation of the absorber then begins to occur, so that, at some time during the pulse's leading edge (time t_1 in Fig. 8.24), absorber loss equals laser gain. For $t > t_1$ the pulse then sees a net gain rather than a net loss. However, if the saturation energy fluence of the gain medium has a suitable value (typically ~ 2 times higher than that of the absorber), gain saturation is produced, so that, at some time during the trailing edge of the pulse (time t_2 in Fig. 8.24), saturated gain becomes equal to saturated loss. For $t > t_2$ the pulse sees a net loss again rather than a net gain. A time window of net gain is thus established for $t_1 < t < t_2$ and, after each pass through the absorber-amplifier combination, the pulse is shortened. A steady-state condition is again established by the balance between this pulse-shortening mechanism and the pulse broadening arising from the finite-gain bandwidth. Thus we expect a pulse duration again comparable to the inverse of the gain bandwidth $\Delta\nu_0$.

The evolution toward mode locking for this slow-saturable-absorber, dynamic-gain-saturation, mechanism can be described by assuming the laser to be initially oscillating with unlocked phases. Saturated gain then equals unsaturated loss; under appropriate circumstances, the most energetic pulse within its noisy time pattern (see Fig. 8.15) begins to produce the time window of net gain as described in Fig. 8.24. This process continues after each passage of noise pulses through the laser cavity until only one laser pulse survives; then the situation described in Fig. 8.24 occurs. It should be noted that, after the mode-locked pulse passes through the absorber-amplifier combination and before the arrival

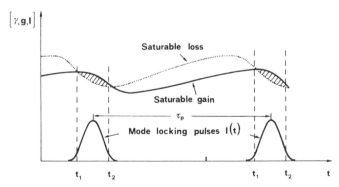

FIG. 8.24. Time-domain picture of slow-saturable-absorber mode locking. Note that the figure is not to scale, since the time duration of a mode-locked pulse is typically in the hundreds femtosecond range while the time interval τ_p between two consecutive pulses, i.e., the cavity round trip time, is typically a few nanoseconds.

of the next one, saturable loss must recover to its unsaturated value by spontaneous (i.e., radiative and nonradiative) decay. The corresponding decay time must then be appreciably shorter than the cavity round-trip time. During the same time interval, the gain medium must recover, due to the pumping process, so as to leave a saturated gain smaller than the loss. This means that the lifetime of the gain medium must be somewhat longer than the cavity round-trip time. We also reiterate that saturation fluences of both amplifier and absorber must be sufficiently low to allow the two media to be saturated by the laser pulse. This type of mode locking can thus be made to occur with short-lifetime ($\sim$ a few nanoseconds) high-cross section ($\sim 10^{16}$ cm^2) gain media, such as dyes or semiconductors. By contrast, this type of mode locking cannot occur with long-lifetime (hundreds of microseconds) gain media, such as crystalline or glass solid-state media, where dynamic gain saturation cannot occur. As saturable absorbers, saturable dyes with a lifetime of a few nanoseconds (determined by spontaneous emission) are often used. When the delicate conditions for this type of mode locking can be met, very short light pulses as low as the inverse of the laser linewidth can, in principle, be obtained. The large gain bandwidths (a few tens of terahertz) available from dye lasers then allow pulses with a duration of some tens of femtosecond to be produced. As discussed in Sect. 8.6.4, cavity dispersion plays a very important role for such short pulses, and its value must be controlled if pulses of the shortest duration are to be obtained.

8.6.4. Role of Cavity Dispersion in Femtosecond Mode-Locked Lasers

As just mentioned in the preceding section, when ultra-broad-band gain media (bandwidths as large as 100 THz) are involved, cavity dispersion plays an important role in establishing the shortest pulse duration achievable in mode locking. We explore this point here, but first we review the concepts of phase velocity, group velocity, and group delay dispersion in a dispersive medium.

8.6.4.1. Phase Velocity, Group Velocity, and Group-Delay Dispersion

Consider first a plane, linearly polarized, monochromatic em wave at frequency ω, propagating in the z-direction in a transparent medium. The electric field $E(t, z)$ of the wave can then be written as $E = A_0 \exp j(\omega t - \beta z)$, where A_0 is a constant and where the propagation constant β is generally a function of the angular frequency ω. The relation $\beta = \beta(\omega)$ is a characteristic of the given medium, referred to as the *dispersion relation* of the medium (see Fig. 8.25). Since the total phase of the wave is now $\phi_t = \omega t - \beta z$, the velocity of a given phase front is such that elemental changes dt and dz of the temporal and spatial coordinates must satisfy the condition $d\phi_t = \omega \, dt - \beta \, dz = 0$. This shows that the phase front moves at a velocity

$$v_{ph} = \frac{dz}{dt} = \frac{\omega}{\beta} \tag{8.6.24}$$

referred to as the *phase velocity* of the wave.

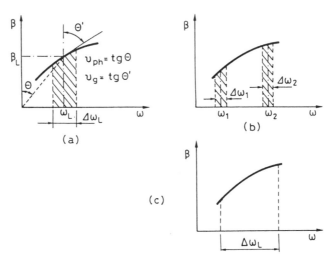

FIG. 8.25. (a) Phase velocity and group velocity in a dispersive medium; (b) dispersion in time delay for two pulses of carrier frequencies ω_1 and ω_2; and (c) group-velocity dispersion for a pulse of large oscillation bandwidth $\Delta\omega_L$

Consider next a light pulse traveling in the medium and let ω_L and $\Delta\omega_L$ be, respectively, the center frequency and the width of the corresponding spectrum (Fig. 8.25a). Assume also that the dispersion relation over the bandwidth $\Delta\omega_L$ can be approximated by a linear law, viz. $\beta = \beta_L + (d\beta/d\omega)_{\omega=\omega_L}(\omega - \omega_L)$, where β_L is the propagation constant corresponding to the frequency ω_L. In this case, upon considering a Fourier expansion of the wave, its electric field can be expressed as (see Appendix G)

$$E(t, z) = A\big[t - (z/v_g)\big] \exp[j(\omega_L t - \beta_L z)] \tag{8.6.25}$$

where A is the pulse amplitude, $\exp[j(\omega_L t - \beta_L z)]$ is the carrier wave, and v_g is given by

$$v_g = \left(\frac{d\omega}{d\beta}\right)_{\beta=\beta_L} \tag{8.6.26}$$

The fact that the pulse amplitude is a function of the variable $t - (z/v_g)$ means that the pulse propagates at a speed v_g without changing its shape. This velocity is referred to as the *group velocity* of the pulse; according to Eq. (8.6.26) it is given by the slope of the ω versus β relation at $\omega = \omega_L$ (i.e., $v_g = \tan\theta'$; see Fig. 8.25a). Note also that, for a general dispersion relation such as that in Fig. 8.25a, the phase-velocity of the carrier wave ($v_{ph} = \tan\theta$; see Fig. 8.25a) is different from the group velocity.

According to the preceding considerations, after traversing the length l of the medium, the pulse is subjected to a time delay:

$$\tau_g = \frac{l}{v_g} = l\left(\frac{d\beta}{d\omega}\right)_{\omega_L} = \phi'(\omega_L) \tag{8.6.27}$$

In this last equation, we have defined a phase ϕ, dependent on the frequency ω, such that

$$\phi(\omega - \omega_L) = \beta(\omega - \omega_L)l \qquad (8.6.28)$$

and we have set $\phi'(\omega_L) = [d\phi(\omega - \omega_L)/d\omega]_{\omega_L}$. For obvious reasons, the quantity $\tau_g = \phi'(\omega_L)$ is referred to as the *group delay* of the medium at frequency ω_L.

Let us see what happens when two pulses with bandwidths $\Delta\omega_1$ and $\Delta\omega_2$ centered at ω_1 and ω_2, respectively, travel in the medium ($\omega_2 > \omega_1$; see Fig. 8.25b). If the slope of the dispersion relation differs at the two frequencies, the two pulses travel at different group velocities v_{g_1} and v_{g_2}. Thus, if the peaks of the two pulses enter the medium at the same time, then, after traversing the length l of the medium, they become separated in time by a delay

$$\Delta\tau_g = \phi'(\omega_2) - \phi'(\omega_1) \cong \phi''(\omega_1) \times (\omega_2 - \omega_1) \qquad (8.6.29)$$

where we have used the symbol $\phi''(\omega_1) = [d^2\phi/d\omega^2]_{\omega_1}$. Note that Eq. (8.6.29) holds exactly if the relation between ϕ and ω, in the frequency range between ω_1 and ω_2, can be approximated by a parabolic law, viz.:

$$\phi = \phi_L + \left(\frac{d\phi}{d\omega}\right)_{\omega_L}(\omega - \omega_L) + \frac{1}{2}\left(\frac{d^2\phi}{d\omega^2}\right)_{\omega_L}(\omega - \omega_L)^2 \qquad (8.6.30)$$

Consider next the case of a light pulse with a bandwidth $\Delta\omega_L$ so large that it is no longer a good approximation to describe the dispersion relation by a linear law (Fig. 8.25c). In this case, different spectral regions of the pulse travel with different group velocities; consequently the pulse broadens as it propagates. Again assuming that the dispersion relation within the bandwidth $\Delta\omega_L$ can be approximated by a parabolic law, then, according to Eq. (8.6.29), pulse broadening due to dispersion, $\Delta\tau_d$, is given approximately by the difference in group delay between the fastest spectral component and the slowest one. According to Eq. (8.6.29) we then have

$$\Delta\tau_d \cong |\phi''(\omega_L)|\Delta\omega_L \qquad (8.6.31)$$

The quantity $\phi''(\omega_L)$ is referred to as the *group-delay dispersion* (GDD) of the medium at frequency ω_L. Its magnitude gives the pulse broadening per unit bandwidth of the pulse. From Eqs. (8.6.28) and (8.6.31) we see that $\Delta\tau_d$ can also be written as:

$$\Delta\tau_d \cong l\left|\left(\frac{d^2\beta}{d\omega^2}\right)_{\omega_L}\right|\Delta\omega_L \qquad (8.6.32)$$

The quantity, GVD, expressed by

$$\text{GVD} = \left(\frac{d^2\beta}{d^2\omega}\right)_{\omega_L} = \left[\frac{d(1/v_g)}{d\omega}\right]_{\omega_L} \qquad (8.6.33)$$

is usually referred to as the *group-velocity dispersion* at frequency ω_L. Its magnitude gives the pulse broadening per unit length of the medium and per unit bandwidth of the pulse. Note that the GVD concept is straightforward, in application, only for a homogeneous medium. For an inhomogeneous or multicomponent medium, such as the two prism pairs in Fig. 8.26 or a multilayer dielectric mirror, GDD is easier to consider.

8.6.4.2. Limitation on Pulse Duration Due to Group-Delay Dispersion

When a dispersive medium is present within a mode-locked laser cavity, an approximate value for the steady-state pulse duration can be obtained from the condition that relative time shortening $(\delta\tau_p/\tau_p)_s$, due to the net gain time window (see Fig. 8.22 or 8.24), must equal the pulse broadening due to both the gain medium $(\delta\tau_p/\tau_p)_g$ and the dispersive medium $(\delta\tau_p/\tau_p)_d$. For the sake of simplicity, we consider here a ring cavity whose pulse passes sequentially through the gain medium, the dispersive medium, and whatever element provides self-amplitude modulation (e.g., a fast saturable absorber). We assume that the light pulse has a Gaussian intensity profile with pulsewidth (FWHM) $\Delta\tau_p$ and that the dispersive medium can be described, at a general frequency ω, by its GDD $\phi'' = \phi''(\omega)$. The gain medium is assumed to be homogeneously broadened; it is then described by its saturated single-pass gain $g_0 = N_0\sigma l$ and its linewidth (FWHM), $\Delta\omega_0$.

For small changes in pulse duration, relative pulse broadenings $(\delta\tau_p/\tau_p)_g$ and $(\delta\tau_p/\tau_p)_d$, after passing through the gain medium and the dispersive medium, are shown in Appendix G to be given, respectively, by

$$\left(\frac{\delta\tau_p}{\tau_p}\right)_g = \left(\frac{2\ln 2}{\pi^2}\right)\left(\frac{1}{\Delta\tau_p^2\Delta v_0^2}\right)g_0 \qquad (8.6.34)$$

where $\Delta v_0 = \Delta\omega_0/2\pi$, and by

$$\left(\frac{\delta\tau_p}{\tau_P}\right)_d = (8\ln^2 2)\frac{\phi''^2}{\Delta\tau_p^4} \qquad (8.6.35)$$

where ϕ'' is calculated at the central frequency ω_L of the laser pulse.

The steady-state pulse duration can now be obtained from the condition that the relative time shortening $(\delta\tau_p/\tau_p)_s$, due to the net gain time window, must equal pulse broadening due to both the gain medium and the dispersive medium. From Eqs. (8.6.34) and (8.6.35) we then obtain

$$\left(\frac{\delta\tau_p}{\tau_p}\right)_s = 0.14\left(\frac{g_0}{\Delta\tau_p^2\Delta v_0^2}\right) + 3.84\left(\frac{\phi''^2}{\Delta\tau_p^4}\right) \qquad (8.6.36)$$

Note that the two terms on the right-hand side of Eq. (8.6.36) are inversely proportional to $\Delta\tau_p^2$ and $\Delta\tau_p^4$, respectively. This means that GDD becomes more important in establishing pulse duration as $\Delta\tau_p$ decreases. To estimate the pulsewidth at which GDD begins to become important, we equate the two terms on the right-hand side of Eq. (8.6.36). We get:

$$\Delta\tau_p = \left(\frac{27.4}{g_0}\right)^{1/2} |\phi''|\Delta v_0 \qquad (8.6.37)$$

Assuming for example $\Delta v_0 \cong 100$ THz (as appropriate for a Ti:sapphire gain medium), $\phi'' = 100$ fs^2 (equivalent, at $\lambda \cong 800$ nm. to the presence of ~ 2 mm of quartz material in the cavity) and $g_0 = 0.1$, from Eq. (8.6.37) we obtain $\Delta\tau_p \cong 162$ fs. Then, to get pulses shorter than ~ 150 fs, down to perhaps the inverse of the gain linewidth Δv_0 ($\Delta\tau_p' = 1/\Delta v_0 \cong 4$ fs), we must reduce GDD by about an order of magnitude.

To obtain the shortest pulses, when second-order dispersion GDD is suitably compensated, we must also compensate higher order dispersion terms in the power expansion (8.6.30). Of course the next term to be considered is the third-order dispersion (TOD), defined as TOD = $\phi''' = \beta'''l$, where the third derivatives are taken at the laser's center frequency ω_L. We do not consider the effects of TOD on an incident pulse and, for a discussion of this topic, we refer the reader to Ref. 24. We just point out that, in the case of, e.g., a Ti:sapphire laser ($\Delta v_0 \cong 100$ THz), TOD begins to play a pulse limiting role for pulses shorter than ~ 30 fs.

8.6.4.3. Dispersion Compensation

According to the preceding discussion, to achieve pulses shorter than ~ 150 fs, the GDD of the laser cavity, contributed by various optical elements (e.g., the active medium itself and the cavity mirrors), must be controlled in broad-band mode-locked lasers. We have also seen from the preceding discussion that fused quartz at $\lambda \cong 800$ nm and more generally in the visible range provides a positive value for ϕ''; this is also the case for all media said to exhibit normal dispersion, i.e., for most common optical media. So, the compensation of cavity GDD requires a suitable element providing a negative ϕ'', i.e., showing anomalous dispersion.

The now classical solution to providing a negative and controllable second-order GDD in a laser cavity uses the four-prism sequence shown in Fig. 8.26.[25] Prisms are generally used at minimum deviation (i.e., with incidence angle equal to the exiting angle) and cut at such an apex angle that rays enter and leave each prism at Brewster's angle. The entrance face of prism II is parallel to the exit face of prism I; the exit face of prism II is parallel to the entrance face of prism I, and so on. The plane MM', normal to rays between prisms II and III and midway between the two prisms, is a plane of symmetry for the ray paths.

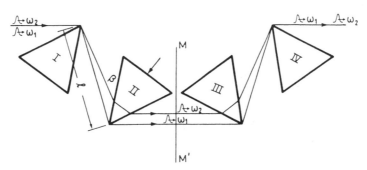

FIG. 8.26. Four-prism sequence having a negative and controllable second-order group-delay dispersion.

According to Eq. (8.6.29) to obtain a negative GDD we must have $\Delta\tau_g < 0$, i.e., $\tau_g(\omega_2) < \tau_g(\omega_1)$ for $\omega_2 > \omega_1$. This is exactly what the four-prism sequence does. In fact, the angular dispersion of the prisms is such that two pulses at ω_2 and ω_1 entering the prism sequence at the same time and in the same direction propagate along two different paths, as indicated in the figure. The path length for the pulse at frequency ω_2 is thus shorter than that for the pulse at frequency ω_1, which means GDD $= \phi'' < 0$. For simplicity we do not present the resulting expression for ϕ'' here; we limit ourselves to observing that ϕ'' depends linearly on the distance l between the two prism pairs. As an example, for quartz prisms, a length $l = 250$ mm gives a negative dispersion, which can compensate for the positive GDD, at $\lambda \cong 800$ nm, of a quartz element with a thickness of 6.6 mm (i.e., $\phi'' \cong -360$ fs^2).

The four-prism sequence in Fig. 8.26 is a convenient way of introducing a negative GDD in a laser cavity for the following reasons: (1) Since all faces are at Brewster's angle to the beam paths, losses introduced by the system are low. (2) The negative value of GDD can be coarsely changed by changing the separation l of the two prism pairs. (3) By translating any one of the prisms along an axis normal to its base (e.g., prism II, see Fig. 8.26), we change the total length of the optical medium traversed by the beam. This motion introduces, in a finely and controlled way, a positive (material) dispersion of adjustable size without altering the ray directions and hence the negative dispersion due to the geometry of the ray paths. (4) The transmitted beam is collinear with the incident beam; this facilitates inserting the four-prism sequence in an already-aligned cavity. Note that, since MM' is a symmetry plane, we can use just the first two prisms in a two-mirror resonator if one cavity mirror is plane and located at the position MM'. In this case the GDD per pass is of course half that of the four-prism sequence.

It should be noted that the four-prism sequence in Fig. 8.26 introduces not only a second order ϕ'' but also a third-order dispersion ϕ'''; this term is the dominant contributor to the overall TOD of a typical femtosecond laser cavity. As with second-order dispersion, the third-order dispersion depends on the ray path geometry in Fig. 8.26; hence its value is proportional to the prism separation. The ratio ϕ'''/ϕ'' is therefore determined only by the prism material and laser wavelength. In this respect fused quartz is one of the best optical materials, and the ratio ϕ'''/ϕ'' has the lowest value [e.g., $\phi'''(\omega_L)/\phi''(\omega_L) = 1.19$ fs at the frequency of the Ti:sapphire laser, i.e., at $\lambda \cong 800$ nm]. Thus, to achieve the smallest value for TOD, we must start with a cavity with the smallest value of positive ϕ'' to ensure the smallest values of both ϕ'' and ϕ''' from the four-prism sequence.

An alternative way of compensating for cavity dispersion involves using a dispersive element that introduces a negative GDD that is wavelength-independent (i.e., such that $\phi''' \cong 0$) instead of the two-prism couple in Fig. 8.26. A very interesting solution uses chirped multilayer dielectric mirrors in the laser cavity.[26] The mirror consists of a large number (~ 40) of alternating low- and high-refractive-index layers whose thickness progressively increases, in a suitable manner, going toward the substrate. In this way, high-frequency components of the laser pulse spectrum are reflected first, and low-frequency components are reflected further on in the multilayer. Hence the group delay of the reflected beam increases with decreasing values of ω, thus giving $\phi'' < 0$. For the appropriate, computer-optimized, design of the spatial-frequency chirp of the layers, one can obtain a value of ϕ'' that, within the bandwidth of interest, is approximately constant with frequency, i.e., $\phi''' \cong 0$. Alternatively, again by computer optimization, the GDD can be designed to exhibit a slight linear variation with frequency and a slope suitable for compensating the TOD of other cavity components (e.g., the gain medium). The main limitation of chirped mirrors arises from the fact that the amount of negative GDD typically obtainable is rather small (~ -50 fs^2). To achieve the required GDD, one must then arrange for the laser beam to undergo many reflections from the same chirped mirror.

8.6.4.4. Soliton-Type Mode Locking

Consider a medium exhibiting optical Kerr effect so that its refractive index can be described as in Eq. (8.6.23) and assume that a light pulse, of uniform transverse intensity profile, is traveling through the medium in the z-direction. After a length z, the carrier wave of the pulse acquires a phase term $\varphi = \omega_L t - \beta_L z$ given by

$$\varphi = \omega_L t - \left[\frac{\omega_L(n_0 + n_2 I)}{c}\right] z \qquad (8.6.38)$$

where I is the light pulse intensity. Since $I = I(t)$, the instantaneous carrier frequency of the pulse is given by:

$$\omega = \frac{\partial(\omega_L t - \beta_L z)}{\partial t} = \omega_L - \frac{\omega_L n_2 z}{c}\frac{\partial I}{\partial t} \qquad (8.6.39)$$

Thus the carrier frequency $\omega = \omega(t)$ is linearly dependent on the negative time derivative of the corresponding light intensity. Then, for a bell-shaped pulse as in Fig. 8.27a, the phase $\varphi = \varphi(t)$ is time-modulated by the beam intensity and the carrier frequency varies with time as indicated in Fig. 8.27b. This phenomenon is referred to as *self-phase modulation* (SPM).

Note that, around the peak of the pulse, i.e., around the region where the pulse's time behavior can be described by a parabolic law, the frequency chirp induced by SPM increases linearly with time. Suppose now that the medium has also a negative GDD. In this case Appendix G shows that, during propagation in this medium, the pulse tends to acquire an instantaneous frequency chirp that decreases linearly with time.* The two effects thus tend

---

* Observe that the pulse spectrum must remain unchanged while the pulse propagates through a passive medium, such as the dispersive medium considered here. In such a medium, however, the pulse broadens on propagation [see Eq. (8.6.32)] and the spectral contribution arising from the finite-pulse duration decreases. It then follows that the pulse must also acquire an appropriate frequency modulation to keep the spectrum unchanged.

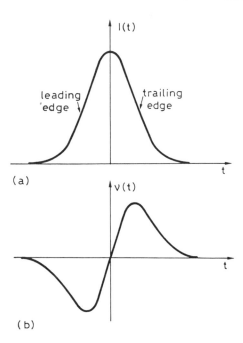

FIG. 8.27. Self-phase-modulation phenomenon. Time behavior of the pulse frequency (b) when a bell-shaped pulse (a) is traversing a medium exhibiting the optical Kerr effect.

to cancel each other, so we expect that, under appropriate conditions, the effect of SPM exactly cancels that due to dispersion, for the whole pulse; this intuitive picture is confirmed by detailed calculation.[27] In fact, if a pulse propagates in a medium showing an optical Kerr effect and a negative GVD (such as a silica optical fiber at $\lambda > 1300$ nm), pulse propagation, due to the presence of SPM, is described by a nonlinear wave equation that admits pulsed solutions that propagate without distortion. For the lowest order solution, the time behavior of the corresponding electric field is described by a hyperbolic secant function [i.e., $A(t') \propto \mathrm{sech}(t'/\tau_p)$], i.e., the pulse is unchirped, and the *whole* pulse propagates without distortion due to the mutual compensation of SPM and GVD, as just discussed. These solutions are called solitary solutions of the nonlinear wave equation or *solitons*. One of the most interesting properties of the lowest order soliton is that its pulse width $\Delta \tau_p$ (FWHM) is related to the pulse peak power P_p by the following equation:[27]

$$\Delta \tau_p^2 = \frac{3.11 |b_2|}{\gamma P_p} \qquad (8.6.40)$$

where $b_2 = (d^2 \beta / d\omega^2)_{\omega_L}$, $\gamma = n_2 \omega_L / c A_{eff}$, and A_{eff} is the effective area of the beam. ($A_{eff} = \pi w^2$ for a Gaussian beam of spot size w.) We can also show that $\gamma/2$ is simply the nonlinear phase shift per unit length per unit peak power [see also Eq. (8.6.38)].

One can now ask whether solitary pulses can be produced in a mode-locked laser cavity with an overall negative GDD and whose medium exhibits the optical Kerr effect. The answer to this question is that solitary solutions alone are unstable in a mode-locking laser cavity[28] but they can be stabilized by some nonlinear loss mechanism producing a self-amplitude modulation (see Fig. 8.22 or Fig. 8.24). In this case, if we neglect pulse broadening due to the finite gain bandwidth and higher order dispersion, we can get an approximate steady-state solution that is equivalent to that of fundamental-soliton propagation in optical fibers with anomalous dispersion. In particular, according to Eq. (8.6.40), the pulse duration turns out to be proportional to the inverse of the pulse energy E ($E \cong 2.27 P_p \Delta \tau_p$) according to the relation[29]

$$\Delta \tau_p = \frac{3.53 |\phi''|}{\delta E} \tag{8.6.41}$$

where ϕ'' is the GDD for a round-trip in the laser cavity and δ is the nonlinear round-trip phase shift per unit power in the Kerr medium (and thus given by $\delta = \gamma l_K$, where l_K is the length of the Kerr medium). From Eq. (8.6.41), taking $\phi'' = -200$ fs^2, $\delta \approx 10^{-6}$ W^{-1}, and $E \approx 50$ nJ (as appropriate for a mode-locked Ti:sapphire laser), we obtain $\Delta \tau_p \cong 14$ fs. Indeed solitary solutions have been observed in both mode-locking Ti:sapphire and dye lasers by carefully adjusting the laser parameters.[29,30] In both of these cases, mode locking exploits self-amplitude modulation, such as that occurring by Kerr lens mode-locking or by a slow-saturable-absorber-, dynamic-gain-saturation mechanism. In both cases, very short (10–20 fs) pulses are generated by either the soliton mechanism alone or the combined action of pulse broadening due to the finite-gain bandwidth. Note that very short mode-locking light pulses ($\sim$100 fs) are produced by this soliton-type mechanism even using mode-locking elements providing a much slower time window of net gain (e.g., semiconductor saturable absorbers with picosecond relaxation time).[31] In this case the mode-locking element helps stabilizing the soliton solution, while the pulse duration is essentially determined by the soliton equation (8.6.41).

8.6.5. Mode-Locking Regimes and Mode-Locking Systems

Mode-locked lasers can be operated with either a pulsed or cw pump; depending on the type of mode-locking element and gain medium used, mode-locking regimes can be rather different. We briefly discuss some examples in this section.

In the pulsed case (Fig. 8.28a), the time duration $\Delta \tau_p'$ of the mode-locked train envelope has a finite value established by the mode-locking method being used. As discussed in Sect. 8.6.3.1, active AM and FM mode locking are commonly achieved by means of a Pockels cell providing electrooptic amplitude or phase modulation, respectively; in this case $\Delta \tau_p'$ is generally determined by the duration of the pump pulse. This occurs for instance in gain media with a rapid recovery time (τ of a few nanoseconds, e.g., dye lasers) that cannot be Q-switched. In this case $\Delta \tau_p'$ may typically be a few tens of microseconds. Passive mode locking is usually achieved by fast (a few tens of picoseconds) dye-solution saturable absorbers; for gain media with a slow recovery time (τ around a few hundreds of microseconds, e.g., solid-state lasers), the presence of the saturable absorber not only results in mode-locked but also in Q-switched operation. In this case, the duration $\Delta \tau_p'$ of the mode-locked train is established by the same consideration that determines the duration $\Delta \tau_p$ of Q-

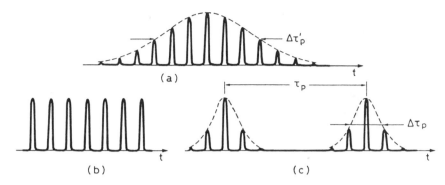

FIG. 8.28. Different mode-locking regimes: (a) ML with a pulsed pump. (b) cw ML with a cw pump. (c) ML with a cw pump and fast saturable absorber, showing the simultaneous occurrence of ML and repetitive Q-switching.

switched pulse behavior (generally a few tens of nanoseconds; see Sect. 8.4.4). Note that, when a slow saturable absorber (τ around a few nanoseconds) is used with a slow-gain medium,* passive Q-switching with single-mode selection rather than mode locking tends to occur, due to the mode-selecting mechanism discussed in Sect. 8.4.2.4.

In the case of active mode locking and a cw pump, the output beam consists of a continuous train of mode-locked pulses (Fig. 8.28b), whose repetition rate depends on whether fundamental or harmonic mode locking is involved (see Fig. 8.19). As discussed in Sect. 8.6.3.1, active mode locking, in this case, is usually achieved by an acoustooptic modulator because of its lower insertion loss compared to a Pockels cell modulator. Continuous-wave passive mode locking, by a slow saturable absorber, can be achieved using slow saturable absorbers combined with fast-gain media (notably dye lasers). Continuous wave passive mode locking can also be achieved by nonlinear elements providing a rapid nonlinear loss (such as a fast saturable absorber or Kerr lens nonlinear element). When the latter is used with a slow-gain medium (a solid-state medium) however, care must be exercised to avoid the simultaneous occurrence of repetitive Q-switching.[32] If this situation is not avoided, the system may operate either repetitively Q-switched with mode-locking (Fig. 8.28c) or repetitively Q-switched without mode locking. In both of these cases, the time duration of the Q-switched pulse $\Delta\tau_p$ and the Q-switching repetition rate $1/\tau_p$ (see Fig. 8.28c) are established by the dynamics of the passive Q-switching process.

A large number of lasers have been made to operate mode-locked, both actively and passively, including many gas lasers (e.g., He-Ne, Ar ion, and CO_2 lasers), all of the commonly used solid-state lasers, many semiconductor lasers, and many dye lasers. As representative examples, Table 8.1 shows the most commonly used media, providing picosecond and femtosecond laser pulses in cw mode locking, and the corresponding values for gain linewidth $\Delta\nu_0$, peak-stimulated-emission cross section σ, and upper-state lifetime τ.

---

* We use the terms fast (rapid) and slow for recovery time differently for absorbers and gain media. Recovery time of a saturable absorber is slow when its value (typically a few nanoseconds) is comparable to typical cavity round-trip time. This lifetime is typical for absorbers whose decay is determined by spontaneous emission by an electric-dipole-allowed transition. Recovery time is fast (a few picoseconds or shorter) when it is comparable to the typical duration of a mode-locked pulse. The lifetime of a gain medium is considered fast when comparable to the cavity round-trip time; this occurs in an electric-dipole-allowed laser transition. The lifetime of a gain medium is considered slow when it corresponds to an electric-dipole-forbidden transition.

TABLE 8.1. Most common media providing picosecond and femtosecond laser pulses together with the corresponding values of: (a) Gain linewidth, $\Delta\nu_0$; (b) peak stimulated-emission cross-section, σ; (c) upper state lifetime, τ; (d) shortest pulse duration so far reported, $\Delta\tau_p$; (e) shortest pulse duration, $\Delta\tau_{mp}$, achievable from the same laser

Laser medium	$\Delta\nu_0$	σ [10^{-20} cm^2]	τ [μs]	$\Delta\tau_p$	$\Delta\tau_{mp}$
Nd:YAG $\lambda = 1.064\ \mu m$	135 GHz	28	230	5 ps	3.3 ps
Nd:YLF $\lambda = 1.047\ \mu m$	390 GHz	19	450	2 ps	1.1 ps
Nd:YVO$_4$ $\lambda = 1.064\ \mu m$	338 GHz	76	98	<10 ps	1.3 ps
Nd:glass $\lambda = 1.054\ \mu m$	8 THz	4.1	350	60 fs	55 fs
Rhodamine 6G $\lambda = 570$ nm	45 THz	2×10^4	5×10^{-3}	27 fs	10 fs
Cr:LiSAF $\lambda = 850$ nm	57 THz	4.8	67	18 fs	8 fs
Ti:sapphire $\lambda = 850$ nm	100 THz	38	3.9	6–8 fs	4.4 fs

In the same table, the shortest pulse duration $\Delta\tau_p$ so far achieved and the minimum pulse duration $\Delta\tau_{mp} \cong 0.44/\Delta\nu_0$, achievable from that particular laser are also shown. Note that, according to Eq. (7.3.12), threshold pump power is inversely proportional to $\sigma\tau$. Thus, for a given gain medium, $1/\sigma\tau$ can be taken as a figure of merit for achieving the lowest threshold while, of course, $1/\Delta\nu_0$ represents a figure of merit for producing the shortest pulses. Note also that, since $\sigma \propto 1/\Delta\nu_0$, lasers capable of the shortest pulses tend to have the highest threshold. Mode-locked lasers are also quite diversified in terms of their configuration, and it is beyond the scope of this book to provide detailed descriptions of these many and varied systems. We therefore limit ourselves to describing just two cases of cw femtosecond lasers, representing particularly relevant and up-to-date examples: These are the colliding pulse mode-locked (CPM) rhodamine 6G laser and the Ti:sapphire Kerr-lens mode-locked laser.

In the CPM rhodamine 6G dye laser (Fig. 8.29), a ring laser cavity is used; the dye active medium, located at the beam waist between the two focusing mirrors M_2 and M_3, consists of a solution of rhodamine 6G in ethylene glycol flowing in a jet stream orthogonal to the plane of the figure. The laser is passively mode-locked by a slow saturable absorber located at the beam waist between the two focusing mirrors M_4 and M_5; it consists of a solution of DODCI in ethylene glycol, again flowing as a jet stream orthogonal to the plane of the figure.[33] The active medium is quasi-longitudinally pumped by a cw Ar ion laser beam, focused in the jet stream by mirror M_1. The ring configuration leads to the generation of two oppositely traveling femtosecond laser pulses that meet each time (i.e., collide) at the saturable absorber jet position. Due to the associated formation of a standing wave pattern in the saturable absorber, absorber saturation is enhanced, which increases the peak net gain generated (Fig. 8.24). The two pulses meet at time intervals separated by L_p/c, where L_p is

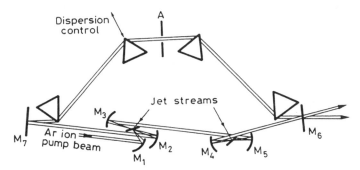

FIG. 8.29. Arrangement for a colliding-pulse mode-locked (CPM) ring dye laser including two-prism pairs for dispersion control.

the length of the ring perimeter. The rhodamine 6G dye jet is positioned at a distance $L_p/4$ from the saturable absorber. As we can see from Fig. 8.30b, this ensures that single pulses passing through the rhodamine 6G gain medium are equally spaced in time by $L_p/2c$. This time symmetry allows the saturated gain of the rhodamine 6G to be the same for the two pulses, thus providing the best mode-locking performances. To control the cavity GDD, the four-prism sequence in Fig. 8.26 is also introduced in the ring. In this way, the overall GDD can be minimized to generate pulses of ~ 50 fs. Under special operating conditions, mode locking can be further enhanced by a soliton-type mechanism, via the SPM phenomenon within the jet streams and with the required amount of negative GDD provided by the four-prism system.[30] In this operating regime, pulses as low as 27 fs have been generated (the shortest for a mode-locked dye laser).

In the Ti:sapphire Kerr-lens mode-locked laser, a 10 mm thick plate of Ti:sapphire is longitudinally pumped, in a z-folded linear laser cavity, by the focused beam of an Ar ion laser (Fig. 8.31). Kerr-lens-type mode locking (see Fig. 8.23) is achieved by exploiting the optical Kerr lens induced in the sapphire plate combined with a suitable aperture at one end of the cavity. To obtain the largest value for Kerr-lens nonlinearity, so that the laser becomes

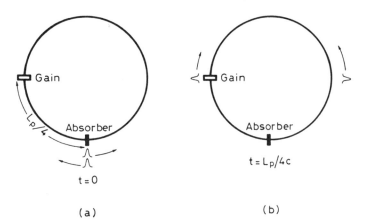

FIG. 8.30. Schematic representation of a CPM ring dye laser: (a) Time $t = 0$ when the two counter-propagating pulses meet (collide) at the location of the saturable absorber; (b) time $t = L_p/4c$, where L_p is the length of the ring perimeter, when one pulse passes through the dye gain medium.

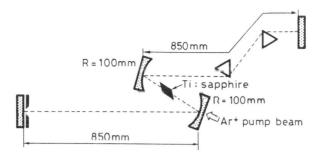

FIG. 8.31. Arrangement for a KLM mode-locked Ti:sapphire laser using a symmetric, *z*-folded, linear cavity and one-prism pair for dispersion control.

self-starting, the two arms of the *z*-shaped cavity must be made equal in length.[34] A two-prism sequence (i.e., the first half of the four-prism sequence in Fig. 8.26) is used for GDD control. Under optimum operating conditions, when the two-prism sequence compensates the positive GDD of sapphire, pulses as short as 30 fs can be obtained from such a laser. By decreasing the sapphire thickness to ~ 2 mm, one can reduce the positive GDD of the active medium to a value sufficiently low to be compensated by the negative GDD of intracavity chirped mirrors in a multipass configuration.[35] In this way the problem of third-order cavity dispersion is greatly reduced and pulses with the record low value of 6–8 fs have been obtained.[35,36]

8.7. CAVITY DUMPING

The cavity dumping technique allows the photon energy, contained in the laser, to be coupled out of the cavity in a time equal to the cavity round-trip time.[37] The principle of this technique is shown in Fig. 8.32, where the laser cavity is assumed to be made of two 100% reflecting mirrors and the output beam is taken from a special kind of output coupler. The reflectivity $R = R(t)$ of this coupler is in fact held at zero until the desired time when it is suddenly switched to 100%. This coupler thus dumps out of the cavity, in a double transit, whatever optical power is circulating in the laser cavity. Alternatively, if the reflectivity R of the output coupler is switched to a value less than 100%, the cavity dumper still works correctly if coupler reflectivity is held, after switching, at its high value for a time equal to the cavity round-trip time and then returned to zero. Cavity dumping is a general technique that can be used to advantage whether the laser is mode-locked, cw, or *Q*-switched. In the

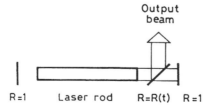

FIG. 8.32. Principle of laser cavity dumping.

discussion that follows we limit ourselves to considering the case of cavity dumping in a mode-locked laser, since this is one of the cases where cavity dumping is most often used.

For pulsed mode-locked lasers, cavity dumping is usually carried out when the intracavity mode-locked pulse reaches its maximum value (see Fig. 8.28a). In this way a single ultrashort pulse of high intensity is coupled out of the laser cavity. This type of dumping is often obtained by a Pockels cell electrooptic modulator used in a configuration similar to that used for Q-switching (see Fig. 8.5a). In this case the reflected beam from the polarizer is taken as output; to obtain switching to 100% reflectivity, the voltage to the Pockels cell is switched from zero to its $\lambda/4$ voltage when cavity dumping is required.

For a cw mode-locked laser, the cavity dumping technique can be used repetitively to produce a train of ultrashort pulses whose repetition rate is now set by the repetition rate of the dumper rather than by the repetition frequency of the mode-locking process. If this rate is low enough (typically from 100 kHz–1 MHz), the corresponding time interval between two successive cavity dumping events (1–10 μs) allows sufficient time for mode locking to reestablish itself fully. Repetitive cavity dumping therefore allows to obtain a train of ultrashort laser pulses of much lower repetition rate and hence much higher peak power that those obtained by ordinary mode locking. These two properties are often a desirable feature when ultrashort pulses are used. Note that, if the output coupler reflectivity is less than 100%, the coupler must be switched on and off so that the time for the on-state equals the cavity round-trip time. In this case, a mode-locked pulse of reduced intensity remains in the cavity after dumping; since mode locking does not have to start again from noise, the system then works more reliably.

Given its lower insertion loss, the acoustooptic cavity dumper is often used for this application. It consists of an acoustooptic modulator working in the Bragg regime and traveling wave mode, where the diffracted beam is taken as the output coupled beam. Figure 8.33 shows its configuration, which differs from that shown for Q-switching in Fig. 8.7a in three ways: (1) The rf oscillator, which drives the piezoelectric transducer, now oscillates at a much higher frequency (e.g., $\nu = 380$ MHz). Its output is gated so that the rf envelope's pulse of duration τ_p equals the cavity round-trip time (e.g., $\tau_p = 10$ ns). Cavity dumping thus occurs when the resulting acoustic pulse interacts with the cavity beam. This pulse must

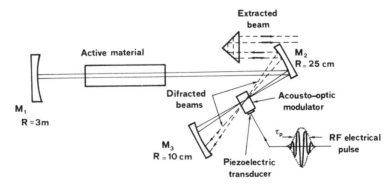

FIG. 8.33. Common arrangement for cavity dumping of a cw-pumped laser (e.g., Nd:YAG or Ar ion). Mirrors M_1–M_3 are nominally 100% reflecting at the laser wavelength. The dashed lines indicate the beam that is diffracted by the modulator. For cavity dumping of a mode-locked laser, a mode locker is also inserted at one end of the cavity (e.g., near mirror M_1.)

therefore be synchronized with the circulating mode-locked pulse, so that the two pulses meet in the modulator. Note that the high carrier frequency serves the double purpose of allowing amplitude modulation by short pulses ($\tau_p = 10$ ns) and of producing a higher diffraction angle θ_d. In fact, since $\theta_d = \lambda/\lambda_a$, where λ_a is the acoustic wavelength, the diffraction angle increases linearly with the carrier frequency. (2) The beam is focused in a very small spot size in the modulator. The time duration of the optical coupling is, in fact, not only established by the duration of the acoustic pulse but also by the pulse transit time through the focused laser beam. Taking as an example a spot diameter of $d = 50$ μm and a sound velocity of $v = 3.76 \times 10^5$ cm/s (shear wave velocity in fused quartz), then $t = d/v = 13.3$ ns. (3) Circulating and diffracted laser pulses interact twice with the acoustic pulse within the modulator. This is achieved with the help of mirror M_3, which also refocuses the scattered beam back into the modulator. In this way higher diffraction efficiency ($\sim 70\%$) is obtained.

8.8. CONCLUDING REMARKS

This chapter considers several cases of transient laser behavior in some detail. Generally speaking these cases fall into two categories: (1) Laser transients occurring on a time scale appreciably longer than the cavity round-trip time. This includes the phenomena of relaxation oscillations, Q-switching, and gain switching. In this case, to first order, we can describe laser light in the cavity in terms of the total number of photons, more or less uniformly filling the cavity, and undergoing a time evolution according to the dynamic situation involved. (2) Laser transients occurring on a time scale appreciably shorter and often much shorter than the cavity round-trip time. This category includes all mode-locking cases of practical interest and some cases of cavity dumping. In this case, laser light in the cavity can be described in terms of a light pulse traveling back and forth in the cavity. For both kinds of transient behavior and in particular for Q-switching and mode-locking, several techniques for inducing the required transient behavior are discussed. In doing so, new physical phenomena are described and characterized, including the interaction of light and sound waves, pulse propagation in dispersive media, and such nonlinear optical phenomena as self-phase modulation, self-focusing, and the formation of solitary waves. Thus we have acquired a new understanding of the light matter interaction, particularly under transient conditions.

The level of treatment is subject to several limitations. All but the simplest analytic treatments are avoided, since we aimed at developing a deeper physical understanding of the complex phenomena involved. In particular, complications arising from spatial variation in the laser beam and pump rate (e.g., variations in the transverse beam profile) are ignored here. Thus the material discussed in this chapter represents, in the author's opinion, the minimum basic knowledge required for a comprehensive and balanced understanding of transient laser behavior.

PROBLEMS

8.1. For the Nd:YAG Q-switched laser considered in Fig. 8.12, calculate the expected threshold energy, output energy, and pulse duration, for $E_{in} = 10$ J, when output coupling is reduced to 20%.

8.2. Consider a Pockels cell in the so-called longitudinal configuration, i.e., with the dc field applied in the direction of the beam passing through the nonlinear crystal. In this case, the induced birefringence $\Delta n = n_x - n_y$ is given by $\Delta n = n_o^3 r_{63} V / L'$, where n_o is the ordinary refractive index, r_{63} is the appropriate nonlinear coefficient of the material, V is the applied voltage, and L' is the crystal length. Derive an expression for the voltage required to keep the polarizer-Pockels-cell combination in Fig. 8.5a in the closed position.

8.3. For a Pockels cell made of KD_2PO_4 (deuterated potassium dihydrogen phosphate, also known as KD*P), the value of the r_{63} coefficient at $\lambda = 1.06 \ \mu m$ is $r_{63} = 26.4 \times 10^{-12}$ m/V, while $n_o = 1.51$. Using results from Problem 8.2 calculate the voltage to be applied for the closed position.

8.4. The Nd:YAG laser in Figs. 7.4 and 7.5 is pumped at a level of $P_{in} = 10$ kW and repetitively Q-switched at a 10-kHz repetition rate by an acoustooptic modulator (whose insertion losses are assumed negligible). Calculate the energy and duration of the output pulses as well as the peak and average powers expected for this case.

8.5. Derive expressions for output energy and pulse duration that apply to a Q-switched quasi-three-level laser.

8.6. A He-Ne laser beam with a wavelength (in air) of $\lambda = 632.8$ nm is deflected by a $LiNbO_3$ acoustooptic deflector operating in the Bragg regime at the acoustic frequency of 1 GHz (the highest acoustic frequency possible in $LiNbO_3$ without introducing excessive losses). Assuming a sound velocity in $LiNbO_3$ of 7.4×10^5 cm/s and a refractive index of $n = 2.3$, calculate the angle through which the beam is deflected.

8.7. Consider a rhodamine 6G dye laser transversely pumped by an atmospheric pressure nitrogen laser (see Chap. 9) with a pulse duration t_p sufficiently short to make the laser operate in gain-switched mode. Assume: (a) The dye is dissolved in ethanol ($n = 1.36$); (b) a length $l = 5$ mm for the dye cell, with the whole length pumped by the focused beam of the nitrogen laser; (c) a circular cross section of the pumped region having diameter $D = 50 \ \mu m$; (d) two end mirrors directly attached to the ends of the dye cell (i.e., $L \cong l$), where one mirror is 100% reflecting and the other has a power transmission of 50%; (e) all other cavity losses are negligible; (f) an effective stimulated emission cross section, at the $\lambda = 570$ nm lasing wavelength, of $\sigma_e = 2 \times 10^{-16}$ cm^2; (g) a square pump pulse of duration t_p much shorter than the upper state lifetime of the rhodamine 6G solution ($\tau \cong 5$ ns). Assume also that the pump amplitude produces a peak inversion four times larger than the threshold inversion. Calculate: (a) The time at which threshold is reached; (b) time behavior of the net gain; (c) time behavior of the photon number, neglecting gain saturation; (d) time duration of the pump pulse t_p such that at the end of the pump pulse, the photon number has reached the value $\phi_p/20$, where the peak value ϕ_p is calculated according to the theory of fast switching.

8.8. Derive Eq. (8.6.6).

8.9. Assuming a phase relation as in Eq. (8.6.13), show that the resulting electric field of the pulse can be written as in Eq. (8.6.14) and calculate the values of α and β as a function of $\Delta\omega_L$ and φ_2.

8.10. The oscillation bandwidth (FWHM) of a mode-locked He-Ne laser is 1 GHz, the spacing between consecutive modes is 150 MHz, and the spectral envelope can be approximately described by a Gaussian function. For fundamental mode locking, calculate the corresponding duration of the output pulses and the pulse repetition rate.

8.11. Assume that the phase relation between consecutive modes is such that $\varphi_{l+1} - \varphi_l = \varphi_l - \varphi_{l-1} + \pi$ and the spectral amplitude is constant over $2n$ modes. Show that the pulse repetition rate is now equal to $2\Delta\nu$, where $\Delta\nu = c/2L$ (case of second harmonic mode locking). Hint: Label the modes with the indices $l = 0$, 1, 2, starting with the mode with the lowest frequency to show that one can write $\varphi_l = l\varphi$ for even l, and $\varphi_l = l\varphi + (\pi/2)$ for odd l where φ is a constant. Then consider separately the even-mode sum and the odd-mode sum and show that they produce two time-intercalated, mode-locked trains.

8.12. Prove Eq. (8.6.16).

8.13. Assuming a uniform laser output spectrum as in Fig. 8.16a, calculate the ratio between the peak power, for fundamental mode-locked operation, and average power, when the modes have random phases.

8.14. For a random-noise intensity pattern as in Fig. 8.15, the probability density p_I (i.e., $p_I \, dI$ gives the elemental probability that beam intensity is between I and $I + dI$) is given by $p_I \propto \exp[-(I/I_0)]$ (see also Fig. 11.11b). Calculate the average beam intensity and the probability that, in the noisy pattern in Fig. 8.15, there is a value exceeding $2I_0$.

8.15. By approximating the sum over all modes in Eq. (8.6.10) with an integral, an important characteristic of the output behavior is lost. What is it?

8.16. Consider a Kerr lens mode-locked Ti:sapphire laser; according to Eq. (8.6.20), assume that total round-trip cavity losses can be written as $2\gamma_l = 2\gamma - kP$, where P is the peak intracavity laser power and the nonlinear loss coefficient k, due to the Kerr lens mode-locking mechanism, can be taken as $\approx 5 \times 10^{-8}$ W^{-1}. Assume a saturated round-trip gain of $2g_0' \cong 0.1$, a gain bandwidth of 100 THz, and an intracavity laser energy of $W = 40$ nJ. Calculate the pulse duration achievable in the limiting case where the effects of cavity dispersion and self-phase modulation can be neglected.

8.17. Assuming a GVD for quartz at $\lambda \cong 800$ nm of 50 fs^2/mm, calculate the maximum thickness of a quartz plane that an initially unchirped 10-fs pulse, of Gaussian intensity profile, can traverse if the output pulse duration is not to exceed input pulse duration by more than 20%.

8.18. Consider a Nd:phosphate glass laser whose linewidth can be taken to be homogeneously broadened to $\Delta\nu_0 \cong 6$ THz. Assuming a saturated round-trip gain of 5% and a peak round-trip loss change of 2%, due to a fast self-amplitude modulation mechanism, calculate the expected width of the mode-locked pulses.

REFERENCES

1. H. Statz and G. de Mars, Transients and Oscillation Pulses in Masers, in *Quantum Electronics* (C. H. Townes, ed.) (Columbia Univ. Press, New York, 1960), p. 530.
2. R. Dunsmuir, Theory of Relaxation Oscillations in Optical Masers, *J. Electron. Control*, **10**, 453 (1961).
3. D. F. Nelson and W. S. Boyle, A Continuously Operating Ruby Optical Maser, *Appl. Optics*, **1**, 181 (1962).
4. N. B. Abraham, P. Mandel, and L. M. Narducci, Dynamical Instabilities and Pulsations in Lasers, in *Progress in Optics Vol. XXV* (Emil Wolf, ed.) (North-Holland, Amsterdam, 1988), pp. 3–167.
5. M. Sargent, M. O. Scully, and W. E. Lamb, *Laser Physics* (Addison-Wesley, London, 1974).
6. C. L. Tang, H. Statz, and G. de Mars, Spectral Output and Spiking Behavior of Solid-State Lasers, *J. Appl. Phys.*, **34**, 2289 (1963).
7. K. Otsuka *et al.*, Alternate Time Scale in Multimode Lasers, *Phys. Rev.*, **46**, 1692 (1992).
8. R. W. Hellwarth, Control of Fluorescent Pulsations, in *Advances in Quantum Electronics* (J. R. Singer, ed.) (Columbia Univ. Press, New York, 1961), pp. 334–341.
9. W. Koechner, *Solid-State Laser Engineering*, 4th ed., Springer Series in Optical Sciences (Springer-Verlag, Berlin, 1996), Chap. 8.

10. A. Yariv, *Optical Electronics*, 4th ed. (Saunders College Publ., Fort Worth, 1991), Chap. 12.
11. W. R. Sooy, The Natural Selection of Modes in a Passive *Q*-Switched Laser, *Appl. Phys. Lett.*, **7**, 36 (1965).
12. W. G. Wagner and B. A. Lengyel, Evolution of the Giant Pulse in a Laser, *J. Appl. Phys.*, **34**, 2040 (1963).
13. W. Koechner, *Solid-State Laser Engineering*, Vol. 1, Springer Series in Optical Sciences (Springer-Verlag, Berlin, 1976), Chap. 11, adapted from Fig. 11.23.
14. L. W. Casperson, Analytical Modeling of Gain-Switched Lasers. 1. Laser Oscillators, *J. Appl. Phys.*, **47**, 4555 (1976).
15. A. E. Siegman, *Lasers* (Oxford University Press, Oxford, 1986), Chap. 27 and 28.
16. Reference 15, Section 27.5.
17. D. J. Kuizenga and A. E. Siegman, FM and AM Mode-Locking of the Homogeneous Laser—Part I: Theory, *IEEE J. Quantum Electron.*, **QE-6**, 694 (1970).
18. H. Haus, A Theory of Forced Mode-Locking, *IEEE J. Quantum Electron.*, **QE-11**, 323 (1975).
19. H. Haus, Theory of Mode Locking with a Fast Saturable Absorber, *J. Appl. Phys.*, **46**, 3049 (1975).
20. U. Keller, Ultrafast All-Solid-State Laser Technology, *Appl. Phys. B*, **58**, 347 (1994).
21. D. E. Spence, P. N. Kean, and W. Sibbett, 60-fs Pulse Generation from a Self-Mode-Locked Ti:Sapphire Laser, *Opt. Letters*, **16**, 42 (1991).
22. M. Piché, Beam Reshaping and Self-Mode-Locking in Nonlinear Laser Resonators, *Opt. Commun.*, **86**, 156 (1991).
23. G. H. C. New, Pulse Evolution in Mode-Locked Quasi-Continuous Lasers, *IEEE J. Quantum Electron.*, **QE-10**, 115 (1974).
24. G. P. Agrawaal, *Nonlinear Fiber Optics*, 2nd ed. (Academic Press, San Diego, 1995), Section 3.3.
25. R. L. Fork, O. E. Martinez, and J. P. Gordon, Negative Dispersion Using Pairs of Prisms, *Opt. Lett.*, **9**, 150 (1984).
26. R. Szipöcs, K. Ferencz, C. Spielmann, and F. Krausz, Chirped Multilayer Coatings for Broadband Dispersion Control in Femtosecond Lasers, *Opt. Lett.*, **19**, 201 (1994).
27. Reference 24, Chapter 5.
28. H. A. Haus, J. G. Fujimoto, and E. P. Ippen, Structures for Additive Pulse Mode Locking, *J. Opt. Soc. Am. B*, **8**, 2068 (1991).
29. C. Spielmann, P. F. Curley, T. Brabec, and F. Krausz, Ultrabroadband Femtosecond Lasers, *IEEE J. Quantum Electron.*, **QE-30**, 1100 (1994).
30. J. A. Valdmanis, R. L. Fork, and J. P. Gordon, Generation of Optical Pulses as Short as 27 Femtosecond Directly from a Laser Balancing Self-Phase Modulation, Group-Velocity Dispersion, Saturable Absorption, and Gain Saturation, *Opt. Lett.*, **10**, 131 (1985).
31. U. Keller *et al.*, Semiconductor Saturable Absorber Mirrors (SESAMs) for Femtosecond to Nanosecond Pulse Generation in Solid-State Lasers, *J. Select. Topics Quantum Electron.*, Dec 1996.
32. H. A. Haus, Parameter Ranges for cw Passive Mode Locking, *IEEE J. Quantum Electron.*, **QE-12**, 169 (1976), F. X. Kärtner *et al.*, Control of Solid-State Laser Dynamics by Semiconductor Devices, *Opt. Engineering*, **34**, 2024 (1995).
33. R. L. Fork, I. Greene, and C. V. Shank, Generation of Optical Pulses Shorter than 0.1 ps by Colliding Pulse Mode Locking, *Appl. Phys. Lett.*, **38**, 671 (1991).
34. G. Cerullo, S. De Silvestri, and V. Magni, Self-Starting Kerr Lens Mode Locking of a Ti:Sapphire Laser, *Opt. Letters*, **19**, 1040 (1994).
35. A. Stingl, M. Lenzner, Ch. Spielmann, F. Krausz, and R. Szipöcs, Sub-10-fs Mirror-Controlled Ti:Sapphire Laser, *Opt. Lett.*, **20**, 602 (1995).
36. I. D. Jung *et al.*, Self-Starting 6.5-fs Pulses from a Ti:Sapphire Laser, *Opt. Lett.*, **22**, 1009 (1997).
37. Reference 9, Section 8.6.
38. D. C. Hanna, B. Luther-Davies, and R. C. Smith, Single Longitudinal Mode Selection of High Power Actively *Q*-Switched Lasers, *Opto-Electronics*, **4**, 249 (1972).

9

Solid-State, Dye, and Semiconductor Lasers

9.1. INTRODUCTION

Chapter 9 considers the most important types of lasers involving high-density active media, namely solid-state, dye, and semiconductor lasers. The chapter concentrates on examples in widest use whose characteristics are representative of a whole class of lasers. The main emphasis is on the laser's physical behavior and how this relates to general concepts developed in previous chapters. Some engineering details are also given with the aim of providing a better insight into the behavior of particular lasers. To complete the picture, data relating to laser performances (e.g., oscillating wavelength(s), output power or energy, wavelength tunability, etc.) are also included to suggest laser applications. The following items are generally covered for each laser: Relevant energy levels, excitation mechanisms, characteristics of the laser transition, engineering details of the laser's structure(s), characteristics of the output beam, and applications.

9.2. SOLID-STATE LASERS

The term solid-state laser is generally reserved for lasers having ions introduced as an impurity in an otherwise transparent dielectric-host material (in crystalline or glass form). Thus semiconductor lasers are not usually included in this category, the mechanisms for pumping and for laser action being in fact quite different. These are considered in a separate section.

Ions belonging to one of the series of transition elements of the Periodic Table, in particular rare earth (RE) or transition-metal ions, are generally used as the active impurities. For host crystals, either oxides, e.g., Al_2O_3, or fluorides, e.g., $YLiF_4$ (abbreviated YLF), are most often used.[1] The Al^{3+} site is too small to accommodate RE ions, so it is generally used for transition-metal ions. A suitable combination of oxides to form synthetic garnets, such as $Y_3Al_5O_{12} = (1/2)(3\,Y_2O_3 + 5\,Al_2O_3)$, are often used and the Al^{3+} site can accommodate transition metal ions while the Y^{3+} site can be used for RE ions. Other oxides include YVO_4 for Nd^{3+} ions and alexandrite for Cr^{3+} ions. Among the fluorides, YLF is used as a host for RE, while $LiSrAlF_6$ (abbreviated LiSAF) or $LiCaAlF_6$ (abbreviated LiCAF) are used for transition metals—most notably for Cr^{3+} ions.

A comparison between oxides and fluorides shows that oxides are harder, with better mechanical and thermo-mechanical (i.e., higher thermal fracture limit) properties. Fluorides, on the other hand, show better thermo-optical properties (i.e., lower thermally induced lensing and birefringence). Glasses from either the silicate (i.e., based on SiO_2) or phosphate (i.e., based on P_2O_5) family have so far been used only for RE ions. Compared to crystals, many glasses have much lower melting temperatures; therefore they are easier and cheaper to fabricate even in large dimensions. On the other hand, glasses have much lower (by approximately an order of magnitude) thermal conductivity, which leads to much worse thermo-mechanical and thermo-optical properties. A comparison between various glasses reveals silicates to have better thermal and mechanical properties, while phosphates show better thermooptical and nonlinear optical properties.

The general electronic structure of a RE is $4f^N 5s^2 5p^6 5d^0 6s^2$ as shown for Nd, Er, Yb, Tm, and Ho in Table 9.1, where for comparison the structure of Xe is also indicated. When a RE is inserted into a host material, the two $6s$ electrons and one of the $4f$ electrons are used for ionic binding, so that the RE presents itself as a triply ionized ion (e.g., $N-1=3$ for Nd^{3+}). The remaining $N-1$ electrons can then arrange themselves in different states of the $4f$ shell, resulting in a large number of energy levels. In fact these states are split by three types of interaction, namely, Coulomb interaction among the $4f^{N-1}$ electrons, spin-orbit coupling, and crystal-field interaction. The Coulomb interaction is the strongest of these three, and it splits the $4f$ states into sublevels typically separated by $\sim 10,000$ cm^{-1}. Spin-orbit coupling then splits each term into manifolds typically separated by ~ 3000 cm^{-1}. The

TABLE 9.1. Electronic configurations of some rare earth and transition metals of interest as laser-active impurities

Atom	Electron Configuration
Xenon, Xe[a]	$(Kr)4d^{10}5s^2 5p^6$
Neodymium, Nd	$(Xe)4f^4 5d^0 6s^2$
Holmium, Ho	$(Xe)4f^{11} 5d^0 6s^2$
Erbium, Er	$(Xe)4f^{12} 5d^0 6s^2$
Thulium, Tm	$(Xe)4f^{13} 5d^0 6s^2$
Ytterbium, Yb	$(Xe)4f^{14} 5d^0 6s^2$
Chromium, Cr	$(Ar)3d^5 4s^1$
Titanium, Ti	$(Ar)3d^2 4s^2$
Cobalt, Co	$(Ar)3d^7 4s^2$
Nickel, Ni	$(Ar)3d^8 4s^2$

[a] For reference, the fundamental configuration of Xe is also shown.

crystal-field interaction produces the weakest perturbation (weakened by the screening effect of $5s^2$ and $5p^6$ orbitals), thus further splitting each manifold into sublevels with an energy separation typically of 200 cm^{-1}.

All relevant absorption and emission features are due to transitions between these $4f$ states ($4f$-$4f$ transitions). Electric dipole transitions within the $4f$ shell are parity-forbidden, so it requires a mixture of wave functions with opposite parity, brought about by the crystal field interaction, to create nonzero, although still weak, transition probabilities. Thus we generally find long (hundreds of microseconds) radiative lifetimes. Furthermore, due to the screening from the $5s^2$ and $5p^6$ orbitals, electron–phonon coupling is very weak. One thus has sharp transition lines and weak nonradiative decay channels for low ion doping. (Ion-ion interaction can lead to nonradiative decay at high RE ion concentrations, see Fig. 2.13). From the preceding considerations we expect large values of the overall lifetime τ and of the product $\sigma\tau$, where σ is the peak cross section. This implies a low threshold pump power for laser action, since, e.g., for a four-level laser, the threshold pump rate is proportional to $1/\sigma\tau$ [see Eq. (7.3.3)].

Electronic configurations for transition metals of interest in laser action are also shown in Table 9.1. Note that the electronic configuration of the most important active species, i.e., Cr, is given by $(Ar)3d^5 4s^1$, while those of Ti, Co, and Ni can be written in the general form $(Ar)3d^N 4s^2$ (with $N = 2$ for Ti, 7 for Co, and 8 for Ni). When a Cr atom is added to an ionic crystal, the one electron belonging to its $4s$ orbital and two $3d$ electrons are used for ionic binding and Cr is found as a triply ionized ion with three electrons left in the $3d$ shell. In titanium, the two $4s$ electrons and one $3d$ electron are used for ionic binding, and Ti is again present as a triply ionized ion with only one electron left in the $3d$ shell. In both Co and Ni, the two $4s$ electrons are used for binding and these elements are present as doubly ionized ions. In all cases, the remaining electrons in the $3d$ orbital can arrange themselves in a large number of states (e.g., 24 for Cr^{3+}), and all absorption and emission features of transition-metal ions arise from $3d$-$3d$ transitions. Lacking the screening that occurs in RE ions, the $3d$ states interact strongly with the crystal field of the host, and, as we see later, this is the fundamental reason for the vibronic character of the corresponding transitions leading to wide absorption and emission bands. Again electric-dipole transitions within the $3d$ shell are parity-forbidden, but, due to the stronger crystal field compared to the RE case, the $3d$-$3d$ transitions are more allowed and thus the lifetimes are significantly shorter (a few microseconds) than those of the $4f$-$4f$ transitions in RE ions. Compared to Nd:YAG, e.g., transition cross sections are somewhat smaller, so that the product $\sigma\tau$ is now typically one order of magnitude smaller.

To conclude this section we observe that ions belonging to the actinide series, notably U^{3+}, were also used in the early days of laser development. (The U^{3+} laser was the second solid-state laser to be developed, i.e., immediately after the ruby laser.) These ions are no longer used, but they deserve a mention here for historic reasons.

9.2.1. Ruby Laser

This was the first type of laser made to operate (T. H. Maiman, June 1960[2,3]), and it continues to be used in some applications.[4] As a naturally occurring precious stone, ruby has been known for at least 2000 years. It consists of a natural crystal of Al_2O_3 (corundum) in which some of the Al^{3+} ions have been replaced by Cr^{3+} ions. For the laser material, one uses artificial crystals obtained by crystal growth from a molten mixture of Al_2O_3 with a small percentage of Cr_2O_3 (0.05% by weight).[4] Without the addition of Cr_2O_3, the colorless crystal that forms is known as sapphire. Due to the strong green and violet absorption bands of Cr^{3+} ions, it needs only that small addition of Cr_2O_3 to give the crystal a pink color (pink ruby). In the case of gem stones, the Cr^{3+} concentration is about an order of magnitude larger, giving them a strong red color (red ruby).

Energy levels in ruby are due to one of the three electrons of the Cr^{3+} ion, in the $3d$ shell, under the influence of the octahedral field at the Al site in the Al_2O_3 lattice. Corresponding levels of interest for laser action are shown in Fig. 9.1. The notation used to label the levels, which is derived from group theory, is not discussed at any length here. We limit ourselves to pointing out that the superscript to the left of each letter indicates multiplicity of the state, while the letter indicates the particular rotational symmetry of the state. As an example, the ground $^4A_2$ state has multiplicity $(2S + 1) = 4$, i.e., $S = 3/2$, where S is the total spin quantum number of the three $3d$ electrons. This means that the spins of these electrons are all parallel in this case.

Ruby has two main pump bands $^4F_1$ and $^4F_2$; transition peaks from the ground $^4A_2$ level to these bands occur at the wavelengths of 550 nm (green) and 420 nm (violet), respectively (see also Fig. 6.7). The two bands are connected by a very fast (picoseconds) nonradiative decay to both $2A$ and $\bar{E}$ states, which together form the ^2E state. The $2A$ and $\bar{E}$ states are connected to each other by a very rapid nonradiative decay, which leads to rapid thermalization of their populations; this results in the $\bar{E}$ level being the most heavily populated. Since the total spin of the ^2E state is $1/2$, the $^2E \to {}^4A_2$ transition is spin-forbidden. The relaxation time of both $2A$ and $\bar{E}$ levels to the ground state is thus very long ($\tau \cong 3$ ms), actually one of the longest among all solid-state laser materials.

From the preceding discussion, one sees that the level $\bar{E}$ accumulates the largest fraction of the pump energy; thus it is a good candidate for the upper laser level. In fact laser action usually occurs on the $\bar{E} \to {}^4A_2$ transition (R_1 line) at the wavelength $\lambda_1 = 694.3$ nm (red). Note that the frequency separation between $2A$ and E levels (~ 29 cm^{-1}) is small compared to kT/h (~ 209 cm^{-1} at $T = 300$ K), so that the $2A$ population is comparable to, although slightly smaller than, the $\bar{E}$-level population. Thus, it is also possible to obtain laser action on the $2A \to {}^4A_2$ transition (R_2 line, $\lambda_1 = 692.8$ nm). Ruby operates as a three-level laser. (Together with Er lasers it represents the most noteworthy example of a three-level laser.) As already discussed in connection with Fig. 2.10, at room temperature the R_1 transition is predominantly homogeneously broadened; the broadening arises from the

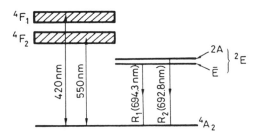

FIG. 9.1. Simplified energy levels of ruby.

interaction of Cr^{3+} ions with lattice phonons. The width of the transition (FWHM) is $\Delta\nu_0 \cong 11$ cm^{-1} (330 GHz) at $T = 300$ K. In summary Table 9.2 lists relevant optical and spectroscopic parameters of ruby at room temperature.

Ruby lasers are usually operated in a pulsed regime. For this, the pump configuration in Fig. 6.1, using a medium-pressure (~ 500 Torr) xenon flashtube, is generally used. Typical rod diameters range from 5–10 mm, with a length from 5–20 cm. Note that a helical flashtube surrounding the active rod was used in the earliest ruby lasers. Since this laser operates on a three-level scheme, threshold pump energy is typically an order of magnitude higher than that of other solid-state lasers operating with four-level schemes (e.g., neodymium lasers). Due to the long upper-state lifetime, ruby lasers lend themselves readily to Q-switched operation. Due to their relatively large laser linewidth, they can also produce short pulses (~ 5–10 ps) in mode-locked operation. Both active and passive methods can be used for Q-switching and mode locking. When slow saturable absorbers are used for Q-switching, the laser tends to operate in a single transverse and longitudinal mode due to the mode-selecting mechanism discussed in Sect. 8.4.2.4. With fast saturable absorbers (usually solutions of cyanine dyes), simultaneous Q-switched and mode-locked operation occurs (see Fig. 8.28a). Peak powers of a few tens of megawatts for Q-switching, and a few gigawatts when also mode-locked, are typical. Since the gain of the R_2 line is somewhat smaller than for the R_1 line, laser action on the R_2 line can be selected by using for instance the dispersive system in Fig. 7.16b. Ruby lasers can also run cw, transversely pumped by a high-pressure mercury lamp or longitudinally pumped by an Ar ion laser.

Once very popular, ruby lasers are now less widely used, since, due to their higher threshold, they have been superseded by such competitors as Nd:YAG or Nd:glass lasers. In fact, ruby lasers were extensively used in the past for the first mass production of military range finders, an application in which this laser is now completely replaced by other solid-state lasers (Nd:YAG, Nd:glass, Yb:Er:glass). However, ruby lasers are still sometimes used for a number of scientific and technical applications where its shorter wavelength compared to Nd:YAG, e.g., represents an important advantage. For example, in the case of pulsed holography, Nd:YAG lasers cannot be used due to the lack of response, in the infrared, of the high-resolution photographic materials that are used.

TABLE 9.2. Optical and spectroscopic parameters of ruby for room temperature operation

Property	Values and Units
Cr_2O_3 doping	0.05 wt.%
Cr^{3+} concentration	1.58×10^{19} ions/cm^3
Ouput wavelengths	694.3 nm (R_1 line)
	692.9 nm (R_2 line)
Upper laser level lifetime	3 ms
Linewidth of R_1 laser transition	11 cm^{-1}
Stimulated emission cross section σ_e	2.5×10^{-20} cm^2
Absorption cross section σ_a	1.22×10^{-20} cm^2
Refractive index ($\lambda = 694.3$ nm)	$n = 1.763$ ($E \perp c$)
	$n = 1.755$ ($E \parallel c$)

9.2.2. Neodymium Lasers

These are the most popular type of solid-state laser. The host medium is often a crystal of $Y_3Al_5O_{12}$ (called YAG, an acronym for yttrium aluminium garnet) in which some of the Y^{3+} ions are replaced by Nd^{3+} ions. Besides this oxide medium, other host media include some fluoride (e.g., $YLiF_4$) or vanadate (e.g., YVO_4) materials as well as some phosphate or silicate glasses. Typical doping levels in Nd:YAG, e.g., are ~ 1 atomic %. Higher doping generally leads to fluorescence quenching and strained crystals, since the radius of the Nd^{3+} ion is somewhat larger (by $\sim 14\%$) than that of the Y^{3+} ion. Doping levels used in Nd:glass are somewhat higher than the value for Nd:YAG ($\sim 4\%$ of Nd_2O_3 by weight). Undoped host materials are usually transparent; when doped, these generally become pale purple because of the Nd^{3+} absorption bands in the red.

9.2.2.1. Nd:YAG Laser

Figure 9.2 shows a simplified energy level scheme for Nd:YAG. As previously discussed, these levels arise from the three inner-shell $4f$ electrons of the Nd^{3+} ion, which are effectively screened by eight outer electrons ($5s^2$ and $5p^6$). Energy levels are only weakly influenced by the crystal field of YAG, so the Russell–Saunders coupling scheme of atomic physics can be used. Level notation is accordingly based on this scheme, and the symbol characterizing each level has the form $^{2S+1}L_J$, where S is the total spin quantum number, J is the total angular momentum quantum number, and L is the orbital quantum number. Note that the allowed values of L, namely, $L = 0, 1, 2, 3, 4, 5, 6, 7, 8, 9, \ldots$ are expressed, for historic reasons, by uppercase letters $S, P, D, F, G, H, I, L, M, N, \ldots$, respectively. Thus the $^4I_{9/2}$ ground level correspond to a state in which $2S + 1 = 4$ (i.e., $S = 3/2$), $L = 6$, and $J = L - S = 9/2$. Each level is $(2J + 1)$-fold degenerate corresponding to the quantum number m_J running from $-J$ to $+J$ in unit steps. In the octahedral symmetry of the YAG crystal-field, states with the same value of $|m_J|$ have the same energy in the presence of the Stark effect, and each $^{2S+1}L_J$ level is split into $(2J + 1)/2$ doubly degenerate sublevels. Thus the $^4I_{11/2}$ and $^4F_{3/2}$ levels are split into six and two sublevels, respectively (see Fig. 9.2). Note that, since the degeneracy of all sublevels is always the same (i.e., $g = 2$), we can disregard this degeneracy and consider each sublevel in Fig. 9.2 as if it were a single nondegenerate level.

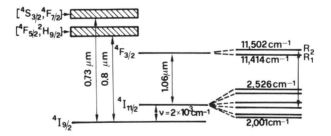

FIG. 9.2. Simplified energy levels of Nd:YAG.

The two main pump bands of Nd:YAG occur at ~ 730 and 800 nm, respectively, although higher lying absorption bands (see Fig. 6.7) also play an important role, notably for flashlamp pumping. These bands are coupled by a fast nonradiative decay to the $^4F_{3/2}$ level from where decay to the lower I levels occurs (to $^4I_{9/2}$, $^4I_{11/2}$, $^4I_{13/2}$, etc., levels; see Fig. 9.2 and Fig. 2.15). The rate of this decay is much slower ($\tau \cong 230$ μs), however, because this transition, in an isolated ion, is forbidden via the electric dipole interaction (the selection rule for electric-dipole-allowed transitions is $\Delta J = 0$ or ± 1) but becomes weakly allowed due to crystal-field interaction. Note that nonradiative decay is not so important because decay due to ion-ion interactions (see Fig. 2.13b) does not play an important role at the stated Nd ion concentrations. Multiphonon decay is also not so effective due both to the screening the of $5s^2$ and $5p^6$ states and the large energy gap between $^4F_{3/2}$ and the nearest level below it. This means that level $^4F_{3/2}$ accumulates a large fraction of the pump power, so that it is a good candidate as the upper level for laser action.

From the preceding discussion, one sees that several laser transitions are possible between $^4F_{3/2}$ and several lower lying I levels; of these transitions, $^4F_{3/2} \rightarrow {}^4I_{11/2}$ is the strongest one. Level $^4I_{11/2}$ is then coupled by a fast (hundreds of picoseconds) nonradiative decay to the $^4I_{9/2}$ ground level, so that thermal equilibrium between these two levels is very rapidly established. Since the energy difference between $^4I_{11/2}$ and $^4I_{9/2}$ levels is almost an order of magnitude larger than kT, then, according to Boltzmann statistics, level $^4I_{11/2}$ may, to a good approximation, be considered empty at all times. Thus laser operation on the $^4F_{3/2} \rightarrow {}^4I_{11/2}$ transition corresponds to a four-level scheme. According to the preceding discussion, the $^4F_{3/2}$ level is split by the Stark effect into two sublevels (R_1 and R_2), while the $^4I_{11/2}$ level is split into six sublevels. Laser action usually occurs from the upper R_2 sublevel to a particular sublevel of the $^4I_{11/2}$ level, since this transition has the highest value for the stimulated emission cross section. The transition occurs at $\lambda = 1.064$ μm (near-infrared), which is the most widely used lasing wavelength for Nd:YAG lasers. Note that laser action can also be obtained on the $^4F_{3/2} \rightarrow {}^4I_{13/2}$ transition (see Fig. 2.15; $\lambda = 1.319$ μm is the strongest transition wavelength in this case) provided the multilayer dielectric coatings of the cavity mirrors have high reflectivity at $\lambda = 1.319$ μm and sufficiently low reflectivity at $\lambda = 1.064$ μm (see Fig. 4.9). With diode laser pumping, laser action has also been made to occur efficiently on the $^4F_{3/2} \rightarrow {}^4I_{9/2}$ transition. In this case the transition, at a wavelength $\lambda = 946$ nm (see Fig. 2.15), is to a sublevel of the $^4I_{9/2}$ state, which, despite being a high-lying sublevel, is still appreciably populated according to Boltzmann statistics; the system thus operates as a quasi-three-level laser. In the case of the usual $\lambda = 1.064$-μm transition, and probably for all other cases, the laser transition is homogeneously broadened at room temperature via interaction with lattice phonons. The corresponding width is $\Delta\nu \cong 4.2$ cm^{-1} $= 126$ GHz at $T = 300$ K (see Fig. 2.10). This makes Nd:YAG a good candidate for mode-locked operation; pulses as short as 5 ps have, indeed, been obtained by passive mode locking (see Example 8.8). The long lifetime of the upper laser level ($\tau \cong 230$ μs) also makes Nd:YAG very suitable for Q-switched operation. Table 9.3 gives a summary of relevant optical and spectroscopic parameters for Nd:YAG at room temperature.

Nd:YAG lasers can operate either cw or pulsed, and can be pumped by either a lamp or an AlGaAs semiconductor laser.[5] For lamp pumping, linear lamps in single-ellipse (Fig. 6.1a), close-coupling (Fig. 6.1b), or multiple-ellipse configurations (Fig. 6.2) are commonly used. Medium pressure (500–1500 Torr) Xe lamps and high-pressure (4–6 atm) Kr lamps

TABLE 9.3. Optical and spectroscopic parameters of Nd:YAG ($\lambda = 1.064$ μm), Nd:YVO$_4$, Nd:YLF ($\lambda = 1.053$ μm), and Nd:glass (phosphate)

	Nd:YAG $\lambda = 1.064$ μm	Nd:YVO$_4$ $\lambda = 1.064$ μm	Nd:YLF $\lambda = 1.053$ μm	Nd:glass $\lambda = 1.054$ μm (Phosphate)
Nd doping	1 atom.%	1 atom.%	1 atom.%	3.8% by weight of Nd$_2$O$_3$
N_t (10^{20} ions/cm^3)[a]	1.38	1.5	1.3	3.2
τ (μs)[b]	230	98	450	300
$\Delta\nu_0$ (cm^{-1})[c]	4.5	11.3	13	180
σ_e (10^{-19} cm^2)[d]	2.8	7.6	1.9	0.4
Refractive index	$n = 1.82$	$n_o = 1.958$ $n_e = 2.168$	$n_o = 1.4481$ $n_e = 1.4704$	$n = 1.54$

[a] N_t is the concentration of the active ions, [b] τ is the fluorescence lifetime, [c] $\Delta\nu_0$ is the transition linewidth (FWHM), [d] σ_e is the effective stimulated emission cross section. Data refer to room temperature operation.

are used for the pulsed and cw cases, respectively. If a rod is used as the active medium, the rod diameter ranges typically from 3–6 mm, with a length from 5–15 cm. To reduce pump-induced thermal lensing and thermal birefringence, a slab configuration (Fig. 6.3a) is also used. The slope efficiency is about 3% for both cw and pulsed operation; average output powers up to a few kW (1–3 kW) are common. Longitudinally diode-pumped lasers (Fig. 6.11) with cw output powers up to $\sim$15 W and transversely diode-pumped lasers (Figs. 6.14 and 6.15) with cw output powers well above 100 W are now available. The slope efficiency for diode pumping is much higher than for lamp pumping, and may exceed 10%.

Nd:YAG lasers are widely used in a variety of applications, including: (1) Material processing, such as drilling and welding. In drilling applications, the beam of a repetitively pulsed laser is focused on the material (average powers of 50–100 W are commonly used with pulse energy $E = 5$–10 J, pulse duration $\Delta\tau_p = 1$–10 ms, and pulse repetition rate $f = 10$–100 Hz). In welding applications, a repetitively pulsed laser beam is conveyed to the working region through a 0.5–2-mm diameter optical fiber (average power as high as 2 kW are now commonly handled in this way). In this application, high-power Nd:YAG lasers supersede their direct competitors (high-power CO$_2$ lasers) due to the system flexibility offered by an optical-fiber delivery. (2) Medical applications. In coagulation and tissue evaporation, cw Nd:YAG lasers with powers as high as 50 W are used; the beam is delivered, through an optical fiber inserted into a conventional endoscope, to the internal organs (lungs, stomach, bladder) of the human body. Repetitively Q-switched Nd:YAG lasers are used for photodisruption of transparent membranes of pathological origin, which can appear in the anterior chamber of the eye (e.g., secondary cataract) or for iridectomy. (3) Laser ranging, in particular for laser range-finders and target designators used in a military context which use Q-switched lasers ($E \approx 100$ mJ, $\Delta\tau_p = 5$–20 ns, $f = 1$–20 Hz). (4) Scientific applications. In this case Q-switched lasers with their second-harmonic ($\lambda = 532$ nm), third-harmonic ($\lambda \cong 355$ nm), and fourth-harmonic beams ($\lambda = 266$ nm), as well as mode-locked lasers are commonly used. It should lastly be noted that diode-pumped Nd:YAG lasers with intracavity second-harmonic generation giving a green ($\lambda = 532$ nm) cw output power up to $\sim$10 W are now available. They provide an all-solid-state alternative to the Ar laser for many of its applications.

9.2.2.2. Nd:Glass Laser

As previously mentioned, relevant transitions of the Nd^{3+} ion involve the three electrons of the $4f$ shell, which are screened by eight outer electrons in the $5s$ and $5p$ configuration.[6] Accordingly, the energy levels of Nd:glass are approximately the same as those of Nd:YAG. Thus the strongest laser transition again occurs at about the same wavelength ($\lambda \cong 1.054\ \mu m$ for phosphate glass; see Table 9.3). Laser transition linewidths are, however, much larger, due to inhomogeneous broadening arising from local field inhomogeneities typical of a glass medium. In particular, the main $\lambda \cong 1.054$-μm laser transition is much broader (by ~ 40 times) while the peak cross section is somewhat smaller (by ~ 7 times) than that of Nd:YAG. A larger bandwidth is of course a desirable feature for mode-locked operation; diode-pumped passively mode-locked Nd:glass lasers have in fact produced femtosecond pulses (~ 100 fs). A smaller cross section is a desirable feature for pulsed high-energy systems, since the threshold inversion for the parasitic process of ASE [see Eq. (2.9.4)] is correspondingly increased. Thus more energy per unit volume can be stored in Nd:glass than in Nd:YAG before the onset of ASE (see Example 2.13). Note that, due to its lower melting temperature, glass can be fabricated more easily than YAG, so that active media of much larger dimensions can be produced. Since the pump absorption bands of Nd:glass are also much broader than those of Nd:YAG and Nd^{3+} concentrations are typically twice as great, the pumping efficiency of a lamp-pumped Nd:glass rod is somewhat larger (~ 1.6 times) than that of a Nd:YAG rod of the same dimensions (see Table 6.1). Against these advantages of Nd:glass compared to Nd:YAG, we must set the disadvantage of its much lower thermal conductivity. (The thermal conductivity of glass is about ten times smaller than that of Nd:YAG.) This has restricted applications of Nd:glass lasers mainly to pulsed laser systems of rather low repetition rate (<5 Hz), so that thermal problems in the active medium (rod or slab) can be avoided.*

As previously discussed, Nd:glass lasers are often used in applications where a pulsed laser with a low-repetition rate is required, for instance in some military range finders and some scientific Nd lasers. A very important application of Nd:glass is in laser amplifiers in the very high-energy systems used in laser-driven fusion experiments. Systems based on Nd:glass amplifiers have indeed been built in several countries, the largest one being in the United States (Nova laser, Lawrence Livermore National Laboratory); it delivers pulses with an energy of ~ 100 kJ and peak power of 100 TW ($\Delta\tau_p = 1$ ns). The laser uses a chain of several Nd:glass amplifiers; the largest consists of Nd:glass disks (see Fig. 6.3b) each disk of ~ 4 cm in thickness and ~ 75 cm in diameter. A national ignition facility delivering pulses of much higher energy (~ 10 MJ, Lawrence Livermore National Laboratory) and a similar system delivering about the same output energy (~ 2 MJ, Limeil Center) are presently being built in the United States and France, respectively.

9.2.2.3. Other Crystalline Hosts

Many other crystal materials have been used as hosts for the Nd^{3+} ion, but we limit ourselves here to mentioning $YLiF_4$ (YLF) and YVO_4.

* An exception is the glass fiber laser, where long length and small transverse dimension eliminate the thermal problem and have allowed cw outputs in excess of 30 W.

Compared to YAG, YLF has better thermooptical (pump-induced thermal lensing and thermal birefringence) properties;[7] lamp-pumped Nd:YLF lasers are thus used to obtain TEM_{00}-mode cw beams of better quality and higher output power. The larger linewidth of Nd:YLF compared to Nd:YAG (approximately three times; see Table 9.3) makes Nd:YLF lasers particularly attractive for mode-locked operation with both lamp and diode pumping (see Example 8.8). The mechanical and thermo-mechanical properties of YLF are however worse than those of YAG, which make YLF rods more difficult to handle and easier to break. Note that the 1053-nm emission wavelength of Nd:YLF provides a good match with the peak gain wavelength of Nd:glass:phosphate lasers (see Table 9.3). Mode-locked Nd:YLF lasers are accordingly used as the first stage in the large energy systems used for laser fusion experiments.

Compared to Nd:YAG, Nd:YVO$_4$ has a much larger peak cross section ($\sigma_e \cong 7.6 \times 10^{-19}$ cm^2) and much shorter fluorescence lifetime ($\tau = 98$ μs). The product $\sigma\tau$ is about the same for the two cases, so we expect approximately the same threshold. For a given inversion, however, the gain coefficient of Nd:YVO$_4$ is about three times larger than that of Nd:YAG, which makes Nd:YVO$_4$ lasers less sensitive to cavity losses. Longitudinally diode-pumped Nd:YVO$_4$ of high-cw power (~ 15 W) are now commercially available; for this kind of laser, Nd:YVO$_4$ seems to be preferred to Nd:YAG.

9.2.3. Yb:YAG Laser

The Yb:YAG laser is the most noteworthy example of a quasi-three-level laser. Since it oscillates at ~ 1.03-μm wavelength, it is a direct competitor to the Nd:YAG laser. Since it operates on a quasi-three-level laser scheme, it is usually pumped by semiconductor laser diodes to provide the intense pumping required.[8]

Figure 9.3 shows a simplified scheme of the energy level diagram of Yb:YAG. The level structure is particularly simple here—only one excited manifold $^2F_{5/2}$ is present because Yb^{3+} is one electron short of a full $4f$ shell (see Table 9.1); this shell thus acts as if it contained one electron hole. The two main absorption lines occur at 968 and 941 nm, respectively; the two lines have approximately the same value of peak absorption cross

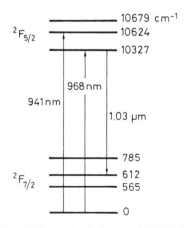

FIG. 9.3. Energy level diagram of Yb:YAG.

section (see Fig. 6.8b). The line at 941 nm is usually preferred for diode pumping due to its larger width. The main gain line occurs at 1.03 μm (quasi-three-level laser). Table 9.4 lists some relevant optical and spectroscopic parameters for Yb:YAG at room temperature. Note the long lifetime $\tau = 1.16$ ms, of essentially radiative origin, which indicates a good storage medium.

Yb:YAG lasers are pumped in a longitudinal pumping configuration using a pump at $\lambda_p - 943$-nm wavelength, usually by InGaAs/GaAs strained quantum well lasers, although they can also be pumped by a Ti:sapphire laser. The optical-to-optical efficiency is very high ($\sim 60\%$), a result mainly due to the high pump quantum efficiency ($\eta_q = h\nu/h\nu_p = \lambda_p/\lambda = 91.5\%$).[9] Average output powers well in excess of 50 W have so far been achieved.[10] Compared to Nd:YAG, the Yb:YAG laser offers the following favorable properties:

1. Very low-quantum defect [$(h\nu_p - h\nu)/h\nu_p \cong 9\%$] and hence very low fractional heating.
2. Long radiative lifetime of the upper state, making Yb:YAG a good medium for Q-switching.
3. High doping levels (due to the simple energy level structure)—6.5 at.% are usually used without incurring fluorescence-quenching due to ion-ion interaction.
4. Broad emission bandwidth (~ 86 cm^{-1}), indicating suitability for mode-locked operation (subpicosecond pulses have indeed been obtained).
5. Low stimulated-emission cross section allowing high energy to be stored before the onset of ASE.

In contrast to these favorable properties, the main limitation of Yb:YAG comes from the high threshold, a result of its quasi-three-level nature and from its low stimulated-emission cross section. The various features of Yb:YAG discussed above suggest that it may be better suited than Nd:YAG for many applications where a diode-pumped laser at a wavelength around 1 μm is needed.

TABLE 9.4. Optical and spectroscopic parameters at room temperature of the most important quasi-three-level laser materials

Active Medium Parameters	Yb:YAG $\lambda = 1.03$ μm	Nd:YAG $\lambda = 946$ μm	Tm:Ho:YAG $\lambda = 2.091$ μm	Yb:Er:Glass[a] $\lambda = 1.54$ μm (Phosphate)
Doping (atom.%)	6.5 atom.	1.1 atom.		
N_t (10^{20} ions/cm^3)	8.97	1.5	8 (Tm)	10 (Yb)
			0.5 (Ho)	1 (Er)
τ (ms)	1.16	0.23	8.5	8
$\Delta\nu_0$ (cm^{-1})	86	9.5	42	120
σ_e (10^{-20} cm^2)	1.8	2.4	0.9	0.8
σ_a (10^{-20} cm^2)	0.12	0.296	0.153	0.8
Refractive index	$n = 1.82$	$n = 1.82$	$n = 1.82$	$n = 1.531$

[a] For Yb:Er:glass, the effective value of the stimulated emission and absorption cross sections are about the same, so the laser can be considered to operate in (almost) a pure three-level scheme.

9.2.4. Er:YAG and Yb:Er:Glass Lasers

Erbium lasers can emit radiation at either $\lambda = 2.94$-μm wavelength (for Er:YAG) or at $\lambda = 1.54$-μm wavelength (for Yb:Er:glass).[11] The former wavelength is particularly interesting for biomedical applications, while the latter wavelength is attractive for application situations where eye safety is important and for optical communications in the third transparency window of optical fibers.

In the case of Er:YAG, the Er^{3+} ion occupies some Y^{3+} ion sites in the lattice; Fig. 9.4a shows relevant energy levels of the laser. Laser oscillation can take place on either the $^4I_{11/2} \rightarrow {}^4I_{13/2}$ transition ($\lambda = 2.94$ μm)[12] or the $^4I_{13/2} \rightarrow {}^4I_{15/2}$ transition ($\lambda \cong 1.64$ μm). Due to their interest for biomedical applications, Er:YAG lasers, oscillating on the $\lambda = 2.94$-μm transition, are the subject of much development. The spectroscopy of Er:YAG indicates that, for flashlamp excitation, the $^4I_{11/2}$ upper laser level is pumped by light absorbed from transitions at wavelengths shorter than 600 nm. For diode laser pumping, diode lasers oscillating at $\lambda = 970$ nm (InGaAs/GaAs strained QW) are used. The lifetime of the upper state (~ 0.1 ms) is much shorter than that of the lower state (~ 2 ms); therefore the laser is usually operated in a pulsed regime. Despite this unfavorable lifetime ratio the laser can also operate cw under the usual operating conditions, particularly when diode pumped. This occurs because, due to the high Er concentrations used (10–50 at.%), a strong Er^{3+}-Er^{3+} interaction occurs, thus leading to an efficient $^4I_{13/2} \rightarrow {}^4I_{9/2}$ up-conversion transition (see Fig. 2.13c). This process leads to energy recycling from the $^4I_{13/2}$ lower level to the $^4I_{11/2}$ upper-laser level.

In a flash-lamp-pumped configuration, the Er:YAG rod dimensions are typically comparable to those of a flash-lamp-pumped Nd:YAG laser (e.g., 6-mm diameter × 7.5 cm length); it is usually pumped using either an elliptical cylinder or a close-coupling pumping chamber (Fig. 6.1). Flash-lamp-pumped Er:YAG lasers with output energies up to 1 J and repetition rates up to 10 Hz are commercially available. Due to the very strong water absorption around $\lambda = 2.94$ μm, flash-lamp-pumped Er:YAG lasers are particularly interesting for biomedical applications and, in particular, for plastic surgery. The human body in fact consists of $\sim 70\%$ water and the skin penetration depth of a 2.94-μm wavelength Er:YAG laser is around 5 μm. More recently, diode-pumped Er:YLF lasers oscillating at $\lambda = 2.8$ μm have produced cw operation with good optical-to-optical slope efficiency ($\sim 35\%$) and sufficiently high output power (> 1 W).

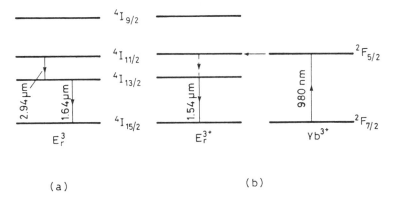

FIG. 9.4. Relevant energy levels of: (a) Er:YAG and (b) Yb:Er:phosphate glass.

Figure 9.4b shows relevant energy levels of the Yb:Er:phosphate-glass laser. The laser can be pumped by either a flashlamp[13] or a cw diode;[14] it usually oscillates on the $^4I_{13/2} \rightarrow {}^4I_{15/2}$ transition ($\lambda = 1.54$ μm). For 1.54-μm Er lasers, the Er concentration must be kept low to avoid the detrimental effect, in this case, of the up-conversion mechanism previously mentioned. For both flashlamp and diode laser pumping, Er absorption coefficients are then too small for efficient laser operation; to increase pump absorption, codoping with Yb^{3+} ions (and Cr^{3+} ions for flashlamp pumping) is used. With diode pumping around the 980-nm wavelength, pump power is mainly absorbed by Yb^{3+} ions ($^2F_{7/2} \rightarrow {}^2F_{5/2}$ transition); excitation is then effectively transferred to the $^4I_{11/2}$ Er level by a Förster-type dipole-dipole interaction (see Sect. 2.6.1). The $^4I_{11/2}$ Er level then decays relatively quickly ($\tau \cong 0.1$ ms), by multiphonon relaxation, to the $^4I_{13/2}$ upper laser level. The lifetime of this level in phosphate glass is particularly long ($\tau \cong 8$ ms), thus making it very suitable for laser action. Note that the peak of the gain spectrum of Er:glass is, only slightly, Stokes shifted to longer wavelengths compared to the peak of the absorption spectrum; both spectra arise, in fact, from several transitions between the $^4I_{13/2}$ and the $^4I_{15/2}$ manifolds. Thus the Yb:Er:glass laser behaves almost like a pure three-level laser. Table 9.4 shows other relevant spectroscopic and optical properties of the Yb:Er:glass laser.

Q-switched flash-lamp-pumped Cr:Yb:Er:glass lasers are used as eye-safe range finders. The 1.5-μm wavelength is in fact particularly safe for the eye.[15] Diode-pumped cw Yb:Er:glass lasers have potential applications in optical communications and for free-space optical measurements where eye safety is of concern.

9.2.5. Tm:Ho:YAG Laser

Figure 9.5 shows relevant energy levels of the Tm:Ho:YAG laser.[16] Both Tm^{3+} and Ho^{3+} ions occupy Y^{3+}-ion sites in the lattice. Typical Tm concentrations are rather high (4–10 at.%), while the concentration of Ho ions is an order of magnitude smaller. In flashlamp pumping, the active medium is also sensitized by Cr^{3+} ions, which are substituted for Al^{3+} ions in the YAG crystal. In this case, the pump energy is absorbed mainly by the $^4A_2 \rightarrow {}^4T_2$ and $^4A_2 \rightarrow {}^4T_1$ transitions* of the Cr^{3+} ions, then efficiently transferred to the $^3F_4$ level of

---

* According to group theory, the $^4T_2$ and $^4T_1$ notations for the Cr^{3+} ion in an octahedral crystal field, used here, are equivalent to the old $^4F_2$ and $^4F_1$ notations used for ruby (see Fig. 9.1).

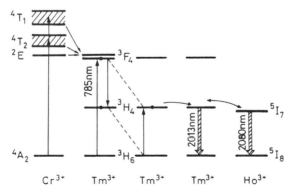

FIG. 9.5. Relevant energy level diagram of Cr:Tm:Ho:YAG system.

the Tm^{3+} ion by a Förster-type ion-ion interaction. For cw diode pumping, the $^3F_4$ level of Tm^{3+} is pumped directly by AlGaAs semiconductor lasers at a 785-nm wavelength, so that codoping with Cr^{3+} ions is not needed. For both flash-lamp and diode-laser pumping, excitation to the $^3F_4$ level of Tm^{3+} ion is then followed by a cross-relaxation process between adjacent ions, of the form $Tm(^3F_4) + Tm(^3H_6) \rightarrow 2Tm(^3H_4)$. This process converts one excited Tm ion in the $^3F_4$ state into two excited Tm ions, both of which are left in the $^3H_4$ state. For the high Tm concentrations used, this cross-relaxation process dominates the $^3F_4$ radiative decay, leading to an overall pump quantum efficiency of nearly 2. A fast spatial migration of the excited energy between Tm ions, again due to Förster-type ion-ion interaction, then occurs until excitation reaches a Tm ion very near a Ho ion. In this case, energy transfer to the $^5I_7$ level of Ho occurs, followed by laser action on the $Ho^{3+}(^5I_7 \rightarrow {}^5I_8)$ transition. Laser action actually occurs between the lowest sublevel of the $^5I_7$ manifold to a sublevel, ~462 cm^{-1} above the ground sublevel, of the $^5I_8$ manifold at $\lambda = 2.08$-μm wavelength (quasi-three-level laser). Without Ho-doping, the crystal can lase on the $^3H_4 \rightarrow {}^3H_6$ Tm transition at $\lambda = 2.02$-μm wavelength.

When flashlamp pumped, the active medium is in the form of a rod of the same typical dimensions as those of the Er:YAG rod considered in Sect. 9.2.4 and again pumped in an elliptical cylinder or close-coupling configuration (see Fig. 6.1). Output energies to 1 J in a ~200 μs long pulse and a slope efficiency to 4% with a repetition rate below 10 Hz are typical laser operating figures. This laser may find interesting applications in the biomedical field, since biological tissue also has a strong absorption around 2 μm (although much less strong than at the 2.94-μm wavelength of the Er laser). When diode pumped, a longitudinal pumping configuration, such as that in Fig. 6.11a, is often used. Given the strong absorption coefficient of Tm^{3+} ions at the pump wavelength ($\alpha_p \cong 6$ cm^{-1}), the thickness of the active medium is now typically 2–3 mm; the medium is generally cooled to low temperatures (-10 to $-40°$C) to reduce the thermal population of the lower laser level.

Eye-safe coherent laser radar systems using Tm:Ho:YAG lasers are used for remote measurement of wind velocity in the atmosphere. This involves using a single-frequency diode-pumped Tm:Ho laser to injection seed a flashlamp-pumped, Pockels-cell-Q-switched, Cr:Tm:Ho:YAG slave oscillator.

9.2.6. Fiber Lasers

In a fiber laser the active medium is the core of the fiber doped with a rare earth.[17] Most commonly, this is a single-mode fiber made of silica. The pump beam is launched longitudinally along the fiber length and it may be guided by either the core itself, as occurs for the laser mode (a conventional single-mode fiber laser), or by an inner cladding around this core (double-clad fiber laser). Note that, although fiber lasers were first demonstrated in the early days of laser development,[18] they have become of practical interest only in recent years after the advent of suitable diode lasers allowing efficient pumping and of techniques for fabricating doped single-mode silica fibers.

In a conventional single-mode fiber, the transverse dimensions of both pump w_p and laser w_0 beams are comparable to the core radius a (typically $a \cong 2.5$ μm). Thus both w_p and w_0 are 10–50 times smaller than corresponding typical values for a bulk device (see Examples 7.4 and 7.5). From Eqs. (6.3.19) and (6.3.24), for a four-level and a quasi-three-level laser, respectively, the threshold pump power P_{th} is seen to be proportional to

($w_0^2 + w_p^2$). Hence, for the same values of laser parameters (e.g., γ, σ_e, η_p, τ for a four-level laser), P_{th} is expected to be smaller in a fiber laser compared to a bulk device by two to three orders of magnitude. Thus, according to Examples 7.4 and 7.5, threshold pump powers well below 1 mW are expected and indeed achieved in fiber lasers. This argument also shows that laser action can be obtained for active media of very low radiative quantum efficiency and hence of very short lifetime τ. On the other hand, the expression for laser slope efficiency for both a four-level and a quasi-three-level laser [the two are identical; see Eqs. (7.3.13) and (7.4.10)] is independent of the upper state lifetime; it depends only on pump efficiency η_p. Thus, even for a transition with a low radiative quantum efficiency, a high laser slope efficiency can be obtained if most of the pump power is absorbed (i.e., $\eta_p \cong 1$). Therefore, transitions that look unpromising in bulk media can still show a high slope efficiency and a low enough threshold in a fiber, when diode-pumped. It should be noted that an interesting effect, occurring at the high pump powers (to ~ 100 mW; see Sect. 6.3.1) available from single-transverse-mode diode lasers, is ground-state depletion. Consider, for instance, a four-level laser, such as Nd:glass, and let F_p be the pump photon flux (assumed for simplicity to be uniform in the core) and N_g and N_2 be the populations of the ground level and upper-laser level, respectively. In the absence of laser action and under cw conditions, we can simply write the following balance equation

$$\sigma_p F_p N_g = (N_2/\tau) \qquad (9.2.1)$$

where σ_p is the pump absorption cross section and τ is the upper state lifetime. Thus, when $N_g = N_2$, one must have $F_p = (I_p/hv_p) = (1/\sigma_p\tau)$, where I_p is the pump intensity and hv_p is the energy of a pump photon. According to Fig. 6.8a and Table 9.3, for a Nd:silica fiber one has $\sigma_p = 2.8 \times 10^{-20}$ cm^2 and $\tau = 300$ μs. From the previous expression, one obtains $I_p = (hv_p/\sigma_p\tau) \cong 25$ kW/cm^2 so that $P_p = I_p A_{core} \cong 0.25$ mW, where A_{core} is the area of the core taken to be $\approx 10^{-7}$ cm^2. Thus, in the example considered, more than half of the ground-state population is raised to the upper laser level at pump powers below 1 mW. Given the ease with which pump-induced depletion of the ground-state population occurs, it follows that typical pump powers can deplete absorption over lengths very much exceeding the small-signal extinction length ($l = 1/\alpha_p$, where α_p is the small-signal absorption coefficient at the pump wavelength). In fact, it can be shown that, if the pump power exceeds this saturation power P_p by a factor x, the pump power penetrates the fiber to a distance roughly x times the extinction length. This circumstance must be taken into account in choosing the optimum fiber length.

Conventional end-pumped single-mode fibers require a diffraction-limited pump source for efficient pumping. So, for diode pumping, only the single-stripe device in Fig. 6.9a meets this requirement; consequently the pump power is limited to ~ 100 mW (~ 1 W in the case of Master Oscillator Power Amplifier diode lasers) and output power is limited correspondingly. High-power diode lasers (see Figs. 6.9b and 6.10) display poor beam quality unsuitable for direct launch into the end of the fiber core. The solution to this problem is provided by *cladding-pumping* and illustrated in Fig. 9.6. The core, which may be monomode, lies within a lower index inner cladding that in turn lies within an outer cladding of a yet lower index. Pump light can be end-launched into the inner cladding with a much less stringent beam quality requirement than into the core. While propagating in the inner cladding, this pump light is progressively absorbed by the core with an effective absorption

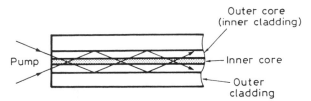

FIG. 9.6. Scheme of cladding pumping.

coefficient smaller than the core's true absorption coefficient by a factor of the order of the ratio of the inner-cladding area to the core area. Thus, for a given core doping, fiber length must be correspondingly increased to allow efficient absorption of pump power. Provided propagation losses of the pump in the inner cladding and of the lasing mode in the core are not excessively increased by an increase in fiber length, then efficient pumping by a multimode diode and efficient monomode lasing can be achieved. Thus the cladding-pumping scheme can provide a very simple means of enhancing the brightness of a (diode) pump source by efficiently converting it to a monomode laser output. Cladding-pumped Nd-doped fibers and Yb-doped fibers with output powers of several watts (4–10 W) are commercially available, and power levels in excess of 30 W have been demonstrated.

As just discussed, the high values of pump intensity available from laser pumping enable a considerable fraction of the ground-state population, in a conventional monomode fiber, to be raised to some upper level of the active ion. Under this condition, a second pump photon of the same or different wavelength can therefore raise this population to a still higher level. From this level, laser action can then take place to a lower level, so that the energy of the emitted photon is actually higher than each pump photon energy (see Fig. 9.7). A laser working on such a scheme where two or more than two pump photons of equal or different wavelength are used, is referred to as an *up-conversion laser*.

While such schemes work with bulk media, they are much more practical with fiber lasers exploiting fiber materials of a special kind. In silica fibers in fact, the main limitations to this up-conversion scheme stem from nonradiative decay of the levels involved, usually occurring by multiphonon deactivation. As explained in Sect. 2.6.1, the probability of such a decay is a very strong function of the number of phonons that must be emitted in the process. In this case the relevant phonon energy is the maximum energy in the phonon spectrum of the host material, since the nonradiative decay rate increases strongly, for a given transition, with increasing values of this energy. For fused silica this energy corresponds to ~ 1150 cm^{-1}, which results in rapid nonradiative decay for energy gaps less than ~ 4500 cm^{-1}. A substantial reduction in the rate of nonradiative decay is obtained by using

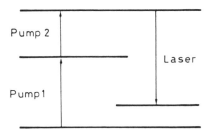

FIG. 9.7. Scheme of an up-conversion laser.

host materials with lower phonon energy. Among the materials that can be fabricated into fibers, the most widely used one consists of a mixture of heavy-metal fluorides, referred to as ZBLAN [an acronym for zirconium, barium, lanthanum, aluminium, and sodium (Na)], which, due to the heavy metals, has a maximum phonon energy of only 590 cm^{-1}. A few years ago, the undoped fiber was available at an advanced stage of development since, due to the correspondingly reduced infrared absorption of heavy-metal fluorides, it was developed as a possible route to ultra-low-loss fibers for communications.

As an example of performance capability, when pumped by three photons of the same wavelength ($\lambda = 1120$–1150 nm), ZBLAN fibers doped with Tm^{3+} have produced a very efficient up-conversion laser in the blue ($\lambda = 480$ nm), giving output powers in excess of 200 mW. When pumped by two photons at ~ 1010 nm and ~ 835 nm, ZBLAN fibers doped with Pr^{3+} have produced laser action on several transitions from blue to red ($\lambda = 491$, 520, 605, 635 nm), giving, e.g., as high as ~ 20 mW output power in the blue. These figures hold promise for a practical all-solid-state blue up-conversion laser source.

9.2.7. Alexandrite Laser

Alexandrite, chromium-doped chrysoberyl, is a crystal of $BeAl_2O_4$ in which Cr^{3+} ions replace some of the Al^{3+} ions (0.04–0.12 at.%).[19] This laser may be considered the archetype of what is now a large class of solid-state lasers, usually referred to as *tunable solid-state lasers*. The emission wavelength of these lasers can in fact be tuned over a wide spectral bandwidth (e.g., $\Delta\lambda \cong 100$ nm around $\lambda = 760$ nm for alexandrite). Tunable solid-state lasers include, among others, Ti:sapphire and Cr:LiSAF, considered in the following sections, as well as Co:MgF_2 ($\Delta\lambda \cong 800$ nm around $\lambda = 1.9$ μm), Cr^{4+}:YAG ($\Delta\lambda \cong 150$ nm around $\lambda = 1.45$ μm), and Cr^{4+}:Forsterite (Cr^{4+}:Mg_2SiO_4, $\Delta\lambda \cong 250$ nm around $\lambda = 1.25$ μm). In this category we can also include color center lasers,[20] which are broadly tunable in the near infrared (at wavelengths from 0.8–4 μm). Once rather popular, color center lasers have declined in popularity and importance due to problems associated with handling and storing the active medium and to the advent of new competitors in the same wavelength range (i.e., other tunable solid-state lasers or parametric oscillators, considered in Chap. 12). For these reasons they are not considered further here.

Energy states of the Cr^{3+} ion in $BeAl_2O_4$ are qualitatively similar to those of Cr^{3+} in other hosts with octahedral crystal field, such as ruby, as already considered. To explain why alexandrite laser is tunable while ruby is not, at least not to the same extent, we refer to a simplified scheme in Fig. 9.8 for energy states as a function of a configuration coordinate of the Cr^{3+} ion (i.e., the distance between this ion and the six surrounding O^{2-} anions of the octahedron; see Fig. 3.3). Figure 9.8 shows that, due to their symmetry, the equilibrium coordinate for both $^4T_2$ and $^4T_1$ states is shifted to a larger value than that of $^4A_2$ and ^2E states.* As in other Cr^{3+}-doped hosts, the decay between the $^4T_2$ and ^2E states is by a fast internal conversion (decay time of less than 1 ps), probably due to the level crossing that occurs between the two states. These two states can therefore be considered in thermal equilibrium at all times; since the energy difference between the lowest vibrational levels of $^4T_2$ and ^2E states in alexandrite ($\Delta E \cong 800$ cm^{-1}) is only a few kT, an appreciable

---

* As already pointed out, the $^4T_2$ and $^4T_1$ notations for the Cr^{3+} ion, considered here, are equivalent to the old $^4F_2$ and $^4F_1$ notations used for ruby (see Fig. 9.1).

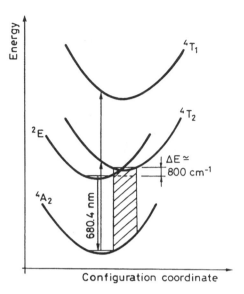

FIG. 9.8. Energy level diagram of alexandrite laser in a configuration coordinate model.

population is present in the vibrational manifold of the $^4T_2$ state when the ^2E state is populated. Invoking the Franck–Condon principle, one sees that vibronic transitions from the $^4T_2$ state end in empty vibrational levels of the $^4A_2$ state, thus becoming the preferred laser transition. Since a very large number of vibrational levels are involved, the resulting emission is in the form of a broad continuous band ($\lambda = 700$–800 nm). Excitation is then terminated by phonon decay to the lowest vibrational level of the $^4A_2$ state. In keeping with the preceding physical description, this type of laser is also referred to as a *phonon-terminated laser* or a *vibronic laser*.

It should be noted that, in ruby laser, laser action takes place between the ^2E and $^4A_2$ states, while phonon-terminated transitions do not occur. This is because the energy difference between the $^4T_2$ (old $^4F_2$) and ^2E states is much larger ($\Delta E \cong 2300$ cm^{-1}) and, hence, there is no appreciable population in the $^4F_2$ level. Note also that, in alexandrite, laser action can occur, as for ruby, on the $^2E \rightarrow {}^4A_2$ transition (compare Fig. 9.8 and Fig. 9.1). In this case, however, alexandrite operates on a three-level scheme, the threshold is much higher, and the emission wavelength occurs at a somewhat different value ($\lambda = 680.4$ nm).

Pumping in alexandrite takes place mostly through its green and blue absorption bands ($^4A_2 \rightarrow {}^4T_2$ and $^4A_2 \rightarrow {}^4T_1$ transitions; see Fig. 9.8), which are very similar to those of ruby. The effective values of the lifetime and stimulated emission cross sections of the $^4T_2$ upper laser state can be roughly calculated by assuming that the upper level consists of two strongly coupled levels with energy spacing of $\Delta E \cong 800$ cm^{-1}. These levels are the lowest vibrational levels of the $^4T_2$ and ^2E states (see Fig. 2.16 and Example 2.11). At $T = 300$ K the upper state lifetime then turns out to be $\tau \cong 200$ μs, which is almost the same as that of Nd:YAG. Note that, although the true lifetime of the $^4T_2$ state is much shorter ($\tau_T \cong 6.6$ μs), the effective lifetime is considerably increased by the presence of the long-lived ^2E state ($\tau_E \cong 1.5$ ms; $^2E \rightarrow {}^4A_2$ is spin-forbidden), which thus acts as a reservoir for the $^4T_2$ state. Due to the coupling of these two states, the effective cross section of the laser transition

($\sigma_e \cong 0.8 \times 10^{-20}$ cm^2) is then considerably smaller than the true value. Note also that both τ and σ_e are temperature dependent because the relative population of the two states depends on temperature. Table 9.5 summarizes some optical and spectroscopic data relevant to the tunable laser transition of alexandrite.

From an engineering point of view, alexandrite lasers are similar to Nd:YAG lasers; in fact alexandrite lasers are usually lamp-pumped in a pumping chamber as in Figs. 6.1 or 6.2. Although they can operate cw, the much smaller cross section than that of, e.g., Nd:YAG makes pulsed operation more practical. The laser can operate in either the free-running regime (output pulse duration ~ 200 μs) or Q-switched regime (output pulse duration ~ 50 ns); it is usually pulsed at a relatively high repetition rate (10–100 Hz). Due to the strong increase in the effective emission cross section with temperature, the laser rod is usually held at an elevated temperature (50–70°C). Performances of a pulsed alexandrite laser, in terms of output versus input energy and slope efficiency, are similar to those of Nd:YAG using a rod of the same dimensions. Average powers, up to 100 W at pulse repetition rates of ~ 250 Hz have been demonstrated. Flash-lamp-pumped alexandrite lasers have proved useful when high average power at $\lambda \cong 700$ nm wavelength is needed (such as in laser annealing of silicon wafers) or tunable radiation is required (as in pollution monitoring).

9.2.8. Titanium Sapphire Laser

The titanium sapphire (Ti:Al$_2$O$_3$) laser is the most widely used tunable solid-state laser.[21–23] It can in fact be operated over a broad tuning range ($\Delta\lambda \cong 400$ nm, corresponding to $\Delta\nu_0 \cong 100$ THz), thus providing the largest bandwidth of any laser. To make Ti:sapphire, Ti$_2$O$_3$ is doped into a crystal of Al$_2$O$_3$ (typical concentrations range between 0.1–0.5% by weight), so that Ti^{3+} ions occupy some of the Al^{3+}-ion sites in the lattice. The Ti^{3+} ion possesses the simplest electronic configuration among transition ions, only one electron being left in the $3d$ shell. The second $3d$ electron and the two $4s$ electrons of the Ti atom (see Table 9.1) are in fact used for ionic binding to oxygen anions. When Ti^{3+} is substituted for an Al^{3+} ion, the Ti ion is situated at the center of an octahedral site whose

TABLE 9.5. Optical and spectroscopic parameters at room temperature of the most important tunable solid-state laser materials

Active Medium Parameters	Alexandrite	Ti:sapphire	Cr:LiSAF	Cr:LiCAF
Doping (at.%)	0.04–0.12	0.1	up to 15	up to 15
N_t (10^{19} ions/cm^3)^a	1.8–5.4	3.3	10	10
Peak wavelength (nm)	760	790	850	780
Tuning range (nm)	700–820	660–1180	780–1010	720–840
σ_e (10^{-20} cm^2)	0.8	28	4.8	1.3
τ (μs)	260	3.2	67	170
$\Delta\nu_0$ (THz)	53	100	83	64
Refractive indices	$n_a = 1.7367$	$n_o = 1.763$		
	$n_b = 1.7421$	$n_e = 1.755$	$n_e = 1.4$	$n_e = 1.39$
	$n_c = 1.7346$			

^{a} The density of active ions N_t for both Cr:LiSAF and Cr:LiCAF is given at $\sim 1\%$ molar concentration of CrF$_3$ in the melt.

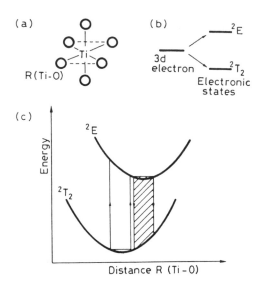

FIG. 9.9. (a) Octahedral configuration of Ti:Al_2O_3, (b) splitting of $3d$ energy states in an octahedral crystal field, and (c) energy states in a configuration coordinate model.

six apexes are occupied by O^{2-} ions (Fig. 9.9a). Assuming for simplicity a field of perfect octahedral symmetry,* the fivefold degenerate (neglecting spin) d-electron levels of an isolated Ti^{3+} ion are split, by the crystal field of the six nearest neighbor oxygen anions, into a triply degenerate $^2T_2$ ground state and a doubly degenerate ^2E upper state (Fig. 9.9b). As usual, the notation for these crystals incorporating a transition metal is derived from group theory. When spin is also taken into account ($S = 1/2$ for this essentially one-electron system), the two states acquire a multiplicity of $2S + 1 = 2$, as denoted by the superscript to the left of each letter. In a configuration coordinate model, where this coordinate is just the Ti-O separation, the two states are represented as in Fig. 9.9c. Note that the rather strong interaction of the $3d$ electron with the crystal field results in a considerably larger equilibrium distance for the upper state than the lower state. This circumstance is particularly relevant because it produces absorption and fluorescence bands that are wide and widely separated, as shown in Fig. 9.10. The Ti^{3+} ion in an octahedral site has only one excited state (i.e., the ^2E state)—a particularly relevant feature. This eliminates the possibility of excited state absorption (e.g., arising from the $^4T_2 \rightarrow \, ^4T_1$ transition in alexandrite), an effect that limits the tuning range and reduces the efficiency of many other transition metal lasers.

Based on the preceding discussion and using the Franck–Condon principle, it follows that laser action takes place from the lowest vibrational level of the ^2E state to some vibrational level of the ground $^2T_2$ state. Table 9.5 lists some relevant optical and spectroscopic properties of this phonon-terminated transition. The upper state lifetime ($\tau \cong 3.2$ μs at $T = 300$ K; the radiative lifetime is $\tau_r \cong 3.85$ μs) is much shorter than for alexandrite because, unlike alexandrite, there is no lengthening effect due to a reservoir of population in another excited state. The stimulated emission cross section, on the other hand,

---

* For a more exact treatment, see Refs. 21 and 22.

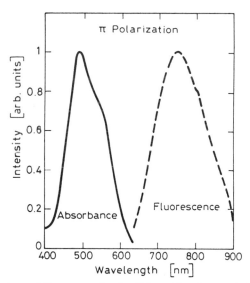

FIG. 9.10. Absorption and fluorescence bands of Ti:sapphire. (By permission from Ref. 55.)

is much ($\sim$40 times) larger than in alexandrite; it is comparable to that of Nd:YAG. Note the large bandwidth of the laser transition which is, in fact, the largest among commonly used solid-state lasers.

Usually, cw Ti:sapphire lasers are pumped by the green output of an Ar laser, while, in pulsed operation, frequency-doubled Nd:YAG or Nd:YLF lasers as well as flashlamps are used. Due to the small value of the $\sigma\tau$ product, flashlamp pumping requires very intense lamps; nonetheless flashlamp-pumped Ti:sapphire lasers are commercially available. Argon-pumped cw lasers provide a convenient source of coherent and high-power (>1 W) light, which is tunable over a wide spectral range (700–1000 nm). Perhaps the most important application of Ti:sapphire lasers is generating (see Sect. 8.6.5) and amplifying (see Chap. 12) femtosecond laser pulses. Sophisticated systems based on Ti:sapphire lasers and Ti:sapphire amplifiers, giving pulses of relatively large energy (20 mJ–1 J) with femtosecond duration (20–100 fs) are now operating in several laboratories and also available commercially.

9.2.9. Cr:LiSAF and Cr:LiCAF Lasers

Two of the most recently developed tunable solid-state materials, based on Cr^{3+} as the active species, are Cr^{3+}:$LiSrAlF_6$ (Cr:LiSAF) and Cr^{3+}:$LiCaAlF_6$ (Cr:LiCAF).[24,25] Both materials offer a wide tuning range and the corresponding lasers can be either flashlamp pumped or diode-laser pumped. In both Cr:LiSAF and Cr:LiCAF, Cr^{3+} ions replace some of the Al^{3+} ions in the lattice, and the impurity ion occupies the center of a (distorted) octahedral site surrounded by six fluorine ions. Thus, to a first approximation, the general energy level picture in the configuration coordinate representation, as presented for alexandrite, also holds in this case (see Fig. 9.8). Figure 9.11 shows the corresponding absorption and fluorescence spectra for the electric field parallel or perpendicular to the c-axis of the crystal (LiSAF and LiCAF are uniaxial crystals) for Cr:LiSAF. Note that the two

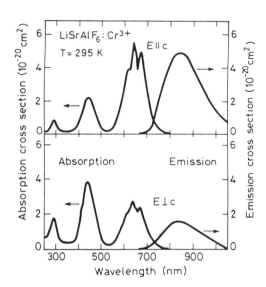

FIG. 9.11. Absorption and fluorescence bands of Cr:LiSAF for polarization parallel and perpendicular to the optical *c*-axis of the crystal. (By permission from Ref. 24.)

main absorption bands centered at 650 nm and 440 nm, respectively, arise from $^4A_2 \rightarrow {^4T_2}$ and $^4A_2 \rightarrow {^4T_1}$ transitions. Note also that the sharp features superimposed on the $^4T_2$ band arise from absorption to the ^2E and $^2T_1$ states. (The latter state is not shown in Fig. 9.8.) Thus the ^2E state is now located within the $^4A_2 \rightarrow {^4T_2}$ absorption band, which implies that the lowest vibrational level of $^4T_2$ must now be located appreciably below the ^2E state. Due to rapid relaxation between the two states, it follows that the most heavily populated state is now $^4T_2$; therefore the ^2E state does not play a role as an energy reservoir, as for alexandrite. This is also evidenced by the fact that the measured lifetime of the $^4T_2$ state is roughly independent of temperature. Table 9.5 lists other relevant optical and spectroscopic parameters of the two laser materials. Among the tunable solid-state laser materials in Table 9.5, Cr:LiSAF exhibits the largest value for the $\sigma\tau$ product. Thus, due to its larger values for both the cross section and $\sigma\tau$ product and its wider tuning range (the tuning range of Cr:LiCAF is limited by excited state absorption), Cr:LiSAF is generally preferred to Cr:LiCAF.

Cr:LiSAF is used as a flashlamp or diode-pumped laser source, providing tunability around 850 nm; the large gain linewidth makes this medium attractive for generating femtosecond pulses. For this application, Kerr lens mode-locked Cr:LiSAF lasers, end pumped by GaInP/AlGaInP QW laser diodes at $\sim$670-nm wavelength in a configuration such as that shown in Fig. 8.31, have been developed. Large flash-lamp-pumped Cr:LiSAF amplifier systems to amplify femtosecond pulses from either a Ti:sapphire or a Cr:LiSAF mode-locked laser have also been developed. Other potential applications of Cr:LiSAF are in tunable systems for pollution monitoring and spectroscopy.

9.3. DYE LASERS

Dye lasers use an active medium consisting of a solution of an organic dye in a liquid solvent, such as ethyl or methyl alcohol, glycerol, or water.[26] Organic dyes constitute a large class of polyatomic molecules containing long chains of conjugated double bonds [c.g., $(-CH=)_n$)]. Laser dyes usually belong to one of the following classes: (1) Polymethine dyes, which provide laser oscillation in the red or near infrared (0.7–1.5 μm); as an example Fig. 9.12a shows the chemical structure of the dye 3,3' diethyl thiatricarbocyanine iodide, which oscillates in the infrared (at a peak wavelength $\lambda_p = 810$ nm). (2) Xanthene dyes, whose laser operation is in the visible; as an example Fig. 9.12b shows the chemical structure of the widely used rhodamine 6G dye ($\lambda_p = 590$ nm). (3) Coumarin dyes, which oscillate in the blue-green region (400–500 nm); as an example Fig. 9.12c shows the chemical structure of coumarin 2, which oscillates in the blue ($\lambda_p = 450$ nm).

9.3.1. Photophysical Properties of Organic Dyes

Organic dyes usually show wide absorption and fluorescence bands without sharp features; the fluorescence is Stokes-shifted to longer wavelengths than the absorption, a feature reminiscent of tunable solid-state laser materials considered in preceding sections. As an example Fig. 9.13 shows relevant absorption and emission characteristics of rhodamine 6G in ethanol solution.

To understand the origin of features shown in Fig. 9.13, we must first consider the energy levels of a dye molecule. A simple understanding of these levels is obtained from the so-called free-electron model,[27] which is illustrated here by considering the case of the cyanine dye shown in Fig. 9.14a. The π-electrons of the carbon atoms are then seen to form two planar distributions, one above and one below the plane of the molecule (dotted regions in Fig. 9.14a–b). The π-electrons are assumed to move freely within their planar

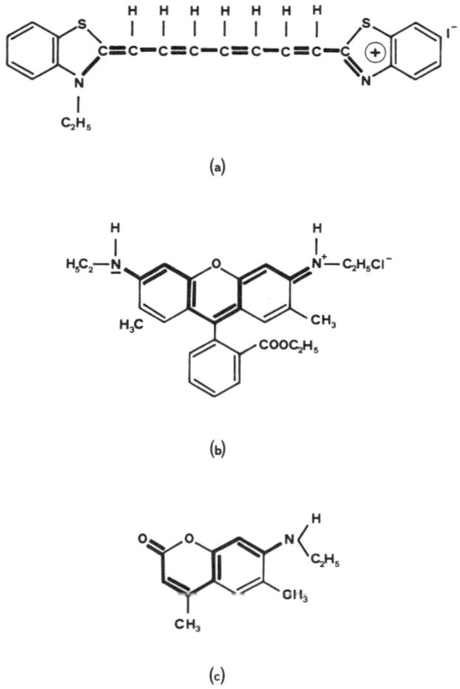

(a)

(b)

(c)

FIG. 9.12. Chemical structure of some common dyes: (a) 3,3' diethyl thiatricarbocyanine iodide, (b) rhodamine 6G, and (c) coumarin 2. In each case the chromophoric region of the dye is indicated by heavier lines.

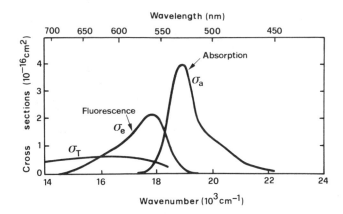

FIG. 9.13. Absorption cross section σ_a, singlet-singlet stimulated emission cross section σ_e, and triplet-triplet absorption cross section σ_T for an ethanol solution of rhodamine 6G.

distributions, limited only by the repulsive potential of the methyl groups at the end of the dye chain; electronic states of the molecule originate from these electrons. To first order, energy levels of the electrons are then simply those of a free electron in a potential well of the form shown in Fig. 9.14c. If this well is approximated by a rectangular one (Fig. 9.14d), the energy levels are known to be given by

$$E_n = \frac{h^2 n^2}{8mL^2} \tag{9.3.1}$$

where n is an integer, m is the electron mass, and L is the length of the well. It should be noted here that stable dye molecules have an even number of electrons in the π-electron

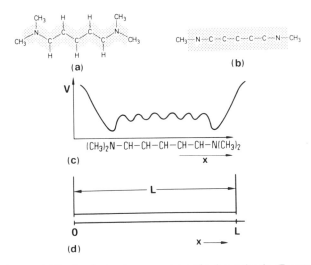

FIG. 9.14. Free electron model for the electronic energy states of a dye molecule. (By permission from Ref. 27.)

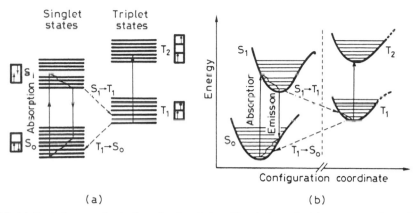

FIG. 9.15. (a) Typical energy levels for a dye in solution. The singlet and triplet states are shown in separate columns. (b) Energy level diagram of a dye in a configuration coordinate representation. (By permission from Ref. 28 with data taken from Ref. 57.)

cloud.* If we then let the number of these electrons be $2N$, the lowest energy state of the molecule corresponds to the situation where these electrons occupy the lowest N energy levels; each level is occupied by two electrons with opposite spin. This molecular state thus has a total spin equal to zero, so it is a singlet state, labeled S_0 in Fig. 9.15. An approximate value for the energy of the uppermost electrons of this state E_N is obtained from Eq. (9.3.1) by letting $n = N$. In Fig. 9.15a the highest occupied level and the next one above it are indicated by two squares, one above the other; the S_0 state thus corresponds to the situation when the lower box is full, having two electrons, and the upper one is empty. The first excited singlet state (labeled S_1 in the figure) corresponds to when one of the two highest lying electrons is promoted without flipping its spin to the next level up. The energy of the uppermost electron of this state E_{N+1} can be roughly calculated from Eq. (9.3.1) by letting $n = N + 1$. The difference of energy between the S_1 and S_0 states is thus seen to be equal to $E_{N+1} - E_N$. According to Eq. (9.3.1) this difference can then be shown to decrease with increasing length L of the chain. If the spin if flipped, the total spin is $S = 1$ and the resulting state is a triplet state, labeled T_1 in the figure. Excited singlet S_2 and triplet T_2 states result when the electron is promoted to the next higher level, and so on. Note in Fig. 9.15a that the corresponding energy levels are indicated by a close set of horizontal lines representing the inclusion of vibrational energy. In Fig. 9.15b, the energy states and the vibrational levels of a dye molecule are represented as a function of a configuration coordinate (i.e., a coordinate describing one of the many vibrational modes that a long-chain dye molecule has). Note that, due to the large number of vibrational and rotational levels involved and the effective line-broadening mechanisms in liquids, the rotational-vibrational structure is in fact unresolved at room temperature.

We now look at what happens when the molecule is subjected to electromagnetic radiation. First we recall that selection rules require $\Delta S = 0$. Hence singlet-singlet as well as triplet-triplet transitions are allowed, while single-triplet transitions are forbidden. Therefore interaction with electromagnetic radiation can raise the molecule from the ground level S_0 to some vibrational levels of the S_1 state, taking into account the Franck–Condon principle (see Fig. 9.15b) or, more precisely, the corresponding Franck–Condon factors (see Sect. 3.1.3).

---

* Molecular systems with unpaired electrons are known as radicals; these tend to react readily, thus forming a more
  stable system with paired electrons.

Since the vibrational and rotational structure is unresolved, the absorption spectrum then shows a broad and featureless transition, as in Fig. 9.13 for the case of rhodamine 6G. Dyes have a very large dipole matrix element μ because the π-electrons are free to move over a distance roughly equal to the chain length L and, since L is quite large, it follows that μ is also large ($\mu \approx eL$). It then follows that the absorption cross section σ_a, which is proportional to μ^2, is also large ($\sim 10^{-16}$ cm^2; see Fig. 9.13).

Once in the excited state, the molecule nonradiatively decays in a very short time ($\tau_{nr} \cong 100$ fs due to collisional deactivation) to the lowest vibrational level of the S_1 state (Fig. 9.15).* From there it decays radiatively to some vibrational level of the S_0 state, taking into account again the Franck–Condon principle (Fig. 9.15b). The fluorescent emission then takes the form of a broad and featureless band, Stokes-shifted to the long wavelength side of the absorption band (see Fig. 9.13). Due to the large value of the dipole moment μ, the stimulated emission cross section is also expected to be quite large again ($\sim 10^{-16}$ cm^2; see Fig. 9.13). Having dropped to an excited vibrational level of the ground S_0 state, the molecule then returns to the lowest vibrational state by another very fast (~ 100 fs) nonradiative decay.

It should be noted that, while the molecule is in the lowest level of S_1, it can also decay to the T_1 state. This process is referred to as *intersystem crossing*; although radiatively forbidden, it may occur rather readily from collisions. Similarly, the transition $T_1 \rightarrow S_0$ takes place mainly by near-resonant energy-transfer collisions with species within the solution (e.g., dissolved oxygen) provided these collisions preserve the total spin of the colliding partners, in accordance with the Wigner rule (see Sect. 6.4.1.1). Note that, while the molecule is in the lowest level of T_1, it can also absorb radiation to undergo the $T_1 \rightarrow T_2$ transition, which is optically allowed. Unfortunately this absorption tends to occur in the same wavelength region where stimulated emission occurs (see again for example Fig. 9.13) and it may represent a serious obstacle to laser action.

The three decay processes just considered, occurring from states S_1 and T_1, can be characterized by the following three constants: (1) τ_{sp}, the spontaneous emission lifetime of the S_1 state; (2) k_{ST}, the intersystem crossing rate (s^{-1}) of the $S_1 \rightarrow T_1$ transition; and (3) τ_T, the lifetime of the T_1 state. If we let τ be the overall lifetime of the S_1 state, then, according to Eq. (2.6.18):

$$\frac{1}{\tau} = \frac{1}{\tau_{sp}} + k_{ST} \qquad (9.3.2)$$

Due to the large value of the dipole matrix element μ, the radiative lifetime falls in the nanosecond range (e.g., $\tau_{sp} \cong 5$ ns for rhodamine 6G). Since k_{ST}^{-1} is usually much longer (e.g., ~ 100 ns for rhodamine 6G), it follows that most of the molecules decay from the S_1 state by fluorescence. The fluorescence quantum yield (number of photons emitted by fluorescence divided by number of molecules raised to the S_1 state) is therefore nearly unity. In fact, according to Eq. (2.6.22) one has

$$\phi = \frac{\tau}{\tau_{sp}} \qquad (9.3.3)$$

---

* More precisely, thermalization among the many rotational-vibrational levels of this state occurs.

TABLE 9.6. Range of optical and spectroscopic parameters of typical dye laser media

Active Medium Parameter	Values
Wavelengths (nm)	320–1500
Concentration (molar)	10^{-3}–10^{-4}
N_t (10^{19} mol/cm^3)	0.1–1
σ_e (10^{-16} cm^2)	1–4
σ_T (10^{-16} cm^2)	0.5–0.8
$\Delta\lambda$ (nm)	25–50
τ (ns)	2–5
k_{ST}^{-1} (ns)	≈ 100
τ_T (s)	10^{-7}–10^{-3}
Refractive index	1.3–1.4

The triplet lifetime τ_T depends on the dye solution and, particularly, on the amount of dissolved oxygen. The lifetime can range from 10^{-7} s in an oxygen-saturated solution to 10^{-3} s or more in a deoxygenated solution.

As a summary, Table 9.6 lists typical ranges of various relevant optical and spectroscopic parameters of dye laser media.

9.3.2. Characteristics of Dye Lasers

From the preceding discussion we see that these materials have the appropriate characteristics for exhibiting laser action, over the wavelength range of fluorescence, in a four-level laser scheme. In fact rapid nonradiative decay within the excited singlet state S_1 populates the upper laser level very effectively, while rapid nonradiative decay within the ground state is effective in depopulating the lower laser level. It was however quite late in the general development of laser devices before the first dye laser was operated (1966),[29,30] and we now look for some reasons for this.

One problem that presents itself is the very short lifetime τ of the S_1 state, since the required pump power is inversely proportional to τ. Although this is to some extent compensated for by the comparatively large value of the stimulated emission cross section, the product $\sigma\tau$ [for a four-level laser, the threshold pump power is inversely proportional to $\sigma\tau$; see Eqs. (7.3.12) and (6.3.20) for the space-independent and space-dependent model, respectively] is still about three orders of magnitude smaller for, e.g., rhodamine 6G than Nd:YAG. A second problem arises from intersystem crossing. In fact, if τ_T is long compared to k_{ST}^{-1}, then molecules accumulate in the triplet state, resulting in absorption at the laser wavelength due to the triplet-triplet transition. In fact a necessary condition for laser action is that τ_T is less than some particular value that depends on other optical parameters of the dye molecule. To obtain this result, let N_2 and N_T be populations of the upper laser state and the triplet state, respectively. A necessary condition for laser action can then be established by requiring the gain coefficient, due to stimulated emission, to exceed the intrinsic loss, due to triplet-triplet absorption, i.e.

$$\sigma_e N_2 > \sigma_T N_T \tag{9.3.4}$$

where σ_T is the cross section for triplet-triplet absorption; the values for both σ_e and σ_T are taken at the wavelength where laser action is considered. In the steady state, the rate of decay of triplet population N_T / τ_T must equal the rate of increase due to intersystem crossing $k_{ST} N_2$, i.e.:

$$N_T = k_{ST} \tau_T N_2 \tag{9.3.5}$$

Combining Eqs. (9.3.4) and (9.3.5) we obtain

$$\tau_T < \frac{\sigma_e}{\sigma_T k_{ST}} \tag{9.3.6}$$

which is a necessary condition for cw laser action [i.e., equivalent to Eq. (7.3.1) for a simple two-level system]. If this condition is not satisfied, the dye laser can operate only in a pulsed regime. In this case the duration of the pump pulse must be short enough to avoid that an excessive population accumulates in the triplet state. Finally, a third crucial problem stems from the presence of thermal gradients produced in the liquid by the pump. These tend to produce refractive index gradients and hence optical distortions that can prevent laser action.

Dye lasers can be operated pulsed or, when Eq. (9.3.6) is satisfied, also cw. Pulsed laser action is obtained from many different dyes by using one of the following pumping schemes: Fast and intense flashlamps, with pulse duration usually less than ~ 100 μs, or short light pulses from another laser. In both cases, the short pulse duration produces laser action before an appreciable population accumulates in the triplet state and before the onset of refractive index gradients in the liquid.

For flashlamp pumping, linear lamps in an elliptical cylinder pumping chamber (see Fig. 6.1a) are used; the liquid containing the active medium flows through a glass tube placed along the second focal line of the ellipse. To achieve better pumping uniformity and hence more symmetric refractive index gradients, annular flashlamps consisting of two concentric glass tubes, with the dye solution in a central glass tube, are also used.

Nitrogen lasers are sometimes used for pulsed laser pumping; the uv output beam is suitable for pumping many dyes that oscillate in the visible range.* To obtain more energy and higher average power, the more efficient excimer lasers (in particular KrF and XeF) are increasingly used as uv pumps; for dyes with emission wavelength longer than ~ 550–600 nm, the second harmonic of a Q-switched Nd:YAG laser ($\lambda = 532$ nm) or the green and yellow emissions of a copper vapor laser are increasingly used. For these visible pump lasers, the conversion efficiency from pump laser to dye laser output is higher (30–40%) than that obtained with uv laser pumping ($\sim 10\%$). Furthermore, dye degradation due to pump light is considerably reduced. For all the cases considered above, involving pulsed laser pumping, a transverse pump configuration (i.e., direction of the pump beam orthogonal to the resonator axis) is generally adopted (Fig. 9.16). The laser pump beam is focused by the lens L, generally a combination of a spherical and cylindrical lens, to a fine line along the axis of the laser cavity. The length of the line focus is made equal to that of the dye cell (a few millimeters), while transverse dimensions are generally less than 1 mm. To tune the output wavelength within the wide emission band of a dye (~ 30–50 nm wide), a grazing-

---

* In this case pump light is usually absorbed by the $S_0 \rightarrow S_2$ transition of the dye, then rapidly transferred to the bottom of the S_1 state.

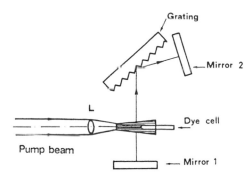

FIG. 9.16. Arrangement for a transversely pumped dye laser. The pumping beam can be a nitrogen, excimer, copper vapor laser, or the second harmonic beam of a Q-switched Nd:YAG laser.

incidence diffraction grating is commonly inserted in the laser cavity (see Fig. 9.16); tuning is achieved by rotating the mirror labeled mirror 2 in the figure. Grazing incidence increases the resolving power of the grating* and hence considerably reduces the bandwidth of the emitted radiation (to ∼0.01–0.02 nm). Smaller bandwidths, down to single-mode operation, are obtained by inserting one or more Fabry–Perot etalons, as discussed in Sect. 7.8.2.1.

For continuous laser pumping, Ar^+ lasers (and sometimes also Kr^+ lasers) are often used. To achieve a much lower threshold, as required for cw pumping, the near-longitudinal pumping configuration in Fig. 9.17 is often used. Liquid dye medium in the form of a thin jet stream (∼200 μm thickness) flows freely in a plane orthogonal to the plane of the figure and inclined at Brewster's angle relative to the dye laser beam direction. Accordingly, this laser beam is linearly polarized with its electric field in the plane of the figure. Both pump and laser beams have their waist in the jet stream with similar, very small, spot sizes (∼10 μm). For laser tuning, a birefringent filter can be inserted within the laser cavity. To achieve single longitudinal mode operation, a birefringent filter and generally two Fabry–Perot etalons in a unidirectional ring cavity are often used (see Fig. 7.25). For femtosecond pulse generation, a colliding pulse mode-locked (CPM) laser configuration is generally used (see

* The resolving power $\nu/\Delta\nu$, where $\Delta\nu$ is the resolved bandwidth, can be shown to just equal the number of lines of the diffraction grating illuminated by the laser beam. At grazing incidence this number increases; thus resolving power also increases.

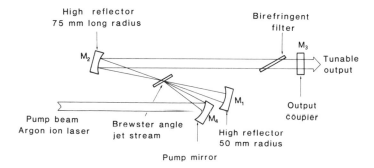

FIG. 9.17. Arrangement for an Ar ion laser-pumped cw dye laser.

Fig. 8.29). To achieve the shortest pulse duration (~ 25 fs in a solution of rhodamine 6G as active medium with DODCI as a saturable absorber), a prism pair is also inserted within the laser cavity for dispersion control.

By virtue of their wavelength tunability, wide spectral coverage, and the possibility of generating femtosecond laser pulses, organic dye lasers play an important role in many fields. In particular, these lasers are widely used in scientific applications, as either a narrow band, single-mode, tunable source of radiation for high-resolution frequency-domain spectroscopy or as femtosecond pulse generators for high-resolution time-domain investigations. Other applications include the biomedical field (e.g., treating diabetic retinopathy or several dermatological diseases) and laser photochemistry. In particular, repetitively pulsed dye laser systems with many dye lasers, each transversely pumped by a copper vapor laser of ~ 100 W average power, have been used for laser isotope separation of ^{235}U.

9.4. SEMICONDUCTOR LASERS

Semiconductor lasers represent one of the most important class of lasers in use today, not only because of the large variety of direct applications in which they are involved, but also because they have found a widespread use as pumps for solid-state lasers.[31,32] These lasers will therefore be considered at some length here.

For the active medium, semiconductor lasers require a direct-gap material, so normal elemental semiconductors (e.g., Si or Ge) cannot be used. The majority of semiconductor laser materials are based on a combination of elements in the third group of the Periodic Table (such as Al, Ga, In) and the fifth group (such as N, P, As, Sb) hence referred to as *III–V compounds*. Examples include the well-known GaAs, as well as some ternary (e.g., AlGaAs, InGaAs) and quaternary (e.g., InGaAsP) alloys. The cw laser emission wavelength of these III–V compounds generally ranges from 630–1600 nm. Quite recently, InGaN semiconductor lasers providing cw room-temperature emission in the blue (~ 410 nm) have been developed; these promise to become the best candidates for semiconductor laser emission in the very important blue-green spectral region. Semiconductor laser materials are not limited to III–V compounds, however. For the blue-green region of the spectrum, we mention the wide-gap semiconductors using a combination between elements of the second group (such as Cd and Zn) and of the sixth group (S, Se) (*II–VI compounds*). For the other end of the e.m. spectrum, we mention semiconductors based on some *IV–VI compounds* such as Pb salts of S, Se, and Te, all oscillating in the mid-infrared (4–29 μm). Due to the small band gap, these last lasers require cryogenic temperatures, however. In the same wavelength range, we thus mention the recent invention of the *quantum cascade* laser,[56] which promises efficient mid-infrared sources without requiring cryogenic temperatures.

9.4.1. Principle of Semiconductor Laser Operation

The operating principle of a semiconductor laser can be explained with the help of Fig. 9.18, which shows the semiconductor valence band V and conduction band C separated by the energy gap E_g. For simplicity we first assume that the semiconductor is held at $T = 0$ K. For a nondegenerate semiconductor, the valence band is completely filled with electrons

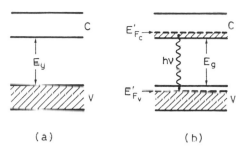

FIG. 9.18. Operation principle of a semiconductor laser.

while the conduction band is completely empty (see Fig. 9.18a; energy states belonging to the hatched area are completely filled by electrons). Suppose now that some electrons are raised from the valence band to the conduction band by a suitable pumping mechanism. After a very short time (~ 1 ps), electrons in the conduction band drop to the lowest unoccupied levels of this band; meanwhile any electron near the top of the valence band also drops to the lowest unoccupied levels of this band, thus leaving holes at the top of the valence band (Fig. 9.18b). This situation is described by introducing the quasi-Fermi levels E'_{F_c} for the conduction band and E'_{F_v} for the valence band (see Sect. 3.2.3). For each band, at $T = 0$ K, they define the energy below which states are fully occupied by electrons and above which states are empty. Light emission can now occur when an electron in the conduction band falls to the valence band to recombine with a hole. This so-called recombination radiation is the process by which radiation is emitted in light-emitting diodes (LED). Given appropriate, conditions, however, stimulated emission from this recombination radiation, leading to laser action, can occur. In Sect. 3.2.5 we showed that the condition for a photon to be amplified rather than absorbed by the semiconductor is simply given by [see Eq. (3.2.39)]:

$$E_g \leq h\nu \leq E'_{F_c} - E'_{F_v} \qquad (9.4.1)$$

In the simple case when $T = 0$ K, this condition is readily understood from Fig. 9.18b, since the nonhatched area in the valence band corresponds to empty states, and a conduction band electron can fall only into an empty state in the valence band. However, the detailed treatment of Sect. 3.2.5 shows that Eq. (9.4.1) holds in fact for any temperature, so that, for the range of transition energy $h\nu$ defined by Eq. (9.4.1), gain from stimulated emission exceeds absorption. To achieve the conditon set by Eq. (9.4.1), one must have $E'_{F_c} - E'_{F_v} \geq E_g$. Values for both E'_{F_c} and E'_{F_v} depend on the intensity of the pumping process, i.e., on the density N of electrons raised to the conduction band (see Fig. 3.15). Actually $E'_{F_c} = E'_{F_c}(N)$ increases while $E'_{F_v} = E'_{F_v}(N)$ decreases as N is increased. Thus, to obtain $E'_{F_c} - E'_{F_v} > E_g$, i.e., gain exceeding absorption losses, the electron density N must exceed some critical value established by the condition:

$$E'_{F_c}(N) - E'_{F_v}(N) = E_g \qquad (9.4.2)$$

---

* Equation (9.4.2) is thus equivalent to the condition $N_2 = N_1$ under which a nondegenerate two-level system becomes transparent.

The value of the injected carrier density that satisfies Eq. (9.4.2) is referred to as the carrier density at transparency,* N_{tr}. If the injected carrier density is now larger than N_{tr}, the semiconductor exhibits a net gain; if this active medium is then placed in a suitable cavity, laser action occurs when this net gain suffices to overcome cavity losses. Thus, to obtain laser action, injected carriers must reach some threshold value N_{th} larger than N_{tr} by a sufficient margin to allow net gain to overcome cavity losses.

Semiconductor laser pumping can in principle be achieved and indeed has been achieved in a number of ways, e.g., by using either the beam of another laser or an auxiliary electron beam to excite, transversely or longitudinally, the bulk semiconductor. By far the most convenient way of excitation, however, involves using the semiconductor laser in the form of a diode, with excitation produced by current flowing in the forward direction of the junction.[33]

Laser action in a semiconductor was in fact first observed in 1962 by using a *p-n* junction diode; the observation was made almost simultaneously by four groups,[34–37] three of which were using GaAs. Devices developed during the early stage of semiconductor laser research used the same material for both the *p* and *n* sides of the junction; therefore these are referred to as homojunction lasers. The homojunction laser is now only of historic importance, since it was superseded by the double-heterostructure (DH) laser, whose active medium is sandwiched between *p* and *n* materials that differ from the active material. Homojunction lasers in fact operate cw only at cryogenic temperatures ($T = 77$ K); only after the invention of the heterojunction laser was it possible to operate semiconductor lasers cw at room temperature. This development occurred 7 years after the invention of the homojunction laser (1969),[38–40] opening the way to the great variety of applications in which semiconductor lasers are nowadays used. Homojunction semiconductor lasers are, nevertheless, discussed briefly in Sect. 9.4.2, since this discussion helps understanding the great advantages offered by DH lasers.

9.4.2. Homojunction Lasers

In the homojunction laser, the pumping process is achieved in a *p-n* junction where both *p*-type and *n*-type regions, of the same material (e.g., GaAs), are in the form of a degenerate semiconductor. Thus donor and acceptor concentrations are so large ($\approx 10^{18}$ atoms/cm^3) that the Fermi level falls in the valence band for the *p*-type, E_{F_p}, and in the conduction band for the *n*-type, E_{F_n}. If no voltage is applied to the *p-n* junction,

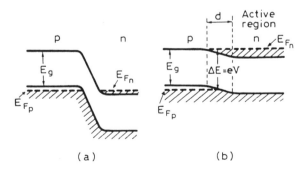

FIG. 9.19. (a) Band structure of a *p-n* junction semiconductor laser with zero voltage, and (b) forward voltage applied to the junction.

the band structure is as shown in Fig. 9.19a, where the two Fermi energies are aligned. Figure 9.19b shows the band structure when a forward-bias voltage V is applied; the two Fermi levels are separated now by $\Delta E = eV$. We see from this figure that, in the junction region, electrons are injected into the conduction band (from the n-type region) while holes are injected into the valence band (from the p-type region). Thus, for appropriate values of the current density, the transparency condition and then the laser threshold condition can be reached. One of the main limitations of this device comes from the very small potential barrier that an electron in the conduction band encounters when it reaches the p-side of the junction. The electron can then penetrate into the p-type material, where it becomes a minority carrier, thus recombining with a hole. The penetration depth d of the electron is then given (according to diffusion theory) by $d = (\sqrt{D\tau})$, where D is the diffusion coefficient and τ is the electron lifetime, as established by electron-hole recombination. In GaAs, one has $D = 10$ cm^2/s and $\tau \cong 3$ ns, so that $d \approx 1$ μm; this shows that the active region is quite thick, being limited by the diffusion length d rather than the thickness of the depletion layer (≈ 0.1 μm).

Figure 9.20 shows a typical configuration of a p-n junction laser; the shaded region corresponds to the active layer. Note that the diode dimensions are very small (some hundreds of microns). To provide feedback for laser action, two parallel end faces are prepared, usually by cleavage, along crystal planes. Often these two surfaces are not provided with reflective coatings; in fact, since the refractive index of a semiconductor is very large (e.g., $n = 3.6$ for GaAs), there is already a sufficiently high reflectivity ($\sim 32\%$ for GaAs) from the Fresnel reflection at the semiconductor-air interface. As mentioned earlier, the thickness of the active region in the direction perpendicular to the junction is $d \approx 1$ μm. Because of diffraction, however, the transverse dimension of the laser beam in this direction (≈ 5 μm) is significantly larger than the active region.

A homojunction laser has a very high threshold current density at room temperature ($J_{th} \cong 10^5$ A/cm^2) that prevents the laser from operating as cw at room temperature (without suffering destruction in a very short time). There are two main reasons for this high threshold: (1) The thickness of the active region ($d \approx 1$ μm) is quite large, so the threshold current, being proportional to the volume of the active medium, is proportionally high. (2) Due to its comparatively large transverse dimensions, the laser beam extends considerably into the p and n regions, where it is strongly absorbed. Given these reasons, homojunction lasers only operate cw at cryogenic temperatures (typically at liquid nitrogen temperature

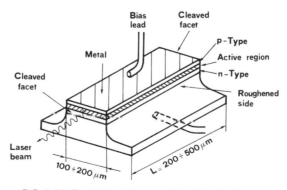

FIG. 9.20. Typical broad-area p-n homojunction laser.

$T = 77$ K). For a given laser transition, in fact, the semiconductor gain, according to Eq. (3.2.37), increases rapidly with decreasing temperature. Contact of the diode with liquid nitrogen also helps providing very efficient cooling.

9.4.3. Double-Heterostructure Lasers

The limitations discussed in Sect. 9.4.2 prevented widespread use of semiconductor lasers until first, single-heterostructure, then, immediately after, double-heterostructure lasers were introduced. We limit our discussion here to the latter structure, since it is the only one now in common use.

Figures 9.21a–b, respectively, show two examples of a double heterostructure, where the active medium is a thin layer (0.1–0.2 μm) of either GaAs or the quaternary alloy InGaAsP. For the two cases considered, p and n sides are made of $Al_{0.3}Ga_{0.7}As$ and InP, respectively. When properly optimized (see Fig. 9.23), the room temperature threshold current density of such a diode structure can be reduced by about two orders of magnitude (i.e., to $\sim 10^3$ A/cm^2) compared to the corresponding homojunction devices, thus making cw room temperature operation feasible. This strong reduction in threshold current density is due to the combined effect of the following three circumstances: (1) The refractive index of the active layer n_1 (e.g., $n_1 = 3.6$ for GaAs) is significantly larger than that, n_2, of the p-side and n-side cladding-layers (e.g., $n_2 = 3.4$ for $Al_{0.3}Ga_{0.7}As$), thus providing a guiding structure (see Fig. 9.22a). This means that the laser beam is now primarily confined to the active layer region, i.e., where the gain exists (*photon confinement*; see Fig. 9.22b). (2) The band gap E_{g_1} of the active layer (e.g., $E_{g_1} \cong 1.5$ eV in GaAs) is significantly smaller than that, E_{g_2} of the cladding layers (e.g., $E_{g_2} \cong 1.8$ eV for $Al_{0.3}Ga_{0.7}As$).* Thus energy barriers form at the two junction planes, thereby effectively confining injected holes and electrons within the active layer (*carrier confinement*; see Fig. 9.22c). For a given current density, hole and electron concentration in the active layer increases, so gain increases. (3) Since E_{g_2} is appreciably larger than E_{g_1}, the laser beam, which has a frequency of $v \cong E_{g_1}/h$, is much less strongly absorbed in its wings (see Fig. 9.22b) by the cladding layers; the loss arises in this case only from free carriers (*reduced absorption*).

To form a double heterostructure, thus taking advantage of all its favorable properties, a very important requirement must be fulfilled, namely, the lattice period of the active layer must equal (within $\sim 0.1\%$) that of the cladding layers.† In fact, if this condition is not

---

* It is a general rule for all III–V compounds that any change in composition producing a change in a given sense, e.g., a *decrease*, in band gap also produces a change in the opposite sense, i.e., an *increase*, in the refractive index.

† All III–V compounds crystallize in the cubic structure.

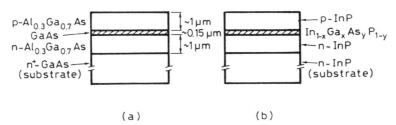

(a)   (b)

FIG. 9.21. Schematic diagram of a double-heterostructure where the active medium (*hatched area*) consists of (a) GaAs and (b) InGaAsP.

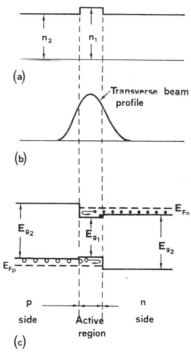

FIG. 9.22. (a) Refractive index profile, (b) transverse beam profile, and (c) band structure (very schematic) of a double-heterostructure diode laser.

fulfilled, the resulting strain at the two interfaces results in misfit dislocations; each dislocation then acts as a rather effective center for electron-hole nonradiative recombination. For the GaAs/AlGaAs structure, the lattice-matching requirement does not constitute a limitation because the lattice periods of GaAs (564 pm) and AlAs (566 pm) are very close in value. (Atomic radii of Ga and Al are in fact almost the same.)

For the quaternary compound $In_{1-x}Ga_xAs_yP_{1-y}$, the alloy can be lattice-matched to InP for a specific y/x ratio, as the following argument shows: Suppose that, starting with InP for the active layer, some fraction x of Ga is added, replacing some In in the lattice (which hence becomes In_{1-x}). Since the Ga radius is *smaller* (by ~ 19 pm) than that of In, the lattice period of the $In_{1-x}Ga_xP$ decreases compared to InP. Suppose now that some fraction y of As (As_y) replaces P (thus becoming P_{1-y}). Since the As radius is now *larger* (by ~ 10 pm) than that of P, this addition tends to increase the lattice period. Therefore, if the y/x ratio of the two substituents has an appropriate value, the two effects cancel each other, thus resulting in $In_{1-x}Ga_xAs_yP_{1-y}$ being lattice-matched to InP. This lattice-matching condition is given by $y \cong 2.2x$. Changing x, while keeping the y/x ratio equal to the lattice-matching value, changes the semiconductor band gap and hence the emission wavelength. In this way, the emission wavelength of $In_{1-x}Ga_xAs_yP_{1-y}$ can be varied from 1150 to 1670 nm for cw room temperature operation, thus encompassing the so-called second (~ 1300 nm) and third (~ 1550 nm) transmission windows of silica optical fibers.

Figure 9.23 shows experimental and theoretical plots of the threshold current density J_{th} versus thickness d of the active layer, for a DH GaAs laser.[41] Note that, as d decreases, J_{th}

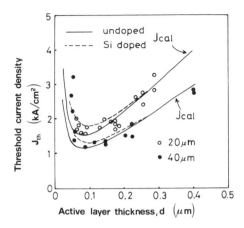

FIG. 9.23. Calculated (*continuous and dashed lines*) and experimental (*open and closed circles*) values of the threshold current density J_{th} versus active layer thickness d for a 300 μm long AlGaAs DH laser. Closed and open circles represent data for a 40-μm and 20-μm stripe width, respectively. Theoretical curves J_{cal} refer to cases of undoped and low Si-doped active layers. (By permission from Ref. 41.)

first decreases, then reaches a minimum value ($J_{th} \cong 1$ kA/cm^2 for $d \cong 0.1$ μm) and thereafter increases. To understand this behavior, we first relate the threshold current density J_{th} to the threshold carrier density N_{th}. We begin by defining R_p as the rate at which electrons (and holes) are injected, for unit volume, into the active layer. We also let η_i, usually referred to as the *internal quantum efficiency*, be the fraction of carriers that recombine radiatively in the layer; the remaining fraction undergoes nonradiative electron-hole recombination primarily at the junction boundaries. We can also view the quantity η_i as the effective fraction of injected carriers and we can consider the remaining fraction as not having been injected into the active region at all. For a given current density J flowing through the junction, R_p is then readily seen to be given by $R_p = \eta_i J/ed$, where e is the electron charge and d is the thickness of the active layer. Under steady-state conditions, a simple balance condition gives the corresponding expression for the carrier density N as $N = R_p \tau_r$, where τ_r is the radiative recombination time (given the previous assumption, all carriers recombine radiatively in the active layer). From the previous two expressions, we get $J = edN/\eta_i \tau_r$ so that at threshold:

$$J_{th} = \left(\frac{ed}{\eta_i \tau_r}\right) N_{th} \qquad (9.4.3)$$

With the help of Eq. (9.4.3) we can now qualitatively understand the relevant features of Fig. 9.23. Note in fact that, for sufficiently large values of d, the threshold carrier density N_{th} is almost the same as the transparency density N_{tr} (see Example 9.1); hence it is a constant. Equation (9.4.3) then predicts a linear relation between J_{th} and d as indeed observed in Fig. 9.23 for sufficiently large values of d (larger than ~0.15 μm). However, when the thickness d becomes sufficiently small, the confinement action of the active layer (see Fig. 9.22b) is no longer so effective; therefore the beam extends considerably into the p and n sides of the junction. This situation results in a reduced effective gain and, at the same time, in increased losses in the cladding layers; both effects lead to a strongly increased N_{th}. Thus, for sufficiently small values of d, J_{th} is expected to increase as d decreases.

Example 9.1. *Carrier and current densities at threshold for a DH GaAs laser.* Since the laser field is space-dependent, the threshold condition must be derived, as in previous examples (see e.g., Sect. 6.3.4), by setting the condition that spatially averaged gain must equal spatially averaged losses. Thus in general we write

$$\langle g \rangle L = \langle \alpha_a \rangle L + \langle \alpha_n \rangle L + \langle \alpha_p \rangle L + \gamma_m \qquad (9.4.1)$$

where: L is the length of the active medium; g is the gain coefficient; α_a is the scattering loss of the active layer; α_n and α_p are the losses in the n and p sides of the cladding, respectively; γ_m is the mirror loss. The average values in Eq. (9.4.4) are taken over the field intensity distribution so that, for example, the average gain is given by

$$\langle g \rangle = \frac{\int_a g|u|^2 \, dV}{\int_c |u|^2 \, dV} \qquad (9.4.5)$$

where $u(x, y, z)$ is the field distribution within the laser cavity, the integral of the numerator is taken over the volume of the active medium, and the integral of the denominator is taken over the whole volume of the cavity. The quantities $\langle g \rangle$ and $\langle g \rangle L$ are usually referred to as the *modal gain coefficient* and the *modal gain*, respectively. Similar expressions hold for the average values on the right-hand side of Eq. (9.4.4); the integral of the numerator is of course always taken over the volume of the material considered. For simplicity we assume that $\alpha_n \cong \alpha_p = \alpha$, and we neglect the spatial variation of the cavity field along the longitudinal z-coordinate (as produced by, e.g., a standing wave pattern) and along the coordinate parallel to the junction. Then, with the help of Eq. (9.4.5) and the corresponding expressions for $\langle \alpha_a \rangle$ and $\langle \alpha_p \rangle$, Eq. (9.4.4) readily gives

$$g\Gamma = \alpha_a \Gamma + \alpha(1 - \Gamma) + [\ln(1/R)/L] \qquad (9.4.6)$$

where R is the power reflectivity of the two end mirrors (assumed to be equal for the two mirrors) and:

$$\Gamma = \frac{\int_{-d/2}^{+d/2} |u|^2 \, dx}{\int_{-\infty}^{+\infty} |u|^2 \, dx} \qquad (9.4.7)$$

where x is the coordinate in the direction orthogonal to the junction. The quantity Γ represents the fraction of the beam power actually in the active layer; it is usually referred to as the *beam confinement factor*. According to the discussion in Sect. 3.2.5, we can now approximate g as $g = \sigma(N - N_{tr})$, where σ is the differential gain coefficient and N_{tr} is the carrier density at transparency. If we now further assume for simplicity $\alpha_a = \alpha$, then Eq. (9.4.6) simplifies to:

$$\sigma \Gamma (N_{th} - N_{tr}) = \alpha + \left[\frac{\ln(1/R)}{L} \right] = \frac{\gamma}{L} \qquad (9.4.8)$$

where N_{th} is the threshold carrier density and $\gamma = \alpha L + \ln(1/R)$ is the total loss per pass. From Eq. (9.4.8) we finally obtain the desired expression for the threshold carrier density as:

$$N_{th} = \left(\frac{\gamma}{\sigma L \Gamma} \right) + N_{tr} \qquad (9.4.9)$$

To proceed with the calculation, we must now evaluate the confinement factor Γ. A fairly accurate and simple expression is given by:[42]

$$\Gamma \cong \frac{D^2}{(2 + D^2)} \tag{9.4.10}$$

where:

$$D = 2\pi(n_1^2 - n_2^2)^{1/2}\frac{d}{\lambda} \tag{9.4.11}$$

is the normalized thickness of the active layer. (Recall that n_1 and n_2 are the refractive indices of the active medium and cladding layers, respectively.) If we now take $n_1 = 3.6$, $n_2 = 3.4$, and $\lambda = 850$ nm (as appropriate for a GaAs laser) and perform the calculation for $d = 0.1$ μm, we obtain $D \cong 0.875$ and hence $\Gamma \cong 0.28$. To obtain a numerical estimate of the corresponding value of N_{th}, we take the reflectivity of the two end faces equal to that of the uncoated surfaces ($R \cong 32\%$) and assume a loss coefficient $\alpha \cong 10$ cm^{-1} and a cavity length of $L = 300$ μm. We obtain $\gamma = \ln(1/R) + \alpha L \cong 1.44$. If we now take (see Table 3.1) $\sigma = 3.6 \times 10^{-16}$ cm^2 and $N_{tr} = 2 \times 10^{18}$ carriers/cm^3, we obtain from Eq. (9.4.9):

$$N_{th} = (0.48 + 2) \times 10^{18} \text{ carriers/cm}^3 \tag{9.4.12}$$

where, for convenience, the numerical values of the two terms on the right-hand side of Eq. (9.4.9) are shown separately. Equation (9.4.12) thus shows that, in this case i.e. for relatively large values of d and hence of the confinement factor Γ, the first term (i.e., the carrier density required to overcome cavity losses) is a relatively small fraction of N_{tr}.

The threshold current density is now readily obtained by substituting Eq. (9.4.9) into Eq. (9.4.3). We obtain

$$J_{th} = \left(\frac{ed}{\eta_i \tau_r}\right)\left[\left(\frac{\gamma}{\sigma L \Gamma}\right) + N_{tr}\right] \tag{9.4.13}$$

We have seen that, for sufficiently large values of d, N_{tr} is the dominant term in the square brackets in Eq. (9.4.13). In this case J_{th} is expected to be proportional to d, as indeed shown in Fig. 9.23 for d larger than ~ 0.15 μm; most of the threshold pump current, in this case, is just used to reach the semiconductor transparency condition. When d becomes very small, however, the confinement factor also becomes very small [according to Eq. (9.4.10), for very small values of d, $\Gamma \propto d^2$]; the first term in brackets eventually dominates, so J_{th} reaches a point where it increases with decreasing d.

To obtain a numerical evaluation of J_{th} from Eq. (9.4.13), we assume $d = 0.1$ μm, we take $\eta_i \cong 1$, $\tau_r = 4$ ns and use the previously calculated value of N_{th}. We obtain $J_{th} \cong 10^3$ A/cm^2 in reasonable agreement with results in Fig. 9.23.

9.4.4. Quantum Well Lasers

If the thickness of the active layer of a DH laser is greatly reduced to a point where the dimension becomes comparable to the deBroglie wavelength, ($\lambda \cong h/p$), a QW double-heterostructure laser is produced.[43,44] Such lasers exploit the more favorable optical properties of a QW or a multiple QW (MQW) structure compared to those of the corresponding bulk material (in particular the increased differential gain, see Example 3.12, and decreased dependence of this gain on temperature). These favorable properties are essentially related to the completely different form for the density of states of QW materials compared to bulk materials, arising from quantum confinement in the well direction (see Sect. 3.3). Single QW and also MQW lasers are however seriously affected by the strong reduction in the confinement factor arising from reduced layer thickness. To limit beam size in the QW direction, one must then use a separate confinement structure.

Figure 9.24a shows a particularly simple example of several structures introduced for this purpose. Everything in this figure is to scale except for the bulk GaAs band gap energy, which is reduced for clarity. At the center of the structure is the thin (~ 10 nm) QW (GaAs) and, on both sides of the well, are two thicker (~ 0.1-μm) inner barrier layers of wider band gap and, hence, lower refractive index material ($Al_{0.2}Ga_{0.8}As$). Outside the inner barrier layers are two much thicker (~ 1-μm) cladding layers of still wider band gap material ($Al_{0.6}Ga_{0.4}As$), constituting the *p*- and *n*-sides of the diode.

Beam confinement is established by the higher refractive index of inner barrier layers compared to cladding layers, while the contribution to confinement by the very thin QW is negligible. The resulting beam intensity profile for this waveguide configuration is also shown as a dashed line in Fig. 9.24a. The full width between the $1/e^2$ points, in this case, is confined to a comparatively small dimension ($\sim 0.8 \mu$m). A somewhat similar and widely used structure is shown in Fig. 9.24b, where the index composition of the inner barrier layer ($Al_xGa_{1-x}As$) gradually changes from, e.g., $x = 0.2$ at the QW interface to the value $x = 0.6$ at the interfaces between the two cladding layers, where it matches the index of the cladding layers. This structure is usually referred to as a graded index separated confinement heterostructure (GRINSCH).

It should be noted that, in both structures in Figs. 9.24, carriers are confined by the QW structure, while the beam is confined by the step index or graded index profile in Figs. 9.24a–b, respectively. Note also that, although the thickness of the QW layers is much smaller than the width of the beam, optical confinement results in a sufficiently high confinement factor to now take advantage of the expected reduction of J_{th} due to the strong reduction in the active layer thickness d [see Eqs. (9.4.3) and (9.4.13)]. In fact, as Example 9.2 shows, we can now obtain values of J_{th}

Example 9.2. *Carrier and current densities at threshold for a GaAs/AlGaAs QW laser.* We assume that a functional relation of the form $g = \sigma(N - N_{th})$ still holds approximately for a QW,* so that relations established in Example 9.1 are still applicable. To compare this case with Example 9.1, we take the same values for the loss coefficient α (10 cm^{-1}) and mirror reflectivity R (32%), hence of the total loss γ ($\gamma = 1.44$). We also take the same value for cavity length (300μm) and carrier density at transparency N_{tr} (2×10^{18} cm^{-3}), while we now take $\sigma \cong 6 \times 10^{-16}$ cm^2 (see Sect. 3.3.5). To calculate the confinement factor, we assume that the field profile can be written as $u \propto \exp[-(x^2/w_\perp^2)]$ where $w_\perp$ is the beam spot size in the direction orthogonal to the junction. From Eq. (9.4.7) we obtain $\Gamma = (d/0.62d_\perp)$, where d is well thickness and $d_\perp = 2w_\perp$, so that, taking $d = 10$ nm and $d_\perp = 1 \mu$m, we obtain $\Gamma = 1.6 \times 10^{-2}$. From Eq. (9.4.9) we now have $N_{th} = (5 + 2) \times 10^{18}$ cm^{-3}, where numerical values for the two terms on the right-hand side of Eq. (9.4.9) are again separated. We see that, due to the much smaller value for the confinement factor, the first term, i.e., the carrier density required to overcome cavity losses, is now

---

* This approximation holds with less accuracy for a QW; a plot of g versus N shows a curve that, due to the essentially two-dimensional structure of the density of states, saturates at sufficiently high values of current injection (see Ref. 45).

appreciably larger than the second term N_{tr}. The threshold current density is now readily obtained by substituting the previously calculated value of N_{th} into Eq. (9.4.3). Assuming again that $\eta_i = 1$ and $\tau = 4$ ns, we obtain $J_{th} \cong 280$ A/cm^2, which is about four times smaller than the value calculated for a DH laser. Note that, in this case, since cavity loss primarily determines the value of N_{th}, a reduction of this loss is helpful in further reducing J_{th}. If we now take, for example, $\alpha = 3$ cm^{-1} and $R = 80\%$, we have $\gamma = 0.28$, hence $N_{th} = (2.3 + 2) \times 10^{18}$ cm^{-3} and $J_{th} \cong 170$ A/cm^2.

$\sim$4–5 times smaller than those of a DH laser (i.e., $\sim$200 A/cm^2). This threshold reduction thus arises from a combination of the following features: (1) Reduction in threshold as expected from the strong reduction in layer thickness once the problem of beam confinement is partially overcome by a separated confinement structure. (2) Increase (by about a factor of 2) in the differential gain, which occurs in a QW compared to the corresponding bulk material.

Separated confinement structures in Fig. 9.24 may include either a single QW, as shown in the figure, or a MQW structure; in this case the structure consists of a number of alternating layers of narrow and wide band gap materials (see Fig. 9.25). The thickness of each QW is 5 nm, while the thickness of the wide band gap barrier (4 nm) must be enough to prevent quantum tunneling of electrons from one well to the next. Compared to the case for a MQW, the threshold for a single QW tends to increase due to the decreased value of the confinement factor; it tends to decrease due to the decreased value of the effective thickness d of the active medium. For long laser cavities ($L > 300$ μm), one has $N_{th} \cong N_{tr}$; thus the second effect dominates the first one, and a single QW exhibits the lowest threshold. For shorter cavity lengths, however, the threshold of a MQW becomes smaller than that of a single QW; the optimum number of wells depends on the actual value of cavity length.[46]

We saw in Sect. 9.4.3 that, in a DH laser, a precise lattice match between the two heterostructures (within better than 0.1%) must be achieved. For the very small thickness of a QW, however, this matching condition can be considerably relaxed, so that a larger lattice mismatch between the QW and the surrounding wider gap material (up to $\sim$1–3%) can be tolerated without creating excessive misfit dislocations at the boundaries between the two materials. Due to the lattice mismatch, a compressive or tensile strain is produced in the QW

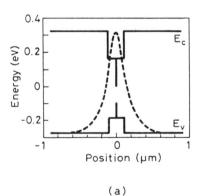
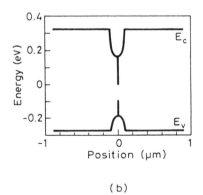

(a)   (b)

FIG. 9.24. (a) Energy bands of a step index Al$_x$Ga$_{1-x}$As-GaAs separate confinement QW heterostructure. The resulting optical mode intensity profile for this waveguide structure is shown as a dashed line. (By permission from Ref. 43) (b) Energy bands of a graded index Al$_x$Ga$_{1-x}$As-GaAs separate confinement QW heterostructure (GRINSCH).

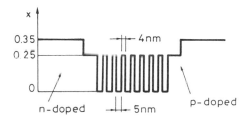

FIG. 9.25. Composition variations in $In_{0.5}Ga_{0.5}P/In_{0.5}(Ga_{0.5-x}Al_x)P$ MQW active layer, where $x = 0.25$ for well barriers and $x = 0.35$ for confinement layers, producing laser emission at 670-nm wavelength.

structure, resulting in a strained QW. Strained QWs present two main advantages: (1) Structures can be grown to produce laser action in wavelength ranges not otherwise covered (e.g., 900–1100 nm for $In_xGa_{1-x}As/GaAs$). (2) Under compressive strain, as discussed in Sect. 3.3.6, the effective mass of the hole in a direction parallel to the junction decreases to a value closer to the effective mass of the electron. This situation lowers the transparency density N_{tr} and increases the differential gain σ compared to an unstrained QW. Thus strained QW lasers allow laser action to be obtained, with very low threshold current density and high efficiency, at wavelengths not previously accessible.

9.4.5. Laser Devices and Performances

Double-heterostructure as well as QW lasers quite often use the so-called stripe geometry configuration in Fig. 9.26, where the active area (*dashed lines*) may be either a double heterostructure or a separated confinement single-QW or MQW structure. We see from both figures that, by introducing a suitable insulating oxide layer, current from the positive electrode is constrained to flow in a stripe of narrow width s ($s = 3$–$10 \ \mu m$). Compared to a broad area device (see Fig. 9.20), this stripe geometry device has the advantage of considerably reducing the area A ($A = Ls$, where L is the semiconductor length) through which current flows. Thus, for a given current density J, the required total current $I = JA$ is correspondingly reduced. Furthermore, since the width of the gain region in the junction plane is also roughly equal to s, this mechanism can be used to confine the beam transverse dimension in the direction parallel to the junction. The corresponding device is

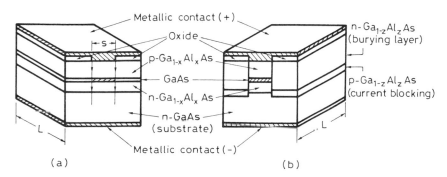

FIG. 9.26. Detail of stripe geometry double-heterostructure semiconductor lasers: (a) gain-guided laser and (b) buried heterostructure index-guided laser.

referred to as a *gain-guided laser* (see Fig. 9.26a). If s is made sufficiently small ($s < 10~\mu$m), gain confinement restricts the beam to the fundamental transverse mode in the direction parallel to the junction. Orthogonal to the junction, the beam is also confined to the fundamental transverse mode by the index-guiding effect produced by the double heterostructure (see Fig. 9.22) or by the separated confinement structure (see Fig. 9.24). The output beam thus consists of a single-transverse mode of elliptical cross section ($\sim 1~\mu$m $\times 5~\mu$m). The gain-guided structure in Fig. 9.26a has the disadvantage that unpumped regions of the active layer are strongly absorbant, so the beam confinement action from these regions inevitably introduces some loss for the beam. A better solution is to provide lateral confinement by a refractive-index-guiding action within the junction plane, as well (*index-guided lasers*). A possible solution involves surrounding the active layer with semiconductor materials of a lower refractive index, such as in the buried heterostructure laser in Fig. 9.26b. With an index-guided laser, the laser beam suffers less absorption by the laterally confining media. In fact, index-guided structures (e.g., buried or ridge waveguide structures) appear to be increasingly favored in commercial devices.

We now consider some properties of the output beam, namely, output power, beam divergence, and spectral content.

Plots of output power versus input current, at two different temperatures, for a gain-guided DH GaAs semiconductor laser are shown in Fig. 9.27. Note that the threshold current I_{th} at room temperature is less than 100 mA as a result of using the stripe geometry. Threshold currents lower than this (~ 15 mA) are now more typical for both gain-guided and index-guided DH GaAs semiconductor lasers; much lower values (~ 1 mA) are obtained with particular QW devices. [Indeed, assuming $J_{th} = 200$ A/cm^2 (see Example 9.2) $s = 4~\mu$m, and $L = 150~\mu$m, one gets $I_{th} = 1.2$ mA.] Figure 9.27 shows a rapid increase in I_{th} with temperature. In most laser diodes, this increase follows the empirical law $I_{th} \propto \exp(T/T_0)$ where T_0 is a characteristic temperature, dependent on the particular diode, whose value is a measure of the quality of the diode. The ratio between threshold values at two temperatures, differing by ΔT, is in fact given by $(I'_{th}/I''_{th}) = \exp(\Delta T/T_0)$. Thus, the

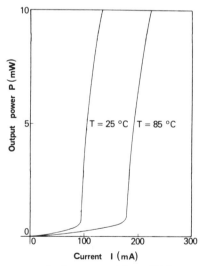

FIG. 9.27. Plot of the output power versus the input current for a DH laser at room temperature and elevated temperature.

greater T_0, the less sensitive I_{th} is to temperature. From Fig. 9.27 we can calculate that $T_0 \cong 91$ K. In DH GaAs lasers T_0 typically ranges from 100–200 K, while T_0 is usually larger (> 270 K) for GaAs QW lasers. Thus the increase in the characteristic temperature of QW lasers is another advantage of QW devices; it results from a weaker dependence of quasi-Fermi energies, and hence of differential gain, on temperature (compare Fig. 3.25 and Fig. 3.15). The characteristic temperature for DH InGaAsP/InP lasers is considerably lower than the preceding values (50 K < T_0 < 70 K), probably due to the rapid increase in the nonradiative decay rate (due to Auger processes) in this narrower bandgap material (see Sect. 3.2.6). Note that the output power in Fig. 9.27 is limited to ~ 10 mW. Higher output powers (typically above 100 mW) can result in beam intensities high enough to damage semiconductor facets. Note also that the slope efficiency of the laser is given by $\eta_s = dP/V\,dI$, where V is the applied voltage. Taking $V \cong 1.8$ V, we obtain $\eta_s = 40\%$. Even higher slope efficiencies than this (up to about 60%) have in fact been reported. Thus semiconductor lasers are currently the most efficient lasers available.

Example 9.3. *Output power and external quantum efficiency of a semiconductor laser.* To calculate the output power, we first observe that, under steady-state conditions, the power emitted by stimulated emission can simply be written as $P_e = (I - I_{th})\eta_i h\nu/e$, where η_i is the internal quantum efficiency, introduced in Sect. 9.4.3, and ν is the frequency of the emitted radiation. Part of this power is dissipated by internal losses (due to scattering and cladding losses), and part is available as output power from the two cavity ends. This power can then be written as:

$$P = \left[\frac{(I - I_{th})\eta_i h\nu}{e}\right]\left(\frac{-\ln R}{\alpha L - \ln R}\right) \qquad (9.4.14)$$

where R is the power reflectivity of the two end mirrors, α is the internal loss coefficient, and L is the cavity length. We can now define the external quantum efficiency η_{ex} as the ratio between the increase in emitted photons and the corresponding increase in injected carriers, i.e., $\eta_{ex} = d(P/h\nu)/d(I/e)$. From Eq. (9.4.14) we then obtain:

$$\eta_{ex} = \eta_{in}\left(\frac{-\ln R}{\alpha L - \ln R}\right) \qquad (9.4.15)$$

This shows that η_{ex} increases by reducing the cavity length. Note also that, according to previous definitions, the relation between external efficiency and slope efficiency is simply $\eta_{ex} = \eta_s(eV/h\nu)$.

Concerning the divergence properties of the output beam, we observe that, due to the small beam dimension in the direction orthogonal to the junction (~ 1 μm), the beam is always diffraction-limited in the plane orthogonal to the junction. Furthermore, as already discussed, if the width of the stripe is smaller than some critical value (~ 10 μm), the beam is also diffraction-limited in the plane parallel to the junction. Now let $d_\perp$ and $d_\parallel$ be the beam dimensions (full width between $1/e$ points of the electric field) in the two directions and let us assume a Gaussian field distribution in both transverse directions. According to Eq. (4.7.19), beam divergences, $\theta_\parallel$ in the plane parallel to the junction and $\theta_\perp$ in the plane orthogonal to the junction, are given by $\theta_\parallel = 2\lambda/\pi d_\parallel$ and $\theta_\perp = 2\lambda/\pi d_\perp$, respectively. For an output beam with an elliptical cross section (e.g., 1 μm $\times$ 5 μm), the divergence in the plane orthogonal to the junction is thus larger than that in the plane parallel to the junction. Then the beam ellipticity rotates by 90° at a distance some tens of microns away from the semiconductor exit face (see Fig. 6.9a). As discussed in Sect. 6.3.2.1, optical systems can be developed to compensate for this astigmatic behavior of the beam.

A typical emission spectrum for a diode laser, in which optical feedback is provided by the two end face reflections, is shown in Fig. 9.28. The equally spaced peaks correspond to different longitudinal modes of the Fabry–Perot cavity. Note two points from this figure: (1) The relative spectral bandwidth $\Delta \nu_L / \nu$ is sufficiently small ($\sim 1.1 \times 10^{-3}$) to justify stating, according to Eq. (9.4.1), that emission frequency is roughly equal to E_g/h. (2) The absolute value of this bandwidth ($\Delta \nu_L \cong 400$ GHz in Fig. 9.28) is sufficiently large, however, to be a problem for optical fiber communications, due to the chromatic dispersion of an optical fiber particularly around $\lambda = 1550$ nm. To obtain much smaller linewidths, the best approach is to use either a distributed feedback (DFB) laser or a laser with distributed Bragg reflectors (DBRs). These lasers are briefly considered in the next section.

9.4.6. Distributed Feedback and Distributed Bragg Reflector Lasers

A distributed feedback laser consists of an active medium in which a periodic thickness variation is produced in one of the cladding layers forming part of the heterostructure.[47] Figure 9.29a shows a schematic of a DFB laser oscillating at 1550 nm, where an InGaAsP active layer ($\lambda = 1550$ nm) is sandwiched between two InGaAsP cladding layers ($\lambda = 1300$ nm); one of the two layers shows this periodic thickness variation. Since the refractive index of the InGaAsP cladding layers is larger than that of the InP p- and n-type layers, the electric field of the oscillating mode will see an effective refractive index $n_{eff}(z) = \langle n(x,z) \rangle_x$ that depends on the longitudinal z-coordinate. In the previous expression $\langle \ \rangle_x$ stands for a weighted spatial average taken over the x-coordinate, orthogonal to the junction; the weight is determined by the transverse distribution of the beam intensity $|u(x)|^2$ [see also Eq. (9.4.5)]. We now assume that $n_{eff}(z)$ is a periodic function of z, i.e.:

$$n_{eff}(z) = n_0 + n_1 \sin[(2\pi z/\Lambda) + \varphi] \qquad (9.4.16)$$

where Λ is the pitch of the periodic thickness change (see Fig. 9.29a). In accordance with Bragg's ideas about scattering from a periodic array of elements, the forward- and backward-propagating beams of the DFB laser are effectively coupled to each other if the free space wavelength of the radiation is such that

$$\lambda = \lambda_B = 2\langle n_{eff} \rangle \Lambda \qquad (9.4.17)$$

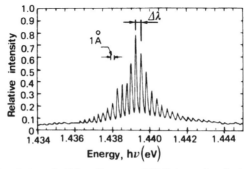

FIG. 9.28. Typical spectral emission of a Fabry–Perot-type DH GaAs semiconductor laser with a cavity length of 250 μm.

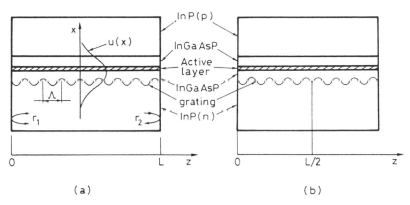

FIG. 9.29. Schematic structure of (a) DFB laser with a uniform grating and (b) $\lambda/4$-shifted DFB laser.

where $\langle n_{eff} \rangle$ is some suitable average value, along the z-coordinate, of n_{eff}, whose value is discussed later. To appreciate the significance of Eq. (9.4.17), we assume that $n_{eff}(z)$ consists of a periodic square-wave function of period Λ. In this case the structure in Fig. 9.29a is equivalent to a periodic sequence of high and low refractive index layers; the thickness of each layer equals $\Lambda/2$. This case is thus rather similar to a periodic sequence of multilayer dielectric mirrors (see Sect. 4.4) and constructive reflection is expected to occur when $(\langle n_{eff} \rangle \Lambda/2) = \lambda/4$. Equation (9.4.17) shows that, for a given pitch Λ, there is only one wavelength, i.e., only one mode satisfying the Bragg condition. Only this mode is expected to oscillate when the appropriate threshold condition is satisfied.

The preceding simple considerations are very approximate and a more precise description of DFB laser behavior would require a detailed analytic treatment. In this analysis, the two oppositely traveling waves are assumed to have an effective gain coefficient, as established by the active medium, and to be coupled by a periodic change in the dielectric constant, i.e., of the refractive index. One also generally assumes some finite values r_1 and r_2 of the electric field reflectivity from the two end faces. We do not go into this analysis here, referring the reader elsewhere for a detailed treatment,[47,48] but discuss only a few important results.

First we consider a rather peculiar result, indicated for the simple case $r_1 = r_2 = 0$ in Fig. 9.30a. There the intensity transmittance T, $T = |E_f(0, L)/E_f(0, 0)|^2$ for, e.g., the forward beam, is plotted against normalized detuning $\delta L = (\beta - \beta_B)L$. In the preceding expressions $E_f(x, z)$ is the electric field of the forward beam, L is the cavity length, $\beta = 2\pi n_0/\lambda$ and $\beta_B = \pi/\Lambda$. The plots shown in the figure are obtained for the value $kL = 2$ of the normalized coupling coefficient k ($k \cong 2\pi n_1/\lambda$) and for several values of the effective gain $\langle g \rangle L$.[49] Figure 9.30a shows that a transmission minimum actually occurs at exact resonance ($\delta = 0$) while several transmission maxima, i.e., several modes, are present, symmetrically located at both sides of exact resonance. Figure 9.31a illustrates the reason for the existence of the first and strongest two resonances; in fact this figure shows the longitudinal variation of the refractive index, the standing wave patterns of the two modes, and the corresponding resonant wavelengths.[49] The mode labeled $+1$ is then seen to be subjected to an effective, i.e., longitudinally averaged refractive index, $\langle n_{eff} \rangle_1$, that is slightly smaller than that, $\langle n_{eff} \rangle_{-1}$, of mode -1. In both preceding expressions, the spatial average $\langle \, \rangle$ is taken over the longitudinal intensity distribution of the cavity mode. According to the general Bragg

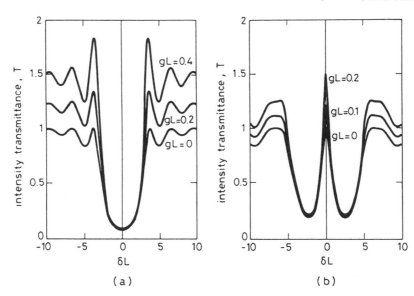

FIG. 9.30. Transmittance T versus normalized detuning δL at several values of the gain gL for zero end face reflectivity and (a) uniform grating, (b) $\lambda/4$-shifted grating. (By permission from Ref. 49.)

condition (9.4.17), the resonant wavelengths of $+1$ and -1 modes are then expected to be slightly smaller and slightly larger than the resonance value $\lambda_B = 2n_0\Lambda$, respectively.

The symmetric situation for DFB modes shown in Fig. 9.30a is obviously not desirable, and several solutions have been considered to ensure that one mode prevails. A commonly used solution makes the device asymmetric by providing end mirror reflectivities, r_1 and r_2, of different values. The best solution, however, appears to be the so-called $\lambda/4$-shifted DFB laser.[50] In this case, the periodic variation in thickness of the inner cladding layer undergoes a shift of $\Lambda/4$ at the center of the active layer (i.e., at $z = L/2$; see Fig. 9.29b). The intensity transmittance T, versus normalized detuning δL, then shows the behavior indicated in Fig. 9.30b, where several plots are made for different values of the gain $\langle g \rangle L$ and for a given value of the normalized coupling constant kL ($kL = 2$). A peak transmission, at exact Bragg resonance $\lambda = \lambda_B$, now occurs. As a further advantage, the difference in transmission between this mode and the two neighboring modes, i.e., mode selectivity, is higher in this case than in the previously considered case of a uniform grating (compare with Fig. 9.30a). The reason for the existence of only one low-loss mode can be understood from Fig. 9.31b, where, due to the $\lambda/4$-shift of the variation in layer thickness, the effective value of the refractive index $n_{eff} = \langle n \rangle_x$ is seen to show a similar shift in its longitudinal variation. The standing wave pattern of the lowest-loss mode is also shown in the figure and the longitudinal spatial average of the effective refractive index $\langle n_{eff} \rangle$ is now seen to equal n_0. The resonance condition is then $(\lambda/2n_0) = \Lambda$, and the wavelength λ of the mode coincides with the Bragg wavelength $\lambda_B = 2n_0\Lambda$.

The fabrication of uniform-grating devices and, even more so, of $\lambda/4$-shifted DFB lasers presents challenging technological problems. In fact, the pitch of the grating Λ typically has to be of submicron dimension [e.g., for a 1550 nm InGaAsP laser, $\langle n_{eff} \rangle \cong 3.4$; from Eq. (9.4.17), $\Lambda \cong 0.23~\mu$m]. It is therefore difficult to make this pitch uniform along the length of the grating and also constant from one grating to the next.

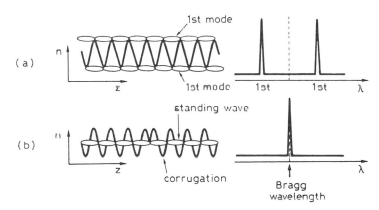

FIG. 9.31. Schematic representation of the refractive index change, mode patterns, and corresponding resonance wavelengths for a DFB laser with a uniform grating and for a DFB laser with a $\lambda/4$-shifted grating. (By permission from Ref. 49.)

In addition to using a DFB laser, another way of ensuring semiconductor laser oscillation on a single line involves the structure shown in Fig. 9.32. In this figure the two cavity ends are made of passive sections where, by appropriate corrugation of a suitable layer, the effective refractive index is modulated with a period Λ in the longitudinal direction. Reflectivity of the two end sections then arises from the constructive interference that occurs in the two sections under the Bragg condition. Since the effect of these sections is somewhat similar to that of a ($\lambda/4$) multilayer dielectric mirror, maximum reflectivity is expected to occur at the wavelength $\lambda = 2\langle n_{eff}\rangle \Lambda$. Compared to DFB lasers, DBR lasers have the advantage that the grating is fabricated in an area separated from the active layer. This simplifies the fabrication process and makes the DBR structure more suitable for integration with other devices, such as separated sections for laser tuning or modulation. The wavelength selectivity of a DBR laser is, however, less than that of a DFB laser, due to the presence of many Fabry–Perot longitudinal modes, as established by the given length of the active medium. Actually, due to the small value of this length, only one mode usually falls within the high-reflectivity bandwidth of the DBR structure. Temperature variations can, however, produce jumping between adjacent modes and, for this reason, DBR lasers are much less widely used than DFB lasers.

9.4.7. Vertical-Cavity Surface-Emitting Lasers

So far, we have considered semiconductor diode lasers that generate light traveling in a direction parallel to the junction plane and hence emitted from one edge of the device (*edge-emitting lasers*). For several applications that will be discussed in Sect. 9.4.8, semiconductor

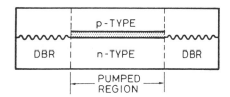

FIG. 9.32. Schematic representation of a DBR semiconductor laser.

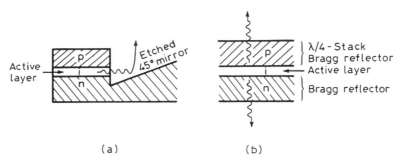

(a)   (b)

FIG. 9.33. Schematic representation of (a) a surface-emitting laser where the light of an edge-emitting laser is deflected vertically by a 45° mirror and (b) a VCSEL.

lasers emitting normal to the junction plane have been developed. These devices are usually referred to as *surface-emitting lasers*, and are made using one of the following approaches: (1) Use of a conventional edge-emitting geometry where some optical element, e.g., a 45° mirror, deflects the output beam vertically (Fig. 9.33a) (2) Use of highly reflective mirrors to clad the active layer, thus resulting in a vertical cavity that produces an output beam propagating normal to the junction plane (vertical-cavity surface-emitting laser, VCSEL, see Fig. 9.33b). Surface-emitting lasers of the type shown in Fig. 9.33a are not different, conceptually, from a conventional edge-emitting laser. A peculiar characteristic of a VCSEL, on the other hand, is the very short length of the active medium and thus the very small gain involved. However, once this low-gain limitation is overcome by using sufficiently high-reflectivity mirrors, low thresholds can be obtained. These lasers then present some distinct advantages over the corresponding edge-emitting devices, due to the inherently high-packaging density and low-threshold currents that can be achieved. In the discussion that follows, we therefore concentrate on vertical-cavity surface-emitting lasers.[51]

Figure 9.34 shows a schematic view of a VCSEL using three $In_{0.2}Ga_{0.8}As/GaAs$-strained QW layers, each 8 nm thick, as an active medium. The three active layers are sandwiched between two $Ga_{0.5}Al_{0.5}As$ spacers to form an overall thickness of one wavelength. The bottom and top mirrors are then made of a 20.5 pairs of n-doped and 16 pairs of p-doped quarter-wave GaAs/AlAs stacks, respectively. Due to the relatively high refractive-index change between the two layers (the refractive index is 3.6 for GaAs and 2.9 for AlAs), high reflectivity ($\sim 99\%$) can be achieved for both end mirrors. Due to the short

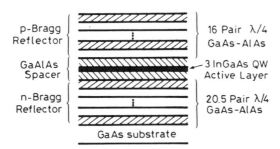

FIG. 9.34. Schematic representation of a bottom-emitting VCSEL design. The top-most GaAs layer is a half-wavelength thick to provide phase matching for the metal contact. (By permission from Ref. 52.)

length of the cavity, scattering and absorption losses in the active layer are very small, so reasonably low threshold-current densities can be obtained ($J_{th} \cong 4 \text{ kA/cm}^2$). Since the diameter of the circular surface through which current flows is usually made very small ($D = 5$–$10~\mu\text{m}$), a very low threshold current is obtained (~ 1 mA). Due to this small cross-sectional area of the active region, VCSELs tend to oscillate on a TEM_{00} mode even at currents well above (e.g., twice) the threshold value. Note that, due to the small length of the laser cavity (1–2 μm), consecutive longitudinal modes are widely spaced in wavelength ($\Delta\lambda \cong 100$ nm). Thus, if one mode coincides with the peak reflectivity of each of the quarter-wave stacks, the two adjacent modes fall outside the high-reflectivity band of the mirrors and single-longitudinal-mode oscillation is also obtained.

Fabrication of VCSELs presents some technological difficulties, one of which is making the cavity length so that one longitudinal mode falls exactly at the center of the high-reflectivity band of the mirrors. Once this difficulty, and that related to fabricating so many layers of exactly $\lambda/4$-thickness, are solved, a low threshold can be achieved. If cavity losses are minimized, the slope efficiency of a VCSEL can be as high as that of an edge-emitting device; slope efficiencies up to 50% have, in fact, been demonstrated. The output power emitted by a single VCSEL is rather limited (~ 1 mW) to avoid the onset of thermal problems due to injecting high pump powers into such a small volume of active layer. Arrays of, e.g., 8×8, independently addressable, lasers as well as matrix-addressable arrays have been made.

Example 9.4. *Threshold carrier density and threshold current for a VCSEL.* According to Eq. (9.4.9), assuming a confinement factor $\Gamma \cong 1$, we can write $N_{th} = (\gamma/\sigma l) + N_{tr}$, where l is the thickness of the active layer. Following Example 9.2, we assume the values $N_{tr} = 2 \times 10^{18}~\text{cm}^{-3}$ and $\sigma \cong 6 \times 10^{-16}~\text{cm}^2$ for the transparency carrier density and the differential gain of the strained layer QWs, respectively. The single-pass loss is then given by $\gamma = -\ln R + \alpha_i L$, where R is the power reflectivity of each of the two mirrors, α_i is the internal loss coefficient, and L is the cavity length. If we take $R = 99\%$, $\alpha_i = 20~\text{cm}^{-1}$ and $L = 2~\mu\text{m}$, we obtain $\gamma = 1.4 \times 10^{-2}$. If we now assume $l = 24$ nm for the overall thickness of the active layers, we obtain $N_{th} \cong 11 \times 10^{18}~\text{cm}^{-3}$, which shows that the value of N_{th} is dominated by the loss term $(\gamma/\sigma l)$. From Eq. (9.4.3), taking $\eta_i \cong 1$ and $\tau = 2$ ns, we obtain $J_{th} = 3.84 \times 10^3~\text{A/cm}^2$. Assuming a diameter $D = 8~\mu\text{m}$ for the active area, the threshold current is given by $I_{th} = (\pi D^2/4)J_{th} \cong 0.44$ mA.

9.4.8. Semiconductor Laser Applications

Semiconductor lasers lend themselves to a wide variety of both low-power and high-power applications, some of which are briefly reviewed here. We first refer to Table 9.7, which lists some characteristics of the most common DH and QW lasers. Since all structures shown are grown on either a GaAs or a InP substrate, the laser material is characterized by the active-layer-substrate combination. In each case, the laser wavelength is largely determined by the composition index of the active layer and, for a QW laser, also by the thickness of this layer. Most recent lasers use separated confinement (e.g., GRINSCH) QW and MQW layers in a gain-guiding, or more often, index-guiding configuration.

In the $\text{Al}_x\text{Ga}_{1-x}\text{As}/\text{Al}_y\text{Ga}_{1-y}\text{As}$ laser structures, the composition index y of the cladding layers must be larger than that, x, of the active layer. Depending on the value of the later composition index, the emission wavelength usually ranges from 720 to 850 nm. Low-power ($P = 5$–20 mW), single-stripe, devices are widely used in compact disk (CD) players

TABLE 9.7. Some characteristic parameters of the most important semiconductor laser diodes

Material/Substrate	InGaAlP/GaAs	AlGaAs/GaAs	InGaAsP/InP	InGaAs/GaAs
Wavelength (nm)	720–850	1200–1650	900–1100	630–700
Internal loss (cm^{-1})	4–15	5–10	2–10	~ 10
Threshold current density J_{th} (A/cm^2)	80–700	200–1500	50–400	200–3000
Characteristic temperature T_0 (K)	120–200	50–70	100–200	60–100

and laser printers. Higher power single-stripe lasers, laser arrays, laser bars, and stacks of laser bars (Figs. 6.9 and 6.10) are used to pump solid-state lasers, such as Nd (pump wavelength $\lambda_p \cong 800$ nm) or Tm:Ho ($\lambda_p \cong 790$ nm). Some of these laser systems and the corresponding fields of applications were discussed in Sect. 9.2 and in previous chapters.

In $In_{1-x}Ga_xAs_yP_{1-y}$/InP lasers, lattice matching is achieved by letting $y \cong 2.2x$, and the oscillation wavelength covers the so-called second (centered at $\lambda = 1310$ nm, corresponding to $x = 0.27$) and third ($\lambda = 1550$ nm center wavelength, corresponding to $x = 0.42$) transmission windows of optical fibers. Thus, these lasers find their widest use in optical communications. Most recent optical communication systems use lasers around 1550-nm wavelength. Due to the relatively large group-delay dispersion of optical fibers around this wavelength, narrow–linewidth ($\Delta\nu_L < 10$ MHz) DFB lasers are now widely used. With these lasers, modulation rates up to a few Gbit/s have been demonstrated by direct modulation of the diode current. Even higher modulation rates (up to a few tens of Gbit/s) have also been demonstrated using external modulators, such as $LiNbO_3$ waveguide modulators. For communication systems operating at even greater bit-rates (from a few hundreds of Gbit/s to the Tbit/s range) wavelength-division-multiplexing (WDM) systems have increasingly been used. For this application, many DFB lasers tuned to distinct wavelengths in the low-loss region (spanning ~ 13 THz) around 1550-nm wavelength are used. Systems based on WDM thus allow exceptionally-high bit-rate capacity to be achieved.

$In_{1-x}Ga_xAs$/GaAs strained-layer QW lasers allow oscillation over a wide range (900–1100 nm) at previously inaccessible wavelengths. Lasers with emission around 980-nm wavelength ($x = 0.8$) are of particular interest as pumps for Er-doped fiber amplifiers and lasers as well as for pumping Yb:Er:glass and Yb:YAG lasers. For these applications, output powers up to ~ 100 mW are available in a diffraction-limited beam (of 1×4 μm area) and up to ~ 1 W from a broader area (1×30 μm) device. To obtain higher output powers (~ 40 W or even greater), diode laser arrays and laser bars are also available. Given the favorable laser properties of strained-layer quantum wells, vertical-cavity surface-emitting lasers based on $In_{1-x}Ga_xAs$/GaAs QW structures have been actively developed. They promise to offer interesting solutions for optical interconnects, optical communications, and optical signal processing.

InGaP/InGaAlP lasers are particularly interesting, since they emit radiation in the visible (red) region of the em spectrum.[53] The $In_{0.5}Ga_{0.5}P/In_{0.5}(Ga_{0.5-x}Al_x)P$ QW or MQW structure (where $x = 0.25$ for well barriers and $x = 0.35$ for confinement layers; see Fig. 9.25), oscillating at 670-nm wavelength, has been especially developed. These lasers are commercially available with sufficient high power (to ~ 20 mW) and long lifetime to be used for CD players or as substitutes for red-emitting He-Ne lasers in such applications as bar code scanners and general alignment.

The development of semiconductor lasers is by no means limited to the laser categories shown in Table 9.7. On the short wavelength side (blue-green region), the most interesting category now appears to be the III–V nitride-based diode lasers, e.g., the $In_{0.2}Ga_{0.8}N/In_{0.05}Ga_{0.95}N$ MQW structure oscillating in the blue (417 nm).[54] In the same wavelength region, wide-gap II–VI lasers, such as ZnCdSe/ZnSSe QW lasers, have also been demonstrated. After a number of years of intense development, however, these lasers still suffer from a few technological limitations, particularly a rather limited operating lifetime (~ 100 h). Although nitride lasers also present similar lifetime problems (less than 100 h), their recent rapid progress suggest that they may become the best candidates for blue-green semiconductor emitters.* Blue-green lasers show potential for, e.g., a new generation of CD players, where, due to the shorter wavelength, substantially higher bit densities can be achieved on the CD. On the long wavelength side, we just mention IV–VI compounds such as the Pb and Sn salts (e.g., PbSSe, PbSnTe, and PbSnSe), oscillating in the middle-far infrared (4–29 μm). These lasers, however, must operate at cryogenic temperatures ($T < 100$ K) to avoid such problems as increased free carrier absorption and increased rate of nonradiative decay arising from the much narrower band gap. Due to the cryogenic requirement, these lasers have found only a limited use (e.g., in spectroscopy). It should be noted, however, that the recent invention of a quantum cascade laser promises efficient mid-infrared sources without requiring cryogenic temperatures.[56]

9.5. CONCLUSIONS

Chapter 9 considers a few of the most notable solid-state, dye, and semiconductor lasers. These lasers use high-density active media and share a few common features. A first feature is that they generally show wide and strong absorption band(s), which indicates that they are generally suitable for optical pumping. This type of pumping is in fact always used for solid-state and dye lasers and sometimes, also, for semiconductor lasers. High absorption coefficients allow lasers with dimensions down to a few microns (microlasers) to be obtained. As a second feature, these media generally show wide fluorescence and hence wide gain bandwidths. This characteristic offers tunability over wide (a few to several nanometers) bandwidths and it also implies that very short pulse durations (femtoseconds) can be obtained in mode-locked operation. A third feature is the large optical-to-optical laser efficiency for solid-state and dye media and the similarly large electrical-to-optical efficiency for semiconductors. Note that laser pumping, involving combination of these three categories of lasers, is increasingly used (e.g., diode-pumped solid-state lasers or solid-state laser-pumped dye lasers), allowing compact and efficient lasers to be achieved. Thus, as a conclusion, high-density laser media appear to present some of the best solutions to requirements for laser radiation in the visible-to-near-infrared range, even at high-power levels.

PROBLEMS

9.1. Make a diagram in which the tuning ranges of all tunable solid-state lasers considered in this chapter are plotted against oscillation wavelength.

9.2. For pollution monitoring, a tunable laser oscillating around 720-nm wavelength is needed. Which kind of solid-state laser would you use?

---

*Note added in proof: Extrapolated lifetimes in excess of 10^4 hours have, quite recently, been demonstrated for nitride-based diode lasers

9.3. For biomedical photocoagulation purposes, using an endoscopic apparatus, a cw laser with power exceeding 50 W must be used. Which laser would you use?

9.4. For material-working applications, a laser with an average power of 2 kW transmitted through a ∼1-mm diameter optical fiber is needed. Which laser would you use?

9.5. Consider a 6-mm diameter, 10 cm long Nd:phosphate glass laser rod. Using data from Table 9.3 and the results discussed in Sect. 2.9.2, calculate the maximum inversion and the corresponding maximum value of stored energy allowed if the onset of amplified spontaneous emission is to be avoided. Compare the results obtained to those for a Nd:YAG rod of the same dimensions.

9.6. Referring to the energy level diagram of alexandrite in Fig. 9.8, assume that the $^4T_2$ and the ^2E states are strongly coupled and take $\tau_T = 1.5$ ms and $\tau_E = 6.6$ μs as the lifetimes of the two states, respectively. From the knowledge that the degeneracy of the two states is the same ($g = 4$), calculate the effective lifetime of the $^4T_2$ state at $T = 300$ K and $T = 400$ K. If the true cross section for the $^4T_2 \rightarrow {}^4A_2$ transition is $\sigma \cong 4 \times 10^{-19}$ cm^2, calculate the effective value of the cross section at the two temperatures. Using these results, consider whether the laser threshold increases or decreases when the crystal temperature is raised from 300 to 400 K.

9.7. Consider a Cr:LiSAF laser longitudinally pumped, at 647.1-nm wavelength, by a Kr ion laser. Assume that the linearly polarized beam of the laser is sent along the c-axis of the LiSAF and assume an active medium with 1 at.% Cr^{3+} concentration and a length of $l = 4$ mm. Also assume that both pump and mode spot sizes are matched at a value of 60 μm, the output coupler transmission is 1%, and the internal loss per pass is 1%. Using data from Fig. 9.11 and Table 9.5 and neglecting both ground-state and excited-state absorption, calculate the expected threshold pump power.

9.8. Referring to Problem 9.7, how would the expression for threshold pump power, given by Eq. (6.3.19), have to be modified if ground-state absorption, characterized by a loss per pass γ_a, and excited-state absorption, characterized by an excited-state absorption cross section σ_{ESA}, were taken into account? Compare the result to that given in Ref. 25.

9.9. Derive an expression for the threshold pump power of a longitudinally pumped dye laser, taking into account triplet-triplet absorption (assume Gaussian transverse profiles for both pump and mode beams). Compare this expression to that obtained for Cr:LiSAF in the previous problem.

9.10. Using the expression for the threshold pump power obtained in Problem 9.9 and using data from Fig. 9.13, calculate the threshold pump power for an Ar$^+$-pumped rhodamine 6G (see Fig. 9.17) laser oscillating at 580-nm wavelength. For this calculation, assume an output coupling of 3%, an internal loss per pass of 1%; assume also that 80% of the pump power is absorbed in the dye jet stream; take the lifetime for the first excited singlet state as 5 ns, the intersystem crossing rate as $k_{ST} \cong 10^7$ s^{-1}, and the triplet lifetime as $\tau_T \cong 0.1$ μs. Compare this value of P_{th} to that obtained for Cr:LiSAF in Problem 9.7 and explain the numerical differences.

9.11. At the very small values of the thickness d corresponding to the minimum of J_{th} in Fig. 9.23, the expression for the beam confinement factor Γ of a DH semiconductor laser, given by Eq. (9.4.10), can be approximated by $\Gamma \cong D^2/2$, where D is expressed by Eq. (9.4.11). Using this approximation, calculate the expression for the thickness d_m that minimizes J_{th}. From data given in Example 9.1, calculate the value of d_m and the corresponding value of J_{th}.

9.12. From the expression for the output power of a semiconductor laser given in Example 9.3, derive an expression for the laser slope efficiency. Using data given in Example 9.1, calculate the predicted slope efficiency of a DH GaAs/AlGaAs laser by using an applied voltage of 1.8 V.

9.13. Assume that the beam, at the exit face of a semiconductor laser, is spatially coherent. Assume that the transverse field distributions have Gaussian profiles in the directions parallel and perpendicular to the junction, with spot sizes $w_\parallel$ and $w_\perp$ respectively. Assume also that, for both field distributions, the location of the beam waists occurs at the exit face. Given these assumptions, derive an expression for the propagation distance at which the beam becomes circular. Taking $w_\perp = 0.5$ μm and $w_\parallel = 2.5$ μm, calculate the value of this distance for $\lambda = 850$ nm.

9.14. Considering that the refractive index of a semiconductor n is a relatively strong function of the wavelength λ, derive an expression for the frequency difference between two consecutive longitudinal modes of a Fabry–Perot semiconductor laser [express this frequency difference in terms of the material group index $n_g = n - \lambda(dn/d\lambda)$].

9.15. The calculations leading to Fig. 9.30a were made assuming $kL = 2$, where k is the coupling constant between forward- and backward-propagating beams in a DFB laser and L is its length. From the definition of k given in Sect. 9.4.6, calculate the value n_1 in Eq. (9.4.16) for $\lambda = 1550$ nm and $L = 600$ μm.

9.16. The two strongest peaks in Fig. 9.30a are seen from the figure to be separated by a normalized frequency difference $\Delta(\delta L) \cong 7.28$. From the definition of the normalized frequency detuning δL given in Sect. 9.4.6, calculate the frequency difference $\Delta \nu$ between the two modes by taking $L = 600$ μm, $n_0 = 3.4$, and $\lambda = 1550$ nm for a InGaAsP DFB laser. Compare this value to the corresponding one obtained for the frequency separation between two consecutive longitudinal modes of a Fabry–Perot semiconductor laser with the same length and wavelength and a group index n_g equal to n_0.

REFERENCES

1. A. A. Kaminskii, *Crystalline Lasers: Physical Processes and Operating Systems* (CRC Press, 1996).
2. T. H. Maiman, Stimulated Optical Radiation in Ruby Masers, *Nature* **187**, 493 (1960).
3. T. H. Maiman, Optical Maser Action in Ruby, *Brit. Commun. Electron.* **7**, 674 (1960).
4. W. Koechner, *Solid-State Laser Engineering*, 4th ed. (Springer Berlin, 1996), Sects. 2.2, 3.6.1.
5. Ref. 4, Sects. 2.3.1., 3.6.3.
6. E. Snitzer and G. C. Young, Glass Lasers, in *Lasers*, vol. 2 (A. K. Levine, ed.) (Marcel Dekker, NY, 1968), Chap. 2.
7. Ref. 4, Sect. 2.3.4.
8. T. Y. Fan, Diode-Pumped Solid-State Lasers, in *Laser Sources and Applications* (A. Miller and D. M. Finlayson, eds.) (Institute of Physics, Bristol, 1996), pp. 163–93.
9. P. Lacovara *et al.*, Room-Temperature Diode-Pumped Yb:YAG Laser, *Opt. Letters* **16**, 1089 (1991).
10. H. Bruesselbach and D. S. Sumida, 69-W-average-power Yb:YAG Laser, *Opt. Letters* **21**, 480 (1996).
11. G. Huber, Solid-State Laser Materials, in *Laser Sources and Applications* (A. Miller and D. M. Finlayson, eds.) (Institute of Physics, Bristol, 1996), pp. 141–62.
12. E. V. Zharikov *et al.*, *Sov. J. Quantum Electron.* **4**, 1039 (1975).
13. S. J. Hamlin, J. D. Myers, and M. J. Myers, High-Repetition Rate *Q*-Switched Erbium Glass Lasers, in *Eyesafe Lasers: Components, Systems, and Applications* (A. M. Johnson, ed.) *SPIE* **1419**, 100 (1991).
14. S. Taccheo, P. Laporta, S. Longhi, O. Svelto, and C. Svelto, Diode-Pumped Bulk Erbium-Ytterbium Lasers, *Appl. Phys.* **B63**, 425 (1996).
15. D. Sliney and M. Wolbarsht, *Safety with Lasers and Other Optical Sources* (Plenum, NY, 1980).
16. T. Y. Fan, G. Huber, R. L. Byer, and P. Mitzscherlich, Spectroscopy and Diode Laser-Pumped Operation of Tm, Ho:YAG, *IEEE J. Quantum Electron.* **QE-24**, 924 (1988).
17. D. C. Hanna, Fibre Lasers, in *Laser Sources and Applications* (A. Miller and D. M. Finlayson, eds.) (Institute of Physics, Bristol, 1996), pp. 195–208.
18. E. Snitzer, Optical Maser Action on Nd^{3+} in a Barium Crown Glass, *Phys. Rev. Letters* **7**, 444 (1961).
19. J. C. Walling, O. G. Peterson, H. P. Jenssen, R. C. Morris, and E. W. O'Dell, Tunable Alexandrite Lasers, *IEEE J. Quantum Electron.* **QE-16**, 1302 (1980).
20. L. F. Mollenauer, Color Center Lasers, in *Laser Handbook*, vol. 4 (M. L. Stitch and M. Bass, eds.) (North Holland, Amsterdam, 1985), pp. 143–228.
21. P. F. Moulton, Spectroscopy and Laser Characteristics of $Ti:Al_2O_3$, *J. Opt. Soc. Am. B* **3**, 125 (1986).

22. G. Hüber, Solid-State Laser Materials: Basic Properties and New Developments, in *Solid-State Lasers: New Developments and Applications* (M. Inguscio and R. Wallenstein, eds.) (Plenum, NY, 1993), pp. 67–81.

23. P. Albers, E. Stark, and G. Huber, Continuous-Wave Laser Operation and Quantum Efficiency of Titanium-Doped Sapphire, *J. Opt. Soc. Am. B* **3**, 134 (1986).

24. S. A. Payne, L. L. Chase, L. K. Smith, W. L. Kway, and H. W. Newkirk, Laser Performance of $LiSrAlF_6:Cr^{3+}$, *J. Appl. Phys.* **66**, 1051 (1989).

25. S. A. Payne L. L. Chase, H. W. Newkirk, L. K. Smith, and W. F. Krupke, $LiCaAlF_6:Cr^{3+}$: A Promising New Solid-State Laser Material, *IEEE J. Quantum Electron.* **QE-24**, 2243 (1988).

26. Dye Lasers, 2d ed (F. P. Schäfer, ed.) (Springer-Verlag, Berlin, 1977).

27. H. D. Försterling and H. Kuhn, *Physikalische Chemie in Experimenten, Ein Praktikum* (Verlag Chemie, Weinheim, Germany 1971).

28. J. T. Verdeyen, *Laser Electronics*, 3d ed. (Prentice-Hall, Englewood Cliffs, NJ, 1995), Fig. 10.19.

29. P. P. Sorokin and J. R. Lankard, Stimulated Emission Observed from an Organic Dye, Chloro-Aluminum Phtalocyanine, *IBM J. Res. Dev.* **10**, 162 (1966).

30. F. P. Schäfer, F. P. W. Schmidt, and J. Volze, Organic Dye Solution Laser, *Appl. Phys. Letters* **9**, 306 (1966).

31. *Semiconductor Lasers: Past, Present, Future* (G. P. Agrawal, ed.) (AIP, Woodbury, NY, 1995).

32. G. P. Agrawal and N. K. Dutta, *Long-Wavelength Semiconductor Lasers* (Chapman and Hall, NY, 1986).

33. N. G. Basov, O. N. Krokhin, and Y. M. Popov, Production of Negative Temperature States in *p*-n Junctions of Degenerate Semiconductors, *Journal Exp. Theoret. Physics* **40**, 1320 (1961).

34. R. N. Hall, G. E. Fenner, J. D. Kinhsley, F. H. Dills, and G. Lasher, Coherent Light Emission from GaAs Junctions, *Phys. Rev. Letters* **9**, 366 (1962).

35. M. I. Nathan, W. P. Dumke, G. Burns, F. H. Dills, and G. Lasher, Stimulated Emission of Radiation from GaAs *p*-n Junctions, *Appl. Phys. Letters* **1**, 62 (1962).

36. N. Holonyak, Jr. and S. F. Bevacqua, Coherent (Visible) Light Emission from $Ga(As_{1-x}P_x)$ Junctions, *Appl. Phys. Letters* **1**, 82 (1962).

37. T. M. Quist, R. J. Keyes, W. E. Krag, B. Lax, A. L. McWhorter, R. H. Rediker, and H. J. Zeiger, Semiconductor Maser of GaAs, *Appl. Phys. Letters* **1**, 91 (1962).

38. Z. I. Alferov, V. M. Andreev, V. I. Korolkov, E. L. Portnoi, and D. N. Tretyakov, Coherent Radiation of Epitaxial Heterojunction Structures in the AlAs-GaAs System, *Soviet. Phys. Semicond.* **2**, 1289 (1969).

39. I. Hayashi, M. B. Panish, and P. W. Foy, A Low-Threshold Room-Temperature Injection Laser, *IEEE J. Quantum Electron.* **QE-5**, 211 (1969).

40. H. Kressel and H. Nelson, Close Confinement Gallium Arsenide *p*-n Junction Laser with Reduced Optical Losses at Room Temperature, *RCA Rev.* **30**, 106 (1969).

41. N. Chinone, H. Nakashima, I. Ikushima, and R. Ito, Semiconductor Lasers with a Thin Active Layer (> 0.1 μm) for Optical Communications, *Appl. Opt.* **17**, 311 (1978).

42. D. Botez, Analytical Approximation of the Radiation Confinement Factor for the TE_0 Mode of a Double-Heterojunction Laser, *IEEE J. Quantum Electron.* **QE-14**, 230 (1978).

43. J. J. Coleman, Quantum-Well Heterostructure Lasers, in *Semiconductor Lasers: Past, Present, Future* (G. P. Agrawal, ed.) (AIP, Woodbury, NY, 1995), Fig. 1.6.

44. *Quantum Well Lasers* (Peter S. Zory, ed.) (Academic Press, Boston, 1993).

45. Ref. 32, Figs. 9.8, 9.10.

46. Ref. 44, Chap. 3.

47. H. Kogelnik and C. V. Shank, Stimulated Emission in a Periodic Structure, *Appl. Phys. Letters* **18**, 152 (1971).

48. Ref. 32, Chap. 7.

49. N. Chinone and M. Okai, Distributed Feed-Back Semiconductor Lasers, in *Semiconductor Lasers: Past, Present, Future* (G. P. Agrawal, ed.) (AIP, Woodbury, NY, 1995), Chap. 2, pp. 28–70.

50. H. A. Haus and C. V. Shank, Antisymmetric Taper of Distributed Feedback Lasers, *IEEE J. Quantum Electron.* **QE-12**, 532 (1976).

51. C. J. Chang-Hasnain, Vertical-Cavity Surface-Emitting Lasers, in *Semiconductor Lasers: Past, Present, Future* (G. P. Agrawal, ed.) (AIP, Woodbury, NY, 1995), Chap. 4, pp. 110–44.

52. C. J. Chang-Hasnain, J. P. Harbison, C.-H. Zah, M. W. Maeda, L. T. Florenz, N. G. Stoffel, and T.-P. Lee, Multiple Wavelength Tunable surface-Emitting Laser Array, *IEEE J. Quantum Electron.* **QE-27**, 1368 (1991).

53. G.-I. Hatakoshi, Visible Semiconductor Lasers, in *Semiconductor Lasers: Past, Present, Future* (G. P. Agrawal, ed.) (AIP, Woodbury, NY, 1995), Chap. 6, pp. 181–207.

54. S. Nakamura *et al.*, *Japn. J. Appl. Phys.* **35**, L74 (1994).

55. P. Moulton, New Developments in Solid-State Lasers, *Laser Focus* **14**, 83 (May 1983).

56. J. Faist, F. Capasso, D. L. Sivco, C. Sirtori, A. L. Hutchinson, and A. Y. Cho, *Science* **264**, 553 (1994).

57. M. Bass, T. F. Deutsch, and M. J. Weber, Dye Lasers, in *Lasers*, Vol. 3 (A. K. Levine and A. De Maria, eds.) (Marcel Dekker, NY, 1971) p. 275.

10

Gas, Chemical, Free-Electron, and X-Ray Lasers

10.1. INTRODUCTION

Chapter 10 considers the most important types of lasers involving low-density active media, namely, gas, chemical, and free-electron lasers, in addition to some aspects of x-ray lasers using highly ionized plasmas. The main emphasis, again, is on the physical behavior of the laser and relating this behavior to general concepts developed in the previous chapters. Some engineering details are also presented with the intention of providing a better insight into the behavior of a particular laser. To complete the picture, some data about laser performances (e.g., oscillation wavelength(s), output power or energy, wavelength tunability, etc.) are also included since they are directly related to the potential applications of the given laser.

10.2. GAS LASERS

In general, for gases, energy levels broadening is rather small (on the order of a few GHz or less), since line-broadening mechanisms are weaker than in solids. For gases at the low pressure often used in lasers (a few tens of Torr), in fact, collision-induced broadening is very small, and the linewidth is essentially determined by Doppler broadening. Thus, no broad absorption bands are present in the active medium, so optical pumping by cw or pulsed lamps is not used. Optical pumping would in fact be very inefficient, since the emission spectrum of these lamps is more or less continuous. Therefore gas lasers are usually excited by electrical means, i.e., by passing a sufficiently large current (which may be continuous, at radio-frequency, or pulsed) through the gas. (The principal pumping mechanisms occurring in gas lasers are discussed in Sect. 6.4.) Note that some lasers can also be pumped by mechanisms other than electrical pumping, in particular gas dynamic expansion, chemical pumping, and optical pumping using another laser (the latter is particularly used for far-infrared lasers).

Once a given species is in its excited state, it can decay to lower states (including the ground state) by four different processes: (1) Collisions between an electron and the excited species, in which the electron takes up the excitation energy as kinetic energy (superelastic collision). (2) Near-resonant collisions between excited species and the same or a different species in the ground state. (3) Collisions with the walls of the container. (4) Spontaneous emission. Regarding this last case, we must always take into account the possibility of radiation trapping, particularly for the, usually very strong, uv or vuv transitions. This process slows down the effective rate of spontaneous emission (see Sect. 2.9.1).

For a given discharge current, these excitation and de-excitation processes lead eventually to some equilibrium distribution of population among the energy levels. Thus, due to the many processes involved, producing a population inversion in a gas is a more complicated matter than, e.g., in a solid-state laser. In general, a population inversion between two given levels occurs when either (or both) of the following circumstances occur: (1) The excitation rate is greater for the upper laser level (level 2) than for the lower laser level (level 1). (2) The decay of level 2 is slower than that of level 1. In this regard, we recall that a necessary condition for cw operation is that the rate of the transition $2 \rightarrow 1$ must be smaller than the decay rate of level 1 [see Eq. (7.3.1)]. If this condition is not satisfied, however, laser action can still occur, under pulsed operation, provided the first condition is satisfied (self-terminating lasers).

10.2.1. Neutral Atom Lasers

These lasers use neutral atoms in either gaseous or vapor form. Neutral atom gas lasers constitute a large class of lasers, including, in particular, most of the noble gases. All these lasers oscillate in the infrared (1–10 μm), apart from the notable exceptions of green and red emission from the He-Ne laser. Metal vapor lasers also constitute a large class of lasers, including, for example, Pb, Cu, Au, Ca, Sr, and Mn. These lasers generally oscillate in the visible; the most important example is the copper vapor laser oscillating on its green (510-nm) and yellow (578.2-nm) transitions. All metal vapor lasers are self-terminating and therefore operate in a pulsed regime.

10.2.1.1. Helium Neon Laser

The He-Ne laser is certainly the most important of the noble gas lasers. Laser action is obtained from transitions of the neon atom, while helium is added to the gas mixture to greatly facilitate the pumping process.[1,2] The laser oscillates on many wavelengths; by far the most popular is $\lambda = 633$ nm (red). Other wavelengths include green (543 nm) and infrared, at $\lambda = 1.15 \, \mu$m and $\lambda = 3.39 \, \mu$m. The He-Ne laser, oscillating on its $\lambda = 1.15 \, \mu$m transition, was the first gas laser and the first cw laser to be operated.[3]

Figure 10.1 shows the energy levels of the He-Ne system relevant for laser action. The level notation for He complies with Russell–Saunders coupling; the principal quantum number of the given level is also indicated as the first number. Thus the 1^1S state

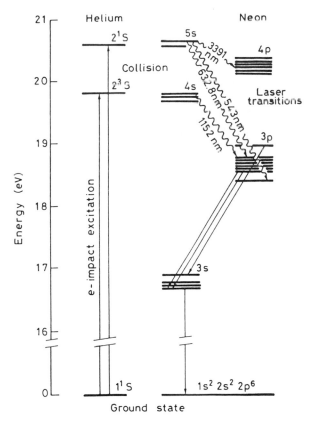

FIG. 10.1. Relevant energy levels of the He-Ne laser.

corresponds to the case when the two electrons of He are in the $1s$ state with opposite spins. The 2^3S and 2^1S states correspond to one of the two electrons being raised to the $2s$ state, with its spin either in the same or opposite direction to that of the other electron, respectively. Neon, on the other hand, has an atomic number of 10 and a number of ways have been used, such as Paschen or Racah notation, to indicate its energy levels. For simplicity, however, we simply indicate here the electron configuration corresponding to each level. Therefore, the ground state is indicated by $1s^2 2s^2 2p^6$, while the excited states in the figure correspond to when one $2p$ electron is raised to excited s states ($3s$, $4s$, and $5s$) or excited p states ($3p$ and $4p$). Note that, due to interaction with the remaining five electrons in the $2p$ orbitals, these s and p states are split into 4 and 10 sublevels, respectively.

It is apparent from Fig. 10.1 that the He levels, 2^3S and 2^1S, are nearly resonant with the Ne $4s$ and $5s$ states. Since the 2^3S and 2^1S levels are metastable ($S \rightarrow S$ transitions are electric-dipole-forbidden, and, furthermore, the $2^3S \rightarrow 2^1S$ transition is also spin-forbidden), He atoms in these states prove very efficient at pumping Ne $4s$ and $5s$ levels by resonant energy transfer. It has been confirmed that this process is dominant in producing population inversion in the He-Ne laser, although direct electron-Ne collisions also contribute to the pumping. Since significant population can then be build-up in the Ne $4s$ and $5s$ states, they prove suitable candidates as upper levels for laser transitions. Taking account of the selection rules, possible transitions are those to the p states. In addition, the decay time of the s states ($\tau_s \cong 100$ ns) is an order of magnitude longer than the decay time of the p states ($\tau_p \cong 10$ ns). Therefore, condition (7.3.1) for operating as a cw laser is satisfied. Note that electron impact excitation rates, from the ground state to the $3p$ and $4p$ levels, are much smaller than the corresponding rates to the $4s$ and $5s$ levels because smaller values of the cross section are involved. However, direct excitation to the $3p$ and $4p$ levels also plays a relevant role in determining laser performance.

The preceding discussion indicates that we can expect laser action in Ne to occur from $5s$ and $4s$ levels, as upper levels, to $3p$ and $4p$ levels, as lower levels. Figure 10.1 indicates some of the most important laser transitions arising from these levels. For transitions differing widely in wavelength ($\Delta\lambda > 0.2\lambda$), the actual oscillating transition depends on the wavelength at which the peak reflectivity of the multilayer dielectric mirror is centered (see Fig. 4.9). Laser transitions are predominantly broadened by the Doppler effect; for instance, in the red He-Ne laser transition ($\lambda = 633$ nm in vacuum and $\lambda = 632.8$ nm in air), Doppler broadening leads to a linewidth of ~ 1.5 GHz (see also Example 2.6). By comparison, natural broadening, according to Eq. (2.5.13), is estimated to be $\Delta\nu_{nat} = 1/2\pi\tau \cong 19$ MHz, where $\tau^{-1} = \tau_s^{-1} + \tau_p^{-1}$ and τ_s, τ_p are the lifetimes of the s and p states, respectively. Collision broadening contributes even less than natural broadening. (For example, for pure Ne, one has $\Delta\nu_c \cong 0.6$ MHz at the pressure of $p \cong 0.5$ torr, see Example 2.2.) Table 10.1 summarizes some spectroscopic properties of the 633-nm laser transition.

Figure 10.2 shows the basic design of a He-Ne laser. The discharge is produced between a ring anode and a large tubular cathode, which can thus withstand collisions from positive ions. The discharge is confined to a capillary for most of the tube length, so high inversion is achieved only where the capillary is present. The large volume of gas, available in the tube surrounding the capillary, acts as a reservoir to replenish the He-Ne mixture in the capillary. When a polarized output is needed, a Brewster angle plate is also inserted inside the laser tube. The laser mirrors are directly sealed at the two tube ends. The most commonly used resonator configuration is nearly hemispherical, since it is easy to adjust, very stable against misalignment, and readily gives TEM$_{00}$ mode operation. The only disadvantage of this configuration is that it does not fully use the volume of the plasma discharge, since the mode spot size is much smaller at the plane mirror than at the concave mirror. If, however, the left-hand mirror in Fig. 10.2 is chosen as the plane mirror, the region

TABLE 10.1. Spectroscopic properties of laser transitions and gas mixture composition in some relevant atomic and ionic gas lasers

Laser Type	He-Ne	Copper Vapor	Argon Ion	He-Cd
Laser wavelength (nm)	633	510.5	514.5	441.6
Cross section (10^{-14} cm^2)	30	9	25	9
Upper state lifetime (ns)	150	500	6	700
Lower state lifetime (ns)	10	$\approx 10^4$	~ 1	1
Transition linewidth (GHz)	1.5	2.5	3.5	1
Partial pressures of gas mixture (Torr)	4 (He)	40 (He)	0.1 (Ar)	10 (He)
	0.8 (Ne)	0.1-1 (Cu)		0.1 (Cd)

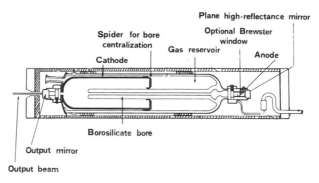

FIG. 10.2. Internal design of a hard-sealed helium neon laser. (Courtesy of Melles–Griot).

of smaller spot size, for the near-hemispherical TEM$_{00}$ mode, is outside the capillary, i.e., in a low-inversion region.

One of the most characteristic features of a He-Ne laser is that the output power does not increase monotonically with the discharge current, but reaches a maximum and thereafter decreases. For this reason, commercially available He-Ne lasers are provided with a power supply designed to give only the optimum current. The fact that there is an optimum value of current, i.e., of current density J within the capillary, is because (at least for 633-nm and 3.39-μm transitions), at high current densities, de-excitation of the He (2^3S and 2^1S) metastable states takes place not only by collision at the walls but also by superelastic collision processes such as:

$$He(2^1S) + e \rightarrow He(1^1S) + e \qquad (10.2.1)$$

Since the rate of this process is proportional to electron density N_e, and hence to J, the overall rate of de-excitation can be written as $k_2 + k_3 J$. In this expression k_2 is a constant that represents de-excitation due to collisions with the walls and $k_3 J$, where k_3 is also a constant, represents the superelastic collision rate of process (10.2.1). The excitation rate can be expressed as $k_1 J$, where k_1 is again a constant. Under steady-state conditions, we can then write $N_t k_1 J = (k_2 + k_3 J)N^*$, where N_t is the ground-state He-atom population and N^* is the excited (2^1S) state population. The equilibrium 2^1S population is then given by

$$N^* = N_t \left(\frac{k_1 J}{k_2 + k_3 J} \right) \qquad (10.2.2)$$

which saturates at high current densities. Since the steady-state population of the $5s$ state of Ne is established by near-resonant energy transfer from the He 2^1S state, the population of the upper $5s$ laser level also shows a similar saturation behavior as J increases (see Fig. 10.3). In the absence of laser action, the population of the lower laser level ($3p$ or $4p$) is produced by direct pumping from ground-state Ne atoms and radiative cascading from higher levels; experimentally, it is found to increase linearly with J (see Fig. 10.3). Therefore, as the discharge current is increased, the population difference, and hence the output power, rises to some optimum value and thereafter falls.

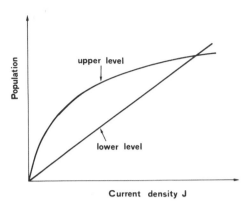

FIG. 10.3. Schematic dependence on current density of upper level and lower level populations in a He-Ne laser.

Besides the optimum value of current density, the He-Ne laser has optimum values for other operational parameters, namely:

- Product of total gas pressure p and capillary diameter D ($pD = 3.6-4\,\text{Torr} \times \text{mm}$); this optimum value of the pD product arises because the electron temperature must be optimized (see Sect. 6.4.5).
- Ratio of the partial pressures of He and Ne ($\sim$5:1 at $\lambda = 632.8\,\text{nm}$ and $\sim$9:1 at $\lambda = 1.15\,\mu\text{m}$).
- Capillary diameter ($D \cong 2\,\text{mm}$); for a constant value of pD, i.e., for a given electron temperature, all electron collision excitation processes scale simply as the number of atoms available for excitation. Since both upper and lower laser levels are ultimately populated by electron collision processes, their populations, and hence also laser gain, are proportional to p, i.e., to D^{-1} at constant pD. Diffraction losses of the laser cavity, on the other hand, increase as D is decreased; a value of the capillary diameter that optimizes the net gain (gain minus diffraction losses) is therefore expected.

According to the behavior in Fig. 10.3, He-Ne lasers are typically low-power devices. (Under optimized conditions, the available output power at the 633-nm transition may range from 1 to 10 mW, for tube lengths ranging from 20–50 cm, while the output power on the green transition is typically an order of magnitude less.) The efficiency of a He-Ne laser, on any of its laser transitions, is always very low ($< 10^{-3}$); a major cause of low efficiency is the low quantum efficiency. In fact, from Fig. 10.1, we readily see that each elementary pumping cycle requires an energy of $\sim$20 eV while laser photon energy is less than 2 eV. The narrow gain linewidth is an advantage when single longitudinal mode is required. In fact, if the cavity length is short enough ($L < 15$–20 cm), single-longitudinal-mode oscillation is readily achieved when the cavity length is tuned (by a piezo-electric translator) to bring a cavity mode into coincidence with the peak of the gain line (see Sect. 7.8.2.1). Single-mode He-Ne lasers can then be frequency stabilized to a high degree [$(\Delta v/v) = 10^{-11}$–10^{-12}] against a frequency reference (such as a high-finesse Fabry–Perot interferometer or, for absolute stabilization, against, e.g., an $^{129}I_2$ absorption line for the 633-nm transition).

He-Ne lasers oscillating on the red transition are widely used for many applications requiring a low-power visible beam (such as alignment or bar code scanners). Most store checkouts use red He-Ne lasers to read information encoded in a product's bar code; for some of these applications, however, He-Ne lasers face now strong competition from red-emitting semiconductor lasers, which are smaller and more efficient. Given the greater visibility of a green beam, green-emitting He-Ne lasers are increasingly used for alignment and cell cytometry. In the latter application, individual cells (e.g., red blood cells), stained by suitable fluorochromes, are rapidly flowed through a capillary onto which an He-Ne laser is focused; the fluorochromes are characterized by subsequent scattering or fluorescent emission. Single-mode He-Ne lasers are also often used in metrological applications (e.g., very precise, interferometric, distance measurements) and holography.

10.2.1.2. Copper Vapor Laser

Figure 10.4 shows relevant energy levels of the copper vapor laser, where Russell–Saunders notation is again used.[4] The $^2S_{1/2}$ ground state of Cu corresponds to the electron configuration $3d^{10}4s^1$, while the excited $^2P_{1/2}$ and $^2P_{3/2}$ levels correspond to the outer $4s$ electron being promoted to the next higher $4p$ orbital. The $^2D_{3/2}$ and $^2D_{5/2}$ levels arise from the electron configuration $3d^94s^2$ in which an electron is promoted from the $3d$ to the $4s$ orbital.

The relative values of the corresponding cross sections are such that the rate of electron impact excitation to the P states is greater than that to the D states; thus the P states are preferentially excited by electron impact. The $^2P \rightarrow {}^2S_{1/2}$ transition is strongly electric-dipole allowed (selection rules for optically allowed transitions require $\Delta J = 0$ or ± 1), so the corresponding absorption cross section is quite large. At the temperature used for Cu lasers ($T = 1500°C$), the vapor pressure is sufficiently high (~ 0.1 torr), however, that the $^2P \rightarrow {}^2S_{1/2}$ transition is completely trapped. Thus, the only effective decay route of the ^2P state is through the ^2D states; the corresponding decay time is rather long ($\sim 0.5 \, \mu s$) since the transition is only weakly allowed.

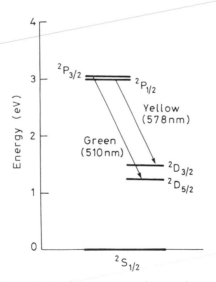

FIG. 10.4. Energy levels of copper atoms relevant to laser operation.

It then follows that, since the ^2P states can accumulate a large population, they are good candidates for upper laser levels. Thus laser action in Cu can occur on both the $^2P_{3/2} \to {}^2D_{5/2}$ (green) and $^2P_{1/2} \to {}^2D_{3/2}$ (yellow) transitions. Note that the $^2D \to {}^2S$ transition is electric-dipole forbidden and the lifetime of the ^2D state is very long (a few tens of microseconds). It then follows that the laser transition is self-terminating, so the laser can operate only on a pulsed basis, with pulse duration on the order of or shorter than the lifetime of the ^2P state. One should also note that the $^2D \to {}^2S$ decay occurs mainly via superelastic collisions, involving cold electrons that remain after the pump pulse; the corresponding decay rate sets an upper limit to the repetition rate of the laser. Table 10.1 indicates relevant spectroscopic properties of the copper vapor green transition as an example.

Metal vapor laser construction is based on the arrangement schematically shown in Figure 10.5, where the vapor is contained in an alumina tube which is thermally isolated in a vacuum chamber. The necessary high temperature is usually maintained by power dissipated in the tube due to repetitively-pulsed pumping current. The anode and cathode, in the form of ring electrodes, are placed at the ends of the alumina tube. A 25–50 Torr neon buffer gas provides enough electron density, after passage of the discharge pulse, to allow for de-excitation of the lower ^2D state by superelastic collisions. The neon gas is also helpful in reducing the diffusion length of the Cu vapor, thus preventing metal vapor deposition on the (cold) end windows. More recently, the so-called copper-HyBrID lasers were introduced, which use HBr in the discharge. Since, in this case, CuBr molecules are formed in the discharge region and these molecules are much more volatile than Cu atoms, lower temperatures are required in the gas discharge.

Copper vapor lasers are commercially available with average output powers in excess of 100 W, short pulse durations (30–50 ns), high-repetition rates (up to ~10 kHz), and relatively high efficiency (~1%). The latter is the result of both the high quantum efficiency of the copper laser (~55%; see Fig. 10.4) and the high electron-impact cross section of the $^2S \to {}^2P$ transition. Even higher output powers (~200 W) and higher efficiencies (~3%) were recently obtained with copper-hybrid lasers.

Copper vapor lasers are used in some industrial applications (such as high-speed photography, resistor trimming, and, more recently, micromachining) and as a pump for dye

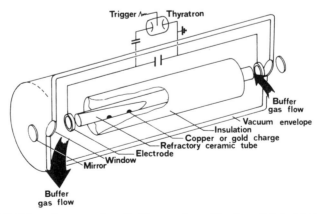

FIG. 10.5. Schematic construction of a Cu vapor laser. (Courtesy of Oxford Lasers, Ltd.)

lasers. In particular, in high-speed flash photography, the short-pulse (tens of nanoseconds) and high-repetition rate (10–20 kHz) are exploited in stroboscopic illumination of various, rapidly moving, objects (e.g., a bullet in flight). A large facility based on copper-laser-pumped dye lasers (using many copper lasers, each with an average power to 100 W) is currently used in a pilot plant, in the United States, for ^{235}U isotope separation.

10.2.2. Ion Lasers

In an ionized atom, the energy levels scale is expanded in comparison with neutral atoms. In this case, in fact, an electron in the outermost orbital(s) experiences the field due to the positive charge Ze of the nucleus (Z is the atomic number and e the electronic charge) screened by the negative charge $(Z - 2)e$ of the remaining electrons. Assuming for simplicity that the screening is complete, the net effective charge is then $2e$, rather than simply e for the corresponding neutral atom. This energy scale expansion means that ion lasers typically operate in the visible or ultraviolet regions. As in the case of neutral atom lasers, ion lasers can be divided into two categories: (1) *Ion gas lasers*, involving most of the noble gases; the most notable example is the Ar$^+$ laser, considered in Sect. 10.2.2.1, and the Kr$^+$ laser. Both lasers oscillate on many transitions; the most common are green and blue (514.5 nm and 488 nm, respectively) for the Ar$^+$ laser and red (647.1 nm) for the Kr$^+$ laser. (2) *Metal-ion vapor lasers*, involving many metals (Sn, Pb, Zn, Cd, and Se); the most notable example is the He-Cd laser, discussed in Sect. 10.2.2.2, and the He-Se laser.

10.2.2.1. Argon Laser

Figure 10.6 shows a simplified scheme for relevant energy levels in an argon laser.[5,6] The Ar$^+$ ground state is obtained by removing one electron from the six electrons in the $3p$ outer shell of Ar. The excited $4s$ and $4p$ states are then obtained by promoting one of the remaining $3p^5$ electrons to the $4s$ or $4p$ state, respectively. As a consequence of interacting with the other $3p^4$ electrons, both the $4s$ and $4p$ levels, indicated as single levels in Fig. 10.6, actually consist of many sublevels.

The Ar ion is excited in a two-step process involving collisions with two distinct electrons. The first collision ionizes Ar, i.e., raises it to the Ar$^+$ ground state, while the second collision excites the Ar ion. Since the lifetime of the $4p$ level ($\sim 10^{-8}$ s, set by the $4p \rightarrow 4s$ radiative transition) is about 10 times longer than the radiative lifetime of the $4s \rightarrow 3p^5$ transition, excited Ar ions accumulate predominantly in the $4p$ level. This means that the $4p$ level can be used as the upper laser level for the $4p \rightarrow 4s$ laser transition and, according to Eq. (7.3.1), cw laser action can be achieved. Note that excitation of the Ar ion can lead to ions in the $4p$ state by three distinct processes (see Fig. 10.6): (a) Direct excitation to the $4p$ level starting from the Ar$^+$ ground level; (b) excitation to higher lying states followed by radiative decay to the $4p$ level; (c) excitation to metastable levels followed by a third collision, leading to excitation to the $4p$ state. Considering, for simplicity, only the first two processes, one sees that the pumping process to the upper state is expected to be proportional to the square of the discharge current density. In fact, since the

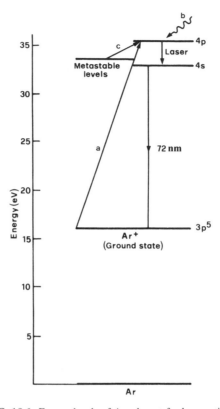

FIG. 10.6. Energy levels of Ar relevant for laser action.

first two processes involve a two-step electron collision, the rate of upper state excitation, $(dN_2/dt)_p$, is expected to take the form

$$\left(\frac{dN_2}{dt}\right)_p \propto N_e N_i \cong N_e^2 \tag{10.2.3}$$

where N_e and N_i are the electron and ion density in the plasma, respectively ($N_e \cong N_i$ in the positive column of the plasma). Since the electric field of the discharge is independent of the discharge current, the drift velocity v_{drift} is also independent of the discharge current. From the standard equation $J = e v_{drift} N_e$ one then sees that the electron density N_e is proportional to the discharge current density and, from Eq. (10.2.3), it follows that $(dN_2/dt)_p \propto J^2$. Laser pumping thus increases rapidly with current density, so high-current densities ($\sim$1 kA/cm^2) are required if the, inherently inefficient, two-step processes are to pump enough ions into the upper state. This may explain why the first operation of an Ar$^+$ laser occurred 3 years after the first He-Ne laser.[7]

From the preceding discussion we expect laser action, in an Ar laser, to occur on the $4p \rightarrow 4s$ transition. Since both the $4s$ and $4p$ levels actually consist of many sublevels, the argon laser oscillates on many lines; the most intense are in the green ($\lambda = 514.5$ nm) and blue ($\lambda = 488$ nm). From spectral measurements of the spontaneously emitted light, the

Doppler linewidth Δv_0^* of, e.g., the green transition, is found to be $\sim 3.5\,\mathrm{GHz}$. From Eq. (2.5.18), this implies an ion temperature of $T \cong 3000\,\mathrm{K}$. The ions are therefore very hot, a result of ion acceleration by the electric field of the discharge. Table 10.1 summarizes relevant spectroscopic properties of the Ar ion green laser transition.

Figure 10.7 shows a schematic diagram of a high-power ($> 1\,\mathrm{W}$) argon laser. Both the plasma current and the laser beam are confined by metal disks (of tungsten) inserted into a larger bore tube of ceramic material (BeO). This, thermally conductive and resistant, metal-ceramic combination is necessary to ensure good thermal conductivity of the tube and, at the same time, to reduce erosion problems arising from the high ion temperature. The diameter of the central holes in the disks is kept small ($\sim 2\,\mathrm{mm}$) to confine oscillation to a TEM_{00} mode (long radius of curvature mirrors are commonly used for the resonator) and to reduce the total current required.

A problem arises in argon lasers from the cataphoresis of argon ions. Due to the high-current density, in fact, a substantial migration of Ar ions occurs toward the cathode, where they are neutralized on combining with electrons emitted by the cathode surface.* Thus neutral atoms tend to accumulate at this electrode, with a corresponding reduction in Ar pressure, in the discharge capillary, below its optimum value. To overcome this problem, off-center holes are also made in the disk to provide return paths for the atoms from cathode to anode, by diffusion. The holes are arranged so that no current flows along this return path because longer path lengths are involved compared to that of the central path.

The inner ceramic tube is water cooled to remove the large amount of heat that it is inevitably dissipated in the tube (some kW/m). Note also that a static magnetic field is also applied in the discharge region, parallel to the tube axis, by a solenoid. With this arrangement, the Lorentz force makes electrons rotate around the tube axis, thus reducing the rate of electron diffusion to the walls. This increases the number of free electrons at the center of the tube, thereby increasing the pump rate. This may explain the observed increase in output power when a magnetic field is applied. By confining charges toward the center of the tube, the magnetic field also alleviates the problem of wall damage (mostly occurring at holes in the tungsten disks). Note that, for high-power lasers ($>1\,\mathrm{W}$), laser mirrors are mounted externally to the laser tube to reduce degradation of mirror coatings due to the intense vuv radiation emitted by the plasma.

---

*Referring to Sect. 6.4.4., we recall that direct electron-ion recombination cannot occur in the discharge volume, since the process cannot simultaneously satisfy both energy and momentum conservation. Electron-ion recombination can therefore occur only in the presence of a third partner, e.g., at the tube walls or at the cathode surface.

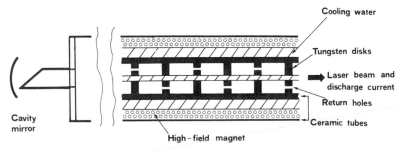

FIG. 10.7. Schematic diagram of a high-power water-cooled Ar^+ laser tube.

For lower power lasers (<1 W), the laser tube is often simply made of a block of ceramic (BeO) provided with a central hole for the discharge current. In this case, no magnetic field is applied, the tube is air cooled, and the mirrors are directly sealed to the ends of the tube as in a He-Ne laser.

Water-cooled argon lasers are commercially available with power ranging from 1 to 20 W, operating on both blue and green transitions simultaneously or, by using the configuration in Fig. 7.16b, operating on a single transition. Air-cooled argon lasers, of lower power (~100 mW) and much simpler design, are also commercially available. In both cases, once above threshold, output increases rapidly with current density ($\propto J^2$), since, in contrast to the behavior of He-Ne lasers, no saturation of the upper-state population occurs. Laser efficiency is however very low ($<10^{-3}$), because the laser quantum efficiency is rather low (~7.5%; see Fig. 10.6) and electron impact excitation involves many levels that are not effectively coupled to the upper laser level. Argon lasers are often operated in a mode-locked regime using an acousto-optic modulator. In this case, fairly short mode-locked laser pulses (~150 ps) are achievable by virtue of the relatively large transition linewidth (~3.5 GHz), which is inhomogeneously broadened.

Argon lasers are widely used in ophthalmology (particularly to treat diabetic retinopathy) and in the field of laser entertainment (laser light shows). Argon lasers are also widely used to study light-matter interaction (particularly in mode-locked operation) and also as a pump for solid-state lasers (particularly Ti:sapphire) and dye lasers. In many of these applications, argon lasers are now superseded by cw diode-pumped Nd:YVO$_4$ lasers, in which a green beam, $\lambda = 532$ nm, is produced by intracavity second harmonic generation. Lower power Ar lasers are extensively used in high-speed laser printers and cell cytometry.

10.2.2.2. He-Cd Laser

Figure 10.8 shows the energy levels in the He-Cd system that are relevant for laser action, where, again, level labeling follows Russell–Saunders notation. Pumping of Cd$^+$ upper laser levels ($^2D_{3/2}$ and $^2D_{5/2}$) is achieved with the help of He in the Penning ionization process. This process can be written in the general form

$$A^* + B \rightarrow A + B^+ + e \qquad (10.2.4)$$

where the ion B^+ may or may not be left in an excited state. Of course, the process occurs only if the excitation energy of the excited species A^* is greater than or equal to the ionization energy of species B (plus the excitation energy of B^+, if the ion is left in an excited state). Note that, unlike near-resonant energy transfer, Penning ionization is a nonresonant process; any surplus energy is, in fact, released as kinetic energy of the ejected electron. In the case of the He-Cd system, the 2^1S and 2^3S metastable states of He act as the species A^* in Eq. (10.2.4); upon collision, this excitation energy is given up to ionize the Cd atom and excite the Cd$^+$ ion. Although the process is not resonant, it has been found that the rate of excitation to the D states is about three times larger than that to the P states.*

* According to, e.g., Eq. (2.6.2), the rate of excitation k_{A^*B} of the general process (10.2.4) can be defined by the relation $(dN/dt)_{AB^+} = k_{A^*B}N_{A^*}N_B$, where $(dN/dt)_{AB^+}$ is the number of species A and B^+ produced, per unit volume per unit time, and N_{A^*} and N_B are the concentrations of the colliding partners.

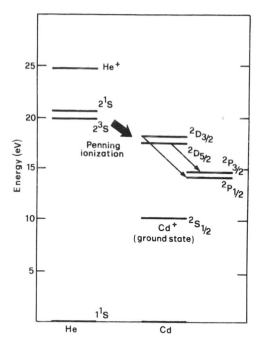

FIG. 10.8. Relevant energy levels in the He-Cd laser.

More importantly, however, the lifetime of the D states ($\sim 0.1\ \mu s$) is much longer than the lifetime of the P states (~ 1 ns). Population inversion between the D and P states can therefore be readily produced, so cw laser action can be achieved. Indeed, in accordance with the selection rule $\Delta J = 0, \pm 1$, laser action is obtained on the $^2D_{3/2} \rightarrow {}^2P_{1/2}$ ($\lambda = 325$ nm, uv) and the $^2D_{5/2} \rightarrow {}^2P_{3/2}$ ($\lambda = 416$ nm, blue) lines. The Cd$^+$ ion then drops to its $^2S_{1/2}$ ground state by radiative decay.

A typical construction for the He-Cd laser is in the form of a tube, terminated by two Brewster's angle windows, with the two laser mirrors mounted separated from the tube. In one possible arrangement, the tube, which is filled with helium, also has a small reservoir containing the Cd metal, located near the anode. The reservoir is raised to a high enough temperature ($\sim 250\ ^\circ$C) to produce the desired vapor pressure of Cd atoms in the tube. The discharge itself then produces enough heat to prevent vapor condensation on the tube walls along the discharge region. Due to ion cataphoresis, however, ions migrate to the cathode where they recombine with electrons emitted by the cathode. Neutral Cd atoms then condense around the cathode region, where there is no discharge and the temperature is low. The net result is a continuous flow of metal vapor from the anode to the cathode. Therefore, a sufficient supply of Cd (~ 1 g per 1000 h) must be provided for long-term laser operation.

He-Cd lasers can give output powers of 50–100 mW, which places these lasers in an intermediate position between red He-Ne lasers (a few milliwatts) and Ar$^+$ lasers (a few watts). Thus, He-Cd lasers are used for many applications where a blue or uv beam of moderate power is required (e.g., high-speed laser printers, holography, cell cytometry, fluorescence analysis of biological specimens).

10.2.3. Molecular Gas Lasers

These lasers exploit transitions between the energy levels of a molecule. Depending on the type of transition involved, molecular gas lasers belong to one of the following categories: (1) *Vibrational-rotational lasers*, which use transitions between vibrational levels of the same electronic state (the ground state), so the energy difference between levels falls in the middle-to far-infrared (2.5–300 μm). By far the most important example of this category is the CO_2 laser oscillating at either 10.6 or 9.6 μm. Other noteworthy examples are the CO laser ($\lambda \cong 5\,\mu$m) and the HF chemical laser ($\lambda \cong 2.7$–3.3 μm). (2) *Vibronic lasers*, which use transitions between vibrational levels of different electronic states. In this case the oscillation wavelength generally falls in the uv region. The most notable example of this laser category is the nitrogen laser ($\lambda = 337$ nm). A special class of lasers, that can be included here is the excimer laser. These lasers involve transitions between different electronic states of special molecules (excimers) with corresponding emission wavelengths generally in the uv. Excimer lasers, however, involve not only transitions between bound states (bound-bound transitions) but also, and actually more often, transitions between a bound upper state and a repulsive ground state (bound-free transitions). It is more appropriate therefore to treat these lasers as a category of their own. (3) *Pure rotational lasers*, which use transitions between different rotational levels of the same vibrational state (usually an excited vibrational level of the ground electronic state). The corresponding wavelength falls in the far infrared (25 μm–1 mm). Since pure rotational lasers are relatively less important than other laser types, we do not discuss them in the sections that follow. We point out here that laser action is more difficult to achieve with this type of laser, since the relaxation between rotational levels is generally very rapid. Therefore, these lasers are usually pumped optically, using the output of another laser as the pump (commonly a CO_2 laser). Optical pumping excites the given molecule (e.g., CH_3F, $\lambda = 496\,\mu$m) to a rotational level belonging to some vibrational state above the ground level. Laser action then takes place between rotational levels of this upper vibrational state.

10.2.3.1. CO_2 Laser

As the active medium, these lasers use a suitable mixture of CO_2, N_2, and He. Oscillation takes place between two vibrational levels of the CO_2 molecule, while N_2 and He greatly improve the efficiency of laser action. The CO_2 laser is actually one of the most powerful lasers (output powers of more than 100 kW have been demonstrated from a CO_2 gas-dynamic laser) and one of the most efficient (15–20% slope efficiency).[8,9]

Figure 10.9 shows the relevant vibrational energy levels for the electronic ground states of the CO_2 and N_2 molecules. The N_2 molecule, being diatomic, has only one vibrational mode; its lowest two energy levels ($v = 0, v = 1$) are indicated in the figure. Energy levels for CO_2 are more complicated, since CO_2 is a linear triatomic molecule. In this case there are three nondegenerate modes of vibration (Figure 10.10): Symmetric stretching, bending, and asymmetric stretching. The oscillation behavior and corresponding energy levels are therefore described by means of three quantum numbers n_1, n_2, and n_3, which give the number of quanta in each vibrational mode. This means that, apart from zero-point energy, the energy of the level is given by $E = n_1 h v_1 + n_2 h v_2 + n_3 h v_3$, where v_1, v_2 and v_3 are the

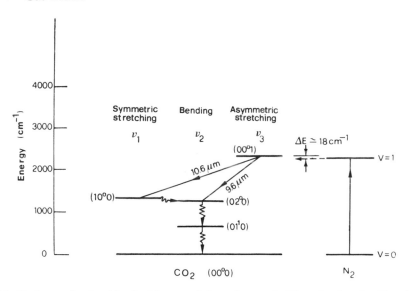

FIG. 10.9. The lowest vibrational levels of the ground electronic state of a CO_2 molecule and an N_2 molecule. (For simplicity, rotational levels are not shown.)

resonance frequencies of the three modes. For example, the 01^10 level* corresponds to an oscillation where there is one vibrational quantum in mode 2. Since mode 2 has the smallest force constant of the three modes (the vibrational motion is transverse), this level has the lowest energy. Laser action occurs between the 00^01 and 10^00 levels ($\lambda \cong 10.6 \, \mu$m), although it is also possible to obtain oscillation between 00^01 and 02^00 ($\lambda \cong 9.6 \, \mu$m).

Pumping the upper 00^01 laser level is efficiently achieved by the following processes: (1) Direct Electron Collisions. The main direct collision considered is obviously as follows: $e + CO_2(000) \rightarrow e + CO_2(001)$. The electron collision cross section for this process is very large and appreciably larger than those for excitation to both the 100 and 020 level— probably because the transition $000 \rightarrow 001$ is optically allowed whereas, for instance, the transition $000 \rightarrow 100$ is not. Note also that direct electron impact can also lead to excitation of upper $(0, 0, n)$ vibrational levels of the CO_2 molecule. However, the CO_2 molecule

---

* The superscript (denoted by l) on the bending quantum number arises because the bending vibration is in this case doubly degenerate. In fact, it can occur in both the plane of Fig. 10.10 and in the orthogonal plane. A bending vibration therefore consists of a suitable combination of these two vibrations. More precisely, $l\hbar$ gives the angular momentum of this vibration about the axis of the CO_2 molecule. For example, in the 02^00 state ($l = 0$), the two degenerate vibrations combine to give an angular momentum $l\hbar = 0$.

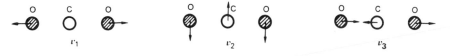

FIG. 10.10. Three fundamental vibration modes for a CO_2 molecule: (v_1) symmetric stretching, (v_2) bending, (v_3) asymmetric stretching.

rapidly relaxes from these upper states to the (001) state by near-resonant collisions of the type:*

$$CO_2(0, 0, n) + CO_2(0, 0, 0) \rightarrow CO_2(0, 0, n - 1) + CO_2(0, 0, 1) \qquad (10.2.5)$$

This process tends to degrade all excited molecules to the $(0, 0, 1)$ state. Note that, since most molecules in a CO_2 laser mixture are in fact in the ground state, collision of an excited with an unexcited molecule constitutes the most likely collisional event. (2) Resonant Energy Transfer from N_2 Molecule. This process is also very efficient due to the small energy difference between excited levels of the two molecules ($\Delta E = 18\,cm^{-1}$). In addition, the excitation of N_2 from the ground level to the $v = 1$ level is a very efficient process, and the $v = 1$ is metastable. Transition $1 \rightarrow 0$ is in fact electric-dipole-forbidden since, by virtue of its symmetry, a N-N molecule, using the same isotope for the two nitrogen atoms, cannot have a net electric dipole moment. The higher vibrational levels of N_2 are also closely resonant ($\Delta E < kT$) with the corresponding CO_2 levels (up to 00^05), and transitions between excited levels, $00n$, and the 001 level of the CO_2 molecule occur rapidly through the process indicated in Eq. (10.2.5).

The next point to consider is the decay of both upper and lower laser levels. Note that, although transitions $00^01 \rightarrow 10^00, 00^01 \rightarrow 02^00, 10^00 \rightarrow 01^00$, and $02^00 \rightarrow 01^00$ are optically allowed, the corresponding decay times τ_{sp}, for spontaneous emission, are very long ($\tau_{sp} \propto 1/v^3$). The decay of these various levels is therefore determined essentially by collisions. Accordingly, the decay time τ_s of the upper laser level can be obtained from a formula of the type

$$\left(\frac{1}{\tau_s}\right) = \sum a_i p_i \qquad (10.2.6)$$

where p_i are partial pressures and a_i are rate constants corresponding to gases in the discharge. Taking for example the case of a total pressure of 15 Torr (in a 1:1:8 CO_2:N_2:He partial-pressure ratio), one finds that the upper level has a lifetime $\tau_s \cong 0.4\,ms$. Concerning the relaxation of the lower level, we begin by noting that the transition $100 \rightarrow 020$ is very rapid and it occurs even in an isolated molecule. In fact, the energy difference between the two levels is much smaller than kT. Furthermore, a coupling between the two states is present (Fermi resonance) because a bending vibration tends to induce a change of distance between the two oxygen atoms (i.e., to induce a symmetric stretching). Levels 10^00 and 02^00 are then effectively coupled to the 01^10 level by the two near-resonant collision processes involving CO_2 molecules in the ground state (VV relaxation), as follows:

$$CO_2(10^00) + CO_2(00^00) \rightarrow CO_2(01^10) + CO_2(01^10) + \Delta E \qquad (10.2.7a)$$

$$CO_2(02^00) + CO_2(00^00) \rightarrow CO_2(01^10) + CO_2(01^10) + \Delta E' \qquad (10.2.7b)$$

---

* Relaxation processes in which vibrational energy is given up as vibrational energy of another like or unlike molecule are usually referred to as *VV relaxations*.

The two processes have a very high probability, since ΔE and $\Delta E'$ are much smaller than kT. It follows, therefore, that the three levels 10^00, 02^00, and 01^10 reach the thermal equilibrium in a very short time.

We are now left with the decay from the 01^10 to the ground level 00^00. If this decay were slow, it would lead to an accumulation of molecules in the 01^10 level during laser action. This would in turn produce an accumulation in the 10^00 and 02^00 levels since these are in thermal equilibrium with the 01^10 level. Thus the decay process would slow down on all three levels, i.e., the $(01^10) \rightarrow (00^00)$ transition would constitute a bottleneck in the overall decay process. It is therefore important to consider the lifetime of the 01^10 level. Note that, since the transition $01^10 \rightarrow 00^00$ is the least energetic transition in any of the molecules in the discharge, relaxation from the 01^10 level can occur only by transferring this vibrational energy to translational energy of the colliding partners (*VT relaxation*). According to the theory of elastic collisions, energy is most likely to be transferred to lighter atoms, i.e., to helium in this case. This means that the lifetime is again given by an expression of the general form of Eq. (10.2.6), where the coefficient a_i, for He, is much larger than that of the other species. For the same partial pressures considered in this example, one obtains a lifetime of about 20 μs. It follows from the preceding discussion that this is also the value of the lifetime of the lower laser level. Therefore, due to the much larger value of the upper state lifetime, population accumulates in the upper laser level, so the condition for cw laser action is also fulfilled. Note that He has another valuable effect: Due to its high thermal conductivity, He helps keeping the CO_2 cool by conducting heat to the walls of the container. A low translational temperature for CO_2 is necessary to avoid populating the lower laser level by thermal excitation; energy separation between levels is in fact comparable to kT.

To summarize most preceding discussions, we can say that nitrogen helps producing a large population in the upper laser level while helium helps removing population from the lower laser level.

From the preceding considerations we see that laser action in a CO_2 laser may occur either on the $(00^01) \rightarrow (10^00)$ ($\lambda = 10.6$ μm) transition or on the $(00^01) \rightarrow (02^00)$ transition ($\lambda = 9.6$ μm). Since the first of these transitions has a greater cross section and both transitions share the same upper level, the $(00^01) \rightarrow (10^00)$ transition usually oscillates. To obtain oscillation on the 9.6-μm line, some appropriate frequency-selective device is placed in the cavity to suppress laser action on the line with highest grain. (The system in Fig. 7.16a is often used.)

So far we have ignored the fact that both upper and lower laser levels consist of many closely spaced rotational levels. Accordingly, laser emission may occur on several equally spaced rotational-vibrational transitions belonging to either P or R branches (see Fig. 3.7), with the P branch exhibiting the largest gain. To complete our discussion we must also consider that, as a consequence of the Boltzmann distribution between rotational levels, the $J' = 21$ rotational level of the upper 00^01 state is the most heavily populated (see Fig. 10.11).* Laser oscillation then occurs on the rotational vibrational transition with the largest gain i.e., originating from the most heavily populated level. This happens because, in a CO_2 laser, the thermalization rate among rotational levels ($\sim 10^7$ s^{-1} Torr^{-1}) is faster than the rate of decrease in population (due to stimulated emission) of the rotational level from which

* By reason of symmetry, only levels with odd values of J are occupied in a CO_2 molecule.

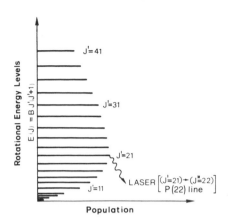

FIG. 10.11. Relative population of rotational levels in the upper laser level of CO_2.

laser emission occurs. Therefore, the entire population of rotational levels contributes to laser action on the rotational level with the highest gain. Consequently, and as a conclusion to this discussion, laser action in a CO_2 laser normally occurs on the $P(22)$ [i.e., $(J' = 21) \rightarrow (J'' = 22)$] line of the $(00^01) \rightarrow (10^00)$ transitions. Other lines of the same transition as well as lines belonging to the $(00^01) \rightarrow (02^00)$ transition (the separation between rotational lines in a CO_2 laser is about $2 \, cm^{-1}$) can be selected with, e.g., the scheme in Fig. 7.16a.

A major contribution to laser linewidth in a CO_2 laser comes from the Doppler effect. Compared with, e.g., a visible gas laser, the corresponding value of the Doppler linewidth is however rather small (about 50 MHz) due to the low frequency ν_0 of the laser transition (see Example 3.2). Collision broadening is, however, not negligible (see Example 3.3), and it actually becomes the dominant line-broadening mechanism for CO_2 lasers operating at high pressures ($p > 100 \, Torr$).

Based on their constructional design, CO_2 lasers fall into the following categories: Lasers with slow axial flow, sealed-off lasers, waveguide lasers, lasers with fast axial flow, diffusion-cooled lasers, transverse-flow lasers, transversely excited atmospheric pressure (TEA) lasers, and gas dynamic lasers. We do not consider the gas dynamic laser here, since its operating principle was described in Sect. 6.1. Before considering the other categories, we point out that, although these lasers differ from one another as far as many of their operating parameters are concerned (e.g., output power), they all share a very important characteristic feature, namely, high slope efficiency (15–25%). This high efficiency results from the high laser quantum efficiency (~40%; see Fig. 10.9) and the highly efficient pumping processes that occur in a CO_2 laser at the optimum electron temperature of the discharge (see Fig. 6.28).

(a) *Lasers with Slow Axial Flow.* Operation of the first CO_2 laser was achieved in a laser of this type (C. K. N. Patel, 1964[10]). The gas mixture is flowed slowly along the laser tube (see Fig. 10.12) simply to remove dissociation products, in particular CO, that otherwise would contaminate the laser. Heat removal is provided by radial heat diffusion to the tube walls (usually made of glass), which are cooled externally by a suitable coolant (usually water). An internal mirror arrangement is often used, and, in the design of Fig. 10.12, one of the metal mounts holding the cavity mirrors must be held at a high voltage.

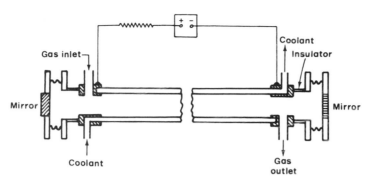

FIG. 10.12. Schematic diagram of a CO_2 laser with longitudinal gas flow.

One of the main limitations of this laser arises from the maximum laser output power, per unit length of the discharge (50–60 W/m), which can be obtained, independently of tube diameter. In fact, from Eq. (6.4.24), we find that the number of molecules pumped into the upper laser level per unit volume and unit time can be written as

$$\left(\frac{dN_2}{dt}\right)_p = N_t \frac{J}{e}\left(\frac{\langle v\sigma\rangle}{v_{drift}}\right) \tag{10.2.8}$$

where: J is the current density; σ is a suitable electron impact cross section, which takes into account both direct and energy transfer excitations; N_t is the total CO_2 ground-state population. For pump rates well above threshold, output power P is proportional to $(dN_2/dt)_p$ and the volume of the active medium V_a. From Eq. (10.2.8) we can therefore write

$$P \propto JN_tV_a \propto JpD^2l \tag{10.2.9}$$

where D is the diameter of the active medium, l its length, and p the gas pressure. Under optimum operating conditions we now have the following: (1) The product pD must be constant (~ 22.5 Torr × cm, e.g., 15 Torr for $D=1.5$ cm) to keep the discharge at the optimum electron temperature. (2) Due to the thermal limitations imposed by the requirement of heat conduction to the tube walls, an optimum value of the current density exists, and this value is inversely proportional to the tube diameter D. The existence of an optimum value of J is understood when we observe that excessive current density leads to excessive heating of the mixture (even with an efficiency of 20%, some 80% of the electrical power is dissipated in the discharge as heat), with resulting thermal population in lower laser levels. The fact that the optimum value of J is inversely proportional to D is understood when we realize that, as diameter is increased, generated heat has more difficulty in escaping to the walls. From this discussion we conclude that, under optimum conditions, both J and p are inversely proportional to D; hence, from Eq. (10.2.9), the optimum value of P is found to be only proportional to the tube length l. CO_2 lasers with slow axial flow and relatively low power (50–100 W) are widely used for laser surgery, resistor trimming, cutting ceramic plates for the electronics industry, and welding thin metal sheets (<1 mm).

(b) *Sealed-off Lasers.* If the flow of the gas mixture is stopped in the arrangement shown in Fig. 10.12, laser action ceases in a few minutes. This arises because chemical reaction products formed in the discharge (CO in particular) are no longer removed but instead are adsorbed on the walls of the tube or react with the electrodes, thus upsetting the CO_2-CO-O_2 equilibrium. Ultimately, this leads to dissociation of a large fraction of CO_2 molecules in the gas mixture. For a non-circulating sealed-off laser, some kind of catalyst must be present in the gas tube to promote regeneration of CO_2 from CO. A simple way of achieving this involves adding a small amount of H_2O (1%) to the gas mixture. This leads to CO_2 regeneration, probably through the reaction

$$CO^* + OH \rightarrow CO_2^* + H \tag{10.2.10}$$

involving vibrationally excited CO and CO_2 molecules. The relatively small amount of water vapor required, may be added in the form of hydrogen and oxygen gas. Actually, since oxygen is produced during the dissociation of CO_2, only hydrogen needs be added. Another way of inducing the recombination reaction uses a hot (300°C) Ni cathode, which acts as a catalyst. With these techniques, lifetimes for sealed-off tubes in excess of 10,000 hours have been demonstrated.

Sealed-off lasers have produced output powers per unit length of $\sim$60 W/m, i.e., comparable to those of longitudinal flow lasers. Low-power ($\sim$1 W) sealed-off lasers of short length (and hence operating on a single mode) are often used as local oscillators in optical heterodyne experiments. Sealed-off CO_2 lasers of somewhat higher power ($\sim$10 W) are attractive for laser microsurgery and micromachining.

(c) Capillary Waveguide Lasers. If the diameter of the laser tube in Fig. 10.12 is reduced to around a few millimeters (2–4 mm), a situation is reached where laser radiation is guided by the inner walls of the tube. Such waveguide CO_2 lasers have a low diffraction loss. Tubes of BeO or SiO_2 were found to give the best performance. Due to the small bore diameter of a waveguide CO_2 laser, the pressure of the gas mixture must be considerably increased (100–200 Torr); with this increase in pressure, the laser gain per unit length is correspondingly increased. Thus, one can manufacture short CO_2 lasers ($L < 50$ cm) without facing the difficult requirement of reducing cavity losses; however, the power available per unit length of the discharge suffers the same limitation discussed earlier for the slow axial flow laser ($\sim$50 W/m). Therefore, waveguide CO_2 lasers are particularly useful when short and compact CO_2 lasers of low power ($P < 30$ W) are needed (e.g., for laser microsurgery).

To fully exploit their compactness, these lasers usually operate as sealed-off devices. The laser configuration may resemble that in Fig. 10.12, where discharge current flows longitudinally along the laser tube, or that in Fig. 10.13, where the current (usually provided by an rf source) flows transversely across the tube. Since $\mathscr{E}/p$ must be constant, then, for a given value of discharge electric field $\mathscr{E}$, the transverse pumping configuration offers a significant advantage over longitudinal pumping due to its much reduced (one to two orders of magnitude) electrode voltage. Radiofrequency ($v \cong 30$ MHz) excitation presents many advantages,[11] the most important of which are perhaps: (1) Avoiding permanent anodes and cathodes, which eliminates the associated plasmo-chemical problems at the cathode. (2) Creating a stable discharge with the help of simple nondissipative elements (e.g., a dielectric slab) in series in the discharge circuit (see, e.g., Fig. 6.21). As a result of these various advantages, rf discharge lasers are increasingly being used not only for waveguide lasers but also for fast axial flow lasers and transverse flow lasers, which we consider next. The laser

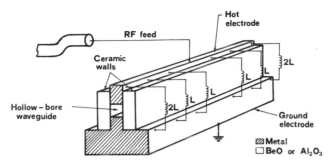

FIG. 10.13. Schematic diagram of an rf-excited waveguide CO_2 laser.

tube of a waveguide CO_2 laser is either not cooled or, for the highest power units, cooled by forced air.

(d) *Lasers with Fast Axial Flow.* To overcome the output power limitation of a CO_2 laser with slow axial flow, as discussed with the help of Eqs. (10.2.8) and (10.2.9), a possible solution, and a very interesting one, involves flowing the gas mixture through the tube at a very-high supersonic speed (about 50 m/s). In this case the heat is removed simply by removing the hot mixture from the discharge region; the mixture is then cooled outside the tube by a suitable heat exchanger and then returned to the discharge region, as shown in Fig. 10.14. A dc longitudinal discharge or, more frequently, an rf transverse discharge across the glass tube (see Fig. 6.21) can be used to provide excitation of each of the two laser tubes indicated in the figure. When operated in this way, there is no optimum for the current density, so the power actually increases linearly with J, resulting in much higher output powers per unit discharge length ($\sim 1\,\text{kW/m}$ or even greater). Besides being cooled, the mixture is also passed, outside the laser tube, over a suitable catalyst to let CO recombine with O_2 and thus achieve the required regeneration of CO_2 molecules. (Some O_2 is already present in the mixture due to dissociation of the CO_2 in the discharge region.) In this way either completely sealed-off operation can be achieved or at least replenishment requirements for the mixture are kept to a minimal level. Fast axial flow CO_2 lasers with high power (1–3 kW) are now commonly used for many material working applications, in particular for cutting metals (with a thickness of up to a few millimeters).

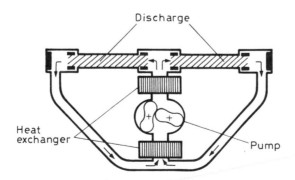

FIG. 10.14. Schematic diagram of a fast axial flow CO_2 laser.

(e) Diffusion-Cooled Lasers. An alternative approach to circumventing the power limitation of a slow axial flow laser is to use a transverse discharge with electrodes spacing d much smaller than the electrode width W (slab geometry, see Fig. 10.15a). In this case, the gas mixture is very effectively cooled by one-dimensional heat flow toward the electrodes, which are water cooled. The laser output power scales, in this case, as $P_{out} = (CWl)/d$, where C is a constant ($C \cong 50\,\text{W/m}$) and l is the electrode length.[12] Thus, for a given electrode spacing, output power scales as the electrode area Wl rather than electrode length, as in, e.g., CO_2 lasers with slow axial flow [see Eq. (10.2.9)]. For sufficiently small electrode spacing, large powers per unit electrode area can then be obtained [e.g., $(P_{out}/Wl) \cong 20\,\text{kW/m}^2$ for $d = 3\,\text{mm}$]. Instead of the slab-geometry configuration in Fig. 10.15a, the annular-geometry configuration in Fig. 10.15b can also be used; the latter configuration is technically more complicated but allows more compact devices to be achieved.

It should be stressed that the preceding results hold if the electrode width is appreciably larger (by approximately an order of magnitude) than the electrode spacing. To produce a good-quality discharge and an output beam with good divergence properties, from a gain medium with such a pronounced elongation, poses some difficult problems. Stable and spatially uniform discharges can however be obtained by exploiting the advantages of rf excitation. For electrode spacings on the order of a few millimeters, on the other hand, the laser beam is guided in the direction normal to the electrode surface, and propagates freely in the direction parallel to this surface. To obtain output beams with good quality, hybrid resonators, which are stable in a direction perpendicular to the electrodes and unstable parallel to the electrodes, have been developed.[13]

Compact, diffusion-cooled slab-geometry CO_2 lasers with output powers of $\sim 2.5\,\text{kW}$ are now commercially available with the potential for a large impact in material working applications.

(f) Transverse-Flow Lasers. Another way of circumventing the power limitations of a slow axial flow laser involves flowing the gas mixture perpendicular to the discharge (Fig. 10.16). If the flow is fast enough, then, as in the case of fast axial flow lasers, heat is removed by convection rather than diffusion to the walls. Saturation of output power versus discharge current does not occur then, and, as in the case of fast axial flow, very high output powers per unit discharge length can be obtained (a few kW/m; see also Fig. 7.7). The optimum total pressure ($\sim 100\,\text{Torr}$) is now typically an order of magnitude greater than that of large-bore longitudinal-flow systems. The increase in total pressure p requires a corresponding increase in the electric field $\mathscr{E}$ in the discharge. In fact, for optimum operating

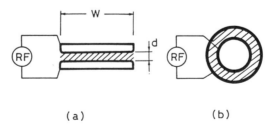

(a)   (b)

FIG. 10.15. Schematic diagram of a diffusion-cooled CO_2 laser using either (a) slab geometry or (b) an annular-geometry configuration.

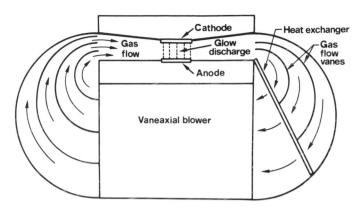

FIG. 10.16. Schematic diagram for a transverse-flow CO$_2$ laser.

conditions, the ratio $\mathscr{E}/p$ must remain approximately the same for all these cases, since this ratio determines the temperature of the discharge electrons [see Eq. (6.4.22)]. With this high value of electric field, a longitudinal-discharge arrangement, such as in Fig. 10.12, is impractical, since it requires very high voltages (100–500 kV for a 1-m discharge). For this reason, the discharge is usually applied perpendicular to the resonator axis (lasers with transverse electric field, i.e., TE lasers).

TE CO$_2$ lasers with fast transverse flow and high-output power (1–20 kW) are used in a great variety of metal-working applications (welding, surface hardening, surface metal alloying). Compared to fast axial flow lasers, these lasers are simpler devices given the reduced flow-speed requirement for transverse flow. However, the beam quality of fast axial flow lasers is considerably better, due to the cylindrical symmetry of their discharge current distribution, and this makes them particularly important for cutting applications.

(g) *Transversely Excited, Atmospheric Pressure, Lasers.* In a cw TE CO$_2$ laser, it is not easy to increase the operating pressure above ∼100 Torr. Above this pressure and at the current densities normally used, glow discharge instabilities set in and cause the formation of arcs within the discharge volume. To overcome this difficulty, the voltage can be applied to the transverse electrodes in the form of a pulse. If the pulse duration is sufficiently short (a fraction of a microsecond), discharge instabilities have no time to develop, so the operating pressure can be increased up to and above atmospheric pressure. These lasers are therefore referred to as transversely excited at atmospheric pressure (TEA). These lasers produce a pulsed output and are capable of large output energies per unit discharge volume (10–50 J/liter). To avoid arc formation, some form of ionization (*preionization*) is also applied; the preionization pulse just precedes the main voltage pulse that produces gas excitation.

Figure 10.17 shows a configuration often used, where ionization is produced by the strong uv emission of a row of sparks running parallel to the tube length. The deep uv emission of these sparks produces the required ionization by both photoionization of gas constituents and uv-induced electron emission from the electrodes (*uv-preionization*). Other preionization techniques include using pulsed e-beam guns (*e-beam preionization*) and ionization by the corona effect (*corona preionization*). Once ionization within the whole volume of the laser discharge is produced, the fast switch (a hydrogen thyratron or a spark

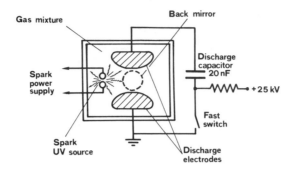

FIG. 10.17. Schematic diagram (*viewed along the laser axis*) of a CO_2 TEA laser. The laser uses uv radiation from several spark sources placed along the tube direction to provide gas preionization.

gap) is closed, so the main discharge pulse is passed through the discharge electrodes. Since the transverse dimensions of the laser discharge are usually large (a few centimeters), the two end mirrors are often chosen to give an unstable resonator configuration (positive-branch unstable confocal resonator; see Fig. 5.18b). For low-pulse repetition rates (~ 1 Hz), it is unnecessary to flow the gas mixture. For higher repetition rates, the gas mixture is flowed transversely to the resonator axis and cooled by a suitable heat exchanger; repetition rates up to a few kilohertz have been achieved in this way. Another interesting characteristic of these lasers is their relatively broad linewidth (~ 4 GHz at $p = 1$ atm, due to collision broadening). Thus optical pulses with less than 1-ns duration can be produced with mode-locked TEA lasers.

Transverse-flow CO_2 TEA lasers of relatively high repetition rate (~ 50 Hz) and relatively high average output power ($\langle P_{out} \rangle \cong 300$ W) are commercially available. Besides being used in scientific applications, these lasers find a number of industrial uses for material working applications where the pulsed nature of the beam presents some advantage (e.g., pulsed laser marking or pulsed ablation of plastic materials).

10.2.3.2. CO Laser

The second example of a gas laser using vibrational-rotational transitions that we will briefly consider is that of the CO laser. This laser has attracted considerable interest on account of its shorter wavelength ($\lambda \cong 5$ μm) than the CO_2 laser, combined with high efficiencies and high power. Output powers in excess of 100 kW and efficiencies in excess of 60% have been demonstrated;[14] however, to achieve this sort of performance, the gas mixture must be kept at cryogenic temperature (77–100 K). Laser action, in the 5-μm region, arises from several rotational-vibrational transitions [e.g., from $v'(11) \rightarrow v(10)$ to $v'(7) \rightarrow v(6)$ at $T = 77$ K] of the highly excited CO molecule.

Pumping of the CO vibrational levels is achieved by electron-impact excitation. Like the isoelectronic N_2 molecule, the CO molecule has an unusually large cross section for electron-impact excitation of its vibrational levels. Thus, nearly 90% of the electron energy in a discharge can be converted into vibrational energy of CO molecules. Another important feature of the CO molecule is that VV relaxation proceeds at a much faster rate than VT relaxation (which is unusually low). As a consequence of this, a non-Boltzmann population

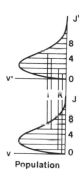

FIG. 10.18. Partial inversion between two vibrational transitions (v and v') with the same total population.

build-up in higher vibrational levels, by a process known as *anharmonic pumping*, plays a very important role.* Although this phenomenon does not allow total inversion in the vibrational population of a CO molecule, a situation known as *partial inversion* may occur. This is illustrated in Fig. 10.18 which indicates the rotational populations of two neighboring vibrational states. Although the total population for the two vibrational states is equal in the figure, an inversion exists for the two P transitions $[(J' = 5) \rightarrow (J = 6), (J' = 4) \rightarrow (J = 5)]$ and also for the two R-branch transitions indicated in the figure.

Under conditions of partial inversion, laser action can take place and a new phenomenon, called cascading, can then play an important role. Laser action, in fact, depopulates a rotational level of the upper state and populates a rotational level of the lower vibrational state. The latter level then accumulates enough population to result in population inversion with respect to a rotational level of a still lower vibrational state. At the same time, the rotational level of the upper state may become sufficiently depopulated to result in population inversion with respect to a rotational level of a still higher vibrational state. This cascading process, coupled with the very-low VT rate, results in most of the vibrational energy being extracted as laser output energy. Combined with the very high excitation efficiency, this accounts for the high efficiency of CO lasers. The low-temperature requirement arises from the need for very efficient anharmonic pumping. In fact, the overpopulation in high vibrational levels, compared to the Boltzman distribution, and hence the degree of partial inversion, increase rapidly with decreasing translational temperature.

As for CO_2 lasers, CO lasers have been operated with longitudinal flow, *e*-beam preionized pulsed TE, and gasdynamic excitation. The requirement of cryogenic temperatures has, so far, limited the commercial development of CO lasers. Recently, however, high-power ($P > 1$ kW) CO lasers, operating at room temperature while retaining a reasonably high slope efficiency ($\sim$10%), have been introduced commercially. These

---

* Anharmonic pumping arises from a collision of the type $CO(v = n) + CO(v = m) \rightarrow CO(v = n + 1) + CO(v = m - 1)$ with $n > m$. Because of anharmonicity (a phenomenon shown by all molecular oscillators), the separation between vibrational levels becomes smaller for levels higher up in the vibrational ladder (see also Fig. 3.1). This means that in a collision process of the type indicated above, with $n > m$, the total vibrational energy of the two CO molecules, after collision, is somewhat smaller than before collision. The collision process therefore has a greater probability of proceeding in this direction than the reverse direction. This means that the hottest CO molecule [$CO(v = n)$] can ascend the vibrational ladder; this leads to a non-Boltzmann distribution of the population among the vibrational levels.

lasers are now under active consideration as a practical source for material-working applications.

10.2.3.3. Nitrogen Laser

As a particularly relevant example of vibronic lasers, we consider the N_2 laser.[15] This laser has its most important oscillation at $\lambda = 337$ nm (uv) and belongs to the category of self-terminating lasers.

Figure 10.19 shows the relevant energy level scheme for the N_2 molecule. Laser action takes place in the so-called second positive system, i.e., in the transition from the $C^3\Pi_u$ state (henceforth called the C state) to the $B^3\Pi_g$ state (B state).* Excitation of the C state is believed to arise from electron-impact collisions with ground-state N_2 molecules. Since both C and B states are triplet states, transitions from the ground state are spin-forbidden. On the basis of the Franck–Condon principle, we can however expect the excitation cross section due to em wave interaction, and hence due to the electron impact, to the $v = 0$ level of the C state, to be larger than that to the $v = 0$ level of the B state. The potential minimum of the B state is in fact shifted (relative to the ground state) to a larger internuclear separation than that of the C state. The lifetime (radiative) of the C state is 40 ns, while the lifetime of the B state is 10 μs. Clearly the laser cannot operate cw, since Eq. (7.3.1) is not satisfied. The laser can however be excited on a pulsed basis provided the electrical pulse is appreciably shorter than 40 ns. Laser action takes place predominantly on several rotational lines of the $v''(0) \to v'(0)$ transition ($\lambda = 337.1$ nm) because this transition exhibits the largest stimulated emission cross section. Oscillation on the $v''(1) \to v'(0)$ ($\lambda = 357.7$ nm) and the $v''(0) \to v'(1)$ ($\lambda = 315.9$ nm) transitions also occurs, although with lower intensity. Table 10.2 summarizes some spectroscopic data for the N_2 laser.

---

* Under different operating conditions, laser action can also take place in the near infrared (0.74–1.2 μm) in the first positive system involving the $B^3\Pi_g \to A^3\Sigma_u^+$ transition.

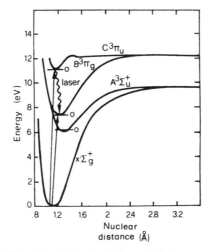

FIG. 10.19. Energy states of the N_2 molecule. For simplicity, only the lowest vibrational level ($v = 0$) is shown for each electronic state.

TABLE 10.2 Spectroscopic properties of uv laser transitions and gas mixture composition in nitrogen and KrF lasers

Laser Type	N_2	$(KrF)^*$
Laser wavelength (nm)	337.1	248
Cross section (10^{-14} cm^2)	40	0.05
Upper state lifetime (ns)	40	10
Lower state lifetime (ns)	10 μs	
Transition linewidth (THz)	0.25	3
Partial pressure of gas mixture (mbar)	40 (N_2)	120 (Kr)
	960 (He)	6 (F_2)
		2400 (He)

Given the high pressure of the gas mixture ($\sim$40 mbar of N_2 and 960 mbar of He) and the corresponding high electric field ($\sim$10 kV/cm), a TEA configuration (as in Fig. 10.17) is normally used for a nitrogen laser. To obtain the required rapid current pulse (5–10 ns), the discharge circuit must have as low an inductance as possible. Due to the high gain of this self-terminating transition, oscillation occurs in the form of ASE, so the laser can be operated even without mirrors. Usually, however, a single mirror is placed at one end of the laser, since this reduces the threshold gain and hence the threshold electrical energy for ASE emission (see Sect. 2.9.2). The mirror also ensures a unidirectional output and reduces the beam divergence. With this type of laser, it is possible to obtain laser pulses of high peak power ($\sim$1 MW) and short duration ($\sim$10 ns) at high repetition rates (to $\sim$100 Hz). Nitrogen lasers, with nitrogen pressure up to atmospheric pressure and without helium, have also been developed. In this case, the problem of arcing is alleviated by further reducing (to $\sim$1 ns) the duration of the voltage pulse. The increased gain per unit length, due to the higher N_2 pressure, and the fast discharge imply that this type of laser is usually operated without mirrors. The length can be kept very short (10–50 cm) and, as a consequence, output pulses of short time duration can be obtained ($\sim$100 ps with 100-kW peak power). Nitrogen lasers of both long ($\sim$10 ns) and short ($\sim$100 ps) pulse duration are still used as pumps for dye lasers, since most dyes absorb strongly in the uv.

10.2.3.4. Excimer Lasers

These lasers represent an interesting and important class of molecular laser involving transitions between different electronic states of special molecules referred to as excimers.[17] Consider a diatomic molecule A_2 with potential energy curves as in Fig. 10.20, for ground and excited states, respectively. Since the ground state is repulsive, the molecule cannot exist in this state; i.e., the species exists only in the monomer form A in the ground state. However, since the potential energy curve shows a minimum for the excited state, the species is bound in this state; i.e., it exists in dimer form. Such a molecule A_2^* is called an *excimer*, a contraction of the words excited dimer. Suppose that a large fraction of excimers are somehow produced in the given volume of the medium. Under appropriate conditions, laser action can then occur on the transition between the upper (bound) state and the lower (free) state (bound-free transition). This is referred to as an excimer laser; a classical example is the Ne_2^* laser, the first excimer laser to be operated ($\lambda = 170$ nm).[16]

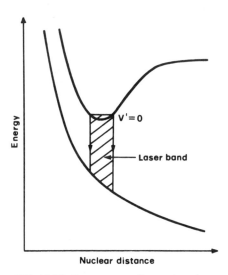

FIG. 10.20. Energy states of an excimer laser.

Excimer lasers have three notable and important properties: (1) Since the transition occurs between different electronic states of a molecule, the corresponding transition wavelength generally falls in the uv spectral region. (2) Once the molecule reaches the ground state, having undergone stimulated emission, it rapidly dissociates due to the repulsive potential of this state. This means that the lower laser level can be considered empty, so the laser operates according to the four-level laser scheme. (3) Due to the lack of energy levels in the ground state, no rotational-vibrational transitions there exist and the transition is observed to be featureless and relatively broad ($\Delta v = 20-100$ cm^{-1}). Note, however, that, in some excimer lasers, the energy curve of the ground state does not correspond to a pure repulsive state, but it features a (shallow) minimum. In this case the transition occurs between an upper bound state and a lower (weakly) bound state (bound-bound transition). However, since the ground state is only weakly bound, a molecule in this state undergoes rapid dissociation either by itself (a process referred to as *predissociation*) or as a result of the first collision with another species of the gas mixture. Thus, in this case also, light emission produces a continuous spectrum.

We now consider a particularly important class of excimer laser in which a rare gas atom (notably Kr, Ar, Xe) is combined, in the excited state, with a halogen atom (notably F, Cl) to form a rare-gas-halide excimer.* Specific examples are ArF ($\lambda = 193$ nm), KrF ($\lambda = 248$ nm), XeF ($\lambda = 351$ nm), and XeCl ($\lambda = 309$ nm); all oscillate in the uv. Rare gas halides are readily formed, in the excited state, when an excited rare gas atom reacts with a ground-state halogen. In fact an excited rare-gas atom becomes chemically similar to an alkali atom, and such an atom is known to react readily with halogens. This analogy also indicates that bonding in the excited state has an ionic character. In the bonding process, the excited electron is transferred from the rare gas atom to the halogen atom. This bound state is also referred to as a charge transfer state.

---

* Strictly speaking these should not be referred to as excimers, since they involve binding between unlike atoms. In fact the word exciplex, a contraction of excited complex, was suggested as more appropriate for this case. However, the word excimer is now widely used in this context, so we follow this usage.

We now consider the KrF laser in some detail, since it represents one of the most important lasers of this category (Fig. 10.21). The upper laser level is an ionically-bound charge-transfer state that, at large internuclear distances ($R \to \infty$), corresponds to the ^2P state of the Kr positive ion and to the ^1S state of the negative F ion. Thus, for large internuclear distances, energy curves of the upper state obey the Coulomb law. The interaction potential between the two ions therefore extends to much greater distances (0.5–1 nm) than those occurring when covalent interactions predominate (compare with, e.g., Fig. 10.19). The lower state is covalently bonded; at large internuclear distances ($R \to \infty$), it corresponds to the ^1S state of the Kr atom and the ^2P state of the F atom. As a result of interaction of the corresponding orbitals, both upper and lower states are split, for short internuclear distances, into the $^2\Sigma$ and $^2\Pi$ states well known from molecular spectroscopy. Laser action occurs on the $B^2\Sigma \to X^2\Sigma$ transition, since it exhibits the largest cross section. Note that, during the transition, the radiating electron transfers from the F$^-$ to the Kr$^+$ ion. Table 10.2 shows some relevant spectroscopic data for this transition.

The two main excitation mechanisms responsible for producing KrF excimers arise from either excited Kr atoms or Kr ions. The route involving excited atoms can be described by the following reactions

$$e + Kr \to e + Kr* \tag{10.2.11a}$$

$$Kr* + F_2 \to (KrF)* + F \tag{10.2.11b}$$

where Kr is first excited by a discharge electron, then reacts with a fluorine molecule. The route involving Kr ions can be described by the following three reactions

$$e + Kr \to 2e + Kr^+ \tag{10.2.12a}$$

$$e + F_2 \to F^- + F \tag{10.2.12b}$$

$$F^- + Kr^+ + M \to (KrF)* + M \tag{10.2.12c}$$

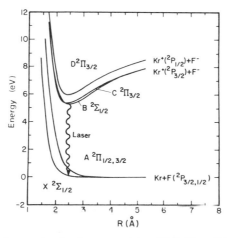

FIG. 10.21. Potential energy diagram showing the energy states of KrF. (From Ref. 17, with kind permission from Elsevier Science-NL, Sara Burgerhartstraat 25, 1055 KV Amsterdam, The Netherlands.)

involving first production of Kr and F ions, then their recombination, in the discharge volume, in the presence of a third partner (usually He, referred to as a buffer gas) to satisfy both energy and momentum conservation. Note that reaction (10.2.12b) is a peculiar one (it is usually referred to as dissociative attachment) and results from the high electron affinity of F atoms. Note also that, due to the long-range interaction between the two reacting ions, reaction (10.2.12c) can proceed very rapidly provided the buffer gas pressure is sufficiently high. Indeed, He partial pressures well above atmospheric pressure are normally used. (A typical gas mixture may contain $\sim$120 mbar of Kr, 6 mbar of F_2, and 2400 mbar of He.) Under this condition, the reaction pathway described by reaction (10.2.12) becomes the dominant mechanism of (KrF)* production.

Since the pressure of the gas mixture is above atmospheric pressure, excimer lasers can be operated only in a pulsed regime, so the general TEA configuration in Fig. 10.17 is used; however, the components of the laser tube and laser flow system must now be compatible with the highly reactive F_2 gas. Furthermore, due to the shorter lifetime of the upper state and to avoid the onset of arc formation, faster pumping is usually provided for excimer lasers compared to TEA CO_2 lasers. (Pump durations of 10–20 ns are typical.) For standard systems, preionization is achieved, as in Fig. 10.17, by a row of sparks. For the largest systems, more complex preionization arrangements are adopted, which use either an auxiliary electron beam or an x-ray source.

Excimer lasers with high repetition rates (up to $\sim$500 Hz) and high average power (up to $\sim$100 W) are commercially available, while larger laboratory systems exist with a higher average power (in excess of a few kilowatts). The efficiency of these lasers is relatively quite high (2–4%) as a result of the high quantum efficiency (see Fig. 10.21) and the high efficiency of the pumping processes.

Excimer lasers are used to ablate plastics as well a biological materials with great precision, since these materials exhibit strong absorption at uv wavelengths. In fact, in some of these materials, the penetration depth for each laser pulse may be only a few micrometers. Due to the strong absorption and short pulse duration, a violent ablation process is produced, wherein the material is directly transformed into volatile components. Applications include drilling very precise holes in thin plastic films (as used, e.g., in the ink jet printer head) and corneal sculpting, to change the refractive power of the eye and hence correct myopia. In the field of lithography, the 248-nm uv light provides a good illumination source for achieving submicron-size features in semiconductor microchips. Excimer lasers can also be used as dye laser pumps, since most dyes absorb strongly in the uv.

10.3. CHEMICAL LASERS

A *chemical laser* is usually defined as one in which the population inversion is *directly* produced by an exothermic chemical reaction.* Chemical lasers usually involve either an associative or a dissociative chemical reaction between gaseous elements.[18,19]

An associative reaction can be described by an equation of the form $A + B \rightarrow AB$. In an exothermic reaction, some of the reaction heat appears as either rotational-vibrational or

---

* According to this definition, the gas dynamic CO_2 laser, briefly considered in Sect. 6.1., is not a chemical laser even though the upper state population arises ultimately from a combustion reaction.

electronic energy of the molecule AB. Thus, if a population inversion is achieved, the associative reaction can, in principle, lead to either a vibrational-rotational or a vibronic transition. In spite of much effort, however, only chemical lasers operating on vibrational-rotational transitions have been demonstrated so far. For this kind of transition, oscillation wavelengths so far range from 3 to 10 μm; the HF and DF lasers considered next are the most notable examples.

A dissociative reaction can be described by an equation of the general form $ABC \rightarrow A + BC$. If the reaction is exothermic, some of the reaction heat can be left as either electronic energy of the atomic species A or as internal energy of the molecular species BC. The most notable example of a laser exploiting this type of reaction is the atomic iodine laser, in which iodine is chemically excited to its $^2P_{1/2}$ state, so laser action occurs between the $^2P_{1/2}$ state and the $^2P_{3/2}$ ground state ($\lambda = 1.315\,\mu$m). Excited atomic iodine can be produced by the exothermic dissociation of CH_3I (or CF_3I, C_3F_7I), the dissociation being produced by means of uv light ($\sim$300 nm) from powerful flash lamps.

More recently, excited iodine was produced by generating excited molecular oxygen from molecular chlorine reacting with hydrogen peroxide. The molecular oxygen, excited to its long-lived singlet state (the ground electronic state of oxygen molecule is a triplet state), in turn transfers its energy to atomic iodine (oxygen-iodine chemical laser).

Chemical lasers are important for two main reasons: (1) They provide an interesting example of direct conversion of chemical energy into electromagnetic energy. (2) They are potentially able to provide either high output power (in cw operation) or high output energy (in pulsed operation). This occurs because the amount of energy available in an exothermic chemical reaction is usually quite large.*

10.3.1 HF Laser

HF chemical lasers can be operated by using either SF_6 or F_2 as compounds to donate the atomic fluorine; from a practical view point, the two lasers are very different.

In commercial devices, the inert SF_6 molecule is used as a fluorine donor, and the gas mixture also contains H_2 and a large amount of He. An electrical discharge is then used to dissociate the SF_6 and excite the reaction. The overall pressure of the mixture is around atmospheric pressure; the laser is pulsed, and the laser configuration is very similar to that of a CO_2 TEA laser. The output energy of this type of device is however appreciably smaller than the input electrical energy. Thus the laser derives only a small part of its output energy from the chemical reaction, so it can be considered only marginally to be a chemical laser.

In an $F_2 + H_2$ chemical laser, a certain amount of atomic fluorine is produced, by dissociation, from fluorine molecules. This atomic fluorine can then react with molecular hydrogen according to the reaction

$$F + H_2 \rightarrow HF^* + H \qquad (10.3.1)$$

which produces atomic hydrogen. This atomic hydrogen can then react with molecular fluorine according to the second reaction:

$$H + F_2 \rightarrow HF^* + F \qquad (10.3.2)$$

---

* For example, a mixture of H_2, F_2, and other substances (16% of H_2 and F_2 in a gas mixture at atmospheric pressure) has a heat of reaction of 2000 J/liter, of which 1000 J is left as vibrational energy of HF (a large value in terms of available laser energy).

Atomic fluorine is then restored, after this second reaction, so this fluorine atom can repeat the same reaction cycle, and so on. We thus have a classical chain reaction that can result in a large production of excited HF molecules. Note that the heats of reaction for Eqs. (10.3.1) and (10.3.2) are 31.6 kcal/mole and 98 kcal/mole, respectively; the two reactions are therefore referred to as the cold and hot reaction, respectively. Note also that, in the case of the cold reaction, the energy released, $\Delta H = 31.6$ kcal/mole, can easily be shown to correspond to an energy of $\Delta H \cong 1.372$ eV for each molecular HF produced. Now, the energy difference between two vibrational levels of HF, corresponding to a transition wavelength of $\lambda \cong 3\,\mu m$, is about $\Delta E \cong 0.414$ eV. So, if all this energy were released as vibrational excitation, vibrationally excited HF molecules, up to the $v = \Delta H/\Delta E \cong 3$ vibrational quantum number, could be produced (see Fig. 10.22a). However, the fraction of the reaction energy that goes into vibrational energy depends on the relative velocity of the colliding partners and on the orientation of this velocity compared to the H-H axis. In a randomly oriented situation, such as in a gas, one can then calculate the fraction of molecules found in the $v = 0, 1, 2,$ or 3 vibrational states, respectively. The relative numbers $N(v)$ of excited HF molecules, in this way obtained, are also indicated in the same figure. For instance, 5 out 18 molecules are found in the $v = 3$ state and thus take up almost all the available energy as vibrational energy. On the other hand, 1 out of 18 molecules are found in the ground ($v = 0$) vibrational state and, in this case, all the reaction enthalpy is found as kinetic energy of the reaction products (mostly H, since this is the lightest product). From the figure we see that, if this were the only reaction, a population inversion would be established, particularly for the $(v = 2) \rightarrow (v = 1)$ transition.

In the case of the hot reaction, excited HF up to the ($v = 10$) vibrational level can be produced (Fig. 10.22b). The relative populations of these vibrational levels $N(v)$ can also be calculated and are shown in the same figure. We see that a strong population inversion exists particularly for the $(v = 5) \rightarrow (v = 4)$ transition. Based on preceding considerations, we can

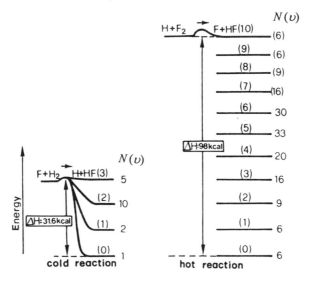

FIG. 10.22. Pumping vibrational levels of the HF molecule by the two reactions (a) $F + H_2 \rightarrow HF^* + H$ and (b) $H + F_2 \rightarrow HF^* + F$. The relative populations $N(v)$ of each vibrational state of quantum number v are also indicated in the two figures.

easily calculate that, e.g., for the cold reaction (10.3.1), more than 60% of the reaction energy is, on average, released as vibrational energy. To understand why the HF molecule is left in an excited vibrational state after a chemical reaction, consider an F atom colliding with a H_2 molecule. As a result of the high electron affinity of a fluorine atom, the interaction is strongly attractive, which leads to considerable polarization of the H_2 charge distribution, even at large F-H_2 distances. As a consequence of the electron's low inertia, an electron can then be transferred to the fluorine atom from the nearest hydrogen atom, hence forming the HF ionic bond, before the spacing between hydrogen and fluorine has adjusted to the internuclear separation corresponding to the HF equilibrium distance. This classical picture indicates that, after a reaction has occurred, the HF molecule is left in an excited vibrational state.

We can see from the preceding discussion that, as a consequence of the combined effect of cold and hot reactions described by Eqs. (10.3.1) and (10.3.2), a population inversion between several vibrational levels of HF occurs. If the active medium is placed in a suitable resonator, laser action on a number of transitions from vibrationally excited HF molecules is thus expected. Laser action has indeed been observed on several rotational lines from the $(v = 1) \rightarrow (v = 0)$ transition up to the $(v = 6) \rightarrow (v = 5)$ transition. In fact, due to the anharmonicity of the interaction potential, the vibrational energy levels in Fig. 10.22 are not equally spaced, so the laser spectrum actually consists of many roto-vibrational lines encompassing a rather wide spectral range ($\lambda = 2.7$–3.3 μm). It should be noted that the number of observed laser transitions is larger than expected according to the population inversion in Fig. 10.22. As discussed in the case of a CO laser, there are two reasons why oscillation can occur on so many lines: (1) The phenomenon of cascading: If, in fact, the $(v = 2) \rightarrow (v = 1)$ transition (usually the strongest one) lases, the population will be depleted from level 2 and will accumulate in level 1. Consequently, laser action on the $(v = 3) \rightarrow (v = 2)$ and $(v = 1) \rightarrow (v = 0)$ transitions now becomes more favored. (2) The phenomenon of partial inversion, according to which there may be a population inversion between some rotational lines even when no inversion exists between overall populations of the corresponding vibrational levels. Finally, it should be noted that, besides laser action in HF, laser action can also be achieved in the analogous compounds DF, HCl, and HBr, thus providing oscillation on a large number of transitions in the 3.5–5 μm range.

Figure 10.23 shows a possible configuration for a high-power cw HF or DF laser. Fluorine is thermally dissociated by an arc jet heater, then expanded to supersonic velocity ($\sim$Mach 4) through appropriate expansion nozzles. Molecular hydrogen is then mixed downstream through appropriate perforated tubes inserted in the nozzles. Downstream in the

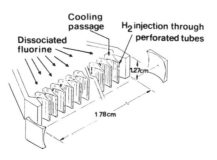

FIG. 10.23. Supersonic diffusion HF (or DF) chemical laser. (By permission from Ref. 18).

expansion regions, excited HF molecules are produced by chain reactions (10.3.1) and (10.3.2); a suitable resonator, with its axis orthogonal to the flow direction, is placed around this region. To handle the large power available in the expanding beam (usually of large diameter) unstable resonators exploiting metallic, water-cooled, mirrors are often used. Chemical lasers of this type can produce very large cw output powers (in the megawatt range!) with good chemical efficiency.

Pulsed TEA-type HF lasers are commercially available, and these have found a limited use when an intense source of middle-infrared radiation is needed (e.g., in spectroscopy). HF and DF chemical lasers of the type shown in Fig. 10.23 are used exclusively for military applications. Safety considerations have prevented the use of these lasers for commercial applications. In fact the F_2 molecule is one of the most corrosive and reactive elements known; its waste products are difficult to dispose of and, under certain conditions, the chain reaction (10.3.1) and (10.3.2) may even become explosive. In the military field, due to the large available output powers, these lasers can be used as directed energy weapons to, e.g., destroy enemy missiles. A military cw device named MIRACL (Mid-Infrared Advanced Chemical Laser), using DF, has given the largest cw power of any laser (2.2 MW). DF rather than HF was used because the system was intended for use from a ground station and DF emission wavelengths fall in a region of relatively good atmospheric transmission. High-power hydrogen-fluoride or, more likely, oxygen-iodine chemical lasers are under active consideration for use from either a high-altitude plane to destroy missiles during ballistic flight or from a space station to destroy missiles during their lift-off phase (when the rocket has lower speed and hence is more vulnerable).

10.4. FREE-ELECTRON LASERS

In free-electron lasers (FELs) an electron beam moving at a speed close to the speed of light passes through the magnetic field generated by a periodic structure (called the wiggler or undulator; see Fig. 10.24).[20] The stimulated emission process occurs through the interaction of the em field of the laser beam with these relativistic electrons moving in the periodic magnetic structure. As in any other laser, two end mirrors provide feedback for laser oscillation. Using suitable bending magnets, the electron beam is first injected into the laser cavity and then deflected out of the cavity.

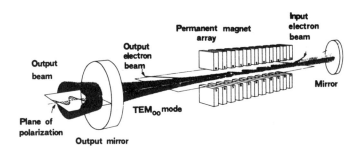

FIG. 10.24. Basic structure of a free-electron laser. (Courtesy of Luis Elias, University of California, at Santa Barbara Quantum Institute.)

To understand how this interaction occurs, we first consider the case of spontaneously emitted radiation, i.e., when no mirrors are used. Once injected into the periodic structure, the electrons acquire a wiggly, or undulatory, motion in the plane orthogonal to the magnetic field (Fig. 10.24). The resulting electron acceleration produces a longitudinal emission of the synchrotron radiation type. The frequency of the emitted radiation can be derived, heuristically, by noting that the electron oscillates in the transverse direction at an angular frequency $\omega_q = (2\pi/\lambda_q)v_z \cong (2\pi/\lambda_q)c$, where λ_q is the magnet period and v_z is the (average) longitudinal velocity of the electron (which is almost equal to the vacuum light velocity c). Let us now consider a reference frame moving longitudinally at velocity v_z. In this frame, the electron oscillates essentially in the transverse direction; thus it looks like an oscillating electric dipole. In this reference frame, due to the Lorentz time contraction, the oscillation frequency is given by

$$\omega' = \frac{\omega_q}{[1 - (v_z/c)^2]^{1/2}} \tag{10.4.1}$$

and this is therefore the frequency of the emitted radiation. If we now go back to the laboratory frame, the radiation frequency undergoes a (relativistic) Doppler shift. The observer frequency ω_0 and the corresponding wavelength λ_0 are then given by

$$\omega_0 = \frac{1 + (v_z/c)}{[1 - (v_z/c)^2]^{1/2}} \omega' \cong \frac{2\omega_q}{1 - (v_z/c)^2} \tag{10.4.2}$$

and

$$\lambda_0 = \frac{\lambda_q}{2}\left[1 - \left(\frac{v_z}{c}\right)^2\right] \tag{10.4.3}$$

respectively. Note that, since $v_z \cong c$, λ_0 is generally much smaller than the magnet period. To calculate the quantity $[1 - (v_z/c)^2]$ in Eqs. (10.4.2) and (10.4.3), we recall that, for a completely free electron moving with velocity v_z along the z-axis, one has $[1 - (v_z/c)^2] = (m_0c^2/E)^2$, where m_0 is the rest mass of the electron and E its energy. However, for a given total energy, the wiggling motion reduces the value of v_z, i.e., it increases the value of $[1 - (v_z/c)^2]$. A detailed calculation then shows that this quantity is given by

$$1 - \left(\frac{v_z}{c}\right)^2 = (1 + K^2)\left(\frac{m_0c^2}{E}\right)^2 \tag{10.4.4}$$

where the numerical constant K, which is usually smaller than 1, is referred to as the undulator parameter. Its value is obtained from the expression $K = e\langle B^2 \rangle^{1/2}\lambda_q/2\pi m_0c^2$, where B is the magnetic field of the undulator and the average is taken along the longitudinal direction. From Eqs. (10.4.2)–(10.4.4), we obtain

$$\omega_0 = \frac{4\pi c}{\lambda_q}\left(\frac{1}{1 + K^2}\right)\left(\frac{E}{m_0c^2}\right)^2 \tag{10.4.5}$$

and

$$\lambda_0 = \frac{\lambda_q}{2} \left(\frac{m_0 c^2}{E} \right)^2 (1 + K^2) \tag{10.4.6}$$

which shows that the emission wavelength can be changed by changing the magnet period λ_q and/or the energy E of the electron beam. Assuming, as an example, $\lambda_q = 10$ cm and $K = 1$, we find that the emitted light can range from the infrared to the ultraviolet by changing the electron energy from 10^2 to 10^3 MeV. Note that, according to our earlier discussion, the emitted radiation is expected to be polarized in the plane orthogonal to the magnetic field direction (see also Fig. 10.24).

To calculate the spectral lineshape and the bandwidth of the emitted radiation, we observe that, in the reference frame previously considered, the electron emission lasts for a time $\Delta t' = (l/c)[1 - (v_z/c)^2]^{1/2}$, where l is the overall length of the wiggler magnet. Using Eq. (10.4.1) we then see that emitted radiation from each electron consists of a square pulse containing a number of cycles $N_{cyc} = \omega' \Delta t' / 2\pi = l/\lambda_q$, i.e., equal to the number of periods $N_w = l/\lambda_q$ of the wiggler. From standard Fourier-transform theory, it then follows that such a pulse has a power spectrum of the $[\sin(x/2)/(x/2)]^2$ form, where $x = 2\pi N_w (v - v_0)/v_0$. The spectral width Δv_0 (FWHM) is then approximately given by the relation:

$$\frac{\Delta v_0}{v_0} = \frac{1}{2N_w} \tag{10.4.7}$$

Figure 10.25a shows this spectrum as a function of the dimensionless quantity x. Since all electrons, injected with the same velocity and in the same direction, show the same lineshape, Eq. (10.4.7) gives the homogeneous linewidth for the FEL laser. Inhomogeneous effects arise from such factors as spread in electron energy, angular divergence of the electron beam, and variation in magnetic field over the beam cross section. Note that, since the number of undulator periods may typically be $N_w \sim 10^2$, from Eq. (10.4.7) one has $\Delta v_0/v_0 \cong 5 \times 10^{-3}$.

It is worthwhile to point out that there is an alternative way of considering the behaviour of the emitted rediation. In the rest frame of the electron, that we considered earlier, the magnetic field of the undulator appears to move at nearly the velocity of light. It can be shown that the static magnetic field then appears to the electrons essentially like a counter-propagating em wave. The synchrotron emission can therefore be considered to arise from Compton backscattering of this virtual em wave from the electron beam. For this reason, the corresponding type of FEL is sometimes referred to as operating in the Compton regime (Compton FEL).

A calculation of the stimulated emission cross section requires a detailed analysis, which we do not consider here, of the interaction between a longitudinally propagating em wave and the electron propagating in the wiggler magnetic field. We merely point out that, unlike the situation previously considered for other lasers, the spectral distribution of this cross section is not the same as that for spontaneously emitted radiation; instead it is proportional to its frequency derivative. Accordingly, its shape is as shown in Fig. 10.25b, with gain on the low-frequency side and loss on the high-frequency side of the transition. This unusual behavior essentially arises because the interaction is based on a light-scattering process rather than absorption or emission from bound states.

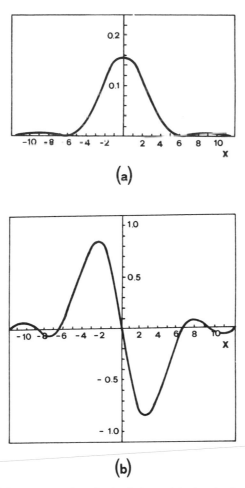

FIG. 10.25. Spectrum (a) of the spontaneously emitted radiation and (b) the stimulated emission cross section in a free-electron laser as a function of the normalized quantity $z = 2\pi N_w(v - v_0)/v_0$.

So far, demonstrations of FEL operation have been made on several devices (more than 10) around the world, with oscillation wavelengths ranging from millimeter waves to the uv region. All of these lasers require large facilities, since they involve using rather large e-beam accelerators. Historically, the first FEL was operated at $\lambda = 3.4$ μm using the Stanford University superconducting linear accelerator.[21] Since the incoming e-beam consisted of 3.2-ps pulses separated by $\tau = 84.7$ ns, the cavity length L was chosen so that τ equals the cavity round-trip time (i.e., $L = c\tau/2 = 12.7$ m); thus, the laser was operated in the synchronously mode-locked regime.

One of the most important issues for a FEL is related to its efficiency. Since the emitted frequency depends on electron energy [see Eq. (10.4.5)], the maximum energy which can be extracted from the electron is that which shifts its energy so that the corresponding operating frequency falls outside the gain curve. Consequently the maximum efficiency η_{max}, defined as the ratio between the maximum energy given to the laser beam and the initial electron energy, is approximately given by $\Delta v_0 / v_0$, i.e., $\eta_{max} = (1/2N_w)$. This means that the efficiency in such a device is rather limited (10^{-2}–10^{-3}). Two ways of obtaining higher efficiency are being actively pursued: (1) The period of the magnet is gradually decreased along the e-beam direction to keep the λ_q / E^2 ratio constant (tapered wiggler). (2) The energy remaining in the electron beam, after leaving the wiggler, is recovered by decelerating the electrons. Much higher efficiencies are predicted to be achievable using these approaches, and indeed they have been achieved to some degree.

As a final comment we point out that all the FELs described so far use high-energy ($E > 10$ MeV) and low-current ($I \sim 1$–100 A) e-beam machines. Under these conditions, as previously discussed, light emission can be described as arising from Compton scattering of the virtual quanta of the magnetic field from individual electrons (*Compton regime FEL*). Free electron lasers using e-beams of lower energy ($E = 1$–2 MeV) and much higher currents ($I \sim 10$–20 kA) have also been operated. In this case the electron-electron interaction becomes so strong that collective oscillatory motions (plasma waves) are induced in the e-beam when interacting with the em wave in the wiggler. The emission can then be considered to arise from scattering of the virtual quanta of the magnetic field from these collective motions, rather than from single electrons. The emitted frequency $v_0 = 2\pi / \omega_0$ is then no longer given by Eq. (10.4.5), but in fact it is downshifted by the frequency of this collective motion. The phenomenon is analogous to Raman scattering of light from molecular vibrations; the corresponding laser is said to operate in the *FEL-Raman regime*. Because of the lower value of electron energy involved, these lasers oscillate in the millimeter wave region.

To conclude this section we list the most attractive properties of FELs: Wide tunability, excellent beam quality (close to the diffraction limit), and potentially very high efficiency and thus very high laser power (average power of the e-beam of the Stanford Linear Accelerator is about 200 kW). Free-electron lasers are however inherently large and expensive machines; interest in their applications is thus likely to be strongest in frequency ranges where more conventional lasers are not so readily available, e.g., the far-infrared (100–400 μm) or vacuum uv ($\lambda < 100$ nm).

10.5. X-RAY LASERS

The achievement of coherent oscillation in the x-ray region was a dream that is slowly but steadily coming true.[22] The potential applications of x-ray lasers are indeed very important. They include, in fact, such possibilities as: (1) X-ray holography or x-ray microscopy of, e.g., living cells or cell constituents, allowing, respectively, three-dimensional or two-dimensional pictures with subnanometer resolution to be obtained. (2) X-ray lithography, where patterns with extremely high resolution could be produced.

Before discussing what has been achieved so far in this wavelength region, we indicate the difficulties that must be overcome to obtain x-ray laser operation. Starting with fundamental considerations, we recall that the threshold pump power of a four-level laser is given by Eq. (7.3.12):

$$P_{th} = \frac{h v_{mp}}{\eta_p} \frac{\gamma A}{\sigma \tau}$$ (10.5.1)

The minimum threshold, P_{mth} is of course attained for $\sigma = \sigma_p$ where σ_p is the cross section at the transition peak. Furthermore, one must also take into account that, in the x-ray region, the upper state lifetime τ is established by the spontaneous lifetime τ_{sp}. From Eqs. (2.4.29) and (2.3.15) we then obtain $1/\sigma_p\tau_{sp} \propto v_0^2/g_t(0)$, independent of the transition matrix element $|\mu|$. For either Eq. (2.4.9a) (homogeneous line) or Eq. (2.4.28) (inhomogeneous line) we find that $g_t(0) \propto 1/\Delta v_0$, where Δv_0 stands for transition linewidth for either a homogeneous or inhomogenous line. Thus, in either case, from Eq. (10.5.1), with $hv_{mp} \cong hv_0$, we obtain $P_{mth} \propto v_0^3\Delta v_0$. At frequencies in the vuv to soft x-ray region and at moderate pressures, we assume that the linewidth is dominated by Doppler broadening. Hence [see Eq. (2.5.18)] one has $\Delta v_0 \propto v_0$ and P_{mth} is expected to increase as v_0^4. At the higher frequencies, corresponding to the x-ray region, linewidth is dominated by natural broadening, since the radiative lifetime becomes very short (down to the femtosecond region). In this case one has $\Delta v_0 \propto 1/\tau_{sp} \propto v_0^3$ and P_{mth} is expected to increase as v_0^6. Thus, if we go from, e.g., the green ($\lambda = 500$ nm) to the soft x-ray region ($\lambda \cong 10$ nm), the wavelength decreases by a factor of 50 and P_{mth} is expected to increases by many orders of magnitude!

From a technical viewpoint, a major difficulty arises from the fact that multilayer dielectric mirrors for the x-ray region are lossy and difficult to make. A basic problem is that the difference in refractive index between various materials becomes very small in this region. Dielectric multilayers with a large number of layers (hundreds) are therefore needed to achieve reasonable reflectivity. Light scattering at the many interfaces then makes the mirrors very lossy; furthermore, the mirrors have difficulty withstanding the high intensity of an x-ray laser beam. For these reasons, so far, x-ray lasers have been operated without mirrors as ASE devices.

As a representative example, we consider the soft x-ray laser based on 24-times ionized selenium (Se^{24+}) as the active medium.[23] In fact, this is the first laser of a kind (the *x-ray recombination laser*) which now includes a large number of other multiply-ionized species as laser media. Pumping was achieved by the powerful second-harmonic beam ($\lambda = 532$ nm) of the Novette laser (pulse energy ~1 kJ, pulse duration ~1 ns), consisting of one arm of the Nova laser at Lawrence Livermore Laboratory in the United States. The beam is focused to a fine line ($d \cong 200$ μm, $l = 1.2$ cm) on a thin stripe (75 nm thick) of selenium, evaporated on a 150-nm-thick foil of Formvar (Fig. 10.26). The foil can be irradiated from one or both sides. Exposed to the high intensity of this pump beam (~5×10^{13} W/cm^2), the foil

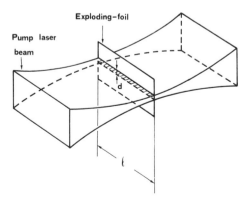

FIG. 10.26 Transverse irradiation geometry of a soft x-ray laser using the exploding-foil technique.

explodes to form a highly ionized, approximately cylindrical, selenium plasma whose diameter is $d \cong 200\,\mu\text{m}$. During the electron-ion recombination process, a particulary long-lived constituent of this plasma, is formed, which consists of $Se^{24\,+}$. This ion has the same ground-state electronic configuration as neutral Ne ($1s^2 2s^2 2p^6$; see Fig. 10.1); accordingly, it is usually referred to as neon-like selenium. Impact collisions with hot-plasma electrons ($T_e \cong 1\,\text{keV}$) then raise $Se^{24\,+}$ from its ground state to excited states, thereby achieving population inversion between the states $2p^5 3p$ and $2p^5 3s$ because the lifetime of the $3s \to$ ground-state transition is much shorter than the lifetime of the $3p \to 3s$ transition. (Both transitions are electric-dipole allowed.) With the pump configuration in Fig. 10.26, a strong longitudinal emission due to ASE is observed on two lines ($\lambda_1 = 20.63\,\text{nm}$ and $\lambda_2 = 20.96\,\text{nm}$) of the $2p^5 3p \to 2p^5 3s$ transition (see Fig. 10.1). Due to the much higher nuclear charge of Se compared to Ne, these lines fall in the soft x-ray region. From the length dependence of the emitted energy, one deduces that a maximum single-pass gain, $G = \exp(\sigma_p N l)$, of about 700 was obtained. Note that this gain is still well below the threshold for ASE, as defined in Sect. 2.9.2. In fact, for the experimental situation described here, the emission solid angle is $\Omega \cong 10^{-4}$ sterad and the linewidth is still expected to be dominated by Doppler broadening. From Eq. (2.9.4b) we then obtain $G_{th} \cong 4.5 \times 10^5$. This means that the emitted intensity, due to ASE, is still much smaller than the saturation intensity of the amplifier. Indeed, the x-ray output energy produced was an extremely small fraction ($\sim 10^{-10}$) of the pump energy.

Since this first laser was demonstrated, research activity in this field has been very strong.[24] Thus many new active media have been operated, namely many more neon-like ions (from $Ag^{37\,+}$ to $Ar^{8\,+}$), many hydrogen-like ions (from $Al^{12\,+}$ to $C^{5\,+}$), lithium-like ions (from $Si^{11\,+}$ to $Al^{10\,+}$), and nickel-like ions (from $Au^{51\,+}$ to $Eu^{35\,+}$). The range of oscillation wavelengths now extends from ~ 3.6 to $47\,\text{nm}$, while the single-pass gain G generally ranges from 10 to 10^3. To achieve the high peak powers required [see Eq. (10.5.1)] and at the same time reduce the required pump energy, picosecond or even femtosecond laser pulses are now often used for pumping. Amplified spontaneous emission (at $\lambda = 46.9\,\text{nm}$) has also been observed in neonlike $Ar^{8\,+}$ by passing a strong current pulse, of short duration, through a 1–$10\,\text{cm}$ long capillary filled with Ar.

10.6. CONCLUDING REMARKS

Chapter 10 considers the most notable examples of lasers involving low-density media. In general these lasers tend to be bulkier and often less efficient than lasers considered in Chapter 9 (notably semiconductor and diode-pumped solid-state lasers). For these reasons, whenever possible, such low-density lasers are tending to be superseded by their solid-state laser counterparts. Such is the case for the Argon laser, which faces strong competition from the green beam emitted by a diode-pumped Nd laser (e.g., Nd:YVO$_4$) with intracavity second-harmonic generation. For quite a few applications, this is also the case of the red-emitting He-Ne laser, which faces strong competition from red-emitting InGaAlP semiconductor lasers. Low-density lasers will however survive in frequency ranges not covered effectively by semiconductor or diode-pumped solid-state lasers. This is, for instance, the case for middle-infrared lasers (the CO_2 lasers being the most notable example), and lasers oscillating in the uv (e.g., excimer lasers) and x-ray spectral regions.

Another field of applications where low-density lasers will continue to perform well is where very high powers are required, with CO_2 lasers, excimer lasers, and chemical lasers as important examples. Thus, as a conclusion, one can foresee that lasers based on low-density media will continue to play an important role in the laser field.

PROBLEMS

10.1. List at least four lasers, using a low-density active medium, whose wavelengths fall in the infrared.

10.2. List at least four lasers, using a low-density active medium, whose wavelengths fall in the uv to soft x-ray region. What problems are faced in achieving laser action in the uv or x-ray region?

10.3. Metal-working applications require a laser with a cw output >1 kW. Which lasers meet this requirement?

10.4. The 514.5-nm transition of an argon ion laser is Doppler-broadened to a linewidth of ~3.5 GHz. The cavity length of the laser is 100 cm; when pumped three times above threshold, the laser emits a power of 4 W in a TEM_{00} mode profile. Assuming that the frequency of one of the oscillating TEM_{00} modes coincides with the center of the gain line, calculate the number of TEM_{00} modes expected to oscillate.

10.5. Consider the argon ion laser described in Problem 10.4 and assume that the laser is mode-locked by an acoustooptic modulator. Calculate: (a) The duration and peak power of the mode-locked pulses; (b) the drive frequency of the rf oscillator.

10.6. Assume that the bond between the two nitrogen atoms of the N_2 molecule can be simulated by a spring of suitable elastic constant. Knowing the vibrational frequency (Fig. 10.9) and the atomic mass, calculate the elastic constant. Compare this constant with that obtainable from the ground-state curve in Fig. 10.19.

10.7. Show that if the elastic constant of the N-N bond is the same as that of the isoelectronic CO molecule, the $(v' = 1) \rightarrow (v = 0)$ transition wavelengh of the N_2 molecule is approximately the same as that of the CO molecule.

10.8. Assume that each of the two oxygen-carbon bonds of the CO_2 molecule can be simulated by a spring with elastic constant k. Assuming there is no interaction between the two oxygen atoms and knowing the v_1 frequency ($v_1 = 1337$ cm^{-1}), calculate this constant.

10.9. Knowing the elastic constant k between the two oxygen carbon bonds obtained in Problem 10.8, calculate the expected frequency v_3 of the asymmetric stretching mode and compare the result with the value shown in Fig. 10.9.

10.10. Show that each C-O bond of the CO_2 molecule cannot be simulated by elastic springs if the harmonic oscillation corresponding to the bending mode of frequency v_2 has to be calculated.

10.11. Knowing that, for a Boltzmann distribution, the maximum population of the upper laser level of a CO_2 molecule occurs for the rotational quantum number $J' = 21$ (see Fig. 10.11), calculate the rotational constant B [assume $T = 400$ K, which corresponds to an energy kT such that $(kT/h) \cong 280$ cm^{-1}]. From this value calculate the equilibrium distance between the C atom and each O atom.

10.12. Using the result from Problem 10.11, calculate the frequency spacing (in cm^{-1}) between the rotational lines of the CO_2 laser transition. (Assume that the rotational constant of the lower laser level is the same as that of the upper laser level and remember that only levels with odd values of J are occupied in a CO_2 molecule.)

10.13. The linewidth due to collision broadening of the CO_2 laser transition is given by $\Delta v_c = 7.58$ $(\psi_{CO_2} + 0.73\psi_{N_2} + 0.6\psi_{He})p(300/T)^{1/2}$ MHz where ψ are fractional partial pressures of the gas mixture, T is the gas temperature, and p is the total pressure (in torr) (see Example 3.3). Taking the ratio of the partial pressures of CO_2, N_2, and He molecules as 1:1:8 and assuming a separation between rotational lines of the CO_2 laser transition of $\Delta v_r \cong 2$ cm^{-1}, calculate the total gas pressure needed to make all rotational lines merge together. What is the width of the gain curve?

10.14. Consider a CO_2 laser with high enough pressure to have all its rotational lines merged together. If this laser is mode-locked, what is the order of magnitude of the corresponding laser pulse width?

10.15. Show that a reaction energy of 31.6 kcal/mole, as in the HF cold reaction [see Eq. (10.3.1)], is equivalent to an energy of 1.372 eV released for each molecular reaction.

10.16. Considering the cold reaction in Fig. 10.22a, take the values shown in the figure for the relative populations of HF molecules that are left in the first three vibrational levels after reaction. Calculate the fraction η of heat released in the reaction that goes into vibrational energy.

10.17. Repeat the previous calculation for the hot reaction in Fig. 10.22b.

10.18. To reach the final end products of the cold reaction [see Eq. (10.3.1)], we can choose a reaction path where we first dissociate the H_2 molecule to obtain the single atoms F, H, and H. Then we let the fluorine and one hydrogen atom recombine together. Similarly, for the hot reaction [see Eq. (10.3.2)], we first dissociate molecular fluorine, then recombine one fluorine with the hydrogen atom. Given the two possibilities, relate the difference in reaction heat from these two reactions to the difference in dissociation energy in the fluorine and hydrogen molecules.

REFERENCES

1. R. Arrathoon, Helium-Neon Lasers and the Positive Column, in *Lasers*, vol. 4 (A. K. Levine and A. J. De Maria, eds.) (Marcel Dekker, New York, 1976), Chap. 3.
2. W. B. Bridges, Atomic and Ionic Gas Lasers, in *Methods of Experimental Physics*, vol. 15 (C. L. Tang, ed.) (Academic, New York, 1979), pp. 33–151.
3. A. Javan, W. R. Bennett, and D. H. Herriott, Population Inversion and Continuous Optical Maser Oscillation in a Gas Discharge Containing a He-Ne Mixture, *Phys. Rev. Lett*, **6** (1961).
4. C. E. Webb, Metal Vapor Lasers, Recent Advances and Applications, in *Gas Flow and Chemical Lasers*, Springer Proceedings in Physics N. 15 (S. Rosenwork ed.), (Springer-Verlag, Berlin, 1987), pp. 481–494.
5. C. C. Davis and T. A. King, Gaseous Ion Lasers, in *Advances in Quantum Electronics*, vol. 3 (D. W. Goodwin, ed.) (Academic, New York, 1975), pp. 170–437.
6. D. H. Dunn and J. N. Ross, Argon Ion Laser, in *Progress in Quantum Electronics*, vol. 4 (J. H. Sanders and S. Steinholm, eds.) (Pergamon, London, 1977), pp. 233–270.
7. W. B. Bridges, Laser Oscillation in Singly Ionized Argon in the Visible Spectrum, *Appl. Phys. Letters* **4**, 128 (1964).
8. P. K. Cheo, CO_2 Lasers, in *Lasers*, Vol. 3 (A. K. Levine and A. J. De Maria, eds.) (Marcel Dekker, New York, 1971), Chap. 2.
9. A. J. De Maria, Review of High-Power CO_2 Lasers, in *Principles of Laser Plasma*, (G. Bekefi, ed.) (Wiley, New York, 1976), Chap. 8.

10. C. K. N. Patel, W. L. Faust, and R. A. McFarlane, CW Laser Action on Rotational Transitions of the $\Sigma_u^+ \rightarrow \Sigma_g^+$ Vibrational Band of CO_2, *Bull. Am. Phys. Soc.* **9**, 500 (1964).

11. D. R. Hall and C. A. Hill, Radiofrequency-Discharge-Excited CO_2 Lasers, in *Handbook of Molecular Lasers* (P. Cheo, ed.) (Marcel Dekker, New York, 1987), Chap. 3.

12. K. M. Abramski, A. D. Colley, H. J. Baker, and D. R. Hall, Power Scaling of Large-Area Transverse Radiofrequency Discharge CO_2 Lasers, *Appl. Phys. Letters* **54**, 1833 (1989).

13. P. E. Jackson, H. J. Baker, and D. R. Hall, CO_2 Large-Area Discharge Laser Using an Unstable Waveguide Hybrid Resonator, *Appl. Phys. Letters* **54**, 1950 (1989).

14. R. E. Center, High-Power, Efficient Electrically Excited CO Laser, in *Laser Handbook*, vol. 3 (M. L. Stitch, ed. (North-Holland, Amsterdam, 1979), pp. 89–133.

15. C. S. Willet, *An Introduction to Gas Lasers: Population Inversion Mechanisms* (Pergamon, Oxford, UK, 1974), Sects. 6.2.1, 6.2.3.

16. N. G. Basov, V. A. Danilychev, and Y. M. Popov, Stimulated Emission in the Vacuum Ultraviolet Region, *Soviet J. Quantum Electrons.* **1**, 18 (1971).

17. J. J. Ewing, Excimer Lasers, in *Laser Handbook*, vol. 3 (M. L. Stitch, ed.) (North-Holland, Amsterdam, 1979), pp. 135–97.

18. A. N. Chester, Chemical Lasers, in *High-Power Gas Lasers* (E. R. Pike, ed.) (Institute of Physics, Bristol and London, 1975), pp. 162–221.

19. C. J. Ultee, Chemical and Gas Dynamic Lasers, in *Laser Handbook*, vol. 3 (M. L. Stitch and M. Bass, eds.) (North-Holland, Amsterdam, 1985), pp. 199–287.

20. G. Dattoli and R. Renieri, Experimental and Theoretical Aspects of the Free-Electron Lasers, in *Laser Handbook*, vol. 4 (M. L. Stitch, ed.) (North-Holland, Amsterdam, 1979), pp. 1–142.

21. D. A. G. Deacon, L. R. Elias, J. M. J. Madey, G. J. Ramian, H. A. Schwettman, and T. I. Smith, First Operation of a Free-Electron Laser, *Phys. Rev. Lett.* **38**, 892 (1977).

22. R. C. Elton, *X-Ray Lasers* (Academic, Boston, 1990).

23. D. L. Matthews *et al.*, Demonstration of a Soft X-Ray Amplifier, *Phys. Rev. Lett.* **54**, 110 (1985).

24. *X-Ray Lasers 1996*, (S. Svanberg and C. G. Wahlstrom, eds.), Institute of Physics Conference Series N. 151 (Institute of Physics, Bristol 1996).

11

Properties of Laser Beams

11.1. INTRODUCTION

In Chapter 1 it was stated that the most characteristic properties of laser beams are: (1) Monochromaticity; (2) coherence (spatial and temporal); (3) directionality; (4) brightness. The material presented in earlier chapters now allows us to examine these properties in more detail and compare them with the properties of conventional light sources (thermal sources).

In most cases of interest to us, the spectral bandwidth of the light source $\Delta\omega$ is much smaller than the mean frequency $\langle\omega\rangle$ of the spectrum (*quasi-monochromatic wave*). In this case, the electric field of the wave, at position $\mathbf{r}$ and time t, can be written as

$$E(\mathbf{r}, t) = A(\mathbf{r}, t) \exp j[\langle\omega\rangle t - \phi(\mathbf{r}, t)] \qquad (11.1.1)$$

where $A(\mathbf{r}, t)$ and $\phi(\mathbf{r}, t)$ are both slowly varying over an optical period, i.e.:

$$\left(\left| \frac{\partial A}{A \partial t} \right|, \left| \frac{\partial \phi}{\partial t} \right| \right) \ll \langle\omega\rangle \qquad (11.1.2)$$

We then define the intensity of the beam as:

$$I(\mathbf{r}, t) = E(\mathbf{r}, t) E^*(\mathbf{r}, t) = |A(\mathbf{r}, t)|^2 \qquad (11.1.3)$$

11.2. MONOCHROMATICITY

We saw in Sects. 7.9–7.11 that frequency fluctuations in a cw single-mode laser mostly arise from phase fluctuations rather than amplitude fluctuations. Amplitude fluctuations arise in fact from pump or cavity loss fluctuations. They are usually very small ($\sim 1\%$) and they can be further reduced by suitable electronically-controlled feedback loops. To first order we

463

can thus take $A(t)$ to be constant for a single-mode laser and consider its degree of monochromaticity to be determined by frequency fluctuations. The theoretical limit to this monochromaticity arises from zero-point fluctuations and is expressed by Eq. (7.9.2). However, this limit generally corresponds to a very low value for the oscillating bandwidth Δv_L, which is seldom reached in practice. In Example 7.9, for instance, the value of Δv_L is calculated to be ~ 0.4 mHz for a He-Ne laser with 1 mW output power. A notable exception occurs for semiconductor lasers, where, due to the short length and high loss of the laser cavity, this limit is much higher ($\Delta v_L \cong 1$ MHz) and actual laser linewidth is often determined by these quantum fluctuations. In most other cases, technical effects, such as vibrations and thermal expansion of the cavity, determine the laser linewidth Δv_L. If a monolithic structure is used for the laser cavity configuration, as in the nonplanar ring oscillator in Fig. 7.26, typical values for Δv_L may fall in the 10–50 kHz range. Much smaller linewidths (down to ~ 0.1 Hz) can be obtained by stabilizing the laser frequency against an external reference, as discussed in Sect. 7.10. In pulsed operation, the minimum linewidth is obviously limited by the inverse of the pulse duration τ_p. For example, in a single-mode Q-switched laser, assuming $\tau_p \cong 10$ ns, one obtains $\Delta v_L \cong 100$ MHz.

In a laser oscillating on many modes, monochromaticity is obviously related to the number of oscillating modes. For example, in mode-locked operation, pulses down to a few tens of femtoseconds can be obtained. In this case, the corresponding laser bandwidth is in the range of a few tens of THz, so the condition for quasi-monochromatic radiation no longer holds well.

The degree of monochromaticity required depends of course on the given application. The narrowest laser linewidths are needed for only the most sophisticated applications in metrology and fundamental measurements in physics (e.g., gravitational wave detection). For more common applications, such as interferometric measurements of distances, coherent laser radar, and coherent optical communications, the required monochromaticity falls in the 10–100 kHz range. A monochromaticity of ~ 1 MHz is typically needed for much of high-resolution spectroscopy, and it certainly suffices for optical communications using WDM. For some applications, of course, laser monochromaticity is not relevant; this is the case in important applications such as laser material working and most applications in the biomedical field.

11.3. FIRST-ORDER COHERENCE

In Chapter 1 we introduced the coherence of an em wave in an intuitive fashion, distinguishing two types of coherence, spatial and temporal.[1] In this section we elaborate on these two concepts. In fact, as it will be appreciated by the end of this chapter, it turns out that spatial and temporal coherence describe the coherence properties of an em wave only to first order.

11.3.1. Degree of Spatial and Temporal Coherence

To describe the coherence properties of a light source, we can introduce a whole class of correlation functions for the corresponding field. For the moment, however, we limit ourselves to looking at the first-order functions.

Suppose that a field is measured at some point $\mathbf{r}_1$ in a time interval from $0-T$. We can then obtain the product $E(\mathbf{r}_1, t_1)E^*(\mathbf{r}_1, t_2)$, where t_1 and t_2 are given time instants within the time interval $0-T$. If the measurement is now repeated a large number of times, we can calculate the average of the preceding product over all the measurements. This average is referred to as the *ensemble average* and written as.

$$\Gamma^{(1)}(\mathbf{r}_1, \mathbf{r}_1, t_1, t_2) = \langle E(\mathbf{r}_1, t_1)E^*(\mathbf{r}_1, t_2)\rangle \qquad (11.3.1)$$

For the remainder of this section, as well as in the next two sections, we consider the case of a stationary beam,[†] as would, for instance, apply to a single-mode cw laser, a cw laser oscillating on many modes not locked in phase, or a cw thermal light source. By definition, in these cases, the ensemble average depends only on the time difference $\tau = t_1 - t_2$, and not on the particular times t_1 and t_2. We can then write

$$\Gamma^{(1)}(\mathbf{r}_1, \mathbf{r}_1, t_1, t_2) = \Gamma^{(1)}(\mathbf{r}_1, \mathbf{r}_1, \tau) = \langle E(\mathbf{r}_1, t + \tau)E^*(\mathbf{r}_1, t)\rangle \qquad (11.3.2)$$

where we have set $t = t_2$ and $\Gamma^{(1)}$ depends only on τ. If, besides being stationary, the field is also ergodic (a condition that usually applies to the preceding cases), then, by definition, the ensemble average is the same as the time average. We can then write

$$\Gamma^{(1)}(\mathbf{r}_1, \mathbf{r}_1, \tau) = \lim_{T \to \infty} \frac{1}{T} \int_0^T E(\mathbf{r}_1, t + \tau)E^*(\mathbf{r}_1, t)dt \qquad (11.3.3)$$

Note that the definition of $\Gamma^{(1)}$ in terms of a time average is perhaps easier to understand than the one based on ensemble average. However, the definition of $\Gamma^{(1)}$ in terms of an ensemble average is more general; in the form given by Eq. (11.3.1), it can also be applied to nonstationary beams, as we will see in Sect. 11.3.4.

Having defined the first-order correlation function $\Gamma^{(1)}$ at a given point $\mathbf{r}_1$, we can define a normalized function $\gamma^{(1)}(\mathbf{r}_1, \mathbf{r}_1, \tau)$ as follows:

$$\gamma^{(1)} = \frac{\langle E(\mathbf{r}_1, t + \tau)E^*(\mathbf{r}_1, t)\rangle}{\langle E(\mathbf{r}_1, t)E^*(\mathbf{r}_1, t)\rangle^{1/2}\langle E(\mathbf{r}_1, t + \tau)E^*(\mathbf{r}_1, t + \tau)\rangle^{1/2}} \qquad (11.3.4)$$

Note that, for a stationary beam, the two ensemble averages in the denominator of Eq. (11.3.4) equal each other, and, according to Eq. (11.1.3), they both equal the average beam intensity $\langle I(\mathbf{r}_1, t)\rangle$. The function $\gamma^{(1)}$, as defined by Eq. (11.3.4), is referred to as the *complex degree of temporal coherence*; its magnitude $|\gamma^{(1)}|$ is the *degree of temporal coherence*. Indeed $\gamma^{(1)}$ gives the degree of correlation between fields, at the same point $\mathbf{r}_1$, at two instants separated by a time τ. The function $\gamma^{(1)}$ has the following main properties: (1) $\gamma^{(1)} = 1$ for $\tau = 0$, as apparent from Eq. (11.3.4); (2) $\gamma^{(1)}(\mathbf{r}_1, \mathbf{r}_1, -\tau) = \gamma^{(1)*}(\mathbf{r}_1, \mathbf{r}_1, \tau)$ as can readily be seen from Eq. (11.3.4) with the help of Eq. (11.1.1); (3) $|\gamma^{(1)}(\mathbf{r}_1, \mathbf{r}_1, \tau)| \leq 1$, which follows upon applying the Schwarz inequality to Eq. (11.3.4).

---

† A process is said to be stationary when the ensemble average of any variable that describes it (e.g., the electric field or the beam intensity) is independent of time.

We can now say that a beam has perfect temporal coherence when $|\gamma^{(1)}| = 1$ for any τ. For a cw beam, this essentially implies that both amplitude and phase fluctuations of the beam are zero, so that the signal is reduced to that of a sinusoidal wave, i.e., $E = A(\mathbf{r}_1) \exp j[\omega t - \phi(\mathbf{r}_1)]$. Indeed, the substitution of this expression into Eq. (11.3.4) shows that $|\gamma^{(1)}| = 1$. The opposite case—the complete absence of temporal coherence—occurs when $\langle E(\mathbf{r}_1, t + \tau)E^*(\mathbf{r}_1, t)\rangle$ and hence $\gamma^{(1)}$ vanish for $\tau > 0$. Such is the case for a thermal light source of very large bandwidth (e.g., a blackbody source; see Fig. 2.3). In more common situations, $|\gamma^{(1)}|$ is generally expected to be a decreasing function of τ, as indicated in Fig. 11.1, where, according to property (2) stated above, $|\gamma^{(1)}|$ is shown as a symmetric function of τ. We can therefore define a characteristic time τ_{co} (referred to as the *coherence time*) as, for instance, the time for which $|\gamma^{(1)}| = 1/2$. For a perfectly coherent wave, one obviously has $\tau_{co} = \infty$ while, for a completely incoherent wave, one has $\tau_{co} = 0$. Note that we can also define a *coherence length* L_c as $L_c = c\tau_{co}$.

Similarly, we can define a first-order correlation function between fields at two points $\mathbf{r}_1$ and $\mathbf{r}_2$, at the same time t, as:

$$\Gamma^{(1)}(\mathbf{r}_1, \mathbf{r}_2, 0) = \langle E(\mathbf{r}_1, t)E^*(\mathbf{r}_2, t)\rangle = \lim_{T \to \infty} \frac{1}{T} \int_0^T E(\mathbf{r}_1, t)E^*(\mathbf{r}_2, t)dt \qquad (11.3.5)$$

We can also define the corresponding normalized function $\gamma^{(1)}(\mathbf{r}_1, \mathbf{r}_2, 0)$ as:

$$\gamma^{(1)} = \frac{\langle E(\mathbf{r}_1, t)E^*(\mathbf{r}_2, t)\rangle}{\langle E(\mathbf{r}_1, t)E^*(\mathbf{r}_1, t)\rangle^{1/2} \langle E(\mathbf{r}_2, t)E^*(\mathbf{r}_2, t)\rangle^{1/2}} \qquad (11.3.6)$$

The quantity $\gamma^{(1)}(\mathbf{r}_1, \mathbf{r}_2, 0)$ is referred to as the *complex degree of spatial coherence*; its magnitude is the *degree of spatial coherence*. Indeed, in this case $\gamma^{(1)}$ gives the degree of correlation between fields, at the two space points $\mathbf{r}_1$ and $\mathbf{r}_2$, at the same time. Note that, from the Schwarz inequality, we again find that $|\gamma^{(1)}| \le 1$. A wave is now said to have perfect spatial coherence if $|\gamma^{(1)}| = 1$ for any two points $\mathbf{r}_1$ and $\mathbf{r}_2$ (if they lie on the same wave front or on wave fronts whose separation is much smaller than the coherence length L_c). However, a beam often has partial spatial coherence. This means that, for a fixed value of $\mathbf{r}_1$, the degree of spatial coherence $|\gamma^{(1)}|$, as a function of $\mathbf{r}_2 - \mathbf{r}_1$, decreases from the value 1 (which occurs for $\mathbf{r}_2 = \mathbf{r}_1$) to zero as $\mathbf{r}_2 - \mathbf{r}_1$ increases. Figure 11.2 shows this situation, where the function $|\gamma^{(1)}(\mathbf{r}_2 - \mathbf{r}_1)|$ is plotted against $\mathbf{r}_2$ for a given position of point P_1 (of

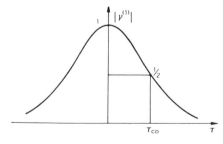

FIG. 11.1. Example of possible behavior of the degree of temporal coherence $|\gamma^{(1)}(\tau)|$. Coherence time τ_{co} can be defined as the half-width of the curve at half-height.

coordinate $\mathbf{r}_1$) on the wave front. We see that $|\gamma^{(1)}|$ will be larger than some prescribed value (e.g., $1/2$) over a certain characteristic area, referred to as the *coherence area* of the beam at point P_1 of the wave front.

The concepts of spatial and temporal coherence can be combined by means of the so-called mutual coherence function, defined as

$$\Gamma^{(1)}(\mathbf{r}_1, \mathbf{r}_2, \tau) = \langle E(\mathbf{r}_1, t + \tau)E^*(\mathbf{r}_2, t)\rangle \tag{11.3.7}$$

which can also be normalized as follows:

$$\gamma^{(1)}(\mathbf{r}_1, \mathbf{r}_2, \tau) = \frac{\langle E(\mathbf{r}_1, t + \tau)E^*(\mathbf{r}_2, t)\rangle}{\langle E(\mathbf{r}_1, t)E^*(\mathbf{r}_1, t)\rangle^{1/2}\langle E(\mathbf{r}_2, t)E^*(\mathbf{r}_2, t)\rangle^{1/2}} \tag{11.3.8}$$

This function, referred to as the *complex degree of coherence*, measures the coherence between two different points on the wave at two different times. For a quasi-monochromatic wave, it follows from Eqs. (11.1.1) and (11.3.8) that we can write

$$\gamma^{(1)}(\tau) = |\gamma^{(1)}|\exp\{j[\langle\omega\rangle\tau - \psi(\tau)]\} \tag{11.3.9}$$

where $|\gamma^{(1)}|$ and $\psi(\tau)$ are both slowly varying functions of τ, i.e.:

$$\left(\frac{d|\gamma^{(1)}|}{|\gamma^{(1)}|d\tau}, \left|\frac{d\psi}{d\tau}\right|\right) \ll \langle\omega\rangle \tag{11.3.10}$$

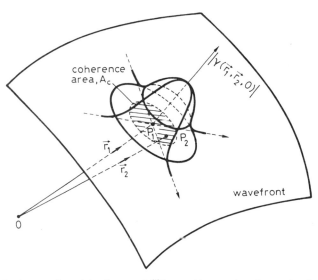

FIG. 11.2. Plot of the degree of spatial coherence $|\gamma^{(1)}(\mathbf{r}_2 - \mathbf{r}_1)|$, around a given point P_1 of a wave front, to illustrate the concept of coherence area.

11.3.2. Measurement of Spatial and Temporal Coherence

One very simple way of measuring the degree of spatial coherence between two points P_1 and P_2 on the wave front of a light wave involves using Young's interferometer (Fig. 11.3). This simply consists of screen 1, which has two small holes at positions P_1 and P_2, and screen 2 on which an interference pattern is produced by light diffracted from the two holes. More precisely interference at point P and time t arises from the superposition of waves emitted from points P_1 and P_2 at times $[t - (L_1/c)]$ and $[t - (L_2/c)]$, respectively. We see therefore that the interference fringes, on screen 2 around point P, will be more distinct the better the correlation between the two light-wave fields $E[\mathbf{r}_1, t - (L_1/c)]$ and $E[\mathbf{r}_2, t - (L_2/c)]$, where $\mathbf{r}_1$ and $\mathbf{r}_2$ are the coordinates of points P_1 and P_2.* If we now let I_{max} and I_{min} represent, respectively, the maximum intensity of a bright fringe and the minimum intensity of a dark fringe in the region of the screen around P, we can define a visibility V_P of the fringes as:

$$V_P = \frac{I_{max} - I_{min}}{I_{max} + I_{min}} \tag{11.3.11}$$

One can now see that, if the diffracted fields, at point P, from holes 1 and 2 have the same amplitude and if the two fields are perfectly coherent, their destructive interference at the point of the dark fringe gives $I_{min} = 0$. From Eq. (11.3.11) one then has $V_P = 1$. If, on the other hand, the two fields are completely uncorrelated, they do not interfere, so $I_{min} = I_{max}$ and $V_P = 0$. Recalling the discussion in Sect. 11.3.1, we see that V_P must be related to the magnitude of the function $\gamma^{(1)}$. To obtain the degree of spatial coherence, however, we must consider the fields $E[\mathbf{r}_1, t - (L_1/c)]$ and $E[\mathbf{r}_2, t - (L_2/c)]$ at the same time. This requires us to choose point P so that $L_1 = L_2$. Example 11.1 then shows that:

$$V_P = |\gamma^{(1)}(\mathbf{r}_1, \mathbf{r}_2, 0)| \tag{11.3.12}$$

On the other hand, if holes 1 and 2 do not produce the same field amplitudes, i.e., the same illumination at point P, then, instead of Eq. (11.3.12), one has

$$V_P = \frac{2(\langle I_1 \rangle \langle I_2 \rangle)^{1/2}}{\langle I_1 \rangle + \langle I_2 \rangle} |\gamma^{(1)}(\mathbf{r}_1, \mathbf{r}_2, 0)| \tag{11.3.13}$$

---

* The integration time T appearing in the correlation function [see Eq. (11.3.5)] now equals the time taken to measure the fringes (e.g., the exposure time of a photographic plate).

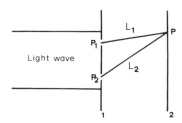

FIG. 11.3. Use of a Young's interferometer to measure the degree of spatial coherence between points P_1 and P_2 of an em wave.

where $\langle I_1 \rangle$ and $\langle I_2 \rangle$ are the average intensities of light diffracted to point P from the two holes. Note also that, for the general point P in Fig. 11.3, the visibility V_P can be shown to equal $|\gamma^{(1)}(\mathbf{r}_1, \mathbf{r}_2, \tau)|$, where $\tau = (L_2 - L_1)/c$.

Example 11.1. *Calculation of the fringe visibility in Young's interferometer.* The field $E(\mathbf{r}_P, t')$, at point P in Fig. 11.3 and time t', can be expressed as a superposition of the fields diffracted from holes 1 and 2 at times $(t' - L_1/c)$ and $(t' - L_2/c)$, respectively. We can then write

$$E(\mathbf{r}_P, t') = K_1 E\left[\mathbf{r}_1, t' - \left(\frac{L_1}{c}\right)\right] + K_2 E\left[\mathbf{r}_2, t' - \left(\frac{L_2}{c}\right)\right] \tag{11.3.14}$$

where $E[\mathbf{r}_1, t' - (L_1/c)]$ and $E[\mathbf{r}_2, t' - (L_2/c)]$ are the fields at points P_1 and P_2, while K_1 and K_2 represent the fractions of these two fields diffracted to point P, respectively. Factors K_1 and K_2 are expected to be inversely proportional to L_1 and L_2 and also to depend on the hole dimensions and the angle between the incident wave and the wave diffracted from P_1 and P_2. Since diffracted secondary wavelets are always a quarter of a period out of phase with the incident wave [see also the discussions of "Huyghens wavelets" in Sect. 4.6], it follows that:

$$K_1 = |K_1| \exp[-(j\pi/2)] \tag{11.3.15a}$$

$$K_2 = |K_2| \exp[-(j\pi/2)] \tag{11.3.15b}$$

If we now define $t = t' - L_2/c$ and $\tau = (L_2/c) - (L_1/c)$, Eq. (11.3.14) can be written as:

$$E = K_1 E(\mathbf{r}_1, t + \tau) + K_2 E(\mathbf{r}_2, t) \tag{11.3.16}$$

According to Eq. (11.1.3), the intensity at point P can be written as $I = EE^* = |K_1 E(\mathbf{r}_1, t + \tau) + K_2 E(\mathbf{r}_2, t)|^2$, where Eq. (11.3.16) has been used. From this expression and Eq. (11.3.15), we have

$$I = I_1(t + \tau) + I_2(t) + 2 \operatorname{Re}[K_1 K_2^* E(\mathbf{r}_1, t + \tau) E^*(\mathbf{r}_2, t)] \tag{11.3.17}$$

where Re stands for real part. In the previous expressions, I_1 and I_2 are the intensities at point P, due to the emission from point P_1 alone and point P_2 alone, respectively. They are given by

$$I_1 = |K_1|^2 |E(\mathbf{r}_1, t + \tau)|^2 = |K_1|^2 I(\mathbf{r}_1, t + \tau) \tag{11.3.18a}$$

$$I_2 = |K_2|^2 |E(\mathbf{r}_2, t)|^2 = |K_2|^2 I(\mathbf{r}_2, t) \tag{11.3.18b}$$

where $I(\mathbf{r}_1, t + \tau)$ and $I(\mathbf{r}_2, t)$ are the intensities at points P_1 and P_2, respectively. Taking the time average of both sides of Eq. (11.3.17) and using Eq. (11.3.7), we find

$$\langle I \rangle = \langle I_1 \rangle + \langle I_2 \rangle + 2|K_1||K_2|\operatorname{Re}[\Gamma^{(1)}(\mathbf{r}_1, \mathbf{r}_2, \tau)] \tag{11.3.19}$$

where Eq. (11.3.15) has also been used. Equation (11.3.19) can be expressed in terms of $\gamma^{(1)}$ by noting that, from Eq. (11.3.8), one has

$$\Gamma^{(1)} = \gamma^{(1)}[\langle I(\mathbf{r}_1, t+\tau)\rangle\langle I(\mathbf{r}_2, t)\rangle]^{1/2} \tag{11.3.20}$$

The substitution of Eq. (11.3.20) into Eq. (11.3.19), with the help of Eq. (11.3.18), gives $\langle I \rangle = \langle I_1 \rangle + \langle I_2 \rangle + 2(\langle I_1 \rangle \langle I_2 \rangle)^{1/2} \, \mathrm{Re}[\gamma^{(1)}(\mathbf{r}_1, \mathbf{r}_2, \tau)]$. From Eq. (11.3.9) we then obtain

$$\langle I \rangle = \langle I_1 \rangle + \langle I_2 \rangle + 2(\langle I_1 \rangle \langle I_2 \rangle)^{1/2}|\gamma^{(1)}|\cos[\langle\omega\rangle\tau - \psi(\tau)] \tag{11.3.21}$$

Since both $|\gamma^{(1)}|$ and $\psi(\tau)$ are slowly varying functions of τ, it follows that the variation of intensity $\langle I \rangle$, as P is changed, i.e., the fringe pattern, is due to the rapid variation of the cosine term with its argument $\langle\omega\rangle\tau$. Therefore, in the region around P, we have

$$I_{max} = \langle I_1 \rangle + \langle I_2 \rangle + 2(\langle I_1 \rangle \langle I_2 \rangle)^{1/2}|\gamma^{(1)}| \tag{11.3.22a}$$

$$I_{min} = \langle I_1 \rangle + \langle I_2 \rangle - 2(\langle I_1 \rangle \langle I_2 \rangle)^{1/2}|\gamma^{(1)}| \tag{11.3.22b}$$

and, from Eq. (11.3.11):

$$V_P = \frac{2(\langle I_1 \rangle \langle I_2 \rangle)^{1/2}}{\langle I_1 \rangle + \langle I_2 \rangle}|\gamma^{(1)}(r_1, r_2, \tau)| \tag{11.3.23}$$

For the case $\tau = (L_2/c) - (L_1/c) = 0$, Eq. (11.3.23) reduces to Eq. (11.3.13).

The degree of temporal coherence is usually measured with a Michelson interferometer (Fig. 11.4a). Let P be the point where the temporal coherence of the wave is to be measured. A combination of a suitably small diaphragm, placed at P, with a lens, having its focus at P, transforms the incident wave into a plane wave (see Fig. 11.12). This wave then falls on a partially reflecting mirror S_1 (of reflectivity $R = 50\%$) that splits the beam into beams A and B. These beams are reflected by mirrors S_2 and S_3 (both of reflectivity $R = 100\%$) and then recombined to form beam C. Since waves A and B interfere, the illumination in the direction of C is either light or dark according to whether $2(L_3 - L_2)$ is an even or odd number of half wavelengths. Obviously this interference is observed only as long as the difference $L_3 - L_2$ does not become so large that beams A and B become uncorrelated in phase. Thus, for a partially coherent wave, the intensity I_c of beam C, as a function of $2(L_3 - L_2)$, behaves as shown in Fig. 11.4b. At a given value of the difference, $L_3 - L_2$, between the lengths of the interferometer arms, i.e., at a given value of the delay $\tau = 2(L_3 - L_2)/c$ between the two reflected waves, we can again define a fringe visibility V_P as in Eq. (11.3.11), where the quantities I_{max} and I_{min} are as shown in Fig. 11.4b. We now expect V_P to be a function of the time delay τ and, likewise the case of Young's interferometer, to be related to the degree of temporal coherence. We can, in fact, show that in this case one has

$$V_P(\tau) = |\gamma^{(1)}(\mathbf{r}, \mathbf{r}, \tau)| \tag{11.3.24}$$

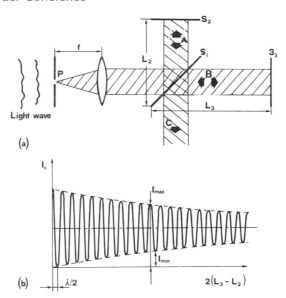

FIG. 11.4. (a) Michelson interferometer for measuring the degree of temporal coherence of an em wave at point P. (b) Behavior of output intensity, along the direction C of the interferometer, as a function of the difference $L_3 - L_2$ between lengths of the interferometer arms.

where $\mathbf{r}$ is the coordinate of point P. Once $V_P(\tau) = |\gamma^{(1)}(\tau)|$ is measured, the value of the coherence time τ_{co}, defined as, e.g., the time delay at which $V_P(\tau_{co}) = 1/2$ (see Fig. 11.1), can be determined. The corresponding coherence length is then given by $L_c = c\tau_{co}$; given the definition of τ_{co} in Fig. 11.1, one sees that L_c equals twice the difference, $L_3 - L_2$, between the interferometer arms for which the fringe visibility decreases to $V_P = 1/2$.

11.3.3. Relation between Temporal Coherence and Monochromaticity

From the preceding discussion about a stationary beam, it is clear that the concept of temporal coherence is intimately connected with monochromaticity. For example, the more monochromatic a wave is, the greater its temporal coherence; i.e., coherence time τ_{co} must depend inversely on the laser oscillation bandwidth $\Delta\nu_L$. In this section we discuss this relationship in more depth.

We start by noting that the spectrum of an em wave, as measured by, e.g., a spectrograph, is proportional to the power spectrum $W(\mathbf{r}, \omega)$ of the field $E(\mathbf{r}, t)$. Since the power spectrum W equals the Fourier transform of the autocorrelation function $\Gamma^{(1)}$, either of these quantities can be obtained once the other is known. To give a precise expression for the relation between τ_{co} and $\Delta\nu_L$ we must redefine these two quantities appropriately. We define τ_{co} as the standard deviation σ_τ of the function $|\Gamma^{(1)}(\tau)|^2$, i.e., such that

$$(\sigma_\tau)^2 = \left[\int_{-\infty}^{+\infty} (\tau - \langle\tau\rangle)^2 |\Gamma^{(1)}(\tau)|^2 \, d\tau \right] \Big/ \left[\int_{-\infty}^{+\infty} |\Gamma^{(1)}(\tau)|^2 \, d\tau \right]$$

where the mean value $\langle \tau \rangle$ is defined by $\langle \tau \rangle = [\int \tau |\Gamma^{(1)}(\tau)|^2 \, d\tau] / [\int |\Gamma^{(1)}(\tau)|^2 \, d\tau]$. As a shorthand notation for this expression, we will write

$$(\sigma_\tau)^2 = \langle (\tau - \langle \tau \rangle)^2 \rangle \qquad (11.3.25)$$

Since $|\Gamma^{(1)}(-\tau)| = |\Gamma^{(1)}(\tau)|$, one has $\langle \tau \rangle = 0$ and Eq. (11.3.25) reduces to:

$$(\sigma_\tau)^2 = \langle \tau^2 \rangle \qquad (11.3.26)$$

Example 11.2. *Coherence time and bandwidth for a sinusoidal wave with random phase jumps.* We assume that the time behavior of the field at a given point is described by a sinusoidal wave with a constant amplitude and a phase undergoing random jumps (see Fig. 2.9). The correlation function $\Gamma^{(1)}(\tau)$ for this wave is calculated in Appendix B as $\Gamma^{(1)}(\tau) \propto \exp[-(|\tau|/\tau_c)]$, where τ_c is the average time between two consecutive phase jumps. According to Eq. (11.3.26), we then obtain

$$\sigma_\tau^2 = \frac{\int_0^\infty \tau^2 \exp[-(2\tau/\tau_c)]d\tau}{\int_0^\infty \exp[-(2\tau/\tau_c)]d\tau} = \frac{\tau_c^2}{4} \frac{\int_0^\infty x^2 \exp(-x)dx}{\int_0^\infty \exp(-x)dx} = \frac{\tau_c^2}{2}$$

where $x = 2\tau/\tau_c$. Thus we obtain $\sigma_\tau = \tau_c/(2)^{1/2}$. The power spectrum $W(v - v_0)$ of this signal is then described by a Lorentzian function (see Appendix B), so that, according to Eq. (11.3.27), we can write

$$\sigma_v^2 = \frac{\int_0^\infty (v - v_0)^2 \{1/[1 + 4\pi^2(v - v_0)^2\tau_c^2]\}^2 d(v - v_0)}{\int_0^\infty \{1/[1 + 4\pi^2(v - v_0)^2\tau_c^2]\}^2 d(v - v_0)}$$

$$= \frac{1}{4\pi^2\tau_c^2} \frac{\int_0^\infty x^2[1/(1 + x^2)]^2 \, dx}{\int_0^\infty [1/(1 + x^2)]^2 \, dx}$$

where $x = 2\pi(v - v_0)\tau_c$. The two integrals on the right-hand side of the preceding equation are both equal to $(\pi/4)$,[11] so that $\sigma_v = (1/2\pi\tau_c)$. From the preceding calculations, we find that the product $\sigma_\tau\sigma_v$ equals $1/(2\sqrt{2}\pi)$ i.e., it is $\sqrt{2}$ times larger than the minimum value, $1/4\pi$, that holds for a Gaussian spectrum.

The coherence time defined in this way is conceptually simpler, although it sometimes involves lengthier calculation, than that defined earlier (i.e., the half-width at half-height of the curve $|\Gamma^{(1)}(\tau)|$, see Fig. 11.1). In fact, if $|\Gamma^{(1)}(\tau)|$ were an oscillatory function of τ, the coherence time τ_{co}, as defined in Fig. 11.1 would not be uniquely determined.

We similarly define the laser oscillation bandwidth Δv_L as the standard deviation of $W^2(v)$, i.e., such that

$$(\Delta v_L)^2 = (\sigma_v)^2 = \langle (v - \langle v \rangle)^2 \rangle \qquad (11.3.27)$$

where $\langle v \rangle$, the mean frequency of the spectrum, is given by $\langle v \rangle = (\int vW^2 \, dv)/\int W^2 \, dv)$. Since W and $\Gamma^{(1)}$ are related by a Fourier transform, one can show that σ_v and σ_τ, as just defined, satisfy the condition:

$$\sigma_\tau\sigma_v \geq (1/4\pi) \qquad (11.3.28)$$

Equation (11.3.28) is closely analogous to the Heisenberg uncertainty relation and can be proved using the same deriving procedure.[2] The equality sign in Eq. (11.3.28) applies when $|\Gamma^{(1)}(\tau)|$ [and hence $W(v)$] are Gaussian functions. This case is obviously the analogue of the minimum uncertainty wave packets of quantum mechanics.[2]

11.3.4. Nonstationary Beams

We briefly consider the case of a nonstationary beam.* By definition the function $\Gamma^{(1)}$ in Eq. (11.3.1) now depends on t_1 and t_2 and not only on their difference $\tau = t_1 - t_2$; examples include an amplitude-modulated laser, an amplitude-modulated thermal light source, a Q-switched or a mode-locked laser. For a nonstationary beam, the correlation function is obtained as the ensemble average of many measurements of the field in a time interval from 0–T, where the origin of the time interval is synchronized to some driving signal (e.g., to the amplitude modulator, for a mode-locked laser, or the Pockels cell driver, for a Q-switched laser). The degree of temporal coherence at a given point $\mathbf{r}$ can then be defined as:

$$\gamma^{(1)}(t_1, t_2) = \frac{\langle E(t_1)E^*(t_2)\rangle}{\langle E(t_1)E^*(t_1)\rangle^{1/2}\langle E(t_2)E^*(t_2)\rangle^{1/2}} \tag{11.3.29}$$

where t_1 and t_2 are two given times, in the interval 0–T, and all signals are measured at point $\mathbf{r}$.

We can now say that the beam has a perfect temporal coherence if $|\gamma^{(1)}(t_1, t_2)| = 1$ for all times t_1 and t_2. According to this definition we can see that a nonstationary beam *without amplitude and phase fluctuations* has a *perfect temporal coherence*. In the absence of fluctuations, in fact, the products $E(t_1)E^*(t_2)$, $E(t_1)E^*(t_1)$, and $E(t_2)E^*(t_2)$ in Eq. (11.3.29) remain the same for all measurements of the ensemble. These products thus equal the corresponding ensemble averages and $\gamma^{(1)}(t_1, t_2)$ reduces to:

$$\gamma^{(1)}(t_1, t_2) = \frac{E(t_1)E^*(t_2)}{|E(t_1)||E(t_2)|} \tag{11.3.30}$$

From Eq. (11.3.30) one immediately sees that $|\gamma^{(1)}| = 1$.

According to this definition of temporal coherence, the coherence time of a nonstationary beam, e.g., a mode-locked laser, is infinite if the beam does not fluctuate in amplitude or phase. This shows that the coherence time of a nonstationary beam is not related to the inverse of the oscillating bandwidth. In a practical situation however, if we correlate, e.g., one pulse of a mode-locked train with some other pulse of the train, i.e., if we choose $t_1 - t_2$ to be larger than the pulse repetition time, some absence of correlation will be found, due to fluctuations. This means that $|\gamma^{(1)}|$ decreases as $t_1 - t_2$ increases beyond the pulse repetition time. Thus coherence time is then expected to be finite, although not related to the inverse of the oscillating bandwidth but to the *inverse of the fluctuation bandwidth*.

11.3.5. Spatial and Temporal Coherence of Single-Mode and Multimode Lasers

Consider first a cw laser oscillating in a single transverse and longitudinal mode. Just above threshold, as explained in Sects. 7.10 and 7.11, the amplitude fluctuations may be neglected, to first order, and the fields of the wave, at points $\mathbf{r}_1$ and $\mathbf{r}_2$, can be written as

$$E(\mathbf{r}_1, t) = a_0 u(\mathbf{r}_1) \exp\{j[\omega t - \phi(t)]\} \tag{11.3.31a}$$

$$E(\mathbf{r}_2, t) = a_0 u(\mathbf{r}_2) \exp\{j[\omega t - \phi(t)]\} \tag{11.3.31b}$$

---

* The author wishes to acknowledge some enlightening discussion on this topic with Prof. V. Degiorgio.

where a_0 is a constant, $u(\mathbf{r})$ describes the mode amplitude, and ω is the angular frequency at band center. The subsitution of Eqs. (11.3.31) into Eq. (11.3.6) then gives $\gamma^{(1)} = u(\mathbf{r}_1)u^*(\mathbf{r}_2)/|u(\mathbf{r}_1)||u(\mathbf{r}_2)|$, which shows that $|\gamma^{(1)}| = 1$. Thus, a laser oscillating in a single mode has perfect spatial coherence. Its temporal coherence, on the other hand, is established by the laser bandwidth $\Delta\nu_L$. As an example, a single-mode monolithic Nd:YAG laser (see Fig. 7.26) may have $\Delta\nu_L \cong 20\,\text{kHz}$. The coherence time is then $\tau_{co} \cong 1/\Delta\nu_L \cong 0.05\,\text{ms}$, and the coherence length $L_c = c\tau_{co} \cong 15\,\text{km}$ (note the very large value of this coherence length).

Let us now consider a laser oscillating in a single transverse mode and in l longitudinal modes. In terms of the cavity mode amplitudes $u(\mathbf{r})$, the fields at two points $\mathbf{r}_1$ and $\mathbf{r}_2$ belonging to the same wave front can generally be written as:

$$E(\mathbf{r}_1, t) = \sum_1^l {}_k a_k u(\mathbf{r}_1) \exp j[\omega_k t - \phi_k(t)] \tag{11.3.32a}$$

$$E(\mathbf{r}_2, t) = \sum_1^l {}_k a_k u(\mathbf{r}_2) \exp j[\omega_k t - \phi_k(t)] \tag{11.3.32b}$$

where a_k are constant factors, ω_k and ϕ_k are, respectively, the frequency and the phase of the kth mode. Note that, since the transverse field configuration is the same for all modes (e.g., that of a TEM$_{00}$ mode), the amplitude u is taken to be independent of the mode index k. The function $u(\mathbf{r})$ can thus be removed from the summation in both Eqs. (11.3.32a–b) and the following result is readily obtained:

$$E(\mathbf{r}_2, t) = \left[\frac{u(\mathbf{r}_2)}{u(\mathbf{r}_1)}\right] E(\mathbf{r}_1, t) \tag{11.3.33}$$

This means that, whatever time variation $E(\mathbf{r}_1, t)$ is observed at $\mathbf{r}_1$, the same time variation is also observed at $\mathbf{r}_2$, except for a proportionality constant. The substitution of Eq. (11.3.33) into Eq. (11.3.6) then yields $|\gamma^{(1)}| = 1$. Thus a laser beam made of many longitudinal modes with the same transverse profile (e.g., corresponding to a TEM$_{00}$ mode) still has a perfect spatial coherence. If all the mode phases are random, the temporal coherence then equals the inverse of the oscillating bandwidth. If no frequency-selecting elements are used in the cavity, the oscillating bandwidth may be comparable to the laser gain bandwidth; hence coherence time may be much shorter than in the example previously considered (generally in the range of nanoseconds to picoseconds). When these modes are locked in phase however, temporal coherence may become very long, as discussed in the previous section. Thus a mode-locked laser can in principle have perfect spatial and temporal coherence.

The last case we should consider is that of a laser oscillating in many transverse modes. The following example shows that, in this case, the laser has only partial spatial coherence.

Example 11.3. *Spatial coherence for a laser oscillating in many transverse modes.* For a laser oscillating in l transverse modes, the fields of the output beam, at two points $\mathbf{r}_1$ and $\mathbf{r}_2$ on the same wave front, can be written as [compare with Eq. (11.3.32)]:

$$E(\mathbf{r}_1, t) = \sum_1^l {}_k a_k u_k(\mathbf{r}_1) \exp j[\omega_k t - \phi_k(t)] \tag{11.3.34a}$$

$$E(\mathbf{r}_2, t) = \sum_1^l {}_k a_k u_k(\mathbf{r}_2) \exp j[\omega_k t - \phi_k(t)] \tag{11.3.34b}$$

If the product $E(\mathbf{r}_1, t)E^*(\mathbf{r}_2, t)$ is performed, one obtains a set of terms proportional to $\exp j[(\omega_k - \omega_{k'})t - (\phi_k - \phi_{k'})]$, with $k' \neq k$. These terms can be neglected because, in taking the time average, they average out to zero. We are thus left only with terms for which $k' = k$, so that we obtain:

$$\langle E(\mathbf{r}_1, t)E^*(\mathbf{r}_2, t)\rangle = \sum_1^l {}_k a_k a_k^* u_k(\mathbf{r}_1) u_k^*(\mathbf{r}_2) \tag{11.3.35}$$

For $\mathbf{r}_2 = \mathbf{r}_1$ one obtains from Eq. (11.3.35):

$$\langle E(\mathbf{r}_1, t)E^*(\mathbf{r}_1, t)\rangle = \sum_1^l {}_k |a_k|^2 |u_k(\mathbf{r}_1)|^2 \tag{11.3.36}$$

Likewise, for $\mathbf{r}_1 = \mathbf{r}_2$, one obtains from Eq. (11.3.35):

$$\langle E(\mathbf{r}_2, t)E^*(\mathbf{r}_2, t)\rangle = \sum_1^l {}_k |a_k|^2 |u_k(\mathbf{r}_2)|^2 \tag{11.3.37}$$

The substitution of Eqs. (11.3.35)–(11.3.37) into Eq. (11.3.6) then gives

$$\gamma^{(1)} = \frac{\sum_1^l {}_k a_k a_k^* u_k(\mathbf{r}_1) u_k^*(\mathbf{r}_2)}{\left[\sum_1^l {}_k |a_k|^2 |u_k(\mathbf{r}_1)|^2\right]^{1/2}\left[\sum_1^l {}_k |a_k|^2 |u_k(\mathbf{r}_2)|^2\right]^{1/2}} \tag{11.3.38}$$

Let now $\mathbf{R}_1$ represent a complex vector, characterized by the components $a_k u_k(\mathbf{r}_1)$ in an l-dimensional space, and similarly for $\mathbf{R}_2$. We can then write their scalar product as:

$$\mathbf{R}_1 \cdot \mathbf{R}_2 = \sum_1^l {}_k a_k a_k^* u_k(\mathbf{r}_1) u_k^*(\mathbf{r}_2) \tag{11.3.39}$$

The magnitude of the two vectors is given by

$$R_1 = |\mathbf{R}_1| = \left[\sum_1^l {}_k |a_k|^2 |u_k(\mathbf{r}_1)|^2\right]^{1/2} \tag{11.3.40}$$

and

$$R_2 = |\mathbf{R}_2| = \left[\sum_1^l {}_k |a_k|^2 |u_k(\mathbf{r}_2)|^2\right]^{1/2} \tag{11.3.41}$$

respectively. The substitution of Eqs. (11.3.39)–(11.3.41) into Eq. (11.3.38) shows that:

$$\gamma^{(1)} = \frac{\mathbf{R}_1 \cdot \mathbf{R}_2}{R_1 R_2} \tag{11.3.42}$$

From the Schwartz inequality, it then follows that, since $\mathbf{R}_1 \neq a\mathbf{R}_2$ where a is a constant, one always has $|\gamma^{(1)}| < 1$.

11.3.6. Spatial and Temporal Coherence of a Thermal Light Source

We now briefly discuss the coherence properties of light emitted by an ordinary lamp, which may be either a filament lamp or a lamp filled with a gas at a suitable pressure. Since the emitted light now arises from spontaneous emission of atoms, under thermal equilibrium conditions, these sources are generally referred to as *thermal sources*.

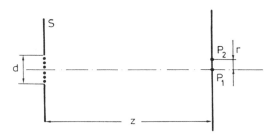

FIG. 11.5. Degree of spatial coherence in a plane at a distance z from independent emitters covering an area of diameter d.

As far as the temporal coherence is concerned, one can observe that light emitted by a cw gas lamp generally consists of several emission lines (see, e.g., Fig. 6.6b); the width of these lines is rather broad ($\Delta v = 1$–10 THz) due to the high pressures generally used. In the case of a flashlamp (see, e.g., Fig. 6.6a) or a filament source, on the other hand, the emission is much broader resembling that of a blackbody radiator (see Fig. 2.3). Thus the coherence time $\tau_{co} \cong 1/\Delta v$ of a thermal light source is generally very short ($\tau_{co} < 1$ ps).

As far as spatial coherence is concerned, since the em wave originates from spontaneous emission by independent emitters, the wave is completely incoherent at a location very near the source. It acquires, however, an increasing degree of spatial coherence as the considered location moves away from the source. This situation can be understood with the help of Fig. 11.5, where uncorrelated emitters (dots) are assumed to be present within the aperture of a hole of diameter d in a screen S. Emission occurs over a 4π solid angle, and the degree of spatial coherence between points P_1 and P_2 is measured at some distance z from the emitters. For simplicity, let P_1 be the point on the symmetry axis of the system and let r be the distance between the two points. It is clear from the figure that, for very small values of z, points P_1 and P_2 ($r < d/2$) primarily see emission from the emitter just facing it, so that fields at the two points are completely uncorrelated. As z is increased, however, each of the two points receives an increasing contribution from all other emitters, so that fields at the two points become more and more correlated. The calculation of the degree of spatial coherence $|\gamma^{(1)}|$ as a function of the coordinates z and r is beyond the scope of this book, so we refer the reader to Ref. 3. We limit ourselves to pointing out that $|\gamma^{(1)}|$ turns out to be a function of the dimensionless quantity $(rd/\lambda z)$; as an example, one has $|\gamma^{(1)}| = 0.88$ for:

$$r \cong 0.16\left(\frac{\lambda z}{d}\right) \tag{11.3.43}$$

This result will be used in Sect. 11.8.

11.4. DIRECTIONALITY

There are usually two ways of measuring the directionality, i.e., the divergence, of a laser beam, or, more generally, of any light source, namely: (1) By measuring the *degree of beam spreading at very large distances from the source*. In fact, if we let W represent a suitably defined radius of the beam at a very large distance z, the half-angle beam divergence can be obtained from the relation:

$$\theta_d = \frac{W}{z} \tag{11.4.1}$$

(2) By measuring the *radial intensity distribution, $I(r)$, of the focused beam in the focal plane of a lens*. To understand the basis of this measurement, let r be the coordinate of a general point, in this focal plane, relative to an origin at the beam center. According to the discussion made in relation to Fig. 1.8, the beam can be regarded as composed of a set of plane waves propagating along slightly different directions.[4] A wave inclined at an angle θ to the propagation axis is then focused, in the focal plane, to a point of radial coordinate r given by (for small θ):

$$r = f\theta \tag{11.4.2}$$

Hence the intensity distribution $I(r)$ in the focal plane of a lens gives the angular distribution of the original beam. In the following discussion, we use either of these methods, as convenient.

The directional properties of a laser beam are strictly related to its spatial coherence. We therefore first discuss the case of an em wave with perfect spatial coherence, then the case of partial spatial coherence.

11.4.1. Beams with Perfect Spatial Coherence

Let us first consider a beam with perfect spatial coherence, consisting of a plane wave front of circular cross section, with diameter D, and constant amplitude over this cross section. Following the preceding discussion, beam divergence can be calculated from the intensity distribution $I(r)$ in the focal plane of a lens. This distribution was derived at the beginning of the nineteenth century by Airy, using diffraction theory.[5] The expression obtained for $I(r)$, known as the *Airy formula*, can be written as

$$I = \left[\frac{2J_1(krD/2f)}{krD/2f}\right]^2 I_0 \tag{11.4.3}$$

where: $k = 2\pi/\lambda$; J_1 is the first-order Bessel function; I_0 is given by

$$I_0 = P_i\left(\frac{\pi D^2}{4\lambda^2 f^2}\right) \tag{11.4.4}$$

where P_i is the power of the beam incident on the lens. Note that, since the value of the expression within square brackets in Eq. (11.4.3) becomes unity when $r = 0$, I_0 represents the beam intensity at the center of the focal spot.

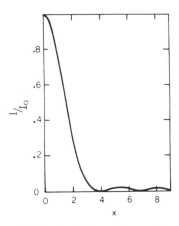

FIG. 11.6. Distribution of light intensity in the focal plane of a lens, of diameter D and focal length f, as a function of the normalized variable $x = krD/2f$, where r is the radial distance.

Figure 11.6 shows the behavior of the intensity I as a function of the dimensionless quantity:

$$x = \frac{krD}{2f} \tag{11.4.5}$$

From this figure one can see that the diffraction pattern consists of a circular central zone (the *Airy disk*) surrounded by a series of rings of rapidly decreasing intensity. The divergence angle θ_d of the original beam can now be defined as corresponding to the radius of the first minimum in Fig. 11.6. While this is an arbitrary definition, it is convenient and is the conventionally accepted definition. Therefore, from the value of x corresponding to this minimum and from Eqs. (11.4.5) and (11.4.2), we obtain

$$\theta_d = \frac{1.22\lambda}{D} \tag{11.4.6}$$

As a second example of propagation of a spatially coherent beam, we consider the case of a Gaussian beam (TEM_{00}) such as can be obtained from a stable laser cavity consisting of two spherical mirrors. If we let w_0 be the spot size at the beam waist, the spot size w and the radius of curvature R of the equiphase surface at a distance z from the waist can be obtained from Eqs. (4.7.17a) and (4.7.17b), respectively. At a large distance from the waist [i.e., for $(\lambda z/\pi w_0^2) \gg 1$], one sees that $w \cong \lambda z/\pi w_0$ and $R \cong z$. Since both w and R increase linearly with distance, the wave can be considered spherical with its origin at the waist. Following the convention of identifying the radius of the beam with the spot size w, the beam divergence is obtained as:

$$\theta_d = \frac{w}{z} = \frac{\lambda}{\pi w_0} \tag{11.4.7}$$

A comparison of Eqs. (11.4.7) and (11.4.6) indicates that, if one sets $D = 2w_0$, a Gaussian beam has a divergence about half that of a plane beam.

We conclude this section by saying that the divergence θ_d of a spatially coherent beam can generally be written as

$$\theta_d = \frac{\beta\lambda}{D} \tag{11.4.8}$$

where D is a suitably defined beam diameter and β is a numerical factor, on the order of unity, whose exact value depends on the field amplitude distribution as well as on the way both θ_d and D are defined. Such a beam is commonly referred to as *diffraction-limited*.

11.4.2. Beams with Partial Spatial Coherence

For an em wave with partial coherence, divergence is greater than for a spatially coherent wave having the same intensity distribution. This can, for example, be understood following the argument used to explain divergence of a beam of uniform amplitude in Fig. 1.6. In fact, if the wave considered in Fig. 1.6 is not spatially coherent, the secondary wavelets emitted over its cross section are no longer in phase. In this case the wave front produced by diffraction has a larger divergence than that given by Eq. (11.4.6). A rigorous treatment of this problem (i.e., the propagation of partially coherent waves) is beyond the scope of this book, so the reader is referred to Ref. 3. We limit ourselves to considering first a particularly simple case of a beam, of diameter D (Fig. 11.7a), that consists of many smaller beams (shaded in the figure) of diameter d. We assume that each of these smaller beams is spatially-coherent and thus diffraction-limited. Now, if the various beams are mutually uncorrelated, the divergence of the beam as a whole equals $\theta_d = \beta\lambda/d$. On the other hand, if the various beams are correlated, the divergence is $\theta_d = \beta\lambda/D$. The latter case is actually equivalent to a number of antennas (the small beams) all emitting in phase with each other.

After this simple case, we consider the general case where the partially coherent beam has a given intensity distribution over its diameter D and a coherence area A_c at a given point P (Fig. 11.7b). By analogy with the previous case, in this case we can write

$$\theta_d = \frac{\beta\lambda}{D_c} \tag{11.4.9}$$

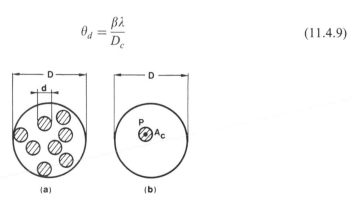

(a)   **(b)**

FIG. 11.7. Examples illustrating the different divergence properties of coherent and partially coherent waves: (a) Beam of diameter D made of the superposition of several smaller and coherent beams of diameter d. (b) Beam of diameter D and coherence area A_c at point P.

where D_c is the diameter of the coherence area and β is a numerical factor, of the order of unity, whose value depends on how θ_d and D_c are defined. The concept of directionality is thus intimately related to spatial coherence.

11.4.3. The M^2 Factor and the Spot Size Parameter of a Multimode Laser Beam

Expressions for beam divergence in Eqs. (11.4.9) and (11.4.8) involve a degree of imprecision that stems from the arbitrary definition of beam diameter. We now present some precise definitions of beam radius and beam divergence that allow us to describe beam propagation, more generally, for both a diffraction-limited laser beam with an arbitrary transverse profile and for a nondiffraction-limited, multi-transverse-mode, partially-coherent laser beam.[6]

Let $I(x, y, z)$ be the time-averaged intensity profile of the laser beam at the longitudinal coordinate z. Note that, for greater generality, we do not restrict ourselves to radially symmetric beams; accordingly, intensity I is written as a function of both transverse coordinates x and y, separately. We can now define a beam standard deviation $\sigma_x(z)$ along, e.g., the x-coordinate, so that

$$\sigma_x^2(z) = \frac{\iint (x - \langle x \rangle)^2 I(x, y, z) dx\, dy}{\iint I(x, y, z) dx\, dy} \tag{11.4.10}$$

where $\langle x \rangle = [\iint x I(x, y, z) dx\, dy]/[\iint I(x, y, z) dx\, dy]$, and similarly for the y-coordinate.

To introduce a similar definition for beam divergence, let $\hat{I}(s_x, s_y)$ be the wave intensity at the normalized angular coordinates $s_x = \theta_x/\lambda$, and $s_y = \theta_y/\lambda$. The s_x and s_y coordinates are usually referred to as the spatial frequency coordinates of the wave, and their use is quite common in diffraction optics.[7] If, for instance, the divergence is obtained by measuring the beam intensity $I(x', y')$ at a plane, x', y', at a large distance z from the source, the angular intensity $\hat{I}(s_x, s_y)$ is obtained from $I(x', y')$ using the relations

$$x' = \theta_x z = s_x \lambda z \tag{11.4.11a}$$

and

$$y' = \theta_y z = s_y \lambda z \tag{11.4.11b}$$

Having defined the intensity $\hat{I}(s_x, s_y)$, the variance of the spatial frequency s_x can now be readily defined as

$$\sigma_{s_x}^2 = \frac{\iint (s_x - \langle s_x \rangle)^2 \hat{I}(s_x, s_y) ds_x\, ds_y}{\iint \hat{I}(s_x, s_y) ds_x\, ds_y} \tag{11.4.12}$$

where $\langle s_x \rangle = [\iint s_x \hat{I}(s_x, s_y) ds_x\, ds_y]/[\iint \hat{I}(s_x, s_y) ds_x\, ds_y]$, and similarly for the y-coordinate.

Let now $u(x, y, z)$ be the transverse amplitude profile of the beam (so that $I \propto |u|^2$) and $\hat{u}(s_x, s_y)$ be the spatial frequency profile (so that $\hat{I} \propto |\hat{u}|^2$). It is known that, for any arbitrary beam, the two functions are related through a Fourier transform.[7] It can then be shown that, for any arbitrary laser beam, $\sigma_x^2(z)$ obeys the free-space propagation equation

$$\sigma_x^2(z) = \sigma_{x0}^2 + \lambda^2 \sigma_{s_x}^2 (z - z_{0x})^2 \tag{11.4.13}$$

where σ_{x0} is the minimum value of σ_x and z_{0x} is the coordinate at which this minimum is attained. One can also show from the same treatment that [compare with Eq. (11.3.28)]

$$\sigma_{x0}\sigma_{s_x} \geq \frac{1}{4\pi} \tag{11.4.14}$$

where the equality holds only for a coherent Gaussian beam. In this case, in fact, one has $I(x, y, z) \propto \exp[-2(x^2 + y^2)/w^2(z)]$. Using the coordinate transformations (11.4.11), we can show that $\hat{I}(s_x, s_y) \propto \exp[-2\pi^2 w_0^2 (s_x^2 + s_y^2)]$. From Eqs. (11.4.10) and (11.4.12) we then obtain

$$\sigma_x(z) = \frac{w(z)}{2} \tag{11.4.15a}$$

$$\sigma_{s_x} = \frac{1}{2\pi w_0} \tag{11.4.15b}$$

and obviously, setting $z = 0$ in Eq. (11.4.15a):

$$\sigma_{x0} = \frac{w_0}{2} \tag{11.4.15c}$$

From Eqs. (11.4.15b–c) we then obtain:

$$(\sigma_{x0}\sigma_{s_x})_G = \frac{1}{4\pi} \tag{11.4.16}$$

Following the preceding argument, we can now define an M_x^2 factor as the ratio of the $(\sigma_{x0}\sigma_{s_x})$ product of the beam to the corresponding product $(\sigma_{x0}\sigma_{s_x})_G$, for a Gaussian beam, i.e.:

$$M_x^2 = \frac{(\sigma_{x0}\sigma_{s_x})}{(\sigma_{x0}\sigma_{s_x})_G} = 4\pi(\sigma_{x0}\sigma_{s_x}) \tag{11.4.17}$$

and similarly for the y-coordinate. Note that, according to Eq. (11.4.14) one has $M_x^2 \geq 1$. The term M_x^2 is usually referred to as the beam quality. Since larger values of M_x^2 correspond to lower beam quality, it has also been referred to (less commonly) as the inverse beam quality factor. Note also that, if a comparison is made with a Gaussian beam having the same variance, i.e., if $(\sigma_{x0})_G = (\sigma_{x0})$, M_x^2 then specifies how much the beam divergence exceeds that of a Gaussian beam.

An alternative way to Eq. (11.4.13) for describing the propagation of a multimode laser beam is obtained by observing that, for a Gaussian beam, according to Eqs. (11.4.15a) and

(11.4.15c), we have $w_x(z) = 2\sigma_x(z)$ and $w_{x0} = 2\sigma_{x0}$. Consequently, for a laser beam of general transverse profile, we can define the spot size parameters $W_x(z)$ and W_{x0} as:

$$W_x(z) = 2\sigma_x(z) \tag{11.4.18a}$$

$$W_{x0} = 2\sigma_{x0} \tag{11.4.18b}$$

Note that we have used the upper case $W_x(z)$ and W_{x0} to indicate spot size parameters of an arbitrary laser beam. The substitution of $\sigma_x(z)$ and σ_{x0}, from Eq. (11.4.18), and σ_{sx} from Eq. (11.4.17), into Eq. (11.4.13) then gives:

$$W_x^2(z) = W_{x0}^2 + M_x^4 \left(\frac{\lambda^2}{\pi^2 W_{x0}^2} \right)(z - z_{0x})^2 \tag{11.4.19}$$

For a Gaussian beam, one obviously has $W_x(z) = w_x(z)$, $W_{x0} = w_{x0}$, and $M_x^2 = 1$ and Eq. (11.4.19) reduces to Eq. (4.7.13a). For a multimode laser beam, Eq. (11.4.19) represents an equation formally similar to that of a Gaussian beam except that the second term on the right-hand side, which expresses spreading due to beam diffraction, is multiplied by the factor M_x^4.

Equation (11.4.19) expresses the propagation of a multimode laser beam as a function of a precisely defined spot size parameter $W_x(z)$. Note that the beam propagation is determined by the three parameters $W_{x0}, M_x^2,$ and z_{0x}. Their values can in principle be determined by measuring the beam size $W_x(z)$ at three different z-coordinates. Note also that, at large distances from the waist position z_{0x}, one obtains from Eq. (11.4.19) $W_x(z) \cong (M_x^2 \lambda / \pi W_{x0})(z - z_{0x})$. For a multimode laser beam, one can then define a beam divergence as:

$$\theta_{dx} = \frac{W_x(z)}{(z - z_{0x})} = M_x^2 \left(\frac{\lambda}{\pi W_{x0}} \right) \tag{11.4.20}$$

The beam divergence of a multimode laser beam is thus M_x^2 times that of a Gaussian beam of the same spot size (i.e., such that $w_{x0} = W_{x0}$). A comparison between Eqs. (11.4.20) and (11.4.9) also allows one to establish a relation between M_x^2, W_{x0}, and the diameter D_c of the coherence area.

Example 11.4. *M^2 factor and spot-size parameters of a broad-area semiconductor laser.* We consider a broad-area AlGaAs/GaAs semiconductor laser with output beam dimensions, at the exit face of the laser, i.e., in the near field, of $d_\perp = 0.8\,\mu m$ and $d_\parallel = 100\,\mu m$, and beam divergences $\theta_\perp = 20°$ and $\theta_\parallel = 10°$. The labels $\perp$ and $\parallel$ represent directions perpendicular and parallel to the junction plane, respectively. Diameters d are measured between the half-intensity points; likewise divergences θ are measured as half-angles between half-maximum-intensity points (HWHM). Since the exit face of the semiconductor is plane, waist positions for both axes can be taken to coincide with this face. The near-field intensity distribution perpendicular to the junction (the fast axis) can be taken as approximately Gaussian. The corresponding spot size $w_{0\perp} = W_{0\perp}$ must then be such that $\exp[-2(d_\perp/2w_{0\perp})^2] = (1/2)$. We obtain $w_{0\perp} = W_{0\perp} = d_\perp/[2\ln 2]^{1/2} \cong 0.68\,\mu m$. The far-field intensity profile along the same direction may also be taken as Gaussian. According to Eq. (11.4.11) its intensity profile, in terms of the transverse coordinate $\theta_\perp z$, can be written for large z as $\propto \exp[-2(\theta_\perp z/W_\perp)^2]$, where $W_\perp(z) = w_\perp(z)$ is the spot size. Its value can be obtained by setting $\exp[-2(\theta_\perp z/W_\perp)^2] = (1/2)$. Since beam divergence is now defined as $\theta_{d\perp} = W_\perp/z$, we obtain $\theta_{d\perp} = (2/\ln 2)^{1/2}\theta_\perp \cong 0.59$ rad, and from Eq. (11.4.20), $M_\perp^2 = \pi W_{0\perp}\theta_{d\perp}/\lambda = \pi \theta_\perp d_\perp/(\ln 2)\lambda = 1.5$, where $\lambda \cong 850$ nm. As expected, the $M_\perp^2$ factor is close to that of a true Gaussian beam. The near-field intensity distribution, in the direction parallel to the junction (the slow axis), can be taken as approximately constant. From Eq. (11.4.10) we then obtain $W_{0\parallel} \cong d_\parallel/2 = 50\,\mu$m. The far-field intensity distribution is a bell-shaped function that can be approximated by a Gaussian function. As before, we then obtain $\theta_{d\parallel} = (2/\ln 2)^{1/2}\theta_\parallel \cong 0.148$ rad and, using Eq. (11.4.20), $M_\parallel^2 = \pi W_{0\parallel}\theta_{d\parallel}/\lambda = \pi\theta_\parallel d_\parallel/(2\ln 2)^{1/2}\lambda = 55$. Thus, in the slow axis direction, the beam divergence is much larger than the diffraction-limit of a Gaussian beam, i.e., the beam is many times the diffraction limit.

11.5. LASER SPECKLE

Following the discussion of first-order coherence in Sect. 11.3, we now briefly consider a striking phenomenon, characteristic of laser light, known as laser speckle.[8,9] It is apparent when one looks at light scattered from a laser beam incident on, for example, the surface of a wall or a transparent diffuser. Scattered light is then seen to consist of a random collection of alternately bright and dark spots (or *speckles*) (Fig. 11.8a). Despite the randomness, one can distinguish an average speckle (or grain) size. This phenomenon was soon recognized, by early workers in the field, as due to constructive and destructive interference of radiation from the small scattering centers within the area where the laser beam is incident. Since the phenomenon depends on a high degree of first-order coherence, it is an inherent feature of laser light.

The physical origin of the observed granularity can be readily understood for both free-space propagation (Fig. 11.8b) and an imaging system (Fig. 11.8c) when we realize that the surfaces of most materials are extremely rough on the scale of an optical wavelength. For

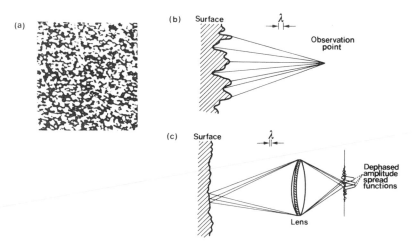

FIG. 11.8. (a) Speckle pattern and its physical origin (b) for free space propagation and (c) an image-forming system.

free space propagation, the resulting optical wave, at any moderately distant point from the scattering surface, consists of many coherent components or wavelets, each arising from a different microscopic element of the surface. Referring to Fig. 11.8b, we see that distances traveled by these various wavelets may differ by many wavelengths. Interference of the phase-shifted, but coherent, wavelets results in the granular intensity (or *speckle pattern*, as it is usually called). When the optical arrangement is that of an imaging system (Fig. 11.8c), an explanation of the observed pattern must take into account diffraction as well as interference. In fact, due to the finite resolving power of even a perfectly corrected imaging system, the intensity at a given image point can result from the coherent addition of contributions from many independent parts of the surface. This situation occurs in practice when the point spread function of the imaging system is broad in comparison to the microscopic surface variations.

We can readily obtain an order-of-magnitude estimate for the grain size d_g (i.e., the average size of spots in the speckle pattern) for the two cases just considered. In the first case (Fig. 11.9a), scattered light is assumed to be recorded on a photographic film at a distance L from the diffuser, with no lens between film and diffuser. Suppose now that a bright speckle is present at some point P in the recording plane. This means that light diffracted by all points of the diffuser interfere, at point P, in a predominantly constructive way to give an overall peak for the field amplitude. Heuristically, we can say that diffractive contributions, at point P, from wavelets scattered from points P_1, P_1', P_1'', etc., add (on the average) in phase with those from points P_2, P_2', P_2'', etc. We now ask how far point P must be moved along the x-axis in the recording plane for this constructive interference to become a destructing interference. This situation occurs when contributions of, e.g., diffracted waves from points P_1 and P_2 interfere, at the new point P', destructively rather than constructively. In this case we can show that contributions from points P_1', and P_2' also interfere destructively, the same situation also occurs for points P_1'' and P_2'', etc., and the overall light intensity has a minimum

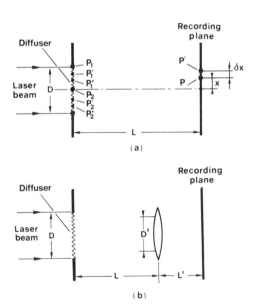

FIG. 11.9. Grain size calculation (a) for free-space propagation and (b) an image-forming system.

value. Taking, e.g., points P_1 and P_2, we then require the change δx, in the x-coordinate of point P, to be such that the corresponding change $\delta(P_2P - P_1P)$ in the path difference $P_2P - P_1P$ equals $\lambda/2$. Since $P_2P = (x^2 + L^2)^{1/2}$ and $P_1P = \{[(D/2) - x]^2 + L^2\}^{1/2}$, we obtain (for $D \ll L$) that $\delta(P_2P - P_1P) \cong (D/2L)\delta x$. The requirement $\delta(P_2P - P_1P) = \lambda/2$ then gives

$$\delta x = \frac{\lambda L}{D} \qquad (11.5.1)$$

Following a similar calculation, we can readily show that the same result is obtained by considering points P_1' and P_2' (or points P_1'' and P_2'', etc.) rather than points P_1 and P_2. All corresponding contributions now (on average) combine destructively rather than constructively, so we can write the following approximate expression for the grain size d_g:

$$d_g \cong 2\delta x = \frac{2\lambda L}{D} \qquad (11.5.2)$$

To obtain an approximate expression for the grain size for the imaging system in Fig. 11.8b, we first observe that a similar argument to that just presented can be used to calculate the beam diameter of the Airy spot in the focal plane of a lens. Consider, in fact, the case when the diffuser in Fig. 11.9a is replaced by a lens of focal length $f = L$. Following the preceding argument, we realize that an intensity maximum must be present at $x = 0$, i.e., at the center of the recording plane. In fact, as a result of the spherical wave front produced by the lens, contributions from points P_1, P_1', P_1'', etc., add in phase, there, with contributions from points P_2, P_2', P_2'', etc. The approximate size of the spot in the focal plane is then expected to be given by Eq. (11.5.2), i.e., equal to $d_g \cong 2\lambda L/D$. This result should be compared with the value $d_g = 2.44\,\lambda L/D$ obtained from the Airy function in Fig. 11.6. From this example, we now understand the following general property of a diffracted wave: Whenever the whole aperture of diameter D, of an optical system, contributes coherently to the diffraction to one or more spots in a plane located at a distance L, the minimum spot size in this recording plane is always approximately given by $2\lambda L/D$.* Note that, in the case of a diffuser, this coherent contribution from the whole aperture D occurs provided that: (1) The size d_s of individual scatterers is much smaller than the aperture D; (2) there is an appreciable overlap, at the recording plane, between wavelets diffracted from various scattering centers. This implies that the dimension of each of these wavelets at the recording plane ($\sim \lambda L/d_s$) is larger than their mean separation ($\sim D$). The length L must therefore be such that $L > d_s D/\lambda$. Thus for instance, when $d_s = 10\,\mu$m and $\lambda = 0.5\,\mu$m, one has $L > 20D$.

We now consider the case when scattered light is recorded on a photographic plate after passing through a lens that images the diffuser onto the plate (Fig. 11.9b). We assume that the diameter D' of the lens aperture is fully illuminated by light diffracted by each individual scatterer, i.e., $(2\lambda L/d_s) \geq D'$. Given this condition, the whole aperture contributes, by

---

* Since, for $D \gg L$, the field distribution in the recording plane is the Fourier transform of that in the input plane,[7] this property emerges as a general property of the Fourier transform.

diffraction, to each spot on the photographic plate and the grain size d_g at the plate is given by:

$$d_g = \frac{2\lambda L'}{D'} \qquad (11.5.3)$$

It should be noted that the arrangement in Fig. 11.9b also corresponds to the case where one looks directly at a scattering surface. In this case the lens and recording plane correspond to the lens of the eye and retina, respectively. Accordingly, d_g, as given by Eq. (11.5.3), can be taken as the expression for the grain size on the retina. Note also that the apparent grain size on the scattering surface d_{ag} is $d_g(L/L')$, so that:

$$d_{ag} = \frac{2\lambda L}{D'} \qquad (11.5.4)$$

This expression, which actually gives the eye's resolution for objects at a distance L (neglecting aberrations of the lens of the eye), shows that d_{ag} is expected to increase with increasing L, i.e., with increasing distance between the observer and the diffuser, and to decrease with an increasing aperture of the iris (i.e., when the eye is dark-adapted). Both these predictions are readily confirmed by experimental observations.

Example 11.5. *Grain size of the speckle pattern as seen by a human observer.* We consider a red, $\lambda = 633$ nm, He-Ne laser beam illuminating, e.g., an area of diameter $D = 2$ cm on a scattering surface for which individual scatterers are taken to have a dimension $d_s = 50\,\mu$m. Scattered light is observed by a human eye at a distance $L = 50$ cm from the diffuser. We take $L' = 2$ cm as the distance between the retina and the lens of the eye and assume a pupil diameter $D' = 2$ mm. Since $(2\lambda L/d_s) \cong 12.7$ mm is much greater than D', the whole aperture of the eye is illuminated by light diffracted by each individual scatterer. The apparent size of the speckle, at the illuminated region of the diffuser, is obtained from Eq. (11.5.4) as $d_{ag} \cong 316\,\mu$m. Note that, if the observer moves to a distance $L = 100$ cm from the diffuser, the apparent grain size, on the $D = 2$ cm illuminated spot of the diffuser, doubles to $\sim 632\,\mu$m.

Speckle noise is often an undesirable feature of coherent light; the spatial resolution of the image of an object made via illumination with laser light is in fact often limited by speckle noise. Speckle noise is also apparent in the reconstructed image of a hologram, which again limits the spatial resolution of this image. Some techniques have therefore been developed to reduce speckle from coherently illuminated objects.[9] Speckle noise is not always a nuisance however; in fact, techniques have been developed to exploit the presence of speckle to indicate, in a rather simple way, the deformation of large objects arising, e.g., from stresses or vibrations (*speckle interferometry*).

11.6. BRIGHTNESS

The brightness B of a light source or a laser source was introduced in Sect. 1.4.4 [see Eqs. (1.4.3) and (1.4.4)]. We note again that the most significant parameter of a laser beam (and in general any light source) is not simply its power or its intensity, but its brightness. This was already pointed out in Sect. 1.4.4. where it was shown that the maximum peak intensity, which can be obtained by focusing a given beam, is proportional to the beam brightness [see Eq. (1.4.6)]. This is further emphasized by the fact that, although beam intensity can be increased, its brightness cannot. In fact, the simple arrangement of confocal

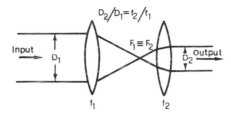

FIG. 11.10. Method for increasing the intensity of a laser beam.

lenses in Fig. 11.10 can be used to decrease the beam diameter, if $f_2 < f_1$, and hence increase its intensity. However, the divergence of the output beam $(\sim \lambda/D_2)$ is also correspondingly increased relative to that $(\sim \lambda/D_1)$ of the input beam, so brightness remains invariant. This property, illustrated here for a particular case, is generally valid (for incoherent sources also): Given some light source and an optical imaging system, the image cannot be brighter than the original source (this is true provided the source and image are in media of the same refractive index).

The brightness of laser sources is typically several orders of magnitude greater than that of the most powerful incoherent sources, due to the extreme directionality of a laser beam. Let us compare, for example, an Ar laser oscillating on its green $\lambda = 514$ nm transition, with power $P = 1$ W, with what is probably the brightest conventional source—a high-pressure mercury vapor lamp (PEK^{TM} Labs type 107/109). The lamp is driven by an electrical power of ~ 100 W and gives an optical output power of $P_{out} \cong 10$ W and a brightness B of ~ 95 W/cm^2 × sr in its most intense green line at $\lambda = 546$ nm wavelength. We assume that the laser oscillates on a TEM$_{00}$ mode, so that we can take $A = (\pi w_0^2/2)$ as the beam area, where w_0 is the spot size at the beam waist. Note that the factor 2 in the denominator of the preceding expression arises from the fact that w_0 actually represents the $(1/e)$ spot size of the laser field rather than of the laser intensity. Likewise, since the $(1/e)$ field divergence is given by $\theta = (\lambda/\pi w_0)$, the emission solid angle can be taken as $\Omega = (\pi \theta^2/2)$. The brightness of this laser source can now be written as $B = (P/A\Omega)$, where P is the power. From the previous two expressions for beam area and the emission solid angle, we then obtain

$$B = \left(\frac{4P}{\lambda^2}\right) \tag{11.6.1}$$

Insertion of the appropriate values for P and λ for this laser gives $B \cong 1.6 \times 10^9$ W/cm^2 × sr. The brightness of the Ar laser is thus more than 7 orders of magnitude larger than that of the lamp. Since this would also be the ratio of the two peak intensities, as obtained by focusing the corresponding sources, we now have a more quantitative understanding of why a focused laser beam can be used, while a focused lamp cannot be used, in applications like, e.g., welding and cutting.

11.7. STATISTICAL PROPERTIES OF LASER LIGHT AND THERMAL LIGHT

The temporal fluctuations of the field generated by a laser source or a thermal light source can be described effectively in terms of the corresponding statistical behavior.[10] Let

$E(t) = A(t) \exp j[\omega t - \phi(t)]$ be the field generated by the source at some given space point. Writing $E(t) = \tilde{E}(t) \exp j(\omega t)$, where $\tilde{E} = A \exp -j(\phi)$, we can then concentrate our considerations on just the slowly varying (over an optical cycle) complex amplitude $\tilde{E}(t)$. Suppose now that a number of measurements are made, at different times t, of $\tilde{E}$, e.g., of its real E_r and imaginary E_i parts. In the limit of a very large number of measurements, we can obtain the bidimensional probability density $p_E(\tilde{E}) = p_E(E_r, E_i)$, defined such that $dp = p_E(\tilde{E}) dE_r\, dE_i$ represents the elemental probability that a field measurement gives a value for the real part between E_r and $E_r + dE_r$ and imaginary part between E_i and $E_i + dE_i$. Alternatively, we can represent $p_E(\tilde{E})$ as a function of the amplitude A and phase ϕ and so write $dp = p_E(\tilde{E}) A\, dA\, d\phi$ as the elemental probability that a measurement gives a value of the amplitude between A and $A + dA$ and phase between ϕ and $\phi + d\phi$. Once $p_E(\tilde{E})$ is known, the average intensity of the wave, according to Eq. (11.1.3), can be written as:

$$\langle I \rangle = \frac{\iint |E|^2 p_E(\tilde{E}) dE_r\, dE_i}{\iint p_E(\tilde{E}) dE_r\, dE_i} = \frac{\iint |A|^2 p_E(\tilde{E}) A\, dA\, d\phi}{\iint p_E(\tilde{E}) A\, dA\, d\phi} \qquad (11.7.1)$$

The probability density can be represented very effectively in three-dimensional space as a function of E_r and E_i.

Figure 11.11a shows the plot of $p_E(\tilde{E})$ versus (E_r, E_i) for a single-mode laser source. As already pointed out in Sect. 7.11, the output intensity and hence the field amplitude of this laser is fixed, for a given pump rate and cavity loss, by the condition that upward transitions due to pumping must be balanced by downward transitions due to both stimulated emission and spontaneous decay. Small amplitude fluctuations may then arise from fluctuations in both pump rate and cavity loss. The phase $\phi(t)$, however, is not controlled by such a balancing process, so it is free to assume any value from 0 to ∞. Now, since one has $A = (E_r^2 + E_i^2)^{1/2}$ and $\phi = -\tan^{-1}(E_i/E_r)$, the expected plot is as shown in Fig. 11.11a. Note that amplitude fluctuations of $A = A(t)$ are greatly exaggerated in the figure; relative amplitude fluctuations of a few percent or less are, in fact, typical in free-running operation for, e.g., a good diode-pumped solid-sate laser source (see Fig. 7.30). To first order we can then assume

$$p_E(\tilde{E}) \propto \delta(A - A_0) \qquad (11.7.2)$$

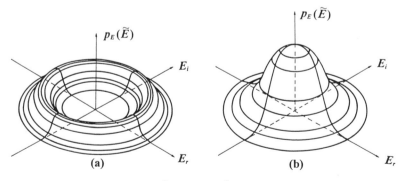

FIG. 11.11. Plot of the probability density $p_E(\tilde{E})$ of the field $\tilde{E}$ versus the real E_r and imaginary E_i, parts of $\tilde{E}$; (a) a coherent signal such as that of a single-mode laser; (b) a thermal light such as that emitted by a conventional light source.

where A_0 is a constant determined by the average beam intensity. The substitution of Eq. (11.7.2) into Eq. (11.7.1) gives in fact $\langle I \rangle = A_0^2$. In the time domain, the point that represents $\tilde{E}(t)$ in the (E_r, E_i) plane travels essentially along the circumference of radius $|\tilde{E}| = A_0$. As a consequence of the statistical nature of phase fluctuations, this movement turns out to be random; the corresponding angular speed, $d\phi/dt$, then establishes the laser bandwidth.

Figure 11.11b shows the plot of $p_E(\tilde{E})$ versus (E_r, E_i) for a thermal light source. In this case, the field is due to the superposition of uncorrelated light emitted, by spontaneous emission, from individual atoms of the light source. Since the number of these emitters is very large, it follows from the central limit theorem of statistics that the probability distribution of both real and imaginary parts of $\tilde{E}$ must follow a Gaussian law. We can thus write

$$p_E(\tilde{E}) \propto \exp\left[-\left(\frac{E_r^2 + E_i^2}{A_0^2}\right)\right] = \exp\left[-\left(\frac{A^2}{A_0^2}\right)\right] \qquad (11.7.3)$$

where A_0 is a constant again determined by the average beam intensity. The substitution of Eq. (11.7.3) into Eq. (11.7.1) again gives, in fact, $\langle I \rangle = A_0^2$. Note that the average values of both E_r and E_i are now zero. Thus, the time domain movement of the point representing $\tilde{E}$ in the (E_r, E_i) plane now consists of a random movement around the origin. The speed of this movement, in terms of both amplitude and phase (i.e., $dA/A\,dt$ and $d\phi/dt$), establishes now the bandwidth of this thermal source.

11.8. COMPARISON BETWEEN LASER LIGHT AND THERMAL LIGHT

We now compare a red-emitting, $\lambda = 633$ nm, He-Ne laser oscillating on a single mode with a modest output power of 1 mW with what is probably the brightest conventional source, already considered in Sect. 11.6. (PEK^{TM} Labs type 107/109), giving an optical output power of $P_{out} \cong 10$ W and a brightness B of ~ 95 W/cm$^2 \times$ sr in its most intense green line at $\lambda = 546$ nm wavelength. To obtain a beam with good spatial coherence from this lamp, one can use the arrangement in Fig. 11.12, where a lens of focal length f and suitable aperture D collects a fraction of the emitted light. The lamp is simulated by individual emitters located within an aperture of diameter d in a screen S. Following the

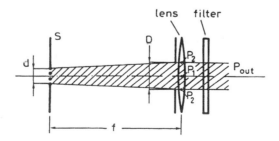

FIG. 11.12. Spatial and frequency filters to obtain a first-order, coherent output beam from an incoherent lamp.

discussion in Sect. 11.3.6, to obtain a beam with a spatial coherence that approaches the ideal, aperture D must be chosen so that [see Eq. (11.3.43)]:

$$D \cong \frac{0.32\lambda f}{d} \tag{11.8.1}$$

This beam, although having a degree of spatial coherence somewhat less than unity $[\gamma^{(1)}(P_1, P_2) \cong 0.88$, in the case considered], may be regarded, generously, as having the same degree of spatial coherence as that of the He-Ne laser (whose value for $\gamma^{(1)}$ may be taken as essentially unity). The output power of the beam obtained after the lens is then given by $P_{out} = BA\Omega$, where B is the brightness of the lamp, A is its emitting area ($A = \pi d^2/4$), and Ω is the acceptance solid angle of the lens ($\Omega = \pi D^2/4f^2$). With the help of Eq. (11.8.1), we obtain [compare with Eq. (11.6.1)]:

$$P_{out} \cong \left(\frac{\lambda}{4}\right)^2 B \tag{11.8.2}$$

Using the value $B = 95\,\text{W/cm}^2 \times \text{sr}$ for lamp brightness, we obtain $P_{out} \cong 1.8 \times 10^{-8}\,\text{W}$. Note that this power is ~ 5 orders of magnitude smaller than that emitted by the He-Ne laser and ~ 9 orders of magnitude smaller than that emitted by the lamp. Note also that, according to Eq. (11.8.2), the power that can be obtained in a spatially coherent beam depends only on the brightness of the lamp; this further illustrates the importance of brightness.

Having paid such a high penalty in terms of output power, we can now say to have a beam with about the same degree of spatial coherence as the He-Ne laser. The degree of temporal coherence, however, is still much worse because the emission bandwidth of the lamp is certainly much larger than that of the He-Ne laser. In fact, the linewidth of the lamp is broadened by the high pressure of its internal gas to a width of $\Delta v \cong 10^{13}\,\text{Hz}$. Even using a modest frequency stabilization scheme, the linewidth of the He-Ne laser may be taken as $\Delta v_L \cong 1\,\text{kHz}$. To achieve the same degree of time coherence, we must then make the bandwidths of the two sources the same. In principle this can be done by inserting a frequency filter with the (exceptionally) narrow linewidth of $1\,\text{kHz}$ in the output beam obtained from the lamp (see Fig. 11.12). However, this filter further reduces the output power of this source by about 10 orders of magnitude [corresponding to $(\Delta v_L/\Delta v) \cong 10^{-10}$], so that the final output power of the spatially- and frequency-filtered lamp is $P_{out} \cong 10^{-18}\,\text{W}$.

Thus, for a penalty of about 19 orders of magnitude on the original green output power emitted by the lamp, we can now say that the He-Ne laser beam and the filtered output beam in Fig. 11.12 show approximately the same degree of spatial and temporal coherence. To compare the two beams at the same output power, we can now place an attenuator with a factor 10^{15} attenuation in front of the He-Ne laser. At this stage the power and degree of coherence for the two beams is the same, so it is natural to ask whether these two sources are now effectively the same, i.e., indistinguishable. The answer is however negative. In fact, a detailed comparison between the two sources shows that the two beams remain basically different and that, notably, the He-Ne laser beam still remains more coherent.

A first comparison can readily be made in terms of the statistical properties of the two light sources. One may notice, in fact, that the filtering operation, applied to the lamp output, and an attenuator, placed in front of the He-Ne laser beam, do not alter the statistical properties of the corresponding light. Thus Figs. 11.11a–b can still be used to describe the statistical properties of the two sources and, from these figures, we can see the basic difference between the two beams. Note that, since the output powers of the two beams have been equalized, the quantity A_0 in Eqs. (11.7.2) and (11.7.3) is the same for the two sources. Note also that, if the degree of temporal coherence of the two beams has been made equal, this means that the speed of movement of the point representing $\tilde{E}$ in the (E_r, E_i) plane is the same for the two cases. If the degree of spatial coherence of the two beams is equalized, this means that, for each beam, the movement in the (E_r, E_i) plane is the same at any point of the wave front. Nonetheless, the statistical properties of the two beams, as represented in Fig. 11.11, still remain fundamentally different.

A second comparison between the two sources can be made in terms of the coherence properties to higher order (see Appendix H). To this purpose we recall that the coherence function $\Gamma^{(1)}$ was introduced in Sect. 11.3 in terms of the product $E(x_1)E^*(x_2)$ between fields taken at two different space-time points $x_i = (\mathbf{r}_i, t_i)$. For this reason, a superscript (1) has been used for Γ as a reminder that one is actually performing a first-order correlation between the two fields. At a higher order, we can then introduce a whole class of correlation functions, e.g., $\langle E(x_1)E(x_2)E^*(x_3)E^*(x_4)\rangle$ involving the four distinct space-time points x_1, x_2, x_3, and x_4, and so on, to yet higher orders.* We can then introduce some suitable definition of higher order coherence $\Gamma^{(n)}$ in terms of these higher order correlation functions. When this is done, the higher order coherence of our monomode laser source turns out to be still greater than that of the filtered beam taken from a lamp (see Appendix H). It also turns out that, at best, we can arrange for the two sources to exhibit the same first-order coherence, i.e., the same degree of spatial and temporal coherence, as indeed achieved by the filtering system in Fig. 11.12.

Thus, despite having paid such a heavy penalty in terms of output power, the filtered thermal source remains basically different from a laser.

PROBLEMS

11.1. Calculate $\Gamma^{(1)}(\mathbf{r}_1, \mathbf{r}_1, \tau)$ for a sinusoidal wave.

11.2. Prove Eq. (11.3.9).

11.3. For the Michelson interferometer in Fig. 11.4a, find the analytical relation between the intensity along the C direction I_c and $\gamma^{(1)}(\mathbf{r}, \mathbf{r}, \tau)$, where $\tau = 2(L_3 - L_2)/c$.

11.4. Assume that the field at point P of the Michelson interferometer in Fig. 11.4a is made of a sinusoidal wave with constant amplitude and random phase jumps (see Fig. 2.9). Using the expression for $\Gamma^{(1)}(\tau)$ calculated in Appendix B for this field, and the relation between I_c and $\Gamma^{(1)}(\tau)$ calculated in Problem 11.3, find an analytical expression for $V_P = V_P(\tau)$.

11.5. The shape of the spectral output of a CO_2 laser beam operating at $\lambda = 10.6\,\mu$m is approximately Gaussian with a bandwidth of 10 kHz [$\Delta\nu_L$ is defined according to Eq. (11.3.27)]. Calculate the coherence length L_c and the distance ΔL between two successive maxima of the intensity curve in Fig. 11.4b.

---

* Actually one can show that a given field $E(x)$ can in principle be completely characterized by either the infinite set of values for E, obtained by changing x, or by the infinite set of its correlation functions.

11.6. A plane wave of circular cross section, uniform intensity, and perfect spatial coherence is focused by a lens. What is the increase in intensity at the focus compared to that of the incident wave?

11.7. A Gaussian beam is focused by a lens of focal length f. Assuming that the waist of the incident beam is located at the lens position and the corresponding spot size w_0 is appreciably smaller than the lens diameter, relate peak intensity at the focal spot to the power P_i of the incident beam. Compare the resulting expression with Eq. (11.4.4).

11.8. The near-field transverse intensity profile of a Nd:YAG laser beam at $\lambda = 1.064\,\mu m$ wavelength is, to a good approximation, Gaussian with a diameter (FWHM) $D \cong 4\,mm$. The half-cone beam divergence, measured at the half-maximum point of the far-field intensity distribution, is $\theta_d \cong 3\,mrad$. Calculate the corresponding M^2 factor.

11.9. The near-field transverse intensity profile of a pulsed TEA CO_2 laser beam at $\lambda = 10.6\,\mu m$ wavelength is, to a good approximation, constant over its $1\,cm \times 4\,cm$ dimension. The laser is advertised to have an M^2 factor of 16 along both axes. Assuming that the waist is located at the position of the output mirror, calculate the spot size parameters at a distance from this mirror of $z = 3\,m$.

REFERENCES

1. M. Born and E. Wolf, *Principles of Optics*, 6th ed. (Pergamon, Oxford, 1980), pp. 491–544.
2. W. H. Louisell, *Radiation and Noise in Quantum Electronics* (McGraw-Hill, New York, 1964), pp. 47–53.
3. Ref. 1, pp. 508–518.
4. J. W. Goodman, *Introduction to Fourier Optics* (McGraw-Hill, New York, 1968).
5. Ref. 1, pp. 395–398.
6. A. E. Siegman, Defining and Measuring Laser Beam Quality, in *Solid-State Lasers—New Developments and Applications* (M. Inguscio and R. Wallenstein, eds.) (Plenum, New York, 1993), pp. 13–28.
7. Ref. 4, Chap. 5.
8. *Laser Speckle and Related Phenomena* (J. C. Dainty, ed.) (Springer-Verlag, Berlin, 1975).
9. M. Françon, *Laser Speckle and Applications in Optics* (Academic, New York, 1979).
10. R. J. Glauber, Optical Coherence and Photon Statistics, in *Quantum Optics and Electronics* (C. De Witt, A. Blandin, and C. Cohen-Tannoudji, eds.) (Gordon and Breach, New York, 1965), pp. 71, 94–98, 103, 151–155.
11. A. Jeffrey, *Handbook of Mathematical Formulas and Integrals* (Academic, San Diego, 1995), p. 244.

12

Laser Beam Transformation: Propagation, Amplification, Frequency Conversion, Pulse Compression, and Pulse Expansion

12.1. Introduction

Before it is put to use, a laser beam is generally transformed in some way. The most common transformation occurs when the beam is made to propagate in free space or through a suitable optical system. Since this produces a change in the spatial distribution of the beam (e.g., the beam may be focused or expanded), we refer to this as a *spatial transformation* of the laser beam. A second, frequently encountered, transformation occurs when the beam is passed through an amplifier or chain of amplifiers. Since the main effect here is to alter the beam amplitude, we refer to this as *amplitude transformation*. A third, rather different, transformation occurs when the wavelength of the beam is changed via propagation through a suitable nonlinear optical material (*wavelength transformation* or *frequency conversion*). The temporal behavior of the laser beam can also be modified by a suitable optical element. For example the amplitude of a cw laser beam can be temporally modulated by an electrooptic or acoustooptic modulator, or the time duration of a laser pulse can be increased (pulse expansion) or decreased (pulse compression) using suitably dispersive optical systems or nonlinear optical elements. This type of transformation is here referred to as *time transformation*. Note that these four types of beam transformation are often interrelated. For instance, amplitude transformation and frequency conversion often result in spatial and time transformations occurring as well.

This chapter briefly discusses these four cases of laser beam transformation. Of the various nonlinear optical effects used[1] to achieve frequency conversion, only the so-called parametric effects are considered here. These in fact provide some of the most useful techniques, so far developed, for producing a new source of coherent light. Time

493

transformation will be considered only for the cases of pulse expansion or pulse compression, while we refer elsewhere for the case of amplitude modulation.[2] We also omit a number of aspects of amplitude and time transformation that arise from the nonlinear phenomenon of self-focusing and the associated phenomenon of self-phase modulation,[3] although these play a very important role in limiting, for instance, the performance of laser amplifiers.

12.2. SPATIAL TRANSFORMATION: PROPAGATION OF A MULTIMODE LASER BEAM

The free-space propagation of a Gaussian beam and of a multi-transverse-mode beam has already been considered in Sects. 4.7.2 and 11.4.3, respectively. In Sect. 4.7.2 a Gaussian beam with, e.g., a circular cross section, is characterized by two parameters—the coordinate z_0 of the beam waist and the corresponding spot size w_0. In contrast, Sect. 11.4.3. shows that a multimode beam with, e.g., a circular cross section, is characterized by three parameters—the coordinate z_0 of the beam waist, the spot size parameter W_0, and the beam quality factor M^2. The propagation of a Gaussian beam through a general optical system characterized by a given *ABCD* matrix, has, on the other hand, been considered in Section 4.7.3. It was shown there that the complex q-parameter of the beam, after passing through the optical system, can be readily obtained from the q-parameter of the input beam once the matrix elements, *A, B, C, D*, are known. To complete this picture, we consider in this section the propagation of a multimode laser beam through a general optical system characterized by a given *ABCD* matrix.[4,5]

Consider first the free-space propagation of a multimode laser beam. The spot size parameter along, e.g., the transverse direction x and at a longitudinal coordinate z is described by Eq. (11.4.19), which we reproduce here:

$$W_x^2(z) = W_{0x}^2 + M_x^4 \left(\frac{\lambda^2}{\pi^2 W_{0x}^2} \right)(z - z_{0x})^2 \qquad (12.2.1)$$

As far as free-space propagation is concerned, the multimode laser beam behaves as if it contained an embedded Gaussian beam having the same waist location z_{0x} as the multimode beam and a spot size, at any coordinate z, given by:

$$w_x(z) = \frac{W_x(z)}{M_x} \qquad (12.2.2)$$

where $M_x = \sqrt{M_x^2}$ is a constant. In fact, the substitution of Eq. (12.2.2) into Eq. (12.2.1) readily gives Eq. (4.7.13a). One can also show that the radius of curvature $R(z)$ of this embedded Gaussian beam, at any z, equals that of the multimode beam.

This notion of an embedded Gaussian beam also holds for propagation through a general optical system described by, e.g., its *ABCD* matrix. Accordingly, propagation of the multimode beam can be obtained by the following procedure:

Example 12.1. *Focusing of a multimode Nd:YAG laser beam by a thin lens.* Consider a multimode beam from a repetitively pulsed Nd:YAG laser at $\lambda \cong 1.06$ μm wavelength, such as used for welding or cutting metallic materials. The near-field transverse intensity profile can

1. Starting with the multimode laser beam characterized by given values of W_{0x}, M_x^2, and z_{0x}, one defines the embedded Gaussian beam such that the beam waist is at the location of the multimode beam waist and $w_{0x} = W_{0x}/M_x$.
2. One then calculates the propagation of the embedded Gaussian beam through the optical system by, e.g., using the *ABCD* law of Gaussian beam propagation.
3. At any location within the optical system, the wave-front radius of curvature of the multimode beam then coincides with that of the embedded Gaussian beam. This means, in particular, that any waist has the same location for the two beams.
4. The spot size parameter W_x of the multimode beam, at any location, is then given by $W_x(z) = M_x w_x(z)$.

be taken as approximately Gaussian with a diameter (FWHM) of $D = 4\,mm$; the M^2 factor can be taken as ~ 40. We want to see what happens when the beam is focused by a spherical lens of $f = 10\,cm$ focal length.

We assume that the waist of this multimode beam coincides with the output mirror, since this is a plane mirror. We also assume that the lens is located very close to this mirror, so that the waist of the multimode beam and hence of the embedded Gaussian beam can be taken to coincide with the lens location. For a Gaussian intensity profile, the spot size parameter of the input beam, $W = W_0$, is then related to the beam diameter, D, by the condition $\exp[-2(D/2W_0)^2] = (1/2)$. We get $W_0 = D/(2\ln 2)^{1/2} \cong 3.4\,mm$, so that $w_0 = W_0/\sqrt{M^2} \cong 0.54\,mm$. According to Eq. (4.7.26), since the Rayleigh range corresponding to this spot size, $z_R = \pi w_0^2/\lambda \cong 85\,cm$, is much larger than the focal length of the lens, the waist formed beyond the lens is located approximately at the lens focus. From Eq. (4.7.28) the spot size of the embedded Gaussian beam at this focus w_{0f} is then given by $w_{0f} \cong \lambda f/\pi w_0 \cong 63\,\mu m$ and the spot size parameter of the multimode beam by $W_{0f} = \sqrt{M^2} w_{0f} \cong 400\,\mu m$.

12.3. AMPLITUDE TRANSFORMATION: LASER AMPLIFICATION

In this section we consider the rate equation treatment of a laser amplifier.[6–8] We assume that a plane wave of uniform intensity I enters (at $z = 0$) a laser amplifier extending for a length l in the z-direction. We limit our consideration to the case when the incoming laser beam is in the form of a pulse (pulse amplification); we refer to Ref. 8 for the amplification of a cw beam (steady-state amplification).

We first consider the case of an amplifier medium working on a four-level scheme and further assume that the pulse duration τ_p is such that $\tau_1 \ll \tau_p \ll \tau$, where τ_1 and τ are the lifetimes of the lower and upper levels of the amplifier medium, respectively. In this case the population of the lower level of the amplifier can be set equal to zero. This is perhaps the most relevant case to consider, since it applies for instance to the case of a Q-switched laser pulse from a Nd:YAG laser that is being amplified. We also assume that pumping to the amplifier upper level and subsequent spontaneous decay can be neglected during passage of the pulse and that the transition is homogeneously broadened. Under these conditions and with the help of Eq. (2.4.17) (in which we set $F = I/hv$), the rate of change of population inversion $N(t, z)$ at a point z within the amplifier can be written as:

$$\frac{\partial N}{\partial t} = -WN = -\frac{NI}{\Gamma_s} \qquad (12.3.1)$$

where

$$\Gamma_s = \left(\frac{h\nu}{\sigma}\right) \tag{12.3.2}$$

is the saturation energy fluence of the amplifier [see Eq. (2.8.29)]. Note that a partial derivative is required in Eq. (12.3.1), since we expect N to be a function of both z and t, i.e., $N = N(t, z)$, because $I = I(t, z)$.

Equation (12.3.1) can be solved for $N(t)$ to yield

$$N(\infty) = N_0 \exp[-(\Gamma/\Gamma_s)] \tag{12.3.3}$$

where $N_0 = N(-\infty)$ is the amplifier's upper level population before the arrival of the pulse, as established by the combination of pumping and spontaneous decay and where

$$\Gamma(z) = \int_{-\infty}^{+\infty} I(z, t)dt \tag{12.3.4}$$

is the total fluence of the laser pulse.

Next we derive a differential equation describing the temporal and spatial variation of intensity I. To do this, we first write an expression for the rate of change of em energy within unit volume of the amplifier. For this we refer to Fig. 12.1, where the shaded area indicates an elemental volume, of the amplifier medium, of length dz and cross section S. We can then write

$$\frac{\partial \rho}{\partial t} = \left(\frac{\partial \rho}{\partial t}\right)_1 + \left(\frac{\partial \rho}{\partial t}\right)_2 + \left(\frac{\partial \rho}{\partial t}\right)_3 \tag{12.3.5}$$

where $(\partial \rho/\partial t)_1$ accounts for stimulated emission and absorption in the amplifier, $(\partial \rho/\partial t)_2$ for amplifier loss (e.g., scattering losses), and $(\partial \rho/\partial t)_3$ for the net photon flux that flows into the volume. With the help again of Eq. (2.4.17) ($F = I/h\nu$), we obtain

$$\left(\frac{\partial \rho}{\partial t}\right)_1 = WNh\nu = \sigma NI \tag{12.3.6}$$

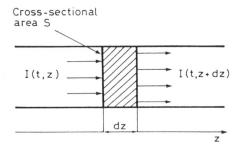

FIG. 12.1. Rate of change of photon energy contained in an elemental volume of length dz and cross-sectional area S of a laser amplifier.

and from Eqs. (2.4.17) and (2.4.32) we obtain

$$\left(\frac{\partial \rho}{\partial t}\right)_2 = -W_a N_a h\nu = -\alpha I \tag{12.3.7}$$

where N_a is the density of loss centers, W_a the absorption rate, and α the absorption coefficient associated with the loss centers. To calculate $(\partial \rho / \partial t)_3$, we refer again to Fig. 12.1 and observe that $(\partial \rho / \partial t)_3 S\, dz$ is the rate of change of em energy in this volume due to the difference between incoming and outgoing laser power. We can then write $(\partial \rho / \partial t)_3 S\, dz = S[I(t, z) - I(t, z + dz)]$, which readily gives

$$\left(\frac{\partial \rho}{\partial t}\right)_3 dz = -\frac{\partial I}{\partial z} dz \tag{12.3.8}$$

Equation (12.3.5), with the help of Eqs. (12.3.6)–(12.3.8) and with the observation that $(\partial \rho / \partial t) = (\partial I / c\partial t)$, gives

$$\frac{1}{c}\frac{\partial I}{\partial t} + \frac{\partial I}{\partial z} = \sigma N I - \alpha I \tag{12.3.9}$$

This equation and Eq. (12.3.1) completely describe the amplification process. Note that Eq. (12.3.9) has the usual form of a time-dependent transport equation.

Equations (12.3.1) and (12.3.9) must be solved with the appropriate boundary and initial conditions. For the initial condition we take $N(0, z) = N_0$, where N_0 is the amplifier upper level population before the arrival of the laser pulse. The boundary condition is obviously established by the intensity $I_0(t)$ of the light pulse injected into the amplifier, i.e., $I(t, 0) = I_0(t)$. For negligible amplifier losses (i.e., neglecting the term $-\alpha I$), the solution to Eqs. (12.3.1) and (12.3.9) can be written as:

$$I(z, \tau) = I_0(\tau)\left\{1 - [1 - \exp(-gz)]\exp\left[-\int_{-\infty}^{\tau} I_0(\tau')d\tau'/\Gamma_s\right]\right\}^{-1} \tag{12.3.10}$$

where $\tau = t - (z/c)$ and $g = \sigma N_0$ is the unsaturated gain coefficient of the amplifier.

From Eqs. (12.3.1) and (12.3.9), we can also obtain a differential equation for the total fluence of the pulse, $\Gamma(z)$, given by Eq. (12.3.4). Thus, we first integrate both sides of Eq. (12.3.1) with respect to time, from $t = -\infty$ to $t = +\infty$, to obtain $(\int_{-\infty}^{+\infty} NI\, dt/\Gamma_s) = N_0 - N(+\infty) = N_0[1 - \exp(-\Gamma/\Gamma_s)]$, where Eq. (12.3.3) has been used. We then integrate both sides of Eq. (12.3.9) with respect to time, over the same time interval, then use the preceding expression for $(\int_{-\infty}^{+\infty} NI\, dt/\Gamma_s)$ and the fact that $I(+\infty, z) = I(-\infty, z) = 0$. We obtain

$$\frac{d\Gamma}{dz} = g\Gamma_s[1 - \exp(-\Gamma/\Gamma_s)] - \alpha\Gamma \tag{12.3.11}$$

Again neglecting amplifier losses, Eq. (12.3.11) gives

$$\Gamma(l) = \Gamma_s \ln\{1 + [\exp(\Gamma_{in}/\Gamma_s) - 1]G_0\} \tag{12.3.12}$$

where $G_0 = \exp(gl)$ is the unsaturated gain of the amplifier and Γ_{in} is the energy fluence of the input beam.

As a representative example, the ratio Γ/Γ_s is plotted in Fig. 12.2 against Γ_{in}/Γ_s for $G_0 = 3$. Note that, for $\Gamma_{in} \ll \Gamma_s$, Eq. (12.3.12) can be approximated as

$$\Gamma(l) = G_0 \Gamma_{in} \tag{12.3.13}$$

and the output fluence increases linearly with the input fluence (linear amplification regime). Equation (12.3.13) is also plotted in Fig. 12.2 as a dashed straight line starting from the origin. At higher input fluences, however, Γ increases with Γ_{in} at a lower rate than that predicted by Eq. (12.3.13); i.e., amplifier saturation begins to occur. For $\Gamma_{in} \gg \Gamma_s$ (deep saturation regime) Eq. (12.3.12) can be approximated to:

$$\Gamma(l) = \Gamma_{in} + gl\Gamma_s \tag{12.3.14}$$

Equation (12.3.14) is also plotted in Fig. 12.2 as a dashed straight line. From Eq. (12.3.14) one sees that, for high input fluences, the output fluence is linearly dependent on the length l of the amplifier. Since $gl\Gamma_s = N_0 lh\nu$, one then realizes that every excited atom undergoes stimulated emission and thus contributes its energy to the beam. Such a condition obviously represents the most efficient conversion of stored energy into beam energy; for this reason, amplifier designs operating in the saturation regime are used wherever practical.

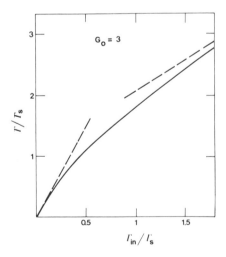

FIG. 12.2. Output laser energy fluence Γ versus input fluence Γ_{in} for a laser amplifier with a small signal gain $G_0 = 3$. Energy fluence is normalized to the laser saturation fluence $\Gamma_s = h\nu/\sigma$.

We point out again that the preceding equations have been derived for an amplifier having an ideal four-level scheme. For a quasi-three-level scheme, on the other hand, one can see from Sect. 7.2.2 that Eq. (12.3.1) still applies provided Γ_s is now given by

$$\Gamma_s = \frac{hv}{(\sigma_e + \sigma_a)} \qquad (12.3.15)$$

where σ_e and σ_a are the effective cross sections for stimulated emission and absorption, respectively. One can also show that Eq. (12.3.9) still applies provided σ is replaced by σ_e. It then follows that Eq. (12.3.12) still remains valid provided Γ_s is given by Eq. (12.3.15) and G_0 is given by $G_0 = \exp \sigma_e N_0 l$. Similar considerations apply to an amplifier operating on a four-level scheme when the pulse duration becomes much shorter than the lifetime of the lower level of the transition. In this case, the population driven to the lower level by stimulated emission remains in this level during the pulse, and we can show that Eq. (12.3.12) is still valid provided σ is replaced by σ_e and Γ_s is given by Eq. (12.3.15), where σ_a is the effective absorption cross section of the lower level of the transition.

If amplifier losses cannot be neglected, the preceding picture must be modified somewhat. In particular the output fluence $\Gamma(l)$ does not continue increasing with input fluence as in Fig. 12.2, but it reaches a maximum, then decreases. This can be understood by noting that, in this case, the output as a function of amplifier length tends to grow linearly due to amplification [at least for high input fluences; see Eq. (12.3.14)] and to decrease exponentially due to loss [due to the term $-\alpha\Gamma$ in Eq. (12.3.11)]. The competition of these two effects then gives a maximum for the output fluence Γ. For $\alpha \ll g$ this maximum value of the output fluence Γ_m is:

$$\Gamma_m \cong \frac{g\Gamma_s}{\alpha} \qquad (12.3.16)$$

It should be noted, however, that, since amplifier losses are typically quite small, other phenomena usually limit the maximum energy fluence that can be extracted from an amplifier. In fact, the limit is usually set by the amplifier damage fluence Γ_d ($10\,\mathrm{J/cm^2}$ is a typical value for a number of solid-state media). From Eq. (12.3.14) we then obtain the condition:

$$\Gamma \cong gl\Gamma_s < \Gamma_d \qquad (12.3.17)$$

Another limitation to amplifier performance arises because unsaturated gain $G_0 = \exp(gl)$ cannot be too high, otherwise two undesirable effects can occur in the amplifier: Parasitic oscillations or ASE. Parasitic oscillation occurs when the amplifier starts lasing due to some internal feedback that will always be present to some degree (e.g., due to amplifier end faces). The phenomenon ASE was discussed in Sect. 2.9.2. Both of these phenomena tend to depopulate the available inversion and hence decrease the gain. To minimize parasitic oscillations, we must avoid elongated amplifiers, and ideally we should use amplifiers with roughly equal dimensions in all directions. Even in this case, however, parasitic oscillations set an upper limit $(gl)_{max}$ to the product of the gain coefficient g with amplifier length l, i.e.

$$gl < (gl)_{max} \qquad (12.3.18)$$

where typically $(gl)_{max}$ ranges from 3–5. The threshold for ASE was given in Sect. 2.9.2. [see Eq. (2.9.4a) for a Lorentzian line]. For an amplifier in the form of a cube (i.e., $\Omega \cong 1$) and for a fluorescence quantum yield of unity, we have $G \cong 8$ [i.e., $gl \cong 2.1$], which is comparable to that established by parasitic oscillations. For smaller values of solid angle Ω, which are more typical, the value of G for the onset of ASE is expected to increase [Eq. (2.9.4a)]. Hence parasitic oscillations rather than ASE usually determine the maximum gain that can be achieved.

When both limits due to damage, Eq. (12.3.17), and parasitic oscillations, Eq. (12.3.18), are taken into account, we can obtain an expression for the maximum energy E_m that can be extracted from an amplifier as

---

Example 12.2. *Maximum energy that can be extracted from an amplifier.* It is assumed that the maximum value of gl is limited by parasitic oscillations such that $(gl)^2_{max} \cong 10$, and the rather low gain coefficient of $g = 10^{-2}$ cm^{-1} is also assumed. For a damage energy-fluence of the amplifier medium of $\Gamma_d = 10$ J/cm^2, we obtain from Eq. (12.3.19) $E_m \cong 1$ MJ. Note that this represents an upper limit to the energy, since it requires a somewhat impractical amplifier dimension on the order of $l_m \cong (gl)_m/g \cong 3$ m.

$$E_m = \Gamma_d l_m^2 = \frac{\Gamma_d (gl)_m^2}{g^2} \qquad (12.3.19)$$

where l_m is the maximum amplifier dimension (for a cubic amplifier) implied by Eq. (12.3.18). Equation (12.3.19) shows that E_m is increased by decreasing the amplifier gain coefficient g. Ultimately, a limit to this reduction in gain coefficient is established by amplifier losses α.

So far, we have primarily concerned ourselves with the change of laser pulse energy as the pulse passes through an amplifier. In the saturation regime, however, important changes in both the temporal and spatial shapes of the input beam also occur. The spatial distortion is readily understood with the help of Fig. 12.2. For an input beam with a bell-shaped transverse intensity profile (e.g., a Gaussian beam), the beam center experiences less gain than that at its periphery, as a result of saturation. Thus, the width of the beam's spatial profile is enlarged as the beam passes through the amplifier. The reason for temporal distortions is also easily understood: Stimulated emission, caused by the leading edge of the pulse, implies that some of the stored energy has been already extracted from the amplifier by the time the trailing edge of the pulse arrives. This edge therefore sees a smaller population inversion and thus experiences a reduced gain. As a result, less energy is added to the trailing edge than to the leading edge of the pulse and this leads to considerable pulse reshaping. The output pulse shape can be calculated from Eq. (12.3.10); one finds that the amplified pulse either broadens or narrows (or even remains unchanged), the outcome depending on the shape of the input pulse.[7]

12.3.1 Examples of Laser Amplifiers: Chirped Pulse Amplification

One of the most important and certainly the most spectacular example of laser pulse amplification is that of Nd:glass amplifiers used to produce high-energy pulses (10–100 kJ) for laser fusion research.[8] Very large Nd:glass laser systems have, in fact, been built and operated at a number of laboratories throughout the world; the one with the largest output energy is operated at the Lawrence Livermore National Laboratory in the United States (the NOVA laser). Most of these Nd:glass laser systems exploit the master oscillator power amplifier (MOPA) scheme, which consists of a master oscillator that generates a well-controlled pulse, of low energy, followed by a series of power amplifiers that amplify the pulse to a high energy. The clear aperture of the power amplifiers is increased along the chain to avoid optical damage as beam energy increases. Figure 12.3 shows a schematic diagram of one of the 10 arms of the NOVA system. The initial amplifiers in the chain consist of phosphate glass rods (of 380-mm length and a diameter of 25 mm for the first amplifiers, 50 mm for the last). The final stage of amplification is achieved by face-pumped disk amplifiers (see Fig. 6.3b) with a large clear aperture diameter (10 cm for the first amplifiers, 50 cm for the last). Note the presence in Fig. 12.3 of Faraday isolators (see Fig. 7.23) to avoid reflected light counter-propagating through the amplifier chain and thus damaging the initial stages of the system. Note also the spatial filters consisting of two lenses in a confocal arrangement (Fig. 11.10) with a pinhole at the common focus. These filters serve the double purpose of removing small-scale spatial irregularities from the beam as well as matching the beam profile between two consecutive amplifiers of different aperture. The laser system in Fig. 12.3 delivers an output energy of $\sim 10\,kJ$ in a pulse duration down to 1 ns, which gives a total energy of the 10-arm NOVA system of $\sim 100\,kJ$. Laser systems based on this layout and delivering an overall output energy of $\sim 1\,MJ$ are now being built in the United States (National Ignition Facility, NIF, Livermore) and in France (Megajoule project, Limeil) (see also Sect. 9.2.2.2).

A second class of laser amplifiers, which has revolutionized the laser field in terms of focusable beam intensity, relies on the chirped pulse amplification (CPA) concept[9] and is used to amplify picosecond or femtosecond laser pulses. At such short pulse durations, in fact, the maximum energy that can be obtained from an amplifier depends on the onset of either self-focusing, which is related to the beam peak power, or multiphoton ionization, which is related to the beam peak intensity.

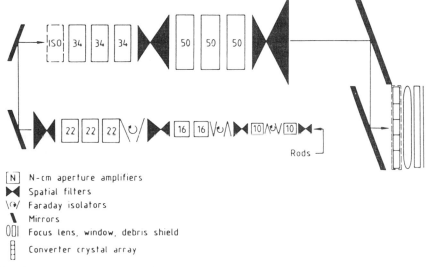

FIG. 12.3. Schematic layout of the amplification system using Nd:glass amplifiers for one arm of the Nova system. (By permission from Ref. 8.)

To overcome these limitations, one can adopt the technique, already used in radar technology, of pulse expansion (or pulse stretching) before amplification, followed by pulse compression to its original shape after the amplification process. In this way the peak power and hence the peak intensity of the pulse, in the amplifier chain, can be reduced by a few orders of magnitude (10^3–10^4). This allows a corresponding increase in the maximum energy that can be safely extracted from a given amplifier. Pulse expansion is achieved via an optical system that provides, e.g., a positive group-delay dispersion (GDD). In this way, the pulse may be considerably expanded in time while acquiring a positive frequency sweep (see Section 12.5.2 and Appendix G). The amplified pulse is then passed through an optical system having negative GDD (see Sect. 12.5.1). This second dispersive element compensates for the frequency sweep introduced by the first and so restores the initial shape of the pulse entering the amplifier chain.

Figure 12.4 illustrates a layout commonly used for a Ti:sapphire CPA. In the figure, P_1, P_2, and P_3 are three polarizers that transmit light whose field is polarized in the plane of the figure (horizontally polarized light) while reflecting light whose field is polarized orthogonal to the figure (vertically polarized light). The combination of the $\lambda/2$-plate and Faraday rotator (FR) is such that it transmits, without rotation, light traveling from right to left and rotates by 90° the polarization of light traveling from left to right (see Fig. 7.24). Low-energy (~ 1 nJ), high-repetition rate ($f \cong 80$ MHz), horizontally polarized femtosecond pulses, emitted by a Ti:sapphire mode-locked oscillator, are sent to the CPA. Since they are transmitted by polarizer P_2, they do not suffer polarization rotation on passing through the $\lambda/2$-plate FR combination. They are then transmitted by polarizer P_1 and thus sent to the pulse stretcher (whose lay out is discussed in Sect. 12.5.2). Typical expansion of the retroreflected pulse from the stretcher can be by a factor of ~ 5000, e.g., from 100 fs to 500 ps. Expanded pulses are then transmitted by polarizer P_1, undergo a 90° polarization rotation in the FR-$\lambda/2$ plate combination, and are then reflected by polarizer P_2. With the

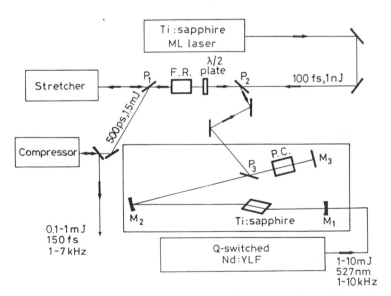

FIG. 12.4. Amplification of femtosecond laser pulses by a Ti:sapphire regenerative amplifier and the chirped pulse amplification technique.

help of polarizer P_3, the expanded pulses are then injected into a so-called regenerative amplifier, which consists of a Ti:sapphire amplifier and a Pockels cell (PC) located in a three mirror (M_1, M_2, and M_3) folded resonator. The PC is oriented to produce a static $\lambda/4$-retardation. The cavity Q is thus low before the pulse arrival and the regenerative amplifier is below the oscillation threshold. In this situation, any injected pulse becomes horizontally polarized, after a double passage through the PC, and is thus transmitted by polarizer P_3 toward the Ti:sapphire amplifier. After returning from the amplifier, the pulse is transmitted by polarizer P_3 and, after again passing through the PC twice, becomes vertically polarized and is thus reflected from the cavity by polarizer P_3. So, in this double transit through the regenerative amplifier, very little amplification is obtained for the output pulse. If, however, a $\lambda/4$-voltage is applied to the PC while the pulse is between the polarizer and mirror M_1, the cell becomes equivalent to a $\lambda/2$-plate and the pulse does not change its polarization state after each double passage through the cell. Therefore, the pulse gets trapped in the regenerative amplifier and, on each pass throught the amplifying medium, it is amplified. After a suitable number of roundtrips in the cavity (typically 15–20), the pulse energy reaches its maximum value and is then extracted from the cavity by applying an additional $\lambda/4$-voltage to the PC. In this case, in fact, after a double pass through the cell, the pulse becomes vertically polarized and is reflected by polarizer P_3 back in the direction of the incoming pulses. This, high-energy, vertically polarized pulse is reflected by polarizer P_2, does not suffer polarization rotation on passing through the $\lambda/2$-plate-FR combination, and is reflected by polarizer P_1 toward the pulse compressor (whose layout is discussed in Section 12.5.1.). The retroflected beam from the compressor then consists of a train of high-energy pulses, each with a duration approximately equal to that of the original pulses emitted by the oscillator and with a repetition rate equal to that at which the Ti:sapphire amplifier is pumped (1–10 kHz, usually by the second-harmonic green beam of a repetitively Q-switched Nd:YLF laser).

Systems of this type, exploiting the CPA technique, have fostered the development of lasers with ultra-high peak power.[10] For instance, using Ti:sapphire active media, table-top CPA systems with a peak power of ~ 20 TW have already been demonstrated, and systems with peak powers approaching 100 TW (e.g., 2 J in a 20-fs pulse) are under construction. The largest peak power so far achieved, by exploiting the CPA technique, is actually ~ 1.25 PW (1 PW $= 10^{15}$ W),[11] obtained by using a chain of amplifiers taken from one arm of the NOVA laser (to obtain an amplified pulse with ~ 580-J energy and ~ 460-fs duration). The peak intensity obtained by focusing these ultra high-power pulses is extremely high (10^{19}–10^{20} W/cm^2), representing an increase of 4 to 5 orders of magnitude compared to that available before introducing the CPA technique. When these ultrahigh-intensity beams interact with a solid target or a gas, a highly ionized plasma is obtained and a completely new class of nonlinear optical phenomena is produced. Applications of these high intensities cover a broad area of science and technology, including ultrafast x-ray and high-energy electron sources, as well as novel fusion concepts and plasma astrophysics.[12]

A third class of amplifier, widely used in optical fiber communications, is represented by the erbium-doped optical-fiber amplifier (EDFA).[13] This amplifier is diode-pumped generally in the 980-nm pump band of the Er^+ ion (see Fig. 9.4); it is used to amplify pulses at wavelengths corresponding to the so-called third transmission window of silica optical fibers ($\lambda \cong 1550$ nm). Since the pulse repetition rate of a communication system is usually very high ($\sim$ GHz) and the upper state lifetime of Er^+ is very long (~ 10 ms; see Table 9.4), the saturation behavior of the Er^+ population is the cumulative result of many laser pulses, i.e., determined by the average beam intensity. The rate-equation treatment of this type of amplification can then be made in terms of average beam intensity and, in principle, is very simple. Complications however arise from several factors, namely: (1) The Er^+ system works on an almost pure three-level scheme (see Sect. 9.2.4), so the effective cross sections of stimulated emission and absorption (both covering a large spectral bandwidth) must be taken into account. (2) Transverse variation, within the fiber, of both the Er^+ population profile and the intensity profile of the propagating mode must be taken into account. (3) Account must also be taken of the simultaneous presence of bidirectional noise from ASE. We therefore make no attempt to cover this subject in detail but instead point out that a vast literature exists[14] and that very high small-signal gain (to ~ 50 dB), relatively large saturated average output powers (~ 100 mW), and low noise are achieved with these amplifiers. Thus, Erbium-doped fiber amplifiers must be considered a major breakthrough in the field of optical fiber communications, with applications in both long-haul systems as well as distribution networks.

12.4. FREQUENCY CONVERSION: SECOND-HARMONIC GENERATION AND PARAMETRIC OSCILLATION

In classical linear optics one assumes that the induced dielectric polarization of a medium is linearly related to the applied electric field, i.e.:

$$\mathbf{P} = \varepsilon_0 \chi \mathbf{E} \tag{12.4.1}$$

where χ is the dielectric susceptibility.[1,15] With the high electric fields involved in laser beams, the above linear relation is no longer a good approximation, and further terms in which $\mathbf{P}$ is related to higher-order powers of $\mathbf{E}$ must also be considered. This nonlinear response can lead to an exchange of energy between em waves at different frequencies.

In this section we consider effects produced by a nonlinear polarization term proportional to the square of the electric field. The two effects that we consider are: (1) Second-harmonic generation (SHG), in which a laser beam at frequency ω is partially converted, in the nonlinear material, to a coherent beam at frequency 2ω [as first shown by Franken et al.[16]]. (2) Optical parameter oscillation (OPO), in which a laser beam at frequency ω_3 causes the simultaneous generation, in the nonlinear material, of two coherent beams at frequency ω_1 and ω_2 such that $\omega_1 + \omega_2 = \omega_3$ [as first shown by Giordmaine and Miller[17]]. With the high electric fields available in laser beams, the conversion efficiency of both these processes can be very high (approaching 100% in SHG). Therefore, these techniques are increasingly used to generate new coherent waves at frequencies different from that of the incoming wave.

12.4.1. Physical Picture

We first introduce some physical concepts using the simplifying assumption that the induced nonlinear polarization P^{NL} is related to the electric field E of the em wave by a scalar equation, i.e.:

$$P^{\mathrm{NL}} = 2\varepsilon_0 \, dE^2 \tag{12.4.2}$$

where d is a coefficient whose dimension is the inverse of an electric field.* The physical origin of Eq. (12.4.2) stems from the nonlinear deformation of the outer, loosely bound, electrons of an atom or atomic system, when subjected to high electric fields. This is analogous to a breakdown of Hooke's law for an extended spring, so that the restoring force is no longer linearly dependent on the displacement from equilibrium. A comparison between Eqs. (12.4.2) and (12.4.1) shows that the nonlinear polarization term becomes comparable to the linear one for an electric field $E \cong \chi/d$. Since $\chi \cong 1$, we see that $(1/d)$ represents the field strength for which linear and nonlinear terms become comparable. At this field strength, a sizable nonlinear deformation of the outer electrons must occur; $(1/d)$ is then expected to be of the order of the electric field that an electronic charge produces at a distance corresponding to a typical atomic dimension a, i.e., $(1/d) \cong e/4\pi\varepsilon_0 a^2$. [Thus $(1/d) \sim 10^{11}$ V/m for $a \cong 0.1$ nm.] Note that d must be zero for a centrosymmetric material, such as a centrosymmetric crystal or the usual liquids and gases. For reasons of symmetry, in fact, if one reverses the sign of E, the sign of the total polarization $P_t = P + P^{NL}$ must also reverse. However, since $P^{NL} \propto dE^2$, this can occur only if $d = 0$. From now on we therefore confine ourselves to considering noncentrosymmetric materials. We will see that the simple Eq. (12.4.2) then accounts for both SHG and OPO.

12.4.1.1. Second Harmonic Generation

We consider a monochromatic plane wave of frequency ω propagating along some direction, denoted as the z-direction, within a nonlinear crystal. The origin of the z-axis is taken at the entrance face of the crystal. For a plane wave of uniform intensity, we can write the following expression for the electric field $E_\omega(z, t)$ of the wave:

$$E_\omega(z, t) = (\tfrac{1}{2})\{E(z, \omega)\exp[j(\omega t - k_\omega z)] + c.c.\} \qquad (12.4.3)$$

where $c.c.$ stands for the complex conjugate of the other term appearing in brackets and:

$$k_\omega = \frac{\omega}{c_\omega} = \frac{n_\omega \omega}{c} \qquad (12.4.4)$$

where c_ω is the phase velocity in the crystal of a wave of frequency ω, n_ω is the refractive index at this frequency, and c is the velocity of light *in vacuum*. The substitution of Eq. (12.4.3) into Eq. (12.4.2) shows that P^{NL} contains a term† oscillating at frequency 2ω, namely:

$$P_{2\omega}^{NL} = \left(\frac{\varepsilon_0 d}{2}\right)\{E^2(z, \omega)\exp[j(2\omega t - 2k_\omega z)] + c.c.\} \qquad (12.4.5)$$

Equation (12.4.5) describes a polarization wave oscillating at frequency 2ω with a propagation constant $2k_\omega$. This wave is expected to radiate at frequency 2ω, i.e., to generate an em wave at the second harmonic (SH) frequency 2ω. The analytic treatment given later

---

* We use $2\varepsilon_0 dE^2$ rather than dE^2 (as often used in other textbooks) to make d conform to the increasingly accepted practice.

† The quantity P^{NL} also contains a term at frequency $\omega = 0$ that leads to the development of a dc voltage across the crystal (optical rectification).

involves substituting this polarization in the wave equation for the em field. The radiated SH field can be written in the form

$$E_{2\omega}(z, t) = \left(\frac{1}{2}\right)\{E(z, 2\omega)\exp[j(2\omega t - k_{2\omega}z)] + c.c.\} \tag{12.4.6}$$

where

$$k_{2\omega} = \frac{2\omega}{c_{2\omega}} = \frac{2n_{2\omega}\omega}{c} \tag{12.4.7}$$

is the propagation constant of a wave of frequency 2ω. The physical origin of SHG can thus be traced back to the fact that, as a result of the nonlinear relation (12.4.2), the em wave at the fundamental frequency ω beats with itself to produce a polarization at 2ω. Comparison between Eqs. (12.4.5) and (12.4.6) reveals a very important condition that must be satisfied if this process is to occur efficiently, viz., the phase velocity of the polarization wave ($v_P = 2\omega/2k_\omega$) must equal that of the generated em wave ($v_E = 2\omega/k_{2\omega}$). This condition can thus be written as:

$$k_{2\omega} = 2k_\omega \tag{12.4.8}$$

If this condition is not satisfied, the phase of the polarization wave at coordinate $z = l$ into the crystal, $2k_\omega l$, differs from that $k_{2\omega}l$, of the wave generated at $z = 0$, which has subsequently propagated to $z = l$. The difference in phase $(2k_\omega - k_{2\omega})l$ then increases with distance l, so the generated wave, being driven by a polarization that does not have the appropriate phase, does not grow cumulatively with distance l. Equation (12.4.8) is therefore referred to as the *phase-matching* condition. Note that, according to Eqs. (12.4.4) and (12.4.7), Eq. (12.4.8) implies that:

$$n_{2\omega} = n_\omega \tag{12.4.9}$$

If the polarization directions of E_ω and P^{NL} (and hence of $E_{2\omega}$) are the same [as implied by Eq. (12.4.2)] condition (12.4.9) cannot be satisfied due to the dispersion ($\Delta n = n_{2\omega} - n_\omega$) of the crystal. This severely limits the crystal length l_c over which P^{NL} can make contributions that add cumulatively to form the second harmonic wave. The length l_c (*the coherence length*) must in fact correspond to the distance over which the polarization wave and the SH wave are out of phase with each other by an amount π. This means that $k_{2\omega}l_c - 2k_\omega l_c = \pi$, from which, with the help of Eqs. (12.4.4) and (12.4.7), we obtain

$$l_c = \frac{\lambda}{4\Delta n} \tag{12.4.10}$$

where $\lambda = 2\pi c/\omega$ is the wavelength in *vacuum* of the fundamental wave. Taking as an example $\lambda \cong 1\ \mu m$ and $\Delta n = 10^{-2}$, we have $l_c = 25\ \mu m$. Note that, at this distance into the crystal, the polarization wave is $180°$ out of phase compared to the SH wave, and the latter begins to decrease with increased distance rather than continuing to grow. Since, as seen in the previous example, l_c is usually very small, only a very small fraction of the incident power can then be transformed into the second harmonic wave.

Another useful way of visualizing the SHG process is in terms of photons rather than fields. First we write the relation between the frequency of the fundamental (ω) and second harmonic (ω_{SH}) wave, viz.:

$$\omega_{SH} = 2\omega \qquad (12.4.11)$$

If we now multiply both sides of Eq. (12.4.11) and (12.4.8) by $\hbar$, we obtain, respectively,

$$\hbar\omega_{SH} = 2\hbar\omega \qquad (12.4.12a)$$

$$\hbar k_{2\omega} = 2\hbar k_{\omega} \qquad (12.4.12b)$$

For energy to be conserved in the SHG process, we must have $dI_{2\omega}/dz = -dI_{\omega}/dz$, where $I_{2\omega}$ and I_{ω} are the wave intensities at the two frequencies. With the help of Eq. (12.4.12a), we have $dF_{2\omega}/dz = -(1/2)dF_{\omega}/dz$, where $F_{2\omega}$ and F_{ω} are the photon fluxes of the two waves. From this equation we can then say that, whenever, in the SHG process, one photon at frequency 2ω is produced, correspondingly two photons at frequency ω disappear. Thus Eq. (12.4.12a) can be regarded as a statement of conservation of photon energy. Remembering that $\hbar k$ is the photon momentum, Eq. (12.4.12b) is then seen to correspond to the requirement that photon momentum is also conserved in the process.

We now reconsider the phase-matching condition (12.4.9) to see how it can be satisfied in a suitable, optically anisotropic, crystal.[18,19] To understand this, we must first explain the propagation behavior of waves in an anisotropic crystal and show how to generalize the simple nonlinear relation (12.4.2) for anisotropic media.

For a given direction of propagation, there are two linearly polarized plane waves, in an anisotropic crystal, that can propagate with different phase velocities. Corresponding to these two different polarizations, we can then associate two different refractive indices; the difference in the refractive index is referred to as the crystal's birefringence. This behavior is usually described in terms of the so-called index ellipsoid, which, for a uniaxial crystal, is an ellipsoid of revolution around the optic axis (the z-axis in Fig. 12.5). Given this ellipsoid, the two allowed directions for linear polarization and their corresponding refractive indices are found as follows: Through the center of the ellipsoid we draw a line in the direction of beam propagation (line OP in Fig. 12.5) and a plane perpendicular to this line. The intersection of this plane with the ellipsoid is an ellipse. The direction of the two axes of the ellipse then give the two polarization directions, and the length of each semiaxis gives the refractive index corresponding to that polarization. One of these directions is necessarily perpendicular to the optic axis; the wave having this polarization is referred to as the *ordinary wave*. Its refractive index n_o is seen from the figure to be independent of the direction of propagation. The wave with the other direction of polarization is referred to as the *extraordinary wave*; the corresponding index $n_e(\theta)$ depends on the angle θ and ranges in value from that of the ordinary wave n_o (when OP is parallel to z) to the value n_e, referred to as the extraordinary index, which occurs when OP is perpendicular to z. Note now that one defines a positive uniaxial crystal as corresponding to the case $n_e > n_o$, while a negative uniaxial crystal corresponds to the case $n_e < n_o$. An equivalent way of describing wave propagation uses the so-called normal (index) surfaces for the ordinary and extraordinary waves (Fig. 12.6). In this case, for a given direction of propagation OP and for either ordinary or extraordinary waves, the length of the segment between the origin O and the point of interception of the

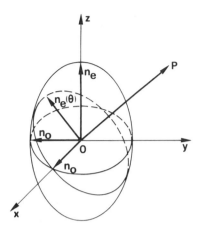

FIG. 12.5. Index ellipsoid for a positive uniaxial crystal.

ray with the surface gives the refractive index of that wave. The normal surface for the ordinary wave is thus a sphere; the normal surface for the extraordinary wave is an ellipsoid of revolution around the z-axis. In Fig. 12.6 the intersections of these two normal surfaces with the y-z plane are indicated for the case of a positive uniaxial crystal.

We now return to the problem of induced nonlinear polarization. In general in an anisotropic medium, the scalar relation (12.4.2) does not hold, so a tensor relation must be introduced. First, we write the electric field $\mathbf{E}^\omega(\mathbf{r}, t)$ of the em wave at frequency ω and a given point $\mathbf{r}$ and the nonlinear polarization vector at frequency 2ω, $\mathbf{P}_{NL}^{2\omega}(\mathbf{r}, t)$ in the form:

$$\mathbf{E}^\omega(\mathbf{r}, t) = \left(\frac{1}{2}\right)[\mathbf{E}^\omega(\mathbf{r}, \omega)\exp(j\omega t) + c.c.] \tag{12.4.13a}$$

$$\mathbf{P}_{NL}^{2\omega}(\mathbf{r}, t) = \left(\frac{1}{2}\right)[\mathbf{P}^{2\omega}(\mathbf{r}, 2\omega)\exp(2j\omega t) + c.c.] \tag{12.4.13b}$$

A tensor relation can then be established between $\mathbf{P}^{2\omega}(\mathbf{r}, 2\omega)$ and $\mathbf{E}^\omega(\mathbf{r}, \omega)$. In fact, the second harmonic polarization component along, e.g., the i-direction of the crystal can be written as:

$$P_i^{2\omega} = \sum_{j,k=1,2,3} \varepsilon_0 d_{ijk}^{2\omega} E_j^\omega E_k^\omega \tag{12.4.14}$$

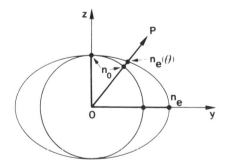

FIG. 12.6. Normal (index) surface for both ordinary and extraordinary waves (for a positive uniaxial crystal).

Note that Eq. (12.4.14) is often written in contracted notation as

$$P_i^{2\omega} = \sum_1^6 {}_m \varepsilon_0 d_{im}^{2\omega} (EE)_m \qquad (12.4.15)$$

where m ranges from 1 to 6. The abbreviated field notation is: $(EE)_1 \equiv E_1^2 \equiv E_x^2$, $(EE)_2 \equiv E_2^2 \equiv E_y^2$, $(EE)_3 \equiv E_3^2 \equiv E_z^2$, $(EE)_4 \equiv 2E_2E_3 \equiv 2E_yE_z$, $(EE)_5 \equiv 2E_1E_3 \equiv 2E_xE_z$ and $(EE)_6 \equiv 2E_1E_2 \equiv 2E_xE_y$, where both the 1, 2, 3 and the x, y, z notations for axes are indicated. Note that, expressed in matrix form, d_{im} is a 3×6 matrix that operates on the column vector $(EE)_m$. Depending on the crystal symmetry, some of the values of the d_{im} matrix may be equal and some may be zero. For the $\bar{4}2m$ point group symmetry, which includes the important nonlinear crystals of KDP type and chalcopyrite semiconductors, only d_{14}, d_{25}, and d_{36} are nonzero, and these three d coefficients are themselves equal. Therefore, only one coefficient, e.g., d_{36}, has to be specified and one can write

$$P_x = 2\varepsilon_0 d_{36} E_y E_z \qquad (12.4.16a)$$

$$P_y = 2\varepsilon_0 d_{36} E_z E_x \qquad (12.4.16b)$$

$$P_z = 2\varepsilon_0 d_{36} E_x E_y \qquad (12.4.16c)$$

where the z-axis is again taken along the optic axis of the uniaxial crystal. The nonlinear optical coefficients, symmetry class, transparency range, and the damage threshold of some selected nonlinear materials are indicated in Table 12.1. Except for cadmium germanium

Table 12.1. Nonlinear Optical Coefficients for Selected Materials

Material	Formula	Nonlinear d Coefficient (Relative to KDP)	Symmetry Class	Transparency Range (μm)	Damage Threshold (GW/cm^2)
KDP	KH$_2$PO$_4$	$d_{36} = d_{14} = 1$	$\bar{4}2m$	0.22–1.5	0.2
KD*P	KD$_2$PO$_4$	$d_{36} = d_{14} = 0.92$	$\bar{4}2m$	0.22–1.5	0.2
ADP	NH$_4$H$_2$PO$_4$	$d_{36} = d_{14} = 1.2$	$\bar{4}2m$	0.2–1.2	0.5
CDA	CsH$_2$AsO$_4$	$d_{36} = d_{14} = 0.92$	$\bar{4}2m$	0.26–1.4	0.5
Lithium iodate	LiIO$_3$	$d_{31} = d_{32} = d_{24}$ $d_{15} = 12.7$	6	0.3–5.5	0.5
Lithium niobate	LiNbO$_3$	$d_{31} = 12.5$ $d_{22} = 6.35$	$3m$	0.4–5	0.05
KTP	KTiOPO$_4$	$d_{31} = 13$ $d_{32} = 10$ $d_{33} = 27.4$ $d_{24} = 15.2$ $d_{15} = 12.2$	$mm2$	0.35–4.5	1
BBO	β-BaB$_2$O$_4$	$d_{22} = 4.1$	$3m$	0.19–3	5
Cadmium germanium arsenide	CdGeAs$_2$	$d_{36} = d_{14} = 538$	$\bar{4}2m$	2.4–20	0.04
Silver gallium selenide	AgGaSe$_2$	$d_{36} = d_{14} = 66$	$\bar{4}2m$	0.73–17	0.05

arsenide and AgGaSe$_2$, which are commonly used around the 10-μm range, all other crystals listed are used in the near-uv to near-infrared range. The table includes the more recent crystals of KTP (potassium titanyl phosphate), and BBO (beta-barium borate); the latter is commonly used for second harmonic generation at, e.g., the Nd:YAG wavelength. Nonlinear d-coefficients are normalized to that of KDP, whose actual value is $d_{36} \cong 0.5 \times 10^{-12}$ m/V.

Following this digression on the properties of anisotropic media, we can now show how phase matching can be achieved for the particular case of a crystal with $\overline{4}2m$ point group symmetry. From Eq. (12.4.16) we observe that, if $E_z = 0$, only P_z is nonvanishing, so it tends to generate a second harmonic wave with a nonzero z-component. We recall (see Fig. 12.5) that a wave with $E_z = 0$ is an ordinary wave, while a wave with $E_z \neq 0$ is an extraordinary wave. Thus an ordinary wave at the fundamental frequency ω, in this case, tends to generate an extra-ordinary wave at 2ω. To satisfy the phase-matching condition, we can then propagate the fundamental wave at an angle θ_m to the optic axis, so that:

$$n_e(2\omega, \theta_m) = n_o(\omega) \qquad (12.4.17)$$

This is better understood with the help of Fig. 12.7, which shows the intercepts of the normal surfaces $n_o(\omega)$ and $n_e(2\omega, \theta)$ with the plane containing the z-axis and the propagation direction. Note that, since crystals usually show a normal dispersion, one has $n_o(\omega) < n_o(2\omega)$; for a negative uniaxial crystal, one then has $n_e(2\omega) < n_o(2\omega)$, where, as a shorthand notation (see Fig. 12.7), we have set $n_e(2\omega) \equiv n_e(2\omega, 90°)$ and $n_o(2\omega) \equiv n_e(2\omega, 0)$. Thus the ordinary circle corresponding to the wave at frequency ω intersects the extraordinary ellipse, corresponding to the wave at frequency 2ω, at some angle θ_m*. For light propagating at this angle θ_m to the optic axis (i.e., for all ray directions lying in a cone around the z-axis with cone angle θ_m), Eq. (12.4.17) is satisfied; hence the phase-matching condition is satisfied.

If $\theta_m \neq 90°$, the phenomenon of double refraction occurs in the crystal,

---

Example 12.3. *Calculation of the phase-matching angle for a negative uniaxial crystal.* Referring to Fig. 12.7, we label the horizontal axis as the y-axis. If we then let z and y represent the cartesian coordinates of a general point of the ellipse describing the extraordinary index, $n_e(2\omega, \theta)$, we can write

$$\frac{z^2}{(n_2^o)^2} + \frac{y^2}{(n_2^e)^2} = 1$$

where, as a shorthand notation, we have set $n_2^o = n_o(2\omega)$ and $n_2^e = n_e(2\omega)$. If the coordinates z and y are now expressed as a function of $n_e(2\omega, \theta)$ and the angle θ, the previous equation can be transformed into:

$$\frac{[n_e(2\omega, \theta)]^2}{(n_2^o)^2} \cos^2 \theta + \frac{[n_e(2\omega, \theta)]^2}{(n_2^e)^2} \sin^2 \theta = 1$$

For $\theta = \theta_m$ Eq. (12.4.17) must hold. Substitution of this equation into the preceding equation then gives

$$\left(\frac{n_1^o}{n_2^o}\right)^2 (1 - \sin^2 \theta_m) + \left(\frac{n_1^o}{n_2^e}\right)^2 \sin^2 \theta_m = 1$$

where, again as a shorthand notation, we have set $n_1^o = n_o(\omega)$. This last equation can be solved for $\sin^2 \theta_m$ to obtain

$$\sin^2 \theta_m = \frac{1 - (n_1^o/n_2^o)^2}{(n_1^o/n_2^e)^2 - (n_1^o/n_2^o)^2} = \frac{(n_2^o/n_1^o)^2 - 1}{(n_2^o/n_2^e)^2 - 1}$$

---

* For this intersection to occur, $n_e(2\omega, 90°)$ must be less than $n_o(\omega)$; otherwise the ellipse for $n_e(2\omega)$ (see Fig. 12.7) lies wholly outside the circle for $n_o(\omega)$. Thus $n_e(2\omega, 90°) = n_e(2\omega) < n_o(\omega) < n_o(2\omega)$, which shows that the crystal birefringence $n_o(2\omega) - n_e(2\omega)$ must be larger than the crystal dispersion $n_o(2\omega) - n_o(\omega)$.

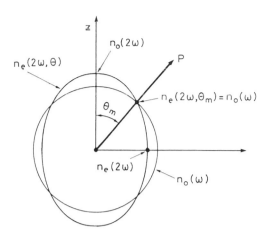

FIG. 12.7. Phase-matching angle θ_m for Type I second harmonic generation in a negative uniaxial crystal.

i.e., the direction of the energy flow for the extraordinary (SH) beam is at an angle slightly different from θ_m. Thus the fundamental and SH beams travel in slightly different directions (although satisfying the phase-matching condition). For a fundamental beam of finite transverse dimensions, this sets an upper limit on the interaction length in the crystal. This limitation can be overcome if it is possible to operate with $\theta_m = 90°$, i.e., if one has $n_e(2\omega, 90°) = n_o(\omega)$. This is referred to as the 90° phase-matching condition. Since n_e and n_o generally undergo different changes with temperature, the 90° phase-matching condition can, in some cases, be reached by changing the crystal temperature. To summarize, phase-matching is possible in a (sufficiently birefringent) negative uniaxial crystal when an ordinary ray at ω [E_x beam of Eq. (12.4.16c)] combines with an ordinary ray at ω [E_y beam of Eq. (12.4.16c)] to give an extraordinary ray at 2ω, or, in symbols, $o_\omega + o_\omega \rightarrow e_{2\omega}$. This is referred to as *type I second harmonic generation*. In a negative uniaxial crystal, another scheme for phase matched SHG, referred to as type II, is also possible. In this case an ordinary wave at ω combines with an extraordinary wave at ω to give an extraordinary wave at 2ω or, in symbols, $o_\omega + e_\omega \rightarrow e_{2\omega}$.*

Second harmonic generation is currently used to provide coherent sources at new wavelengths. The nonlinear crystal can be placed either outside or inside the cavity of the laser producing the fundamental beam. In the latter case one takes advantage of the greater em field strength inside the resonator to increase the conversion efficiency. Very high conversion efficiencies (approaching 100%) have been obtained with both arrangements. Among the most frequent applications of SHG are frequency doubling the output of a Nd:YAG laser (thus producing a green beam, $\lambda = 532$ nm, from an infrared one, $\lambda = 1.064\,\mu$m) and generation of tunable uv radiation (down to $\lambda \cong 205$ nm) by frequency doubling a tunable dye laser. In both of these cases either cw or pulsed laser sources are used. The nonlinear crystals most commonly used as frequency doublers for Nd:YAG lasers are KTP and β-BaB$_2$O$_4$ (BBO); due to its more extended transparency into the uv, BBO is used particularly when a SH beam at uv wavelengths down to ~ 200 nm has to be generated.

---

* More generally, interactions where polarizations of the two fundamental waves are the same are termed type I (e.g., also $e_\omega + e_\omega \rightarrow o_{2\omega}$), and interactions where the polarizations of the fundamental waves are orthogonal are termed type II.

Efficient frequency conversion of infrared radiation from CO_2 or CO lasers is often produced in chalcopyrite semiconductors (e.g., $CdGeAs_2$).

12.4.1.2. Parametric Oscillation

To describe parametric oscillation, we first note that the preceding ideas, introduced in the context of SHG, can be readily extended to the case of two incoming waves at frequencies ω_1 and ω_2 combining to give a wave at frequency $\omega_3 = \omega_1 + \omega_2$ (sum-frequency generation). Harmonic generation can in fact be thought of as a special case of sum-frequency generation with $\omega_1 = \omega_2 = \omega$ and $\omega_3 = 2\omega$. The physical picture is again very similar to the SHG case: By virtue of the nonlinear relation (12.4.2) between P^{NL} and the total field E [$E = E_{\omega_1}(z, t) + E_{\omega_2}(z, t)$], the wave at ω_1 will beat with that at ω_2, to give a polarization component at $\omega_3 = \omega_1 + \omega_2$. This will then radiate an em wave at ω_3. Thus, for sum frequency generation, we can write

$$\hbar\omega_1 + \hbar\omega_2 = \hbar\omega_3 \qquad (12.4.18a)$$

which, according to a description in terms of photons rather than fields, implies that one photon at ω_1 and one photon at ω_2 disappear when a photon at ω_3 is created. We therefore expect the photon momentum also to be conserved in the process, i.e.

$$\hbar\mathbf{k}_1 + \hbar\mathbf{k}_2 = \hbar\mathbf{k}_3 \qquad (12.4.18b)$$

where the relationship is in its general form, with $\mathbf{k}$ denoted by vectors. Equation (12.4.18b), which expresses the phase-matching condition for sum-frequency generation, is a straightforward generalization of that for SHG [compare with Eq. (12.4.12b)].

Optical parametric generation is, now, just the reverse of sum frequency generation: A wave at frequency ω_3 (the pump frequency) generates two waves (called the idler and signal waves) at frequencies ω_1 and ω_2, so that the total photon energy and momentum is conserved, i.e.:

$$\hbar\omega_3 = \hbar\omega_1 + \hbar\omega_2 \qquad (12.4.19a)$$

$$\hbar\mathbf{k}_3 = \hbar\mathbf{k}_1 + \hbar\mathbf{k}_2 \qquad (12.4.19b)$$

The physical process in this case can be visualized as follows. First imagine that a strong wave at ω_3 and a weak wave at ω_1 are both present in the nonlinear crystal. As a result of the nonlinear relation (12.4.2), the wave at ω_3 will beat with the wave at ω_1 to give a polarization component at $\omega_3 - \omega_1 = \omega_2$. If the phase-matching condition (12.4.19b) is satisfied, a wave at ω_2 thus builds up as it travels through the crystal. Then, the total E field will in fact be the sum of three fields [$E = E_{\omega_1}(z, t) + E_{\omega_2}(z, t) + E_{\omega_3}(z, t)$] and the wave at ω_2 will in turn beat with the wave at ω_3 to produce a polarization component at $\omega_3 - \omega_2 = \omega_1$. This polarization also causes the ω_1 wave to grow. Thus power is transferred from the beam at ω_3 to those at ω_1 and ω_2, and the weak wave at ω_1, which was assumed initially to be present, is amplified.

From this picture one sees a fundamental difference between parametric generation and SHG. In the latter case only a strong beam at the fundamental frequency is needed for the SHG process to occur. In the former case, however, a weak beam at ω_1 is also needed and the system behaves like an amplifier at frequency ω_1 (and ω_2). In practice, however, the weak beam need not be supplied by an external source (such as another laser), since it is generated internally to the crystal as a form of noise (so-called parametric noise). We can then generate coherent beams from this noise in a way analogous to that used in a laser oscillator. Thus the nonlinear crystal, which is pumped by an appropriately focused pump beam, is placed in an optical resonator (Fig. 12.8). The two mirrors (1 and 2) of this parametric oscillator have high reflectivity (e.g., $R_1 = 1$ and $R_2 \cong 1$) either at ω_1 only [singly resonant oscillator (SRO)] or at both ω_1 and ω_2 [doubly resonant oscillator (DRO)]. The mirrors are, ideally, transparent to the pump beam. Oscillation starts when the gain arising from the parametric effect just exceeds losses from the optical resonator. Some threshold power of the input pump beam is therefore required before oscillation begins. When this threshold is reached, oscillation occurs at both ω_1 and ω_2; the particular pair of values of ω_1 and ω_2 is determined by Eq. (12.4.19). For instance, with type I phase matching involving an extraordinary wave at ω_3 and ordinary waves at ω_1 and ω_2 (i.e., $e_{\omega_3} \to o_{\omega_1} + o_{\omega_2}$), Eq. (12.4.19b) yields:

$$\omega_3 n_e(\omega_3, \theta) = \omega_1 n_o(\omega_1) + \omega_2 n_o(\omega_2) \tag{12.4.20}$$

For a given θ, i.e., for a given inclination of the nonlinear crystal with respect to the cavity axis, Eq. (12.4.20) provides a relation between ω_1 and ω_2, which, together with relation (12.4.19a), determines the values of both ω_1 and ω_2. Type I and type II phase-matching schemes (e.g., $e_{\omega_3} \to o_{\omega_1} + e_{\omega_2}$ for a negative uniaxial crystal) are both possible, and tuning can be achieved by changing either the crystal inclination (*angle tuning*) or its temperature (*temperature tuning*). As a final comment, we note that, if the gain from the parametric effect is large enough, one can dispense with the mirrors altogether; an intense emission at ω_1 and ω_2 grows from parametric noise in a single pass or just in a few passes through the crystal. This behavior is often referred to as *superfluorescent parametric emission*, and such a device is referred to as an *optical parametric generator* (OPG).

Both singly resonant and doubly resonant optical parametric oscillators are used. Doubly resonant parametric oscillation has been achieved with both cw and pulsed pump lasers. For cw pumping, threshold powers as low as a few milliwatts have been demonstrated. The requirement for simultaneous resonance of both parametric waves in the same cavity generally leads to poor amplitude and frequency stability of the output beams. Singly resonant parametric oscillation had, until recently, been achieved only using

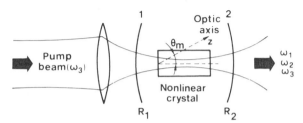

FIG. 12.8. Schematic diagram of an optical parametric oscillator.

pulsed pump lasers, since the threshold pump power for the singly resonant case is much higher (by as much as 2 orders of magnitude) than that of the doubly resonant case. However, with improved nonlinear crystals, cw oscillation is now readily achieved. Singly resonant oscillators produce a much more stable output, and, for this reason, is the configuration most frequently used.

Optical parametric oscillators producing coherent radiation from the visible to the near infrared (0.5–5 μm) are now well-developed; the most successful devices are based on BBO, LBO, and lithium niobate (LiNbO$_3$). Optical parametric oscillators can also generate coherent radiation at longer infrared wavelengths (to $\sim$14 μm) using such crystals as silver gallium selenide (AgGaSe$_2$) and cadmium selenide (CdSe). Synchronous pumping of OPOs, using a mode-locked pump, is also proving very attractive as a means of generating short pulses with very wide tunability. A notable feature of these devices is that their gain is determined by the peak power of the pump pulse, so that thresholds corresponding to very low average powers (a few milliwatts) can be achieved even for a singly resonant oscillator. Note lastly that the efficiency of an OPO can be very high, approaching the theoretical 100% photon efficiency.

12.4.2. Analytical Treatment

To arrive at an analytic description of both SHG and parametric processes, we must see how the nonlinear polarization [e.g., Eq. (12.4.2)], which acts as the source term to drive the generated waves, is introduced into the wave equation. The fields within the material obey Maxwell's equations:

$$\nabla \times \mathbf{E} = -\frac{\partial \mathbf{B}}{\partial t} \qquad (12.4.21a)$$

$$\nabla \times \mathbf{H} = \mathbf{J} + \frac{\partial \mathbf{D}}{\partial t} \qquad (12.4.21b)$$

$$\nabla \cdot \mathbf{D} = \rho \qquad (12.4.21c)$$

$$\nabla \cdot \mathbf{B} = 0 \qquad (12.4.21d)$$

where ρ is the free-charge density. For the media of interest here we can assume the magnetization $\mathbf{M}$ to be zero; thus:

$$\mathbf{B} = \mu_0 \mathbf{H} + \mu_0 \mathbf{M} = \mu_0 \mathbf{H} \qquad (12.4.22)$$

Losses within the material (e.g., scattering losses) can be simulated by introducing a fictitious conductivity σ_s such that:

$$\mathbf{J} = \sigma_s \mathbf{E} \qquad (12.4.23)$$

Finally we can write

$$\mathbf{D} = \varepsilon_0 \mathbf{E} + \mathbf{P}^L + \mathbf{P}^{NL} = \varepsilon \mathbf{E} + \mathbf{P}^{NL} \qquad (12.4.24)$$

where $\mathbf{P}^L$, the linear polarization of the medium, is accounted for, in the usual way, by introducing the dielectric constant ε. Applying the $\nabla\times$ operator to both sides of Eq. (12.4.21a), interchanging the order of $\nabla\times$ and $\partial/\partial t$ operators on the right-hand side of the resulting equation, then using Eqs. (12.4.22), (12.4.21b), (12.4.23), and (12.4.24), we obtain

$$\nabla \times \nabla \times \mathbf{E} = -\mu_0 \left(\sigma_s \frac{\partial \mathbf{E}}{\partial t} + \varepsilon \frac{\partial^2 \mathbf{E}}{\partial t^2} + \frac{\partial^2 \mathbf{P}^{NL}}{\partial t^2} \right) \tag{12.4.25}$$

Using the identity $\nabla \times \nabla \times \mathbf{E} = \nabla(\nabla \cdot \mathbf{E}) - \nabla^2 \mathbf{E}$ and given the assumption that $\nabla \cdot \mathbf{E} \cong 0$, we obtain from Eq. (12.4.25)

$$\nabla^2 \mathbf{E} - \frac{\sigma_s}{\varepsilon c^2} \frac{\partial \mathbf{E}}{\partial t} - \frac{1}{c^2} \frac{\partial^2 \mathbf{E}}{\partial t^2} = \frac{1}{\varepsilon c^2} \frac{\partial^2 \mathbf{P}^{NL}}{\partial t^2} \tag{12.4.26a}$$

where $c = (\varepsilon\mu_0)^{-1/2}$ is the phase velocity of the wave in the material. Equation (12.4.26a) is the wave equation with the nonlinear polarization term included. Note that the linear part of the medium polarization has been transferred to the left-hand side of Eq. (12.4.26a), and its effect is included in the dielectric constant ε. The nonlinear part $\mathbf{P}^{NL}$ has been kept on the right-hand side, since it will be shown to act as a source term for waves generated at new frequencies as well as a loss term for the incoming wave. Confining ourselves to the simple scalar case of plane waves propagating in the z-direction, we see that Eq. (12.4.26a) reduces to:

$$\frac{\partial^2 E}{\partial z^2} - \frac{\sigma_s}{\varepsilon c^2} \frac{\partial E}{\partial t} - \frac{1}{c^2} \frac{\partial^2 E}{\partial t^2} = \frac{1}{\varepsilon c^2} \frac{\partial^2 P^{NL}}{\partial t^2} \tag{12.4.26b}$$

We now write the field of the wave at frequency ω_i as

$$E_{\omega_i}(z, t) = \left(\frac{1}{2}\right)\{E_i(z)\exp[\,j(\omega_i t - k_i z)] + c.c.\} \tag{12.4.27a}$$

where E_i is taken to be complex in general. Likewise, the amplitude of the nonlinear polarization at frequency ω_i is written as:

$$P_{\omega_i}^{NL} = \left(\frac{1}{2}\right)\{P_i^{NL}(z)\exp[\,j(\omega_i t - k_i z)] + c.c.\} \tag{12.4.27b}$$

Since Eq. (12.4.26b) must hold separately for each frequency component of the waves present in the medium, Eq. (12.4.27a) and (12.4.27b) can be substituted into the left- and right-hand sides of Eq. (12.4.26b), respectively. Within the slowly varying amplitude approximation, we can neglect the second derivative of $E_i(z)$, i.e., assume that $d^2 E_i/dz^2 \ll k_i (dE_i/dz)$. Equation (12.4.26b) then yields

$$2\frac{dE_i}{dz} + \frac{\sigma_i}{n_i \varepsilon_0 c} E_i = -j\left(\frac{\omega_i}{n_i \varepsilon_0 c}\right)P_i^{NL} \tag{12.4.28}$$

where σ_i, n_i, and ε_i are, respectively, the loss, refractive index, and dielectric constant of the medium at frequency ω_i, and where the relations $k_i = n_i\omega_i/c$ and $\varepsilon_i = n_i^2\varepsilon_0$ have been used.

Equation (12.4.28) is the basic equation used in the following sections. Note that this equation assumes a scalar relation between $\mathbf{P}^{NL}$ and $\mathbf{E}$ [see Eq. (12.4.2)]. This assumption is not correct and actually a tensor relation should be used [see Eq. (12.4.15)]. However, it can be shown that one can still use this scalar equation if E_i now refers to the field component along an appropriate axis and an effective coefficient d_{eff} is substituted for d in Eq. (12.4.2). In general, d_{eff} is a combination of one or more of the d_{im} coefficients appearing in Eq. (12.4.15) multiplied by appropriate trigonometric functions of the angles θ and ϕ that define the direction of wave propagation in the crystal[20] (θ is the angle between the propagation vector and the z-axis, ϕ is the angle that the projection of the propagation vector in the x-y plane makes with the x-axis of the crystal). For example, in a crystal with $\bar{4}2m$ point group symmetry and type I phase-matching, one obtains $d_{eff} = d_{36}\sin 2\phi\sin\theta$. As a shorthand notation, however, we retain the symbol d in Eq. (12.4.2), keeping in mind that it means the effective value of the d coefficient, d_{eff}.

12.4.2.1. Parametric Oscillation

We now consider three waves at frequencies ω_1, ω_2, and ω_3 (where $\omega_3 = \omega_1 + \omega_2$) interacting in the crystal. Thus we write the total field $E(z, t)$ as:

$$E(z, t) = E_{\omega_1}(z, t) + E_{\omega_2}(z, t) + E_{\omega_3}(z, t) \tag{12.4.29}$$

where each of the fields can be written in the form of Eq. (12.4.27a). Substituting Eq. (12.4.29) into Eq. (12.4.2) and using Eq. (12.4.27a), we obtain an expression for the components $P_i^{NL}(z)$ [as defined by Eq. (12.4.27b)] of the nonlinear polarization at frequency ω_i. After some lengthy but straightforward algebra, we find that the component P_i^{NL} at frequency ω_1, for instance, is given by:

$$P_1^{NL} = 2\varepsilon_0\, d E_3(z)E_2^*(z)\exp[j(k_2 - k_3)z] \tag{12.4.30}$$

The components of P^{NL} at ω_2 and ω_3 are obtained similarly. For each of the three frequencies, we obtain the field equation by substituting the expression for P^{NL} corresponding to the appropriate frequency into Eq. (12.4.28). We thus arrive at the following three equations:

$$\frac{dE_1}{dz} = -\left(\frac{\sigma_1}{2n_1\varepsilon_0 c}\right)E_1 - j\left(\frac{\omega_1}{n_1 c}\right)dE_3 E_2^*\exp[-j(k_3 - k_2 - k_1)z] \tag{12.4.31a}$$

$$\frac{dE_2}{dz} = -\left(\frac{\sigma_2}{2n_2\varepsilon_0 c}\right)E_2 - j\left(\frac{\omega_2}{n_2 c}\right)dE_3 E_1^*\exp[-j(k_3 - k_1 - k_2)z] \tag{12.4.31b}$$

$$\frac{dE_3}{dz} = -\left(\frac{\sigma_3}{2n_3\varepsilon_0 c}\right)E_3 - j\left(\frac{\omega_3}{n_3 c}\right)dE_1 E_2\exp[-j(k_1 + k_2 - k_3)z] \tag{12.4.31c}$$

These are the basic equations describing the nonlinear parametric interaction. One sees that they are coupled to each other by the nonlinear coefficient d.

It is convenient at this point to define new field variables A_i as:

$$A_i = \left(\frac{n_i}{\omega_i}\right)^{1/2} E_i \tag{12.4.32}$$

Since the intensity of the wave is $I_i = n_i \varepsilon_0 c |E_i|^2/2$, the corresponding photon flux F_i is given by $F_i = I_i/\hbar\omega_i = (\varepsilon_0 c/2\hbar)|A_i|^2$. Thus $|A_i|^2$ is proportional to the photon flux F_i with a proportionality constant independent of both n_i and ω_i. When reexpressed in terms of these new field variables, Eq. (12.4.31) can be transformed into

$$\frac{dA_1}{dz} = -\frac{\alpha_1 A_1}{2} - j\delta A_3 A_2^* \exp[-j(\Delta kz)] \tag{12.4.33a}$$

$$\frac{dA_2}{dz} = -\frac{\alpha_2 A_2}{2} - j\delta A_3 A_1^* \exp[-j(\Delta kz)] \tag{12.4.33b}$$

$$\frac{dA_3}{dz} = -\frac{\alpha_3 A_3}{2} - j\delta A_1 A_2 \exp[j(\Delta kz)] \tag{12.4.33c}$$

where we have set $\alpha_i = \sigma_i/n_i\varepsilon_0 c$, $\Delta k = k_3 - k_2 - k_1$ and:

$$\delta = \frac{d}{c}\left(\frac{\omega_1 \omega_2 \omega_3}{n_1 n_2 n_3}\right)^{1/2} \tag{12.4.34}$$

The advantage of using A_i instead of E_i is now apparent, since, unlike Eq. (12.4.31), relations (12.4.33) now involve a single coupling parameter δ.

If losses are neglected (i.e., when $\alpha_i = 0$), we can obtain from Eqs. (12.4.33) some useful conservation laws. For instance, multiplying both sides of Eq. (12.4.33a) by A_1^* and both sides of Eq. (12.4.33b) by A_2^*, then comparing the results, we arrive at the following relation: $d|A_1|^2/dz = d|A_2|^2/dz$. Similarly, from Eqs. (12.4.33b–c), we obtain $d|A_2|^2/dz = -d|A_3|^2/dz$. We can therefore write

$$\frac{d|A_1|^2}{dz} = \frac{d|A_2|^2}{dz} = -\frac{d|A_3|^2}{dz} \tag{12.4.35}$$

which are known as the Manley–Rowe relations. Note that, since $|A_i|^2$ is proportional to the corresponding photon flux, Eq. (12.4.35) implies that whenever a photon at ω_3 is destroyed, a photon at ω_1 and a photon at ω_2 are created. This is consistent with the photon model for the parametric process which was discussed in Section 12.4.1.2. Note also that Eq. (12.4.35) means, for instance, that $(dP_1/dz) = -(\omega_1/\omega_3)(dP_3/dz)$, where P_1 and P_3 are the powers of the two waves. Thus, only a fraction (ω_1/ω_3) of the power at frequency ω_3 can be converted into power at frequency ω_1.

Strictly speaking, Eq. (12.4.33) applies to a traveling-wave situation in which an arbitrarily long crystal is traversed by the three waves at $\omega_1, \omega_2, \omega_3$. We now want to see how to apply these equations to the case of an optical parametric oscillator, as in Fig. 12.8. We first consider the DRO scheme. The waves at ω_1 and ω_2 therefore travel back and forth within the cavity, so the parametric process occurs only when their propagation direction is the same as that of the pump wave (only under these circumstances can phase matching be satisfied). If we unfold the optical path, it will look like that in Fig. 12.9a; one sees that loss occurs on every pass while parametric gain occurs only once in every second pass. This situation can be reduced to that of Fig. 12.9b if we appropriately define the effective loss coefficient $\alpha_i (i = 1, 2)$. The loss due to a crystal of length l in Fig. 12.9b must in fact equal the losses incurred in a double pass in Fig. 12.9a. The latter losses must account for the actual losses in the crystal as well as mirror and diffraction losses. Thus coefficients α_1 and α_2 in Eq. (12.4.33) must be appropriately defined to incorporate these various losses. From Eq. (12.4.33), neglecting the parametric interaction (i.e., setting $\delta = 0$), we see that, after the beam at frequency ω_i ($i = 1, 2$) traverses the length l of the crystal, its power is reduced by a fraction $\exp(-\alpha_i l)$. This reduction must then account for roundtrip losses of the cavity, which requires that

$$\exp(-\alpha_i l) = R_{1i} R_{2i} (1 - L_i)^2 \qquad (12.4.35a)$$

where R_{1i} and R_{2i} are the power reflectivities of the two mirrors at frequency ω_i, and L_i is the crystal loss (and diffraction loss) per pass. If we now define [compare to Eq. (1.2.4)] $\gamma_{1i} = -\ln R_{1i}$, $\gamma_{2i} = -\ln R_{2i}$, $\gamma_i' = -\ln(1 - L_i)$, and $\gamma_i = [(\gamma_{1i} + \gamma_{2i})/2] + \gamma_i'$, we can rewrite Eq. (12.4.35a) as

$$\alpha_i l = 2\gamma_i \qquad (12.4.36)$$

where γ_i is the overall cavity loss per pass at frequency ω_i. Note that this amounts to simulating mirror losses by losses distributed through the crystal and then including them in an effective absorption coefficient of the crystal $\alpha_i (i = 1, 2)$. The loss α_3, on the other hand, involves only crystal losses, so it can in general be neglected. Thus we can say that Eq. (12.4.33) can be applied to a DRO if α_1 and α_2 are given by Eq. (12.4.36) and we set $\alpha_3 \cong 0$.

To obtain the threshold condition of a DRO, Eq. (12.4.33) can be further simplified if we neglect pump wave depletion by the parametric process. This assumption and the

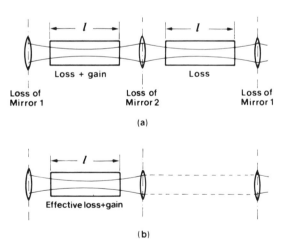

FIG. 12.9. (a) Unfolded path for an optical parametric oscillator and (b) reduction to a single-pass scheme with mirror losses incorporated into the distributed losses of the crystal.

assumption $\alpha_3 = 0$ means that we can take $A_3(z) \simeq A_3(0)$, where $A_3(0)$, the field amplitude of the incoming pump wave, is taken as real. With the further assumption of $\Delta k = 0$ (perfect phase matching), Eq. (12.4.33) is considerably simplified and becomes

$$\frac{dA_1}{dz} = -\frac{\alpha_1 A_1}{2} - j\frac{g}{2} A_2^* \qquad (12.4.37a)$$

$$\frac{dA_2}{dz} = -\frac{\alpha_2 A_2}{2} - j\frac{g}{2} A_1^* \qquad (12.4.37b)$$

where we have set

$$g = 2\delta A_3(0) = 2d\frac{E_3(0)}{c}\left(\frac{\omega_1 \omega_2}{n_1 n_2}\right)^{1/2} \qquad (12.4.38)$$

The threshold condition for a DRO is then readily obtained from Eq. (12.4.37) by imposing the condition $dA_1/dz = dA_2/dz = 0$. This leads to:

$$\alpha_1 A_1 + jgA_2^* = 0 \quad (12.4.39a)$$

$$jgA_1 - \alpha_2 A_2^* = 0 \quad (12.4.39b)$$

where the complex conjugate of Eq. (12.4.37b) has been taken. The solution of this homogeneous system of equations yields nonzero values for A_1 and A_2 only if

$$g^2 = \alpha_1 \alpha_2 = 4\left(\frac{\gamma_1 \gamma_2}{l^2}\right) \quad (12.4.40)$$

where Eq. (12.4.36) has been used. Note that, according to Eq. (12.4.38), g^2 is proportional to $E_3^2(0)$, i.e., to the intensity of the pump wave. Thus Eq. (12.4.40) means that a certain threshold intensity of the pump wave is needed for parametric oscillation to start. As example 12.4. shows, the threshold intensity is proportional to the product of the single-pass (power) losses γ_1 and γ_2 of the two waves at ω_1 and ω_2 and inversely proportional to d^2 and l^2.

Example 12.4. *Calculation of threshold intensity for the pump beam in a doubly resonant optical parametric oscillator.* From Eqs. (12.4.38) and (12.4.40), we obtain the following expression for $E_3^2(0)$, the square of the pump beam amplitude at threshold:

$$E_{3th}^2(0) = \frac{c^2}{d^2} \frac{n_1 n_2}{\omega_1 \omega_2} \frac{\gamma_1 \gamma_2}{l^2}$$

Since the pump beam intensity is given by $I_3 = n_3 \varepsilon_0 c |E_3|^2/2$, we then obtain the following expression for the threshold pump intensity

$$I_{3th} = \left(\frac{n_3}{2Zd^2}\right)\left[\frac{n_1 n_2 \lambda_1 \lambda_2}{(2\pi l)^2}\right](\gamma_1 \gamma_2)$$

where $Z = 1/\varepsilon_0 c = 377\,\text{ohm}$ is the free-space impedance while λ_1 and λ_2 are the wavelengths of the signal and idler waves, respectively. Note that the term in the first bracket on the right-hand side of the preceding equation has the dimension of an intensity, while factors in the other two brackets are dimensionless.

The SRO case is somewhat more involved. If the laser cavity is resonant only at ω_1, then α_1 can again be written as in Eq. (12.4.36). Since the wave at ω_2 is no longer fed back, α_2, which involves only crystal losses, can be neglected. Again, neglecting pump wave depletion and assuming perfect phase matching, we find that Eq. (12.4.37) is still applicable if we now set $\alpha_2 = 0$. For a small parametric conversion, we can put $A_1^*(z) \cong A_1^*(0)$ on the right-hand side of Eq. (12.4.37b) to obtain

$$A_2(z) = -j\frac{gA_1^*(0)z}{2} \tag{12.4.41}$$

where the condition $A_2(0) = 0$ is assumed (i.e., no field at frequency ω_2 is fed back into the crystal by the resonator). If we substitute Eq. (12.4.41) into Eq. (12.4.37a) and put $A_1(z) \cong A_1(0)$ in the right-hand side of the latter equation, we obtain

$$\frac{dA_1}{dz} = \left(-\frac{\alpha_1}{2} + \frac{g^2 z}{4}\right)A_1(0) \tag{12.4.42}$$

Integration of Eq. (12.4.42) over the length l of the crystal gives the following expression for the field at frequency ω_1:

$$A_1(l) = A_1(0)\left(1 - \frac{\alpha_1 l}{2} + \frac{g^2 l^2}{8}\right) \tag{12.4.43}$$

The threshold condition is then reached when $A_1(l) = A_1(0)$, i.e., when:

$$g^2 = \frac{4\alpha_1}{l} = \frac{8\gamma_1}{l^2} \tag{12.4.44}$$

Since g^2 is proportional to the intensity I of the pump wave, a comparison between Eqs. (12.2.44) and (12.4.40) gives the ratio of the threshold pump intensities as:

$$\frac{I_{\text{SRO}}}{I_{\text{DRO}}} = \frac{2}{\gamma_2} \tag{12.4.45}$$

For example, if one takes a loss per pass of $\gamma_2 = 0.02$ (i.e., 2% loss), one finds from Eq. (12.4.45) that the threshold power for the SRO is 100 times larger than for the DRO case.

12.4.2.2. Second-Harmonic Generation

In the case of SHG we write

$$E(z, t) = \left(\tfrac{1}{2}\right)\{E_\omega \exp[j(\omega t - k_\omega z)] + E_{2\omega} \exp[j(2\omega t - k_{2\omega}z)] + c.c.\} \tag{12.4.46}$$

$$P^{\text{NL}}(z, t) = \left(\tfrac{1}{2}\right)\{P_\omega^{\text{NL}} \exp[j(\omega t - k_\omega z)] + P_{2\omega}^{\text{NL}} \exp[j(2\omega t - k_{2\omega}z)] + c.c.\} \tag{12.4.47}$$

Substitution of Eqs. (12.4.46) and (12.4.47) into Eq. (12.4.2) gives

$$P_{2\omega}^{\text{NL}} = \varepsilon_0 \, dE_\omega^2 \exp[-j(2k_\omega - k_{2\omega})z] \tag{12.4.48a}$$

$$P_\omega^{\text{NL}} = 2\varepsilon_0 \, dE_{2\omega}E_\omega^* \exp[-j(k_{2\omega} - 2k_\omega)z] \tag{12.4.48b}$$

If one then substitutes Eq. (12.4.48) into Eq. (12.4.28) and neglect crystal losses (i.e., one takes $\sigma_i = 0$), one gets

$$\frac{dE_{2\omega}}{dz} = -j\frac{\omega}{n_{2\omega}c}dE_\omega^2 \exp(j\Delta kz) \tag{12.4.49a}$$

$$\frac{dE_\omega}{dz} = -j\frac{\omega}{n_\omega c}dE_{2\omega}E_\omega^* \exp(-j\Delta kz) \tag{12.4.49b}$$

where $\Delta k = k_{2\omega} - 2k_\omega$. These are the basic equations describing SHG.

To solve these equations, it is convenient to define new field variables E_ω' and $E_{2\omega}'$ as:

$$E_\omega' = (n_\omega)^{1/2}E_\omega \tag{12.4.50a}$$

$$E_{2\omega}' = (n_{2\omega})^{1/2}E_{2\omega} \tag{12.4.50b}$$

Since the intensity I_ω of the wave at ω is proportional to $n_\omega|E_\omega|^2$, the quantity $|E_\omega'|^2$ is proportional to I_ω with a proportionality constant independent of refractive index. Substitution of Eq. (12.4.50) into Eq. (12.4.49) then gives

$$\frac{dE_{2\omega}'}{dz} = -\frac{j}{l_{SH}}\frac{E_\omega'^2}{E_\omega'(0)}\exp[j(\Delta kz)] \tag{12.4.51a}$$

$$\frac{dE_\omega'}{dz} = -\frac{j}{l_{SH}}\frac{E_{2\omega}'E_\omega'^*}{E_\omega'(0)}\exp[-j(\Delta kz)] \tag{12.4.51b}$$

where: $E_\omega'(0)$ is the value of E_ω' at $z = 0$, assumed to be a real quantity; l_{SH} is a characteristic length, for second-harmonic interaction, given by

$$l_{SH} = \frac{\lambda(n_\omega n_{2\omega})^{1/2}}{2\pi\, dE_\omega(0)} \tag{12.4.52}$$

where λ is the wavelength and $E_\omega(0)$ is the field amplitude of the incident wave at frequency ω, also assumed to be a real quantity. Note that the advantage of using the new field variables E_ω' and $E_{2\omega}'$ is now apparent from Eqs. (12.4.51), since they involve a single coupling parameter l_{SH}. From Eqs. (12.4.51) we find

$$\frac{d|E_{2\omega}'|^2}{dz} = -\frac{d|E_\omega'|^2}{dz} \tag{12.4.53}$$

which represents the Manley–Rowe relation for SHG. Equation (12.4.53) shows that, e.g., a decrease of beam power, or intensity, at frequency ω must correspond to an increase, by the same amount, of power or intensity at frequency 2ω. Thus, 100% conversion of fundamental beam power into second harmonic power is possible in this case.

As a first example of the solution of Eq. (12.4.51), we consider the case when there is an appreciable phase mismatch (i.e., $l_{SH}\Delta k \gg 1$), so that little conversion of fundamental beam power into SH is expected to occur. We therefore set $E'_\omega(z) \cong E'_\omega(0)$ on the right-hand side of Eq. (12.4.51a). The resulting equation can then be integrated; using the boundary condition $E'_{2\omega}(0) = 0$, we obtain:

$$E'_{2\omega}(l) = -\frac{E'_\omega(0)}{l_{SH}}\left[\frac{\exp(-j\Delta kl) - 1}{\Delta k}\right] \qquad (12.4.54)$$

We can then readily show that:

$$\left|\frac{E'_{2\omega}(l)}{E'_\omega(0)}\right|^2 = \frac{\sin^2(\Delta kl/2)}{(\Delta kl_{SH}/2)^2} \qquad (12.4.55)$$

Since $|E'_{2\omega}|^2$ is proportional to the SH intensity $I_{2\omega}$, the variation of this intensity with crystal length is immediately obtained from Eq. (12.4.55). The corresponding behavior of I_ω versus l can then be obtained from the condition $I_\omega + I_{2\omega} = I_\omega(0)$; this condition is an immediate consequence of Eq. (12.4.53). The plots of $[I_\omega/I_\omega(0)]$ and $[I_{2\omega}/I_\omega(0)]$ versus (l/l_{SH}), obtained in this way for $l_{SH}\Delta k = 10$, are shown as dashed curves in Fig. 12.10. Note that, due to the large phase mismatch, only a small conversion into second harmonic occurs. Note also that, as can be shown from Eq. (12.4.55), the first maximum of $[I_{2\omega}/I_\omega(0)]$ occurs at $l = l_c$, where l_c, the coherence length, is given by Eq. (12.4.10).

As a second example of a solution to Eq. (12.4.51), we consider the case of perfect phase matching ($\Delta k = 0$). In this case appreciable conversion into second harmonic may occur, so fundamental-beam depletion must be considered. This means that Eq. (12.4.51)

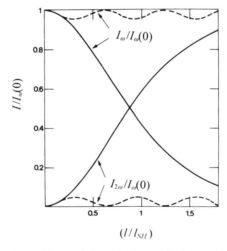

FIG. 12.10. Normalized plots of second harmonic intensity $I_{2\omega}$ and fundamental intensity I_ω versus crystal length l for perfect phase matching (continuous curves) and a finite-phase mismatch (dashed curves).

must now be solved without the approximation $E'_\omega(z) \cong E'_\omega(0)$. If however $\Delta k = 0$, we can show from Eq. (12.4.51) that, if $E'_\omega(0)$ is taken to be real, then $E'_\omega(z)$ and $E'_{2\omega}(z)$ are real and imaginary, respectively. We can therefore write

$$E'_\omega = |E'_\omega| \tag{12.4.56a}$$

$$E'_{2\omega} = -j|E'_{2\omega}| \tag{12.4.56b}$$

and Eq. (12.4.51) gives

$$\frac{d|E'_\omega|}{dz} = -\frac{1}{l_{SH}} \frac{|E'_{2\omega}||E'_\omega|}{E'_\omega(0)} \tag{12.4.57a}$$

$$\frac{d|E'_{2\omega}|}{dz} = \frac{1}{l_{SH}} \frac{|E'_\omega|^2}{E'_\omega(0)} \tag{12.4.57b}$$

The solution of Eqs. (12.4.57) with the boundary conditions $E'_\omega(z = 0) = E'_\omega(0)$ and $E'_{2\omega}(0) = 0$ is:

$$|E'_{2\omega}| = E'_\omega(0) \tanh(z/l_{SH}) \tag{12.4.58a}$$

$$|E'_\omega| = E'_\omega(0) \operatorname{sech}(z/l_{SH}) \tag{12.4.58b}$$

Since the intensity of the wave at a given frequency is proportional to $|E'|^2$, one has $I_{2\omega}/I_\omega(0) = |E'_{2\omega}|^2/E'^2_{2\omega}(0)$ and $I_\omega/I_\omega(0) = |E'_\omega|^2/E'^2_\omega(0)$. The dependence of $I_{2\omega}/I_\omega(0)$ and $I_\omega/I_\omega(0)$ on crystal length, as predicted by Eqs. (12.4.58), is shown by solid curves in Fig. 12.9. Note that, for $l = l_{SH}$, an appreciable fraction ($\sim 59\%$) of the incident wave is converted into a SH wave. This illustrates the role of l_{SH} as a characteristic length for the second harmonic interaction; according to Eq. (12.4.52), the value of l_{SH} is inversely proportional to the square root of the beam intensity at the fundamental frequency ω. Note also that, for $l \gg l_{SH}$, the radiation at the fundamental frequency can be completely converted into second harmonic radiation, in agreement with the Manley–Rowe relation (12.4.53).

12.5. TRANSFORMATION IN TIME

The phenomena of pulse compression and pulse expansion, now widely used in the field of ultrashort laser pulses, are considered in this section. Before discussing these phenomena, we recall that, given a homogeneous medium, such as a piece of glass or an optical fiber, characterized by the dispersion relation $\beta = \beta(\omega)$, one can define a group velocity, v_g, and a group-velocity-dispersion, GVD, as $v_g = (d\omega/d\beta)_{\omega_L}$ [see Eq. (8.6.26)] and GVD $= (d^2\beta/d\omega^2)_{\omega_L}$ [see Eq. (8.6.33)], respectively, where ω_L is the central frequency of the beam spectrum. For an inhomogeneous medium such as the prism pair described in Sect. 8.6.4.4 or the grating pair described in Sect. 12.5.1, a more useful approach is obtained upon letting $E_{in} \propto \exp j(\omega t)$ be the field of a monochromatic input beam at frequency ω and $E_{out} \propto \exp j[\omega t - \phi(\omega)]$ the field of the corresponding output beam. For a pulsed input beam, one can then define a group delay, τ_g, and a group-delay-dispersion, GDD, of the medium as $\tau_g = (d\phi/d\omega)_{\omega_L}$ [see Eq. (8.6.27)], and GDD $= (d^2\phi/d\omega^2)_{\omega_L}$, respectively.

12.5.1. Pulse Compression

Figure 12.11 shows a commonly used arrangement to compress ultrashort laser pulses. The pulse of a mode-locked laser of sufficient power (in practice a relatively modest peak power of, e.g., $P_p = 2\,\text{kW}$) and long time duration (e.g., $\tau_p = 6\,\text{ps}$) is sent through a single-mode silica optical fiber of suitable length (e.g., $L = 3\,\text{m}$). The pulse wavelength (e.g., $\lambda \cong 590\,\text{nm}$) falls in the positive GVD region of the fiber ($\lambda < 1.32\,\mu\text{m}$, for nondispersion-shifted fibers). After leaving the fiber, the output is collimated and passed through an optical system consisting of two identical gratings, parallel to each other, with appropriately chosen tilt and spacing, as described below. Under appropriate conditions, the output beam then consists of a light pulse with a much shorter duration (e.g., $\tau_p = 200\,\text{fs}$) than that of the input pulse and hence of much higher peak power (e.g., $P_p = 20\,\text{kW}$). Thus the arrangement in Fig. 12.11 provides a large compression factor (e.g., ~ 30 in the illustrated case) of the input pulse. The rather subtle phenomena involved in this pulse compression scheme are now discussed.[21,22]

We start by considering what happens when a pulse propagates in the optical fiber. Due to self-phase modulation, a light pulse of uniform intensity profile, that travels a distance z in a material exhibiting the optical Kerr effect, acquires a nonlinear phase term given by Eq. (8.6.38). In an optical fiber, however, the situation is somewhat more complicated due to the nonuniform transverse intensity profile of its fundamental mode (EH_{11}). In this case, one can show that the whole mode profile acquires a phase term given by[22]

$$\phi(t, z) = \omega_L t - \frac{\omega_L n_0}{c} z - \frac{\omega_L n_2 P}{c A_{eff}} z \qquad (12.5.1)$$

where n_0 is the low-intensity refractive index, n_2 is the coefficient of the nonlinear index of the medium [see Eq. (8.6.23)], $P = P(t, z)$ is the power of the beam traveling in the fiber, and A_{eff} is a suitably defined effective area of the beam in the fiber. The instantaneous carrier frequency of the light pulse is then obtained from Eq. (12.5.1) as:

$$\omega(t, z) = \frac{\partial \phi}{\partial t} = \omega_L - \frac{\omega_L}{c A_{eff}} z n_2 \frac{\partial P}{\partial t} \qquad (12.5.2)$$

It is linearly dependent on the negative time derivative of the corresponding power P. Thus, for a bell-shaped pulse as in Fig. 12.12a, the carrier frequency varies with time as indicated

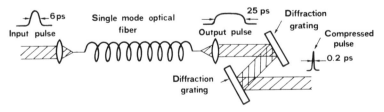

FIG. 12.11. Experimental setup for pulse compression.

by the solid curve in Fig. 12.12b. Note that, around the peak of the pulse, the time behavior of the power can be described by a parabolic law. The instantaneous carrier frequency then increases linearly with time (i.e., the pulse is said to show a positive frequency chirp). However the frequency chirp becomes negative after the pulse inflection points, i.e., for $t < t_A$ or $t > t_B$ in Fig. 12.12b.

It should be noted that the physical situation described so far has neglected the presence of GVD in the fiber. In the absence of GVD, in fact, the pulse shape does not change with propagation, i.e., the field amplitude remains a function of the variable $(z - v_g t)$, where v_g is the group velocity (see Appendix G). The z-dependence of the pulse at any given time is then obtained from the corresponding time dependence by reversing the positive direction of the axis and multiplying the time scale by v_g (see Fig. 12.12). This means that a point such as A in Fig. 12.12a is actually on the leading edge, while a point such as B is on the trailing edge of the pulse. According to Fig. 12.12b, the carrier frequency ω of the pulse, around point A, is smaller than around C, where it roughly equals ω_L. The carrier frequency of the pulse around point B, on the other hand will be higher than around C.

Assume now that the fiber has a positive GVD. The part of the pulse in the vicinity of point A then moves faster than that corresponding to point C, which moves faster than the region around point B. This means that, while traveling in the fiber, the central part of the pulse is expanded. Following a similar argument, one sees that the outer parts of the pulse are compressed rather than expanded because, there, the frequency chirp is negative. Thus, when a positive GVD is considered, the actual shape of the pulse intensity, as a function of time at a given z-position, will look like the dashed curve in Fig. 12.12a. The corresponding behavior for the frequency change is then as shown by dashed curve in Fig. 12.12b. Note in Fig. 12.12a that, due to pulse broadening produced by GVD, the peak intensity of the dashed curve is lower than that of the solid curve. Note also that, since the parabolic part of the pulse now extends over a wider region around the peak, the linear part of the positive frequency chirp extends over a larger fraction of the pulse.

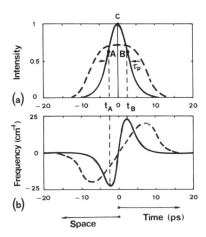

FIG. 12.12. Time behavior (a) of the pulse intensity and (b) the pulse frequency, when propagating through a single-mode fiber of suitable length. Solid and dashed curves refer, respectively, to the cases of no group-velocity dispersion and positive group-velocity dispersion in the fiber.

Having established these general features of the interplay between self-phase modulation (SPM) and GVD, we are now able to understand how, for a long enough fiber, the time behavior of the pulse amplitude and the pulse frequency, at the exit of the fiber, can actually develop into those shown in Figs. 12.13a–b. One sees that the pulse is squared and expanded to a duration of $\tau'_p \cong 23$ ps, while the positive frequency chirp is linear with time over most of the light-pulse duration. The corresponding pulse spectrum is shown in Fig. 12.13c, where one sees that, due to the strong SPM for such a small beam in the fiber (the core diameter for the condition depicted in Fig. 12.13 is $d \cong 4$ μm), the bandwidth of the output pulse, $\Delta v'_L = 50$ cm^{-1}, is considerably larger than that of the input pulse to the fiber. The latter is, in fact, established by the inverse of the pulse duration and, for the case considered of $\tau_p \cong 6$ ps, it corresponds to $\Delta v_L \cong 0.45/\tau_p \cong 2.5$ cm^{-1}. This means that the bandwidth of the output pulse is predominantly determined by the phase modulation of the pulse rather than by the duration of its envelope.

Suppose now that the pulse in Fig. 12.13a–b is passed through a medium of negative GDD. With the help of an argument similar to that used in relation to Fig. 12.12, we see that the region of the pulse around point A moves more slowly than that around C, which in turn moves more slowly than that around B. This implies that the pulse will now be compressed. Let us next suppose that besides being negative, the medium GDD is also independent of frequency. According to Eq. (8.6.27), the dispersion in group delay $d\tau_g/d\omega$ is also negative and independent of frequency. Thus τ_g decreases linearly with frequency and, since the frequency chirp of the pulse increases linearly with time (see Fig. 12.13b), all points of the pulse in Fig. 12.13a will tend to be compressed together at the same time if the GDD has the appropriate value. According to Eq. (8.6.31) this optimum value of GDD must be such that

$$\left(\frac{d^2\phi}{d\omega^2}\right)_{\omega_L} \Delta\omega'_L = \tau'_p \tag{12.5.3}$$

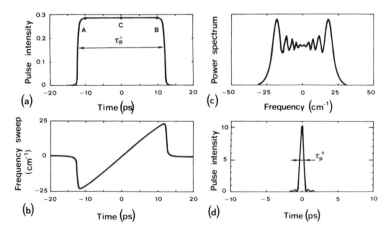

FIG. 12.13. Plots of calculated values of (a) self-broadening and (b) self-phase modulation, in an initial 6-ps pulse, after propagating through 30 m of single-mode fiber with positive group-velocity dispersion; (c) output pulse spectrum; (d) compressed pulse produced by an optical system having a negative group-velocity dispersion with linear dispersion. (By permission from Ref. 21.)

where $\Delta\omega'_L = 2\pi\Delta v'_L$ is the total frequency sweep of the pulse in Fig. 12.13b and τ'_p is the duration of the expanded pulse in Fig. 12.13a. Note that this compression mechanism does not produce an indefinitely sharp pulse as, at first sight, one may be led to believe. In fact, the system providing the negative GDD is a linear medium, which implies that the pulse spectrum must remain unchanged on passing through such a system, so the spectrum of the compressed pulse still remains as shown in Fig. 12.13c. Therefore, even under optimal compression conditions, the duration of the compressed pulse τ''_p cannot be shorter than approximately the inverse of the spectral bandwidth; i.e., $\tau''_p \cong 1/\Delta v'_L \cong 0.75$ ps. Since the time duration τ_p of the pulse originally entering the optical fiber was ~ 6 ps (see Fig. 12.11), the preceding result indicates that a sizable compression of the incoming light pulse has been achieved.*

The preceding heuristic discussion is based on the assumption that a chirped pulse can be subdivided into different temporal regions with different carrier frequencies. Although this idea is basically correct and allows a description of the phenomena in simple physical terms, a more critical and detailed examination of this approach reveals some conceptual difficulties. To validate this analysis, however, the analytical treatment of the problem can be performed in a straightforward way although the intuitive picture of the phenomenon becomes, in this way, somewhat obscured. For this analytical treatment, in fact, we take the Fourier transform $E_\omega(\omega)$ of the pulse in Figs. 12.13a–b, then multiply $E_\omega(\omega)$ by the transmission $t(\omega)$ of the medium exhibiting negative GDD. The resulting pulse in the time domain is then obtained by taking the inverse Fourier transform of $E(\omega)t(\omega)$. Note that, in a lossless medium, $t(\omega)$ must be represented by a pure phase term, i.e., it can be written as

$$t(\omega) = \exp(-j\phi) \tag{12.5.4}$$

where $\phi = \phi(\omega)$. If the medium has a constant GDD, the Taylor series expansion of $\phi(\omega)$ around the central carrier frequency ω_L gives

$$\phi(\omega) = \phi(\omega_L) + \left(\frac{d\phi}{d\omega}\right)_{\omega_L}(\omega - \omega_L) + \frac{1}{2}\left(\frac{d^2\phi}{d\omega^2}\right)_{\omega_L}(\omega - \omega_L)^2 \tag{12.5.5}$$

where $(d\phi/d\omega)_{\omega_L}$ is the group delay and $(d^2\phi/d\omega^2)_{\omega_L}$ is the group-delay dispersion. By substituting Eq. (12.5.5) into Eq. (12.5.4) and taking the inverse Fourier transform of $E(\omega)t(\omega)$, we then find that if $(d^2\phi/d\omega^2)_{\omega_L}$ is negative and satisfies Eq. (12.5.3), optimum pulse compression occurs. The optimally compressed pulse calculated in this way is shown in Fig. 12.13d. The resulting pulse duration is $\tau''_p \cong 0.6$ ps rather than the approximate value (0.75 ps) estimated before.

We are now left with the problem of finding a suitable optical system to provide the required negative GDD. Note that, since we can write GDD $= d\tau_g/d\omega$, a negative GDD implies that group delay must decrease with increasing ω. As discussed in Sect. 8.6.4.4, one such system consists of the two-prism pair shown in Fig. 8.26; another such system is the pair of parallel and identical gratings already shown in Fig. 12.11.[23] To understand the main properties of the latter system, we refer to Fig. 12.14 which shows a plane wave,

---

* Techniques of this type to produce shorter pulses, by first imposing a linear frequency chirp followed by pulse compression, have been used extensively in the field of radar (chirped radars) since World War II.

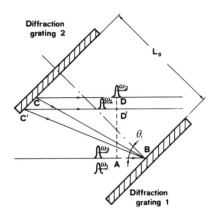

FIG. 12.14. Grating-pair for pulse compression.

represented by the ray AB, incident on grating 1 with a propagation direction at an angle θ_i to the grating normal. We now assume that the incident wave consists of two synchronous pulses at frequencies ω_2 and ω_1, with $\omega_2 > \omega_1$. As a result of the grating dispersion, the two pulses then follow paths $ABCD$ and $ABC'D'$, respectively. We see that the delay suffered by the pulse at frequency ω_2, $\tau_{d_2} = ABCD/v_g$, is smaller than that for the pulse at frequency ω_1, $\tau_{d_1} = ABC'D'/v_g$. Since $\omega_2 > \omega_1$, the pulse-delay dispersion is then negative. Detailed calculation then shows that the GDD can be expressed as[23]

$$\text{GDD} = \frac{d^2\phi}{d\omega^2} = -\frac{4\pi^2 c}{\omega^3 d^2} \frac{1}{\{1 - [\sin\theta_i - (\lambda/d)]^2\}^{3/2}} L_g \qquad (12.5.6)$$

where ω is the frequency of the wave, λ its wavelength, d the grating period, and L_g the distance between the two gratings. Note the minus sign on the right-hand side of Eq. (12.5.6), which indicates a negative GDD. Note also that the value of the dispersion can be changed by changing L_g and/or the incidence angle θ_i.

The two-grating system shown in Fig. 12.14 has the drawback that a lateral walk-off is present in the output beam (as occurs, e.g., between rays CD and $C'D'$); the amount of walk-off depends on the difference in frequency between beam components. For beams of finite size, this walk-off can represent a problem; however this problem can be circumvented by retroreflecting the output beam, back onto itself, with a plane mirror. In this case the overall dispersion resulting from the double pass through the diffraction-grating pair is obviously twice that given by Eq. (12.5.6).

The system in Fig. 12.11 is used to compress both picosecond and femtosecond laser pulses over a wide range of conditions.[24] For example, pulses of $\sim6\,\text{ps}$ duration (and $\sim2\,\text{kW}$ peak power) from a synchronously pumped mode-locked dye laser were compressed, using a 3-m-long fiber, to about 200 fs ($P_p = 20\,\text{kW}$). These pulses were again compressed, by a second system as in Fig. 12.11 using a 55-cm-long fiber, to obtain optical pulses of 90-fs duration. One of the most interesting results involves compressing 50-fs pulses from a colliding pulse, mode-locked, dye laser to $\sim6\,\text{fs}$ by using a 10-mm-long fiber.[25] To achieve this record value of pulse duration for such a configuration, second-order group-delay dispersion [GDD $= (d^2\phi/d\omega^2)_{\omega_L}$] and third-order group-delay dispersion [TOD $= (d^3\phi/d\omega^3)_{\omega_L}$] were compensated via the use of two consecutive grating pairs (each pair as in Fig. 12.14) and a four-prism sequence as in Fig. 8.26. In fact, the TOD of the two compression systems could be arranged to have opposite signs, so as to cancel each other.

A limitation of the optical fiber compression scheme in Fig. 12.11 arises from the small diameter ($d \cong 5\ \mu m$) of the fiber core. Accordingly, the pulse energy that can be transmitted through the fiber is necessarily limited to a low value ($\sim 10\ nJ$). A recently introduced guiding configuration to produce wide bandwidth SPM spectra, uses a hollow silica fiber filled with noble gases (Kr, Ar) at high pressures (1–3 atm).[26] With an inner diameter for the hollow fibre of 150–300 μm, a much higher energy input ($\sim 2\ mJ$) could be launched into the fibre. Using a fiber length of $\sim 1\ m$, wide SPM spectra ($\sim 200\ nm$) have been obtained starting with input pulses of femtosecond duration (20–150 fs). With the help of a specially designed two-prism sequence, in a double-pass configuration, and also using two reflections from a specially designed chirped mirror,[27] 20-fs pulses from the amplified beam of a mode-locked Ti:sapphire laser were compressed to ~ 4.5 fs.[28] These pulses, containing ~ 1.5 cycle of the carrier frequency, are the shortest pulses generated to date and have a relatively large amount of energy.

12.5.2. Pulse Expansion

As pointed out in Sect. 12.3, to obtain chirped-pulse amplification, we must first subject the pulse to a large expansion in time. In principle this expansion can be achieved by a single-mode fiber of suitable length (see Figs. 12.11 and 12.12a). However, the linear chirp produced in this way (see Fig. 12.13b), cannot be exactly compensated by a grating-pair compressor (Fig. 12.14) due to the higher order dispersion exhibited by this compressor. For pulses of short duration (subpicosecond), this system thus provides only partial compression of the expanded pulse to its original shape. A much better solution[29] involves using an expander that also consists of a grating pair but in an antiparallel configuration and with a 1:1 inverting telescope between the two gratings, as shown in Fig. 12.15.[30] To achieve the desired positive GDD, the two gratings must be located outside the telescope but within a focal length of the lens, i.e., one must have $(s_1, s_2) < f$, where f is the focal length of each lens. In this case, under the ideal paraxial-wave-propagation conditions and negligible dispersion of the lens material, the GDD is given by[30]

$$\text{GDD} = \frac{d^2\phi}{d\omega^2} = \frac{4\pi^2 c}{\omega^3 d^2 \cos^2\theta}(2f - s_1 - s_2) \qquad (12.5.7)$$

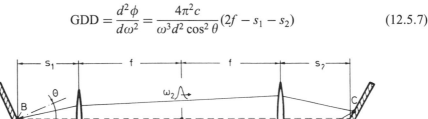
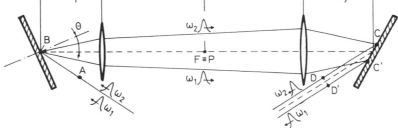

FIG. 12.15. Pulse expander consisting of two gratings, in an antiparallel configuration, with a 1-to-1 inverting telescope between them.

where ω is the frequency of the wave, d is the grating period, and θ is the angle shown in Fig. 12.15. Equation (12.5.7) shows that we have a positive value of GDD in this case. To understand this result we refer again to Fig. 12.15, where the plane wave incident on the first grating, represented by the ray AB, is assumed to consist of two synchronous pulses at frequencies ω_2 and ω_1, with $\omega_2 > \omega_1$. Due to grating dispersion, the two pulses follow paths $ABCD$ and $ABC'D'$, respectively. The delay suffered by the pulse at frequency ω_2, $\tau_{d_2} = ABCD/v_g$, is now larger than that of the pulse at frequency ω_1, $\tau_{d_1} = ABC'D'/v_g$. Since $\omega_2 > \omega_1$, the pulse-delay dispersion is now positive. Note that the two-grating telescopic system in Fig. 12.15 has the drawback that a lateral walk-off is present in the output beam; the amount of walk-off depends on the difference in frequency between beam components (e.g. between rays CD and $C'D'$). For beams of finite size, this walk-off can represent a problem; however, this problem can be circumvented by retroreflecting the output beam back onto itself by a plane mirror. In this case, the overall dispersion resulting from the double pass through the system in Fig. 12.15 is twice that given by Eq. (12.5.7).

To compare the positive GDD of this pulse expander with the negative GDD of the grating pair in Fig. 12.14, we recall that, according to the grating equation, one has $\sin\theta_i - (\lambda/d) = \sin\theta'$, where θ_i is the angle of incidence at the grating and θ' is the corresponding diffraction angle. We can substitute this grating equation into Eq. (12.5.6) and compare the resulting expression to Eq. (12.5.7). We see that, if $\theta' = \theta$, the two expressions become identical, apart from having opposite sign, provided that

$$\left(\frac{L_g}{\cos\theta}\right) = 2f - s_1 - s_2 \tag{12.5.8}$$

It should be stressed that this equivalence holds only under the ideal conditions considered previously. In this case, and when Eq. (12.5.8) applies, the expander in Fig. 12.15 is said to be conjugate to the compressor in Fig. 12.14. Physically, the conjugate nature of this expander arises because the telescope produces an image of the first grating located beyond the second grating and parallel to it. The expander in Fig. 12.15 is thus equivalent to a two-parallel-grating system with negative separation and, under condition (12.5.8), this system has the exact opposite dispersion, for all orders, to that of the compressor in Fig. 12.14.

In practice, due to lens aberrations and dispersion, the expander in Fig. 12.15 works well for pulse durations longer than $\sim 100\,\text{fs}$ and expansion ratios less than a few thousands. For shorter pulses and larger expansion ratios, the one-to-one telescope of the expander is usually realized by a suitably designed cylindrical[31] or spherical mirror configuration.[32] In particular, the cylindrical-mirror configuration has resulted in an expander with an expansion ratio greater than 10^4 and second-, third-, and fourth-order dispersion matched, for a suitable choice of material dispersion in the amplification chain, to that of the compressor.[31]

PROBLEMS

12.1. The Nd:YAG laser beam in Example 12.1. is first propagated in free space for a distance of 1 m starting from its waist, then focused by a positive lens with $f = 10\,\text{cm}$ focal length. Calculate the waist position after the lens and the spot size parameter at the waist. [Hint: To calculate this waist position, consider the lens to consist of two positive lenses f_1, and f_2 ($f_1^{-1} + f_2^{-1} = f^{-1}$); the first lens compensates for the curvature of the incoming wave front, thus producing a plane wave front, while the second focuses the beam.]

12.2. The output of a Q-switched Nd:YAG laser ($E = 100$ mJ, $\tau_p = 20$ ns) is amplified by a 6.3-mm diameter Nd:YAG amplifier having a small signal gain of $G_0 = 100$. Assume that: (a) The lifetime of the lower level of the transition is much shorter than τ_p; (b) the beam transverse intensity profile is uniform; (c) the effective peak cross section for stimulated emission is $\sigma \cong 2.8 \times 10^{-19}$ cm^2. Calculate the energy of the amplified pulse, the corresponding amplification, and the fraction of the energy stored in the amplifier that is extracted by the incident pulse.

12.3. A large Nd:glass amplifier, used to amplify 1-ns laser pulses for fusion experiments, consists of a disk amplifier with a disk clear aperture of $D = 9$ cm and overall disk length of 15 cm. Assume: (a) A measured small signal gain G_0 for this amplifier of ~ 4; (b) an effective stimulated-emission peak cross-section for Nd:glass of $\sigma = 4 \times 10^{-20}$ cm^2 (see Table 9.3); (c) that the lifetime of the lower transition level is much shorter than the laser pulse. Calculate the total energy available in the amplifier and the required energy of the input pulse to generate an output energy of $E_{out} = 450$ J.

12.4. Following the analysis used in deriving Eqs. (12.3.1) and (12.3.9) (assume $\alpha = 0$) as well as the rate equation calculation for a quasi-three-level laser [see Eqs. (7.2.21)–(7.2.24)], prove Eq. (12.3.15).

12.5. Referring to Problem 12.2., assume now that the input pulse duration is much shorter than the lifetime τ_1 of the lower laser level ($\tau_1 \cong 100$ ps). Using data obtained in Example 2.10 and knowing that the fractional population of the lower laser sublevel of the $^4I_{11/2}$ state is $f_{13} \cong 0.187$, calculate the energy of the amplified pulse and the corresponding amplification. Compare results with those obtained in Problem 12.2.

12.6. A large CO_2 TEA amplifier (with a gas mixture CO_2:N_2:He in the proportion 3:1.4:1) has dimensions of $10 \times 10 \times 100$ cm. The small signal gain coefficient for the $P(22)$ transition is measured to be $g = 4 \times 10^{-2}$ cm^{-1}. The duration of the input light pulse is 200 ns, which can therefore be assumed to be much longer than the thermalization time of the rotational levels of both the upper and lower vibrational states. The laser pulse, however, is much shorter than the decay time of the lower laser level. The true peak cross section for the $P(22)$ transition is $\sigma \cong 1.54 \times 10^{-18}$ cm^2; for $T = 300$ K, the fractional population of both initial and final rotational states is calculated to be $f = 0.07$. Calculate the output energy and the gain available from this amplifier for an input energy of 17 J. Also calculate the energy, per unit volume, available in the amplifier.

12.7. Prove Eq. (12.3.12).

12.8. The frequency of a Nd:YAG laser beam ($\lambda = 1.06$ μm) is doubled in a KDP crystal. For KDP one has $n_o(\lambda = 1.06$ μm$) = 1.507$, $n_o(\lambda = 532$ nm$) = 1.5283$, and $n_e(\lambda = 532$ nm$) = 1.48222$; calculate the phase-matching angle θ_m.

12.9. Prove Eq. (12.4.30).

12.10. Calculate the threshold pump intensity for parametric oscillation at $\lambda_1 \cong \lambda_2 = 1$ μm in a 5-cm-long LiNbO$_3$ crystal pumped at $\lambda_3 \cong 0.5$ μm ($n_1 = n_2 = 2.16$, $n_3 = 2.24$, $d \cong 6 \times 10^{-12}$ m/V, $\gamma_1 = \gamma_2 = 2 \times 10^{-2}$). If the pumping beam is focused in the crystal to a spot with a diameter of ~ 100 μm, calculate the resulting threshold pump power.

12.11. Calculate the second harmonic conversion efficiency by type I second harmonic generation, in a perfectly phase-matched 2.5-cm-long KDP crystal, for an incident beam at $\lambda = 1.06\ \mu$m having an intensity of $100\ \text{MW/cm}^2$ (For KDP one has $n \cong 1.5$, and $d_{\mathit{eff}} = d_{36} \sin \theta_m = 0.28 \times 10^{-12}\ \text{m/V}$ where $\theta_m \cong 50°$ is the phase-matching angle).

REFERENCES

1. R. W. Boyd, *Nonlinear Optics* (Academic, New York, 1992).
2. A. Yariv, *Optical Electronics*, 4th ed. (Holt Rinehart and Winston, New York, 1991). Chaps. 9, 12.
3. O. Svelto, Self-Focusing, Self-Steepening, and Self-Phase Modulation of Laser Beams, in *Progress in Optics*, vol. 12, (E. Wolf ed.) (North-Holland, Amsterdam, 1974), pp. 3–50.
4. A. E. Siegman, New Developments in Laser Resonators, in *Laser Resonators* (D. A. Holmes, ed.) *Proc. SPIE* **1224**, 2 (1990).
5. A. E. Siegman, Defining and Measuring Laser Beam Quality, in *Solid State Lasers—New Developments and Applications* (M. Inguscio and R. Wallenstein, eds.) (Plenum, New York, 1993), pp. 13–28.
6. L. M. Franz and J. S. Nodvick, Theory of Pulse Propagation in a Laser Amplifier, *J. Appl. Phys.* **34**, 2346 (1963).
7. P. G. Kriukov and V. S. Letokhov, Techniques of High-Power Light Pulse Amplification, in *Laser Handbook*, vol. 1 (F. T. Arecchi and E. O. Schultz-Dubois, eds.) (North-Holland, Amsterdam, 1972), pp. 561–595.
8. Walter Koechner, *Solid-State Laser Engineering*, 4th ed. (Springer, Berlin, 1996), Chap. 4.
9. D. Strickland and G. Mourou, Compression of Amplified Chirped Optical Pulses, *Opt. Commun.* **56**, 219 (1985).
10. G. Mourou, Ultra High Peak Power Laser: Present and Future, *Appl. Phys.* B **65**, 205 (1997).
11. M. D. Perry *et al.*, Petawatt Laser and Its Application to Inertial Confinement Fusion, *CLEO '96 Conference Digest* (Optical Society of America, Washington, DC., 1996), Paper CW14.
12. *Superstrong Fields in Plasmas, First International Conference* (M. Lontano, G. Mourou, F. Pegoraro, and E. Sindoni, eds.) (American Institute of Physics Series, New York, 1998).
13. R. J. Mears, L. Reekie, I. M. Jauncey, and D. N. Payne, Low-Noise Erbium-Doped Fiber Amplifier at 1.54 μm, *Electron. Lett.* **23**, 1026 (1987).
14. Emmanuel Desurvire, *Erbium-Doped Fiber Amplifiers* (Wiley, New York, 1994).
15. R. L. Byer, Optical Parametric Oscillators, in *Quantum Electronics*, vol. 1 (H. Rabin and C. L. Tang eds.) (Academic, New York, 1975). Part B, pp. 588–694.
16. P. A. Franken, A. E. Hill, C. W. Peters, and G. Weinreich, Generation of Optical Harmonics, *Phys. Rev. Lett.* **7**, 118 (1961).
17. J. A. Giordmaine and R. C. Miller, Tunable Optical Parametric Oscillation in $LiNbO_3$ at Optical Frequencies, *Phys. Rev. Lett.* **14**, 973 (1965).
18. J. A. Giordmaine, Mixing of Light Beams in Crystals, *Phys. Rev. Lett.* **8**, 19 (1962).
19. P. D. Maker, R. W. Terhune, N. Nisenhoff, and C. M. Savage, Effects of Dispersion and Focusing on the Production of Optical Harmonics, *Phys. Rev. Lett.* **8**, 21 (1962).
20. F. Zernike and J. E. Midwinter, *Applied Nonlinear Optics* (Wiley, New York, 1973), Sect. 3.7.
21. D. Grischkowsky and A. C. Balant, Optical Pulse Compression Based on Enhanced Frequency Chirping, *Appl. Phys. Lett.* **41**, 1 (1982).
22. G. P. Agrawal, *Nonlinear Fiber Optics*, 2nd ed. (Academic, San Diego, CA, 1995), Chap. 2.
23. E. B. Treacy, Optical Pulse Compression with Diffraction Gratings, *IEEE J. Quantum Electron.* **QE-5**, 454 (1969).
24. Ref. 22, Chap. 6.
25. R. L. Fork *et al.*, Compression of Optical Pulses to 6 Femtosecond by Using Cubic Phase Compensation, *Opt. Lett.* **12**, 483 (1987).
26. M. Nisoli, S. De Silvestri, and O. Svelto, Generation of High-Energy 10-fs Pulses by a New Compression Technique, *Appl. Phys. Letters* **68**, 2793 (1996).
27. R. Szipöcs, K. Ferencz, C. Spielmann and F. Krausz, Chirped Multilayer Coatings for Broadband Dispersion Control in Femtosecond Lasers, *Opt. Letters* **19**, 201 (1994).
28. M. Nisoli *et al.*, Compression of High-Energy Laser Pulses below 5 fs, *Opt. Letters* **22**, 522 (1997).

29. M. Pessot, P. Maine, and G. Mourou, 1000 Times Expansion-Compression of Optical Pulses for Chirped Pulse Amplification, *Opt. Comm.* **62**, 419 (1987).

30. O. E. Martinez, 3000 Times Grating Compressor with Positive Group Velocity Dispersion: Application to Fiber Compensation in 1.3–1.6 μm Region, *IEEE J. Quantum Electron.* **QE-23**, 59 (1987).

31. B. E. Lemoff and C. P. J. Barty, Quintic Phase-Limited, Spatially Uniform Expansion, and Recompression of Ultrashort Optical Pulse, *Opt. Letters* **18**, 1651 (1993).

32. Detau Du *et al.*, Terawatt Ti:Sapphire Laser with a Spherical Reflective Optic Pulse Expander, *Opt. Letters* **20**, 2114 (1995).

A

Semiclassical Treatment of the Interaction of Radiation and Matter

The calculation that follows uses the so-called semiclassical treatment of the interaction of radiation with matter. In this treatment the atomic system is assumed to be quantized and it is therefore described quantum mechanically while the em radiation is treated classically, i.e., by using Maxwell's equations.

We will first examine the phenomenon of absorption. We therefore consider the usual two-level system where we assume that, at time $t = 0$, the atom is in its ground state 1 and a monochromatic em wave at frequency ω interacts with it. Classically, the atom aquires an additional energy H' when interacting with the em wave, which may be due, for instance, to the interaction of the electric dipole moment of the atom $\boldsymbol{\mu}_e$ with the electric field $\mathbf{E}$ of the em wave ($H' = \boldsymbol{\mu}_e \cdot \mathbf{E}$), referred to as an electric-dipole interaction. This is not the only type of interaction by which a transition can occur, however. For instance, a transition may result from the interaction of the magnetic dipole moment of the atom $\boldsymbol{\mu}_m$ and the magnetic field $\mathbf{B}$ of the em wave ($\boldsymbol{\mu}_m \cdot \mathbf{B}$, magnetic-dipole interaction).

To describe the time evolution of this two-level system, we resort to quantum mechanics. As the classic treatment involves an interaction energy H', the quantum mechanical approach introduces an interaction Hamiltonian $\mathscr{H}'$. This Hamiltonian can be obtained from the classical expression for H' according to the well-known rules of quantum mechanics. The precise expression for $\mathscr{H}'$ need not concern us here, however. We need only note that $\mathscr{H}'$ is a sinusoidal function of time, with frequency ω equal to that of the incident wave. Accordingly we set

$$\mathscr{H}' = \mathscr{H}'_0 \sin \omega t \tag{A.1}$$

The total Hamiltonian $\mathscr{H}$ for the atom can then be written as

$$\mathscr{H} = \mathscr{H}_0 + \mathscr{H}' \tag{A.2}$$

where $\mathcal{H}_0$ is the atomic Hamiltonian in the absence of the em wave. Once the total Hamiltonian $\mathcal{H}$ for $t > 0$ is known, the time evolution of the wave function ψ of the atom is obtained from the time-dependent Schrödinger equation:

$$\mathcal{H}\psi = j\hbar \frac{\partial \psi}{\partial t} \tag{A.3}$$

To solve Eq. (A.3) for the wave function $\psi(t)$, we begin by introducing, according to Eq. (2.3.1), $\psi_1 = u_1 \exp[-(jE_1 t/\hbar)]$ and $\psi_2 = u_2 \exp[-(jE_2 t/\hbar)]$ as the unperturbed eigenfunctions of levels 1 and 2, respectively. Thus u_1 and u_2 satisfy the time-independent Schrödinger wave equation:

$$\mathcal{H}_0 u_i = E_i u_i \qquad (i = 1, 2) \tag{A.4}$$

Under the influence of the em wave, the wave function of the atom can be written as

$$\psi = a_1(t)\psi_1 + a_2(t)\psi_2 \tag{A.5}$$

where a_1 and a_2 are time-dependent complex numbers that, according to quantum mechanics, obey the relation:

$$|a_1|^2 + |a_2|^2 = 1 \tag{A.6}$$

According to Eq. (1.1.6) one has $W_{12} = -d|a_1(t)|^2/dt = d|a_2(t)|^2/dt$. To calculate W_{12}, we must then calculate the function $|a_2(t)|^2$. To do this we first generalize Eq. (A.5) as

$$\psi = \sum_{1}^{m} a_k \psi_k = \sum_{1}^{m} a_k u_k \exp[-j(E_k/\hbar)t] \tag{A.7}$$

where k denotes a general state of the atom and m gives the number of these states. By substituting Eq. (A.7) into Eq. (A.3), we obtain

$$\sum_k (\mathcal{H}_0 + \mathcal{H}') a_k u_k \exp[-j(E_k/\hbar)t] = \sum_k \{ j\hbar (da_k/dt) u_k \exp[-j(E_k/\hbar)t]$$
$$+ a_k u_k E_k \exp[-j(E_k/\hbar)t] \} \tag{A.8}$$

With the help of Eq. (A.4), this equation reduces to:

$$\sum j\hbar (da_k/dt) u_k \exp[-j(E_k/\hbar)t] = \sum a_k \mathcal{H}' u_k \exp[-j(E_k/\hbar)t] \tag{A.9}$$

Multiplying each side of this equation by the arbitrary eigenfunction u_n^*, then integrating over the whole of space, we obtain

$$\sum j\hbar (da_k/dt) \exp[-j(E_k/\hbar)t] \int u_k u_n^* \, dV =$$
$$= \sum a_k \exp[-j(E_k/\hbar)t] \int u_n^* \mathcal{H}' u_k \, dV \tag{A.10}$$

Since the wave functions u_k are orthogonal (i.e., $\int u_n^* u_k dV = \delta_{kn}$), Eq. (A.10) gives

$$\left(\frac{da_n}{dt}\right) = \frac{1}{(j\hbar)} \sum_1^m H'_{nk} a_k \exp\left[-j\frac{(E_k - E_n)t}{\hbar}\right] \tag{A.11}$$

where $H'_{nk} = H'_{nk}(t)$ is given by:

$$H'_{nk}(t) = \int u_n^* \mathcal{H}' u_k \, dV \tag{A.12}$$

Equation (A.11) comprises a set of m differential equations for the m variables $a_k(t)$, and these equations can be solved once the initial conditions are known. For the simpler case of a two-level system, the wave function ψ is given by Eq. (A.5), and Eq. (A.11) reduces to the following equations:

$$\left(\frac{da_1}{dt}\right) = \left(\frac{1}{j\hbar}\right)\left\{H'_{11} a_1 + H'_{12} a_2 \exp\left[-j\frac{(E_2 - E_1)t}{\hbar}\right]\right\} \tag{A.13a}$$

$$\left(\frac{da_2}{dt}\right) = \left(\frac{1}{j\hbar}\right)\left\{H'_{21} a_1 \exp\left[-j\frac{(E_1 - E_2)t}{\hbar}\right] + H'_{22} a_2\right\} \tag{A.13b}$$

which must be solved with initial conditions $a_1(0) = 1, a_2(0) = 0$.

So far no approximations have been made. To simplify the solution of Eq. (A.13), we now use a perturbation method. We assume that, on the right-hand side of Eq. (A.13), we can make the approximations that $a_1(t) \cong 1$ and $a_2(t) \cong 0$. By solving Eq. (A.13) subject to this approximation, we obtain first-order solutions for $a_1(t)$ and $a_2(t)$. For this reason the theory that follows is known as *first-order perturbation theory*. The solutions $a_1(t)$ and $a_2(t)$ obtained in this way can then be substituted into the right-hand side of Eq. (A.13) to obtain a solution which would then be a second-order approximation, and so on to higher orders. To first order, therefore, Eq. (A.13) gives

$$\left(\frac{da_1}{dt}\right) = \left(\frac{1}{j\hbar}\right)H'_{11} \tag{A.14a}$$

$$\left(\frac{da_2}{dt}\right) = \left(\frac{1}{j\hbar}\right)H'_{21} \exp(j\omega_0 t) \tag{A.14b}$$

where we have written $\omega_0 - (E_2 - E_1)/\hbar$ for the transition frequency of the atom. To calculate the transition probability, we need only solve Eq. (A.14b). Using Eqs. (A.1) and (A.12), we write

$$H'_{21} = H'^0_{21} \sin \omega t = \frac{H'^0_{21}[\exp(j\omega t) - \exp(-j\omega t)]}{2j} \tag{A.15}$$

where H'^0_{21} is given by

$$H'^0_{21} = \int u_2^* \mathcal{H}'_0 u_1 \, dV \tag{A.16}$$

and, in general, it is a complex constant. By substituting Eq. (A.15) into Eq. (A.14b) and integrating with the initial condition $a_2(0) = 0$, we obtain:

$$a_2(t) = \frac{H_{21}^{r0}}{2j\hbar} \left[\frac{\exp[j(\omega_0 - \omega)t] - 1}{\omega_0 - \omega} - \frac{\exp[j(\omega_0 + \omega)t] - 1}{\omega_0 + \omega} \right] \qquad (A.17)$$

If we now assume that $\omega \cong \omega_0$, we see that the first term in the square brackets is much larger than the second. We can then write

$$a_2(t) \cong -\frac{H_{21}^{r0}}{2j\hbar} \left[\frac{\exp(-j\Delta\omega t) - 1}{\Delta\omega} \right] \qquad (A.18)$$

where $\Delta\omega = \omega - \omega_0$, so that:

$$|a_2(t)|^2 = \frac{|H_{21}^{r0}|^2}{\hbar^2} \left[\frac{\sin(\Delta\omega t/2)}{\Delta\omega} \right]^2 \qquad (A.19)$$

The function, $y = [\sin \Delta\omega t/2)/\Delta\omega]^2$, is plotted in Fig. A.1 against $\Delta\omega$. One sees that the peak value of the function becomes greater and its width narrower as t increases. Furthermore one can show that:

$$\int_{-\infty}^{+\infty} \left[\frac{\sin(\Delta\omega t/2)}{\Delta\omega} \right]^2 d\Delta\omega = \frac{\pi t}{2} \qquad (A.20)$$

For large enough values of t, we can then set

$$\left[\frac{\sin(\Delta\omega t/2)}{\Delta\omega} \right]^2 \cong \frac{\pi t}{2} \delta(\Delta\omega) \qquad (A.21)$$

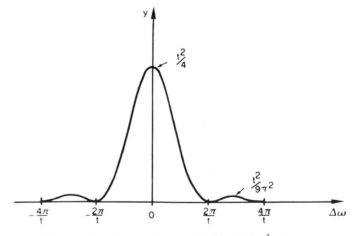

FIG. A.1. Plot of the function $y = [\sin(\Delta\omega t/2)/\Delta\omega]^2$ versus $\Delta\omega$.

where δ is the Dirac δ function. Within this approximation, Eq. (A.19) gives

$$|a_2(t)|^2 = \frac{|H_{21}'^0|^2}{\hbar^2} \frac{\pi}{2} t\delta(\Delta\omega) \tag{A.22}$$

which shows that, for long enough times, the probability $|a_2(t)|^2$ of finding the atom in level 2 is proportional to time t. Consequently, the transition probability W_{12} is obtained as:

$$W_{12} = \frac{d|a_2(t)|^2}{dt} = \frac{\pi}{2} \frac{|H_{21}'^0|^2}{\hbar^2} \delta(\Delta\omega) \tag{A.23}$$

To calculate W_{12} explicitly, we must calculate the quantity $|H_{21}'^0|^2$. Let us assume that the interaction responsible for the transition occurs between the electric field of the em wave and the electric dipole moment of the atom (electric dipole interaction). Classically, if we let $\mathbf{r}$ be the vector that specifies the position of the electron with respect to the nucleus and e the magnitude of the electron charge, the corresponding dipole moment of the atom is $\boldsymbol{\mu}_e = -e\mathbf{r}$. The classical interaction energy H' is then given by $H' = \boldsymbol{\mu}_e \cdot \mathbf{E} = -e\mathbf{E}(\mathbf{r}, t) \cdot \mathbf{r}$, where $\mathbf{E}$ is the electric field of the incident em wave at the electron position. Following the well-known rules of quantum mechanics, the interaction Hamiltonian is then simply given by:

$$\mathscr{H}' = -e\mathbf{E}(\mathbf{r}, t) \cdot \mathbf{r} \tag{A.24}$$

Substitution of Eq. (A.24) into Eq. (A.12) with $n = 2$ and $k = 1$ gives

$$H_{21}' = -e \int u_2^* \mathbf{E} \cdot \mathbf{r} u_1 \, dV \tag{A.25}$$

Let us now assume that the wavelength of the em wave is much larger than the atomic dimension. This is satisfied very well for em waves in the visible. In fact, one has $\lambda = 500 \, \text{nm}$ for green light while typical atomic dimensions are $\sim 0.1 \, \text{nm}$. In this case, we assume that $\mathbf{E}(\mathbf{r}, t)$ does not change appreciably over an atomic dimension and thus remains equal to its value $\mathbf{E}(0, t)$ at $\mathbf{r} = 0$, i.e., at the center of the nucleus (*electric-dipole approximation*). We can thus write

$$\mathbf{E}(\mathbf{r}, t) \cong \mathbf{E}(0, t) = \mathbf{E}_0 \sin \omega t \tag{A.26}$$

where $\mathbf{E}_0$ is a constant. If Eq. (A.26) is now substituted into Eq. (A.25) and the resulting expression for H_{21}' compared with that given in Eq. (A.15), $H_{21}'^0$ can be expressed as:

$$H_{21}'^0 = \mathbf{E}_0 \cdot \boldsymbol{\mu}_{21} \tag{A.27}$$

where $\boldsymbol{\mu}_{21}$ is given by:

$$\boldsymbol{\mu}_{21} = -\int u_2^* e\mathbf{r} u_1 \, dV \tag{A.28}$$

It is called the matrix element of the electric dipole moment. If we now let θ be the angle between $\boldsymbol{\mu}_{21}$ and $\mathbf{E}_0$, we obtain from Eq. (A.27)

$$|H_{21}'^0|^2 = E_0^2 |\mu_{21}|^2 \cos^2 \theta \tag{A.29}$$

where $|\mu_{21}|$ is the magnitude of the complex vector $\boldsymbol{\mu}_{21}$. If we now assume that the em wave interacts with several atoms whose vectors $\boldsymbol{\mu}_{21}$ are randomly oriented with respect to $\mathbf{E}_0$, the average value of $|H_{21}'^0|^2$ is obtained by averaging Eq. (A.29) over all possible angles θ and ϕ (in two dimensions). To carry this out, let $p(\theta)$ represent the probability density for the orientation of atomic dipoles, so that $p(\theta)d\Omega$ gives the elemental probability that the vector $\boldsymbol{\mu}_{21}$ is within the solid angle $d\Omega$ making an angle θ with the direction of $\mathbf{E}_0$. For randomly oriented dipoles, one has $p(\theta) = $ const. Accordingly it follows that $\langle \cos^2 \theta \rangle = 1/3$, where the brackets $\langle \, \rangle$ indicate the average value over all dipole orientations. From Eq. (A.29) we then get:

$$\langle |H_{21}'^0|^2 \rangle = \frac{E_0^2 |\mu_{21}|^2}{3} \tag{A.30}$$

Substitution of the latter expression into Eq. (A.23) then gives:

$$W_{12} = \frac{\pi}{6} \frac{(2\pi)^2 E_0^2 |\mu_{21}|^2}{h^2} \delta(\omega - \omega_0) \tag{A.31}$$

If the function $\delta(\nu - \nu_0)$ is used instead of $\delta(\omega - \omega_0)$ in Eq. (A.31), then, since $\delta(\nu - \nu_0) = 2\pi\delta(\omega - \omega_0)$, Eq. (A.31) can be transformed into Eq. (2.4.5).

Having calculated the rate of absorption, we can now go on to calculate the rate of stimulated emission. We again start with Eq. (A.13), this time using the initial conditions $a_1(0) = 0$ and $a_2(0) = 1$. We immediately see, however, that the required equations in this case are obtained from Eqs. (A.13)–(A.31) for the case of absorption simply by interchanging indices 1 and 2. Since we see from Eq. (A.28) that $|\mu_{12}| = |\mu_{21}|$, it follows from Eq. (A.31) that $W_{12} = W_{21}$, which shows that the probabilities for absorption and stimulated emission are equal.

B

Lineshape Calculation for Collision Broadening

As explained in Sect. 2.5.1, we can calculate the lineshape for collision broadening from the normalized spectral density $g(v' - v)$ of the sinusoidal waveform with random phase jumps as shown in Fig. 2.9. The signal wave of Fig. 2.9 is written as

$$E(t) = E_0 \exp j[\omega t - \phi(t)] \tag{B.1}$$

where E_0 is taken to be a real constant, $\omega = 2\pi v$ is the angular frequency of the radiation, and the phase $\phi(t)$ is assumed to undergo random jumps at each collision of the atom. We assume that the probability density $p_\tau(\tau)$ of the time τ between two consecutive collisions can be described by Eq. (2.5.7). The calculation of $g(v' - v)$ is best performed in terms of angular frequency ω, i.e., in terms of the distribution $g(\omega' - \omega)$. Obviously one has $g(v' - v)dv' = g(\omega' - \omega)d\omega'$ and it follows that $g(v' - v) = 2\pi g(\omega' - \omega)$. Apart from a proportionality constant, the function $g(\omega' - \omega)$ is then given by the power spectrum $W(\omega' - \omega)$ of the waveform in Fig. 2.9. For this proportionality constant to be unity, according to Eq. (2.5.4), $W(\omega' - \omega)$ must be such that $\int W(\omega' - \omega)d\omega' = 1$. From Parseval's theorem one has:

$$\int_{-\infty}^{+\infty} W(\omega' - \omega)d\omega' = \lim_{T \to \infty} \frac{\pi}{T} \int_{-T}^{+T} |E(t)|^2 \, dt = 2\pi E_0^2 \tag{B.2}$$

The condition $\int W(\omega' - \omega)d\omega' = 1$ then leads to the result $2\pi E_0^2 = 1$. This means that $g(\omega' - \omega)$ can be obtained as the power spectrum of the signal $E(t)$ given by Eq. (B.1) and with amplitude:

$$E_0 = (2\pi)^{-1/2} \tag{B.3}$$

To calculate the power spectrum $W(\omega' - \omega)$ of the signal given in Eq. (B.1) we use the Wiener–Kintchine theorem, so that $W(\omega' - \omega)$ is obtained as the Fourier transform of the signal autocorrelation fuction $\Gamma(\tau)$. We can thus write

$$W(\omega' - \omega) = \int_{-\infty}^{+\infty} \Gamma(\tau) \exp[-(j\omega'\tau)]d\tau \qquad (B.4)$$

where the autocorrelation function $\Gamma(\tau)$ is given by:

$$\Gamma(\tau) = \lim_{T \to \infty} \frac{1}{2T} \int_{-T}^{+T} E^*(t)E(t+\tau)dt \qquad (B.5)$$

For the waveform in Fig. 2.9, we can then write

$$\Gamma(\tau) = \lim_{T \to \infty} \frac{1}{2T} E_0^2 \exp(j\omega\tau) \left\{ \int_{corr.} dt + \int_{uncorr.} \exp[-j(\Delta\phi)]dt \right\} \qquad (B.6)$$

The first integral, on the right-hand side of Eq. (B.6), is calculated over the time intervals, between $-T$ and $+T$, in which no phase-jumping collision has occurred and thus the signals $E(t)$ and $E(t+\tau)$ have the same phase (*correlated intervals*). The second integral, on the right-hand side of Eq. (B.6), is calculated over the time intervals in which a collision has occurred and thus the two signals $E(t)$ and $E(t+\tau)$ have a random phase difference $\Delta\phi$ (*uncorrelated intervals*). This situation is illustrated in Fig. B.1, where vertical bars (continuous line) indicate phase jumps for $E(t)$ and $E(t+\tau)$. Projecting the vertical bars of, e.g., $E(t+\tau)$ onto the plot of $E(t)$ (dashed vertical lines), the correlated time intervals, shown as hatched areas, are obtained.

To calculate $\Gamma(\tau)$ from Eq. (B.6), we first notice that the contribution of the second integral on the right-hand side of Eq. (B.6) vanishes as T tends to infinity because the integrand $\exp[-j(\Delta\phi)]$ is a random number with zero average. Equation (B.6) then reduces to:

$$\Gamma(\tau) = \lim_{T \to \infty} \frac{E_0^2}{2T} \exp(j\omega\tau) \left(\int_{corr.} dt \right) \qquad (B.7)$$

The integral in Eq. (B.7) equals the total time of phase correlation, i.e., the sums of the time intervals corresponding to hatched areas in Fig. B.1. If we let τ'_n represent the time interval between the nth phase jump and the next one (see Fig. B.1), this sum can be expressed as $\sum_n (\tau'_n - \tau)$, where summation is extended over the values τ'_n for which $\tau'_n > \tau$. Equation (B.7) can then be written as:

$$\Gamma(\tau) = E_0^2 \exp(j\omega\tau) \lim_{T \to \infty} \frac{\sum_n (\tau'_n - \tau)}{2T} \qquad (B.8)$$

If we now let N be the number of phase jumps between $-T$ and $+T$, we can write

$$\lim_{T \to \infty} \frac{\sum_n (\tau'_n - \tau)}{2T} = \lim_{T \to \infty} \frac{[\sum_n (\tau'_n - \tau)]/N}{2T/N} = \frac{\langle (\tau' - \tau) \rangle}{\tau_c} \qquad (B9)$$

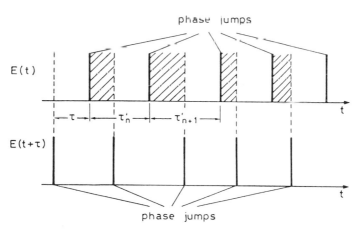

FIG. B.1. Plots of the phase jumps of the two functions $E(t)$ and $E(t + \tau)$ vs time. Hatched areas indicate time intervals where the two functions are correlated.

where $\tau_c = 2T/N$ is the average time between two consecutive phase jumps and $\langle(\tau' - \tau)\rangle$ is the average value of $(\tau' - \tau)$ (with the constraint $\tau' > \tau$). Using the probability density of time intervals between consecutive jumps, given by Eq. (2.5.7), the value $\langle(\tau' - \tau)\rangle$ can be obtained as:

$$\langle \tau' - \tau \rangle = (1/\tau_c) \int_{\tau}^{+\infty} (\tau' - \tau) \exp[-(\tau'/\tau_c)]d\tau' = \tau_c \exp[-(\tau/\tau_c)] \qquad (B.10)$$

From Eqs. (B.8)–(B.10) we then obtain the expression for the correlation function as $\Gamma(\tau) = E_0^2 \exp[j\omega\tau - (\tau/\tau_c)]$. If $\Gamma(\tau)$ is now also extended to the case $\tau < 0$, then, since $\Gamma(-\tau) = \Gamma(\tau)$, we can write our final result as:

$$\Gamma(\tau) = E_0^2 \exp[j\omega\tau - (|\tau|/\tau_c)] \qquad (B.11)$$

Using the Wiener–Kintchine theorem [see Eq. (B.4)] we can now calculate $W(\omega' - \omega)$ and hence, using Eq. (B.3) for E_0, we can also calculate $g(\omega' - \omega)$. We obtain:

$$g(\omega' - \omega) = \frac{\tau_c}{\pi} \frac{1}{[1 + (\omega' - \omega)^2 \tau_c^2]} \qquad (B.12)$$

Since $g(v' - v) = 2\pi g(\omega' - \omega)$, we can also write [see Eq. (2.5.9)]

$$g(v' - v) = 2\tau_c \frac{1}{[1 + (v' - v)^2 4\pi^2 \tau_c^2]} \qquad (B.13)$$

from which the lineshape function given by Eq. (2.5.10) is readily obtained.

C

Simplified Treatment of Amplified Spontaneous Emission

We will assume that ASE emission occurs in both directions within the active medium and therefore refer to the geometry shown in Fig. C.1. We further assume that the amplifier behaves as an ideal four-level system, so that the lower level population can be neglected. We consider both homogeneously and inhomogeneously broadened transitions. A detailed theory that holds for these conditions is developed in Ref. 1. The theory is rather complicated, however, and, as a result, one cannot readily obtain the main aspects of the physical behavior of ASE from this approach. Following some very recent work,[2] we present here a simplified treatment of ASE designed to obtain asymptotic expressions for ASE behavior in the low-saturation regime.

In the low-saturation regime, we assume that the upper state population N_2, and hence the population inversion $N \cong N_2$, are not appreciably saturated by the ASE intensity. With reference to Fig. C.1, we let $I_\nu(z, \nu)$ be the spectral ASE intensity at coordinate z for the beam propagating in the positive z-direction. The elemental variation dI_ν along the z-coordinate must not only account for stimulated emission but also for the spontaneous emission contribution from the element dz. We can thus write

$$\frac{\partial I_\nu}{\partial z} = \sigma N I_\nu + N A_\nu \frac{\Omega(z)}{4\pi} \tag{C.1}$$

where σ is the transition cross section at frequency ν, A_ν is the rate of spontaneous emission at frequency ν, and $\Omega(z)$ is the solid angle subtended by the exit face as seen from the element dz. Note that the factor $\Omega(z)/4\pi$, on the right-hand side of Eq. (C.1), accounts for the fact that spontaneous radiation from the element dz is emitted uniformly over the whole solid angle 4π, while we are interested only in that fraction emitted within the solid angle $\Omega(z)$.

To calculate the ASE spectral intensity $I_\nu(l, \nu)$ at the $z = l$ exit face of the medium, we must integrate Eq. (C.1) over the z-coordinate. Since most of the ASE emission arises from

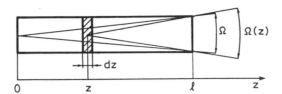

FIG. C.1. Calculation of the ASE spectral emission along the z-direction.

emitting elements near $z = 0$, which experience the largest gain, we can assume $\Omega(z) \cong \Omega$, where Ω is the solid angle subtended by one face of the active medium as seen from the other face (see Fig. C.1). Equation (C.1) can then be readily integrated; using the boundary condition $I_v(0, v) = 0$, we obtain:

$$I_v(l, v) = \frac{\Omega}{4\pi} A_v \frac{hv[\exp(\sigma Nl) - 1]}{\sigma} \tag{C.2}$$

Equation (C.2) can be rearranged by observing that $A_v = Ag(v - v_0)$ and $\sigma = \sigma_p g(v - v_0)/g_p$, where A is the rate of spontaneous emission, $g(v - v_0)$ is the transition lineshape function, g_p is its peak value, σ_p is the peak cross section and v_0 is the frequency of the transition peak. Equation (C.2) can then be written as

$$I_v(l, v) = \phi I_s \frac{\Omega}{4\pi} g_p[\exp(\sigma Nl) - 1] \tag{C.3}$$

where $\phi = \tau/\tau_r = \tau A$ is the fluorescence quantum yield [see Eq. (2.6.22)] and $I_s = hv/\sigma_p \tau$ is the peak saturation intensity of the amplifier [see Eq. (2.8.24)].

The ASE emission at a general frequency v, normalized to its peak value, is readily obtained from Eq. (C.3) as $[I_v(l, v)/I_v(l, v_0)] = [\exp(\sigma Nl) - 1]/[\exp(\sigma_p Nl) - 1]$. For both Lorentzian and Gaussian lines, this spectral emission can be readily computed for a given value of the peak gain $G = \exp(\sigma_p Nl)$. As an example, Fig. 2.24 shows (as solid lines) the computed spectral profiles versus the normalized frequency offset $2(v - v_0)/\Delta v_0$ for a Lorentzian line with peak gains of 10^3 and 10^6, respectively. An approximate expression for the ASE linewidth, Δv_{ASE}, can be obtained from Eq. (C.3) if we assume that the ASE spectrum can be approximated by a Gaussian function. We accordingly write

$$[\exp(\sigma Nl) - 1] \cong [\exp(\sigma_p Nl) - 1] \times \exp(-kx^2) \tag{C.4}$$

where: k is a constant; x represents the normalized frequency offset, i.e.

$$x = \frac{2(v - v_0)}{\Delta v_0} \tag{C.5}$$

where Δv_0 is the transition linewidth (for either a homogeneously or inhomogeneously broadened transition). From Eq. (C.4) we readily obtain the expression for kx^2 as:

$$kx^2 \cong \ln[\exp(\sigma_p Nl) - 1] - \ln[\exp(\sigma Nl) - 1] \tag{C.6}$$

If we now let $f(x)$ represent the function on the right-hand side of Eq. (C.6), the constant k can be readily obtained from the relation:

$$k = \frac{1}{2}\left(\frac{d^2 f}{dx^2}\right)_{x=0} \tag{C.7}$$

For a Lorentzian line, we now write $\sigma = \sigma_p/(1+x^2)$ in Eq. (C.6). After some lengthy but straightforward calculations, we obtain from Eq. (C.7):

$$k = \frac{G \ln G}{(G-1)} \tag{C.8}$$

Similarly, for the case of a Gaussian line, we write $\sigma = \sigma_p \exp -(x^2 \ln 2)$ in Eq. (C.6) to obtain from Eq. (C.7):

$$k = (\ln 2)\frac{G \ln G}{(G-1)} \tag{C.9}$$

With the preceding Gaussian approximation, the normalized ASE linewidth, in terms of the normalized frequency offset x, is now readily obtained as $\Delta x_{ASE} = 2(\ln 2/k)^{1/2}$. Since, from Eq. (C.5), one has $\Delta v_{ASE} = \Delta x_{ASE} \times (\Delta v_0/2)$, the ASE linewidth can be found from the two previous expressions once the value for k given by Eqs. (C.8) or (C.9) is used. We thus obtain

$$\Delta v_{ASE} = (\ln 2)^{1/2}\left(\frac{G-1}{G \ln G}\right)^{1/2}\Delta v_0 \tag{C.10}$$

for a Lorentzian line and

$$\Delta v_{ASE} = \left(\frac{G-1}{G \ln G}\right)^{1/2}\Delta v_0 \tag{C.11}$$

for a Gaussian line. As an example, Eq. (C.10) is used to obtain the normalized ASE linewidth $\Delta v_{ASE}/\Delta v_0$ versus peak gain G, which is plotted in Fig. 2.25 as a dashed line.

With the help of these considerations on the spectral behavior of ASE, we can now also obtain an approximate expression for the total ASE intensity, $I = \int I_v \, dv$, at the exit face of the active medium. To do so, we first integrate both sides of Eq. (C.3) over frequency v to obtain

$$I = \phi I_s\left(\frac{\Omega}{4\pi}\right)g_p\int_{-\infty}^{+\infty}[\exp(\sigma Nl) - 1]dv \tag{C.12}$$

With the help of the Gaussian approximation for the ASE spectrum given by Eq. (C.4), the integral in Eq. (C.12) is readily calculated, using Eqs. (C.8) and (C.9), to give [see Eq. (2.9.3)]

$$I = \phi I_s \left(\frac{\Omega}{4\pi^{3/2}} \right) \frac{(G-1)^{3/2}}{(G \ln G)^{1/2}} \tag{C.13}$$

for a Lorentzian and

$$I = \phi I_s \left(\frac{\Omega}{4\pi} \right) \frac{(G-1)^{3/2}}{(G \ln G)^{1/2}} \tag{C.14}$$

for a Gaussian line.

REFERENCES

1. L. W. Casperson, Threshold Characteristics of Mirrorless Lasers, *J. Appl. Phys.* **48**, 256 (1977).
2. O. Svelto, S. Taccheo, and C. Svelto, Analysis of Amplified Spontaneous Emission: Some Corrections to the Lyndford Formula, *Optic. Comm.* **149** 277–282 (1998).

D

Calculation of the Radiative Transition Rates of Molecular Transitions

We consider a simplified approach here, merely to show how selection rules arise for a molecular transition.

The transition probability can be expressed in the form of Eq. (2.4.9) if the appropriate value for the amplitude of the oscillating dipole moment $|\mu|$ is used. We begin by recalling that, for an ensemble of negative charges (electrons of the molecule), each of value e (with the sign included) and of positive charges of value e_h (nuclei of the molecule), the classical electric-dipole moment is given by $\boldsymbol{\mu} = \sum_i e\mathbf{r}_i + \sum_j e_h\mathbf{R}_j$. Here $\mathbf{r}_i$ and $\mathbf{R}_j$ specify positions of electrons and nuclei, respectively, relative to some given reference point; the sum is taken over all electrons and nuclei in the molecule. If the reference point is taken as the center of the positive charges, then $\sum_j e_h\mathbf{R}_j = 0$ and $\boldsymbol{\mu}$ reduces to:

$$\boldsymbol{\mu} = \sum_i e\mathbf{r}_i \tag{D.1}$$

To simplify matters we now consider a diatomic molecule. In this case, the nuclear coordinates may be reduced to the magnitude R of the internuclear spacing $\mathbf{R}$ and to the angular coordinates θ and ϕ of $\mathbf{R}$ relative to a given reference system. According to quantum mechanics, the oscillating dipole moment of the molecule is then given by [see also Eq. (2.3.6)]

$$\boldsymbol{\mu}_{osc} = 2\,\mathrm{Re}\int \psi_2^*(\mathbf{r}_i, R, \mathbf{r}_r)\boldsymbol{\mu}\psi_1(\mathbf{r}_i, R, \mathbf{r}_r)\,d\mathbf{r}_i\,dR\,d\mathbf{r}_r \tag{D.2}$$

where ψ_2 and ψ_1 are, respectively, the wave functions of the final and initial states of the transition. Note that both ψ_1 and ψ_2 are taken as functions of the positions of all electrons, of the internuclear distance R, and the rotational coordinates $\mathbf{r}_r$ (shorthand notation for θ and ϕ); the integral is taken over all these coordinates.

Following the Born–Oppenheimer approximation, the molecular wave functions ψ can now be written as

$$\psi(\mathbf{r}_i, R, \mathbf{r}_r) = u_e(\mathbf{r}_i, R)u_v(R)u_r(\mathbf{r}_r)\exp[-j(E/\hbar)t] \tag{D.3}$$

where u_e, u_v, and u_r are the electronic, vibrational, and rotational wave functions, respectively, and $E = E_e + E_v + E_r$ is the total energy of the given state. Substitution of Eq. (D.3) into Eq. (D.2) shows that $\boldsymbol{\mu}_{osc}$ oscillates at the frequency $v_{21} = (E_2 - E_1)/h$ with a complex amplitude $\boldsymbol{\mu}_{21}$ given by [compare with Eq. (2.3.7)]:

$$\boldsymbol{\mu}_{21} = \left(\int u_{v2}^* \boldsymbol{\mu}_e u_{v1}\, dR\right)\left(\int u_{r2}^* u_{r1}\, d\mathbf{r}_r\right) \tag{D.4}$$

In Eq. (D.4) we have set

$$\boldsymbol{\mu}_e = \boldsymbol{\mu}_e(R) = \int u_{e2}^*(\mathbf{r}_i, R)\boldsymbol{\mu} u_{e1}(\mathbf{r}_i, R)d\mathbf{r}_i \tag{D.5}$$

where $\boldsymbol{\mu}$ is the dipole moment given by Eq. (D.1). Since electronic wave functions are slowly varying functions of R, $\boldsymbol{\mu}_e(R)$ can be expanded in a power series around the equilibrium internuclear distance R_0 in the form:

$$\boldsymbol{\mu}_e(R) = \boldsymbol{\mu}_e(R_0) + \frac{d\boldsymbol{\mu}_e}{dR}(R - R_0) + \cdots \tag{D.6}$$

Let us first consider pure rotational transitions; in this case one has $u_{e2} = u_{e1}$ and $u_{v2} = u_{v1}$. From Eq. (D.5) the dipole moment $\boldsymbol{\mu}_e(R_0)$ is then seen to be given by

$$\boldsymbol{\mu}_e(R_0) = \int \boldsymbol{\mu}|u_{e1}(\mathbf{r}_i, R_0)|^2\, d\mathbf{r}_i \tag{D.7}$$

which is the permanent electric dipole moment, $\boldsymbol{\mu}_{ep}$, of the molecule. If we assume $\boldsymbol{\mu}_e \cong \boldsymbol{\mu}_e(R_0)$ in the first integral on the right-hand side of Eq. (D.4) and recall that

$$\int u_{v2}^* u_{v1}\, dR = \int |u_{v1}|^2\, dR = 1$$

we obtain the following expression for $|\mu_{21}|^2 = |\boldsymbol{\mu}_{21}|^2$, to be used in Eq. (2.4.9):

$$|\mu_{21}|^2 = |\boldsymbol{\mu}_{ep}|^2\left|\int u_{r2}^* u_{r1}\, d\mathbf{r}_r\right|^2 \tag{D.8}$$

The first factor in the right-hand side of Eq. (D.8) indicates that pure rotational transitions are possible only in molecules possessing a permanent dipole moment $\boldsymbol{\mu}_{ep}$. This is easily understood because the stimulated emission process can be considered to arise from the interaction of the incident em wave with this rotating dipole moment. For molecules with a permanent dipole moment, $|\mu_{21}|^2$ is then proportional to the second factor on the right-hand side of Eq. (D.8). From the symmetry properties of rotational wave functions, it then follows that this factor is nonzero only if the quantum jump between the rotational numbers of the two states obeys the selection rule $\Delta J = \pm 1$.

Let us next consider rotational-vibrational transitions. One has again $u_{e2} = u_{e1}$ and, to first order, we again put $\boldsymbol{\mu}_e(R) \cong \boldsymbol{\mu}_e(R_0) = \boldsymbol{\mu}_{ep}$ into Eq. (D.4). We then readily see that μ_{21} reduces to $(\boldsymbol{\mu}_{ep} \int u_{v2}^* u_{v1} \, dR)(\int u_{r2}^* u_{r1} \, d\mathbf{r}_r)$ which is zero due to the orthogonality of vibrational wave functions belongins to the same electronic state. To calculate the transition rate, we must therefore consider the second term in Eq. (D.6). Substitution of this term into Eq. (D.4) gives the following expression for $|\mu_{21}|^2$:

$$|\mu_{21}|^2 = \left|\frac{d\boldsymbol{\mu}_e}{dR}\right|^2 \left|\int u_{v2}^* (R - R_0) u_{v1} \, dR\right|^2 \left|\int u_{r2}^* u_{r1} d\mathbf{r}_r\right|^2 \tag{D.9}$$

The third factor on the right-hand side of Eq. (D.9) again gives the selection rule $\Delta J = \pm 1$ for the rotational quantum jump. As for the second factor, we recall that, if the potential energy curve $U(R - R_0)$ of the molecule is approximated by a parabola, i.e, for a harmonic restoring force, the wave functions u_v are given by the well-known harmonic-oscillator functions, i.e., the product of Hermite polynomials with a Gaussian function. Due to the symmetry properties of these functions, $|\mu_{21}|^2$ is then nonzero only if $\Delta v = \pm 1$. Overtones arise if this parabolic assumption is relaxed (i.e., for anharmonicity of the potential energy) or if higher order terms in Eq. (D.6) are taken into account (electrical anharmonicity). Note that, under certain symmetry conditions for the ground-state electronic wave function, the first factor in Eq. (D.9) is zero, and the transition is said to be infrared inactive. An obvious case occurs when two atoms of the molecule are identical, which occurs when, e.g., an N_2 molecule involves the same isotopic species for the two atoms. In this case, for symmetry reasons, the molecule cannot have a net dipole moment $\boldsymbol{\mu}_e(R)$ for any value of the internuclear distance R, and $|\mu_{21}|^2$ in Eq. (D.9) is always zero.

Lastly we consider vibronic transitions. If we again take only the first term in Eq. (D.6), then $|\mu_{21}|^2$ in Eq. (D.4) is given by:

$$|\mu_{21}|^2 = |\boldsymbol{\mu}_e(R_0)|^2 \left|\int u_{v2}^* u_{v1} \, dR\right|^2 \left|\int u_{r2}^* u_{r1} \, d\mathbf{r}_r\right|^2 \tag{D.10}$$

As a result of the symmetry properties of the electronic wave functions of the two states, the first factor on the right-hand side of Eq. (D.10) may turn out to be zero. In this case the vibronic transition is said to be electric-dipole forbidden. For a dipole-allowed transition, the third factor on the right-hand side of (D.10) again leads to the selection rule $\Delta J = \pm 1$. Within this selection rule and again for a dipole-allowed transition, Eq. (D.10) shows that $|\mu_{21}|^2$ is proportional to the second factor on the right-hand side of the equation, known as the Franck–Condon factor. Note that, in this case, this factor is nonzero because u_{v2} and u_{v1} belong to different electronic states. The transition probability W is thus determined by the degree of overlap between nuclear wave functions, as discussed in Sect. 3.1.3.

E

Space-Dependent Rate Equations

The purpose of this appendix is to develop a rate-equation treatment and to solve these equations for the cw case, when the spatial variation of both the pump rate and the cavity field are taken into account. As a result of these spatial variations, the population inversion is also space-dependent. In all cases we assume that the laser oscillates in a single mode.

E.1. FOUR-LEVEL LASERS

For an ideal four-level laser, we can neglect the population N_1 of the lower laser level and thus let $N \cong N_2$ represent the population inversion. We then write

$$\frac{\partial N}{\partial t} = R_p - WN - \frac{N}{\tau} \tag{E.1.1a}$$

$$\frac{d\phi}{dt} = \int_a WN \, dV - \frac{\phi}{\tau_c} \tag{E.1.1.b}$$

where the integral in Eq. (E.1.1b) is taken over the volume of the active medium and where the meaning of other symbols is given in Chap. 7. Equation (E.1.1a) expresses a local balance between pumping, stimulated emission, and spontaneous decay processes. Note that a partial derivative is used on the left-hand side of the equation because of the expected spatial variation of N. The integral term on the right-hand side of Eq. (E.1.1b) accounts for the contribution of stimulated processes to the total number of cavity photons ϕ. This term is written on the basis of a simple balance, using the fact that each individual stimulated process produces a photon. For a plane wave, we can now write $W = \sigma F = \sigma I / h\nu$ and $I = c\rho / n$, where σ is the stimulated emission cross section, F is the photon flux, I is the intensity of the wave, ρ is the energy density in the active medium, and n is its refractive index. From the preceding expressions, we can relate W to the energy density of the wave as:

$$W = \frac{c\sigma}{nh\nu} \rho \tag{E.1.2}$$

Although this equation has been derived, for simplicity, for a plane wave, we recognize that it merely establishes a local relation between transition rate and energy density of the em field. It follows that this equation has a general validity and thus represents the relation between W and ρ for, e.g., the em field of the cavity. In this case ρ is expected to depend on both space $\mathbf{r}$, and, in a transient case, also on time t, the spatial dependence accounting for the spatial variation of the cavity mode. Equations (E.1.1) with the help of (E.1.2) give:

$$\frac{\partial N}{\partial t} = R_p - \frac{c\sigma}{nh\nu}\rho N - \frac{N}{\tau} \tag{E.1.3a}$$

$$\frac{d\phi}{dt} = \frac{c\sigma}{nh\nu}\int_a \rho N \, dV - \frac{\phi}{\tau_c} \tag{E.1.3b}$$

Note that, since R_p and ρ are assumed to depend on position (and time, for a transient case), this will also apply to N, which cannot therefore be removed from the integral in Eq. (E.1.3b).

The total number of cavity photons ϕ can now be related to the energy density of the em wave by

$$\phi = \frac{1}{h\nu}\int_c \rho \, dV \tag{E.1.4}$$

where the integral is taken over the whole volume of the cavity. We consider a laser cavity of length L in which an active medium of length l and refractive index n is inserted; we assume that the beam waist is located somewhere in the active medium. Under these conditions, the energy density of the mode outside, ρ_{out}, and inside, ρ_{in}, the active medium can be written, respectively, as

$$\rho_{out} = \rho_0|u(\mathbf{r})|^2 \tag{E.1.5a}$$

$$\rho_{in} = n\rho_0|u(\mathbf{r})|^2 \tag{E.1.5b}$$

where $u(\mathbf{r})$ is the field amplitude at the general coordinate $\mathbf{r}$, normalized to its peak value (occurring at the waist), and $n\rho_0$ is the energy density at the waist. From Eqs. (E.1.4) and (E.1.5) we obtain

$$\phi = \frac{\rho_0}{h\nu}\left(n\int_a |u|^2 \, dV + \int_r |u|^2 \, dV\right) \tag{E.1.6}$$

where the two integrals are taken over the active medium and the remaining volume of the cavity, respectively. The form of Eq. (E.1.6) suggests that we define an effective volume of the mode in the cavity, V, as

$$V = \left(n\int_a |u|^2 \, dV + \int_r |u|^2 \, dV\right) \tag{E.1.7}$$

and a volume, V_a, of the mode in the active medium, as:

$$V_a = \int_a |u|^2 \, dV \tag{E.1.8}$$

With the help of Eqs. (E.1.5b), (E.1.6), and (E.1.7), Eq. (E.1.3) can be transformed into

$$\frac{\partial N}{\partial t} = R_p - \frac{c\sigma}{V}\phi N|u|^2 - \frac{N}{\tau} \tag{E.1.9a}$$

$$\frac{d\phi}{dt} = \left(\frac{c\sigma}{V}\int_a N|u|^2\,dV - \frac{1}{\tau_c}\right)\phi \tag{E.1.9b}$$

which represent our final result describing the space-dependent rate equations for a four-level laser.

We now solve Eq. (E.1.9) for the case of a cw laser oscillating in a single TEM_{00} mode. For simplicity we further assume that the em cavity field $u(\mathbf{r})$ is independent of the longitudinal coordinate z. This implies that we are neglecting both spot size variation and the mode standing-wave pattern along the laser cavity. We also take the cladded rod model discussed in Sect. 6.3.3, so that we need not worry about the aperturing effect caused by a finite rod diameter. Under these assumptions, we write the following simple expression for $|u(r)|$, which holds for any value of the longitudinal coordinate z in the active medium and for any value of the radial coordinate r from 0 to ∞:

$$|u| = \exp[-(r/w_0)^2] \tag{E.1.10}$$

where w_0 is the spot size at the beam waist. Equation (E.1.7) then gives

$$V = \frac{\pi w_0^2}{2}L_e \tag{E.1.11}$$

where the equivalent length of the cavity L_e can be expressed as [see Eq. (7.2.11)]:

$$L_e = L + (n-1)l \tag{E.1.12}$$

Likewise, from Eq. (E.1.8), we obtain

$$V_a = \frac{\pi w_0^2}{2}l \tag{E.1.13}$$

The threshold condition for the population inversion is now obtained from Eq. (E.1.9b) by letting $(d\phi/dt) = 0$. We can now define a spatially averaged population inversion as [see Eq. (7.3.20)]:

$$\langle N\rangle = \frac{\int_a N|u|^2\,dV}{\int_a |u|^2\,dV} = \frac{\int_a N|u|^2\,dV}{V_a} \tag{E.1.14}$$

Equation (E.1.9b) then gives [see Eq. (7.3.19)]

$$\langle N\rangle_c = \frac{1}{c\sigma\tau_c}\frac{V}{V_a} = \frac{\gamma}{\sigma l} \tag{E.1.15}$$

where Eqs. (E.1.11), (E.1.13), and (7.2.14) have been used. The threshold expression for the pump rate can be obtained from Eq. (E.1.9a) by letting $(\partial N/\partial t) = 0$ and $\phi = 0$. We obtain

$$R_p(r, z) = \frac{N(r, z)}{\tau} \qquad (E.1.16)$$

A spatially averaged pump rate $\langle R_p \rangle$ can now be defined as:

$$\langle R_p \rangle = \frac{\int_a R_p |u|^2 \, dV}{\int_a |u|^2 \, dV} = \frac{\int_a R_p |u|^2 \, dV}{V_a} \qquad (E.1.17)$$

Substitution of Eq. (E.1.16) on the right-hand side of Eq. (E.1.17), with the further help of Eq. (E.1.15), gives

$$\langle R_p \rangle_c = \frac{\langle N \rangle_c}{\tau} = \frac{\gamma}{\sigma l \tau} \qquad (E.1.18)$$

Above threshold, the steady-state average population $\langle N \rangle_0$ is obtained from Eq. (E.1.9b) by letting $(d\phi/dt) = 0$. This gives

$$\langle N \rangle_0 = \langle N \rangle_c = \frac{\gamma}{\sigma l} \qquad (E.1.19)$$

The steady-state photon number ϕ_0 is obtained from Eq. (E.1.9a) by letting $(\partial N/\partial t) = 0$. We obtain

$$N\left(1 + \frac{c\sigma\tau}{V} \phi_0 |u|^2\right) = R_p \tau \qquad (E.1.20)$$

To proceed further we relate ϕ_0, in Eq. (E.1.20), to the output power, P_{out}, with the help of Eq. (7.2.18). Equation (E.1.20) can then be transformed into

$$N = \frac{R_p \tau}{[1 + (P_{out}/P_s)|u|^2]} \qquad (E.1.21)$$

where we have used Eq. (E.1.11) for the cavity volume and Eq. (7.2.14) for the photon decay time and where the saturation power, P_s, has been defined as [see Eq. (7.3.28)]:

$$P_s = \frac{\gamma_2}{2} \frac{\pi w_0^2}{2} \frac{h\nu}{\sigma\tau} \qquad (E.1.22)$$

Multiplying both sides of Eq. (E.1.21) by $|u|^2$ and integrating over the volume of the active medium, we obtain

$$\langle N \rangle_0 = \frac{1}{V_a} \int \frac{R_p |u|^2 \tau}{[1 + (P_{out}/P_s)|u|^2]} dV \qquad (E.1.23)$$

where Eq. (E.1.14) has been used and where the spatially-averaged inversion is denoted by $\langle N \rangle_0$ since the laser is operating cw. Using Eqs. (E.1.19) and (E.1.13), the preceding expression gives:

$$\frac{\gamma}{\sigma} = \frac{2}{\pi w_0^2} \int \frac{R_p |u|^2 \tau}{[1 + (P_{out}/P_s)|u|^2]} dV \qquad (E.1.24)$$

To proceed further we need to specify the spatial variation of R_p and relate its value to the pump power P_p. This will be done for a pumping profile that is either uniform or has a Gaussian transverse distribution.

In the case of uniform pumping, R_p is a constant and, for both lamp pumping and electrical pumping, it is given by [see Eqs. (6.2.6) and (6.4.26)]:

$$R_p = \eta_p \frac{P_p}{\pi a^2 l h \nu_{mp}} \qquad (E.1.25)$$

Note that Eq. (E.1.25) holds only for $0 \leq r \leq a$, where a is the radius of the medium, while one has $R_p(r) = 0$ for $r > a$. Note also that, according to Sect. 6.3.3, the expression holding for diode pumping with uniform illumination is simply obtained from Eq. (E.1.25) by replacing ν_{mp}, the minimum pump frequency described in Fig. 6.18, with ν_p, the frequency of the pumping diode. Substitution of Eq. (E.1.25) into the integral in Eq. (E.1.24) then gives

$$\frac{\gamma}{\sigma} = \eta_p \left[\frac{P_p \tau}{\pi a^2 h \nu_{mp}} \right] \left[\frac{2}{\pi w_0^2} \right] \int_0^a \frac{|u|^2}{[1 + (P_{out}/P_s)|u|^2]} 2\pi r \, dr \qquad (E.1.26)$$

where integration along the longitudinal coordinate z of the active medium has already been performed. We now define a minimum pump threshold P_{mth} and the dimensionless variables x and y as in Eqs. (7.3.26), (7.3.25), and (7.3.27), respectively. Equation (E.1.26) can then be transformed into

$$\frac{1}{x} = \left(\frac{2}{\pi w_0^2} \right) \int_0^a \frac{\exp[-2(r/w_0)^2]}{\{1 + y \exp[-2(r/w_0)^2]\}} 2\pi r \, dr \qquad (E.1.27)$$

where Eq. (E.1.10) has been used for $|u|^2$. Equation (E.1.27) can be readily integrated using the subsitution:

$$t = \exp[-2(r/w_0)^2] \qquad (E.1.28)$$

We obtain [compare with Eq. (7.3.30)]:

$$\frac{1}{x} = \int_\beta^1 \frac{dt}{1 + yt}$$

$$= \frac{1}{y} \ln \left(\frac{1+y}{1+\beta y} \right) \qquad (E.1.29)$$

where

$$\beta = \exp[-2(a/w_0)^2] \tag{E.1.30}$$

In the case of a Gaussian distribution for the transverse pumping profile, as can be produced by, e.g., longitudinal diode pumping, R_p can be related to the pump power P_p by Eq. (6.3.7). Using the expression for $|u|$ given by Eq. (E.1.10) in Eq. (E.1.24), we obtain

$$\frac{\gamma}{\sigma} = \eta_r \eta_t \left(\frac{2}{\pi w_p^2}\right) \left(\frac{P_p \tau}{h v_p}\right) \left(\frac{2}{\pi w_0^2}\right) \left\{\int_0^\infty \frac{\exp[(-2r^2)(w_0^2 + w_p^2)/w_0^2 w_p^2]}{1 + (P_{out}/P_s)\exp[-(2r^2/w_0^2)]} 2\pi r dr\right\} \times$$

$$\times \int_0^l \alpha \exp[-(\alpha z)]dz \tag{E.1.31}$$

According to Eq. (6.3.11), the second integral on the right-hand side of Eq. (E.1.31) gives the pump absorption efficiency η_a. We now define a minimum pump threshold as [see Eq. (7.3.32)]

$$P_{mth} = \left(\frac{\gamma}{\eta_p}\right)\left(\frac{h v_p}{\tau}\right)\left(\frac{\pi w_p^2}{2\sigma_e}\right) \tag{E.1.32}$$

where $\eta_p = \eta_r \eta_t \eta_a$ is the pump efficiency. We also define the dimensionless variables x and y as in Eqs. (7.3.25) and (7.3.27), respectively. Equation (E.1.31) then gives

$$\frac{1}{x} = \left(\frac{2}{\pi w_0^2}\right)\left\{\int_0^\infty \frac{\exp[(-2r^2)(w_0^2 + w_p^2)/w_0^2 w_p^2]}{1 + y\exp[-(2r^2/w_0^2)]} 2\pi r dr\right\} \tag{E.1.33}$$

If we now define a variable t as in Eq. (E.1.28) and a quantity δ as $\delta = (w_0/w_p)^2$, the preceding equation simplifies to:

$$\frac{1}{x} = \int_0^1 \frac{t^\delta}{1 + yt}dt \tag{E.1.34}$$

The integral in Eq. (E.1.34) can be calculated analytically for integer values of δ. In particular, for $\delta = 1$, we obtain

$$\frac{1}{x} = \left[\frac{t}{y} - \frac{1}{y^2}\ln(1 + yt)\right]_0^1 \tag{E.1.35}$$

from which we immediately obtain Eq. (7.3.34).

E.2. QUASI-THREE-LEVEL LASERS

The procedure to establish the space-dependent rate equations for a quasi-three-level laser and to solve them for the cw case follows the same path as that for a four-level laser. According to Eq. (7.2.19) we now write

$$(N_1 + N_2) = N_t \tag{E.2.1a}$$

$$\frac{\partial N_2}{\partial t} = R_p - (W_e N_2 - W_a N_1) - \frac{N_2}{\tau} \tag{E.2.1b}$$

$$\frac{d\phi}{dt} = \int_a (W_e N_2 - W_a N_1) dV - \frac{\phi}{\tau_c} \tag{E.2.1c}$$

According to Eq. (E.1.2), we can write the rates for stimulated emission W_e and for absorption W_a, appearing in Eqs. (E.2.1b) and (E.2.1c), as:

$$W_e = \frac{c\sigma_e}{nh\nu} \rho \tag{E.2.2a}$$

and

$$W_a = \frac{c\sigma_a}{nh\nu} \rho \tag{E2.2.b}$$

where σ_e and σ_a are the effective cross sections for emission and absorption, respectively. We now follow the same steps, from Eq. (E.1.4) to (E.1.8), as for a four-level laser, to obtain [compare with Eq. (7.2.24)]:

$$\frac{\partial N}{\partial t} = R_p(1+f) - \frac{c(\sigma_e + \sigma_a)}{V} \phi N|u|^2 - \frac{fN_t + N}{\tau} \tag{E.2.3a}$$

$$\frac{d\phi}{dt} = \left(\frac{c\sigma_e}{V} \int_a N|u|^2 \, dV - \frac{1}{\tau_c} \right) \phi \tag{E.2.3b}$$

where $N = N_2 - fN_1$ [see Eq. (7.2.23)] and $f = \sigma_a/\sigma_e$ [see Eq. (7.2.22)]. These equations represent our final result describing the space-dependent rate equations of a quasi-three-level laser.

We now solve Eq. (E.2.3) for the case of a cw laser oscillating in a single TEM$_{00}$ mode. Under the assumption that $|u(r)|$ is described by Eq. (E.1.10) for $0 < r < \infty$, we again obtain Eqs. (E.1.11), (E.1.12), and (E.1.13) for V, L_e, and V_a, respectively. We also define the spatially averaged values, $\langle N \rangle$, and $\langle R_p \rangle$ as in Eqs. (E.1.14) and (E.1.17), respectively.

The threshold value of $\langle N \rangle$ is obtained from Eq. (E.2.3b) by letting $(d\phi/dt) = 0$. This gives:

$$\langle N \rangle_c = \frac{\gamma}{\sigma_e l} \tag{E.2.4}$$

The threshold value of $\langle R_p \rangle$ is obtained from Eq. (E.2.3a) by letting $(\partial N / \partial t) = 0$ and $\phi = 0$. We have

$$\langle R_p \rangle_c = \frac{f \langle N_t \rangle + \langle N \rangle_c}{(1+f)\tau} = \frac{\sigma_a \langle N_t \rangle l + \gamma}{(\sigma_e + \sigma_a)l\tau} \qquad (E.2.5)$$

Above threshold, the steady-state value $\langle N \rangle_0$ is obtained from Eq. (E.2.3b) by letting $(d\phi/dt) = 0$. We get

$$\langle N \rangle_0 = \langle N \rangle_c = \frac{\gamma}{\sigma_e l} \qquad (E.2.6)$$

From Eq. (E.2.3a), with the condition $(\partial N / \partial t) = 0$, we obtain

$$N = \frac{R_p(1+f)\tau - fN_t}{1 + [c(\sigma_e + \sigma_a)\tau / V]\phi_0 |u|^2} \qquad (E.2.7)$$

The steady-state number of photons ϕ_0 can again be related to the output power P_{out} by Eq. (7.2.18). Equation (E.2.7) can then be transformed into:

$$N = \frac{R_p(1+f)\tau - fN_t}{1 + y|u|^2} \qquad (E.2.8)$$

In Eq. (E.2.8) we have again defined $y = P_{out}/P_s$, where the saturation power P_s is now given by:

$$P_s = \frac{\gamma_2}{2} \frac{\pi w_0^2}{2} \left[\frac{h\nu}{(\sigma_e + \sigma_a)\tau} \right] \qquad (E.2.9)$$

We now multiply both sides of Eq. (E.2.8) by $|u|^2$, then integrate over the volume of the active medium. With the help of Eqs. (E.1.13), (E.1.14), and (E.2.6), we obtain [compare with Eq. (E.1.24)]:

$$\frac{\gamma}{\sigma} = \frac{2}{\pi w_0^2} \int \frac{[R_p(1+f)\tau - fN_t]|u|^2}{[1 + y|u|^2]} dV \qquad (E.2.10)$$

To proceed further we must specify the spatial dependence of R_p. In the case of longitudinal pumping with a pump beam of Gaussian radial profile, we use Eq. (6.3.7). Using Eq. (E.1.10) for $|u|$, Eq. (E.2.10) can be written as

$$\frac{\gamma}{\sigma_e l} = \eta_r \eta_t (1+f) \left(\frac{P_p \tau}{h\nu_p} \right) \left(\frac{2}{\pi w_p^2 l} \right) \int_0^1 \frac{t^\delta}{1 + yt} dt \int_0^1 \alpha \exp[-(\alpha z)] dz - fN_t \int_\beta^1 \frac{dt}{1 + yt} \qquad (E.2.11)$$

where t and β are expressed by Eqs. (E.1.28) and (E.1.30), respectively, and

$$\delta - \left(\frac{w_0}{w_p}\right)^2 \tag{E.2.12}$$

Again recognizing that the integral over the longitudinal coordinate z is the absorption efficiency η_a and given the simplifying assumption $w_0 \ll a$ (i.e. $\beta \cong 0$), Eq. (E.2.11) yields

$$\gamma = \eta_p(\sigma_e + \sigma_a)\left(\frac{P_p\tau}{hv_p}\right)\left(\frac{2}{\pi w_p^2}\right)\int_0^1 \frac{t^\delta}{1+yt}dt - \sigma_a N_t l\frac{\ln(1+y)}{y} \tag{E.2.13}$$

This equation can be solved for P_p to obtain

$$P_p = \gamma\left(\frac{hv_p}{\eta_p\tau}\right)\left[\frac{\pi w_p^2}{2(\sigma_e + \sigma_a)}\right]\frac{1}{\int_0^1 [t^\delta/(1+yt)]dt}\left[1 + B\frac{\ln(1+y)}{y}\right] \tag{E.2.14}$$

where $B = \sigma_a N_t l/\gamma$. We obtain the minimum pump threshold from Eq. (6.3.25) when $w_0 \ll w_p$ and $\sigma_a N_t l \ll \gamma$. We obtain [see Eq. (7.4.16)]:

$$P_{mth} = \gamma\left(\frac{hv_p}{\eta_p\tau}\right)\left[\frac{\pi w_p^2}{2(\sigma_e + \sigma_a)}\right] \tag{E.2.15}$$

Taking the ratio of Eq. (E.2.14) to Eq. (E.2.15), we obtain our final result [see Eq. (7.4.18)]:

$$x = \frac{1 + B\dfrac{\ln(1+y)}{y}}{\displaystyle\int_0^1 \frac{t^\delta}{1+yt}dt} \tag{E.2.16}$$

F

Mode Locking Theory: Homogeneous Line

According to Sect. 8.6.3, a theory of mode locking can be developed, in the time domain, by requiring the pulse to reproduce itself after each cavity round trip. We limit our discussion here to the case of a homogeneous line, and we further assume that the lifetime of the upper laser level is much longer than the cavity round-trip time. Under these conditions, the saturated single-pass power gain of the amplifier, at the transition peak, is given by $g_0 = \sigma_p N_0 l$, where σ_p is the peak cross section, l is the length of the active medium, and N_0 is the steady-state inversion, as established by the cumulative effect of the passage of many pulses. This means that this saturated gain is determined by the average intracavity beam intensity $\langle I \rangle$; it can thus be related to the unsaturated gain g by

$$g_0 = \frac{g}{1 + (\langle I \rangle / I_s)} \tag{F.1}$$

where $I_s = h\nu_0 / \sigma_p \tau$ is the saturation intensity of the amplifier at the transition peak.

F.1. ACTIVE MODE LOCKING

This theory was developed by Kuizenga and Siegman[1] and, later on, put in a more general framework by Haus.[2,3] We follow the latter treatment, and, for brevity, we limit our considerations to mode locking by an amplitude modulator. Thus we consider the laser configuration in Fig. F.1 and assume that the modulator is very thin and placed as close as possible to mirror 2. Under these conditions, a single light pulse is expected to travel back and forth within the cavity (see Fig. 8.19). At any given position within the cavity, the electric field of the pulse can be written as

$$E(t) = A(t) \exp j(\omega_0 t - \phi) \tag{F.1.1}$$

563

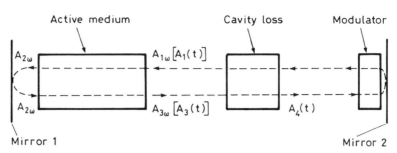

FIG. F.1. Schematic diagram of the laser cavity considered for the theoretical analysis of mode locking.

where t represents a suitable local time to account for pulse propagation. The spectral amplitude $A_\omega(\omega - \omega_0)$ of the pulse is then obtained by taking the Fourier transform of $E(t)$ i.e.:

$$A_\omega(\omega - \omega_0) = \int_{-\infty}^{+\infty} E(t) \exp(-j\omega t)dt = \int_{-\infty}^{+\infty} A(t) \exp[-j(\omega - \omega_0)t]dt \qquad (F.1.2)$$

The field amplitude $A(t)$ is then related to $A_\omega(\omega - \omega_0)$ by the inverse Fourier transform:

$$A(t) = \int_{-\infty}^{+\infty} A_\omega(\omega - \omega_0) \exp[j(\omega - \omega_0)t]d(\omega - \omega_0) \qquad (F.1.3)$$

We first consider the passage of the light pulse through the amplifier. If we let $A_{1\omega}$ and $A_{2\omega}$ represent the spectral amplitude of the light pulse before and after a single passage (see Fig. F.1), we can write $A_{2\omega} = t_g A_{1\omega}$, where the single-pass electric field transmission t_g through the amplifier is given by[4]

$$t_g = \frac{A_{2\omega}}{A_{1\omega}} = \{\exp[-j(\omega nl/c)]\} \times \exp\left\{\frac{(g_0/2)}{1 + [2j(\omega - \omega_0)/\Delta\omega_0]}\right\} \qquad (F.1.4)$$

where n is the refractive index of the active medium and $\Delta\omega_0$ is the width (FWHM) of the laser line. Note that, according to Eq. (F.1.4), the power gain is given by

$$G(\omega) = |t_g|^2 = \exp[g_0(\omega)] \qquad (F.1.5)$$

where the frequency-dependent gain $g_0(\omega)$ is given by

$$g_0(\omega) = \frac{g_0}{\{1 + [2(\omega - \omega_0)/\Delta\omega_0]^2\}} \qquad (F.1.6)$$

i.e., it shows the expected Lorentzian shape of a homogeneous line. If we assume that the spectral width of the light pulse is appreciably narrower than $\Delta\omega_0$, the argument of the second exponential function in Eq. (F.1.4) can be expanded as a power series in $(\omega - \omega_0)$. To first order, this gives:

$$t_g = \exp(-j\{(\omega nl/c) + [g_0(\omega - \omega_0)/\Delta\omega_0]\})$$
$$\times \exp(g_0/2)\{1 - [2(\omega - \omega_0)/\Delta\omega_0]^2\} \tag{F.1.7}$$

The imaginary terms in the first exponential function correspond to a phase delay $\phi = (\omega nl/c) + [g_0(\omega - \omega_0)/\Delta\omega_0]$, from which, according to Eq. (8.6.27), the time delay τ_d experienced by the pulse after traveling in the active medium is obtained as:

$$\tau_d = \frac{d\phi}{d\omega} = \frac{nl}{c} + \frac{g_0}{\Delta\omega_0} \tag{F.1.8}$$

Note that this delay is not simply nl/c because the gain line makes an additional finite contribution. This delay must be taken into account when considering the requirement that the round-trip pulse propagation time must equal the period of the amplitude modulator. For simplicity we will not consider any further the effect of this delay or that due to other cavity elements; the pulse amplitude refers, in any case, to a local time where these delays are taken into account. We therefore ignore the phase term in Eq. (F.1.7) to write:

$$t_g = \exp\{(g_0/2)\{1 - [2(\omega - \omega_0)/\Delta\omega_0]^2\}\} \tag{F.1.9}$$

We also ignore the loss introduced by mirror 1, since this is taken into account in the overall cavity loss. After passing once more through the active medium, the spectral amplitude of the light pulse experiences another transmission factor t_g as given by Eq. (F.1.9). Round-trip transmission through the amplifier is then given by $t_g^2 = (A_{3\omega}/A_{1\omega}) = \exp\{(g_0)\{1 - [2(\omega - \omega_0)/\Delta\omega_0]^2\}\}$, where $A_{3\omega}$ is the spectral amplitude of the light pulse after one round trip through the amplifier. Under the assumption $g_0 \ll 1$, the latter equation gives:

$$A_{3\omega} = t_g^2 A_{1\omega} = A_{1\omega}\{1 + (g_0)\{1 - [2(\omega - \omega_0)/\Delta\omega_0]^2\}\} \tag{F.1.10}$$

To proceed further we must calculate the effect of this transmission in the time domain rather than the frequency domain. To do this, we observe the following property of a Fourier transform, FT:

$$FT\left[\frac{d^n A(t)}{dt^n}\right] = [j(\omega - \omega_0)]^n A_\omega(\omega - \omega_0) \tag{F.1.11}$$

This relation is proved by first taking the nth derivative, then the Fourier transform of both sides of Eq. (F.1.3). Equation (F.1.11) shows that multiplication of the spectral amplitude A_ω by $k(\omega - \omega_0)^n$, where k is a constant, is equivalent, in the time domain, to taking (k/j^n)

times the nth derivative of the amplitude $A(t)$. We apply this rule to each individual term on the right-hand side of Eq. (F.1.10) to obtain

$$A_3(t) = \left\{1 + g_0\left[1 + \left(\frac{2}{\Delta\omega_0}\right)^2\frac{d^2}{dt^2}\right]\right\}A_1(t) \tag{F.1.12}$$

where $A_1(t)$ and $A_3(t)$ are, respectively, the amplitudes of the light pulse entering the amplifier and after one round-trip (see Fig. F.1). Equation (F.1.12) shows that the effect, on the light-pulse amplitude, of a round-trip passage through the gain medium, can be described by a round-trip operator:

$$\hat{T}_g = \left\{1 + g_0\left[1 + \left(\frac{2}{\Delta\omega_0}\right)^2\frac{d^2}{dt^2}\right]\right\} \tag{F.1.13}$$

We next consider the effect of cavity losses arising from finite mirror reflectivities and internal losses. These losses are represented by the central box in Fig. F.1. If we then let γ be the logarithmic power loss per pass, we can write for a single passage in the cavity,

$$A_4(t) = [\exp(-\gamma/2)]A_3(t) \tag{F.1.14}$$

In fact, according to Eq. (F.1.14), the ratio of corresponding intensities $(I_4/I_3) = (A_4/A_3)^2$ shows the expected value $\exp(-\gamma)$. From Eq. (F.1.14) we then find that the transmission accounting for round-trip losses is given by $\exp(-\gamma)$ and, for $\gamma \ll 1$, this expression can be approximated by $1 - \gamma$. This means that the operator corresponding to the round-trip loss in the cavity is simply:

$$\hat{T}_l = 1 - \gamma \tag{F.1.15}$$

Lastly we consider the effect of the amplitude modulator. We let $\gamma_m[1 - \cos\omega_m t]$ represent the single-pass logarithmic power loss introduced by the modulator. In this expression ω_m is the modulator frequency; it is assumed to be such that the modulator period equals the round-trip time of the light pulse in the laser cavity. The single-pass transmission of the field amplitude through the modulator is then given by:

$$t_m = \exp[-(\gamma_m/2)(1 - \cos\omega_m t)] \tag{F.1.16}$$

Transmission for a double pass through the modulator is then $t_m^2 = \exp[-(\gamma_m)(1 - \cos\omega_m t)]$. For $\gamma_m \ll 1$ this expression can be approximated to $t_m^2 \cong 1 - (\gamma_m)(1 - \cos\omega_m t)$. We now assume that the pulse passes through the modulator when modulator loss is zero (see Fig. 8.20), i.e., at time $t = 0$, and that the pulse width is much smaller than the modulator period $2\pi/\omega_m$. Under these conditions, the round-trip transmission can be further approximated to $t_m^2 = 1 - (\gamma_m/2)(\omega_m t)^2$. The operator corresponding to a double passage of the pulse through the modulator is then simply given by:

$$\hat{T}_m = 1 - \frac{\gamma_m}{2}(\omega_m t)^2 \tag{F.1.17}$$

Having established the operators that describe the time domain evolution of the pulse on a double passage through the three components considered, we now require the pulse amplitude $A(t)$, in the steady-state regime, to reproduce itself after a round-trip. We thus write

$$\hat{T}_m\hat{T}_l\hat{T}_g A(t) = A(t) \tag{F.1.18}$$

Using the previous expressions for $\hat{T}_m$, $\hat{T}_l$, and $\hat{T}_g$ and the condition that $(g_0, \gamma, \gamma_m) \ll 1$, we obtain the following differential equation

$$\left\{ g_0 \left[1 + \left(\frac{2}{\Delta\omega_0} \right)^2 \frac{d^2}{dt^2} \right] - \gamma - \frac{\gamma_m}{2} \omega_m^2 t^2 \right\} A(t) = 0 \tag{F.1.19}$$

which is the final result of our calculation. The preceding equation is formally equivalent to the Schrödinger equation for a particle in a parabolic potential (harmonic oscillator). From the well-known solution of this problem, we can write, in our case,

$$A(t) = H_n(\omega_p t) \exp\left[-\left(\frac{\omega_p^2 t^2}{2} \right) \right] \tag{F.1.20}$$

where H_n is the Hermite polynomial of order n, and where:

$$\omega_p = \left(\frac{\gamma_m}{2g_0} \right)^{1/4} \left(\frac{\omega_m\Delta\omega_0}{2} \right)^{1/2} \tag{F.1.21}$$

with g_0 being such that

$$1 - \frac{\gamma}{g_0} = \frac{4\omega_p^2}{\Delta\omega_0^2}(2n + 1) \tag{F.1.22}$$

One can show, however, that, of all these solutions, only the first-order Gaussian solution $(n = 0)$ is stable.

Equations (F.1.21) and (F.1.22) represent a pair of relations for the unknown parameters ω_p and g_0. From the knowledge of ω_p we then obtain the width of the mode-locked pulse. The full width at half-maximum intensity of the pulse, $\Delta\tau_p$, is in fact given by $\Delta\tau_p = 2(\ln 2)^{1/2}/\omega_p$; from Eq. (F.1.21) we then obtain

$$\Delta\tau_p = \left(\frac{2\sqrt{2}\ln 2}{\pi^2} \right)^{1/2} \left(\frac{g_0}{\gamma_m} \right)^{1/4} \left(\frac{1}{\nu_m\Delta\nu_0} \right)^{1/2} \tag{F.1.23}$$

where $\nu_m = \omega_m/2\pi$ and $\Delta\nu_0 = \Delta\omega_0/2\pi$. We observe that the first factor on the right-hand side of Eq. (F.1.23) is approximately equal to 0.45; as a result of the (1/4)th power, the second factor is approximately equal to unity. The values of $\Delta\tau_p$ and, hence, ω_p then depend only weakly on g_0 and from Eq. (F.1.23) we obtain the following approximate expression for $\Delta\tau_p$ [see Eq. (8.6.19)]: $\Delta\tau_p \cong 0.45/(\nu_m\Delta\nu_0)^{1/2}$. From Eq. (F.1.22), with $n = 0$, we can then obtain the value of g_0. Note that, according to Eq. (F.1.22), g_0 turns out to be larger than γ, due to modulator loss. Once the value of g_0 is calculated, the average intracavity laser intensity $\langle I \rangle$ is obtained from Eq. (F.1), since one has $g = \sigma_p N l = x\gamma$, where $x = N/N_c = R_p/R_{pc}$ is the amount by which threshold is exceeded. Knowing the average intracavity laser intensity, the laser pulse duration, and the pulse repetition rate, one then obtains the laser peak intensity.

F.2. PASSIVE MODE LOCKING

We now consider the theory of passive mode locking by a saturable absorber with a lifetime much shorter than the pulse duration (fast saturable absorber).[5] We again refer to Fig. F.1, where the modulator is now replaced by this fast absorber.

According to Eq. (F.1.13), the effect on the pulse amplitude of a round-trip passage through the gain medium can be described by the operator

$$\hat{T}_g = \left\{ 1 + g_0' \left[1 + \left(\frac{2}{\Delta\omega_0} \right)^2 \frac{d^2}{dt^2} \right] \right\} \tag{F.2.1}$$

where, conforming to the discussion in Sect. 8.6.3.2, saturated gain is now denoted by g_0'. According to Eq. (F.1.15), the effect of unsaturable cavity losses can be described by

$$\hat{T}_l = 1 - \gamma_c \tag{F.2.2}$$

where γ_c is the cavity loss without the saturable absorber. According to Eq. (F.1.14), the single-pass amplitude transmission of the saturable absorber is then written as:

$$t_{sa} = \exp[-(\gamma_{sa}/2)] \tag{F.2.3}$$

In the preceding equation, γ_{sa} represents the saturated single-pass power loss of the absorber, and is given by

$$\gamma_{sa} = \frac{\gamma'}{1 + (I/I_s)} \tag{F.2.4}$$

where γ' is the unsaturated loss, $I = I(t)$ is the pulse intensity, and I_s is the saturation intensity of the absorber. The amplitude transmission, for a double passage through the saturable absorber is then given by $t_{sa}^2 = \exp[-(\gamma_{sa})]$; under the assumptions $\gamma' \ll 1$ and $(I/I_s) \ll 1$, the latter equation, with the help of Eq. (F.2.4), gives $t_{sa}^2 \cong 1 - \gamma_{sa} \cong 1 - \gamma'[1 - (I/I_s)] = 1 - \gamma'[1 - (|A|^2/I_s)]$, where the amplitude $A(t)$ is now normalized, so that $|A|^2$ is the beam intensity. From the previous expression for t_{sa}^2, we obtain the operator corresponding to a double pass through the saturable absorber as:

$$\hat{T}_{sa} = 1 - \gamma' + \gamma' \frac{|A|^2}{I_s} \tag{F.2.5}$$

Self-consistency now requires that:

$$\hat{T}_m \hat{T}_l \hat{T}_{sa} A(t) = A(t) \tag{F.2.6}$$

Assuming $(g_0', \gamma, \gamma') \ll 1$, from Eqs. (F.2.1), (F.2.2), and (F.2.5), we obtain:

$$\left\{ g_0' \left[1 + \left(\frac{2}{\Delta \omega_0} \right)^2 \frac{d^2}{dt^2} \right] - \gamma_c - \gamma' + \gamma' \frac{|A|^2}{I_s} \right\} A(t) = 0 \tag{F.2.7}$$

The solution to Eq. (F.2.7) can be written as

$$A(t) = \frac{A_0}{\cosh(t/\tau_p)} \tag{F.2.8}$$

where

$$\tau_p = \left(\frac{2g_0'}{\gamma'} \right)^{1/2} \left(\frac{2}{\Delta \omega_0} \right) \left(\frac{I_s}{|A_0|^2} \right)^{1/2} \tag{F.2.9}$$

with g_0'; being such that

$$\gamma_c + \gamma' - g_0' = \frac{4g_0'}{\Delta \omega_0^2 \tau_p^2} \tag{F.2.10}$$

Since the width of the pulse intensity $\Delta \tau_p$ (FWHM) is given by $\Delta \tau_p = 1.76 \tau_p$ and $\Delta \omega_0 = 2\pi \Delta \nu_0$, from Eq. (F.2.9) we obtain Eq. (8.6.22) by recognizing that $|A_0|^2$ is the peak laser intensity. Equation (F.2.10) then shows that $g_0' < \gamma_c + \gamma' = \gamma$, where γ is the overall unsaturated loss of the cavity. This means that, in the absence of the pulse, the laser has a net loss while a time window of net gain exists only during the passage of the mode-locked pulse (see Fig. 8.22).

REFERENCES

1. D. J. Kuizenga and A. E. Siegman, FM and AM Mode Locking of the Homogeneous Laser—Part I: Theory, *IEEE J. Quantum Electron.* **QE-6**, 694 (1970).
2. H. A. Haus, A Theory of Forced Mode Locking, *IEEE J. Quantum Electron.* **QE-11**, 323 (1975).
3. H. A. Haus, *Waves and Fields in Optoelectronics* (Prentice-Hall, Englewood Cliffs, NJ, 1984), Sect. 9.3.
4. A. E. Siegman, *Lasers* (Oxford University Press, Oxford, UK, 1986), Sect. 7.4.
5. H. A. Haus, Theory of Mode Locking with a Fast Saturable Absorber, *J. Appl. Phys.* **46**, 3049 (1975); see also Ref. 3, Sect. 10.3.

G

Propagation of a Laser Pulse through a Dispersive Medium or a Gain Medium

Consider first a light pulse traveling in a dispersive medium, and let ω_L and $\Delta\omega_L$ be, respectively, the center frequency and the width of the corresponding spectrum (Fig. 8.25a). The electric field $E(t, z)$ of the corresponding waveform, at coordinate z along the propagation direction, can generally be expressed in terms of a Fourier expansion as

$$E(t, z) = \int_{-\infty}^{+\infty} A_\omega(\omega - \omega_L) \exp[j(\omega t - \beta z)] d\omega \tag{G.1}$$

where $A_\omega = A_\omega(\omega - \omega_L)$ is the complex amplitude of each field component and $\beta = \beta(\omega - \omega_L)$ describes the dispersion relation of the medium.

Let us now assume that this dispersion relation, over the bandwidth $\Delta\omega_L$, can be approximated by a linear relation, i.e.

$$\beta = \beta_L + \left(\frac{d\beta}{d\omega}\right)_{\omega_L} (\omega - \omega_L) \tag{G.2}$$

where β_L is the propagation constant corresponding to the frequency ω_L. Substituting Eq. (G.2) into Eq. (G.1), one can see the latter equation can be written as

$$E(t, z) = \exp[j(\omega_L t - \beta_L z)] \times \int_{-\infty}^{+\infty} A_\omega(\Delta\omega) \exp\left\{ j\Delta\omega \left[t - \left(\frac{d\beta}{d\omega}\right)_{\omega_L} z \right] \right\} d\Delta\omega \tag{G.3}$$

where $\Delta\omega = \omega - \omega_L$. From Eq. (G.3) we see that integration over $\Delta\omega$ leads to a function in the single variable $[t - (d\beta/d\omega)_{\omega_L} z]$. Equation (G.3) can therefore be put in the form

$$E(t, z) = A\left[t - \left(\frac{z}{v_g}\right) \right] \exp[j(\omega_L t - \beta_L z)] \tag{G.4}$$

571

where A is the pulse amplitude, $\exp[j(\omega_L t - \beta_L z)]$ is the carrier wave, and v_g is given by:

$$v_g = \left(\frac{d\omega}{d\beta}\right)_{\beta=\beta_L} \tag{G.5}$$

Since the pulse amplitude is a function of the variable $t - (z/v_g)$, the pulse propagates without changing its shape and at a speed v_g. This velocity is called the *group velocity* of the pulse; according to Eq. (G.5), it is given by the slope of the ω versus β relation at $\omega = \omega_L$ (i.e. $v_g = \tan \theta'$; see Fig. 8.25a).

Consider next the case of a light pulse with a bandwidth $\Delta\omega_L$ so large that it is no longer a good approximation to describe the dispersion relation by a linear law (Fig. 8.25c). In this case, different spectral regions of the pulse travel with different group velocities; consequently, the pulse broadens as it propagates. After traversing a length l of the medium, the broadening of the pulse, $\Delta\tau_d$, is then given approximately by the difference in group delay between the slowest and the fastest spectral components. We can then write

$$\Delta\tau_d = \left(\frac{l}{v_g'} - \frac{l}{v_g''}\right) = l\left[\left(\frac{d\beta}{d\omega}\right)_{\omega'} - \left(\frac{d\beta}{d\omega}\right)_{\omega''}\right] \tag{G.6}$$

where v_g' and v_g'' are two group velocities of these components, respectively, and ω' and ω'' are the corresponding frequencies. Let us now assume that the dispersion relation, within the bandwidth $\Delta\omega_L$, can be approximated by a parabolic (or quadratic) law, i.e.:

$$\beta = \beta_L + \left(\frac{d\beta}{d\omega}\right)_{\omega_L}(\omega - \omega_L) + \frac{1}{2}\left(\frac{d^2\beta}{d\omega^2}\right)_{\omega_L}(\omega - \omega_L)^2 \tag{G.7}$$

From Eqs. (G.6) and (G.7) we then have

$$\Delta\tau_d \cong l\left|\left(\frac{d^2\beta}{d\omega^2}\right)_{\omega_L}\right|\Delta\omega_L = |\phi''(\omega_L)|\Delta\omega_L \tag{G.8}$$

where we have defined $\phi = \beta l$, $\phi'' = d^2\phi/d\omega^2$ and ϕ'' is taken at the central laser frequency ω_L. Given the form of Eq. (G.8), the quantity $\phi''(\omega_L)$ is referred to as the *group-delay dispersion* (GDD) while the quantity

$$\left(\frac{d^2\beta}{d\omega^2}\right)_{\omega_L} = \text{GVD} = \left[\frac{d(1/v_g)}{d\omega}\right]_{\omega_L} \tag{G.9}$$

is referred to as the *group-velocity dispersion* (GVD) at frequency ω_L.

The calculation leading to Eq. (G.8) is open to criticism, because, to obtain Eq. (G.8), we have considered the propagation of limited spectral components of the pulse, each of which actually corresponds to a different, and, in fact, longer pulse than the original one. A more precise and instructive calculation can be performed by assuming that the pulse, upon entering the medium at coordinate $z = 0$, has a Gaussian amplitude profile, i.e.

$$E(t) = A_0 \exp[-(t^2/2\tau_p^2)] \exp(j\omega_L t) \tag{G.10}$$

where τ_p is the $1/e$ half-width of the pulse intensity. Since the spectral amplitude $A_\omega(\omega - \omega_L)$ is a Gaussian function of $(\omega - \omega_L)$, the electric field, after a distance z in the medium, is easily calculated from Eq. (G.1) if we assume that the dispersion relation can be developed in a Taylor expansion to second order in $(\omega - \omega_L)$, as shown in Eq. (G.7). In this case, the integrand in Eq. (G.1) can be expressed in terms of the inverse Fourier transform of a generalized Gaussian function, of complex argument, whose integral is well-known. To show the final result for the pulse amplitude $A(t, z)$, we refer to a new coordinate system given by

$$t' = t - \left(\frac{z}{v_g}\right) \tag{G.11a}$$

$$z' = z \tag{G.11b}$$

where v_g is the group velocity. This means that the pulse amplitude is referred to a local time that takes into account the group delay of the pulse. Using this new coordinate system, the pulse amplitude turns out to be given by[1]

$$A(t', z) = \frac{A_0 \tau_p}{(\tau_p^2 + jb_2 z)^{1/2}} \exp\left[-\frac{(t')^2}{2(\tau_p^2 + jb_2 z)}\right] \tag{G.12}$$

where, for simplicity, we have written $b_2 = (d^2\beta/d\omega^2)_{\omega_L} = \text{GVD}$, and where, on account of Eq. (G.11b), we have written, again for simplicity, $z' = z$. According to Eq. (G.12), $A(t', z)$ is given by a Gaussian function of t' with complex argument, so that we can write

$$A(t', z) = |A(t', z)| \exp[-j\varphi(t', z)] \tag{G.13}$$

From Eq. (G.12) the pulse magnitude $|A(t', z)|$ is calculated as:

$$|A(t', z)| = \frac{A_0 \tau_p}{(\tau_p^4 + b_2^2 z^2)^{1/4}} \exp\left\{-\left[\frac{t'^2 \tau_p^2}{2(\tau_p^4 + b_2^2 z^2)}\right]\right\} \tag{G.14}$$

Equation (G.14) shows that the Gaussian pulse maintains its shape on propagation; comparison with Eq. (G.10) shows that the width of the pulse $\tau_p(z)$, at coordinate z, is such that $\tau_p^2(z) = (\tau_p^4 + b_2^2 z^2)/\tau_p^2$. This expression can then be transformed into

$$\tau_p(z) = \tau_p\left[1 + \left(\frac{z}{L_D}\right)^2\right]^{1/2} \tag{G.15}$$

where $L_D = \tau_p^2/|b_2|$ is referred to as the *dispersion length* of the pulse in the medium. Note from Eq. (G.15) the analogy of time broadening of a Gaussian pulse in a dispersive medium and a spot-size increase of a Gaussian beam due to diffraction [compare with Eq. (4.7.17a)], the dispersive length, in the former case, being equivalent to the Rayleigh range in the latter case. This analogy can be traced back to a formal analogy between the diffraction equation in the paraxial approximation and the differential equation describing pulse propagation in a quadratic dispersive medium.[1]

From Eq. (G.12), the phase $\varphi(t', z)$ can also be calculated as

$$\varphi(t', z) = -\frac{\mathrm{sgn}(b_2)(z/L_D)}{1 + (z/L_D)^2} \frac{t'^2}{\tau_p^2} + \frac{1}{2}\tan^{-1}\left(\frac{z}{L_D}\right) \tag{G.16}$$

where $\mathrm{sgn}(b_2)$ stands for the sign of b_2, i.e., of GVD. Equation (G.16) shows that, besides the constant term $(1/2)\tan^{-1}(z/L_D)$, $\varphi(t', z)$ contains another term that is quadratic in t'. This means that the instantaneous frequency of the pulse, $\omega(t') = \partial[(\omega_L t') - \varphi(t', z)]/\partial t'$, given by

$$\omega = \omega_L + \mathrm{sgn}(b_2)\frac{(z/L_D)}{1 + (z/L_D)^2}\frac{2t'}{\tau_p^2} \tag{G.17}$$

now has a term that changes linearly with time. Thus the pulse acquires a linear frequency chirp whose sign depends on the sign of b_2. In particular the instantaneous frequency decreases in time for a negative GVD.

If the length l of the medium is much smaller than the dispersion length L_D, the relative pulse broadening, $(\delta\tau_p/\tau_p)_D = [\tau_p(l) - \tau_p]/\tau_p$, is obtained from Eq. (G.15) as:

$$\left(\frac{\delta\tau_p}{\tau_p}\right)_D \simeq \frac{1}{2}\left(\frac{l}{L_D}\right)^2 = \frac{1}{2}\left(\frac{\phi''}{\tau_p^2}\right)^2 \tag{G.18}$$

where $\phi'' = \phi''(\omega_L) = (d^2\phi/d\omega^2)_{\omega_L}$. The width of a Gaussian-pulse intensity profile (FWHM) is then related to the quantity τ_p in Eq. (G.10) by $\Delta\tau_p = 2(\ln 2)^{1/2}\tau_p$. From Eq. (G.18) we then obtain

$$\left(\frac{\delta\tau_p}{\tau_p}\right)_D = (8\ln^2 2)\frac{\phi''^2}{\Delta\tau_p^4} \tag{G.19}$$

Let us now consider a Gaussian pulse, as in Eq. (G.10), entering a homogeneously broadened gain medium. The spectral amplitude of the pulse $A_\omega(\omega - \omega_0)$, while entering the medium, is obtained by taking the inverse Fourier transform of Eq. (G.10); it is then given by

$$A_\omega(\omega - \omega_L) \propto \exp\left\{-\left[\frac{(\omega - \omega_L)^2\tau_p^2}{2}\right]\right\} \tag{G.20}$$

where ω_L is the central laser frequency. If the spectral width of the pulse is much smaller than the gain linewidth, the gain for the electric field amplitude can be approximated as (see Appendix F)

$$G_e(\omega - \omega_0) = \exp\left\{\left(\frac{g_0}{2}\right)\{1 - [2(\omega - \omega_0)/\Delta\omega_0]^2\}\right\} \tag{G.21}$$

where $g_0 = N_0\sigma l$ is the saturated single-pass power gain through the amplifier and $\Delta\omega_0$ is the transition linewidth. The spectral amplitude of the pulse, after passing through the gain medium, is then given by:

$$A_{g\omega}(\omega - \omega_0) = G_e(\omega - \omega_0) \times A_\omega(\omega - \omega_L) \tag{G.22}$$

From Eq. (G.22) with the help of Eqs. (G.20) and (G.21) and assuming $\omega_L = \omega_0$, we obtain

$$A_{g\omega}(\omega - \omega_0) \propto \exp\left\{-(\omega - \omega_0)^2\left[\frac{\tau_p^2}{2} + \frac{g_0}{2}\left(\frac{2}{\Delta\omega_0}\right)^2\right]\right\} \tag{G.23}$$

Equation (G.23) shows that, within the approximation made, the spectrum remains Gaussian after the pulse traverses the gain medium. A comparison between Eqs. (G.23) and (G.20) then indicates that the pulse duration is broadened to a value τ_p', where $\tau_p'^2/2$ corresponds to the term in the square brackets in Eq. (G.23). We thus obtain

$$\tau_p' = \tau_p\left[1 + g_0\left(\frac{2}{\tau_p\Delta\omega_0}\right)^2\right]^{1/2} \tag{G.24}$$

For small changes in pulse duration, the relative pulse broadening, $(\delta\tau_p/\tau_p)_g = (\tau_p' - \tau_p)/\tau_p$, after the gain medium, is obtained from Eq. (G.24) as:

$$\left(\frac{\delta\tau_p}{\tau_p}\right)_g = \frac{1}{2}\left(\frac{2}{\tau_p\Delta\omega_0}\right)^2 g_0 \tag{G.25}$$

Equation (G.25) can be recast in terms of the gain linewidth $\Delta\nu_0 = \Delta\omega_0/2\pi$ and laser pulse width $\Delta\tau_p = 2(\ln 2)^{1/2}\tau_p$. We obtain

$$\left(\frac{\delta\tau_p}{\tau_p}\right)_g = \left(\frac{2\ln 2}{\pi^2}\right)\left(\frac{1}{\Delta\tau_p^2\Delta\nu_0^2}\right)g_0 \tag{G.26}$$

REFERENCE

1. Hermann A. Haus, *Waves and Fields in Optoelectronics* (Prentice-Hall, Englewood Cliffs, NJ, 1984), Sect. 6.6.

H

Higher-Order Coherence

The degree of coherence $\Gamma^{(1)}$ introduced in Sect. 11.3 involves the first-order correlation function $\langle E(x_1)E^*(x_2)\rangle$ [see Eq. (11.3.7)], where $x_i = (\mathbf{r}_i, t_i)$ denotes both space and time coordinates of the field. Likewise we can define

$$\Gamma^{(n)}(x_1, x_2, \ldots, x_{2n}) = \langle E(x_1)\cdots E(x_n)E^*(x_{n+1})\cdots E^*(x_{2n})\rangle \tag{H.1}$$

which involves the product of $2n$ terms, these being the functions E evaluated at the $2n$ space-time points $x_1, x_2, \ldots, x_{2n}$. The corresponding normalized quantity $\gamma^{(n)}$ can be defined as

$$\gamma^{(n)}(x_1, x_2, \ldots, x_{2n}) = \frac{\langle E(x_1)\cdots E(X_n)E^*(x_{n+1})\cdots E^*(x_{2n})\rangle}{\prod_1^{2n}{}_r \langle E(x_r)E^*(x_r)\rangle^{1/2}} \tag{H.2}$$

where $\prod$ is the symbol for product. These expressions can be reduced to Eqs. (11.3.7) and (11.3.8) for the case $n = 1$.

We now need to define, in terms of these higher-order correlation functions, what we mean by a completely coherent beam. We begin by noting that, if a wave is perfectly coherent to first order (i.e., if $|\gamma^{(1)}(x_1, x_2)| = 1$), then one must have

$$\Gamma^{(1)}(x_1, x_2) = E(x_1)E^*(x_2) \tag{H.3}$$

i.e., $\Gamma^{(1)}$ must factorize into a product of the fields at x_1 and x_2. Indeed, if field fluctuations are completely absent, the ensemble averages of, e.g., Eq. (11.3.7) or (11.3.8) simply reduce to the product of corresponding signals. By analogy, we can define a perfectly coherent em wave as one where $\Gamma^{(n)}$ factorizes to all orders n; this means that:

$$\Gamma^{(n)}(x_1, x_2, \ldots, x_{2n}) = \prod_1^n{}_r E(x_r) \prod_{n+1}^{2n}{}_k E^*(x_k) \tag{H.4}$$

Indeed, when field fluctuations are completely absent, the ensemble average of Eq. (H.1) is simply the product of corresponding fields. If we now substitute Eq. (H.4) in the numerator on the right-hand side of Eq. (H.2), we readily find that

$$|\gamma^{(n)}(x_1, x_2 \ldots, x_{2n})| = 1 \tag{H.5}$$

for all orders n. It should be noted that the field of a cw laser oscillating in single mode with a narrow linewidth can be considered, for all practical purposes, to satisfy Eq. (H.4) to all orders. In fact, as discussed in Sect. 11.7, this field can be considered to show only phase fluctuations. For a narrow linewidth laser, the rate of change of this phase is rather slow however. For example, in the case of the He-Ne laser considered in Sect. 11.8, which has a bandwidth of $\Delta v_L \cong 1\,\mathrm{kHz}$, the phase change occurs in a time $\tau_{co} \cong 1/\Delta v_L = 1\,\mathrm{ms}$. This means that, for time intervals much smaller than τ_{co}, i.e., for separations between equiphase surfaces of the $2n$ space time points much smaller than $c\tau_{co} = 300\,\mathrm{km}$, phase fluctuations are the same for all $2n$ space-time points. We then readily obtain Eqs. (H.4) and (H.5).

The difference, to the nth order, between a completely coherent beam, e.g., the single-mode He-Ne laser just considered, and a thermal light source, is easily illustrated in the case where $x_1 = x_2 \cdots = x_{2n} = x$, i.e., by considering field correlations at the same point and at the same time. The correlation function $\Gamma^{(n)}(x, x, \ldots, x)$ can then be obtained from Eq. (H.1) as

$$\Gamma^{(n)} = \langle |E|^{2n} \rangle = \frac{\iint A^{2n} p_E(\tilde{E}) A\,dA\,d\phi}{\iint p_E(\tilde{E}) A\,dA\,d\phi} \tag{H.6}$$

where Eq. (11.1.1) gives the field amplitude, $A = A(x)$, and $p_E(\tilde{E})$ is the probability density introduced in Sect. 11.7. In particular, for $n = 1$, one has:

$$\Gamma^{(1)}(x, x) = \langle |E|^2 \rangle = \langle I \rangle = \frac{\iint A^2 p_E(\tilde{E}) A\,dA\,d\phi}{\iint p_E(\tilde{E}) A\,dA\,d\phi} \tag{H.7}$$

For a coherent field, we use Eq. (11.7.2) for $p_E(\tilde{E})$. From Eq. (H.6) we then obtain $\Gamma^{(n)} = A_0^{2n}$ while, from Eq. (H.7), we have $\Gamma^{(1)} = A_0^2$. Then we can write:

$$\Gamma^{(n)}(x, x, \ldots, x) = [\Gamma^{(1)}(x, x)]^n \tag{H.8}$$

For a thermal light source, Eqs. (H.6) and (H.7), with the help of Eq. (11.7.3) for $p_E(\tilde{E})$, give:

$$\Gamma^{(n)}(x, x, \ldots, x) = n! [\Gamma^{(1)}(x, x)]^n \tag{H.9}$$

To obtain the normalized nth order coherence function $\gamma^{(n)}$, using Eq. (H.2), we observe that the denominator of the expression on the right-hand side of this equation is, in any case, equal to $[\Gamma^{(1)}(x, x)]^n$. From Eqs. (H.8) and (H.9) we then obtain

$$\gamma^{(n)}(x, x, \ldots, x) = 1 \tag{H.10}$$

and

$$\gamma^{(n)} = n! \tag{H.11}$$

for the single-mode laser source and the thermal source, respectively. Equation (H.10) shows that, as already pointed out, the laser beam satisfies the general coherent condition (H.5). Equation (H.11), on the other hand, shows that a thermal source can satisfy the coherence condition only for $n = 1$, i.e., only to first order. It then follows that one can, at best, arrange for a thermal light source to have perfect first-order coherence, i.e., perfect spatial and temporal coherence, as indeed shown in Sect. 11.8.

I

Physical Constants and Useful Conversion Factors

Planck constant (h)	$6.6260755(40) \times 10^{-34}$	$J \cdot s$
($\hbar = h/2\pi$)	$1.05457266(63) \times 10^{-34}$	$J \cdot s$
Electronic charge (e)	$1.60211733(49) \times 10^{-19}$	C
Electron rest mass (m_e)	$9.1093897 \times 10^{-31}$	kg
Proton rest mass (m_p)	$1.6726231(10) \times 10^{-27}$	kg
Neutron rest mass (m_n)	$1.6749286(10) \times 10^{-27}$	kg
Velocity of light in vacuum (c)	2.99792458×10^{8}	m/s
Boltzmann constant (k)	$1.380658(12) \times 10^{-23}$	$J \cdot K^{-1}$
Bohr magneton (β)	$9.2740154(30) \times 10^{-24}$	$A \cdot m^2$
Permittivity of vacuum (ϵ_0)	$8.854187817\ldots \times 10^{-12}$	F/m
Permeability of vacuum (μ_0)	$4\pi \times 10^{-7}$	H/m
Avogadro's number (N_A)	$6.0221367(36) \times 10^{23}$	mol^{-1}
Ideal gas constant ($R = N_A \cdot k$)	8.31451	$J \cdot K^{-1} \cdot mol^{-1}$
Radius of first Bohr orbit [$a_o = (4\pi^2\epsilon_0/me^2)$]	$0.529177249(24) \times 10^{-10}$	m
Stefan–Boltzmann constant (σ_{SB})	$5.67051(19) \times 10^{-8}$	$W \cdot m^{-2} \cdot K^{-4}$
Free space impedance ($Z = 1/\epsilon_0 c$)	$376.73\ldots$	Ω
Ratio of the mass of the proton to the mass of the electron (m_p/m_e)	$1836.152\ldots$	
Energy corresponding to 1 eV	$1.602\ldots \times 10^{-19}$	J
Energy of a photon with wavelength $\lambda = 1\,\mu m$	$1.986\ldots \times 10^{-19}$	J
Wavenumbers corresponding to an energy spacing of kT; ($T = 300$ K) ($\tilde{v} = kT/hc$)	$208.512\ldots$	cm^{-1}
Atmospheric pressure (1 atm $= 760$ torr $\cong 1,013\ldots$ mbar)	$1.013\ldots \times 10^{5}$	Pa

Answers to Selected Problems

Chapter 1

1.1. Far infrared: 1 mm–50 μm, medium infrared: 50–2.5 μm, near infrared: 2.5 μm–750 nm, visible: 750–380 nm, ultraviolet: 380–180 nm, vacuum ultraviolet 180–40 nm, soft x-ray: 40 – 1 nm, x-ray: 1–0.01 nm.

1.4. For $g_1 = g_2$ one gets from Eq. (1.2.2) $E_2 - E_1 = kT = 208.5 \text{ cm}^{-1}$, so that $\lambda = (1/208.5) \text{ cm} \cong 48 \, \mu m$, falling in the medium infrared.

1.5. $\gamma_1 = 1$, $\gamma_2 = -\ln R_2 \cong 0.693$, $\gamma_i \cong 0.01$, $\gamma = \gamma_i + (\gamma_1 + \gamma_2)/2 \cong 0.357$, $N_c = \gamma/\sigma l \cong 1.7 \times 10^{17} \text{ cm}^{-3}$.

1.6. $D_m = (2\lambda/D)L \cong 533$ m, where D_m is the beam diameter on the moon, D is the telescope aperture, and L is the distance between earth and moon. The first earth moon ranging experiment was achieved under these conditions using a Q-switched ruby laser. Due to the large beam diameter on the moon and due to surface variations over this diameter, the precision of this ranging experiment was rather limited (~ 1 m). Using special mirrors as beam reflectors, placed on the moon surface by visiting astronauts, the earth–moon distance can now be measured with an accuracy of a few millimeters.

Chapter 2

1.1. $N(\Delta\lambda) = 8\pi V \Delta\lambda/\lambda^4 \cong 1.9 \times 10^{12}$ modes!

2.2. $\rho_\lambda = \rho_\nu |d\nu/d\lambda| - (c_n/\lambda^2)\rho_\nu$, where the relation $\lambda\nu = c_n$ (c_n is the light velocity in the medium filling the black body cavity) has been used. From Eq. (2.2.22), with the substitution $\nu = c_n/\lambda$, we obtain

$$\rho_\lambda = \frac{8\pi c_n}{\lambda^5} \frac{1}{\exp(hc_n/\lambda kT) - 1}$$

2.3. By imposing the condition $(d\rho_\lambda/d\lambda) = 0$ and using the expression for ρ_λ given in the answer to Problem 2.2, we obtain $5 \times [\exp(hc_n/\lambda kT) - 1] - (hc_n/\lambda kT)\exp(hc_n/\lambda kT) = 0$. If we write $y = (hc_n/\lambda kT)$ in the previous expression, we see that the value of y corresponding to the peak of ρ_λ must satisfy the equation $5 \times [1 - \exp(-y_M)] = y_M$. The solution of this equation can be obtained, by a fast converging iterative procedure, as $y_M \cong 4.965$. For $c_n = c$ (c is the light

583

velocity in vacuum) the wavelength λ_M where the maximum value of ρ_λ occurs must then satisfy the relation (Wien's law): $\lambda_M T = hc_n/y_M k \cong 2.3 \times 10^{-3}$ m × K.

2.6. The density of the Nd^{3+} ions N expressed in ions/cm^3, and hence the Nd^{3+} concentration in the $^4I_{9/2}$ manifold, is given by: $N = 1 \times 10^{-2} \times 3(\rho/\text{M.W.})N_A$ where ρ is the density expressed in g/cm^3, M.W. is the molecular weight of YAG, and N_A is Avogadro's number. The factor 3 in the preceding expression accounts for the presence of three yttrium atoms per molecule. Since the YAG molecular weight is 594 g/mol, we obtain $N \cong 1.38 \times 10^{20}$ ions/cm^3. According to Eq. (1.2.2), the fraction f of this population belonging to the lowest sublevel of the $^4I_{9/2}$ state is then given by:

$$f = \frac{1}{1 + \sum_{i=1}^{4} \exp[-(E_i/kT)]}$$

where E_i ($i = 1$–4) is the energy seperation between the higher sublevels and the ground sublevel. Given the values of E_i for these sublevels, we have $f = 46\%$.

2.7. From Eqs. (2.4.25) and (2.3.15) we have $\sigma_{in} = (\lambda_n^2/8\pi)[g_t(v - v_0)/\tau_{sp}]$ where $\lambda_n = c/nv_0$ is the wavelength, in a medium of refractive index n, of an em wave of frequency v_0. For $v = v_0$ and a pure inhomogeneous broadening, using Eq. (2.4.28), we obtain the following expression for the peak cross section: $\sigma_p = 0.939(\lambda_n^2/8\pi)(1/\Delta v_0^* \tau_{sp})$. For $\lambda_n = 1.15\ \mu m$ ($n \cong 1$), $\Delta v_0^* = 9 \times 10^8$ Hz, and $\tau_{sp} \cong 10^{-7}$ s, we then have $\sigma_p \cong 5.5 \times 10^{-12}$ cm^2.

2.8. Consider a plane wave, of uniform intensity I, crossing a surface of area S in a medium of refractive index n. The em energy flux through the surface S in a time Δt is $E = IS\Delta t$; this energy is uniformly distributed in a volume $V = S(c/n)\Delta t$. The energy density in the medium is $\rho_n = (E/V) = (n/c)I$.

2.11. The answer is readily obtained from Example 2.13: One gets $(\Omega/4\pi) = (D/4l)^2 \cong 4.4 \times 10^{-4}$ and from Eq. (2.9.4a), by a fast iterative procedure, $G = 1.24 \times 10^4$. The threshold inversion is then $N_{th} = \ln G/\sigma_p l = 4.49 \times 10^{18}$ cm^{-3} and the maximum stored energy $E_M = N_{th}(\pi D^2 l/4)hv = 1.96$ J.

2.13. Under thermal equilibrium, the two processes (2.6.9) and (2.6.10) must balance each other. Thus the relation $\kappa_{B^*A} N_{B^*} N_A = \kappa_{BA^*} N_B N_{A^*}$ must hold. At exact resonance and again at thermal equilibrium, we have $(N_{A^*}/N_A) = (N_{B^*}/N_B) = \exp[(E/kT)]$, where E is the energy-level separation between either one of the two-level system. We then obtain $\kappa_{B^*A} = \kappa_{BA^*}$.

2.14. For a Lorentzian line we have

$$I_s = I_{so}\left\{1 + \left[\frac{2(v - v_0)}{\Delta v_0}\right]^2\right\}$$

We then obtain

$$
\begin{aligned}
\alpha(v - v_0) &= \frac{\alpha_0(0)}{1 + [2(v - v_0)/\Delta v_0]^2} \cdot \frac{1}{1 + (I/I_s)} \\
&= \frac{\alpha_0(0)}{1 + [2(v - v_0)/\Delta v_0]^2} \cdot \frac{1}{1 + \dfrac{I}{I_{so}}\dfrac{1}{1 + [2(v - v_0)/\Delta v_0]^2}} \\
&= \frac{\alpha_0(0)}{1 + [2(v - v_0)/\Delta v_0]^2 + (I/I_{so})}
\end{aligned}
$$

2.15. On setting $v = v_0$ in the expression obtained in Problem 2.14, we obtain

$$\alpha_s(0) = \frac{\alpha_0(0)}{1 + (I/I_{s0})}$$

According to this equation, the saturation intensity I_{s0} is the intensity of a resonant ($v \cong v_0$) em wave where the peak-saturated absorption coefficient $\alpha_s(0)$ is half the corresponding unsaturated value $\alpha_0(0)$. The (1/2)-power points of $\alpha = \alpha(v - v_0)$ in the expression obtained in Problem 2.14 occur at frequency v' so that: $[2(v' - v_0)/\Delta v_0]^2 = 1 + (I/I_{s0})$. The saturated linewidth (FWHM) is then readily obtained as $\Delta v_s = \Delta v_0[1 + (I/I_{s0})]^{1/2}$.

Chapter 3

3.1. The center of mass is midway between the two atoms. Considering the x-axis to be along the vibration direction, with the origin at the center of mass, the restoring force on each atom is given by $F = -2k_0(x - x_0)$, where x_0 is the equilibrium coordinate of each atom. The equation of motion can then be written as: $[Md^2(x - x_0)/dt^2] = -2k_0(x - x_0)$ so that the resonance frequency is $\omega = (2k_0/M)^{1/2}$.

3.2. Using the result of Problem 3.1, we obtain $k_0 = (2\pi\tilde{v}c)^2(M/2) = 2314\,\text{J} \cdot \text{m}^{-2}$ where the atomic weight of the N atom is taken as $\cong 14$. The potential energy of the system is then given by $U = k_0(R - R_0)^2/2$ where R is the internuclear separation and R_0 is the equilibrium value. For $R - R_0 = 0.03\,\text{nm}$ one obtains $U \cong 6.5\,\text{eV}$.

3.6. $B = \hbar^2/2I$. In this expression, $I = 2M_o R_0^2$ where M_o is the oxygen mass and R_0 is the equilibrium distance between oxygen and carbon. For $B = 0.37\,\text{cm}^{-1}$ and $M_o = 16\,\text{g/mol}$ (atomic weight) one gets $R_0 = \hbar/2(M_o B)^{1/2} = 0.0515\,\text{nm}$.

3.8. From Fig. 3.15b, for $N = 1.6 \times 10^{-18}\,\text{cm}^{-3}$, we obtain $E_{F_c} = 2.35\,kT$ and $E_{F_v} = -1.45\,kT$. The overall gain bandwidth is then $\Delta\tilde{v} = (E_{F_c} + E_{F_v})/hc = 0.9(kT/hc) = 187.65\,\text{cm}^{-1}$. [We recall that $(kT/hc) = 208.5\,\text{cm}^{-1}$ at $T = 300\,\text{K}$.]

3.9. $E_2 + E_1 = 0.45\,kT$. From Eq. (3.2.2) one gets $(E_2/E_1) = m_v/m_c = 6.865$, where m_v is the hole mass and m_c is the electron mass in the conduction band. From the preceding two equations we obtain $E_2 = 0.392\,kT$ and $E_1 = 0.0572\,kT$.

3.10. From Fig. 3.16, for $E - E_g = 0.45\,kT \cong 12\,\text{meV}$, we obtain $\alpha = \alpha_0 = 1.8 \times 10^3\,\text{cm}^{-1}$. Since the probabilities of occupation of the upper and lower laser level are given by [see Eq. (3.2.10)]:

$$f_c(E_2) = \frac{1}{1 + \exp[(E_2 - E_{F_c})/kT]}$$

$$f_v(E_1) = \frac{1}{1 + \exp[(E_{F_v} - E_1)/kT]}$$

from the results of Problems 3.8 and 3.9, we obtain $f_c(E_2) = 0.877$ and $f_v(E_1) = 0.8186$. From Eq. (3.2.37) we then get $g = \alpha_0[f_c(E_2) - f_v(E_1)] \cong 104\,\text{cm}^{-1}$.

3.11. $\sigma = (dg/dN) = [g/(N - N_{tr})] = 2.6 \times 10^{-16}\,\text{cm}^2$ where $g = 104\,\text{cm}^{-1}$, $N = 1.6 \times 10^{18}\,\text{cm}^{-3}$, and (see Example 3.7) $N_{tr} = 1.2 \times 10^{18}\,\text{cm}^{-3}$.

3.13. From Fig. 3.26, for $N = 2 \times 10^{18}\,\text{cm}^{-3}$, we have $E_{F_c} - E_{1_c} \cong 2.8\,kT \cong 72\,\text{meV}$ and $E_{F_v} - E_{1_v} \cong -1.1\,kT \cong -28.6\,\text{meV}$. Using the results of Example 3.9 for E_{1_c} and E_{1_v}, we then obtain $E_{F_c} = 128.2\,\text{meV}$ and $E_{F_v} = -20.6\,\text{meV}$. According to Eq. (3.3.26) the gain bandwidth occurs for $E - E_g$ ranging from $\Delta E_1 = E_{1_c} + E_{1_v} = 64.2\,\text{meV}$ to $E_{F_c} + E_{F_v} = 107.6\,\text{meV}$.

Chaper 4

4.3. $T = 1 - R - A = 5 \times 10^{-3}$. From Eq. (4.5.6a) the peak transmission is obtained as $(T_{FP})_p = [T/(1-R)]^2 = 25\%$. Note the strong reduction in peak transmission even for such a small mirror loss. Comparing Eqs. (4.5.6a) and (4.5.6) (with $R_1 = R_2 = R$) show that the expression for the finesse still remains that given by Eq. (4.5.14); thus $F = \pi R^{1/2}/(1-R) \cong 312.4$.

4.4. From Eqs. (4.5.8) and (4.5.3), assuming normal incidence ($\theta = 0$) and unit refractive index ($n_r = 1$), we obtain $L = c/2\Delta\nu_{fsr} = 5\,\text{cm}$. From Eq. (4.5.13) the finesse is then obtained as $F = \Delta\nu_{fsr}/\Delta\nu_c = 50$. From Eq. (4.5.14), with $R_1 = R_2 = R$, we obtain the value of the mirror reflectivity as $R \cong 94\%$. From Eq. (4.5.6a), for a peak transmission of 50%, we then obtain $T = 4.24 \times 10^{-2}$ and hence $A = 1 - R - T = 1.76 \times 10^{-2}$.

4.7. The wavefront radius of curvature at the lens position is given by [see Eq. (4.7.17b)] $R = d[1 + (z_R/d)^2]$, where $z_R = \pi w_0^2/\lambda$ is the Rayleigh length. To compensate for this curvature, the focal length of the lens must just equal R.

4.9. From Eq. (4.7.19) we have $w_0 = \lambda/\pi\theta_d \cong 201\,\mu\text{m}$. If the output beam forms a waist, the peak intensity occurs at this waist; the peak intensity is given by $I_p = P/(\pi w_0^2) \cong 7.85\,\text{W/cm}^2$. From Eqs. (2.4.10) and (2.4.6), the intensity of an em wave is seen to be related to the field amplitude E_0 by $I = nE_0^2/2Z$ where $Z = 1/\varepsilon_0 c \cong 377\,\Omega$ is the impedance of free space. For $n = 1$, one gets $E_0 = (2IZ)^{1/2} \cong 77\,\text{V/cm}$.

4.11. From Eq. (4.7.27) we readily obtain $w_{02} = \lambda f/D$. The numerical aperture of a lens is defined as (see Sect. 1.4.4) $N.A. = \sin(\theta)$, where θ is the half-angle of the cone formed by the aperture D seen from the central point in the focal plane. Since $\tan(\theta) = D/2f$, we have $w_{02} = \lambda/2\tan(\sin^{-1} N.A.)$. For a small numerical aperture, $w_{02} \cong \lambda/2N.A.$.

4.13. According to Eq. (4.7.4), the beam parameter q, after the plate, is related to the input beam parameter q_0 by

$$\frac{1}{q} = \frac{C + (D/q_0)}{A + (B/q_0)}$$

In our case $q_0 = jz_R$, where $z_R = \pi w_0^2/\lambda$ is the Rayleigh length. Using the $ABCD$ matrix elements of a plate of length L and refractive index n (see Table 4.1), we obtain

$$\frac{1}{q} = \frac{-j/z_R}{1 - j(L/nz_R)}$$

From the real and imaginary parts of the preceding equation and with the help of Eq. (4.7.8), we obtain

$$w^2 = w_0^2[1 + (L'/z_R)^2]$$
$$R = L'[1 + (z_R/L')^2]$$

where $L' = L/n$. This proves the statement of Problem 4.13. After the plate, the spot size at distance z from the waist then equals that which would occur, without the plate, at a distance $z' = (z - L) + (L/n) = z - [(n - 1)/n]L$. At large distances from the waist, i.e., for $z \gg z_R$, then $w(z) = (\lambda/\pi w_0)\{z - [(n - 1)/n]L\}$. The beam divergence is then $\theta_d = w(z)/z$ and, if $L \leq z_R$, one has $z \gg L$ and the beam divergence remains equal to $(\lambda/\pi w_0)$, i.e., it is unaffected by the presence of the plate.

4.14. Equation (4.7.26) can be written as

$$z_m = \frac{1}{(1/f) + (f/z_{R_1}^2)}$$

For a given value of z_{R_1}, the denominator on the right-hand side of this equation has a minimum for $f = z_{R_1}$. At this value of focal length, z_m then reaches its maximum value given by $z_m = z_{R_1}/2$.

Chapter 5

5.2. $w_o = (L\lambda/2\pi)^{1/2} = 0.29\,\text{mm}$, $w_s = (2)^{1/2}w_o \cong 0.4\,\text{mm}$, $\Delta\nu_c = c/2L = 150\,\text{MHz}$. The number of modes is $N = \Delta\nu_o^*/(c/4L) = 47$, where $(c/4L)$ is the frequency spacing between two consecutive nondegenerate modes of a confocal resonator (see Fig. 5.10a).

5.4. The curvature of the wave front must coincide with that of the mirror, at the mirror's location. From Eq. (4.7.13b), setting $z = L/2$ where L is the cavity length, we then obtain

$$\frac{\pi w_0^2}{\lambda} = \frac{L}{2}\left(\frac{2R}{L} - 1\right)^{1/2}$$

The preceding expression gives $z_R = \pi w_0^2/\lambda \cong 1.32\,\text{m}$ and $w_0 \cong 0.466\,\text{mm}$. The spot size at the mirror is then obtained from Eq. (4.7.13a) as $w = w_0[1 + (L/2z_R)^2)]^{1/2} = 0.498\,\text{mm}$.

5.6. Since the equiphase surfaces at the two mirror positions must coincide with the mirror surfaces, from Eq. (4.7.17b) we write

$$-R_1 = z_1 + \left(\frac{z_R^2}{z_1}\right)$$

$$R_2 = z_2 + \left(\frac{z_R^2}{z_2}\right)$$

$$L = z_2 - z_1$$

where z_1 and z_2 are the coordinates of the mirrors as measured from the waist. Note the minus sign in front of the term R_1 in the first equation. It arises because, if, e.g., mirror 1 is a concave mirror, R_1 is positive while the sign of the wave front is negative because the center of curvature is to the right of the wavefront. From the preceding three equations one can then eliminate z_2 to obtain

$$-R_1 z_1 = z_1^2 + z_R^2$$

$$R_2(L + z_1) = (L + z_1)^2 + z_R^2$$

From these two equations we can eliminate z_R^2 to obtain $z_1(2L - R_1 - R_2) = L(R_2 - L)$. Note that, for $R_1 = R_2 = R$, this equation gives $z_1 = -L/2$, which shows that, in this case, the waist is located at the cavity center. From Eq. (5.4.10) we now find $R_1 = L/(1 - g_1)$ and $R_2 = L/(1 - g_2)$. Substitution of these two expressions for R_1 and R_2 in the preceding equation then readily gives $-z_1(g_1 + g_2 - 2g_1g_2) = g_2(1 - g_1)L$.

5.7. $g_1 = 0.333, g_2 = 0.75$. From the expression for z_1 obtained in Problem 5.6 we then find $z_1 = -0.857$ m. From Eqs. (5.5.8) and (5.5.9) we also find $w_1 = 0.533$ mm, $w_2 = 0.355$ mm, and $w_0 = 0.349$ mm.

5.9. For symmetry reasons the waist must be located at a distance $L_p/2$ from the lens. From Eq. (4.7.17b) we then find that the absolute value of the wave-front radius of curvature, on both sides of the lens, is given by $R = (L_p/2)[1 + (2z_R/L_p)^2]$. The lens must then transform one wave front into the other. We must then have $f = R/2$. Since $z_R = \pi w_0^2/\lambda$ one obtains from the preceding expressions $w_0^2 = (\lambda/2\pi)[L(4f - L_p)]^{1/2}$. Note that the latter expression gives a real value of w_0^2 only when $L_p \leq 4f$, which represents the stability condition for our case.

5.11. $g_1 = 1$, $g_2 = 1 - L/(L + \Delta) \cong \Delta/L$. From Eq. (5.5.8b) we then have $w_2 = w_m = (L\lambda/\pi)^{1/2}\{1/(\Delta/L)[1 - (\Delta/L)]\}^{1/4}$. For $w_m = 0.5$ mm, $L = 30$ cm, and $\lambda = 633$ nm, the preceding expression gives $\Delta = 1.85$ cm. Then $R_2 = L + \Delta = 31.85$ cm, $g_2 = 0.058$, and, from Eq. (5.5.8a), the spot size at the plane mirror is $w_1 \cong 0.122$ mm.

5.13. In this case, from Eqs. (5.5.5) and (5.4.6), the stability condition is given by $-2 < 2(2A_1D_1 - 1) < 2$, i.e., $0 < A_1D_1 < 1$. Since $A_1D_1 - B_1C_1 = 1$, then $-1 < B_1C_1 < 0$.

5.17. (a) $g_1 = 1, g_2 = 1.25$; (b) from Eq. (5.6.1), $r_1 = 2.24, r_2 = 1.24$; (c) $a_1 > M_{21}a_2 = 1.8\ a_2$; (d) $M = M_{12}M_{21} = 2.62$, so that $\gamma = (M^2 - 1)/M^2 = 0.85$.

5.18. Positive branch confocal unstable resonator. From Fig. 5.22, for $\gamma = 0.2$ and $N_{eq} = 7.5$, we find $M = 1.35$. It then follows that $2a_2 = 2[2L\lambda N_{eq}/(M - 1)]^{1/2} = 4.26$ cm. To achieve a single-ended resonator, one must have $a_1 > 2Ma_2 = 5.75$ cm. The radii of the two mirrors must then be such that $L = (R_1 + R_2)/2$ and $M = -R_1/R_2$ (Note: $R_2 < 0$). We obtain $R_1 = 7.7$ m and $R_2 = -5.7$ m.

5.20. (a) From Eq. (5.6.20), $\exp[-2(a/w)^6] = 2 \times 10^{-2}$, i.e., $w = 2.94$ mm. (b) From Eq. (5.6.21), $w_m = w/(M^6 - 1)^{1/6} = 2.32$ mm. (c) $\gamma = 1 - (R_0/M_2) \cong 0.744$. (d) Since $g_2 = 1$, from Eqs. (5.6.3) and (5.6.1) one gets $M = g_1\{1 + [1 + (1/g_1)]^{1/2}\}^2$ from which one obtains $g_1 = 1.0285$. The radius of curvature of the convex mirror is then $R_1 = L/(1 - g_1) \cong -17.5$ m.

Chapter 6

6.1. For radial propagation, the power absorbed in the laser rod can be written as $P_a = \int S\{1 - \exp[-(2\alpha R)]\}I_{e\lambda}\, d\lambda$, where S is the rod's lateral surface area. Since the power entering the rod is $P_e = \int SI_{e\lambda}\, d\lambda$, we obtain $\eta_a = P_a/P_e = \int\{1 - \exp[-(2\alpha R)]\}I_{e\lambda}\, d\lambda / \int I_{e\lambda}\, d\lambda$.

6.4. From Eq. (6.2.6) with $h\nu_{mp} \cong 2.11 \times 10^{-19}$ J ($\lambda_{mp} = 940$ nm), one readily gets $R_{cp} \cong 2.01 \times 10^{20}$ cm^{-3}s^{-1}.

6.6. The pump efficiency in this case equals $\eta_p = \eta_t\eta_a\eta_{pq} = 5.3\%$. The laser in Problem 6.4, which has a pump efficiency of $\eta_p' = 4.5\%$, has a threshold of $P_{th}' = 2$ kW. With the present pump configuration, the threshold pump power is then $P_{th} = (\eta_p/\eta_p')P_{th}' = 2.36$ kW. To pump the laser two-times above threshold, we need a pump power of $P_p \cong 4.72$ kW. The area of the collecting optics is given by $A = P_p/I \cong 4.72$ m^2, where I is the solar intensity. If we let D and l be the rod diameter and rod length, respectively, the focal lengths of the two lenses must be such that $f_1\alpha = D$ and $f_2\alpha = l$, where α is the angle that the sun's disk subtends at the earth. We obtain $f_1 = 0.64$ m and $f_2 = 8.05$ m. A cheaper focusing scheme can be made from a spherical mirror with focal length $f_2 = 8.05$ m and diameter $D = (4A/\pi)^{1/2} = 2.45$ m followed by a cylindrical lens.

6.7. $R_{cp} = \eta_p E_{th}/Vh\nu_{mp}\Delta t$, where Δt is the pump duration. Since $h\nu_{mp} = 2.11 \times 10^{-19}$ J, we obtain $R_{cp} = 5.75 \times 10^{21}$ cm^{-3}s^{-1}. The rate equation involving pumping and spontaneous decay is $(dN_2/dt) = R_p - (N_2/\tau)$ whose solution, for $R_p = $ const and $t \geq 0$, is $N_2(t) = R_p\tau\{1 - \exp[-(t/\tau)]\}$. Assuming $\tau = 230$ μs, $t = \Delta t = 100$ μs, and with the help of the previously calculated value of R_{cp}, we obtain critical the inversion as $N_{2c} = R_{cp}\tau\{1 - \exp[-(\Delta t/\tau)]\} = 4.66 \times 10^{17}$ cm^{-3}. If the pulse duration is increased to $\Delta t' = 300$ μs, to achieve the same inversion the pump rate R'_{cp} must be such that $R'_{cp}\{1 - \exp[-(\Delta t'/\tau)]\} = R_{cp}\{1 - \exp[-(\Delta t/\tau)]\}$. Then $R'_{cp} \cong 0.48 R_{cp} \cong 2.78 \times 10^{21}$ cm^{-3} s^{-1}. The new threshold pump energy is then $E'_p = (R'_{cp}\Delta t'/R_{cp}\Delta t)E_p \cong 1.44E_p \cong 4.9$ J.

6.10. From Table 6.2 one gets $N_t = 9 \times 10^{20}$ cm^{-3}, so that $\sigma_a N_t l + \gamma = 0.169$. Assuming an $\sim 80\%$ efficiency for the transfer optics, the pump efficiency can be taken as $\eta_p = \eta_t\eta_a = \eta_t[1 - \exp(\alpha l)] = 0.424$, where $\alpha = 5$ cm^{-1} is the absorption coefficient at the pump wavelength (see Table 6.2). From Eq. (6.3.25), with $w_p = w_0$, we then readily obtain $P_{th} \cong 177$ mW.

6.12. For a Maxwell–Boltzmann distribution we have $kT_e = (2/3)(mv_{th}^2/2)$. Since $(mv_{th}^2/2) = 10$ eV, we then have $kT_e = 6.67$ eV.

6.15. Using Gauss theorem, the radially oriented electric field, at the radial coordinate r within the medium, can be expressed as $E(r) = N_i er/2\varepsilon_0$. By integration, the potential drop between wall and center is obtained as $V = N_i eR^2/4\varepsilon_0 \cong 4.56 \times 10^6$ V, where R is the tube radius. The very high value of the voltage obtained shows that there is a negligible probability for electrons to disappear at a different rate from that of the ions.

6.18. The thermal velocity is given by $v_{th} = (2E/m)^{1/2} = 1.87 \times 10^8$ cm/s, where m is the mass of the electron. For an ideal gas, the molar volume is given by $V = RT/p$, where $R = 8.314$ J mol^{-1} K^{-1} is the gas constant. For $p = 1.3$ torr $\cong 1.73 \times 10^2$ Pa and $T = 400$ K, we have $V = 19.2 \times 10^6$ cm^3. Atomic density of the gas is then given by $N = N_A/V = 3.14 \times 10^{16}$ cm^{-3} where N_A is Avogadro's number (see Appendix I). We then obtain the electron mean free path as $l = 1/N\sigma = 638$ μm. From Eqs. (6.4.14) and (6.4.15), the drift velocity can be obtained as $v_{drift} = e\mathscr{E}l/mv_{th} \cong 1.8 \times 107$ cm/s $(v_{drift}/v_{th} \cong 9.6 \times 10^{-2})$.

Chapter 7

7.3. $L_e = L + (n - 1)l = 56.15$ cm, $\gamma = 0.12$, $\tau_c = 15.6$ ns.

7.4. The overall lifetime of the upper laser level is such that $(1/\tau) = (1/\tau_{21}) + (1/\tau')$, where $(1/\tau_{21})$ is the rate of transition $2 \rightarrow 1$ and $(1/\tau')$ is the rate of all other spontaneous transitions originating from level 2. Since the branching ratio β is given by $\beta = (1/\tau_{21})/(1/\tau) = \tau/\tau_{21}$, we have $\tau_{21} = \tau/\beta = 451$ μs. Below threshold, at steady state, we have $(N_1/\tau_1) = (N_2/\tau_{21})$, where τ_1 is the lifetime of the lower laser level. Thus, for $(N_1/N_2) < 1\%$, we must have $\tau_1 < 10^{-2}\tau_{21} \cong 4.5$ μs. For an output power of $P_{out} = 200$ W, the emission rate of photons from the active medium is $(d\phi/dt) = P_{out}(2\gamma/\gamma_2)/h\nu = 1.58 \times 10^{21}$ photons/s. The population, ending up in the lower laser level per unit time, is then $dN'/dt = (d\phi/dt)/A_b l = 9.16 \times 10^{20}$ cm^{-3} s^{-1}, where A_b is the effective beam area in the rod and l is the rod length. The steady-state population N_1 is given by $N_1 = \tau_1(dN'/dt)$; the upper state population N_2 now equals the threshold population, i.e., $N_2 = N_c \cong 5.7 \times 10^{16}$ cm^{-3}. For $(N_1/N_2) < 1\%$ one must then have $\tau_1 < 10^2 \, N_c/(dN'/dt) \cong 0.6 \, \mu$s. (The actual lifetime τ_1 for Nd:YAG is ~ 100 ps.)

7.7. The minimum threshold power is given by $P_{mth} = P_{th}\gamma_i/\gamma = 2.75$ kW, where γ_i is the internal loss ($\gamma_i = 0.02$) and γ is the total loss ($\gamma = 0.32$). At a pump power of $P_p = 140$ kW, one then has $x_m = P_p/P_{mth} = 50.9$. From Eq. (7.5.5) one finds $S_{op} = 6.135$, so that $\gamma_{2op} = 2S_{op}\gamma_i = 0.25$. The corresponding optimum output power is obtained from Eq. (7.5.6) as $P_{op} = 16.78$ kW. We have $T_{2op} = 1 - \exp(-\gamma_{2op}) \cong 0.22$. Since the peak intensity in the focal plane of a lens is proportional to $(M^2 - 1)/M^2$ (see Sect. 5.6.3), the ratio between the two intensities is $(I_{op}/I) = P_{op}T_{2op}/PT_2 = (16.78 \times 0.22)/(12 \times 0.45) = 0.68$.

7.8. The lens, of focal length f, can be divided into two closely spaced lenses, each of focal length $2f$. The radius of curvature of the wave front at the lens position is given by $R = (L/2)\{1 + [z_R/(L/2)]^2\}$, with $z_R = \pi w_0^2/\lambda$, where w_0 is the spot size at each of the two mirrors. For symmetry reasons, the wave front between the two lenses must be plane. Thus we must have $R = 2f = 50$ cm. Using this value of R in the preceding expression, we find $z_R = 25$ cm, i.e., $w_0 = 290 \, \mu$m. The spot size at the lens position is then $w = \sqrt{2} \, w_0 = 410 \, \mu$m.

7.10. The minimum threshold pump power, for zero output coupling, is $P'_{mth} = P_{th}\gamma_i/\gamma = 12.5$ mW. Note that P'_{mth} should not be confused with P_{mth} given by Eq. (7.3.32). At $P_p = 1.14$ W, we have $x_m = 91.2$. From Eq. (7.5.5) we obtain $S_{op} = 8.54$, i.e., $\gamma_{2op} = 2S_{op}\gamma_i = 8.5\%$. To calculate the output power we note that the total loss is now given by $\gamma' = (\gamma'_{2op}/2) + \gamma_i = 4.75\%$ and the amount by which threshold is exceeded is $x' = x\gamma/\gamma'$, where x and γ are the corresponding values obtained in Example 7.4 ($x = 30$ and $\gamma = 3\%$). We obtain $x' = 19$, so that, from Eq. (7.3.34), we find $y' = 15.6$. The expected power is then $P'_{out} = P_{out}(y'/y)(\gamma'_{2op}/\gamma_2) = 510$ mW where P_{out}, y, and γ_2 are the corresponding values obtained in Example 7.4 ($P_{out} = 500$ mW, $y = 26$, $\gamma_2 = 5\%$). Note that our optimization procedure results in a value of P'_{out} (slightly) larger than P_{out}.

7.12. The lens of focal length f can be divided into two closely spaced lenses, each of focal length $f' = 2f$. The spot size between the two lenses is w_a and, for symmetry reasons, the wave front is plane. The position of the two plane mirrors must also correspond to a beam waist. From Eq. (4.7.26) we find that the distance between each plane mirror and the corresponding lens f' is given by $z_m = f'/[1 + (f'/z_R)^2]$, where, in our case, $z_R = \pi w_a^2/\lambda \cong 581$ cm. For $f' = 2f = 42$ cm we find $z_m \cong 41.8$ cm. The spot size on each mirror is then obtained from Eq. (4.7.28) as $w_0 = (\lambda/\pi w_a)f' \cong 100 \, \mu$m.

7.15. $\gamma_2 = -\ln(1 - T_2) \cong 5.1\%$. To avoid oscillation in the TEM$_{01}$ mode, the required total loss per pass for this mode must be $\gamma \geq 7.45\%$. From Fig. 5.13b, with $g = 1 - (L/R) = 0.8$, we have $N = a^2/\lambda L \leq 2$, i.e., $a \leq (2\lambda L)^{1/2} \cong 1$ mm.

7.17. $N_c = \gamma/\sigma_e l \cong 4 \times 10^9$ ions/cm^3, $R_{cp} = N_c/\tau \cong 8 \times 10^{17}$ cm^{-3} s^{-1}, $\Delta \nu = c/2L \cong 1.5 \times 10^8$ Hz, $(R_p/R_{cp}) = \exp[(2\Delta\nu/\Delta\nu_0^*)^2 \ln 2] = 1.005$.

7.19. One has $L = n(\lambda/2)$, where n is an integer number. According to this equation, if L is increased by $\lambda/2$, the oscillation wavelength increases by $\Delta\lambda = \lambda/n$. Since $\lambda\nu = c$, then $\Delta\nu \cong -(\Delta\lambda/\lambda)\nu = -\nu/n$. From the relation $\nu = n(c/2L)$, we then obtain $\Delta\nu \cong -(c/2L)$.

7.21. Assume that a transmission peak of the FP etalon is coincident with the central mode. The two adjacent longitudinal modes, which are frequency spaced by $\Delta\nu = c/2L$, do not oscillate if the FP transmission at these two frequencies $T(\Delta\nu)$ is such that $T(\Delta\nu)\exp(\sigma_p Nl - \gamma) \leq 1$. We obtain $T(\Delta\nu) \leq 0.8$. From Eqs. (4.5.6) and (4.5.14), with $R_1 = R_2 = R$, one obtains $T(\Delta\nu) = 1/[1 + (2F/\pi)^2 \sin^2 \phi]$ where $\phi = 2\pi L'\Delta\nu/c$ with $L' \cong n_r L_{et} = 2.9$ cm (L_{et} is the thickness and n_r the refraction index of the etalon). We obtain $\phi = \pi L'/L = 9.1 \times 10^{-1}$ rad; since

$T(\Delta v) \le 0.8$, we obtain $(2F/\pi)^2 \sin^2\phi \cong (2F/\pi)^2\phi^2 \ge 0.25$, i.e. $F \ge (0.5\pi/2\phi) = 8.63$. For $R_1 = R_2 = R$, Eq. (4.5.14) can then be written as $(1 - R) = \pi R^{1/2}/F$, from which, by an iterative procedure, we have $R \cong 0.7$. We must also ensure that the mode near the next peak of the FP etalon is below threshold. This occurs if $\exp[-(2\Delta v_{fsr}/\Delta v_0^*)^2 \ln 2] \times \exp(\sigma_p Nl - \gamma) \le 1$ i.e., for $(2\Delta v_{fsr}/\Delta v_0^*)^2 \ln 2 \ge 0.223$, where $\Delta v_{fsr} \cong (c/2n_r L_{et})$ is the FP free spectral range. We obtain $l_{et} \le (\ln 2/0.223)^{1/2}c/n_r\Delta v_0^* \cong 10.4\,\text{cm}$, which shows that the preceding condition is satisfied in our case.

Chapter 8

8.2. Since $k\Delta nL' = (\pi/2)$, one finds $V = \lambda/4n_0^3 r_{63}$.

8.4. From Fig. 8.14, for $f^* = f\tau = 2.3$ and $x = 10\,\text{kW}/2.2\,\text{kW} = 4.55$, we obtain $N_i/N_p \cong 1.89$, so that, from Fig. 8.11, we find $\eta_E \cong 0.76$. Since $\gamma_2 = 0.162$ and $A_b = 0.23\,\text{cm}^2$ (see Example 7.2), from Eq. (8.4.20) we find $E \cong 18\,\text{mJ}$, which gives an average output power of $\langle P \rangle = Ef = 180\,\text{W}$, i.e., very close to the cw value (202 W; see Fig. 7.5). Since $\gamma = 0.12$ (see Example 7.2) and $L_e = L + (n - 1)l \cong 56\,\text{cm}$ (where $n = 1.8$ is the refractive index of the YAG crystal), we obtain $\tau_c = L/c\gamma = 15.6\,\text{ns}$ and, from Eq. (8.4.21), $\Delta\tau_p \cong 90\,\text{ns}$.

8.7. (a) Since t_p is much shorter than the upper state lifetime, one has $(dN/dt) = R_p$, i.e., $N = R_p t$. Since $R_p t_p = 4N_{th}$, where N_{th} is the threshold inversion, the time when threshold is reached is $t_{th} = t_p/4$. (b) The time behavior of the net gain is then given by $g_{net} = \sigma(N - N_{th})l = (4\sigma N_{th}l)(t - t_{th})/t_p$ where $t_{th} = t_p/4$. (c) Neglecting saturation, one has $(d\phi/dt) = g_{net}\phi/t_T$, where t_T is the transit time. Using the expression for net gain, just derived, we find by integration $\phi(t') = \phi_i \exp[(4\sigma N_{th}l)(t'^2/2t_p t_T)]$ where $t' = t - t_{th}$ and $\phi_i \cong 1$. (d) At the end of the pump pulse one has $t' = 3t_p/4$; from the preceding expression for $\phi(t')$ one then gets $(9\sigma N_{th}l/8)(t_p/t_T) = \ln(\phi_p/20)$, where ϕ_p is given by Eq. (8.4.14). Using the latter expression, we find $t_p = t_T(8/9\gamma)\ln(\phi_p/20)$, where $\gamma = (-\ln T_2)/2 = 0.35$. To calculate ϕ_p from Eq. (8.4.14), one notes that $N_i/N_p = 4$ and $V_a N_p = \gamma A/\sigma$, where A is the beam area. Thus one finds $\phi_p = 5.54 \times 10^{10}$ and, since $t_T = 22.7\,\text{ps}$, it follows that $t_p = 1.25\,\text{ns}$.

8.12. Equation (8.6.14) can be expressed more conveniently as $E(t) \propto \exp(-\Gamma t^2)\exp(j\omega_0 t)$ where $\Gamma = \alpha - j\beta$; i.e., it can be transformed into a Gaussian pulse with a complex Gaussian parameter Γ. Its Fourier transform can then be written as $E(\omega - \omega_0) \propto \exp[-(\omega - \omega_0)^2]/4\Gamma = \exp(-\{[(\omega - \omega_0)^2/4(\alpha^2 + \beta^2)](\alpha + j\beta)\})$. The power spectrum is then $|E(\omega - \omega_0)|^2 \propto \exp\{-[(\omega - \omega_0)^2\alpha/2(\alpha^2 + \beta^2)]\}$. If we now write $|E(\omega - \omega_0)|^2 \propto \exp\{-[4(\omega - \omega_0)^2 \ln 2/\Delta\omega_L^2]\}$, where $\Delta\omega_L$ is the bandwidth, a comparison between the two preceding expressions gives $\Delta\omega_L^2 = (8\ln 2)\alpha[1 + (\beta^2/\alpha^2)]$. This equation, with the help of expression (8.16.15) for α and using the relation $\Delta v_L = \Delta\omega_L/2\pi$, then leads to (8.6.16).

8.14. The average intensity is $\langle I \rangle = \int_0^\infty I p_I\, dI = I_0$. The required probability is given by $p = \int_{2I_0}^\infty p_I\, dI / \int_0^\infty p_I\, dI = \exp(-2) = 0.135$.

8.16. $2\gamma_t = 2\gamma - kP$, in the fast saturable absorber case, we can write [see Eq. (8.6.20)]: $2\gamma_t = 2\gamma - 2\gamma'(P/P_s)$. Comparison of these two expressions shows that k is equivalent to $2\gamma'/P_s$. According to Eq. (8.6.22), the pulse duration can then be written as $\Delta\tau_p \cong (0.79/\Delta v_0)(2g_0'/kP_p)^{1/2}$ where P_p is the peak power. For a hyperbolic secant function, the peak power is related to the pulse energy by $E = 1.13 P_p \Delta\tau_p$. From the two preceding expressions we obtain $\Delta\tau_p \cong (0.79/\Delta v_0)^2(2g_0'/k)(1.13/E) \cong 3.5\,\text{fs}$.

8.17. From Eq. (8.6.35) with $\phi'' = \beta''l$ one gets $l = (\delta\tau_p/\tau_p)^{1/2} \times (\Delta\tau_p^2/\beta'')/2\sqrt{2}\ln 2 \cong 0.46\,\text{mm}$, where β'' is the GVD.

Chapter 9

9.5. The emission solid angle is $\Omega = \pi D^2/4l^2 \cong 2.83 \times 10^{-3}$ sr. Assuming a Gaussian line, from Eq. (2.9.4b), with $\phi \cong 1$, one gets $G \cong 1.37 \times 10^4$, i.e., $N_{th} = \ln G/\sigma_p l \cong 2.38 \times 10^{19}$ cm^{-3}, where $\sigma_p \cong 4 \times 10^{-20}$ cm^2 for Nd:glass (see Table 9.3). We then obtain $E = N_{th} Vhv \cong 12.7$ J, where $V = 2.83$ cm^3. For Nd:YAG, at the same value of solid angle and assuming a Lorentzian line, we have, from Eq. (2.9.4a), with $\phi = 1$, $G \cong 2.5 \times 10^4$, i.e. $N_{th} = 3.61 \times 10^{18}$ cm^{-3}, so that $E \cong 1.91$ J.

9.7. Neglecting ground-state and excited absorption and under mode-matching conditions ($w_0 \cong w_p$), the threshold pump power from Eq. (6.3.20) is given by: $P_{th} = (\gamma/\eta_p)(hv_p/\tau)(\pi w_0^2/\sigma_e)$. We have $\gamma = (\gamma_2/2) + \gamma_i = 1.5 \times 10^{-2}$, $\eta_p \cong 1 - \exp[-(\alpha_p l)] \cong 0.86$ where $\alpha_p = 5$ cm^{-1} is the pump absorption coefficient, $hv_p \cong 3.1 \times^{-19}$ J, $\tau = 67$ μs, $w_0 = 60$ μm, and $\sigma_e = 4.8 \times 10^{-20}$ cm^2. We thus obtain $P_{th} = 190$ mW.

9.8. $P_{th} = [(\gamma + \gamma_a)/\eta_p](hv_p/\tau)[\pi(w_0^2 + w_p^2)/2(\sigma_e - \sigma_{ESA})]$.

9.9. $P_{th} = (\gamma/\eta_p)(hv_p/\tau)[\pi(w_0^2 + w_p^2)/2(\sigma_e - k_{ST}\tau_T\sigma_T)]$.

9.12. $\eta_s = dP/V \, dI = (hv/eV)[-\ln R/(\alpha L - \ln R)]$. For $\lambda = 850$ nm, $hv/e = 1.46$ eV, and $V = 1.8$ V we get $\eta_s \cong 64\%$.

9.13. One must have $w_{0\|}^2[1 + (z\lambda/\pi w_{0\|}^2)^2]) = w_{0\perp}^2[1 + (z\lambda/\pi w_{0\perp}^2)^2]$, which gives $z = \pi w_{0\|} w_{0\perp}/\lambda$. For the given values of $w_{0\|}$, $w_{0\perp}$, and λ, we obtain $z = 4.6$ μm (note the very short distance).

9.15. $(2\pi n_1 L/\lambda) = 2$, i.e., $n_1 = \lambda/\pi L \cong 8.22 \times 10^{-4}$.

Chaper 10

10.4. If we let $\sigma(v - v_0)$ be the unsaturated cross section of Ar$^+$, oscillation will occur up to the nth mode, away from the central mode, when $\sigma(n\Delta v)Nl \geq \gamma$, where Δv is the frequency spacing between consecutive longitudinal modes, N is the unsaturated inversion, l is the length of the active medium, and γ is the cavity loss. The unsaturated cross section is then given by $\sigma(n\Delta v) = \sigma_p \exp\{-[(2n\Delta v/\Delta v_0^*)^2 \ln 2]\}$. In this expression σ_p is the peak cross section and, with the laser pumped 3 times above threshold, one has $\sigma_p Nl = 3\gamma$. From the preceding three expressions, one finds $3 \exp\{-[(2n\Delta v/\Delta v_0^*)^2 \ln 2]\} \geq 1$, from which one obtains $n \leq (\ln 3/\ln 2)^{1/2}(\Delta v_0^*/2\Delta v)$. Since $\Delta v_0^* = 3.5$ GHz and $\Delta v = c/2L = 150$ MHz (L is the cavity length), we find that $n \leq 14.7$. The number of oscillating modes is then $N_{osc} = 2n + 1 \cong 30$.

10.6. For a homonuclear molecule, consisting of two atoms of mass M, the vibrational frequency, according to Eq. (3.1.3), is given by $v_0 = (1/2\pi)(2k_0/M)^{1/2}$ where k_0 is the elastic constant. For $M \cong 14$ a.u. $\cong 2.32 \times 10^{-26}$ kg and $\tilde{v}_0 = 2300$ cm^{-1}, we find $k_0 = 2180$ Nm^{-1}.

10.8. For the symmetric stretching mode, the carbon position is fixed, and the force acting on each oxygen atom is $F = -k(x - x_0)$, where k is the elastic constant and x_0 is the equilibrium separation between carbon and oxygen. The resonance frequency of this mode is $\omega_1 = (k/M_0)^{1/2}$, where M_0 is the mass of the oxygen atom. For $\tilde{v}_1 = 1337$ cm^{-1} and $M_0 = 16$ a.u. $\cong 2.65 \times 10^{-26}$ kg we obtain $k = 1683$ Nm^{-1}.

10.10. Let x_0 be the equilibrium distance between one of the oxygen atoms and the carbon atom. A transverse displacement of the carbon atom by Δy corresponds to an elongation Δd of the spring given by $\Delta d = (x_0^2 + \Delta y^2)^{1/2} - x_0$. For $\Delta y \ll x_0$, one then gets $\Delta d \cong \Delta y^2/2x_0$. Thus the force produced by the spring is proportional to Δy^2. This implies that the harmonic oscillator model, for oscillation in the y-direction, cannot be derived from the simpified spring model considered in this problem.

10.13. All roto-vibrational lines merge when the collision-broadened linewidth Δv_o becomes comparable to the frequency separation between rotational lines. Assuming $\Delta v_o = \Delta v_r = 60\,\text{GHz}$, from the given value of Δv_c we obtain a total pressure of $p \cong 13{,}997\,\text{Torr} = 18.4\,\text{atm}$. From Fig. 10.11 we see that the width of the gain curve, Δv_0, corresponds to J'-values ranging from $J' \cong 11$ to $J' \cong 41$, i.e., $\Delta J' \cong 30$. From the solution of Problem 10.11, we find that the rotational constant B of a CO_2 molecule is $B \cong 0.3\,\text{cm}^{-1}$. The width Δv_0 of the gain curve is then given by $\Delta v_0 = 2B\Delta J' \cong 60B \cong 18\,\text{cm}^{-1}$.

10.16. The energy left, after a reaction, as vibrational energy is $E_v = \overset{3}{\underset{0}{\sum}}_v vN(v)\Delta E$ where $N(v)$ is the population of the vibrational level with vibrational quantum number v, and ΔE is the energy spacing between vibrational levels (assumed the same for all levels). On the other hand, the total energy of reaction, E_t, is given by $E_t = \Delta H \overset{3}{\underset{0}{\sum}} vN(v)$ where $\Delta H \cong 3\Delta E$ is the reaction energy. From the preceding equations: $\eta = (E_v/E_t) = \overset{3}{\underset{0}{\sum}}_v vN(v)/\overset{3}{\underset{0}{\sum}}_v 3N(v) = 68.5\%$.

Chapter 11

11.3. The field of the beam along the C direction, due to the superposition of the two beams of the interferometer, can be written as: $E_c = K_A E(t) + K_B E(t + \tau)$. If the power reflectivity of mirror S_1 is 50% and neglecting, for simplicity, any phase shift arising from reflections at mirrors S_1, S_2, and S_3, we can assume $K_A = K_B = K$. Then $\langle I_c(t) \rangle = \langle E_c(t)E_c^*(t) \rangle = 2|K|^2\{\langle I \rangle +\text{Re}[\Gamma^{(1)}(\tau)]\}$ where $\langle I \rangle = \langle E(t)E^*(t) \rangle = \langle E(t + \tau)E^*(t + \tau) \rangle$ and Re stands for the real part. From Eqs. (11.3.4) and (11.3.9) one then gets $\langle I_c(t) \rangle = 2|K|^2\langle I \rangle\{1 + |\gamma^{(1)}| \cos[\langle \omega \rangle \tau - \psi(\tau)]\}$. Around a given time delay τ, since both $|\gamma^{(1)}|$ and ψ are slowly varying functions of τ, one then has $I_{max} = \langle I_c(\tau) \rangle_{max} = 2|K|^2\langle I \rangle[1 + |\gamma^{(1)}(\tau)|]$, $I_{min} = \langle I_c(\tau) \rangle_{min} = 2|K|^2\langle I \rangle[1 - |\gamma^{(1)}(\tau)|]$, so that $V_P = |\gamma^{(1)}(\tau)|$.

11.5. For a Gaussian spectral output, $\gamma^{(1)}(\tau)$ is also a Gaussian function, i.e., it can be written as $\gamma^{(1)} = \exp\{-(\tau/\tau_{co})^2 \ln 2\}$, where τ_{co}, the coherence time, is defined as in Fig. 11.1. According to Eq. (11.3.28) one then has $\sigma_\tau = 1/4\pi\sigma_v$. In our case we have $\sigma_v = \Delta v_L$ while the standard deviation σ_τ of the function $(\gamma^{(1)})^2 = \exp\{-[2(\tau/\tau_{co})^2 \ln 2]\}$ is $\sigma_\tau = \tau_{co}/2(\ln 2)^{1/2}$. From the preceding expressions we obtain $\tau_{co} = \sqrt{\ln 2}/2\pi\sigma_v \cong 13.25\,\mu\text{s}$, and $L_{co} = c\tau_{co} \cong 3.98\,\text{km}$.

11.7. $I_0 = 2P_i/[\pi(\lambda f/\pi w_0)^2]$. To avoid excessive diffraction losses and the creation of diffraction rings from beam truncation by the finite lens aperture D_L, we choose a large enough D_L, typically $D_L = \pi w_0$ [see (5.5.31)]. From the preceding expressions, we then find $I_0 = (2/\pi)P_i D_L^2/(\lambda f)^2$ while, from Eq. (11.4.4) with $D = D_L$, we find $I_0 = (\pi/4)P_i D_L^2/(\lambda f)^2$.

11.9. If we let x and y be the coordinates along the smaller and larger dimensions, respectively, of the near-field pattern, one has $W_{x0} = 0.5\,\text{cm}$ and $W_{y0} = 2\,\text{cm}$. From Eq. (11.4.19), one then has $W_x(z = 3\,\text{m}) \cong 3.28\,\text{cm}$ while, from the equivalent equation in the y-direction, one gets $W_y(z = 3\,\text{m}) \cong 2.16\,\text{cm}$.

Chapter 12

12.1. Since $w_0 = 0.54\,\text{mm}$, one has $w(z = 1\,\text{m}) = w_0[1 + (z/z_R)^2]^{1/2} = 0.83\,\text{mm}$ and $R(z = 1\,\text{m}) = z[1 + (z_R/z)^2] \cong 1.74\,\text{m}$, where $z_R = \pi w_0^2/\lambda \cong 86.1\,\text{cm}$. The lens of focal length f can be divided into a first lens, of focal length $f_1 = R = 1.74\,\text{m}$, to compensate for the wavefront curvature, and a second lens, of focal length $f_2 = f_1 f / (f_1 - f) \cong 10.61\,\text{cm}$, to focus the beam. To a good approximation, the waist position then occurs at a distance of $z_m \cong f_2 \cong 10.61\,\text{cm}$ from the original lens. The spot size of the embedded Gaussian beam is $w_0' \cong (\lambda / \pi w) f_2 \cong 0.043\,\text{mm}$, and the corresponding spot size parameter is $W_0' = (M^2)^{1/2} w_0' \cong 0.274\,\text{mm}$.

12.3. One has $\Gamma_s = h\nu/\sigma \cong 4.71\,\text{J/cm}^2$ and $S = \pi D^2 / 4 \cong 63.6\,\text{cm}^2$, so that $\Gamma_{out} = E_{out}/S \cong 7.07\,\text{J/cm}^2$. The total energy available in the amplifier is $E_{av} = h\nu N V = S\Gamma_s \ln G_0 = 415\,\text{J}$, where N is the initial inversion and V is the volume of the amplifier. To calculate the required input energy, Eq. (12.3.12) can be solved for Γ_{in} to give $\Gamma_{in} = [\{[\exp(\Gamma_{out}/\Gamma_s) - 1]/G_0\} + 1] \cong 2.95\,\text{J/cm}^2$ which results in $E_{in} = \Gamma_{in} S = 187.8\,\text{J}$. Thus, out of an available energy of 415 J, the energy extracted from the amplifier is $E_{ex} = E_{out} - E_{in} \cong 262.2\,\text{J}$. Note that the length of the amplifier does not enter into this calculation.

12.9. With the help of Eq. (12.4.27a), substitution of Eq. (12.4.29) into Eq. (12.4.2) gives $P^{NL} = (\varepsilon_0 d / 2)\{\sum_i E_i(z) \exp[j(\omega_i t - k_i z)] + c.c.\}^2$. After manipulating the right-hand side of the preceding equation, since $\omega_1 = \omega_3 - \omega_2$, the only term at frequency ω_1 is found to be $P^{NL}_{\omega_1} = (\varepsilon_0 d / 2)\{E_2^*(z)E_3(z) \exp[j(\omega_3 - \omega_2)t - j(k_3 - k_2)z] + c.c.\}$. Using the relation $\omega_1 = \omega_3 - \omega_2$ and Eq. (12.4.27b), we then obtain Eq. (12.4.30).

12.11. From Eq. (12.4.58a) the second harmonic conversion efficiency is obtained as $\eta = I_{2\omega}/I_\omega(0) = |E_{2\omega}'|^2 / |E_\omega'(0)|^2 = [\tanh(z/l_{SH})]^2$. From Eq. (12.4.52), since $E_\omega(0)$ is related to the incident intensity $I = I_\omega(0)$ by $E_\omega(0) = (2ZI)^{1/2}$, where $Z = 1/\varepsilon_0 c \cong 377\,\Omega$ is the free-space impedance, one gets $l_{SH} \cong \lambda n_o / [2\pi d_{eff}(2ZI)^{1/2}] = 2.75\,\text{cm}$, where n_o is the ordinary refractive index of KDP at frequency ω. Substituting this value of l_{SH} into the preceding expression for η and assuming $z = 2.5\,\text{cm}$, we obtain $\eta = 51.9\%$.

Index

Orazio Svelto is professor of Quantum Electronics at the Polytechnic Institute of Milan and Director of the Quantum Electronics Center of the Italian National Research Council. His research has covered a wide range of activity in the field of laser physics and quantum electronics, starting from the very beginning of these disciplines. This activity includes ultrashort-pulse generation and applications, development of laser resonators and mode-selection techniques, laser applications in biology and medicine, and development of solid-state lasers. Professor Svelto is the author of more than 150 scientific papers and his researches have been the subject of more than 50 invited papers at international conferences. He has served as a program chair of the IX International Quantum Electronics Conference (1976), as a chair of the European program committee for CLEO '85 and CLEO '90, and he was general co-chair for the first CLEO–Europe Conference (1994). He is an elected member of the Italian "Accademia dei XL" and a Fellow of the IEEE.